Wireless Receiver Design
for
Digital Communications

Wireless Receiver Design for Digital Communications

Second Edition

Kevin McClaning

Raleigh, NC
scitechpub.com

Published by SciTech Publishing, Inc.
911 Paverstone Drive, Suite B
Raleigh, NC 27615
(919) 847-2434, fax (919) 847-2568
scitechpublishing.com

Editor: Dudley R. Kay
Production Manager: Robert Lawless
Typesetting: MPS Limited, a Macmillan Company
Cover Design: Brent Beckley and Jennifer McClaning
Printer: Sheridan Books, Inc., Chelsea, MI

This book is available at special quantity discounts to use as premiums and sales promotions, or for use in corporate training programs. For more information and quotes, please contact the publisher.

Printed in the United States of America
10 9 8 7 6 5 4 3 2 1

ISBN: 978-1-891121-80-7

Library of Congress Cataloging-in-Publication Data

McClaning, Kevin, 1959-
Wireless receiver design for digital communications / Kevin McClaning. – 2nd ed.
p. cm.
Rev. ed. of Radio receiver design / Kevin McClaning, Tom Vito. 2000.
Includes bibliographical references and index.
ISBN 978-1-891121-80-7 (hardcover : alk. paper)
1. Radio–Receivers and reception–Design and construction. 2. Digital communications. I. McClaning, Kevin, 1959- Radio receiver design. II. Title.
TK6563.M38 2012
384–dc23

2011027275

Contents

Preface to the Second Edition

Avoid studies of which the result dies with the worker.

Leonardo da Vinci

There have been two paradigm shifts since we wrote the first edition of this book. The first major shift is that the world is now full of digital signals. Commercial television and radio have changed from analog to digital formats. Cellular telephone signals are digital, and even video baby monitors transmit digital signals. I'm not sure you can even find an analog signal on the air anymore. In a way, this is sad. In the "either you get a perfect signal or you get nothing" world of the digital broadcast, our children will never know the distinct pleasure of watching a snowy ultrahigh frequency (UHF) broadcast or listening to a crackling AM radio station after it has traveled from the other side of the nation.

The second paradigm shift since the publication of the first edition is the availability of inexpensive digital processing power. As of this writing (early 2011), processing power is essentially free. In the past, radio receivers performed filtering and demodulation in the analog world. Many old-school receivers performed demodulation using analog techniques that were finicky and required large amounts of circuit board real estate. Control functions such as automatic gain control (AGC) and automatic frequency control (AFC) were realized in the analog world using individual diodes, operational (op) amps, resistors and capacitors. Today, it makes technical and economic sense to perform these functions in the digital domain.

We find ourselves in a world where the receiver is primarily a downconverter whose sole purpose is to translate the signal of interest to a frequency and power level that is suitable to be sampled by an analog-to-digital converter (ADC). Similarly, transmitters convert the user's information into complex waveforms using digital signal processing (DSP) techniques. A digital-to-analog converter (DAC) then converts these signals to the analog domain. The transmitter's remaining task is to convert the modulated signal to its final frequency and power level.

However, it is still an analog world, and we must address analog concepts. Our receivers require analog filters, and, although filter realizations have changed, filtering concepts of previous years have remained valid and useful. Oscillator phase noise is essentially low-level, accidental analog phase modulation, and it still limits receiver performance in many areas. Linearity, as measured by component compression points and second- and third-order intercept points, is very important in a world that contains many signals existing in proximity.

The format of this book reflects the state of receiver design as it exists in early 2011. I have only lightly updated many chapters from the previous edition as the material there continues to be relevant and useful. I added chapters on ADCs and included an overview of the demodulation of digital signals as it is performed in the digital domain.

As in the first edition, I leave you with a quote from Groucho Marx: "Why, a 4-year-old child could understand this report. Run out and find me a 4-year-old child. I can't make head or tail out of it" (from *Duck Soup*).

Kevin McClaning
August 2011

Preface to the First Edition

My sister, the lawyer, said that, if we wanted this book to sell, we needed to put in a bloodthirsty clown who lives in the sewers. So far, though, we haven't been able to work him in.

We write as engineers. We don't pretend to be heavy theoretical types, but we do write in the hope that this book will be useful. We didn't concentrate on technology but on useful and proven concepts. Some of the most useful books we own were written in the 1940s by fellows named Frederick Terman, John Kraus, and Mischa Schwartz. Sure, the books contain a lot of information on electron tubes and are a little light on modern filter design, statistical decision theory, and quadrature modulation, but they are clear and well written. They were formulated by engineers for engineers, and as such they are still useful. We hope that in 20 years people will say, "The McClaning/Vito book is a little dated, but it's clear and well written."

We made up our minds early to follow one cardinal rule: to be clear. If we have succeeded, errors should be easy to detect. If we hear about errors in this book (and I hope we do), I consider it a good thing. It means we have been clear enough to bring doubts into the reader's mind. The reader is thinking about the material and understands it enough to find inconsistencies.

We'd like to thank our normally noisy children, Chris and Jenny (McClaning) and Mandy, Nick, Steve, and James (Vito) for being quiet while we wrote this. We'd also like to thank our wives, Kitty and Terri, for putting up with us (and not just while we were writing this book).

We'll leave you with a quote from Groucho Marx: "From the moment I picked up your book until I laid it down, I was convulsed with laughter. Someday, I intend reading it" (quoted in *Life,* February 9, 1962).

Enjoy.

Kevin McClaning and Tom Vito
August 1998

Bibliography

Terman, Frederick Emmons, *Radio Engineering*, Third Edition, McGraw-Hill Book Company, Inc., 1947.

Kraus, John D., *Antennas*, McGraw-Hill Book Company, Inc., 1950.

Schwartz, Mischa, *Information Transmission, Modulation, and Noise,* McGraw-Hill Book Company, Inc., 1959.

Acknowledgments

No book is an island, to horribly paraphrase the poet John Donne. I would like to thank the following individuals for their contributions to this ponderous tome:

James McPherson and Tom Vito, who both contributed mightily in terms of scheduling, reviews, and occasional material

Jeff Houser, Roger Kaul, and Ron Tobin, from Johns Hopkins University, friends with whom I voyaged through the teaching experience

Edward Sheriff and Brian Sherlock, buddies from work and two of the smartest people I know.

Christopher and Jennifer McClaning, my two children and my world

Kathy Sessions, the true love of my life who is always there with support and laughter

I would also like to thank the following brilliant and eccentric people:

Mark Bennett, Roger Gilbert, Paul Hughes, Dan Loveday, and Philip Pring, from Detica, United Kingdom

Dr. David Waymont and Philip Morris of Waymont Consulting in the United Kingdom;

Terry Fry, of Zeus Technology Systems

CHAPTER 1

Radio Frequency Basics

Begin at the beginning, and go on 'til you come to the end; then stop.

The King of Hearts, *Alice in Wonderland*

The journey of a thousand miles begins with a single step.

Loa Tsu

Chapter Outline

1.1 INTRODUCTION

There are entire books that deal with statistics, logarithms, significant figures, transmission lines, and s-parameters. We are interested in understanding and designing radio receivers and systems, which requires at least a passing knowledge of all these topics and more. This chapter serves as an introduction to many of the basic concepts we will need in later chapters.

This is an optional chapter. Many readers will find the information presented here very basic and may choose to pass this chapter over. For others, it will be required reading.

Some of the purists out there may argue that the information presented here is imprecise, incomplete, and greatly simplified. Again, the purpose of this chapter is to acquaint the reader with some of the fundamental material we'll need later on. For a more complete treatment of any of the topics listed here, refer to the bibliography at the end of the chapter.

1.2 NOMENCLATURE

Throughout this chapter and the following chapters, we will use quantities commonly expressed in both linear and logarithmic (or dB) formats. In this book, anytime we express a quantity in dB, we will give the quantity a *dB* subscript. For example, noise figure expressed in decibels is

$$F_{dB} \tag{1.1}$$

while noise figure expressed as a linear quantity is

$$F \tag{1.2}$$

If we really want to make the point, we may write a linear quantity as

$$F_{Lin} \tag{1.3}$$

A major source of mistakes in receiver design is improperly swapping decibel and linear quantities.

1.3 | DECIBELS

Engineers originally used decibels to make measurements of human hearing more understandable. Human hearing covers a 1:10,000,000,000,000 ($1{:}10^{13}$) range from the threshold of hearing to the threshold of pain. This enormous range makes it hard to plot and analyze data. Expressing quantities in decibel units makes data comparisons easier and graphical data more manageable.

Figure 1-1 shows the range of powers that exist on a typical INMARSAT satellite link. The power levels fall over a 10^{26} range. Since Figure 1-1 uses decibels and a logarithmic scale, the data are very readable.

Other examples of the large ranges system designers must accommodate are as follows:

- A typical communications receiver can easily process signals whose input powers vary by a factor of 10^{12} or 1 trillion.
- On its journey to a geosynchronous satellite, a signal might be attenuated by a factor of 10^{20}.
- Filters routinely provide attenuation factors of 10^6 or 1 million.

1.3.1 Definitions

The Bel is defined as

$$Bel = B = \log\left(\frac{P_2}{P_1}\right) \tag{1.4}$$

The decibel (or dB) is ten Bels or

$$dB = 10\log\left(\frac{P_2}{P_1}\right) \tag{1.5}$$

where P_2 and P_1 are signal *powers*.

These are the fundamental definitions of the Bel and decibel. All other decibel-like quantities are derived from these two definitions. These are power ratios (not voltage or current ratios—we'll derive those shortly). A quantity described in decibels is fundamentally a dimensionless number, although we'll often label the quantity with a "reminder" subscript.

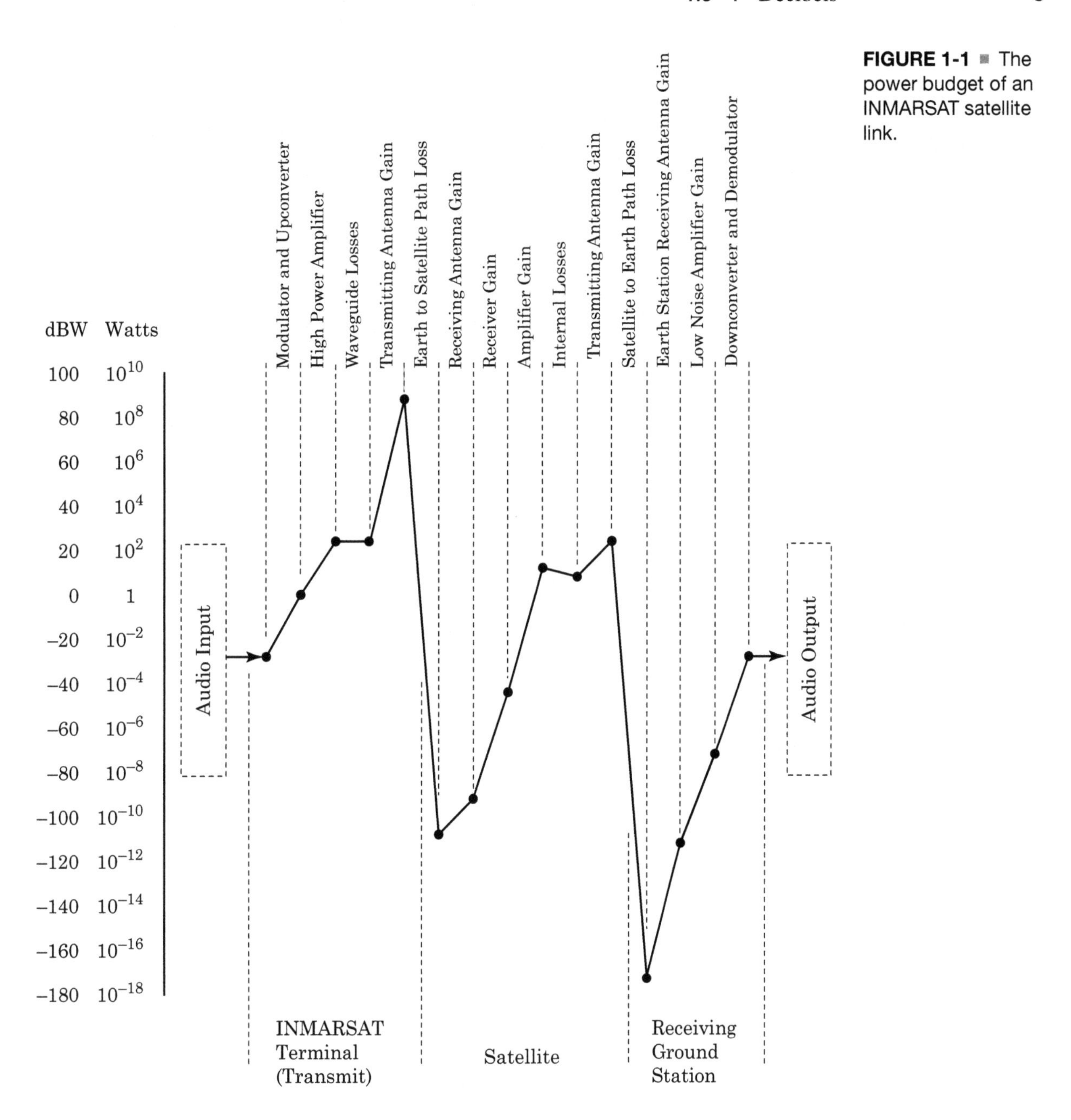

FIGURE 1-1 ■ The power budget of an INMARSAT satellite link.

EXAMPLE

dB Power Conversions

A radio frequency (RF) amplifier accepts 3.16E-6 watts from a signal source and supplies its load resistor with 2.2E-3 watts. What's the power gain of the RF amplifier? Give your answer in dB.

Solution

Using equation (1.5), we set P_2 equal to the amplifier's output power and P_1 equal to the amplifier's input power. So, $P_1 = 3.16\text{E-}6$ watts, and $P_2 = 2.2\text{E-}3$ watts and

$$\begin{aligned} Gain_{dB} &= 10\log\left(\frac{P_2}{P_1}\right) \\ &= 10\log\left(\frac{2.2\cdot 10^{-3}}{3.16\cdot 10^{-6}}\right) \\ &= 10\log(696.2) \\ &= 28.4\ dB \end{aligned} \tag{1.6}$$

The *power gain* or simply the *gain* of the amplifier is about 28.4 dB.

1.3.2 dB Math

The logarithmic nature of the decibel allows us to calculate quickly. For example:

- A doubling of power is a 3 dB increase:

$$\begin{aligned} 10\log\left(\frac{2P_{in}}{P_{in}}\right) &= 3.01\ dB \\ &\approx 3\ dB \end{aligned} \tag{1.7}$$

- Similarly, halving the power is a 3 dB decrease:

$$\begin{aligned} 10\log\left(\frac{P_{in}}{2P_{in}}\right) &= -3.01\ dB \\ &\approx -3\ dB \end{aligned} \tag{1.8}$$

- A 4-times increase in power:

$$\begin{aligned} 10\log\left(\frac{4P_{in}}{P_{in}}\right) &= 6.02\ dB \\ &\approx 6\ dB \end{aligned} \tag{1.9}$$

  A 4-times power increase is like doubling the power twice and 3 dB + 3 dB = 6 dB.

- A 4-times decrease in power:

$$\begin{aligned} 10\log\left(\frac{P_{in}}{4P_{in}}\right) &= -6.02\ dB \\ &\approx -6\ dB \end{aligned} \tag{1.10}$$

  A 4-times power decrease is like halving the power twice and (−3 dB) + (−3 dB) = −6 dB.

- A 10-times increase in power:

$$10\log\left(\frac{10P_{in}}{P_{in}}\right) = 10\ dB \tag{1.11}$$

  Table 1-1 summarizes these results.

TABLE 1-1 ■ Power ratios and their decibel equivalents.

Power Ratio	Decibels
10^{-6}	−60.0 dB
0.001	−30.0 dB
0.01	−20.0 dB
0.1	−10.0 dB
0.5	−3.0 dB
1.0	0.0 dB
2	3.0 dB
3	4.8 dB
4	6.0 dB
5	7.0 dB
8	9.0 dB
10	10.0 dB
100	20.0 dB
1,000	30.0 dB
10^{6}	60.0 dB

TABLE 1-2 ■ Power levels in dBm and their linear equivalents.

Power Level in dBm	Linear Power Level
−30 dBm	0.001 mW = 1 μW
−20 dBm	0.01 mW = 10 μW
−10 dBm	0.1 mW = 100 μW
0 dBm	1.0 mW = 1000 μW
10 dBm	10 mW
20 dBm	100 mW
30 dBm	1,000 mW = 1 W
40 dBm	10 W
50 dBm	20 W

1.3.2.1 Orders of Magnitude[1]

When we compare a power level to a 1 milliwatt reference, we use the symbol "dBm" to mean "dB above or below a milliwatt." Similarly, we use dBW to describe a power level that is reference to 1 watt or "dB above or below a watt." Table 1-2 shows that every 10 dB increase in a quantity represents a linear factor of 10 increase in that quantity (in other words, you gain an order of magnitude for every 10 dB increase). Similarly, a decrease of 10 dB represents multiplying the quantity by 1/10 (or you lose an order of magnitude).

Given a number in the format of

ABCD.EF dB

[1]Herein lies a story. The term *decade* undisputedly refers to a factor of 10. The term *order of magnitude* was first used in the world of astronomy, where it refers to the brightness of a star. An astronomical order of magnitude works out to be the fifth root of 100, and five orders of magnitude add up to 20 dB. In the 1960s, engineers stole the term order of magnitude for their own nefarious uses. In the world of RF and in this book, the term *order of magnitude* means a factor of 10.

the numbers represented by the "ABC" tell you the order of magnitude, while the "D.EF" tells you the position in that order of magnitude.

1.3.2.2 Amplifiers, Attenuators, and Decibels

In wireless system design, we often encounter situations where signals pass through amplifiers and attenuators. These calculations are easy to perform if we express all our quantities in decibels.

When we multiply in the linear domain, we add in the decibel (or logarithmic) domain. When we divide in the linear domain, we subtract in the decibel domain. Mathematically, the relationship

$$\begin{aligned} C &= A \cdot B \\ \Rightarrow C_{dB} &= A_{dB} + B_{dB} \end{aligned} \tag{1.12}$$

is always true.

For example, if a signal with –82 dBm of power passes through an amplifier with 15 dB of power gain, the output signal power is

$$\begin{aligned} P_{out,\,dBm} &= P_{in,\,dBm} + G_{p,\,dB} \\ &= -82 + 15 \\ &= -67\ dBm \end{aligned} \tag{1.13}$$

Using the linear quantities, we first convert the input signal power and gain from decibel to linear terms:

$$\begin{aligned} -82\text{ dBm} &= 10\log\left(P_{in,\,mW}\right) & 15\text{ dB} &= 10\log\left(G_P\right) \\ \Rightarrow P_{in,\,mW} &= 6.31\cdot 10^{-9}\text{ mW} & \Rightarrow G_p &= 31.6 \end{aligned} \tag{1.14}$$

We multiply and convert back to decibels:

$$\begin{aligned} P_{out} &= P_{in}\cdot G_P \\ P_{out,\,mW} &= \left(6.31\cdot 10^{-9}\text{mW}\right)(31.6) \\ &= 199.5\cdot 10^{-9}\text{mW} \\ P_{out,\,dBm} &= 10\log\left(199.5\cdot 10^{-9}\right) \\ &= -67.0\text{ dBm} \end{aligned} \tag{1.15}$$

1.3.2.3 Gains and Losses

The terms gain and loss come up quite a bit when we discuss signals and systems. It's worth taking a little time to clarify these terms and their concepts.

For example, when a device has a 6 dB gain, the device will accept a signal, multiply its power by 4 times (or add 6 dB), and present that power to the outside world. If the input signal power is P_{in}, then the output signal power is

$$P_{out} = 4P_{in} \tag{1.16}$$

or, in decibels,

$$P_{out,\,dBm} = P_{in,\,dBm} + 6\text{ dB} \tag{1.17}$$

If a devices has a 6 dB loss, the device accepts a signal, divides the signal power by 4 (or subtracts 6 dB from the input signal power) and then presents that power to the outside world. If the input signal power is P_{in}, the output signal power will be

$$P_{out} = \frac{P_{in}}{4} \tag{1.18}$$

or

$$P_{out,\,dBm} = P_{in,\,dBm} - 6 \text{ dB} \tag{1.19}$$

So, a loss of x dB is equivalent to a gain of $-x$ dB. A negative gain is equivalent to a positive loss. We relate gains and losses in equation form as

$$Gain_{dB} = -Loss_{dB} \tag{1.20}$$

and

$$Gain_{Lin} = \frac{1}{Loss_{Lin}} \tag{1.21}$$

EXAMPLE

Gains and Losses in Cascades

Figure 1-2 shows a cascade made up of blocks with gain and losses. Find the cascade gain.

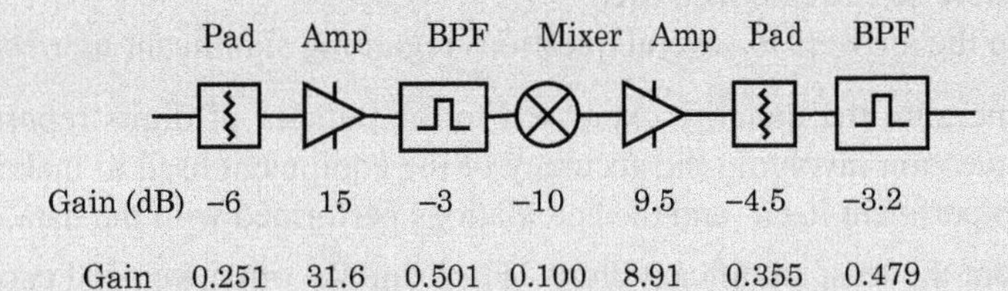

FIGURE 1-2 ■ Calculating cascade gain.

Solution

The cascade gain is

$$\begin{aligned} G_{p,\,cas,\,dB} &= G_{1,dB} + G_{2,dB} + G_{3,dB} + G_{4,dB} + G_{5,dB} + G_{6,dB} + G_{7,dB} \\ &= (-6) + 15 + (-3) + (-10) + 9.5 + (-4.5) + (-3.2) \\ &= -2.2 \text{ dB} \end{aligned} \tag{1.22}$$

The cascade has a gain of −2.2 dB or a loss of 2.2 dB. We can find the cascade gain using only the linear terms:

$$\begin{aligned} G_{p,cas} &= G_1 G_2 G_3 G_4 G_5 G_6 G_7 \\ &= \left(\frac{1}{3.98}\right)(31.6)\left(\frac{1}{2}\right)\left(\frac{1}{10}\right)(8.91)\left(\frac{1}{2.82}\right)\left(\frac{1}{2.09}\right) \\ &= 0.60 \\ &= -2.2 \text{ dB} \end{aligned} \tag{1.23}$$

1.3.2.4 Measurement Accuracy, Significant Figures, and Decibels

Suppose we make a power measurement and our meter reads 0.000000 dBm. Is this number of decimal places reasonable? If not, then how many are reasonable?

Assume that the power level is 0.000000 dBm (or 1 mW) exactly. If the power level increases by 1 dB to +1 dBm, then

$$\begin{aligned} 1 \text{ dBm} &= 10 \log(P_{Measured,\,mW}) \\ \Rightarrow P_{Measured,\,mW} &= 1.2589 \text{ mW} \end{aligned} \tag{1.24}$$

which is a 26% change. This is a universal truth—a change of 1 dB in any quantity represents a change of about 26%. Any reasonable meter would have a good chance of measuring this amount of change accurately.

A 0.1 dB increase in a measured quantity results in

$$\begin{aligned} 0.1 \text{ dB} &= 10 \log \left(P_{Measured,\, mW} \right) \\ \Rightarrow P_{Measured,\, mW} &= 1.0233 \text{ mW} \end{aligned} \tag{1.25}$$

which is about a 2% change. This is measurable with care. The 10th dB digit is meaningful in a carefully controlled environment.

Looking at the 100th (or $0.0x$) dB digit, we find

$$\begin{aligned} 0.01 \text{ dB} &= 10 \log \left(P_{Measured,\, mW} \right) \\ \Rightarrow P_{Measured,\, mW} &= 1.00233 \text{ mW} \end{aligned} \tag{1.26}$$

which is a 0.2% change. Under normal engineering circumstances, this amount of accuracy is unusual unless we exercise extreme measures.

We're interested in the answers to several questions regarding significant figures:

- Can I reasonably measure the quantity expressed to the number of digits reported? This is a complex question involving the accuracy of the equipment used to make the measurement, the experiment itself, and the calculations performed with the data.
- Do I really care about the least significant digits? For example, often we need enough gain to overcome the system noise or enough noise figure to see a -125 dBm signal. We do not care about the precise value. The situation is a little like the amount of money you have—it doesn't matter exactly how much you have as long as you have enough.
- Components with very tight tolerances tend to be expensive. Testing to tight tolerances is also expensive. It is important to understand how many significant digits are really needed to avoid requiring more accuracy than is necessary.
- Given the component specifications with which I have built my system, can I really justify the number of significant digits I'm using? For example, the typical gain and noise figure specifications for a typical RF amplifier might be

  Power Gain: 11.0 dB (± 0.5 dB)

  Noise Figure: 2.5 dB (± 0.5 dB)

- Given these specifications (along with the topology of the system), it might be hard to justify stating that the power gain of the system is 24.22 dB when the gain variation of just one of the amplifiers is ± 0.5 dB.

1.3.3 Decibels, Current, and Voltage

Figure 1-3 shows a simplified schematic diagram of an amplifier. We apply a voltage V_{in} to the amplifier. The power being dissipated in the input resistor R_{sys} is

$$P_{in} = \frac{V_{in}^2}{R_{sys}} \tag{1.27}$$

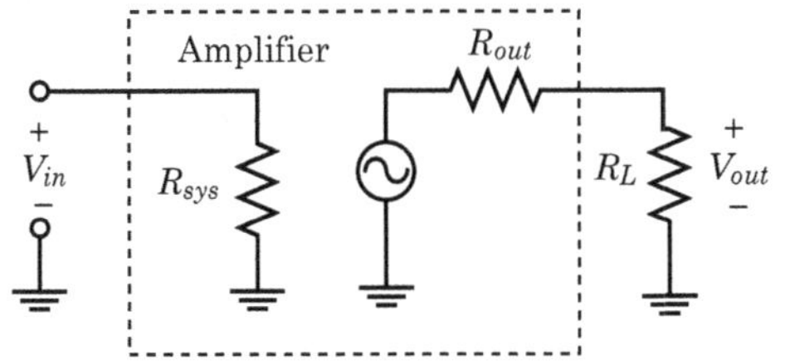

FIGURE 1-3 ■ Simplified schematic of an RF amplifier.

The power delivered by the amplifier to the load resistor R_L is

$$P_{out} = \frac{V_{out}^2}{R_L} \tag{1.28}$$

Using equation (1.5), the gain of the amplifier in decibels (dB) is

$$\begin{aligned} Gain_{dB} &= 10 \log \left(\frac{P_{out}}{P_{in}} \right) \\ &= 10 \log \left(\frac{V_{out}^2 / R_L}{V_{in}^2 / R_{sys}} \right) \\ &= 20 \log \left(\frac{V_{out}}{V_{in}} \right) + 10 \log \left(\frac{R_{sys}}{R_L} \right) \end{aligned} \tag{1.29}$$

Equation (1.29) is exactly correct. However, electrical engineers aren't known especially for their rigor, and last term of equation (1.29) involving R_{sys} and R_L is often ignored, even in systems with wildly differing impedance levels.

Most RF systems have the same impedances on the input and on the output. In Figure 1-3, for example, R_{sys}, R_{out}, and R_L are the same value (usually 50 or 75 ohms). If $R_{sys} = R_L$, then

$$10 \log \left(\frac{R_L}{R_{sys}} \right) = 0 \text{ when} R_L = R_{sys} \tag{1.30}$$

and

$$G_{dB} = 20 \log \left(\frac{V_{out}}{V_{in}} \right) \text{ when } R_L = R_{sys} \tag{1.31}$$

Similarly, we can show

$$G_{dB} = 20 \log \left(\frac{I_{out}}{I_{in}} \right) \text{ when } R_L = R_{sys} \tag{1.32}$$

where I_{in} and I_{out} are the signal currents flowing in R_{sys} and R_L of Figure 1-3.

Equations (1.31) and (1.32) are true only if the source and load resistors are the same.

1.4 SIGNAL STANDARDS

Engineers commonly use the decibel format to express absolute power or voltage levels. We replace P_1 in equation (1.5) with some standard, agreed on value. For example, assume we are building a system where milliwatt (or mW) power levels are common. We might choose P_1 to be 1 mW, and equation (1.5) would become

$$\begin{aligned} P_{Referenced\ to\ 1mW} &= 10 \log\left(\frac{P_{Measured,\ mW}}{1\ \text{mW}}\right) \\ &= 10 \log\left(P_{Measured,\ mW}\right) \\ &= 10 \log\left(\frac{P_{Measured,\ W}}{0.001}\right) \end{aligned} \tag{1.33}$$

where

$P_{Measured,\ mW}$ = the power level measured in mW
$P_{Measured,\ W}$ = the same power level measured in watts

Equation (1.33) defines the dBm standard.

Here are several of the most common standard ways of expressing various RF quantities:

1.4.1 dBm

This is the oldest standard. The telephone company originally set this one up to measure signal levels on their lines, but now almost everyone has adapted it.

The reference power for dBm is 1 milliwatt (that's where the *m* in *dBm* comes from—it stands for power referenced to 1 mW). A quantity expressed in dBm represents an absolute power level (e.g., $P_{out} = 20$ dBm).

$$\begin{aligned} 0\ \text{dBm} &= 1\ \text{mW} \\ P_{dBm} &= 10 \log\left(\frac{P_{watts}}{0.001}\right) \\ &= 10 \log(P_{mW}) \end{aligned} \tag{1.34}$$

The impedance of a telephone line is about 600 Ω, so 1 mW (or 0 dBm) measures a little less than 0.775 V_{RMS}. In a 50 Ω system, a 1 mW or 0 dBm sine wave has a root mean square (RMS) voltage of 0.224 V_{RMS}.

1.4.2 dBW

This is a power measurement referenced to 1 watt. Quantities expressed in dBW are useful in the world of transmitting equipment. dBW is an absolute unit for expressing power level (e.g., The transmitter was putting out +30 dBW).

$$\begin{aligned} 0\ \text{dBW} &= 1\ \text{watt} \\ P_{dBW} &= 10 \log(P_{watts}) \end{aligned} \tag{1.35}$$

EXAMPLE

dBm and dBW

Find an expression relating dBm and dBW.

Solution

Solving equation (1.34) for P_{watts} produces

$$P_{watts} = (0.001)(10^{P_{dBm}/10}) \tag{1.36}$$

Equating this to equation (1.35) produces

$$\begin{aligned} P_{dBW} &= 10\log\left[0.001(10^{P_{dBm}/10})\right] \\ &= 10\log(0.001) + 10\log(10^{P_{dBm}/10}) \\ &\Rightarrow P_{dBw} = P_{dBm} - 30 \end{aligned} \tag{1.37}$$

So to convert dBm to dBW, we add 30 dB. For example, 0 dBW is +30 dBm—both equal 1 watt.

1.4.3 dBf

This standard often refers to the sensitivity of consumer receiving equipment. The reference power is 1 fW = 1 femtowatt = 10^{-15} watts. dBf is an absolute power level (usually a rather small one).

$$\begin{aligned} 0\text{ dBf} &= 1\text{ fW} = 1\text{ femtowatt} = 10^{-15}\text{ watt} \\ P_{dBf} &= 10\log(P_{fW}) \end{aligned} \tag{1.38}$$

1.4.4 dBV

This standard is dB referenced to 1 volt_{RMS}. Quantities expressed as dBV are not power references unless we specify some impedance (or assumed one from the syntax).

$$\begin{aligned} 0\text{ dBV} &= 1\text{ volt}_{RMS} \\ V_{dBV} &= 20\log(V_{volts,RMS}) \end{aligned} \tag{1.39}$$

Note the multiplier of 20 (which arises because voltage is squared to produce power) and the missing

$$10\log\left(\frac{R_L}{R_{sys}}\right) \tag{1.40}$$

term. A quantity expressed in dBV represents an absolute voltage level in a system (for example, 2 Volts_{RMS} = 6 dBV) and can represent an absolute power level if the system impedance is specified or understood.

1.4.5 dBmV

This is another voltage standard. dBmV is decibels referenced to one millivolt_{RMS}. This standard is common in the video and cable TV industries.

$$\begin{aligned} 0\text{ dBmV} &= 1\text{ mV}_{RMS} \\ V_{dBmV} &= 20\log\left(\frac{V_{volts,RMS}}{0.001}\right) \\ &= 20\log\left(V_{mV,RMS}\right) \end{aligned} \tag{1.41}$$

Like dBV, dBmV represents an exact voltage measurement (1 volt_{RMS} = 60 dBmV). dBmV can express power if we know the system impedance.

Unfortunately, the previous definitions are often written simply as dB without identifying subscripts. This leaves the reader to ascertain just what reference the author had in mind.

EXAMPLE

Power Measurements

Express 24 mW in dBm, dBW, dBf, dBV. and dBmV. Assume a 50 Ω system.

Solution

dBm: Equation (1.34) produces

$$\begin{aligned} P_{dBm} &= 10\log(P_{mW}) \\ &= 10\log(24) = 13.8 \text{ dBm} \end{aligned} \tag{1.42}$$

dBW: Using equation (1.35), $PdBW = 10\log(P_{watts})$. Since 24 mW $= 24 \cdot 10^{-3}$ watts then $P_{dBW} = 10\log(24 \cdot 10^{-3}) = -16.2$ dBW.

dBf: Using equation (1.38), we know

$$\begin{aligned} 24\text{ mW} &= 24 \cdot 10^{-3} \text{ watts} \\ &= 24 \cdot 10^{12} \text{ fW} = 133.8 \text{ dBf} \end{aligned} \tag{1.43}$$

To find the dBV and dBmV solutions, we must determine the voltage across a 50 Ω resistor that's dissipating 24 mW. Since $P = V_{RMS}^2/R$ with $P = 24 \cdot 10^{-3}$ watts and $R = 50\ \Omega$, solving for $V_{RMS} = 1.1\ V_{RMS} = 1100\ mV_{RMS}$.

dBV: Using equation (1.39), we know

$$\begin{aligned} V_{dBV} &= 20\log(V_{volts,\,RMS}) \\ &= 20\log(1.1) = 0.83 \text{ dBV} \end{aligned} \tag{1.44}$$

dBmV: Using equation (1.41), we know

$$\begin{aligned} V_{dBmV} &= 20\log(V_{mV,\,RMS}) \\ &= 20\log(1100) = 60.8 \text{ dBmV} \end{aligned} \tag{1.45}$$

EXAMPLE

Power Measurements

A Yamaha RX-700U stereo receiver has a usable sensitivity specification of 9.3 dBf or 0.8 μV_{RMS} at the antenna terminals. What is the input impedance of this stereo receiver?

Solution

Using equation (1.38), we know $P_{dBf} = 9.3$ dBf, which implies

$$\begin{aligned} 9.3\text{ dBf} &= 10\log(P_{fW}) \\ P_{dBf} &= 8.51 \text{ fW} \\ P_{Watt} &= 8.51 \cdot 10^{-15} \text{ watts} \end{aligned} \tag{1.46}$$

Since

$$P_{watts} = \frac{V_{in,RMS}^2}{R_{Rcvr,in}} \tag{1.47}$$

and we're given that $V^2_{in,RMS} = 0.8\text{E-}6$ volts, then

$$\begin{aligned} 8.51 \cdot 10^{-6} \text{ watts} &= \frac{(0.8 \cdot 10^{-6}\, V_{RMS})^2}{R_{Rcvr,in}} \\ R_{Rcvr,in} &= 75\Omega \end{aligned} \tag{1.48}$$

EXAMPLE

Power Amplifiers and Decibels

a. An amplifier is generating an output signal of −30 dBm. If we increase the output power by 3 dB, how much more power (in linear terms) does the amplifier produce?

b. A power amplifier is generating an output signal of +40 dBm. If we increase the output power by 3 dB, how much more power (in linear terms) does the amplifier produce?

Solution

a. Equation (1.34) relates the amplifier's output power in dBm to the power in mW:

$$\begin{aligned} P_{dBm} &= 10\log(P_{mW}) \\ -30\text{ dBm} &= 10\log(P_{mW}) \\ \Rightarrow P_{out,mW} &= 10^{-3}\text{ mW} \end{aligned} \tag{1.49}$$

Increasing this power by 3 dB is equivalent to doubling the power, so the increase is 10^{-3} mW (a small number).

b. Using equation (1.34) again produces

$$\begin{aligned} +40\text{ dBm} &= 10\log(P_{out,mW}) \\ \Rightarrow P_{out,mW} &= 10^{4}\text{ mW} \\ \Rightarrow P_{out,W} &= 10\text{ watts} \end{aligned} \tag{1.50}$$

Doubling the output power represents a 10 watt increase in the amplifier's output power (a big number).

The lesson is that a 3 dB power increase is easy to accomplish at relatively small power levels because it doesn't represent much power. At high output power levels, unit decibel increases amount to large increases in output power.

1.4.6 Other Standards

Later, we will need other quantities expressed as decibels. These quantities don't fit the definition of a decibel because they aren't power ratios, but they are still useful.

1.4.6.1 dBK

This is a temperature measurement. The reference is 1°K.[2]

$$\begin{aligned} 0\text{ dBK} &= 1^\circ\text{ kelvin} \\ \text{dBK} &= 10\log(\text{Temperature in}^\circ\text{K}) \end{aligned} \tag{1.51}$$

[2]A temperature expressed in kelvin is unitless, and 1K = −273.15°C. However, there is the opportunity for confusion as 100K may refer to 100,000 or a temperature of 100 Kelvin in the absence of context. To address possible confusion, we will refer to temperature expressed in Kelvin as °K, while the kilo prefix will be expressed as *k* (e.g., 100 kHz).

We will use dBK when we discuss amplifier noise, satellite communications and link budgets.

EXAMPLE

Convert 430°K to dBK.

Solution

Using equation (1.51)

$$\begin{aligned}\text{dBK} &= 10\log(\text{Temperature in K})\\ &= 10\log(430)\\ &= 26.3\text{ dBK}\end{aligned} \tag{1.52}$$

1.4.6.2 dBHz

This is a measure of bandwidth. The reference is 1 hertz.

$$\begin{aligned}0\text{ dBHz} &= 1\text{ hertz}\\ \text{dBHz} &= 10\log(\text{Bandwidth in hertz})\end{aligned} \tag{1.53}$$

We will need dBHz for discussions of noise bandwidth and receiver sensitivity.

EXAMPLE

Convert 30 kHz to dBHz.

Solution

Using equation (1.53)

$$\begin{aligned}\text{dBHz} &= 10\log(\text{Frequency in hertz})\\ &= 10\log(30,000)\\ &= 44.8\text{ dBHz}\end{aligned} \tag{1.54}$$

1.5 | FREQUENCY, WAVELENGTH, AND PROPAGATION VELOCITY

1.5.1 General Case

We relate frequency, wavelength, and propagation velocity by

$$\lambda f = v \tag{1.55}$$

where

λ = the wavelength of the signal
v = the velocity of propagation
f = the frequency of the signal

Equation (1.55) is the most general form of the equations relating wavelength to propagation velocity. The period T of a wave is

$$T = \frac{1}{f} \tag{1.56}$$

1.5.2 Free Space

When a wave is traveling through empty space or through air, the velocity of propagation [v in equation (2.36)] is the speed of light, commonly denoted as c and

$$\begin{aligned} c &= 2.9979 \cdot 10^8 \frac{\text{meters}}{\text{second}} \\ &\approx 3 \cdot 10^8 \frac{\text{meters}}{\text{second}} \end{aligned} \tag{1.57}$$

so equation (1.56) becomes

$$\lambda_0 f = c \tag{1.58}$$

where

λ_0 = the wavelength GHz in free space.

EXAMPLE

Wavelength and Frequency

Find the wavelengths of the following frequencies in the atmosphere: 1 MHz (low high-frequency [HF] band), 20 MHz (upper HF band), 100 MHz (middle very high frequency [VHF] band), 150 MHz (the 2 meter band), 300 MHz, 600 MHz, 1 GHz, 3 GHz, 10 GHz, 30 GHz, and 100 GHz.

Solution

Since we are discussing signals that are propagating through the atmosphere, we can assume the wave is moving at the speed of light and that equation (1.58) applies.

TABLE 1-3 The relationship between frequency and wavelength.

	Wavelength			
Frequency (MHz)	Meters	Yards	Feet	Inches
1	300.0	328.0	984.0	11,800.0
20	15.0	16.4	49.2	591.0
100	3.0	3.28	9.84	118.0
150	2.0	2.19	6.56	78.7
300	1.0	1.10	3.28	39.4
600	0.50	0.547	1.64	19.7
1,000	0.30	0.328	0.948	11.8
3,000	0.10	0.109	0.328	3.94
10,000	0.03	0.0328	0.0948	1.18
30,000	0.01	0.0109	0.0328	0.394
100,000	0.003	0.00328	0.00948	0.118

It is convenient to remember that 300 MHz equates to 1 meter of wavelength.

1.5.3 The Speed of Light

The speed of light is 2.998E8 meters/second, and there are 39.37 inches in a meter; thus, we can write

$$c = \left(2.998 \cdot 10^8 \frac{\text{meters}}{\text{second}}\right)\left(39.37 \frac{\text{inches}}{\text{meter}}\right) = 11.80 \frac{\text{inches}}{\text{nanosecond}} \tag{1.59}$$

or

$$c \approx 1 \frac{\text{ft}}{\text{nanosecond}} \tag{1.60}$$

With less than a 2% error, we can say that light travels about 1 foot in 1 nanosecond (or conversely, it takes about 1 nanosecond for light to travel 1 foot).

EXAMPLE

Satellite Delay Time

A geostationary satellite is 35.863E6 meters (= 22,284 miles or $117.66 \cdot 10^6$ feet) above the earth. How long does it take a signal to travel from the surface of the earth to the satellite and back to the earth again?

Solution

Since the speed of light = $c = 2.9979 \cdot 10^8$ meters/second, it takes

$$\frac{(2)(35.863 \cdot 10^6)\text{ meters}}{2.9979 \cdot 10^8 \text{ meters/second}} = 0.239 \text{ seconds} \tag{1.61}$$

to travel the distance. Using our 1 nsec/foot rule of thumb, we find

$$(2)(35.863 \cdot 10^6 \text{ meters})\left(\frac{1 \text{ foot}}{0.304 \text{ meters}}\right)\left(\frac{1 \text{ nsec}}{\text{foot}}\right) = 0.236 \text{ sec} \tag{1.62}$$

1.5.4 Physical Size

The physical size of a system or device expressed in wavelengths is an important quantity. For example:

- Wires begin to act like antennas when their physical sizes approach wavelength dimensions (typically when their physical dimensions get larger than $\lambda/10$ or so).
- Wires begin to look like transmission lines when the physical length of the wire is of the same order of magnitude as a wavelength ($\lambda/15$ is the rule of thumb).
- If two wires carrying a signal are physically separated by $\lambda/5$, they begin to act like an antenna and will radiate energy into space.
- When the manufacturing tolerances of a device approach a tenth of a wavelength, the device will begin to misbehave. Connectors will exhibit excessive losses; antennas will exhibit gain and sidelobe variations.

We will encounter this effect many times as we discuss wireless systems. The behavior of an antenna depends strongly on its physical size in wavelengths. Table 1-4 relates frequency to wavelength and physical size:

TABLE 1-4 ■ The relationship between frequency and physical dimensions of an object.

f	λ	λ/2π	λ/20
10 Hz	30,000 km	4,800 km	1,500 km
60 Hz	5,000 km	800 km	250 km
100 Hz	3,000 km	480 km	150 km
400 Hz	750 km	120 km	38 km
1 kHz	300 km	48 km	15 km
10 kHz	30 km	4.8 km	1.5 km
100 kHz	3 km	480 m	150 m
1 MHz	300 m	48 m	15 m
10 MHz	30 m	4.8 m	1.5 m
100 MHz	3.0 m	0.48 m	15 cm
1 GHz	30 cm	4.8 cm	1.5 cm
10 GHz	3.0 cm	4.8 mm	1.5 mm

where

f = frequency
λ = wavelength
$\lambda/2\pi$ = the boundary between antenna near and far fields
$\lambda/20$ = antenna effects begin to occur in wires and slots

1.6 TRANSMISSION LINES

Transmission line effects occur when the physical size of a system approaches a wavelength at the frequency of operation. An equivalent description is that the time a signal takes to travel between components is nonzero.

Figure 1-4 illustrates the point. We drive a 300 MHz signal to two receivers: one located 1 inch away from the driver and the second located 7 inches away.

Signals travel more slowly in a transmission line than they travel in free space. We will assume the signal travels at 2E8 meters/second so it requires 0.127 ns to travel 1 inch. The signal at point B arrives at $t = 0.127$ ns. The signal at point C arrives at $t = 0.889$ ns. The difference in propagation time between A-B and A-C is 0.762 ns. This time is an appreciable fraction of the 3.33 ns period of the 300 MHz clock. The digital logic at point A and point B would not see the rising clock edge at the same instant. This propagation delay problem (termed *clock skew*) is very important in high-speed computers.

War Story—Transmission Line Propagation Delay

Open up an old computer and examine the memory printed circuit boards (PCBs). Most memory PCBs contain several integrated circuits (ICs) that must be clocked at the same time. The time-critical signals will enter the memory PCB from the motherboard and then will be run to the individual memory ICs. Some lines run in a serpentine, back-and-forth manner so that the propagation delay to each IC is identical.

When they were first building radars early in World War 2, the designers needed to generate precisely timed, very high-voltage pulses to power their transmitters. For example, a system might need a 20 kV, 1 μsec pulse. The engineers built special high-voltage transmission lines whose propagation delay times were 1 μsec. They would charge

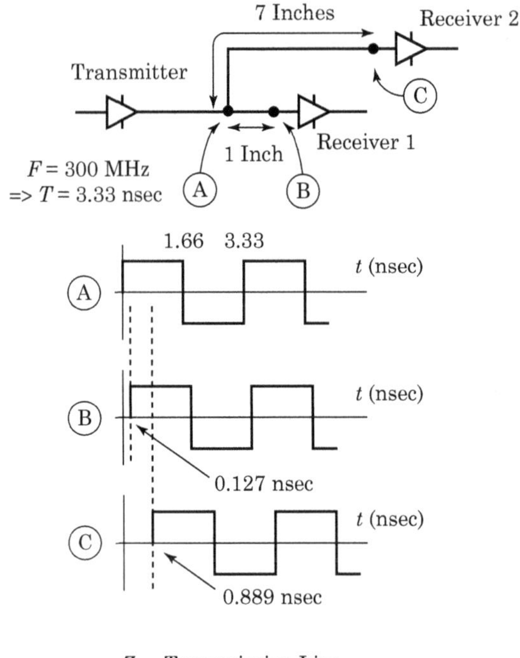

FIGURE 1-4 Clock skew in digital logic caused by unequal line lengths.

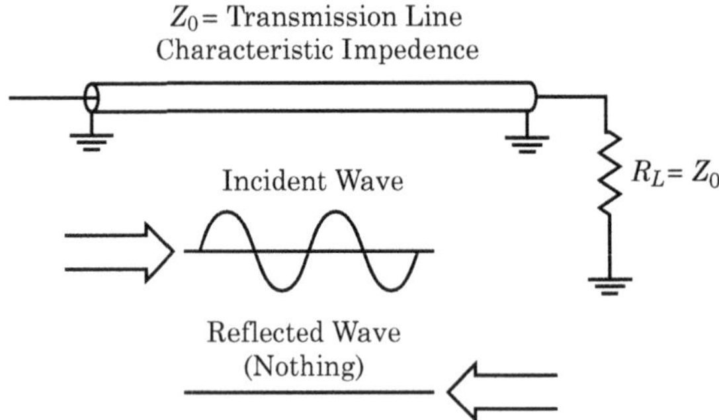

FIGURE 1-5 Transmission line reflection for a matched load. The signal incident on the matched load is completely absorbed by the matched load.

the line up to 20 kV and then connect the line to the high-voltage sink at the instant that they needed the pulse. The sink would pull the energy from the transmission line until the line dissipated, that is, for exactly 1 μsec.

1.6.1 Characteristic Impedance

Characteristic impedance, denoted Z_0, is the most important property of a transmission line. We'll come to think of characteristic impedance as the impedance at which our RF systems operate.

Terminating a transmission line with a resistor whose value is the characteristic impedance of the line allows the resistor to absorb completely any signal we push into the line, as in Figure 1-5. If the transmission line is terminated with a resistor whose value is not Z_0, then the load resistor will not dissipate the entire signal. The resistor will absorb some of the energy, and the rest will be reflected back into the line, as shown in Figure 1-6. These reflections have profound effects on system performance.

A transmission line's characteristic impedance is a function of its physical geometry and the materials used to build it. We'll look at two common examples.

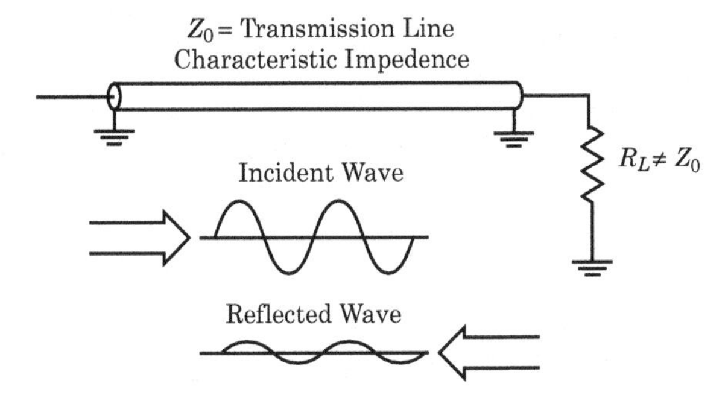

FIGURE 1-6 ■ Transmission line reflection for a mismatched load. Part of the signal incident on the mismatched load is absorbed by the load, but some of the signal reflects from the mismatch and returns to the source.

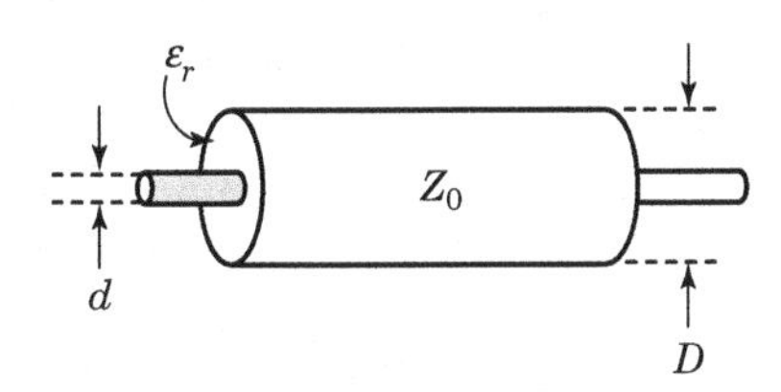

FIGURE 1-7 ■ The construction details of a coaxial cable.

1.6.1.1 Coaxial Cable

Figure 1-7 shows the physical configuration of a coaxial cable. The characteristic impedance is

$$Z_0 = \frac{138}{\sqrt{\varepsilon_R}} \log\left(\frac{D}{d}\right) \tag{1.63}$$

where

ε_R is the permittivity or dielectric constant of the insulator
d is the outer diameter of the inner conductor
D is the inner diameter of the outer conductor

The characteristic impedance of the cable depends on the permittivity of the insulator and the relative diameters of the outer conductor and the center conductor.

War Story

Equation (1.63) tells us that the ratio of diameters of the outer conductor to the inner conductor strongly determines the characteristic impedance of a coaxial cable. The characteristic impedance of a cable will change if someone steps, kinks, or otherwise physically damages the cable. Small changes in the D/d ratio result in large changes in Z_0.

1.6.1.2 Twin Lead Cable

Figure 1-8 shows the physical configuration of a twin lead cable. The characteristic impedance is

$$Z_0 = \frac{276}{\sqrt{\varepsilon_R}} \log\left(\frac{2D}{d}\right) \tag{1.64}$$

where

ε_R is the permittivity or dielectric constant of the insulator
d is the diameter of each wire
D is the center-to-center distance between the conductors

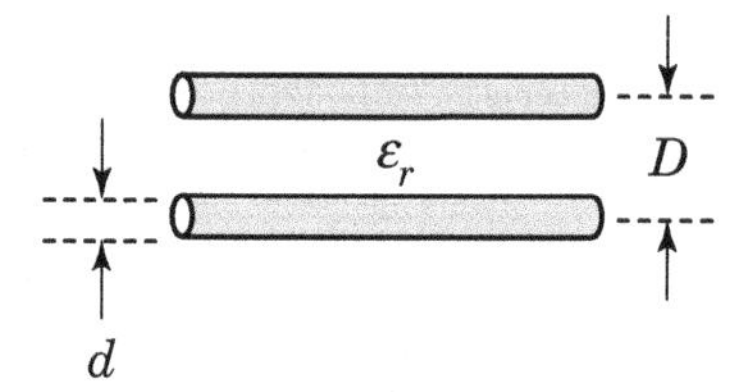

FIGURE 1-8 ■ The construction details of a twin lead cable.

Again, the characteristic impedance of the cable depends on the permittivity of the insulator and the physical configuration of the two wires.

War Story

Equation (1.64) produces the Z_0 of the cable shown in Figure 1-8. The conductors in twin lead are not shielded from the outside world, as is the inner conductor of coaxial cable. Equation (1.64) is entirely different if the cable is near a metal sheet.

One of the authors worked in a popular electronics store in college, where they sold both coaxial and twin lead TV antenna cable. The return rate on the twin lead cable was much higher than the return rate on the coaxial cable. We eventually reasoned that people would run the cable along their aluminum siding and that the proximity of the metal would change the characteristic impedance of the twin lead more than it would change the Z_0 of the coaxial cable. Since the antenna was operating into an unmatched characteristic impedance, signal loss occurred, and the cable was judged to be of poor quality.

1.6.2 Transmission Lines and Pulsed Input Signals

Figure 1-9 shows a transmission line experiment. We are feeding a lossless transmission line, whose characteristic impedance is Z_0, with a very narrow pulse while we vary the load resistor R_L. We will plot V_{in}, the voltage at the input of the transmission line, over time.

1.6.2.1 Open-Circuit Line

Plot (b) of Figure 1-9 shows V_{in} when the load is an open circuit ($R_L = \infty$). We observe two pulses at point A: the source-generated pulse at $t = 0$ and a second pulse. The second pulse has the same magnitude as the first pulse but appears at $t = t_d$.

We explain the second pulse by considering the transmission line as a simple time delay. The initial pulse travels through the cable until it encounters the open circuit at the load end. The pulse reflects off the impedance discontinuity and travels back in the direction it came. The second pulse is traveling from the load end of the cable toward the generator.

The time t_d is the time required to travel down the cable and back again. We can use this technique to determine the cable's propagation velocity.

When the pulse first enters the transmission line, but before the return pulse has had time to return from the open circuit, the transmission line appears to be a resistor whose value is its characteristic impedance. Voltage division between R_S and the cable's characteristic impedance Z_0 causes the initial input voltage to be one-half the input voltage.

1.6.2.2 Short-Circuit Line

Plot (c) of Figure 1-9 shows V_{in} when we short-circuit the load end of the transmission line. Everything is the same except that the return pulse is in the opposite polarity. The short circuit at the end of the cable causes the pulse to come back inverted.

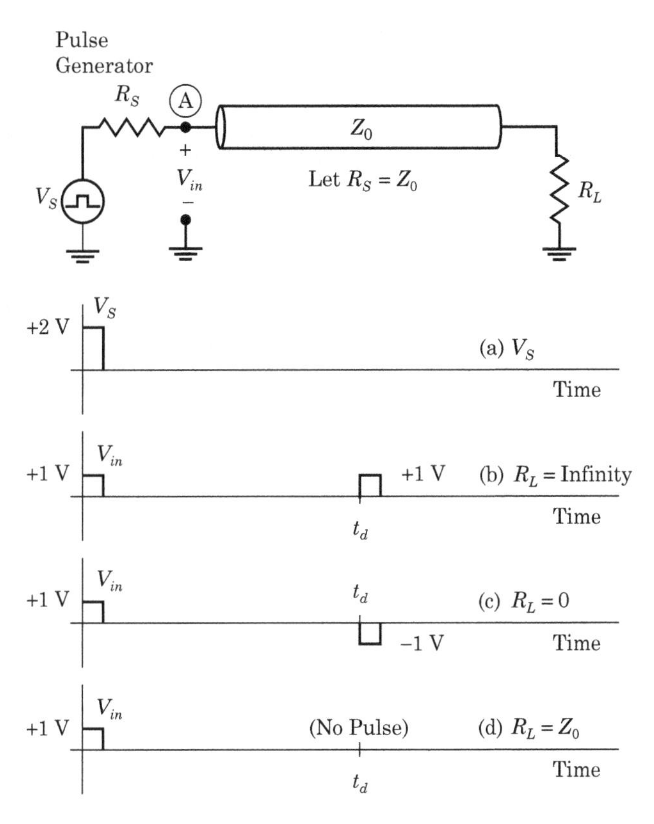

FIGURE 1-9 Transmission line under pulsed conditions with various values of loading. We will set V_S equal to 2 volts, and we will set the source impedance, R_S, equal to the characteristic impedance of the cable.

1.6.2.3 Z_0 Terminated Line

Plot (d) of Figure 1-9 shows the effect of terminating the cable with a resistor whose value equals the characteristic impedance of the cable. No pulse ever returns because the matched load resistor absorbs all of the energy in the pulse. This is a desirable situation and is one of the reasons we like to keep our system impedances matched.

1.6.3 Propagation Velocity or Velocity Factor

Waves travel slower in a transmission line than they travel in free space. The propagation velocity in any medium is v.

Velocity factor is the ratio of the propagation velocity of a wave in a transmission line to the wave's velocity in free space or

$$\begin{aligned} v_f &= \frac{v}{c} \\ &= \frac{t_0}{t_d} \\ &\leq 1 \end{aligned} \tag{1.65}$$

where

v_f = the velocity factor of the transmission line ($0 \leq v_f \leq 1$)
v = the propagation velocity in the transmission line
c = the speed of light
t_0 = the propagation time in free space
t_d = the propagation time in the transmission line

EXAMPLE

Measuring Velocity Factor

We launch an 8 nsec long pulse into a 6 foot long piece of RG-223 coaxial cable. The load end of the cable is an open circuit. The return pulse came back 19.1 nsec after we launched the first pulse into the cable. What is the velocity factor of this piece of cable?

Solution

The 19.1 nsec time between pulses is the time required for the pulse to travel down the cable and return. The one-way travel time is 19.1/2 = 9.55 ns. Our pulse traveled 6 feet in 9.55 ns. Using equation (1.60), our pulse would travel the 6 feet in 6 ns if it were traveling in free space, so the velocity factor of the cable is

$$\begin{aligned} v_f &= \frac{t_0}{t_d} = \frac{6 \text{ nsec}}{9.55 \text{ nsec}} \\ &= 0.63 \end{aligned} \tag{1.66}$$

1.6.4 Guide Wavelength

Waves move slower than the speed of light inside a transmission line, so the wavelength inside the transmission line (the guide wavelength $= \lambda_g$) will be shorter than the wavelength in free space. The guide wavelength is

$$v_f = \frac{\lambda_g}{\lambda_0} \tag{1.67}$$

where

v_f = the velocity factor of the transmission line ($0 \le v_f \le 1$)
λ_g = the guide wavelength in the transmission line
λ_0 = the wavelength in free space

Equation (1.58) relates λ_0 to the frequency.

1.7 DESCRIPTIONS OF IMPEDANCE

The notion of impedance permeates radio frequency design. Many of the properties of our systems are related directly to the terminal impedances of our components, referenced to the system impedance. The environment in which we're operating determines the system impedance, and it is always some standard—often 50 Ω or 75 Ω.

The system impedance is set by the impedances of the test gear we have available and by the characteristic impedance of the transmission lines with which we connect the components.

We have many ways to describe impedance.

1.7.1 Reflection Coefficient

Open and short circuits are extreme conditions. If we place an arbitrary complex load at the load end of a transmission line, the magnitude of the reflected pulse will be

$$V_{Reflected} = \rho V_{Incident} \tag{1.68}$$

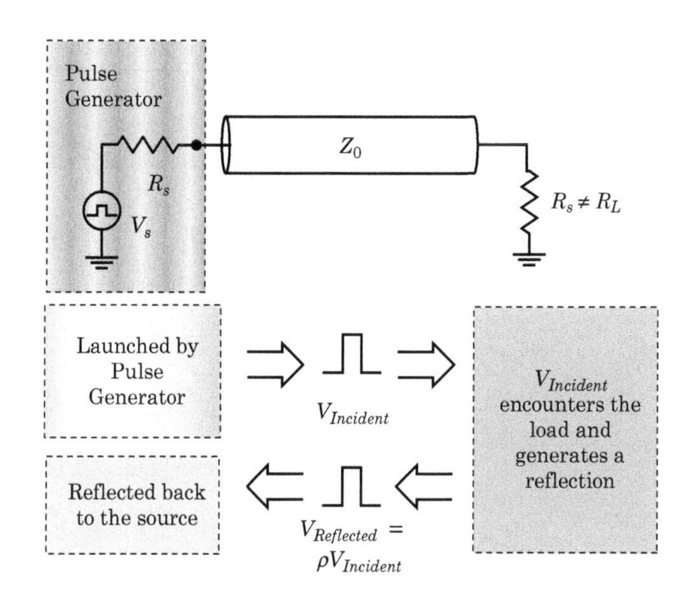

FIGURE 1-10 ■ Transmission line reflection coefficient.

where

ρ = complex reflection coefficient. The reflection coefficient will be a complex number if the load impedance is complex.

$V_{Incident}$ = the complex voltage incident on the load. In other words, it's the magnitude of the pulse we sent into the transmission line. $V_{Incident}$ travels from the source to the load.

$V_{Reflected}$ = the complex voltage reflected off the load and back into the transmission line. $V_{Reflected}$ travels from the load to the source.

See Figure 1-10. We can show

$$\rho = \frac{Z_L - Z_0}{Z_L + Z_0} \qquad -1 \leq |\rho| \leq +1 \tag{1.69}$$

where

ρ = the complex reflection coefficient

Z_L = the complex impedance that we are comparing to the characteristic impedance Z_0

Z_0 = the characteristic impedance of the system in which we are operating

Figure 1-11 shows a graph of real load impedance vs. the reflection coefficient in a 50 Ω system. At low values of R_L, the reflection coefficient approaches –1; most of the voltage we send down the cable reflects back toward the source (although the magnitude of the pulse is reversed). When R_L is large, ρ approaches +1, and most of the voltage reflects off the load but the sign stays the same.

When the value of the load resistor equals the transmission line's characteristic impedance, the reflection coefficient equals zero. None of the energy we send into the cable comes back—all of the energy dissipates in the load resistor.

1.7.1.1 Transmitting Systems

In transmitting systems, we send power into an antenna so the power will radiate away into free space. If the impedance of the transmitting antenna isn't matched to the impedance of the transmission line, some of the energy we are sending into the cable reflects off antenna and is not radiated into space. We've lost efficiency.

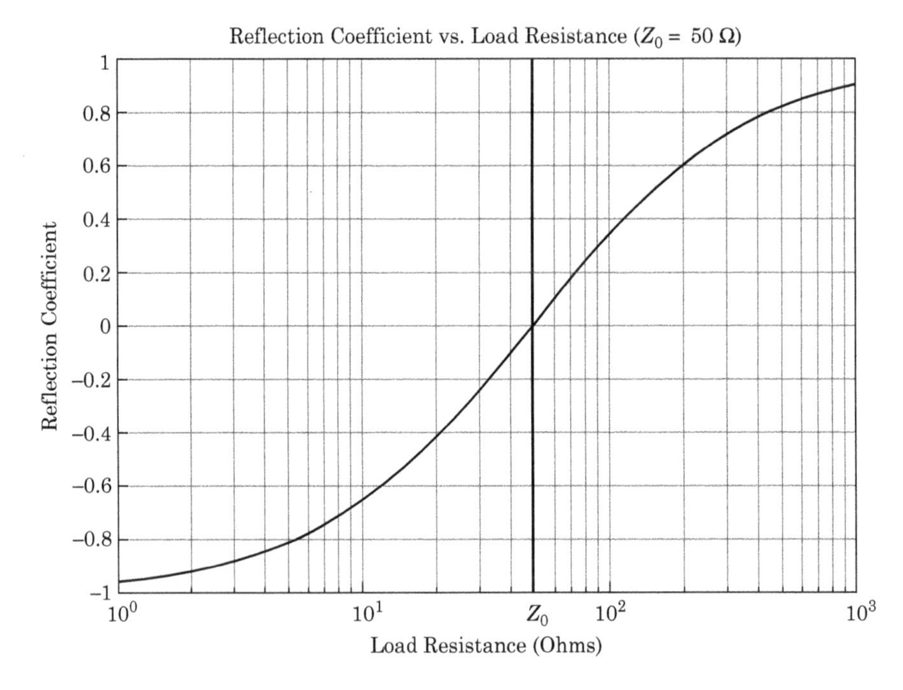

FIGURE 1-11 ■ Transmission line reflection coefficient versus load resistance for a 50 Ω system. Load resistance is real valued load impedance.

1.7.1.2 Receiving Systems

In receiving systems, we feed a receiver with an antenna. Any impedance mismatch between the antenna and the receiver will cause energy to reflect off the mismatch. Signal energy will be lost, and our receiving system will lose sensitivity.

EXAMPLE

Reflection Coefficient

Find the reflection coefficient for the following load impedances in a 50 Ω system:

a. 10 Ω
b. 125 Ω
c. 50 Ω
d. $20 - j40\ \Omega$
e. $50 + j20\ \Omega$

Solution

Liberally applying equation (1.69) produces

a.
$$\rho = \frac{10 - 50}{10 + 50} = -0.667 \tag{1.70}$$

A real impedance less than Z_0 produces a real, negative reflection coefficient.

b.
$$\rho = \frac{125 - 50}{125 + 50} = 0.429 \tag{1.71}$$

A real impedance greater than Z_0 produces a real, positive reflection coefficient.

c.

$$\rho = \frac{50 - 50}{50 + 50} = 0 \tag{1.72}$$

A real impedance equal to Z_0 produces a zero valued reflection coefficient.

d.

$$\begin{aligned}\rho &= \frac{(20 - j40) - 50}{(20 - j40) + 50} \\ &= \frac{-30 - j40}{70 - j40} \\ &= 0.077 - j0.615 \\ &= 0.620\angle -97^\circ\end{aligned} \tag{1.73}$$

A complex impedance produces a complex reflection coefficient.

e.

$$\begin{aligned}\rho &= \frac{(50 + j20) - 50}{(50 + j20) + 50} \\ &= \frac{j20}{100 + j20} \\ &= 0.039 + j0.192 \\ &= 0.196\angle -79^\circ\end{aligned} \tag{1.74}$$

EXAMPLE

Mismatched Transmission Line

Figure 1-12 shows a signal source driving a transmission line.

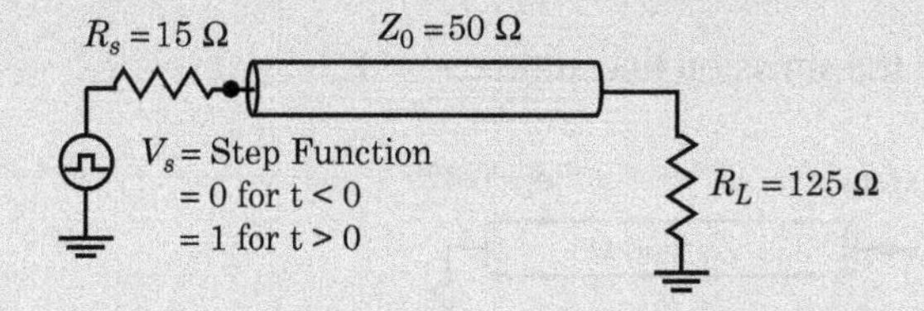

FIGURE 1-12 ■ Example of a mismatched transmission line. The transmission line is mismatch on both the source and load ends.

The signal source is a step function, that is,

$$\begin{aligned}V_S &= 0 \text{ volts for } t < 0 \\ V_S &= 1 \text{ volt for } t \geq 0\end{aligned} \tag{1.75}$$

and $Z_0 = 50\ \Omega$, $R_S = 15\ \Omega$, and $R_L = 125\ \Omega$. The time for a wave to propagate from one end of the transmission line to the other is τ seconds. Plot the voltage versus distance over the length of the cable for $t = 0.3\tau, 1.3\tau, 2.3\tau$, etc.

Solution

We use equation (1.69) to find the source and load reflection coefficients. The source reflection coefficient is

$$\rho_S = \frac{R_L - Z_0}{R_L + Z_0} = \frac{15 - 50}{15 + 50} = -0.538 \tag{1.76}$$

The load reflection coefficient is

$$\begin{aligned}\rho_L &= \frac{R_L - Z_0}{R_L + Z_0} \\ &= \frac{125 - 50}{125 + 50} \\ &= 0.429\end{aligned} \tag{1.77}$$

Before any reflections propagate down the cable and back again, the transmission line looks like a resistive element with a value of Z_0. So when the generator turns on at $t = 0$, the transmission line looks like a 50 Ω resistor to the source. Using voltage division, the magnitude of the incident wave propagating down the transmission line is

$$\begin{aligned}V_{Incident} &= V_S \frac{Z_0}{R_S + Z_0} \\ &= (1 \text{ volt}) \frac{50}{15 + 50} \\ &= 0.770 \text{ volts}\end{aligned} \tag{1.78}$$

The graph labeled $t = 0.3\tau$ in Figure 1-13 shows the initial pulse traveling from the source to the load.

At $t = \tau$, the incident pulse encounters the load resistor. The load resistor absorbs some of the incident pulse's energy and reflects the rest. Equation (1.68) relates the incident and reflected voltage, and we know the magnitude of the incident voltage is 0.770 V while the load reflection coefficient is 0.429. The reflected voltage is

$$\begin{aligned}V_{Reflected} &= \rho_L V_{Incident} \\ &= (0.429)(0.770V) = 0.330 \text{ V}\end{aligned} \tag{1.79}$$

Figure 1-13 shows the state of the transmission line at time $t = 1.3\tau$.

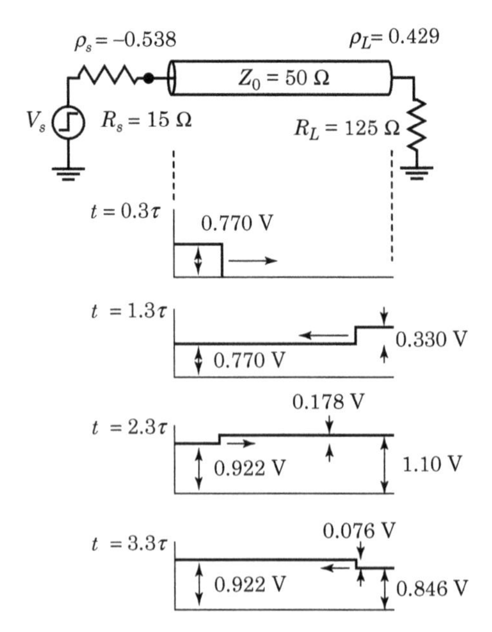

FIGURE 1-13 ■ Snapshots of voltages present on a mismatched transmission line at times $t = 0.3\tau$, $t = 1.3\tau$, $t = 2.3\tau$ and $t = 3.3\tau$.

The next interesting event occurs at $t = 2\tau$. The 0.330 V wave traveling from the load encounters the source. At this discontinuity, the incident voltage is 0.330 volts, and the reflection coefficient is −0.538. Equation (1.68) gives us the reflected voltage

$$\begin{aligned} V_{Reflected} &= \rho_S V_{Incident} \\ &= (-0.538)(0.330V) \\ &= -0.178 \text{ V} \end{aligned} \tag{1.80}$$

Figure 1-13 shows the state of the transmission line at time $t = 2.3\tau$.

At $t = 3\tau$, the −0.178 volt waveform transient is incident on the load. The value of the reflected wave is

$$\begin{aligned} V_{Reflected} &= \rho_L V_{Incident} \\ &= (0.429)(-0.178V) \\ &= -0.076 \text{ V} \end{aligned} \tag{1.81}$$

Figure 1-13 shows the state of the transmission line at time $t = 3.3\tau$.

If we continue this process, we will find that the voltage on the transmission line eventually settles down to what we would expect from a simple direct current (DC) analysis. In other words, after all the transients have died down, the final voltage on the transmission line will be

$$\begin{aligned} V_{Final} = V_{DC} &= V_S \frac{R_L}{R_L + R_S} \\ &= (1 \text{ V}) \left(\frac{125}{15 + 125} \right) \\ &= 0.893 \text{ V} \end{aligned} \tag{1.82}$$

We can arrive at this result either through simple DC analysis or by summing up the reflections on the transmission line until they die out. The result is the same.

1.7.1.3 Complex Reflection Coefficient

We have spent our time examining resistive cable terminations. The astute reader may wonder what happens if we decided to terminate a cable with an inductor or capacitor.

The answer is nothing special. Everything we have discussed still applies, except that the reflection coefficient has now become a complex number; it now has a magnitude and a phase angle. With a resistive load, the reflection coefficient is always real. With a complex load, the time-domain examples have certainly become harder to visualize because of the frequency-dependent nature of the load, but the effects are identical.

Since the impedance of a capacitor or inductor changes with frequency, the complex reflection coefficient will also change with frequency. Different frequency components of a signal will experience different phase and amplitude changes as they reflect off the complex load.

For a complex load, $Z_L = R_L + jX_L$, the reflection coefficient is

$$\begin{aligned} \rho &= |\rho| \angle \theta_\rho \\ &= \frac{Z_L - Z_0}{Z_L + Z_0} \end{aligned} \tag{1.83}$$

where

$Z_L = R_L + jX_L =$ the complex load (resistance and reactance)
$\rho =$ the complex reflection coefficient

$|\rho|$ = the magnitude of the complex reflection coefficient
θ_ρ = the angle of the complex reflection coefficient

A complex ρ means that the incident wave suffers a phase change as well as a magnitude change when it encounters the mismatch.

The equations we are developing for return loss and voltage standing wave ratio (VSWR), for example, will apply for complex reflection coefficient. We have been careful to specify the magnitude wherever it was required. If the magnitude is not specified, use the complex reflection coefficient and be prepared for a complex result.

1.7.2 VSWR

The notion of VSWR as a description of impedance arises directly from the effects of mismatched impedances on signals in transmission lines. The term evolved into a common expression of impedance even when there is no transmission line immediately present.

1.7.2.1 Transmission Lines and Sine Wave Input Signals

We have discussed the behavior of transmission lines under pulsed input signal conditions. Our discussions have centered on the behavior of pulses as they move up and down the transmission line and when they encounter impedance discontinuities.

Since a pulse is made up of a series of sine waves, we can argue that we will observe the same effects if we launch a continuous RF carrier (i.e., a sine wave) down the transmission line. Reflections occur, and the mathematical description is identical to the pulsed conditions. However, because of the continuous nature of the sine wave, the effects will not be as intuitive as with the pulsed case.

Figure 1-14 shows an example. The voltage generator V_S is a sine wave source that turns on at time $t = 0$. The system is completely mismatched (i.e. $R_S \neq Z_0 \neq R_L$). When we turn the signal generator on at $t = 0$, the sine wave first encounters the mismatch between the signal generator's source impedance and the impedance of the transmission line. This effect acts to reduce the voltage incident on the line.

The sine wave travels down the line until it encounters the load resistor. The load absorbs some of the signal energy and reflects the rest according to the equation

$$V_{Reflected} = \rho V_{Incident} \tag{1.84}$$

The energy that reflects off the load travels back along the line toward the source. If the load is complex, then ρ is complex and the sine wave experiences a phase shift and a magnitude change.

At the source end of the line, the signal from the load splits into two pieces again as it encounters the mismatched source impedance. Some energy is absorbed by the source while the rest is reflected and sent back again toward the load.

This process of absorption and reflection at each end of the transmission line continues until the transients die out and the line reaches a steady-state condition. The voltage present at any particular point on the transmission line is the sum of the initial incident wave and all of the reflections.

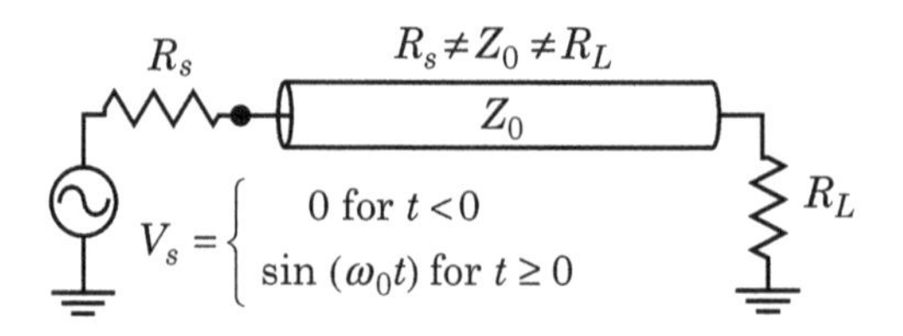

FIGURE 1-14 ■ A mismatched transmission line with a sinusoidal input voltage.

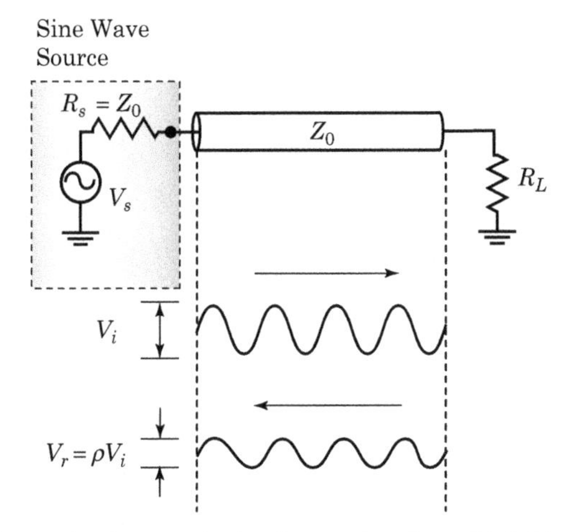

FIGURE 1-15 ■ A mismatched transmission line under sinusoidal drive showing the incident and reflected voltages. The incident sine wave, V_i, travels toward the load and reflects from the mismatched load resistor, forming $V_r = \rho V_i$. The reflected signal, V_r, is completely absorbed by the matched source resistor R_S.

The mathematical expression describing this process is quite complicated, involving an infinite series. To simplify our analysis, we will examine a transmission line with an arbitrary load resistance but with a matched source resistance (Figure 1-15). When we assume $R_S = Z_0$, we eliminate the reflections caused by a mismatch at the source end of the cable. The voltage at any physical point on the line is due to only the incident wave and one reflection from the load.

1.7.2.2 Voltage Minimums, Voltage Maximums

Figure 1-16 and Figure 1-17 show the magnitudes of the sine waves we will observe along a length of transmission lines for several load mismatch conditions. Figure 1-16 describes loads that are greater than Z_0 (or 50 Ω, in this case), while Figure 1-17 describes the situation when the loads are less than 50 Ω.

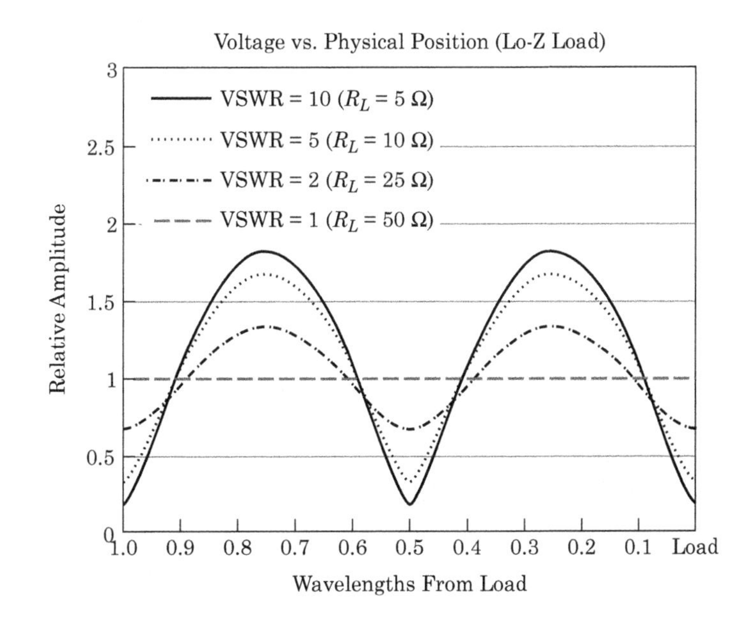

FIGURE 1-16 ■ VSWR. The magnitude of the voltages present on a mismatched transmission line when $Z_L > Z_0$.

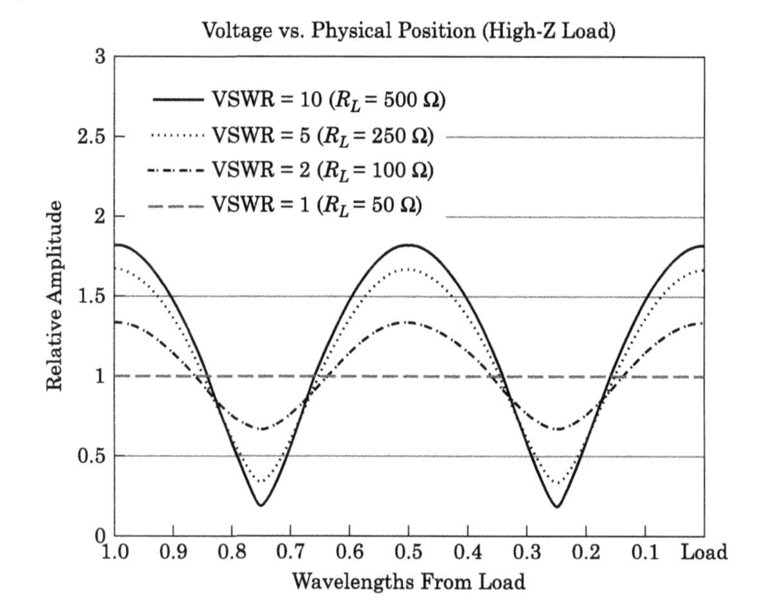

FIGURE 1-17 ■ VSWR. The magnitude of the voltages present on a mismatched transmission line when $Z_L < Z_0$.

EXAMPLE

Transmission Line Voltages (Hi-*Z* Load)

Assume we are operating a 50 Ω transmission line with a 250 Ω load resistance. What are the time-domain voltages we would observe at the following distances from the load?

a. $0.10\ \lambda_G$
b. $0.25\ \lambda_G$
c. $0.40\ \lambda_G$
d. $0.50\ \lambda_G$
e. $0.60\ \lambda_G$
f. $0.65\ \lambda_G$

Solution

Reading from the $R_L = 250\ \Omega$ (or VSWR = 5) curve of Figure 1-16, we find

a. $1.37 \cos(\omega t + \phi_1)$
b. $0.37 \cos(\omega t + \phi_2)$
c. $1.37 \cos(\omega t + \phi_3)$
d. $1.65 \cos(\omega t + \phi_4)$
e. $1.37 \cos(\omega t + \phi_5)$
f. $1.00 \cos(\omega t + \phi_6)$

where

$\omega = 2\pi f$ = the frequency of operation
ϕ_n = a phase constant depending on the frequency and upon the distance from the source to the load

EXAMPLE

Transmission Line Voltages (Low-*Z* Load)

Assume we are operating a 50 Ω transmission line with a 10 Ω load resistance. What are the time-domain voltages we would observe at the following distances from the load?

a. $0.10\ \lambda_G$
b. $0.25\ \lambda_G$
c. $0.40\ \lambda_G$
d. $0.50\ \lambda_G$
e. $0.60\ \lambda_G$
f. $0.65\ \lambda_G$

Solution

Reading from the $R_L = 10\ \Omega$ (or VSWR = 5) curve of Figure 1-17, we find

a. $0.90\cos(\omega t + \phi_1)$
b. $1.35\cos(\omega t + \phi_2)$
c. $0.90\cos(\omega t + \phi_3)$
d. $0.65\cos(\omega t + \phi_4)$
e. $0.90\cos(\omega t + \phi_5)$
f. $1.12\cos(\omega t + \phi_6)$

where

$\omega = 2\pi f$ = the frequency of operation
ϕ_n = a phase constant depending on the frequency and upon the distance from the source to load

Given a line with a mismatched load, we'll define the largest voltage present on the line as V_{max} and the smallest voltage present on the line as V_{min} or

$$\begin{array}{c}\text{Largest voltage on the T-Line} \Rightarrow V_{max}\cos(\omega t + \phi_1) \\ \text{Smallest voltage on the T-Line} \Rightarrow V_{min}\cos(\omega t + \phi_2)\end{array} \tag{1.85}$$

The values of V_{max} and V_{min} change with the value of the load resistor.

When the line is matched (when $R_L = Z_0$), then $V_{max} = V_{min}$. When the line is not matched, $V_{max} \neq V_{min}$. As the load's reflection coefficient increases (i.e., as the match gets worse), V_{max} and V_{min} become increasingly different.

1.7.2.3 VSWR

We refer to the ratio of the voltage maximum to the voltage minimum as the voltage standing wave ratio of the line:

$$VSWR = \frac{V_{max}}{V_{min}} \geq 1 \tag{1.86}$$

The concept of VSWR comes about naturally from our observations of physical transmission lines and from theory. At one time, VSWR was a very common unit of impedance

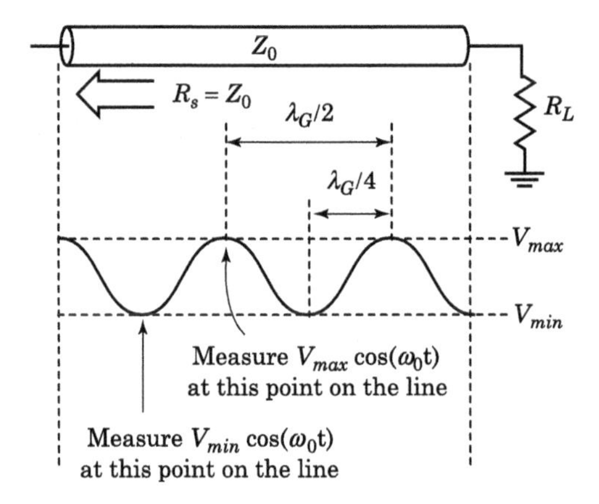

FIGURE 1-18 Measuring VSWR on a transmission line.

measurement because it was relatively easy to determine with simple equipment and without disturbing the system under test. Although it is derived from transmission line effects, VSWR is simply a measure of impedance—just like the reflection coefficient (ρ).

1.7.2.4 VSWR Relationships

A little algebra with V_{max}, V_{min}, ρ, Z_0, and R_L reveals the following relationships:

$$VSWR = \frac{1+|\rho|}{1-|\rho|} 1 \leq VSWR \leq \infty \quad 0 \leq |\rho| \leq 1 \tag{1.87}$$

and

$$|\rho| = \frac{VSWR - 1}{VSWR + 1} \tag{1.88}$$

Figure 1-16 and Figure 1-17 show another feature of the voltage peaks and valleys. As Figure 1-18 shows, the distance between two adjacent voltage maximums or two adjacent voltage minimum is $\lambda_G/2$ where λ_G is the wavelength in the transmission line or the guide wavelength. Also, the distance between a voltage minimum and its closest voltage maximum is $\lambda_G/4$.

War Story: VSWR and Microwave Ovens

We can think of a microwave oven as a transmission line system. The energy source is the magnetron tube inside of the microwave. The transmission line is the cavity in which we place the food to be heated, and the load is the water in the food.

As long as we have a load on the transmission line (i.e., we have food in the oven), energy transfers from the magnetron tube through the waveguide transmission line and into the food. We have a low VSWR inside of the microwave oven, and all is well.

However, if we operate the oven without an adequate load (i.e., empty), then the oven operates under high VSWR conditions. Areas of high voltages (large electric fields) inside the cavity will stress the magnetron tube. Also, since the energy from the magnetron meets a load with a high VSWR (or, equivalently, a high reflection coefficient), all of the energy sent out by the magnetron will bounce off the poor load and travel back into the tube. The magnetron now has to dissipate a large amount of heat and it can fail.

Even when we operate a microwave oven with food in the cavity, the load is still poorly characterized. In other words, if we think of a microwave oven and its load as a transmission line with a source and a load, the load is not exactly Z_0. After all, what is the impedance of a bologna sandwich? The system exhibits a nonunity VSWR.

From a practical point of view, this effect causes hot and cold spots in the oven. The hot spots are areas of high field energy, whereas the cold spots are areas of low field energy. Food that lies in a hot spot heats quickly, while food in a cold spots heats slowly.

To combat this problem, manufacturers often arrange to rotate thc food in the oven. Since the rotating food alternately passes through hot and cold spots in the cavity, the food heats more evenly. Another solution is to install a "mode stirrer" in the cavity. The mode stirrer is a metal propeller-like structure that rotates slowly in the cavity. It reflects the energy from the magnetron in different directions as it moves. This changes the positions of the hot and a cold spot in the oven, and the effect is to heat the food more evenly.

War Story: VSWR and Screen Room Testing

We must often perform sensitive electronic tests in a screen room. A screen room usually takes the form of a large metal room that shields the inside from the ambient electromagnetic fields on the outside of the room. The room has metal walls, doors, and ceilings with the appropriate fixtures to get power and air into the room.

Screen rooms are commonly used to test antennas or to measure the electromagnetic radiation emanating from a piece of equipment. Such tests are necessary to make sure one piece of equipment won't interfere with another.

One of the first questions we ask after constructing a screen room is how much shielding the room supplies. There are many such tests to verify the performance.

In one test, the operator places a transmitter inside the room, closes all the doors, then checks for radiation from the transmitter on the outside of the room.

Like the microwave oven example, we have to be concerned with the load that the transmitter experiences and with the VSWR inside the room. In an empty metal room, the transmitter effectively has no load, and we will experience a high VSWR condition inside the screen room. The effect of the high VSWR is to place large electromagnetic fields in some areas of the room and small electromagnetic fields in other areas of the room. If an area of small electromagnetic field happens to fall in an area where the screen room is leaky, then you may not detect the leak. If the area of strong electromagnetic field falls in the neighborhood of a good joint, you may detect that the joint is leaky due to the excess field placed across the joint by the high VSWR condition in the room.

Commercial gear gets around this problem by sweeping the transmitter slightly in frequency. The positions of the voltage maximums and minimums on a transmission line depend on the operating frequency (remember that the distances between voltage maxima and minima were $\lambda_G/2$). When we change the frequency, we change the position of the voltage maxima and minima, so we hope to obtain a more realistic measurement of the attenuation of the screen room.

1.7.2.5 Return Loss

Like VSWR and reflection coefficient, return loss is a measure of impedance with respect to the characteristic impedance of a transmission line.

We have discussed launching pulses down a transmission line. The load absorbed some of the pulse's energy, while some of the energy was reflected back toward the source. Return loss characterizes this energy loss in a very direct manner (see Figure 1-19).

When a known voltage is incident on a load ($V_{Incident}$), we can calculate an equivalent power incident on the load ($P_{Incident}$). Similarly, when voltage reflects from a load ($V_{Reflected}$), that voltage represents power reflected from the load ($P_{Reflected}$).

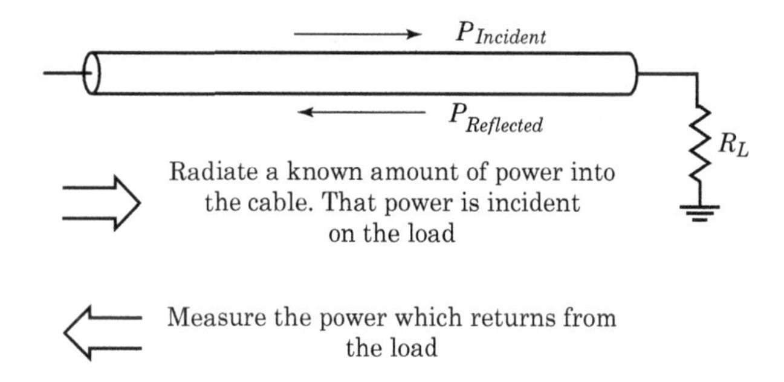

FIGURE 1-19 ■ Measuring return loss.

The definition for return loss is

$$Return\ Loss_{dB} = 10\log\left(\frac{P_{Reflected}}{P_{Incident}}\right) \leq 0\ dB \tag{1.89}$$

We often measure return loss directly by launching a known amount of power into a transmission line ($P_{Incident}$) and measuring the amount of power that reflects off the load ($P_{Reflected}$). Return loss measures the power lost when a signal is launched into a transmission line.

A little algebraic manipulation reveals

$$\begin{aligned} Return\ Loss_{dB} &= 20\log\left|\frac{V_{Reflected}}{V_{Incident}}\right| \\ &= 20\log|\rho_L| \\ &= 20\log\left|\frac{Z_L - Z_0}{Z_L + Z_0}\right| \end{aligned} \tag{1.90}$$

Figure 1-20 shows a plot of return loss versus load resistance for a 50 Ω system. When the load resistance is very small or very large, most of the power we launch into the transmission line reflects off the load. Since $P_{Reflected} \approx P_{Incident}$, the return loss is near 0 dB. This is a poorly matched condition.

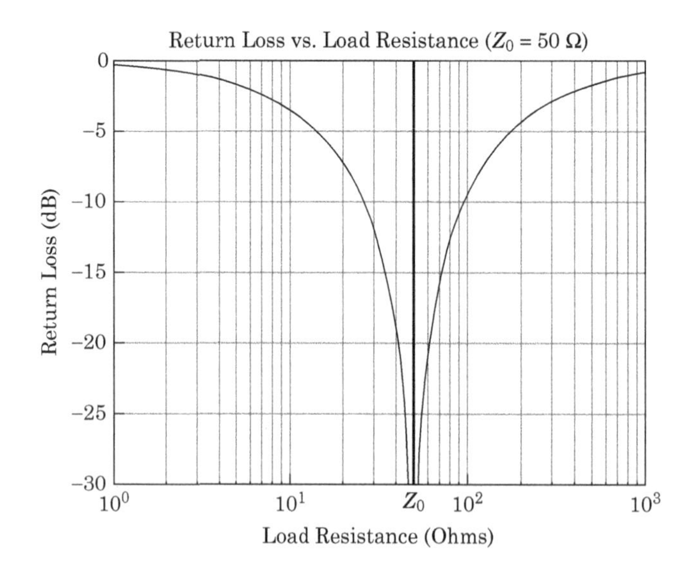

FIGURE 1-20 ■ Return loss versus load resistance in a 50 Ω system.

When the load resistance is approximately equal to the system characteristic impedance (Z_0, which is 50 Ω in this case), most of the power we launch into the line is absorbed by the load; very little returns. The return loss is a large negative number. A large negative number indicates a good match.

EXAMPLE

Reactive Terminations

Figure 1-21 shows a transmission line with a reactive termination. Find the reflection coefficient, return loss, and VSWR for this load. Assume a 50 Ω system and that the frequency of operation is 4 GHz.

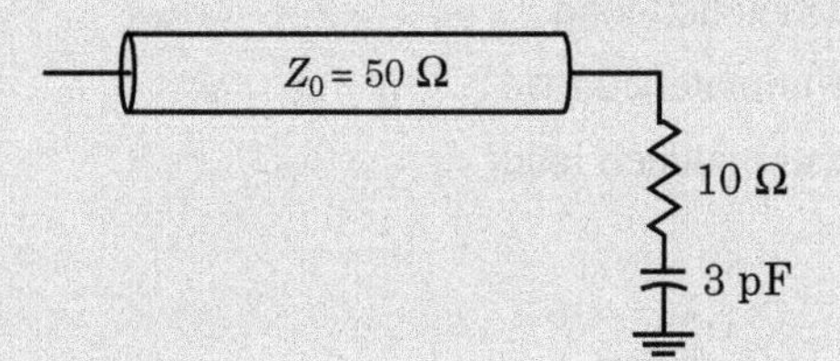

FIGURE 1-21 ■ Transmission line with a complex load.

Solution

The reactance of the load at 4 GHz is $10 - j13.3\Omega$ or $16.6\angle 53°$. Using equation (1.69) with $Z_L = 10 - j13.3\ \Omega$ and $Z_0 = 50\ \Omega$ produces

$$\begin{aligned} \rho &= \frac{(10 - j13.3) - 50}{(10 - j13.3) + 50} \\ &= \frac{-40 - j13.3}{60 - j13.3} \\ &= -0.589 - j0.352 \\ &= 0.686\angle -149° \end{aligned} \tag{1.91}$$

Note that the reflection coefficient is complex because of the complex load. Since the load impedance changes with frequency, the complex reflection coefficient changes with frequency.

Equations (1.88) and (1.90) relate the complex reflection coefficient to the VSWR and return loss:

$$\begin{aligned} \rho &= \frac{(10 - j13.3) - 50}{(10 - j13.3) + 50} \\ &= \frac{-40 - j13.3}{60 - j13.3} \\ &= -0.589 - j0.352 \\ &= 0.686\angle -149° \end{aligned} \tag{1.92}$$

and

$$\begin{aligned} \mathit{Return\ Loss}_{dB} &= 20\log(|\rho|) \\ &= 20\log(|0.686|) \\ &= -3.2\text{ dB} \end{aligned} \tag{1.93}$$

1.7.2.6 Mismatch Loss

The voltage reflecting off a mismatched load at the end of a transmission line represents power loss. If the load were perfectly matched to the transmission line, we would transfer all of the available incident signal power to the load resistor. However, since our load isn't matched to the transmission line's characteristic impedance, we don't transfer all the available power to the load. This is mismatch loss, which we define as

$$\begin{aligned} \textit{Mismatch Loss} &= \frac{P_{Available} - P_{Reflected}}{P_{Available}} \\ &= \frac{P_{Delivered}}{P_{Available}} \end{aligned} \tag{1.94}$$

where

$P_{Available}$ = the power delivered to the matched load
$P_{Delivered}$ = the power delivered to the unmatched load
$P_{Reflected}$ = the power reflected off the unmatched load

We know

$$P_{Available} = \frac{V_{Incident}^2}{Z_0} \tag{1.95}$$

and

$$P_{Reflected} = \frac{V_{Reflected}^2}{Z_0} \tag{1.96}$$

Combining equations (1.94) through (1.96) with equation (1.68) produces

$$\begin{aligned} \textit{Mismatch Loss}_{Load} &= \frac{V_{Incident}^2/Z_0 - V_{Reflected}^2/Z_0}{V_{Incident}^2/Z_0} \\ &= \frac{V_{Incident}^2 - V_{Reflected}^2}{V_{Incident}^2} \\ &= \frac{V_{Incident}^2 - (|\rho_L| V_{Incident})^2}{V_{Incident}^2} \\ &= 1 - |\rho_L|^2 \end{aligned} \tag{1.97}$$

Mismatch loss is yet another way of describing the relationship between the characteristic impedance of a transmission line and its load.

We usually specify mismatch loss in decibels

$$\textit{Mismatch Loss}_{Load,dB} = 10\log(1 - |\rho_L|^2) \tag{1.98}$$

1.7.2.7 Source and Load Mismatches

We can experience mismatch at both the source and load ends of a transmission line. When we have a source mismatch, we will not transfer the maximum amount of power into the line. The mismatch loss due to source mismatch is

$$\textit{Mismatch Loss}_{Source,dB} = 10\log(1 - |\rho_S|^2) \tag{1.99}$$

EXAMPLE

Mismatch Loss

Find the mismatch loss (in decibels) for the following resistors in a 50 Ω system:

a. 10 Ω ($\rho = -0.667$),
b. 25 Ω ($\rho = -0.333$),
c. 50 Ω ($\rho = 0$),
d. 100 Ω($\rho = 0.333$)
e. 250 Ω($\rho = 0.667$)
f. $20 - j100$ Ω($\rho = 0.530 - j0.671 = 0.855\angle 51.7°$).

Solution

Liberally applying equation (1.98) produces

a.

$$\begin{aligned} Mismatch\ Loss_{dB} &= 10\log(1 - |\rho|^2) \\ &= 10\log(1 - |-0.667|^2) \\ &= -2.6 \text{ dB} \end{aligned} \tag{1.100}$$

b.

$$\begin{aligned} Mismatch\ Loss_{dB} &= 10\log(1 - |\rho|^2) \\ &= 10\log(1 - |-0.333|^2) \\ &= -0.5 \text{ dB} \end{aligned} \tag{1.101}$$

c.

$$\begin{aligned} Mismatch\ Loss_{dB} &= 10\log(1 - |\rho|^2) \\ &= 10\log(1 - |0|^2) \\ &= 0.0 \text{ dB} \end{aligned} \tag{1.102}$$

d.

$$\begin{aligned} Mismatch\ Loss_{dB} &= 10\log(1 - |\rho|^2) \\ &= 10\log(1 - |0.333|^2) \\ &= -0.5 \text{ dB} \end{aligned} \tag{1.103}$$

e.

$$\begin{aligned} Mismatch\ Loss_{dB} &= 10\log(1 - |\rho|^2) \\ &= 10\log(1 - |0.667|^2) \\ &= -2.6 \text{ dB} \end{aligned} \tag{1.104}$$

f.

$$\begin{aligned} Mismatch\ Loss_{dB} &= 10\log(1 - |\rho|^2) \\ &= 10\log(1 - |0.855|^2) \\ &= -5.7 \text{ dB} \end{aligned} \tag{1.105}$$

The angle of the reflection coefficient is irrelevant as far as mismatch loss goes.

1.7.2.8 Transmission Line Summary

The major tool we have used to analyze transmission lines is to launch energy at something then observe how the energy scatters. This is a common concept in radio frequency design.

VSWR, reflection coefficient and return loss all measure one single thing: how well our load is matched to our system's characteristic impedance. In the examples, we have used 50 Ω as Z_0, but there are many other candidates: 23 Ω, 75 Ω, 93 − 125 Ω, and 300 Ω all immediately come to mind.

We will be seeing VSWR, reflection coefficient, and return loss again. Although their definitions are rooted in transmission line theory, we will use them in many other places.

1.7.2.9 Lossy Transmission Lines

Real-world transmission lines all exhibit some power loss. The equations describing lossy transmission lines are long; however, we can easily adjust our mental models.

When a signal propagates down a lossy transmission line, the signal will experience loss. In the pulsed-input case, the pulses we apply to the source end of the cable will suffer loss as they travel toward the load. This means that the power (and voltage) incident on the load will not be as large as we'd expect it in the lossless case. The reflected signal will also experience signal loss as it travels from the load to the source.

If we were to measure the power of the returning pulse at the sending end of the transmission line, it will be less than it would have been had the cable lossless. In other words, both $P_{Reflected}$ and $V_{Reflected}$ will be smaller because of the line loss. Since

$$\rho = \frac{V_{Reflected}}{V_{Incident}} \tag{1.106}$$

the loss in the cable makes the reflection coefficient smaller, and the load appears to be closer to Z_0. In other words, a lossy cable tends to make our loads look like they are a better match than they really are.

This concept applies to return loss measurements (1.90). We know how much power we initially launched into the source end of the cable ($P_{Incident}$), and we can measure how much returns from the load ($P_{Reflected}$). However, we do not know how much power was absorbed by the load and how much power was absorbed by the lossy line. Since the reflected power is less than we would see if we had a lossless cable, the return loss looks better (i.e., the load will appear to be more closely matched) than if the cable was lossless.

In a receiving situation, the cable loss represents signal loss. The loss eats directly into your system's sensitivity. In a transmitting situation, cable loss represents power that is being dissipated in the cable between the power amplifier and the transmitting antenna.

1.8 S-PARAMETERS

1.8.1 Philosophy

In the radio receiver arena, we are building systems and circuits that exist in some system characteristic impedance. When our systems are operating normally, they will be working in an environment with a Z_0 source resistance and a Z_0 load resistance. It is natural and useful to characterize our systems under these conditions.

When we discussed transmission lines, we launched a voltage down the line and watched as the incident voltage reflected off the load. We can use the very same technique to characterize RF devices such as amplifiers and mixers as they work in a Z_0 environment.

Figure 1-22 shows the measurement environment we will use. We always terminate the input and the output of the device with a Z_0 load. Then we launch an incident wave (V_{i1}) at our two-port. The incident wave will encounter the two-port device, and some of

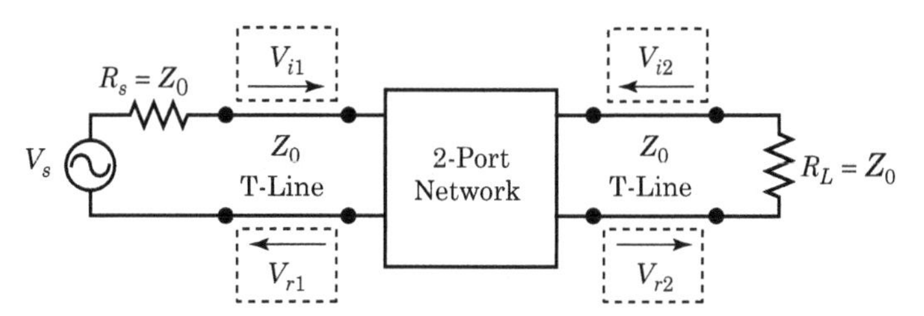

FIGURE 1-22 ■ Measurement of the four s-parameters of a general two-port device. This measurement requires only matched loads and sources.

the voltage will reflect off port 1 (this is V_{r1}). Some of the incident signal will enter the two-port and find its way to port 2 where it will make itself known as V_{r2} and V_{i2}.

Terminating the device on both ports with a matched load causes any signals that are incident upon the termination to be completely absorbed. In Figure 1-22, for example, the matched load placed on the output port of the device will force the voltage V_{i2} to be zero.

By measuring the s-parameters of a network, we are radiating energy at the device and watching how the energy scatters off. The s-parameters of a network are the *scattering parameters*.

1.8.2 Definition

The s-parameters are

$$\begin{aligned} V_{r1} &= s_{11}V_{i1} + s_{12}V_{i2} \\ V_{r2} &= s_{21}V_{i1} + s_{22}V_{i2} \end{aligned} \quad or \quad \begin{bmatrix} V_{r1} \\ V_{r2} \end{bmatrix} = \begin{bmatrix} s_{11} & s_{12} \\ s_{21} & s_{22} \end{bmatrix} \begin{bmatrix} V_{i1} \\ V_{i2} \end{bmatrix} \tag{1.107}$$

To calculate the various s-parameters from measured data, we perform

$$\begin{aligned} s_{11} &= \left.\frac{V_{r1}}{V_{i1}}\right|_{V_{i2}=0} \qquad s_{12} = \left.\frac{V_{r1}}{V_{i2}}\right|_{V_{i1}=0} \\ s_{21} &= \left.\frac{V_{r2}}{V_{i1}}\right|_{V_{i2}=0} \qquad s_{22} = \left.\frac{V_{r2}}{V_{i2}}\right|_{V_{i1}=0} \end{aligned} \tag{1.108}$$

1.8.3 Measurement Technique

It is difficult to measure voltage directly at high frequencies, but it is easy to measure power. It is also easy to measure the phase difference between two voltages. The measurement of s-parameters considers these issues.

To measure the s-parameters of a network, we drive one port with a matched source (a source whose series impedance is Z_0), and we terminate the other port with a Z_0 load (Figure 1-23).

The source launches energy at the two-port (represented by the voltage V_{i1}); some of that energy reflects off the two-port's input and travels back toward the source, becoming V_{r1}. Some of the energy that entered the input port will travel through the two-port and find its way to the right-hand side of Figure 1-23.

Some of this energy will exit the right-hand side of the two-port becoming V_{r2}. Since the load resistor R_L is to equal Z_0, R_L absorbs all of the energy and reflects none (in other words, setting R_L to equal Z_0 sets V_{i2} to zero). The same effects occur in the bottom half of Figure 1-23 when we reverse the source and load sides.

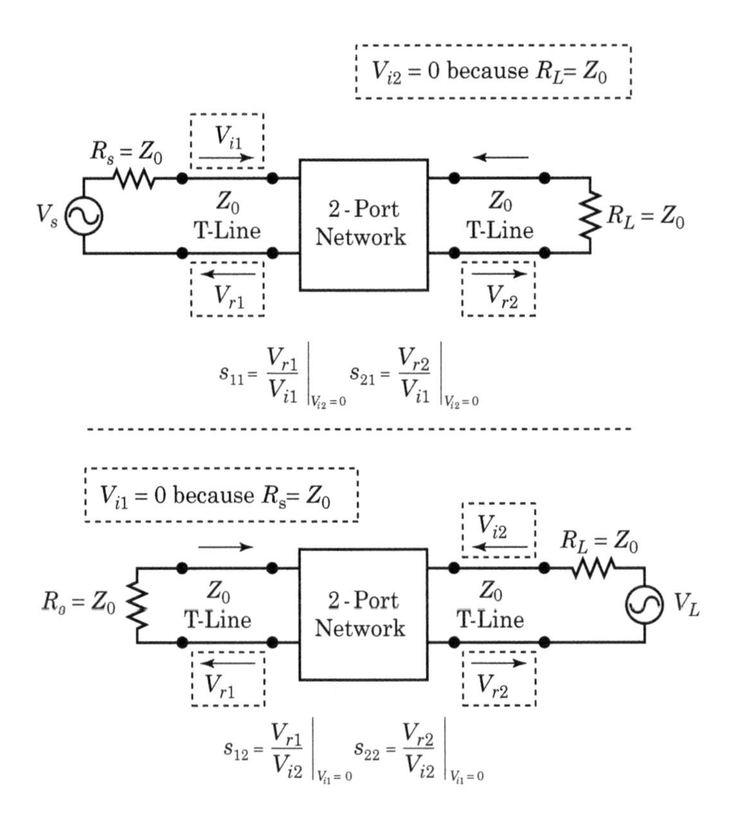

FIGURE 1-23 ■ Detailed measurement of the s-parameters of a two-port device indicating incident and reflected voltages.

At this point, we must make a small change to our s-parameter defining equations:

$$\begin{aligned} b_1 &= s_{11}a_1 + s_{12}a_2 \\ b_2 &= s_{21}a_1 + s_{22}a_2 \end{aligned} \quad or \quad \begin{bmatrix} b_1 \\ b_2 \end{bmatrix} = \begin{bmatrix} s_{11} & s_{12} \\ s_{21} & s_{22} \end{bmatrix} \begin{bmatrix} a_1 \\ a_2 \end{bmatrix} \tag{1.109}$$

where

$$\begin{aligned} b_1 &= \frac{V_{r1}}{\sqrt{Z_0}} \quad a_1 = \frac{V_{i1}}{\sqrt{Z_0}} \\ b_2 &= \frac{V_{r2}}{\sqrt{Z_0}} \quad a_2 = \frac{V_{i2}}{\sqrt{Z_0}} \end{aligned} \tag{1.110}$$

These equations are identical to equation (1.107) except that we've divided everything by $\sqrt{Z_0}$. Note that

$|a_1|^2$ = Power incident on the input of the network
= Power available from a source of impedance Z_0

$|a_2|^2$ = Power incident on the output of the network
= Power reflected from the load

$|b_1|^2$ = Power reflected from the input port of the network
= Power available from a Z_0 source minus the power delivered to the input of the network

$|b_2|^2$ = Power reflected or emanating from the output of the network
= Power incident on the load
= Power that would be delivered to a Z_0 load

Since we can easily measure power and phase angles, we can measure these quantities and arrive at the s-parameters of a given network.

1.8.4 S-Parameter Relationships

The s-parameters relate simply to power gain and mismatch loss, which are usually more interesting than the voltage transfer function:

$$\begin{aligned}
|s_{11}|^2 &= \frac{\text{Power reflected from the network input}}{\text{Power incident on the network input}} \\
|s_{22}|^2 &= \frac{\text{Power reflected from the network output}}{\text{Power incident on the network output}} \\
|s_{21}|^2 &= \frac{\text{Power delivered to a } Z_0 \text{ load}}{\text{Power available from a } Z_0 \text{ load}} \\
&= \text{Transducer power gain with } Z_0 \text{ load and source} \\
|s_{12}|^2 &= \text{Reverse transducer power gain with } Z_0 \text{ load and source} \\
&= \text{Reverse isolation}
\end{aligned} \tag{1.111}$$

These equations are valid only when the network is driven by a source with Z_0 characteristic impedance and when the network is terminated with a Z_0 load. We are interested in these very operating conditions.

The s-parameters relate to the reflection coefficient and the return loss of the two-port:

$$\begin{aligned}
s_{11} &= \text{Reflection Coefficent} \\
&= \rho \\
\textit{Return Loss}_{dB} &= 20\log(|\rho|) \\
&= 20\log(|s_{11}|)
\end{aligned} \tag{1.112}$$

1.9 | MATCHING AND MAXIMUM POWER TRANSFER

We can model any practical signal source as Figure 1-24. The combination of V_S and R_S can be a signal generator, antenna, RF amplifier, or whatever. For a variety of reasons, we usually want to set R_S equal to Z_0, the system's characteristic impedance.

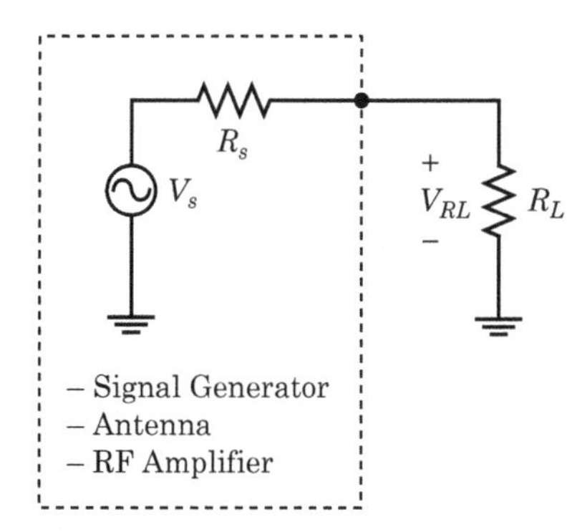

FIGURE 1-24 ■ We can model any signal source or signal sink combination using this simple model. We'll use it in this section to derive the conditions for maximum power transfer.

Assume the source in Figure 1-24 is really an antenna. Then V_S represents the signal energy the antenna receives, and R_S represents the radiation resistance of the antenna. If the load resistor R_L represents a receiving system, we would like the receiver to absorb the maximum amount of signal power from the antenna. We want to maximize the power dissipation in R_L.

1.9.1 Resistive Loads

What value of R_L will produce the maximum power transfer from V_S to R_L? The power dissipated in R_L is

$$P_{RL} = \frac{V_{RL}^2}{R_L} \tag{1.113}$$

We know

$$V_{RL} = V_S \frac{R_L}{R_L + R_S} \tag{1.114}$$

So

$$P_{RL} = V_S^2 \frac{R_L}{(R_L + R_S)^2} \tag{1.115}$$

Then the maximum power transfer will occur when

$$\begin{gathered}\frac{\partial P_{RL}}{\partial R_L} = 0 \\ \Rightarrow V_S^2 \frac{(R_L + R_S)^2 - 2R_L(R_L + R_S)}{(R_L + R_S)^2} = 0\end{gathered} \tag{1.116}$$

which simplifies to

$$R_S = R_L \tag{1.117}$$

The receiver will pull the maximum amount of power from the antenna if we set the input impedance of the receiver equal to the output impedance of the antenna (the matched condition). Figure 1-25 shows a graph of the power dissipated in R_L, P_{RL}, versus R_L with $R_S = 100\ \Omega$ and $V_S = 1$ volt. As advertised, the curve peaks when $R_S = 100\ \Omega = R_L$.

Assume we are operating in a 50 Ω system and the combination of V_S and R_S is an antenna. To pull the maximum amount of signal energy from the antenna, we must set $R_S = 50\ \Omega$. The farther R_L is from 50 Ω, the less power we will be able to draw out of the antenna.

1.9.2 Complex Loads

If the source impedance is complex (and equal to Z_S), then we will achieve maximum power transfer from the source to the load when the load impedance (Z_L) satisfies the relationship

$$Z_L = Z_S^* \tag{1.118}$$

where Z_S^* is the complex conjugate of the source impedance (Figure 1-26). If $Z_L = Z_S^*$, then $R_L = R_S$ and $X_L = -X_S$. The reactances cancel, and we are left with two matched resistors.

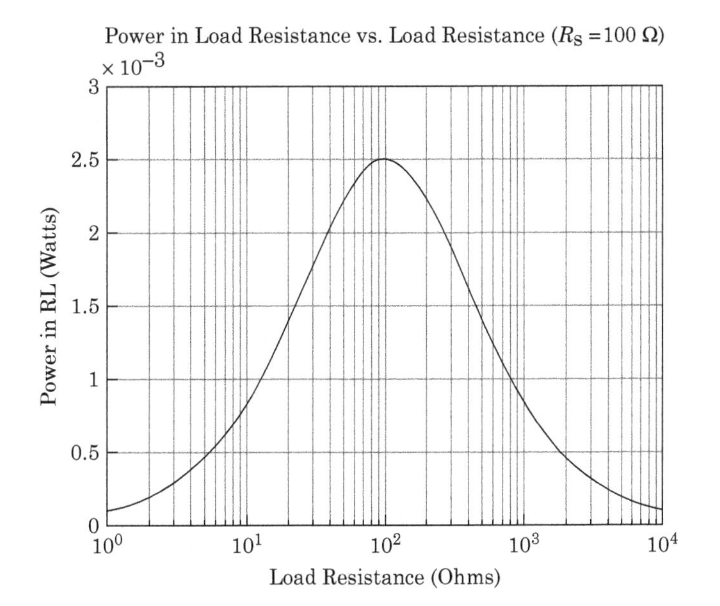

FIGURE 1-25 Power in a 100 Ω load resistor versus the load resistor value. Note that the power dissipated in the load resistor is maximum when $R_S = R_L$.

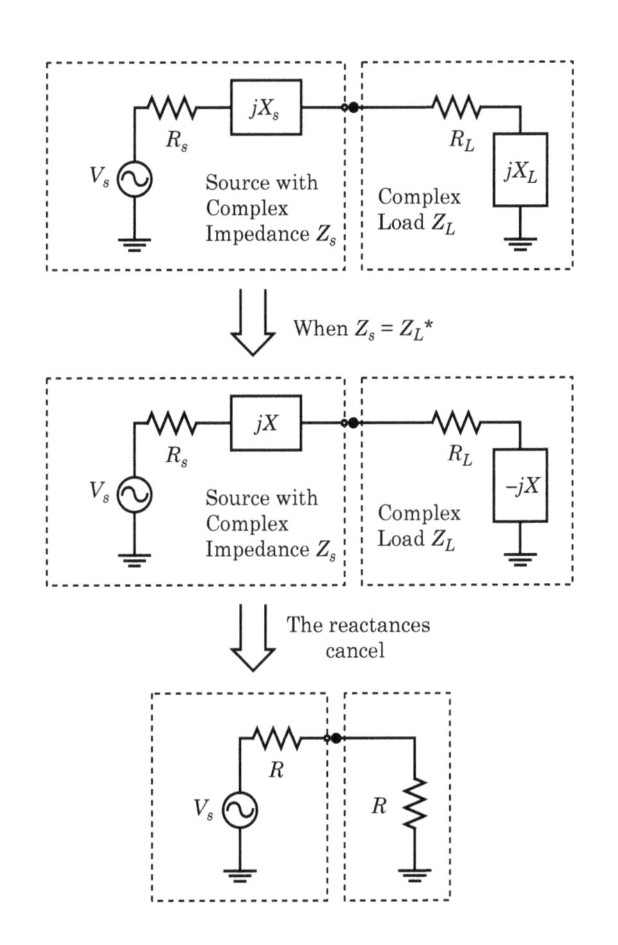

FIGURE 1-26 Maximum power transfer for a complex load and source impedances. When $Z_S = Z_l^*$, the reactances cancel and we're left with a purely resistively matched system.

1.9.3 When to Match

Matching usually means setting one impedance equal to another. If you were to ask an engineer, "Why do we match?" the likely answer will be, "For maximum power transfer." This answer is correct as far as it goes; however, there are other important reasons we match. There are also important situations where matching is not a good idea.

1.9.3.1 Antennas

We match to achieve maximum power transfer from a source to a load. This situation is most common when we are interfacing to an antenna.

In a receiving situation, we want to pull all the signal energy we possibly can out of the antenna. In a transmitting situation, we want to dump as much power as we can into the antenna. Matching allows such power transfers.

1.9.3.2 Terminations

We often match to present termination-sensitive devices with a fixed, known impedance. For example, mixers and filters usually require a broadband termination on their input and output ports. If they do not see a broadband match, mixer performance will suffer.

Some devices are unstable or change their mode of operation radically if we present them with an improper termination. Amplifiers might break into oscillation, or an oscillator's frequency may change.

1.9.3.3 Transmission Lines

Mismatches on the source or load end of a cable will produce signal reflections. These reflections distort the signal and produce voltage maximums and minimums on the line. Matching allows us to avoid these problems.

Using matched loads, we can use arbitrary lengths of transmission line to connect the different pieces of our systems together. If a transmission line is not matched, it will cause the performance of our system to vary as we change the length of the cables connecting the pieces together.

1.9.3.4 Calculations

Matching impedances makes calculations and their associated mathematics easier and more intuitive. We do not need to factor the changing impedance levels into our equations if all the source and load impedances are equal. This is the difference between doing a calculation in your head and using a computer.

1.9.3.5 Time-Domain Effects

In digital and other time-domain systems, a poorly terminated transmission line will cause signal reflections as pulses bounce off mismatched loads. These reflections will distort the pulses and their edges to the point of ruining a system's performance.

1.9.4 When Not to Match

In some cases, we absolutely do not want to match the source impedance to our load impedance.

1.9.4.1 Efficiency

Matching is a poor idea when we are interested in efficiency. Figure 1-27 shows a model for an RF amplifier. If the load resistor R_L is dissipating power P_{RL} and if $R_{out} = R_L$

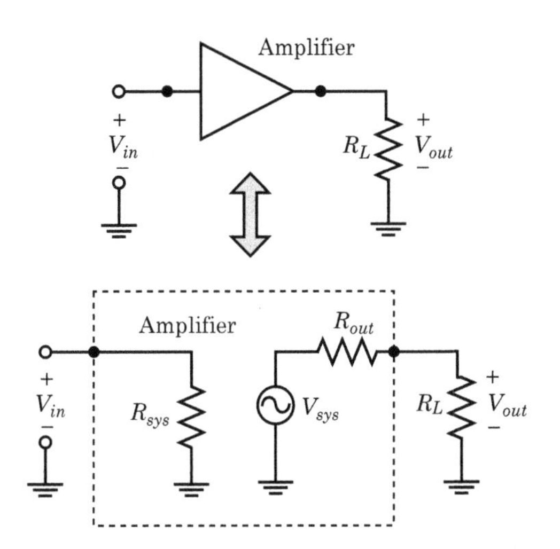

FIGURE 1-27 ■ Model of an RF amplifier delivering power to a load resistor.

(i.e. the matched condition), then R_{out} must also be dissipating P_{RL}. We are losing one-half (or 3 dB) of the amplifier's power in the source resistor.

The output efficiency of the system is

$$\eta = \frac{\text{Power dissipated in R}_\text{L}}{\text{Power generated by V}_\text{sys}} = \frac{\dfrac{\left(V_{sys}\dfrac{R_L}{R_L} + R_{out}\right)^2}{R_L}}{\dfrac{V_{sys}^2}{R_L + R_{out}}} = \frac{R_L}{R_L + R_{out}} \tag{1.119}$$

Figure 1-28 shows a graph of efficiency versus R_{out}. We achieve maximum efficiency when the amplifier's output impedance is zero.

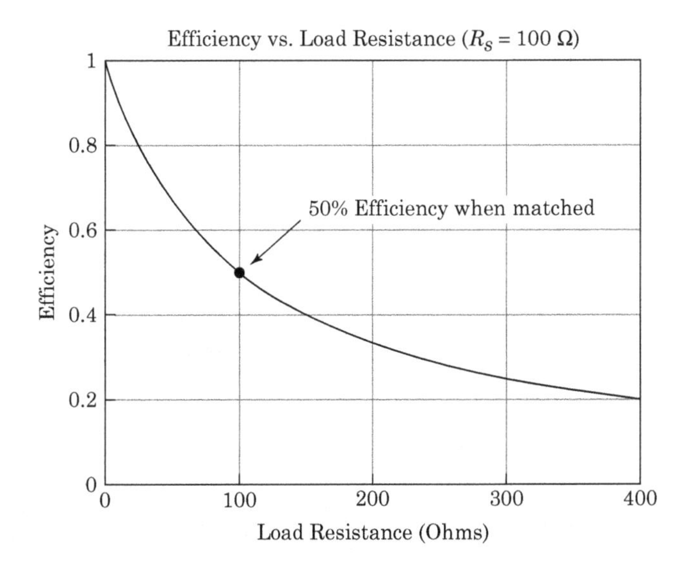

FIGURE 1-28 ■ Efficiency versus load resistance for a 100 Ω load resistance. Note that efficiency is only 50% when the system is matched. High efficiency occurs when $R_S \ll R_L$.

Consider the power company's 60 hertz wall outlets. Are you interested in maximum power transfer in that situation? Do you want to match into the power grid? Obviously not. In the case of the power company, we are more interested in having a constant voltage than we are in maximum power transfer. Making the source impedance as small as possible will allow the voltage to stay almost constant despite what type of load we plug in.

1.10 | INTRODUCTION TO RADIO FREQUENCY COMPONENTS

1.10.1 Amplifiers

As part of our introduction to radio frequency systems, we are now going to discuss the basic characteristics of amplifiers. In later chapters, we will expand on this material (especially regarding noise and linearity).

1.10.1.1 Power Gain

We place an amplifier in our system because the amplifier provides signal power gain. Figure 1-29 shows the model for an RF amplifier, its source, and its load.

The signal source can be a signal generator, antenna, or another RF amplifier. We will model the signal source as a voltage source V_S in series with a resistor R_S.

The amplifier in Figure 1-29 accepts power from some external source, and that power is dissipated in R_{sys}. The amplifier measures the power being dissipated in R_{sys} and adjusts its internal voltage source V_{sys} so that a fixed multiple of the input power is delivered to R_L. The equations in various forms are

$$\begin{aligned} P_{RL} &= G_P P_{in} \\ P_{RL,\,dBm} &= G_{P,dB} + P_{in,\,dBm} \\ P_{RL,\,dBW} &= P_{in,\,dBW} + G_{P,\,dB} \end{aligned} \tag{1.120}$$

where

P_{in} = the power delivered to the amplifier's input resistor R_{sys} by some external source (in linear units such as watts or mW).

P_{RL} = the power delivered to the load resistor R_L by the RF amplifier (in linear units).

G_P = the power gain of the amplifier; usually > 1 (in linear units)

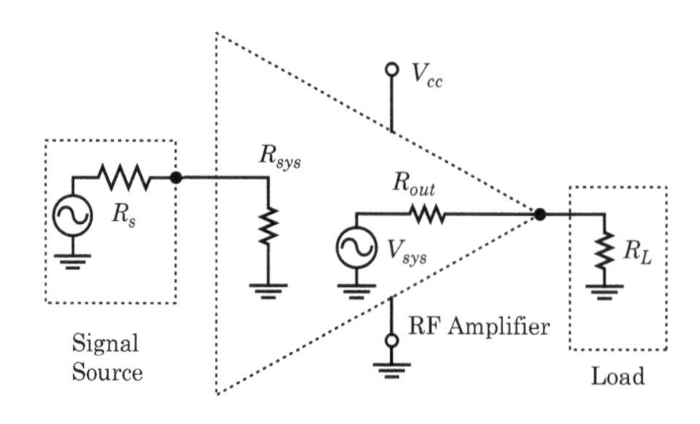

FIGURE 1-29 ■ General model for a signal source, an RF amplifier, and a load under matched conditions. We force $R_S = R_{SYS} = R_{OUT} = R_L$.

$P_{in,\,dBm}$ = the power delivered to the amplifier's input resistor by some external source in dBm

$P_{RL,\,dBm}$ = the power delivered to the load resistor R_L by the RF amplifier in dBm.

$G_{P,\,dB}$ = the power gain of the amplifier; usually > 0 dB in dB

$P_{in,\,dBW}$ = the power delivered to the amplifier's input resistor by some external source in dBW

$P_{RL,\,dBW}$ = the power delivered to the load resistor R_L by the RF amplifier in dBW

The power gain of an amplifier (or attenuator) does not change with how we choose to describe the input and output power. The power gain is the same when we describe the power in dBW, dBm, dBf, or any other units.

However, the power gain of an amplifier will change with frequency, temperature, and power supply voltage. A typical specification for the power gain of an amplifier is

> Power Gain = 15 dB $\pm$ 1 dB over a 20–500 MHz frequency range and a -20 to $+55$°C temperature range.

This specification describes the nominal power gain (15 dB) but also that the gain may vary from 14 to 16 dB over a –20 to +55°C temperature range and over the amplifier's frequency range. This specification does not tell us how the power gain will vary with temperature or frequency or from unit to unit.

The equations of (1.120) are valid for lossy devices such as attenuators and passive filters. In the lossy case, the linear gain is less than unity, or and the decibel gain is negative.

1.10.1.2 Mental Model

A convenient mental model of an amplifier is as follows:

> The amplifier measures the input power delivered to R_{sys} by some external source. It multiplies the signal power by some power gain G_P and delivers the multiplied power to the load resistor R_L.

This model works for both signals and noise. However, the amplifier will always generate and add noise power of its own to the input signal.

1.10.1.3 Reverse Isolation

Amplifiers are not unilateral devices. In other words, if we present an amplifier with a signal on its *output* port, some of the signal will leak through to the amplifier's *input* port. This is the amplifier's reverse isolation (Figure 1-30).

This effect is not intentional. Through one mechanism or another, signals will end up traveling backward through the amplifier (or any device). The reverse isolation specification simply quantifies this effect.

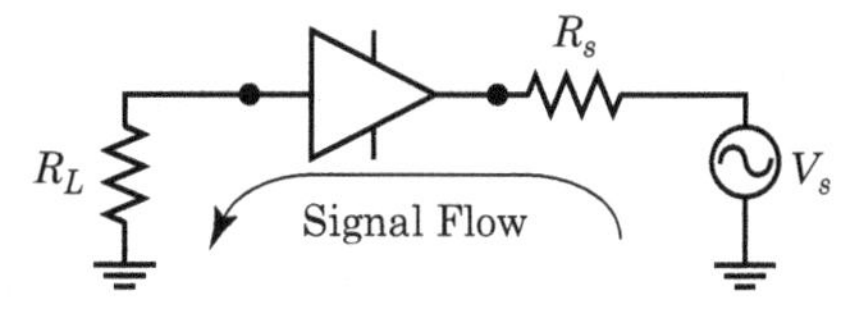

FIGURE 1-30 ▪ Reverse isolation in an RF amplifier. The signal flows backward through a component.

Referring to Figure 1-30, the amplifier's reverse isolation is

$$\textit{Reverse Isolation} = \frac{\text{Power available from the input port}}{\text{Power delivered to the output port}} \tag{1.121}$$

We are concerned about this problem when we work with amplifiers and mixers. The reverse isolation can cause problems when we connected devices together in cascade.

1.10.1.4 Frequency Coverage

Amplifiers will provide their guaranteed specifications only over some given frequency range. Outside the specified frequency range, the amplifier may still operate, but its performance specifications are not guaranteed over time and temperature.

We will often find it convenient to divide frequency ranges into *wideband* and *narrowband*. Generally, a wideband system has a frequency coverage that is greater than an octave (2:1 range). A narrowband system has a frequency coverage is less than an octave.

1.10.1.5 Signals, Noise, and Power Gain

An amplifier will add noise to any signal it processes. This noise may be problematic depending on the relative power levels of the noise and signal being applied to the amplifier.

An amplifier processes signal power and noise power equally. The external source (e.g., an antenna) will deliver both signal power and noise power to the amplifier. The amplifier will add noise of its own to the input and then will amplify the sum by its power gain.

1.10.1.6 Linearity

We have drawn the amplifier of Figure 1-29 with connections to both a power supply (V_{cc}) and ground. These connections are meant to emphasize the linearity constraints of the amplifier.

An amplifier will distort any signal it processes, no matter now small the signal. The distortion could be problematic depending on the final system architecture.

There are many linearity specifications such as third-order intercept, second-order intercept, and compression point. We will discuss their definitions and effects in Chapter 7.

1.10.1.7 Input and Output Impedances

We would like our amplifiers to exhibit carefully controlled input and output impedances. If we intend to operate the amplifier in a system whose characteristic impedance is Z_0, then we almost universally match the amplifier so that $R_S = R_{sys} = R_{out} = R_L = Z_0$.

The input and output impedance of an amplifier will change with frequency and temperature. A typical specification is

> Input VSWR is $< 2.0{:}1$ over a -55 to $85°$C temperature range and over a 5 to 500 MHz frequency range.

This specification tells us that the input impedance of the amplifier will never create a VSWR of more than 2.0:1 over the temperature and frequency range. We are on our own if we operate the amplifier outside of these ranges.

There are similar specifications for the output impedance in terms of VSWR.

In practice, we have found two distinct types of RF devices: Some devices present a wideband match to the outside world; other devices demand that the outside world present a wideband match to them.

For example, the source and load terminations are strong players in a filter's performance. The filter designer assumes that we will provide a Z_0 resistor on the input and output terminals of the filter. If we do not meet this requirement, the filter will misbehave.

Alternately, we often ask an attenuator to provide something close to a Z_0 impedance to the outside world. We often use attenuators to quiet down a widely varying impedance at some interface.

War Story: Frequency Synthesizer Design

When you change the impedance an amplifier sees on its output, the amplifier's input impedance changes. Likewise, when you change the load presented to the input of an amplifier, the output impedance of the amplifier changes. This is different effect from the reverse isolation of the amplifier.

When designing a frequency synthesizer with a voltage-controlled oscillator (VCO), you find that the VCO's output frequency changes when you change the load impedance. The VCO often drives a frequency divider (a modulo-2 prescaler). The prescaler input impedance changes slightly depending on its output state. This puts an unwanted frequency modulation (FM) on the VCO.

One solution is to isolate devices as much as possible. It is common practice to place amplifiers and attenuators between the VCO and the prescaler. This has the effect of quieting the impedance changes presented to the VCO.

1.10.1.8 Output Power and Efficiency

An amplifier can produce only so much output power before it falls out of specification. The output power specification is important in transmitter and some mixer applications.

The amount of signal power you can draw from an amplifier depends on the amount of distortion you can tolerate. Generally, higher-power waveforms contain higher levels of distortion.

We can build very linear amplifiers that produce a lot of output power. However, the trade-off is that these amplifiers are inefficient and will require a lot of DC power.

As a rule of thumb, we have found that a wideband amplifier (greater than an octave frequency coverage) will deliver a maximum of about 10% of its DC power as RF energy; that is,

$$P_{out,\,max} \approx \frac{P_{DC}}{10} \tag{1.122}$$

1.10.1.9 Stability

Any device with power gain has the ability to become unstable and break into self-sustaining oscillation. Amplifier designers are aware of this effect and usually make sure their amplifiers are unconditionally stable. This means that the amplifier will not oscillate no matter what impedance we place on the input and output ports.

For example, we test amplifiers and other RF devices with laboratory-grade signal generators and spectrum analyzers. The manufacturers of the test equipment ensure that the ports on their equipment are a nonreactive Z_0 impedance. The designers will quench any oscillations under these circumstances.

However, realized filters, antennas, oscillators, and mixers do not present a wideband Z_0 match to the outside world. The terminal impedances of a filter, for example, vary wildly with frequency. In the passband, the filter presents a Z_0 impedance to the outside world. In the stopband, a filter usually presents either a short or open circuit to the world.

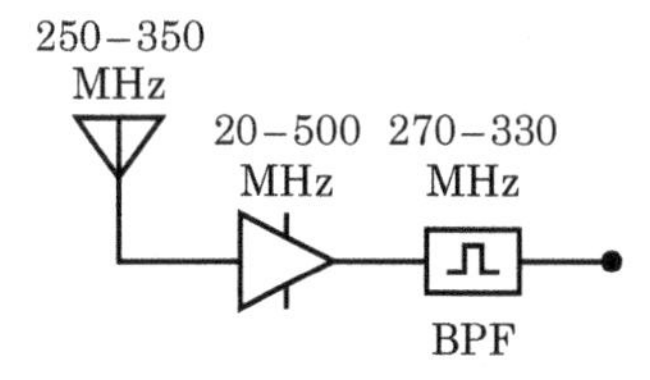

FIGURE 1-31 ■ A common system configuration consisting of an antenna, amplifier, and filter. Each component will present a non-Z_0 impedance to the outside world at frequencies outside of their specified operating ranges.

Similarly, an antenna may look like a voltage source with a Z_0 series impedance in its operating band, but there are no guarantees once we operate outside of the frequency range of the antenna.

Figure 1-31 shows a simple RF system. The amplifier will provide gain over its specified frequency range of 20–500 MHz. The antenna covers a range of 250–350 MHz, whereas the filter following the amplifier has a passband of only 270–330 MHz. The impedance of each device will be in the neighborhood of 50 Ω in the 270–330 MHz frequency range.

The impedance of the antenna changes rapidly once we leave the 250–350 MHz range. The amplifier's input sees a non-Z_0 termination on its input port when the frequency is below 250 MHz and when the frequency is greater than 350 MHz. The amplifier's output sees a non-Z_0 termination when the frequency is outside of the filter's 270–350 MHz range. If the amplifier is designed improperly, these non-Z_0 terminations can cause the amplifier to oscillate.

If the amplifier of Figure 1-31 oscillates, it will often do so at frequencies corresponding to the band edges of the filters or antenna. The input impedance of these devices change rapidly as they transverse from their passband into their stopbands.

Finally, as the following war story illustrates, an amplifier can be oscillating yet can still perform all of its design functions.

War Story: Amplifier Stability

A colleague of ours was performing some work on an intermediate frequency (IF) amplifier with AGC. This was an old design that was deployed successfully in the field for years. The company had just purchased a new, wideband oscilloscope they were using to characterize the IF amplifier.

When our colleague placed the probe on one of the points in the circuit, he noticed that the trace got just a little fatter. Further investigation revealed that the amplifier was oscillating. Still further investigation showed that every amplifier he could find in the factory was also oscillating. For one final test, our colleague dug up some of the first units ever built (and in use for some 10 years). They were also oscillating.

Despite the oscillation, the amplifiers performed every function they were designed to perform. Their gain was flat, the AGC functioned beautifully, and the amplifiers did not introduce additional noise or distortion in the system.

This is not to say that you should not remove unwanted oscillations. They represent nonlinear behavior, which is bad. In addition, this company was lucky. Normally this sort of undiscovered design flaw comes back later (and often at a very high cost).

War Story: Specialized Amplifiers

Several times in the past, we have purchased amplifiers with special design characteristics (i.e., very low noise figure or very wide bandwidth). We used these custom amplifiers for several years before we realized they were oscillating when we connected them to antennas.

War Story: More on Amplifier Stability

Often, an oscillating amplifier makes itself known in unusual ways. We were developing a system with a steerable antenna, and we noticed that some of the signals we saw changed frequency depending on the position of the antenna. As you might expect, we were baffled for quite a while until we discovered that one of the amplifiers connected directly to the antenna feed was oscillating.

Apparently, the impedance seen by the amplifier was a strong function of the antenna direction. You might imagine that the amplifier would see a non-Z_0 impedance when the antenna was pointed toward a large metal plate and a Z_0 impedance when the antenna was pointed directly at the sky.

Since the impedance seen by the amplifier changed depending on what the antenna would look at, the frequency of oscillation changed also. Incoming signals mixed with the oscillation and found their way into the system's passband. Finally, we saw the signals of interest move as we moved the antenna.

This example is on the very edge of weirdness and helps to foster the "RF is magic" belief. We spent much time on this problem before we realized what was happening. The initial clue came when an engineer accidentally walked in front of the antenna as another engineer was viewing the signals on a spectrum analyzer.

The lesson is that you should specify any amplifiers you buy as being unconditionally stable for all values and phases of source and load impedances.

The frequency of oscillation is not always constrained to be within the frequency of operation of the amplifier. An amplifier designed to operate in the 10 to 1,000 MHz range can oscillate at 2 MHz or at 4,560 MHz. As long as the device exhibits power gain, you have the possibility of oscillation.

1.10.2 Resistive Attenuators

Resistive attenuators are radio components designed to produce signal loss. Resistive attenuators also have the effect of making a non-Z_0 load look more like a resistive Z_0 load. We often refer to attenuators as *pads*. A 6 dB attenuator is a 6 dB pad.

1.10.2.1 Characteristic Impedance

The two design parameters for a resistive attenuator are the characteristic impedance of the system in which the attenuator will be operating and the desired signal attenuation. Figure 1-32 shows two 6 dB attenuators—the top attenuator works in a 50 Ω system, whereas the bottom attenuator is designed for a 300 Ω system. Note that the values of the resistors are different.

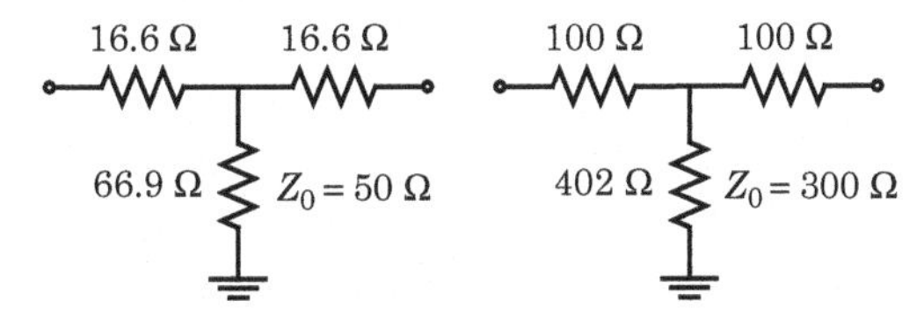

FIGURE 1-32 ■ Two different 6 dB attenuators. The top attenuator operates in a 50 Ω system, whereas the bottom attenuator operates in a 300 Ω system.

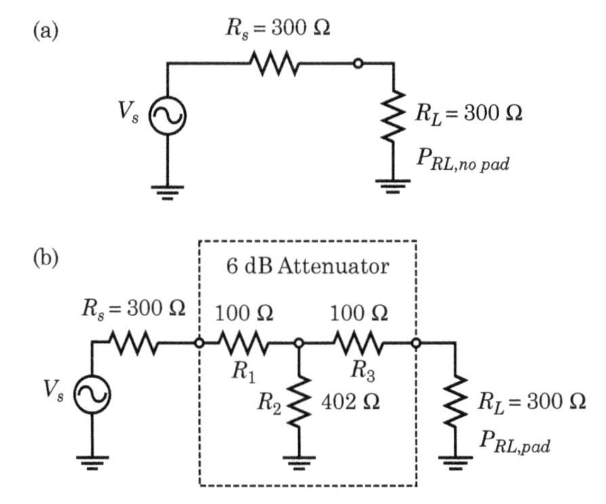

FIGURE 1-33 (a) A 300 Ω system without an attenuator. (b) The same system with a 6 dB attenuator.

1.10.2.2 Signal Attenuation

Let's look at the 300 Ω attenuator in detail. We designed the attenuator to provide a precise amount of signal loss. Figure 1-33 shows a 300 Ω system without and with a 6 dB attenuator in the signal path.

We will set V_S of Figure 1-33 to be 1.10 V_{RMS}. The power dissipated in the load resistor of circuit (a) is

$$\begin{aligned} P_{RL,no\,pad} &= \frac{V_{S,RMS}^2}{Z_0} \\ &= \frac{(1.10/2)^2}{300} \\ &= 1\ \text{mW} \\ &= 0\ \text{dBm} \end{aligned} \tag{1.123}$$

Tedious math reveals the current traveling through the load resistor of circuit (b) is 0.918 mA_{RMS}. The power dissipated in R_L is

$$\begin{aligned} P_{RL,\,pad} &= i_{RL,\,RMS}^2 R_L \\ &= (0.918E-3)^2(300) \\ &= 0.25\ \text{mW} \\ &= -6.0\ \text{dBm} \end{aligned} \tag{1.124}$$

The power gain of the attenuator is

$$\begin{aligned} G_p &= \frac{P_{RL,\,pad}}{P_{RL,no\,pad}} \\ &= \frac{0.250}{1.00} \\ &= 0.25 \\ &= -6\ \text{dB} \end{aligned} \tag{1.125}$$

The power gain of the attenuator is –6 dB. Equivalently, the power loss of the attenuator is 6 dB.

1.10.2.3 Input Impedance

Resistive attenuators have the effect of making the input impedance of follow-on devices look closer to the system characteristic impedance. This change of effective impedance comes at the cost of signal loss.

1.10.2.4 Z_0 Terminated

If we terminate the output of an attenuator in its characteristic impedance, the input impedance of the attenuator will be the characteristic impedance. Let's look again at circuit (b) of Figure 1-33.

The input impedance of the attenuator of Figure 1-34, Z_{in}, is

$$\begin{aligned} Z_{in} &= 100 + \frac{(402)(400)}{402+400} \\ &= 100 + 200 \\ &= 300\ \Omega \end{aligned} \tag{1.126}$$

The input impedance of *any* resistive attenuator terminated in its characteristic impedance will be the characteristic impedance.

1.10.2.5 Open- and Short-Circuit Terminations

A resistive attenuator exhibits a desirable trait when terminated in impedances other than the characteristic impedance. Figure 1-35 shows our 6 dB, 300 Ω resistive attenuator terminated in an open circuit (a) and in a short circuit (b).

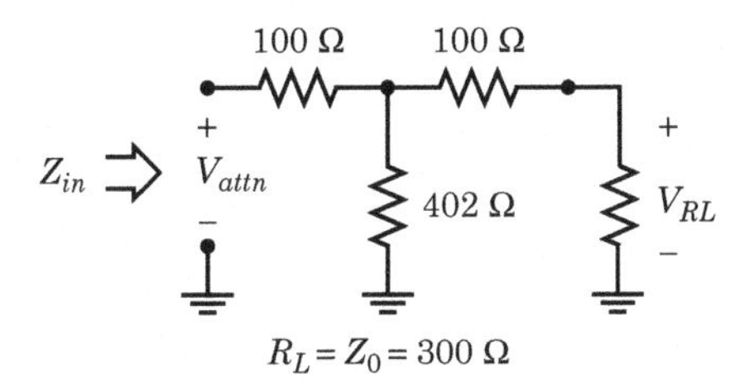

FIGURE 1-34 ■ The input impedance of a 300 Ω, 6 dB attenuator terminated in its characteristic impedance.

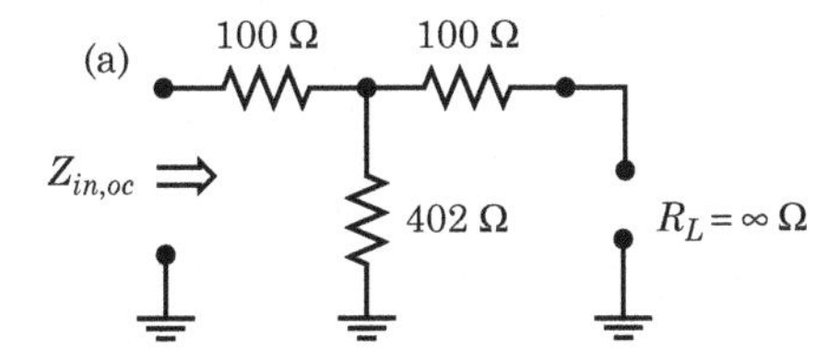

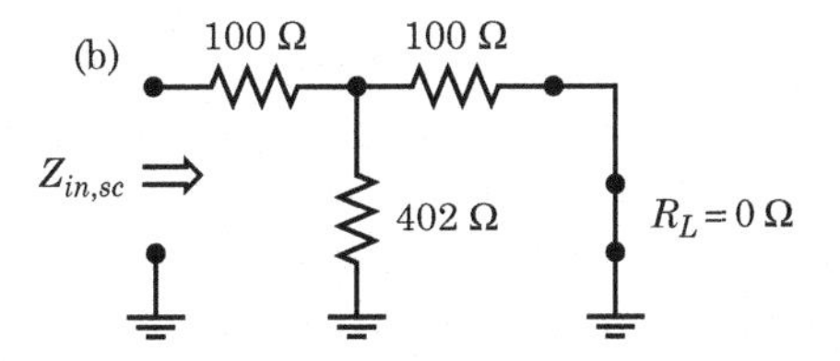

FIGURE 1-35 ■ The input impedance of a 300 Ω, 6 dB attenuator terminated in (a) an open circuit and (b) a short circuit.

The respective attenuator input impedances are

$$\begin{aligned} Z_{in,OC} &= 100 + 402 \\ &= 500\ \Omega \end{aligned} \qquad \begin{aligned} Z_{in,SC} &= 100 + \frac{(402)(100)}{402 + 100} \\ &= 100 + 80 \\ &= 180\ \Omega \end{aligned} \tag{1.127}$$

The attenuator has the effect of bringing both the open-circuit and short-circuit input impedances closer to the characteristic impedance. The equivalent return losses are

$$\begin{aligned} RL_{OC} &= 20 \log \left| \frac{Z_{in,OC} - Z_0}{Z_{in,OC} + Z_0} \right| \\ &= 20 \log \left| \frac{500 - 300}{500 + 300} \right| \\ &= 20 \log |0.25| \\ &= -12\,\text{dB} \end{aligned} \qquad \begin{aligned} RL_{SC} &= 20 \log \left| \frac{Z_{in,SC} - Z_0}{Z_{in,SC} + Z_0} \right| \\ &= 20 \log \left| \frac{180 - 300}{180 + 300} \right| \\ &= 20 \log |-0.25| \\ &= -12\,\text{dB} \end{aligned} \tag{1.128}$$

Both the open- and short-circuit cases produce a return loss of 12 dB or twice the attenuator value. We can find a general rule for the return loss of a resistive attenuator by examining the attenuator and load as a two-element cascade and using the s-parameter technique of watching the signal power move through the cascade.

We launch a known amount of signal power, $P_{incident}$, into the cascade of Figure 1-36. The signal encounters the attenuator and loses A dB. Leaving the attenuator, the signal strikes the load; some power is absorbed, and some power is reflected. The reflected power is given by

$$P_{L,\,Reflected} = \rho_L^2 P_{L,\,Incident} \tag{1.129}$$

and travels again through the attenuator, losing another A dB. A complete description of the power loss through the cascade is

$$\begin{aligned} P_{Reflected,\,dBm} &= P_{Incident,\,dBm} - A_{dB} - RL_{L,\,dB} - A_{dB} \\ &= P_{Incident,\,dBm} - 2A_{dB} \end{aligned} \tag{1.130}$$

In the worst case, the load is either an open or short circuit, and $R_{L,\,dB}$ is 0 dB (i.e., the load reflects all of the power) so the worst-case return loss of the attenuator-load cascade is 2A dB, where A_{dB} is the value of the attenuator in dB.

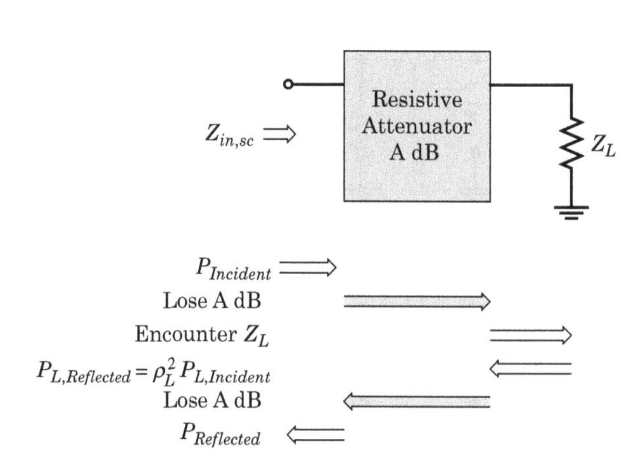

FIGURE 1-36 ■ A resistive attenuator and load as a cascade. Analysis of this cascade reveals the worst-case input impedance of the attenuator-load cascade is 2*A* dB, where *A* is the value of the attenuator.

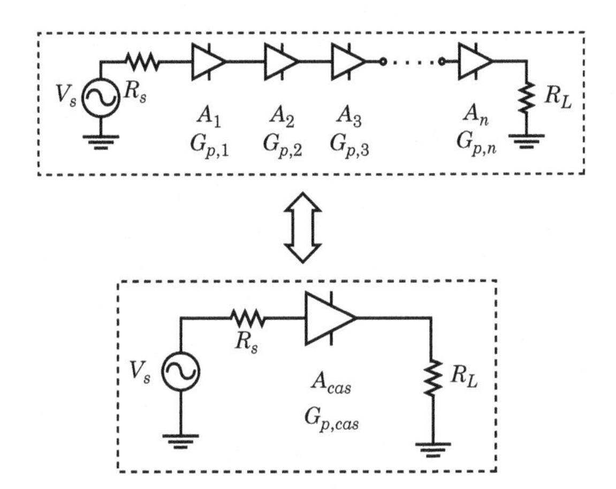

FIGURE 1-37 ■ An *n*-element cascade and its single-element equivalent.

This is a general rule. The worst-case return loss looking into a resistive attenuator of A dB is $2A$ dB.

1.10.3 Gain of Cascaded Devices

The RF system (Figure 1-37) contains several amplifiers in cascade (i.e., the output of one amplifier feeds the input of the next). Given the power gain characteristics of all the devices in a cascade, we'd like to find the power gain of the cascade.

Let

$G_{p,1}$ = the power gain of the first amplifier in linear terms

$G_{p,2}$ = the power gain of the second amplifier in linear terms

$G_{p,n}$ = the power gain of the n-th amplifier in linear terms

$G_{p,\,n,\,dB}$ = the power gain of the n-th amplifier in decibels

$G_{p,\,cas}$ = the power gain of the cascade in linear terms

$G_{p,\,cas,\,dB}$ = the power gain of the cascade in decibels

We will assume that our entire system is exactly matched to the system's characteristic impedance Z_0. This means that $R_S = R_L = Z_0$ and that the input and output impedances of all the amplifiers are exactly Z_0. This assumption is not always true in the real world, but we'll make corrections to our analysis later on.

The goal is to describe the cascade as a single amplifier with a power gain of $G_{p,cas}$.

First, we'll present amplifier A_1 with an input signal power of $S_{in,A1}$. The signal power present at the output of A_1 is

$$S_{out,A1} = S_{in,A1} G_{p,1} \tag{1.131}$$

The output of amplifier A_1 is connected to the input of amplifier A_2. The input signal applied to A_2 is $S_{out,A1}$ and the signal present at the output of A_2 is

$$\begin{aligned} S_{out,A2} &= S_{out,A1} G_{P,2} \\ &= S_{in,A1} G_{P,1} G_{P,2} \end{aligned} \tag{1.132}$$

Finally, amplifier A_3 sees an input signal of $S_{out,A2}$ to the input of amplifier A_3. The output of A_3 is

$$\begin{aligned} S_{out,A3} &= S_{out,A2} G_{P,3} \\ &= S_{in,A1} G_{P,1} G_{P,2} G_{P,3} \end{aligned} \tag{1.133}$$

We continue this process until we reach the end of the cascade (i.e., the output of the n-th amplifier). The signal out of the n-th amplifier is

$$S_{out,n} = G_{p,1} G_{p,2} G_{p,3} \cdots G_{p,n} S_{in,A1} \tag{1.134}$$

The power gain of the cascade is

$$\begin{aligned} G_{p,cas} &= G_{p,1} G_{p,2} G_{p,3} \cdots G_{p,n} \\ G_{p,cas,dB} &= G_{p,1,dB} + G_{p,2,dB} + G_{p,3,dB} + \ldots + G_{p,n,dB} \end{aligned} \tag{1.135}$$

EXAMPLE

Cascade Gain Calculation

Find the gain of the cascade shown in Figure 1-38.

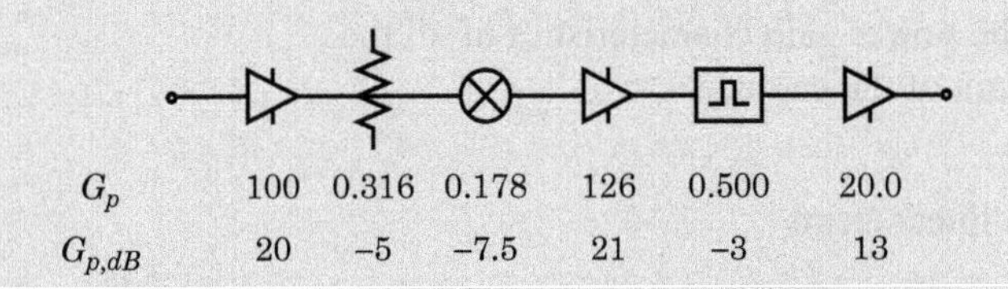

FIGURE 1-38 ■ Cascade gain example.

Solution

Using equation (1.135), we find

$$\begin{aligned} G_{p,cas} &= G_{p,1} G_{p,2} G_{p,3} \cdots G_{p,n} \\ &= (100)\left(\frac{1}{3.16}\right)\left(\frac{1}{5.62}\right)(126)\left(\frac{1}{2}\right)(20) \\ &= 7.09 \end{aligned} \tag{1.136}$$

and, in dB,

$$\begin{aligned} G_{p,cas,dB} &= G_{p,1,dB} + G_{p,2,dB} + G_{p,3,dB} + \ldots + G_{p,n,dB} \\ &= 20 + (-5) + (-7.5) + 21 + (-3) + 13 \\ &= 38.5 \text{ dB} \\ &= 7.09 \end{aligned} \tag{1.137}$$

1.10.3.1 Real-World Effects

As usual, if we build up a cascade and measure its power gain, we'll find that the real-world results deviate slightly from equation (1.135). There are several reasons for the deviation.

One of the implicit assumptions we made as we derived the cascade gain equations was that all of the amplifiers had input and output impedances that were matched exactly to the system impedance. In other words, the terminal impedances of all the amplifiers were exactly equal to Z_0. If this is the case, then each amplifier accepts all of the signal power available from the amplifier preceding it. Equation (1.135) is exactly accurate when all of the amplifiers are matched.

In the real world, however, amplifiers are not exactly matched to the system's characteristic impedance. This is especially true for wideband amplifiers (amplifiers whose frequency range is greater than an octave or so). For example, one vendor specifies a wideband amplifier with a maximum input/output VSWR of 2.5 over the entire band of operation (some 6 octaves). Our earlier discussion of VSWR relationships revealed

$$R_L = Z_0(VSWR) \quad \text{or} \quad R_L = \frac{Z_0}{VSWR} \tag{1.138}$$

The quantity R_L must be

$$\begin{aligned} R_L &= Z_0(VSWR) \\ &= 50(2.5) \\ &= 125\ \Omega \end{aligned} \qquad \begin{aligned} R_L &= \frac{Z_0}{VSWR} \\ &= \frac{50}{2.5} \\ &= 20\ \Omega \end{aligned} \tag{1.139}$$

so the terminal impedance on this amplifier can vary from 20 Ω to 125 Ω (in a 50 Ω system). The terminal impedances usually change with frequency. For example, the input impedance may be 120 Ω at one frequency, 75 Ω at another frequency, and 25 Ω at a third frequency.

Another factor we didn't account for when we derived equation (1.135) is that the gain of a garden-variety amplifier will not always be the same over time, temperature, and frequency. For example, one manufacturer sells an amplifier with a gain specification of 25 dB $\pm$ 1 dB. The $\pm$1 dB span addresses the change in amplifier gain with time, temperature, and frequency. We normally run through equation (1.135) only once, using the nominal gain for each amplifier.

The effect is that the cascade will, on average, exhibit the gain given by equation (1.135). However, the actual gain of the cascade may be higher or lower than expected. The gain may also vary with frequency.

1.11 | BIBLIOGRAPHY

Anderson, Richard W., "S-Parameter Techniques for Faster, More Accurate Network Design," Hewlett-Packard Application Note 95-1, February 1967.

Bowick, Chris, *RF Circuit Design,* Howard W. Sams and Co., 1982.

Hewlett-Packard, Inc., "S-Parameter Design," Hewlett-Packard Application Note 154, April 1972.

Kraus, Herbert L., Bostian, Charles W., and Raab, Frederick H., *Solid State Radio Engineering,* John Wiley and Sons, 1980.

Royle, David, "Rules Tell Whether Interconnections Act Like Transmission Lines—Designer's Guide to Transmission Line and Interconnections Part One," *EDN Magazine,* June 23, 1988.

Royle, David, "Correct Signal Faults by Implementing Line-Analysis Theory—Designer's Guide to Transmission Line and Interconnections Part Two," *EDN Magazine,* June 23, 1988.

Royle, David, "Quiz Answers Show How to Handle Connection Problems—Designer's Guide to Transmission Line and Interconnections Part Three," *EDN Magazine,* June 23, 1988.

Amp Incorporated, Guide to RF Connectors, Catalog 80-570, Streamlined May 1990.

Montgomery, David, "Borrowing RF Techniques for Digital Design," *Computer Design Magazine,* May 1982.

Johnson, Walter C., *Transmission Lines and Networks,* McGraw-Hill Book Company, Inc., 1950.

Westman, H.P., Karsh, M., Perugini, M.M., and Fujii, W.S., Eds., *Reference Data for Radio Engineers,* 5th ed., Howard W. Sams Inc., 1968.

Wolff, Edward A., and Kaul, Roger, *Microwave Engineering and Systems Applications,* Wiley-Interscience, 1988.

Terman, Frederick E., *Electronic and Radio Engineering,* McGraw-Hill Book Company, Inc., 1955.

Schroenbeck, Robert J., *Electronic Communications—Modulation and Transmission,* Macmillan Publishing Co., 1992.

Sklar, Bernard, *Digital Communications—Fundamentals and Applications,* Prentice-Hall, 1988.

Taub, Herbert and Schilling, Donald L., *Principles of Communications Systems,* McGraw-Hill, 1971.

Hewlett-Packard Corp., Application Note 150-1—Spectrum Analysis—Amplitude and Frequency Modulation, November 1971.

1.12 PROBLEMS

1. **FM Booster** An FM radio booster has a power gain of 10 dB. What is the voltage gain?
2. **Mismatched Amplifier** An amplifier has an input and output impedance of 50 Ω. When measured in a 50 Ω system, the amplifier exhibits a power gain of 15 dB. If the amplifier is connected into a 30 Ω system, what is the power gain?
3. **Cascaded Amplifiers** Three cascaded amplifier have power gains of 10, 15, and 20 dB. What is the overall power gain? What is the overall voltage gain?
4. **dBm Conversions**
   a. Convert 2 watts into dBm.
   b. What is the voltage of this 2 watt signal in a 600 Ω telephone line?
5. **Splitter** A four-set TV splitter advertises a loss of 5 dB per output. Can this specification be true?
6. **Filter Slope** Filter slope characteristics and control system Bode plots are often given as 6 dB per octave or 20 dB per decade, referring to the slope of the amplitude curve versus frequency. Show that these slopes are equivalent.
7. **Signal-to-Noise Ratio** A geostationary satellite has an output power of 10 watts into its transmitting antenna. The satellite antenna has a gain of 6 dB. The path loss between the transmitting and receiving antenna is 200 dB. The ground station receive antenna has a gain of 40 dB. The received noise is 2 dBf. What is the received signal-to-noise ratio (SNR)?
8. **Amplifier Ripple** A 25 Watt stereo amplifier is advertised as "flat" from 20 to 20000 Hertz, ± 2 dB. What is the maximum and minimum output power over this range of frequencies? Answer in Watts.
9. **FM Receiver Gain** An FM radio receiver can deliver 50 watts of sound to the speakers with an input signal of 10 dBf. What is the overall power gain in dB?

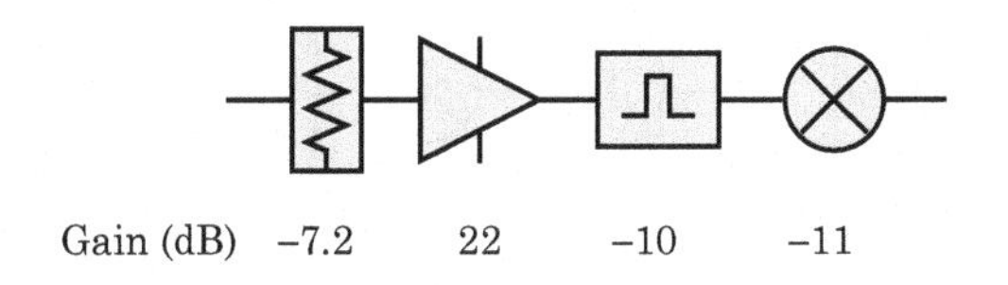

FIGURE 1-39 ■ Simple cascade.

10. **Cable Loss** A coaxial cable has a loss of 3 dB per 100 feet. What is the power loss of 500 feet of cable?
11. **Noise Reduction** The Dolby tape-noise reduction system reduces tape hiss by 15 dB. How much is the noise voltage reduced in linear terms?
12. **Cable Diameter** What change in the quantity D/d will cause the characteristic impedance of a coaxial cable to change from 50 Ω to 55 Ω?
13. **Channel Separation** A stereo amplifier has a channel separation of 30 dB. How much of the left channel signal voltage appears in the right channel?
14. **Input and Output Impedances** An RF amplifier has a maximum input voltage standing wave ratio (VSWR) specification of 2.5.
    a. What is the corresponding range in input reflection coefficient?
    b. What are the minimum and maximum terminal impedances of this amplifier?
15. **Manual Cascade Calculations** For the cascade shown in Figure 1-39, calculate
    a. The gain of each component in linear terms
    b. The cascade gain using the component gains expressed in dB
    c. The power present at the output of each component, in dBm, if the input power is −80 dBm
16. **Manual Cascade Calculation with Mismatch** Given the cascade shown in Figure 1-40:
    a. Find the cascade gain assuming the mismatch loss of every component is zero (i.e., each component is perfectly matched to 50 Ω).
    b. If each component has a maximum input VSWR of 2.0, what are the minimum and maximum reflection coefficients present on the input port of each device? What are the allowable angles of the reflection coefficient?
    c. If each component has a maximum output VSWR of 1.5, what are the minimum and maximum output impedances present on the output port of each device? What are the allowable angles of the reflection coefficient?
    d. What is the maximum possible input mismatch loss of each component? What is the maximum possible output mismatch loss of each component?
    e. What is the minimum and maximum gain of the cascade if each component has a maximum input VSWR specification of 2.0 and a maximum output VSWR specification of 1.5?
    f. Describe, in words, the condition in which the gain variation varies the most from its nominal or matched value.

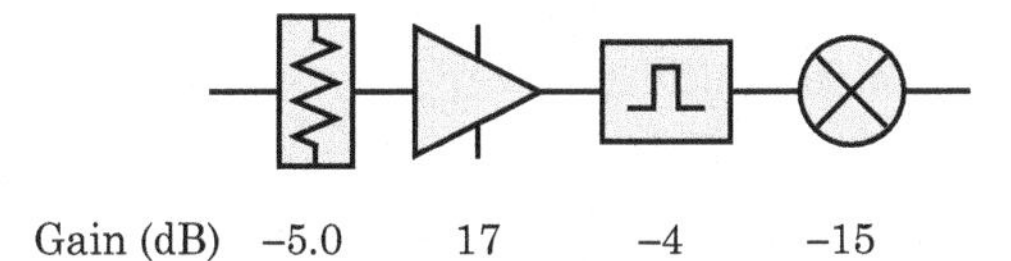

FIGURE 1-40 ■ A cascade with mismatch.

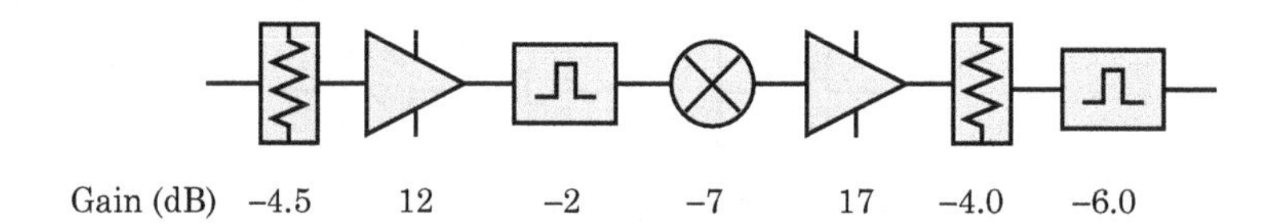

FIGURE 1-41 ▪ A simple cascade.

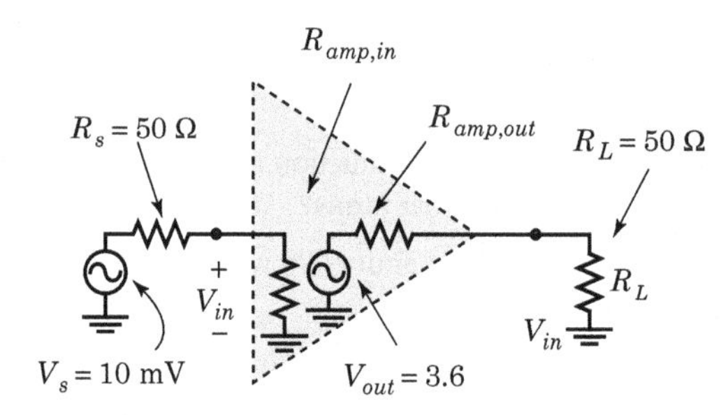

FIGURE 1-42 ▪ An amplifier whose input and output impedances are not matched.

17. **Manual Cascade Calculation** For the cascade of Figure 1-41:
    a. Find the power gain of each component in linear terms.
    b. Find the cascade gain using dB.
    c. Find the cascade gain using the linear numbers.
    d. Find the output power if the input power is 27 dBf. Answer in mW and dBm.
18. **Amplifier Modeling and Mismatching** Find the signal power delivered to the load resistor for the amplifier of Figure 1-42 when
    a. $R_{amp,in} = R_{amp,out} = 50\ \Omega$.
    b. $R_{amp,in} = 40\ \Omega$ and $R_{amp,out} = 60\ \Omega$.
19. **Essay: Why 50 Ω?** Find the impedances most commonly used in the following industries:
    - Cable TV
    - Military systems
    - Commercial radio and television transmitting systems

    Why were these characteristic impedances chosen? Why 75 Ω instead of 32 Ω, for example?
20. **Multiple Choice: Decibels** The reason we commonly use decibels (or dBs) in radio calculations is:
    a. The quantities in radio frequency systems exhibit a large range and expressing the quantities logarithmically is more convenient.
    b. It's easier to perform common RF-oriented calculations in decibels.
    c. It's possible and convenient to represent absolute quantities as in decibels.
    d. All of the above.
21. **Multiple Choice: Impedance Matching** Which statement regarding impedance matching is false?
    a. Impedance matching makes calculations easier.
    b. We match for maximum power transfer.
    c. We match to suppress reflections on transmission lines.

d. We can use selective matching to remove unwanted signals.

e. We must match only at the very input of a receiver, before the first amplifier.

22. Units of Power and Decibels

a. Convert the following power measurements to watts: 27 dBm, 30 dBW.

b. Convert the following power measurements to dBm: 20 mW, 500 W, 10 nW.

c. Convert 17 dBV to volts.

d. Convert 63 dBHz into hertz. We often use this notation for noise calculations.

e. If the alternate current (AC) current gain of a bipolar transistor is 40, to how many dB does this correspond?

f. Assume for the following that AC power is being dissipated in a 50 Ω load.

- If the power dissipated in the load is 1 watt, what is the root mean square (RMS) voltage across the load?
- If the power dissipated in the load is 0 dBm, what is the peak-to-peak voltage across the load?
- If the dissipated in the load is −174 dBm, what is the peak-to-peak voltage across the load?
- If the peak-to-peak voltage across the load is 1 V, what is the power dissipated in the load in dBm?

CHAPTER 2

Signals, Noise, and Modulation

Modulation is good.

Ron Tobin

Chapter Outline

2.1 INTRODUCTION

This chapter is an introduction to waveforms and the various expressions we can use to describe them. In upcoming chapters, we will describe the effects that various receiver specifications have on those waveforms.

2.2 A REAL-VALUED, IDEAL COSINE WAVE

The simplest waveform we will encounter is a mathematically pure cosine wave, free from noise. The ratio of signal power to noise power (i.e., the signal-to-noise ratio, or SNR) is infinite.

2.2.1 The Time Domain

Figure 2-1 shows the time-domain representation of a 1 MHz cosine wave whose initial phase is 30°.

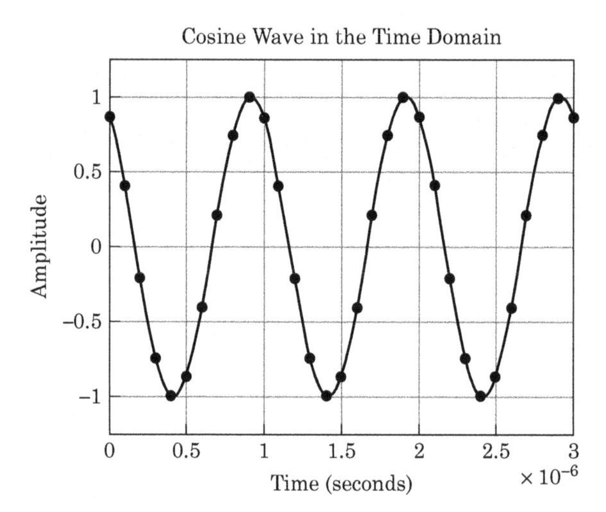

FIGURE 2-1 ■ The time-domain representation of a 1 MHz cosine wave exhibiting a 30° initial phase. In a digital signal processor (DSP), we view only the discrete samples (marked here by •). It's difficult to measure the instantaneous amplitude, frequency, or phase of the signal without considering many samples. If we wish to perform such measurements (e.g., when we want to demodulate the signal), there are better ways to represent the signal.

The equation of the pure waveform is

$$f(t) = V_{pk} \cos(\omega_0 t + \theta) \qquad \omega_0 = 2\pi f_0 \tag{2.1}$$

The period of the cosine wave is

$$T_0 = \frac{1}{f_0} \tag{2.2}$$

There are a few problems with this simple representation. Consider measuring various signal parameters such as amplitude, frequency, and phase of this simple cosine. To measure the amplitude of the cosine, we must observe the signal over many samples and then search for the minimum and maximum values. To measure the frequency, we might measure the zero crossings and derive the period, but the measurement would require a lot of samples and some "interesting" algorithms to figure out exactly when the waveform crosses zero. Measuring the instantaneous phase presents similar problems.

2.2.1.1 Zero Crossings

We are often interested in the zero crossings of a waveform. In a mixer, for example, we use a cosine wave to control a radio frequency (RF) switch. When the instantaneous value of the waveform is greater than zero, the switch will be in one position. When the waveform is less than zero, the switch will be in the other position.

We label the zero crossing waveform of Figure 2-2 as V_{Limit}. The quantity V_{Limit} describes exactly when and how often the waveform passes through 0 volts. Figure 2-2 shows a cosine and its zero crossings. We can write

$$V_{Limit}(t) = \begin{cases} +1 \text{ when } \cos(\omega_0 t + \theta) \geq 0 \\ -1 \text{ when } \cos(\omega_0 t + \theta) < 0 \end{cases} \tag{2.3}$$

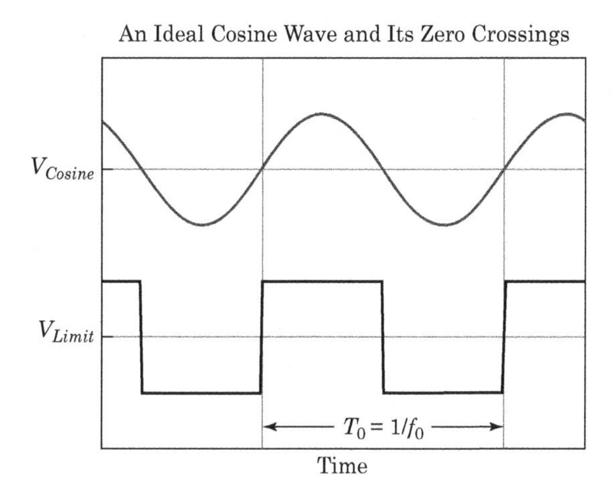

FIGURE 2-2 ▪ The time-domain representation of a cosine wave exhibiting a 30° initial phase, emphasizing its zero crossings. We label the zero crossing waveform as V_{Limit}.

The quantity V_{Limit} retains the waveform's phase and frequency information in the zero crossings. However, we have lost any information regarding the amplitude of the original waveform, V_{pk}. If V_{pk} is a function of time, we have lost all of the information about that function. Limiting preserves phase and frequency modulation but discards amplitude modulation.

2.2.2 The Frequency Domain

The Fourier transform relates the time- and frequency-domain representations of a signal:

$$V(\omega) = \int_{t=-\infty}^{t=\infty} v(t)e^{-j\omega t}dt \tag{2.4}$$

This equation isn't very intuitive, but Euler's identities help. Euler's identities relate various combinations of sines and cosines to complex exponentials. Various forms of Euler's equation are

$$e^{-j\omega t} = \cos(\omega t) - j\sin(\omega t) \qquad e^{j\omega t} = \cos(\omega t) + j\sin(\omega t) \tag{2.5}$$

and

$$\cos(\omega t) = \frac{e^{j\omega t} + e^{-j\omega t}}{2} \sin(\omega t) = \frac{e^{j\omega t} - e^{-j\omega t}}{2j} \tag{2.6}$$

We substitute

$$e^{-j\omega t} \tag{2.7}$$

in the right-hand side of equation (2.24) with

$$\cos(\omega t) - j\sin(\omega t) \tag{2.8}$$

to produce

$$\begin{aligned} V(\omega) &= \int_{t=-\infty}^{t=\infty} v(t)e^{-j\omega t}dt \\ &= \int_{t=-\infty}^{t=\infty} v(t)[\cos(\omega t) - j\sin(\omega t)]dt \\ &= \int_{t=-\infty}^{t=\infty} v(t)\cos(\omega t)dt - j\int_{t=-\infty}^{t=\infty} v(t)\sin(\omega t)dt \end{aligned} \tag{2.9}$$

We define

$$I(\omega) = \int_{t=-\infty}^{t=\infty} v(t)\cos(\omega t)dt \qquad Q(\omega) = \int_{t=-\infty}^{t=\infty} v(t)\sin(\omega t)dt \tag{2.10}$$

and we reexpress equation (2.24) as

$$\begin{aligned} V(\omega) &= \int_{t=-\infty}^{t=\infty} v(t)\cos(\omega t)dt - j\int_{t=-\infty}^{t=\infty} v(t)\sin(\omega t)dt \\ &= I(\omega) - jQ(\omega) \end{aligned} \tag{2.11}$$

Every waveform has a real component, $I(\omega)$, and an imaginary or quadrature component, $Q(\omega)$.

2.2.2.1 A Correlation

Let's look at $I(\omega)$, the component of equation (2.11) that involves the cosine:

$$I(\omega) = \int_{t=-\infty}^{t=\infty} v(t)\cos(\omega t)dt \tag{2.12}$$

The left side of equation (2.12) is $I(\omega)$, a function of radian frequency, ω. The result tells us something about $v(t)$ with respect to frequency.

The right side of equation (2.12) is a correlation of an arbitrary waveform, $v(t)$, with a cosine of various frequencies. Loosely speaking, a correlation describes how well one waveform matches another so $I(\omega)$ describes how well an arbitrary waveform $v(t)$ matches a cosine of a particular frequency. A similar argument holds for the quantity $Q(\omega)$ of equation (2.11). $Q(\omega)$ describes how well an arbitrary waveform $v(t)$ matches a sine of a particular frequency.

2.2.2.2 Example: A Simple Cosine

Let us work through equation (2.4) when $v(t)$ is a cosine whose amplitude is A_0 and whose radian frequency is ω_0.

$$\begin{aligned} V(\omega) &= \int_{t=-\infty}^{t=\infty} v(t)e^{-j\omega t}dt \\ &= \int_{t=-\infty}^{t=\infty} A_0\cos(\omega_0 t)\, e^{-j\omega t}dt \end{aligned}$$

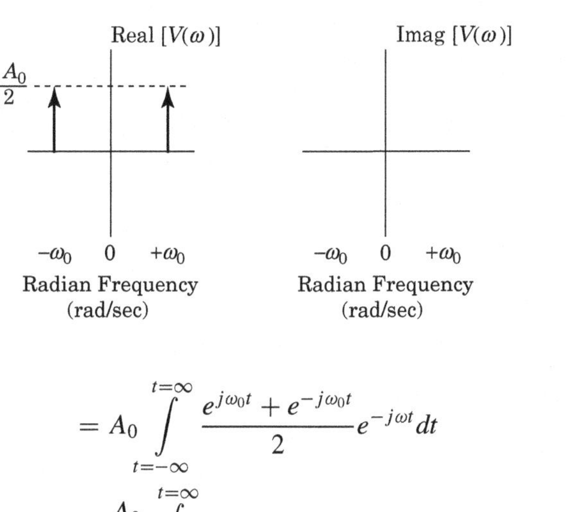

FIGURE 2-3 ■ The frequency-domain representation of a cosine wave showing both the positive and negative frequency components. There is no imaginary component.

$$
\begin{aligned}
&= A_0 \int_{t=-\infty}^{t=\infty} \frac{e^{j\omega_0 t} + e^{-j\omega_0 t}}{2} e^{-j\omega t} dt \\
&= \frac{A_0}{2} \int_{t=-\infty}^{t=\infty} \left(e^{-j(\omega+\omega_0)t} + e^{-j(\omega-\omega_0)t}\right) dt \\
&= \frac{A_0}{2}[\delta(\omega+\omega_0) + \delta(\omega-\omega_0)]
\end{aligned}
\tag{2.13}
$$

where $\delta(\omega)$ is the Dirac delta function.

Figure 2-3 shows the frequency-domain representation of a cosine wave of radian frequency ω_0. The plot shows equal-valued real peaks at $+\omega_0$ and $-\omega_0$. The Fourier components of a cosine wave are real—there is no imaginary component.

2.2.2.3 A Second Example: A Simple Sine

Let's work through equation (2.4) when $v(t)$ is a sine with amplitude A_0 and radian frequency of ω_0.

$$
\begin{aligned}
V(\omega) &= \int_{t=-\infty}^{t=\infty} v(t) e^{-j\omega t} dt \\
&= \int_{t=-\infty}^{t=\infty} A_0 \sin(\omega_0 t) e^{-j\omega t} dt \\
&= A_0 \int_{t=-\infty}^{t=\infty} \frac{e^{j\omega_0 t} - e^{-j\omega_0 t}}{2j} e^{-j\omega t} dt \\
&= \frac{A_0}{2j} \int_{t=-\infty}^{t=\infty} \left(e^{-j(\omega+\omega_0)t} - e^{-j(\omega-\omega_0)t}\right) dt \\
&= \frac{A_0}{2j}[\delta(\omega+\omega_0) - \delta(\omega-\omega_0)] \\
&= j\frac{A_0}{2}[\delta(\omega-\omega_0) - \delta(\omega+\omega_0)]
\end{aligned}
\tag{2.14}
$$

where $\delta(\omega)$ is the Dirac delta function.

Figure 2-4 shows the frequency-domain representation of the sine wave of radian frequency ω_0. Like the cosine wave, the Fourier components of a sine wave are two Dirac

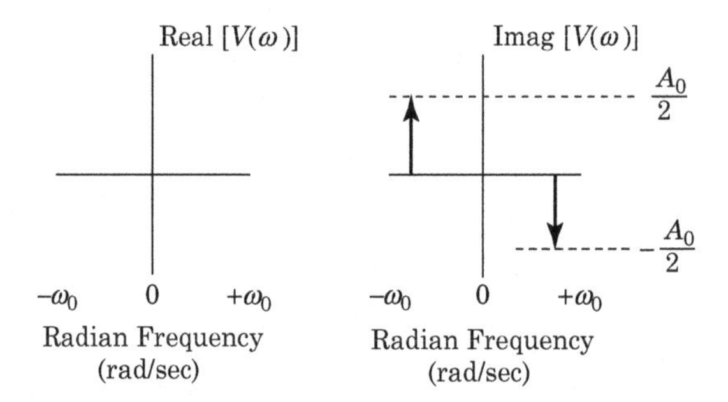

FIGURE 2-4 ■ The frequency-domain representation of a sine wave showing the positive and negative frequency components and showing the real and imaginary parts. In the case of the sine wave, the Fourier components are completely imaginary.

functions at $+\omega_0$ and $-\omega_0$. Unlike the cosine wave, the two components are completely imaginary.

2.2.2.4 A Final Example: A Cosine with a Nonzero Phase Angle

Let's work through equation (2.4) when $v(t)$ is a cosine with amplitude A_0, radian frequency ω_0 and phase angle of θ. We'll make use of the trigonometric relationship

$$\cos(\alpha \pm \beta) = \cos(\alpha)\cos(\beta) \mp \sin(\alpha)\sin(\beta) \tag{2.15}$$

to break the problem down into simple cosine and sine problems. We can write

$$\cos(\omega_0 t + \theta) = \cos(\omega_0 t)\cos(\theta) - \sin(\omega_0 t)\sin(\theta) \tag{2.16}$$

Our Fourier transform is

$$\begin{aligned}
V(\omega) &= \int_{t=-\infty}^{t=\infty} v(t)e^{-j\omega t}dt \\
&= \int_{t=-\infty}^{t=\infty} A_0\cos(\omega_0 t + \theta)e^{-j\omega t}dt \\
&= A_0 \int_{t=-\infty}^{t=\infty} \cos(\theta)\cos(\omega_0 t)e^{-j\omega t}dt - A_0 \int_{t=-\infty}^{t=\infty} \sin(\theta)\sin(\omega_0 t)e^{-j\omega t}dt \qquad (2.17) \\
&= A_0\cos(\theta) \int_{t=-\infty}^{t=\infty} \cos(\omega_0 t)e^{-j\omega t}dt - A_0\sin(\theta) \int_{t=-\infty}^{t=\infty} \sin(\omega_0 t)e^{-j\omega t}dt \\
&= \left\{\frac{A_0\cos(\theta)}{2}[\delta(\omega+\omega_0) + \delta(\omega-\omega_0)]\right\} - j\left\{\frac{A_0\sin(\theta)}{2}[\delta(\omega-\omega_0) - \delta(\omega+\omega_0)]\right\}
\end{aligned}$$

where $\delta(\omega)$ is the Dirac delta function.

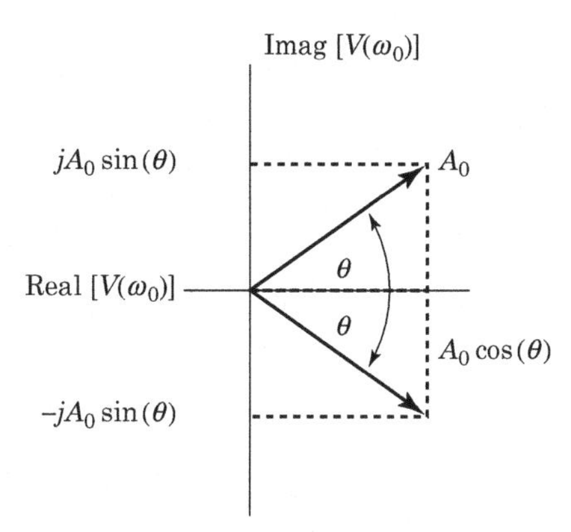

FIGURE 2-5 ■ The phasor diagram representation of a cosine wave of amplitude A_0, radian frequency ω_0, and phase angle θ. This figure is a geometrical interpretation of equation (2.18) as two vectors whose amplitudes are A_0 and whose angles with the real axis are θ and $-\theta$. The real part is $A_0 \cos(\theta)$, and the imaginary part is $A_0 \sin(\theta)$.

2.2.3 The Phasor Domain

Equation (2.17) is made up of components from equations (2.13) and (2.14). We regroup equation (2.17) into its positive- and negative-frequency components.

$$
\begin{aligned}
V(\omega) = &\left[\frac{A_0 \cos(\theta)}{2} \partial(\omega + \omega_0) + j\frac{A_0 \sin(\theta)}{2} \partial(\omega + \omega_0)\right] \\
&+ \left[\frac{A_0 \cos(\theta)}{2} \partial(\omega - \omega_0) - j\frac{A_0 \sin(\theta)}{2} \partial(\omega - \omega_0)\right]
\end{aligned}
\tag{2.18}
$$

It's useful to interpret equation (2.18) geometrically. If we ignore the common factor of ½, we can graphically interpret equation (2.18) as the phasor diagram of Figure 2-5.

2.2.3.1 Single-Sided Spectra

Analysis of equation (2.4) reveals that the positive- and negative-frequency components of any arbitrary real-valued signal are related in the following ways:

- The amplitude of any component present at a positive frequency, f_{POS}, always equals the amplitude of the component present at the corresponding negative frequency, f_{NEG}.
- The phase of the negative-frequency component always equals the negative of the phase of the corresponding positive-frequency component. In other words, for any positive-frequency component f, the phase of f, $\varphi_{POS} = -\varphi_{NEG}$, the phase of the component at $-f$.

More concisely, we can say that if a frequency-domain component at $+f$ has a value of $a + jb = M\angle\varphi$, then the component at $-f$ will have a value of $a - jb = M\angle - \varphi$. The components are complex conjugates of each other. The complex conjugate relationship is true only for real-valued signals, or signals that we can express in the time domain using real-valued samples (as opposed to signals we must express using complex samples).

Figure 2-6 shows the double-sided spectrum of a modulated, real-valued signal. The major implication of the spectral symmetry of real-valued signals is that the positive- and negative-frequency components contain identical information. We need only the positive- or the negative-frequency components to completely describe the signal.

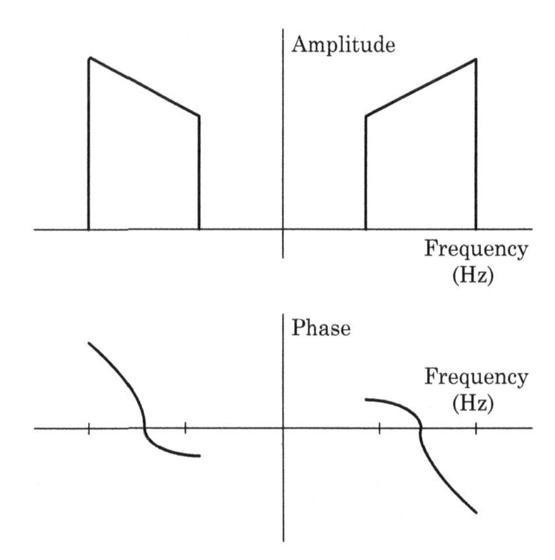

FIGURE 2-6 ■ The frequency-domain spectrum of a real-valued modulated signal, showing both positive and negative frequency components. The amplitude of every component at negative frequency $-f$ equals the amplitude of the component at positive frequency $+f$. In other words, the spectra are mirror images of each other. The phases of the components at $-f$ is the negative of the component at $+f$.

EXAMPLE

Double-Sided Components of a Sawtooth Wave

Figure 2-7 shows the positive-frequency Fourier components of a sawtooth wave whose period is 1 μsec. Find the negative-frequency Fourier components.

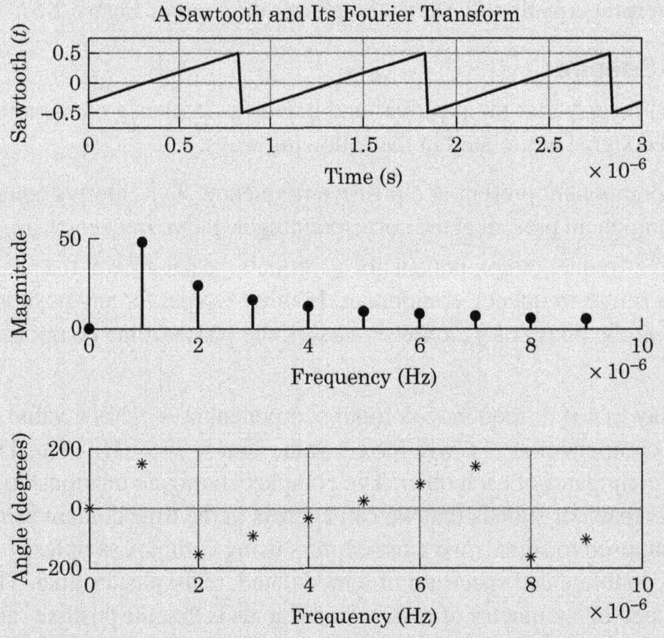

FIGURE 2-7 ■ The positive-frequency components of a sawtooth wave.

Solution

We know that if a frequency-domain component at $+f$ has a value of $a+jb=M\angle\varphi$, then the component at $-f$ will have a value of $a-jb=M\angle-\varphi$. The positive and negative Fourier components are given in Table 2-1.

TABLE 2-1 The positive and negative Fourier components of a triangle wave.

	Positive-Frequency Fourier Components		Negative-Frequency Fourier Components	
Frequency (MHz)	**Rectangular**	**(Magnitude∠Phase)**	**Rectangular**	**(Magnitude∠Phase)**
0	$0.0+j0.0$	$0\angle 0$	$0.0+j0.0$	$0\angle 0$
1	$-40.5+j24.9$	$47.6\angle 148$	$-40.5-j24.9$	$47.6\angle-148$
2	$-21.2-j10.7$	$23.8\angle-153$	$-21.2+j10.7$	$23.8\angle 153$
3	$-1.3-j15.8$	$15.9\angle-95$	$-1.3+j15.8$	$15.9\angle 95$
4	$9.6-j7.1$	$11.9\angle-36$	$9.6+j7.1$	$11.9\angle 36$
5	$8.8+j3.6$	$9.5\angle 22$	$8.8-j3.6$	$9.5\angle-22$
6	$1.3+j7.8$	$7.9\angle 80$	$1.3-j7.8$	$7.9\angle-80$
7	$-5.1+j4.5$	$6.8\angle 139$	$-5.1-j4.5$	$6.8\angle-139$
8	$-5.7-j1.8$	$6.0\angle-163$	$-5.7+j1.8$	$6.0\angle 163$
9	$-1.3-j5.1$	$5.3\angle-104$	$-1.3+j5.1$	$5.3\angle 104$

Note: The positive and negative components are complex conjugates of each other. The magnitudes of the components are identical, but the angles are negatives of one another.

EXAMPLE

Fourier Components of the Sawtooth on the Complex Plane

We can completely describe any signal via its complex Fourier components. For example, the +2 MHz component of our triangle wave is $-21.2-j10.7$. Figure 2-8 shows this quantity plotted on the complex plane.

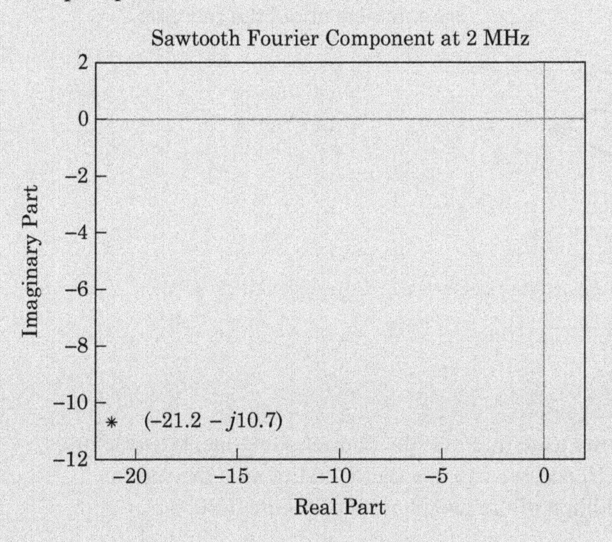

FIGURE 2-8 ■ The 2 MHz component of the 1 MHz sawtooth waveform plotted on the complex plane.

We can interpret the individual Fourier components of a signal as phasors. Figure 2-9 shows the 2 MHz component of the sawtooth expressed as a vector or phasor.

Figure 2-10 shows the sawtooth's positive and negative 2 MHz components plotted as phasors.

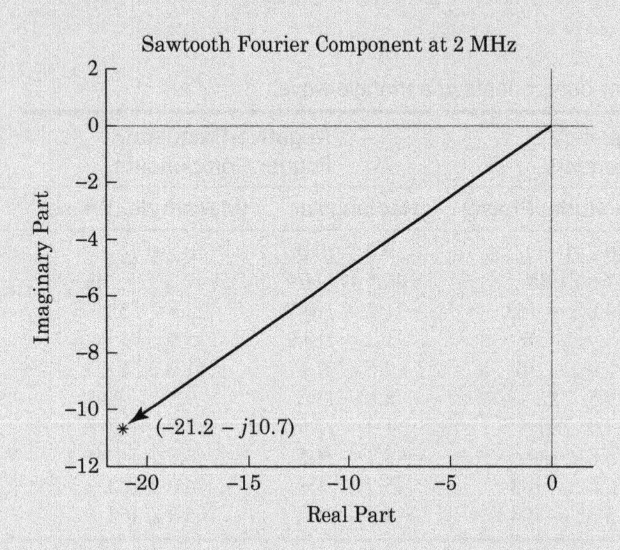

FIGURE 2-9 ■ The 2 MHz component of the 1 MHz sawtooth interpreted as a vector or phasor. This interpretation implies both magnitude and direction. It also implies that we can perform vector mathematics on multiple components. This figure shows only one positive-frequency component.

As Figure 2-10 shows, the +2 and −2 MHz Fourier components are equal in magnitude and are symmetrical about the real axis. This symmetry holds for any two Fourier components at f_{pos} and f_{neg} where $f_{pos} = -f_{neg}$. All the information we need to describe the signal is present on one side of the spectrum—either the positive or the negative band will describe the signal completely.

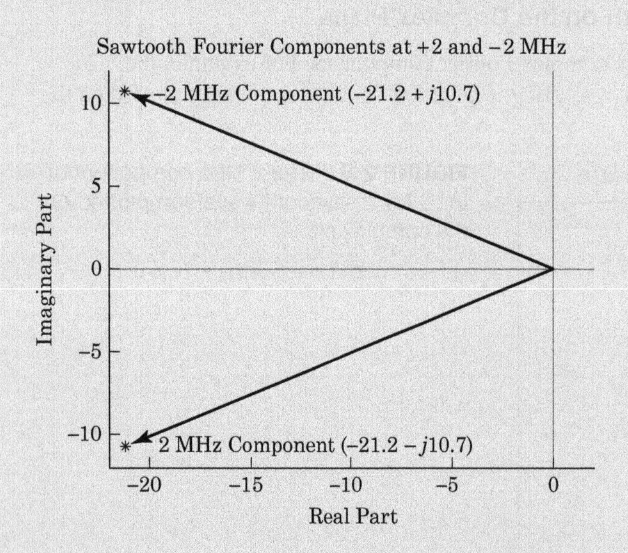

FIGURE 2-10 ■ The +2 MHz and −2 MHz Fourier components of the 1 MHz sawtooth plotted as vectors or phasors. The two Fourier components are symmetrical about the real axis.

The phasor representation of a signal allows us to interpret the Fourier components as vectors, which possess both magnitude and direction. Hence, we can use vector addition on the various components. Figure 2-11 shows the vector addition of the two phasors of Figure 2-10.

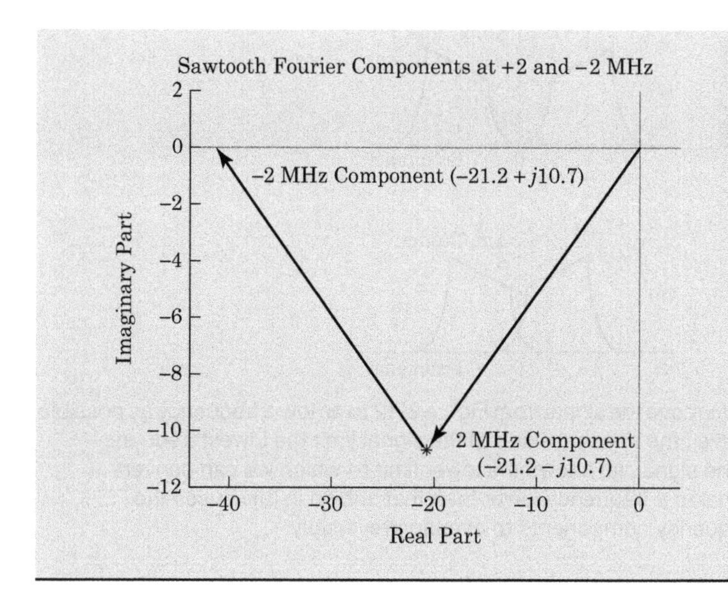

FIGURE 2-11 ■ The +2 MHz and −2 MHz Fourier components of the 1 MHz sawtooth, plotted as vectors placed head to tail. Since the original sawtooth signal was a real-valued signal, any two spectrally symmetrical components always vectorially add to produce a real-valued resultant.

2.3 | SINGLE-SIDED SPECTRA AND COMPLEX BASEBANDING

2.3.1 Complex Signals

We'd like to remove (or never produce) the redundant spectrum present in real-valued signals. Mathematical analysis reveals the spectral redundancy is a consequence of the signal being real-valued. Figure 2-12 shows the double-sided spectrum of a signal.

To digitize the signal shown in Figure 2-12, we'd first like to move it down to a low frequency before applying the signal to an analog-to-digital converter (ADC). Reducing the center frequency allows us to use less expensive components and to consume less power. However, the skirts of the signal in Figure 2-12 do not drop immediately off to zero amplitude, and, as Figure 2-13 shows, the skirts limit the lowest frequency to which we can convert the signal.

Figure 2-14 shows the results of removing the negative frequencies from our signal. We could move the signal to any center frequency, including 0 hertz, without worrying about the upper and lower components overlapping.

The signal shown in Figure 2-14 is a complex-valued signal and exhibits a single-sided spectrum. The process by which we convert a signal from the double-side representation of Figure 2-12 to its single-sided representation of Figure 2-14 is complex basebanding. Representing a signal as its complex equivalent describes the signal very efficiently. For example, the complex representation allows us to measure the instantaneous amplitude

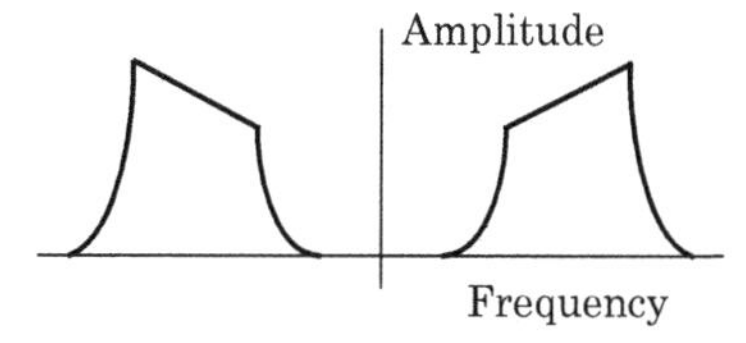

FIGURE 2-12 ■ The double-sided spectrum of a signal. Note that the skirts of the signal do not immediately drop to zero amplitude.

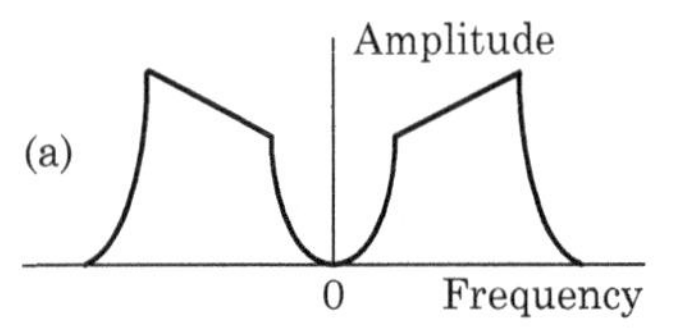

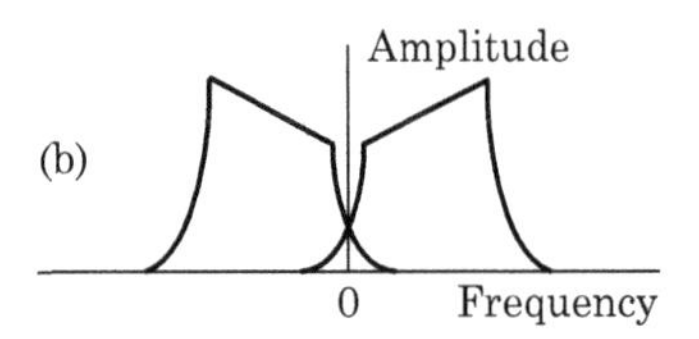

FIGURE 2-13 ■ We'd like to move the signal from Figure 2-12 to as low a frequency as possible before we digitize it. However, the spectral skirts of the signal limit the lowest frequency to which we can convert the signal. (a) shows the lower limit to which we can convert the signal. Moving the signal to a frequency lower than that shown in (b) causes the positive- and negative-frequency components to overlap irreversibly.

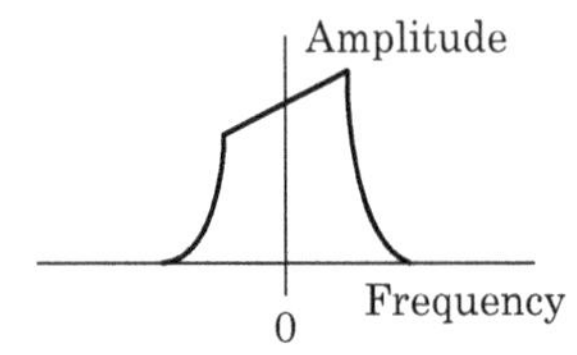

FIGURE 2-14 ■ By removing the negative-frequency elements from the signal shown in Figure 2-12 and Figure 2-13, we could move the signal to any arbitrary center frequency, including 0 hertz.

and phase of a signal on a sample-by-sample basis. We can measure the instantaneous frequency using only two samples.

To remove the negative frequencies from the double-sided signal of Figure 2-12 to produce the single-sided signal of Figure 2-14, we turn to the inverse Fourier transform (IFT). The IFT converts a signal from its frequency-domain representation, $V(\omega)$, into its time-domain representation, $v(t)$. The equation of the IFT is

$$\begin{aligned} v(t) &= \int_{\omega=-\infty}^{\omega=\infty} V(\omega)e^{j\omega t}d\omega \\ &= \int_{\omega=-\infty}^{\omega=\infty} V(\omega)[\cos(\omega t) + j\sin(\omega t)]d\omega \end{aligned} \tag{2.19}$$

We want to find the time-domain representation of a signal whose spectrum consists only of positive frequencies. Figure 2-15 shows the spectrum of a single-sided signal.

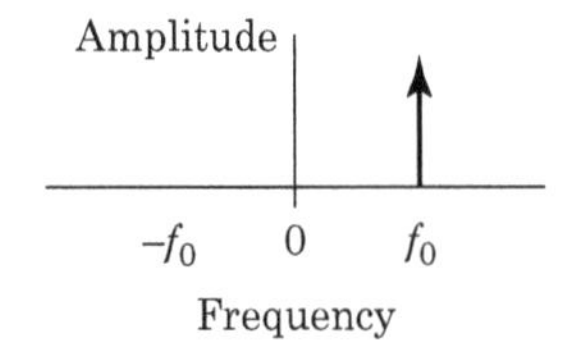

FIGURE 2-15 ■ The spectrum of a simple single-sided signal consisting of a single impulse at f_0. We seek the time-domain representation of this signal.

We use equation (2.19) to convert the spectrum of Figure 2-15 into its time-domain representation:

$$\begin{aligned} v(t) &= \int_{\omega=-\infty}^{\omega=\infty} V(\omega) e^{j\omega t} d\omega \\ &= \int_{\omega=-\infty}^{\omega=\infty} V(\omega)\, [\cos(\omega t) + j \sin(\omega t)] d\omega \\ &= \int_{\omega=-\infty}^{\omega=\infty} \partial(\omega_0)\, [\cos(\omega t) + j \sin(\omega t)] d\omega \\ &= \int_{\omega=\omega_0} [\cos(\omega t) + j \sin(\omega t)] d\omega \\ &= \cos(\omega_0 t) + j \sin(\omega_0 t) \\ &= I(t) + jQ(t) \end{aligned} \tag{2.20}$$

Equation (2.20) reveals that the time-domain waveform is complex-valued. The quantity $I(t)$ is the real component of the complex signal, whereas the quantity $Q(t)$ is the signal's imaginary component. We must keep track of the two separate waveforms, $I(t)$ and $Q(t)$.

2.3.2 Examples of Complex Signals

Figure 2-16 shows a 1 MHz, single-sided sinusoid expressed in the frequency domain.

We use the IFT of equation (2.19) to convert +1 MHz frequency-domain signal shown in Figure 2-16 into its time-domain representation:

$$\begin{aligned} v(t) &= \int_{\omega=-\infty}^{\omega=\infty} V(\omega) e^{j\omega t} d\omega \\ &= \int_{\omega=-\infty}^{\omega=\infty} V(\omega)\, [\cos(\omega t) + j \sin(\omega t)] d\omega \\ &= \int_{\omega=-\infty}^{\omega=\infty} \partial(\omega_{1\,MHz})\, [\cos(\omega t) + j \sin(\omega t)] d\omega \\ &= \int_{\omega=\omega_{1\,\mathrm{MHz}}} [\cos(\omega t) + j \sin(\omega t)] d\omega \\ &= \cos(\omega_{1\,\mathrm{MHz}} t) + j \sin(\omega_{1\,\mathrm{MHz}} t) \end{aligned} \tag{2.21}$$

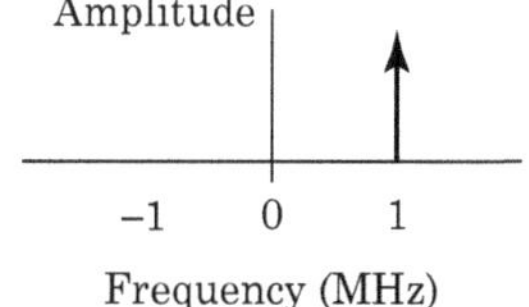

FIGURE 2-16 ■ The spectrum (i.e., frequency-domain representation) of a +1 MHz sinusoid. This signal contains only positive frequencies.

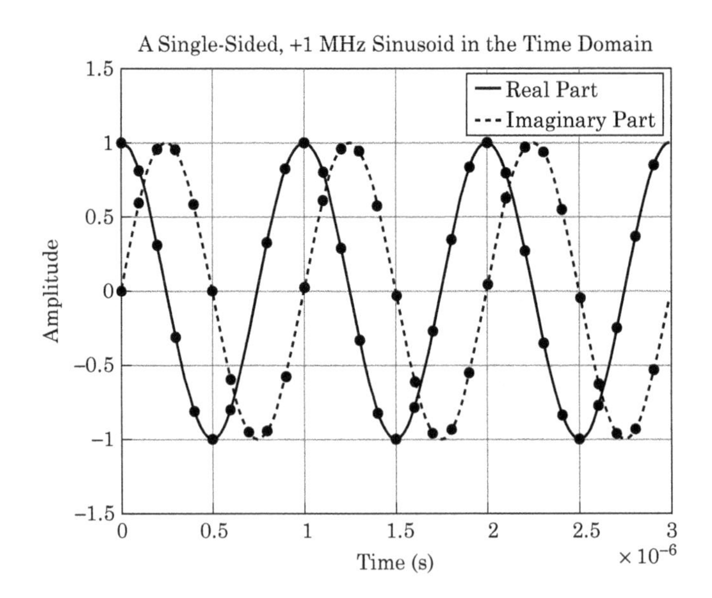

FIGURE 2-17 ■ The time-domain representation of the single-sided spectrum of Figure 2-16. The solid trace is the real part of the signal, whereas the dashed trace is the imaginary part of the signal. The black • indicates a sample point. Note that we require two real numbers, or one complex number, per sample point.

The time-domain equivalent of the spectrum shown in Figure 2-16 is a signal with both real and imaginary parts. Equation (2.21) shows that the real part of the +1 MHz sinusoid is a 1 MHz cosine. The imaginary part of the +1 MHz sinusoid is a 1 MHz sine. Figure 2-17 shows the complex sinusoid in the time domain. The time-domain representation requires two separate signals (or plots) to express completely the signal.

The complex samples of the signal of Figure 2-17 are shown in Table 2-2.

Table 2-2 shows the discrete samples from Figure 2-17, taken every 0.1 μsecond. We can view the two sample streams as separate channels (or wires or PCB traces), or we can view the two samples as a complex number $I(t) + jQ(t)$. These two representations are interchangeable.

2.3.3 Complex Signals as Phasors

Figure 2-18 shows the time-domain samples from Figure 2-17 and from Table 2-2 plotted on the complex plane. We interpret each complex point as a phasor.

2.3.4 Positive and Negative Frequencies

Figure 2-19 shows a −1 MHz sinusoid in the frequency domain.

We use the IFT of equation (2.19) to convert the spectrum of Figure 2-19 into its time-domain representation:

$$\begin{aligned} v(t) &= \int_{\omega=-\infty}^{\omega=\infty} V(\omega) e^{j\omega t} d\omega \\ &= \int_{\omega=-\infty}^{\omega=\infty} V(\omega) [\cos(\omega t) + j \sin(\omega t)] d\omega \\ &= \int_{\omega=-\infty}^{\omega=\infty} \partial(\omega_{-1\,MHz}) [\cos(\omega t) + j \sin(\omega t)] d\omega \end{aligned} \tag{2.22}$$

$$= \int_{\omega=\omega_{-1\,\text{MHz}}} [\cos(\omega t) + j\sin(\omega t)]d\omega$$
$$= \cos(\omega_{-1\,\text{MHz}} t) + j\sin(\omega_{-1\,\text{MHz}} t)$$
$$= \cos(\omega_{1\,\text{MHz}} t) - j\sin(\omega_{1\,\text{MHz}} t)$$

Figure 2-20 shows the time-domain representation of a −1 MHz sinusoid.

Recall that the time-domain expression for the +1 MHz sinusoid is

$$v(t) = \cos(\omega_{1\text{MHz}} t) + j\sin(\omega_{1\text{MHz}} t) \tag{2.23}$$

The time-domain representation of a −1 MHz sinusoid differs from that of the +1 MHz sinusoid only by sign of the complex term. Figure 2-21 shows a +1 MHz sinusoid plotted together with a −1 MHz sinusoid.

TABLE 2-2 ■ The time-domain signal of Figure 2-17 represented as a series of discrete samples.

t (μ second)	$I(t)$ – The Solid Trace of Figure 2-17	$Q(t)$ – The Dashed Trace of Figure 2-17	$I(t) + jQ(t)$
0.0	1.0000	0.0000	1.0000
0.10	0.8078	0.5895	0.8078 + j0.5895
0.20	0.3050	0.9523	0.3050 + j0.9523
0.30	−0.3150	0.9491	−0.3150 + j0.9491
0.40	−0.8139	0.5810	−0.8139 + j0.5810
0.50	−0.9999	−0.0105	−0.9999 − j0.0105
0.60	−0.8015	−0.5979	−0.8015 − j0.5979
0.70	−0.2950	−0.9555	−0.2950 − j0.9555
0.80	0.3250	−0.9457	0.3250 − j0.9457
0.90	0.8200	−0.5724	0.8200 − j0.5724
1.00	0.9998	0.0210	0.9998 + j0.0210
1.10	0.7952	0.6063	0.7952 + j0.6063
1.20	0.2849	0.9585	0.2849 + j0.9585
1.30	−0.3349	0.9423	−0.3349 + j0.9423
1.40	−0.8260	0.5637	−0.8260 + j0.5637
1.50	−0.9995	−0.0315	−0.9995 − j0.0315
1.60	−0.7888	−0.6146	−0.7888 − j0.6146
1.70	−0.2749	−0.9615	−0.2749 − j0.9615
1.80	0.3448	−0.9387	0.3448 − j0.9387
1.90	0.8318	−0.5550	0.8318 − j0.5550
2.00	0.9991	0.0420	0.9991 + j0.0420
2.10	0.7823	0.6229	0.7823 + j0.6229
2.20	0.2647	0.9643	0.2647 + j0.9643
2.30	−0.3546	0.9350	−0.3546 + j0.9350
2.40	−0.8376	0.5463	−0.8376 + j0.5463
2.50	−0.9986	−0.0525	−0.9986 − j0.0525
2.60	−0.7757	−0.6311	−0.7757 − j0.6311
2.70	−0.2546	−0.9670	−0.2546 − j0.9670
2.80	0.3644	−0.9312	0.3644 − j0.9312
2.90	0.8433	−0.5374	0.8433 − j0.5374

Note: We take a sample of the complex signal every 0.1 μsecond, and each sample is two numbers: one number describes the $I(t)$ waveform, whereas the second number describes the $Q(t)$ waveform. We can interpret these two sample streams as separate $I(t)$ and $Q(t)$ signals, or we can interpret the stream as a series of complex numbers.

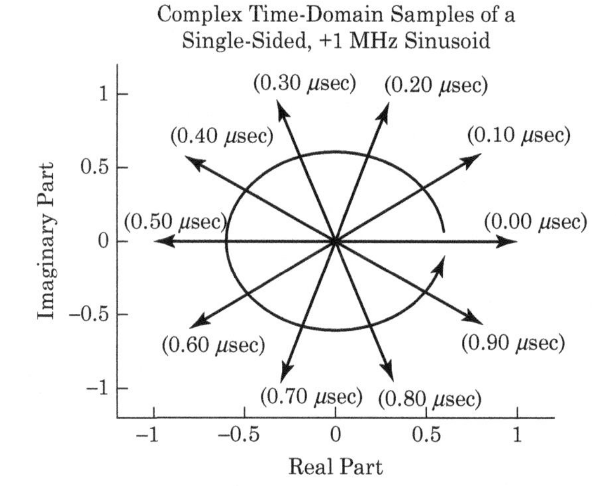

FIGURE 2-18 ■ The complex time-domain samples of the +1 MHz sinusoid from Figure 2-17 and Table 2-2 plotted on the complex plane as a series of phasors. Each phasor is labeled according to its sample time. The +1 MHz phasor rotates counterclockwise (CCW) once around the origin in 1 microsecond.

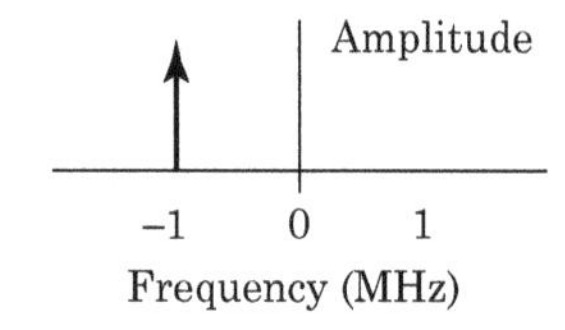

FIGURE 2-19 ■ The spectrum of a −1 MHz sinusoid. This signal contains only a single impulse at −1 MHz.

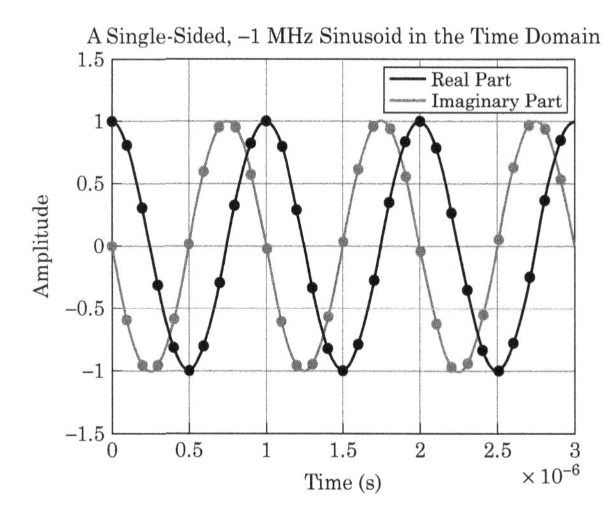

FIGURE 2-20 ■ The time-domain representation of a −1 MHz sinusoid. The black trace is the real part of the signal, whereas the gray trace is the imaginary part of the signal. The • indicates a sample point.

Figure 2-22 shows the phasor representation of a −1 MHz sinusoid.

The +1 MHz and the −1 MHz phasors rotate at the same speed, but they rotate in opposite directions.

2.3.5 Phasors and Frequencies

Figure 2-23 shows the phasor representation of sinusoids of different frequencies sampled at the same rate.

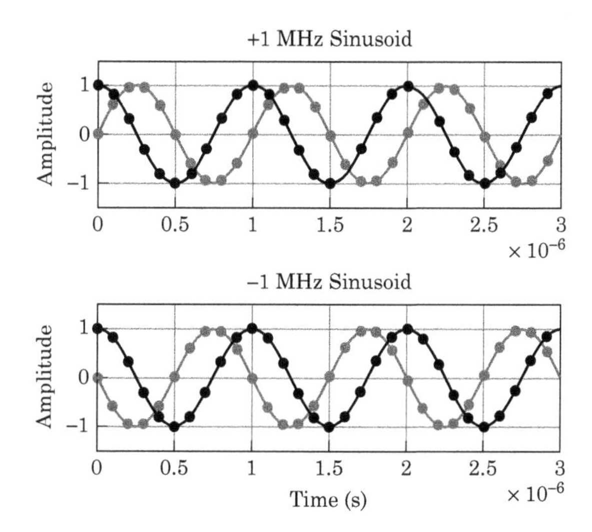

FIGURE 2-21 ■ The time-domain representation of a +1 MHz (top) and −1 MHz (bottom) sinusoid. The two complex waveforms differ only in the sign of the imaginary part. The black trace is the real part of the signal, whereas the gray trace is the imaginary part of the signal.

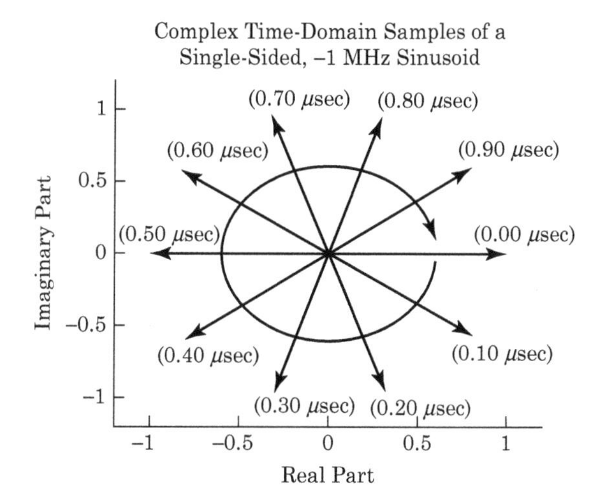

FIGURE 2-22 ■ The complex time-domain samples of the −1 MHz sinusoid, from Figure 2-20, plotted on the complex plane as a series of phasors. The −1 MHz phasor rotates clockwise. Like the +1 MHz phasor, the −1 MHz phasor makes it once around the origin in a microsecond.

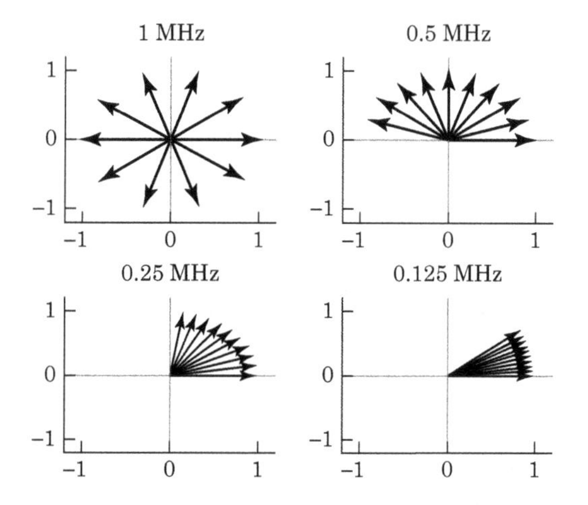

FIGURE 2-23 ■ The complex time-domain samples of sinusoids of +1 MHz, +0.5 MHz, +0.25 MHz, and +0.125 MHz. The time duration of each plot is 1 μsecond, and the sample rate is 0.1 μsecond.

The sinusoids in Figure 2-23 are positive frequencies so they rotate counterclockwise (CCW) about the origin. The +1 MHz sinusoid rotates once around the origin in 1 μsecond. The +0.5 MHz sinusoid makes it only halfway around the origin in the 1 μsec plot time. As you might expect, the 0.25 and 0.125 MHz sinusoids rotate only 0.25 and 0.125 times around the origin in the 1 μsecond plot time.

2.3.6 The Amplitude, Phase, and Frequency of a Complex Signal

Let us calculate the instantaneous amplitude, phase, and frequency of the −1 MHz sinusoid shown in Figure 2-20. We will arbitrarily select the 0.6 μsecond sample instant.

2.3.6.1 Measuring Amplitude

The instantaneous amplitude of the complex signal is the magnitude of the phasor at the sample instant:

$$\begin{aligned} \text{Amplitude} &= \sqrt{(\text{Real})^2 + (\text{Imag})^2} \\ \Rightarrow A(t) &= \sqrt{I^2(t) + Q^2(t)} \end{aligned} \tag{2.24}$$

We calculate the sample-by-sample amplitude of a complex signal by performing the square root of the sum of the squares on each sample. For the specific case of the −1 MHz sinusoid of Figure 2.20 at the 0.6 μsec sample instant, the amplitude is

$$\begin{aligned} A(0.6\mu\,\text{sec}) &= \sqrt{I^2(0.6\mu\,\text{sec}) + Q^2(0.6\mu\,\text{sec})} \\ &= \sqrt{(-0.8015)^2 + (0.5979)^2} \\ &= 1.0 \end{aligned} \tag{2.25}$$

2.3.6.2 Measuring Phase

The instantaneous phase of a complex number is

$$\begin{aligned} \phi &= \tan^{-1}\left(\frac{\text{Imag}}{\text{Real}}\right) \\ \Rightarrow \phi(t) &= \tan^{-1}\left[\frac{Q(t)}{I(t)}\right] \end{aligned} \tag{2.26}$$

For the specific case of the −1 MHz sinusoid of Figure 2-20 at the 0.6 μsec sample instant, the phase is

$$\begin{aligned} \phi(0.6\mu\,\text{sec}) &= \tan^{-1}\left[\frac{Q(0.6\mu\,\text{sec})}{I(0.6\mu\,\text{sec})}\right] \\ &= \tan^{-1}\left[\frac{0.5979}{-0.8015}\right] \\ &= \tan^{-1}[-0.7460] \\ &= -36.7^\circ \end{aligned} \tag{2.27}$$

2.3.6.3 Measuring Frequency

Frequency is the rate of change of the phase between two complex samples:

$$\omega = \frac{d\phi}{dt} \tag{2.28}$$

Practically, we find the phases of two adjacent complex samples then approximate equation (2.28) as a difference equation:

$$\begin{aligned} \omega &\approx \frac{\Delta\phi}{\Delta t} \\ &= \frac{\phi_1 - \phi_0}{t_1 - t_0} \end{aligned} \tag{2.29}$$

where

φ_0 and φ_1 are the phases of two adjacent complex samples and φ_1 occurs later than φ_0
Δt is the time between samples and $\Delta t = t_1 - t_0$

We'll use the two samples at 0.6 and 0.7 μsec to approximate the frequency at 0.65 μsec. The phase at 0.6 μsec is -36.7° or -0.641 radians. The complex sample at 0.7 μsec is $-0.2950 - j0.9555$, so the phase is -72.8° or -1.27 radians. The instantaneous frequency is

$$\begin{aligned} \omega &\approx \frac{\Delta\phi}{\Delta t} \\ &= \frac{\phi_1 - \phi_0}{\Delta t} \\ &= \frac{-1.27 - (-0.641)}{0.7\mu - 0.6\mu} \\ &= \frac{-0.630}{0.1\mu} \\ &= -6.30 \frac{\text{Mrad}}{\text{sec}} \\ f &= -1.0\,\text{MHz} \end{aligned} \tag{2.30}$$

2.3.7 Example: An Upchirp

Figure 2-24 shows a waveform whose frequency is a linear function of time.

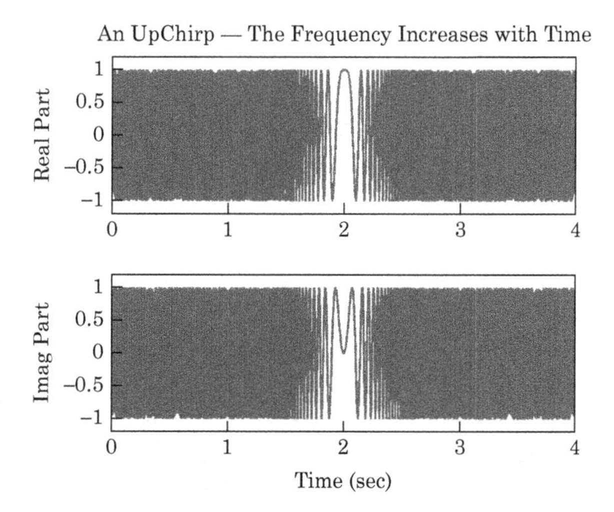

FIGURE 2-24 ■ The complex time-domain samples of an upchirp. The frequency of this signal increases linearly with time. At time $t = 0$ seconds, the frequency is -200 Hz. The frequency at $t = 4$ seconds is $+200$ Hz.

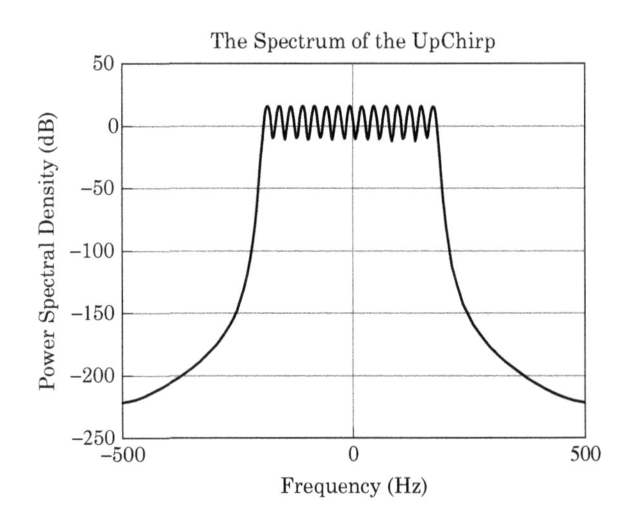

FIGURE 2-25 ■ A plot of the power spectral density (PSD) of the signal from Figure 2-24. Over the 4 second signal duration, the signal moves from −200 Hz to +200 Hz.

Figure 2-25 shows the power spectral density of the signal from Figure 2-24. Over the 4 second signal duration, the signal moves from −200 Hz to +200 Hz. Figure 2-25 does not express a sense of the movement of the frequency over time.

Figure 2-26 shows a spectrogram, or a time versus frequency plot, of the signal of Figure 2-24 and Figure 2-25. A spectrogram gives a sense of how the frequency of a signal changes with time.

Figure 2-27 shows an expanded view of the time-domain plot of Figure 2-24. We'll refer to this figure for the following discussion.

2.3.7.1 Amplitude Detection of the Upchirp

Equation (2.24) gives the amplitude of a complex sample. Table 2-3 shows the amplitude of the sample from Figure 2-27. By design, the amplitude of every sample of this signal is unity.

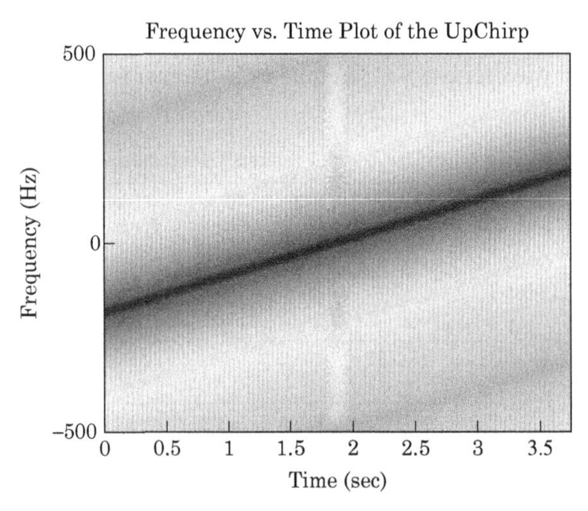

FIGURE 2-26 ■ The spectrogram, or a time versus frequency plot, of the signal of Figure 2-24. The x-axis is time, the y-axis is frequency, and darker colors represent stronger signal amplitudes. The frequency of the signal begins at −200 Hz at $t = 0$ seconds and ends at +200 Hz at $t = 4$ seconds.

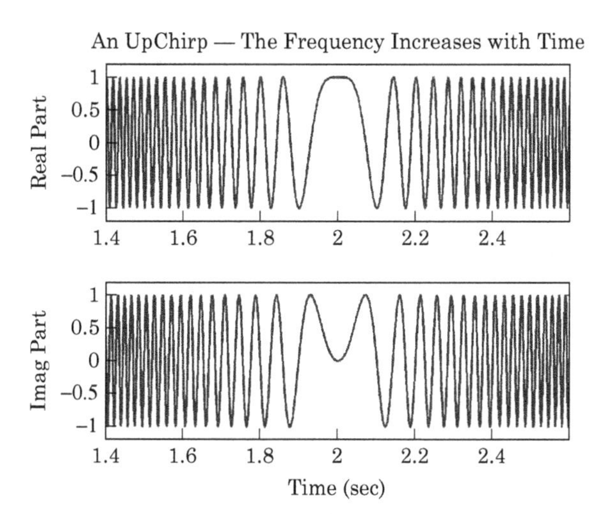

FIGURE 2-27 ■ A close-up view of the time-domain plot of Figure 2-24. The scale is expanded around time $t = 2$ seconds, where the signal frequency passes through 0 hertz.

TABLE 2-3 ■ Several samples of the upchirp signal from Figure 2-27.

Time (sec)	Complex Sample $I(t) + jQ(t)$	Amplitude via Equation (2.24)
1.9000	$-1.0000 - j0.0000$	1.0000
1.9100	$-0.8271 + j0.5621$	1.0000
1.9200	$-0.4258 + j0.9048$	1.0000
1.9300	$0.0314 + j0.9995$	1.0000
1.9400	$0.4258 + j0.9048$	1.0000
1.9500	$0.7071 + j0.7071$	1.0000
1.9600	$0.8763 + j0.4818$	1.0000
1.9700	$0.9603 + j0.2790$	1.0000
1.9800	$0.9921 + j0.1253$	1.0000
1.9900	$0.9995 + j0.0314$	1.0000
2.0000	$1.0000 + j0.0000$	1.0000
2.0100	$0.9997 + j0.0254$	1.0000
2.0200	$0.9936 + j0.1132$	1.0000
2.0300	$0.9653 + j0.2611$	1.0000
2.0400	$0.8880 + j0.4599$	1.0000
2.0500	$0.7288 + j0.6848$	1.0000
2.0600	$0.4593 + j0.8883$	1.0000
2.0700	$0.0750 + j0.9972$	1.0000
2.0800	$-0.3801 + j0.9250$	1.0000
2.0900	$-0.7942 + j0.6077$	1.0000
2.1000	$-0.9980 + j0.0625$	1.0000

Note: The rightmost column shows the amplitude of each complex sample. The amplitude of every sample is unity by design.

2.3.7.2 Phase Detection of the UpChirp

Equations (2.28) through (2.29) describe the instantaneous phase of a complex number as the arc tangent of the imaginary part over the real part. As a function of time, the relationship is

$$\phi(t) = \tan^{-1}\left[\frac{Q(t)}{I(t)}\right] \tag{2.31}$$

TABLE 2-4 ■ The phase of several samples of the upchirp signal of Figure 2-27.

Time (sec)	Complex Sample $I(t) + jQ(t)$	Phase (Radians) Via Equation (2.31)	Cumulative Phase (radians) Via Equation (2.31)
1.9000	$-1.0000 - j0.0000$	-3.1416	-9.4248
1.9100	$-0.8271 + j0.5621$	2.5447	-10.0217
1.9200	$-0.4258 + j0.9048$	2.0106	-10.5558
1.9300	$0.0314 + j0.9995$	1.5394	-11.0270
1.9400	$0.4258 + j0.9048$	1.1310	-11.4354
1.9500	$0.7071 + j0.7071$	0.7854	-11.7810
1.9600	$0.8763 + j0.4818$	0.5027	-12.0637
1.9700	$0.9603 + j0.2790$	0.2827	-12.2836
1.9800	$0.9921 + j0.1253$	0.1257	-12.4407
1.9900	$0.9995 + j0.0314$	0.0314	-12.5350
2.0000	$1.0000 + j0.0000$	0.0000	-12.5664
2.0100	$0.9997 + j0.0254$	0.0254	-12.5409
2.0200	$0.9936 + j0.1132$	0.1134	-12.4530
2.0300	$0.9653 + j0.2611$	0.2642	-12.3022
2.0400	$0.8880 + j0.4599$	0.4778	-12.0885
2.0500	$0.7288 + j0.6848$	0.7543	-11.8121
2.0600	$0.4593 + j0.8883$	1.0936	-11.4728
2.0700	$0.0750 + j0.9972$	1.4957	-11.0707
2.0800	$-0.3801 + j0.9250$	1.9607	-10.6057
2.0900	$-0.7942 + j0.6077$	2.4885	-10.0779
2.1000	$-0.9980 + j0.0625$	3.0791	-9.4873

Note: The third column shows the phase of each complex sample, restricted to lie between $-\pi$ and $+\pi$. The fourth column is the unwrapped, or cumulative phase, of each complex sample.

Table 2-4 shows several samples of the complex signal and the phase of each signal. Figure 2-28 shows the phase as a function of time.

For example, the phase of the sample at $t = 1.9100$ seconds is

$$\begin{aligned}\phi(1.9100\text{ sec}) &= \tan^{-1}\left[\frac{Q(1.9100\text{ sec})}{I(1.9100\text{ sec})}\right]\\ &= \tan^{-1}\left[\frac{0.5621}{-0.8271}\right]\\ &= \tan^{-1}[-0.6796]\\ &= 145^\circ\\ &= 2.547\text{ rad}\end{aligned} \tag{2.32}$$

Figure 2-28 shows the phase of the upchirp signal from Figure 2-27 and Table 2-4. While accurate, the plot isn't very intuitive because the discontinuities obscure the true nature of the phase. The sudden jumps in phase of Figure 2-28 are artifacts of the phase changing abruptly from $-\pi$ to $+\pi$ or from $+\pi$ to $-\pi$.

Figure 2-29 is a phasor diagram of the two complex sample points at 1.8200 and 1.8300 seconds. There are two ways to interpret the phases of the samples. We can allow the phase to abruptly change signs from -2.3876 radians at 1.82 seconds to $+2.7960$ radians at 1.83 seconds. This abrupt phase changes results in the sawtooth behavior of

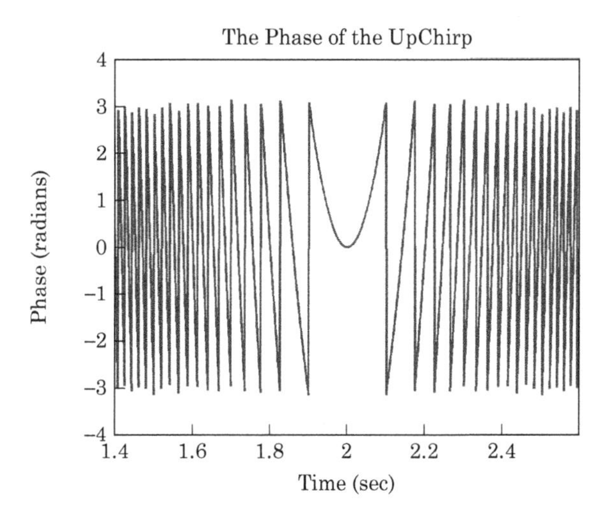

FIGURE 2-28 ■ The phase of the upchirp of Figure 2-27 and *Table 2-4*. This plot was made from the data in the third column of *Table 2-4*.

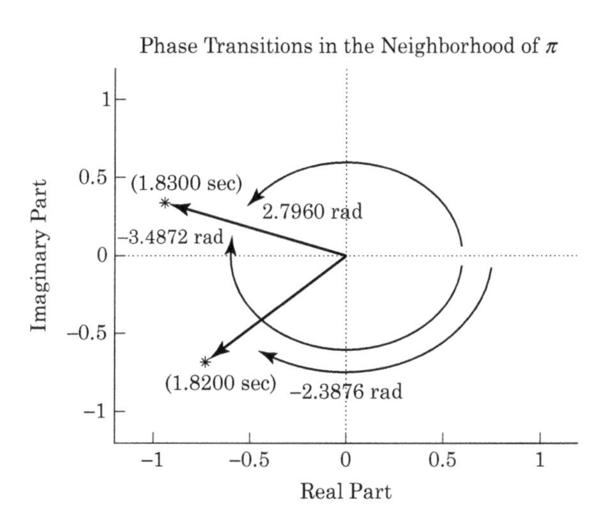

FIGURE 2-29 ■ The phasor diagram of the two samples at 1.82 and 1.83 seconds. We can interpret the phase of the complex sample at 1.83 seconds as either 2.7960 radians or −3.4872 radians. The second assumption (of $\varphi = -3.4872$) produces a more intuitive phase plot (see Figure 2-30).

Figure 2-28. Alternately, we can assume that the phasor continues to rotate in the clockwise direction and assume that the phase changed from −2.3876 radians at 1.82 seconds to −3.4872 radians at 1.83 seconds. This assumption produces the smoother, more intuitive phase plot of Figure 2-30.

Figure 2-30 shows the phase plot of the upchirp assuming that the phase never makes a change of more than π radians. If we see a phase change that's larger than π radians, we add or subtract 2π to remove the apparent phase discontinuities.

The phase plot of Figure 2-30 is a paraboloid. We built the upchirp to exhibit a linear change in frequency with time. Phase is the integral of frequency; hence a linear frequency change with time results in a parabolic phase change over time.

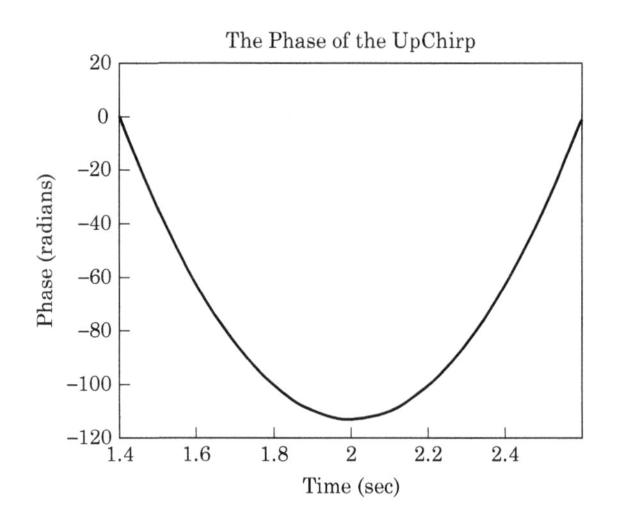

FIGURE 2-30 The unwrapped or cumulative phase of the upchirp of Figure 2-27 and *Table 2-4*. This plot shows the same phase data as Figure 2-28, but we've removed the large discontinuities associated with the rotation of the phase through π radians.

2.3.7.3 Frequency Detection of the Upchirp

Equation (2.29), repeated and expanded here, is an approximation for the instantaneous frequency of the signal

$$\begin{aligned} \omega &\approx \frac{\Delta\phi}{\Delta t} \\ &= \frac{\phi_1 - \phi_0}{t_1 - t_0} \\ &= \frac{\phi_1 - \phi_0}{\Delta t} \\ &= (\phi_1 - \phi_0)(SR) \end{aligned} \tag{2.33}$$

where

φ_0 and φ_1 are the phases of two adjacent complex samples and φ_1 occurs later than φ_0

Δt is the time between samples and the sample rate, *SR*, $= 1/\Delta t$

The frequency detection algorithm of equation (2.33) is the derivative of the phase plot. Applying equation (2.33) using the complex samples at 1.82 sec and 1.83 sec produces

$$\begin{aligned} \omega &\approx \frac{\phi_1 - \phi_0}{t_1 - t_0} \\ &= \frac{\phi_1 - \phi_0}{\Delta t} \\ &= \frac{-2.3876-(-3.4872)}{0.01} \\ &= 110\,\frac{\text{rad}}{\text{sec}} \\ &= 17.5\,\text{Hz} \end{aligned} \tag{2.34}$$

The instantaneous frequency of the upchirp 1.825 seconds is about 17.5 Hz. Figure 2-31 shows a plot of the frequency of the upchirp.

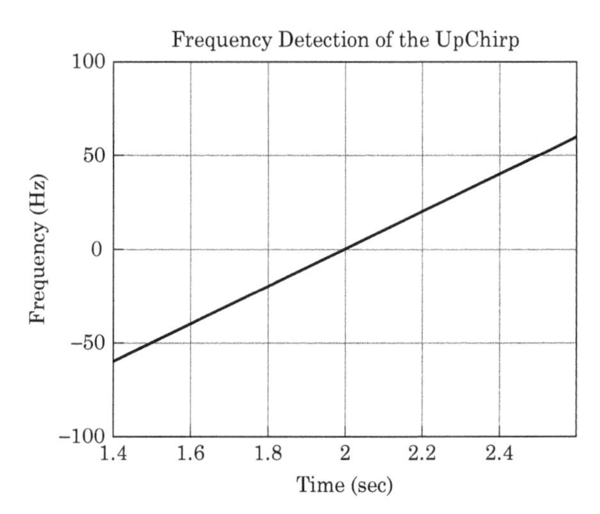

FIGURE 2-31 ■ The results of applying the frequency-detection algorithm of equation (2.33) the upchirp signal. We designed the signal to exhibit a linear frequency change over time. The frequency detection algorithm reveals that the frequency of the signal is, in fact, a linear function of time. Since frequency is the rate of change of phase over time, this plot is the derivative of Figure 2-30.

2.4 TWO NOISELESS SINE WAVES

As a step toward examining modulation and noisy waveforms, we now look at the sum of two sine waves.

2.4.1 Time Domain

Figure 2-32 shows the time-domain sum of two sine waveforms. The equation is

$$\begin{aligned} V_1(t) &= V_{pk,1}\cos(\omega_1 t + \theta_1) \\ V_2(t) &= V_{pk,2}\cos(\omega_2 t + \theta_2) \\ f(t) &= V_1(t) + V_2(t) \\ &= V_{pk,1}\cos(\omega_1 t + \theta_1) + V_{pk,2}\cos(\omega_2 t + \theta_2) \end{aligned} \tag{2.35}$$

where

ω_1 = radian frequency of the first sine wave
ω_2 = radian frequency of the second sine wave and $\omega_1 \neq \omega_2$

2.4.1.1 Zero Crossings

Figure 2-33 shows the zero crossings of the waveform described by equation (2.35).

Due to the noiseless nature of this waveform, the zero crossings occur at exact, mathematically deterministic times. The zero crossings occur when

$$V_{pk,1}\cos(\omega_1 t + \theta_1) = -V_{pk,2}\cos(\omega_2 t + \theta_2) = 0 \tag{2.36}$$

If the amplitude of $V_{1,pk}$ is much greater than the amplitude of $V_{2,pk}$, then we can view the zero crossings of the composite signal as being mostly controlled by the larger signal. The smaller signal causes the zero crossings to deviate slightly from those of the larger signal. Figure 2-34 shows the effect.

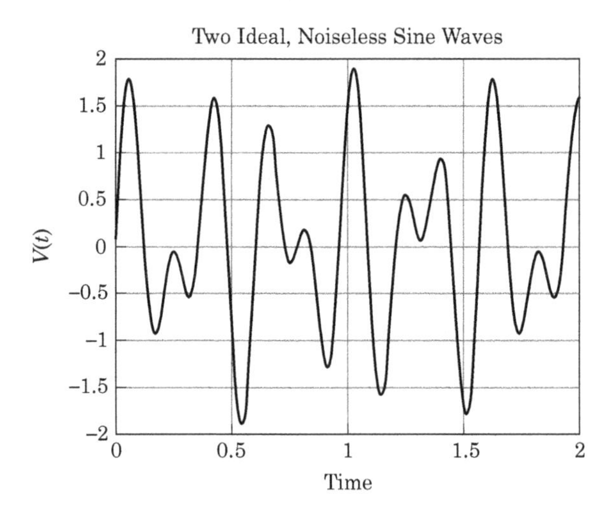

FIGURE 2-32 ■ Two sine waves, at different frequencies, added together. The frequencies of the two sine waves are f_1 and f_2.

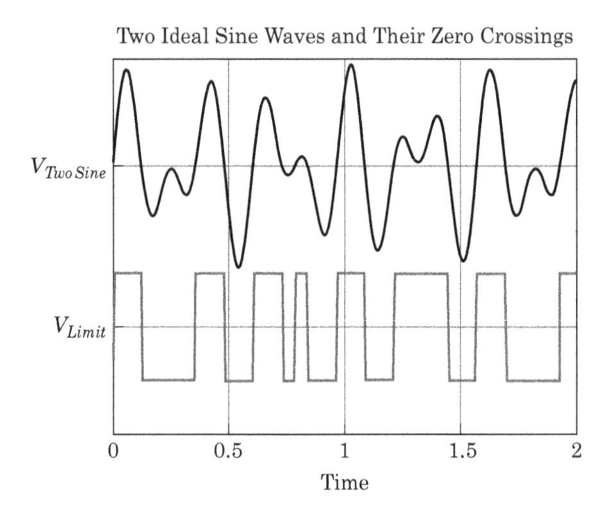

FIGURE 2-33 ■ The zero crossings of the waveform shown in Figure 2-32.

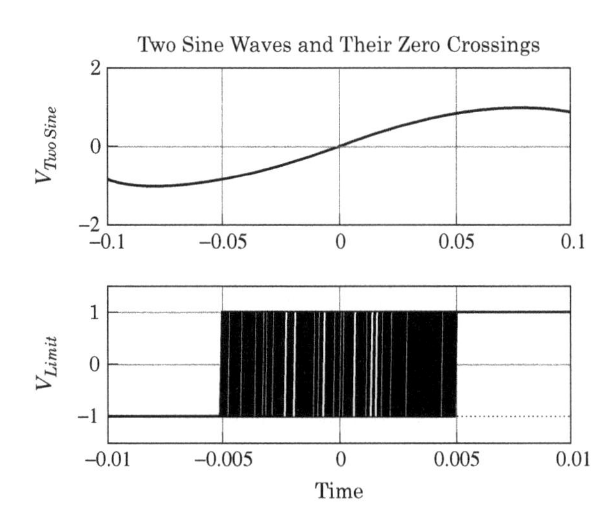

FIGURE 2-34 ■ The zero crossings of two sine waves when one sine wave is much larger than the other. The zero crossings occur in the same neighborhood as the larger sine wave, but the exact time of zero crossing is affected by the smaller sine. The time scale of the lower plot is expanded for clarity.

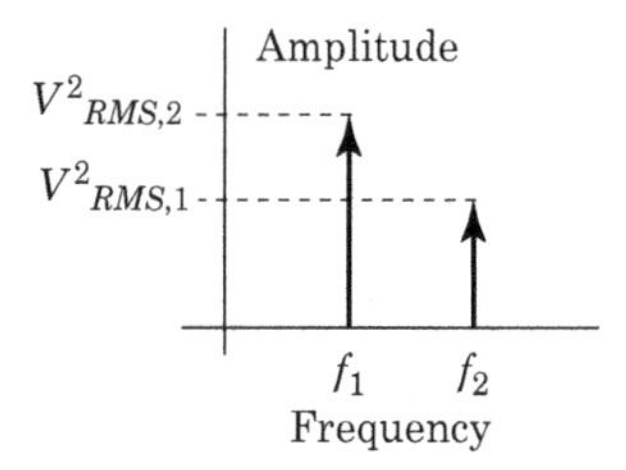

FIGURE 2-35 ■ The frequency-domain representation of two noiseless sine waves.

This view of the zero crossings will be helpful when we discuss mixers and phase noise in later chapters.

2.4.2 Frequency Domain

Figure 2-35 shows the frequency-domain representation of equation (2.35). We see two impulse functions at frequencies f_1 and f_2. Note that the magnitude of each impulse function is a function of $V_{pk,1}$ and $V_{pk,2}$.

2.4.3 Phasor

Figure 2-36 shows the phasor representation of the two sine waves. There are two vectors: $\mathbf{V}_1$ and $\mathbf{V}_2$. The length of the phasors $\mathbf{V}_1$ and $\mathbf{V}_2$ are $V_{pk,1}$ and $V_{pk,2}$, respectively.

Each phasor rotates about the origin. Phasor $\mathbf{V}_1$ rotates with a radian velocity of ω_1 while phasor $\mathbf{V}_2$ rotates with a radian velocity of ω_2.

The two phasor are rotating at different rates. It's often useful to use one of the phasors as a reference. The effect is to freeze one of the phasors so it doesn't rotate. The rest of the phasors then rotate with respect to the reference phasor (see Figure 2-37). We usually choose the vector with the largest magnitude to be the reference phasor ($\mathbf{V}_1$ in this case).

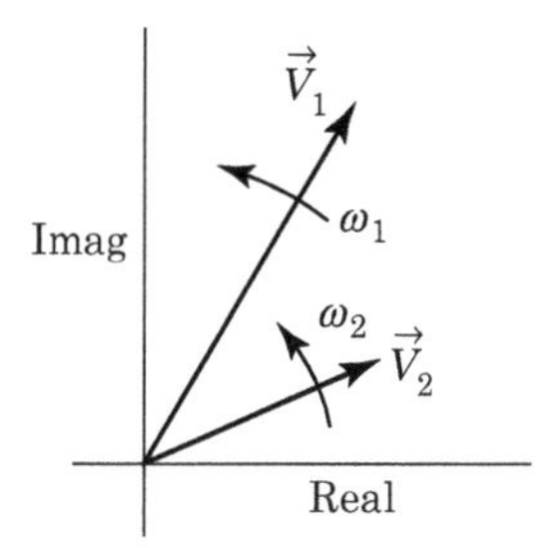

FIGURE 2-36 ■ The phasor-domain representation of two sine waves. Each vector rotates counterclockwise. One vector, V_1, rotates at a radian frequency of ω_1, while the second vector, V_2, rotates at ω_2.

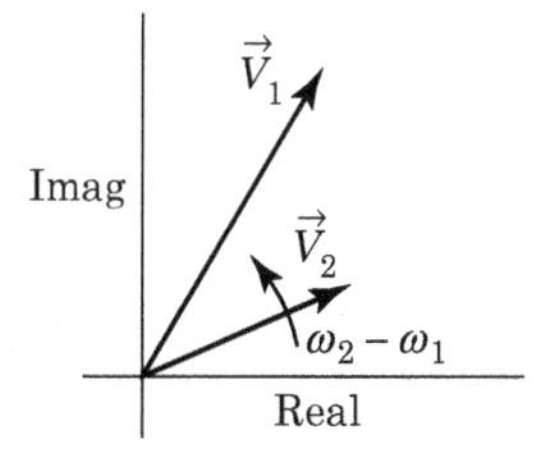

FIGURE 2-37 ■ The phasor-domain representation of two sine waves. The phasor V_1 is the reference phasor and does not rotate. The phasor V_2 rotates relative to phasor V_1 at a radian frequency of $\omega_2 - \omega_1$.

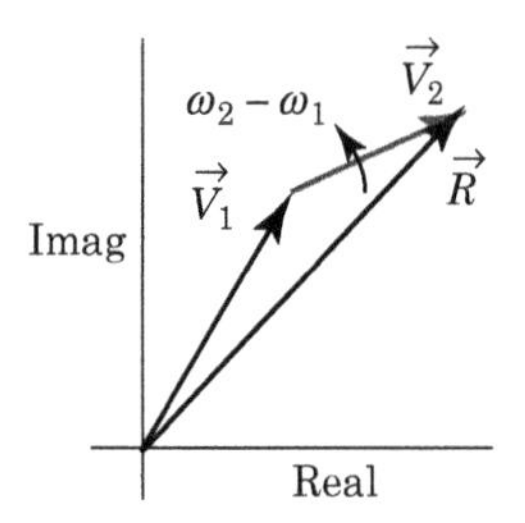

FIGURE 2-38 ■ The vectorial sum of two phasors. Phasor V_1 is the reference phasor and does not rotate. Phasor V_2 rotates about the end of phasor V_1 at a radian frequency of $\omega_1 - \omega_2$.

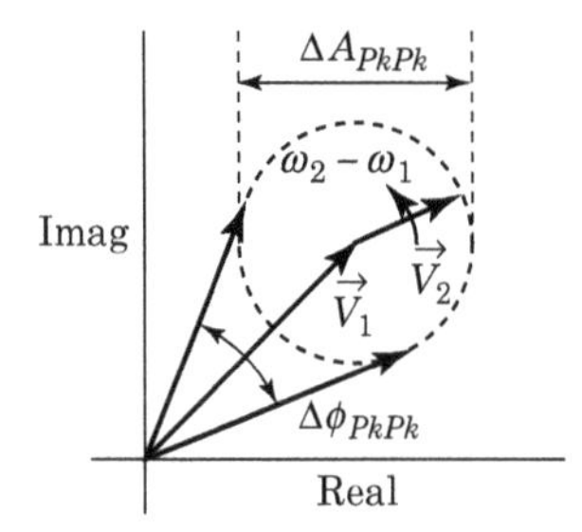

FIGURE 2-39 ■ The locus of the resultant vector R over time. The summation of the two vectors (V_1 and V_2) causes the resultant to exhibit a phase deviation of $\Delta\phi_{PkPk}$ and an amplitude deviation of ΔA_{PkPk}.

Since we're using phasor $\mathbf{V}_1$ as the reference, phasor $\mathbf{V}_2$ will appear to rotate around the origin counterclockwise with an angular velocity of $\omega_2 - \omega_1$. Phasor $\mathbf{V}_2$ rotates counterclockwise because $\omega_2 > \omega_1$. If $\omega_2 < \omega_1$, then the smaller phasor would rotate clockwise.

It takes one more interpretation of Figure 2-36 to produce a mathematically interesting interpretation. We've vectorially added the two phasors together in Figure 2-38.

The resultant or vector sum of the two phasors $\mathbf{V}_1$ and $\mathbf{V}_2$ is

$$\vec{R} = \vec{V}_1 + \vec{V}_2 \tag{2.37}$$

We can make some interesting observations that will be useful when we discuss oscillator phase noise. Figure 2-39 shows the locus of the resultant vector **R** over time. The vector addition of two sine waves produces amplitude and phase modulation of the resultant.

The resultant is phase modulated because the resultant oscillates between the extremes shown in Figure 2-39 at a rate of $\omega_2 - \omega_1$. The peak-to-peak change in phase is $\Delta\phi_{PkPk}$ and

$$\sin\left(\frac{\Delta\phi_{PkPk}}{2}\right) = \frac{V_{Pk,2}}{V_{Pk,1}} \tag{2.38}$$

For small $\Delta\phi_{PkPk}$,

$$\Delta\phi_{PkPk} \approx 2\frac{V_{pk,2}}{V_{pk,1}} \text{ for small}\Delta\phi_{PkPk} \tag{2.39}$$

Small $\Delta\phi_{pk}$ occurs when $V_{pk,1} \gg V_{pk,2}$.

We can derive the amplitude variation of the resultant from Figure 2-39. We can write

$$\Delta A_{PkPk} = 2V_{pk,2} \tag{2.40}$$

EXAMPLE

Two Additive Sine Waves

Figure 2-40 shows two complex sinusoids. The frequency and the amplitude of the top sinusoid are 1 Hz and 1.0 volts, respectively. The frequency and amplitude of the bottom sinusoid are 1.5 Hz and 0.7 volts, respectively.

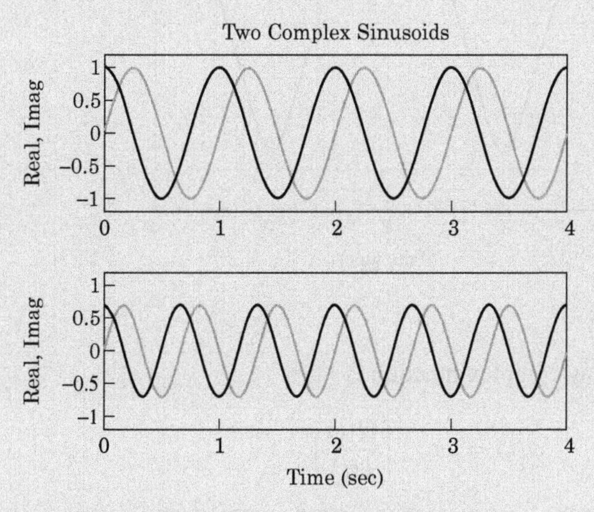

FIGURE 2-40 ■ Two complex sinusoids. The top plot shows the real and imaginary parts (in black and gray, respectively) of the first sinusoid. The frequency of the top sinusoid is 1.0 hertz, and the amplitude is 1.0 volts. The frequency of the bottom sinusoid is 1.5 hertz, and its amplitude is 0.7 volts.

Figure 2-41 shows the complex samples of the arithmetic sum of the two sinusoids from Figure 2-40.

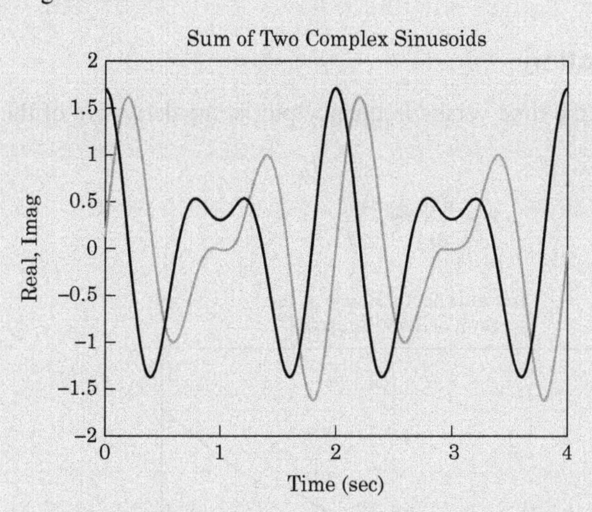

FIGURE 2-41 ■ The complex sum of the two sinusoids from Figure 2-40. Each point of this function is the arithmetic sum of the two sinusoids from Figure 2-40. We'll find the amplitude, phase, and frequency of this complex function.

The waveform of Figure 2-41 is periodic with a period of 2 seconds or with a frequency of $1/2$ hertz.

2.4.4 Amplitude Detection

The amplitude of any complex signal is

$$A(t) = \sqrt{I^2(t) + Q^2(t)} \tag{2.41}$$

Figure 2-42 shows the results of applying equation (2.41) to the function of Figure 2-41.

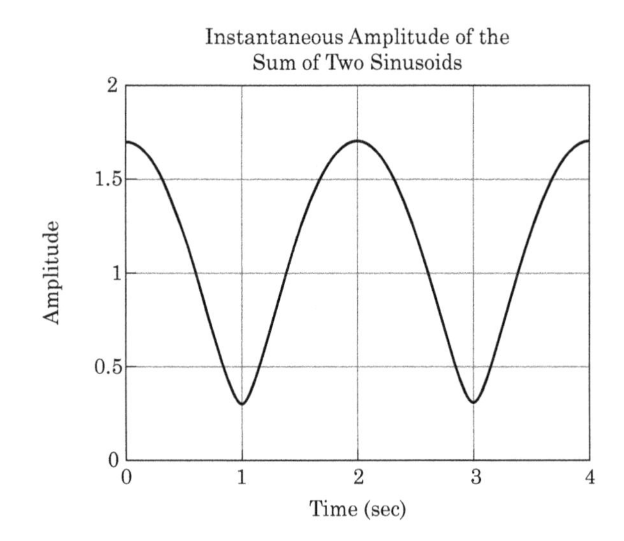

FIGURE 2-42 ■ The instantaneous amplitude of the function of Figure 2-41, derived using equation (2.41) The amplitude is periodic with a frequency of 0.5 Hz.

2.4.5 Phase Detection

The instantaneous phase of any complex function is

$$\phi(t) = \tan^{-1}\left[\frac{Q(t)}{I(t)}\right] \tag{2.42}$$

Figure 2-43 shows the phase of the complex function of Figure 2-41. This is the unwrapped or cumulative phase, ignoring the modulo 2π nature of phase.

2.4.6 Frequency Detection

To within a constant or two, the time versus frequency plot is the derivative of the time versus phase plot:

$$\omega = \frac{\partial \phi}{\partial t} \approx \frac{\Delta \phi}{\Delta t} \tag{2.43}$$

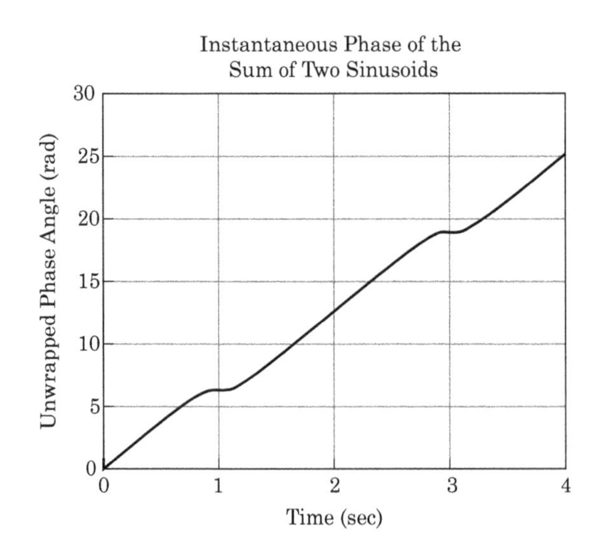

FIGURE 2-43 ■ The instantaneous phase of the function of Figure 2-41, derived using equation (2.42). This is the unwrapped or cumulative phase, ignoring the modulo 2π nature of phase. Like the amplitude plot, the phase plot exhibits a periodicity of 0.5 Hz.

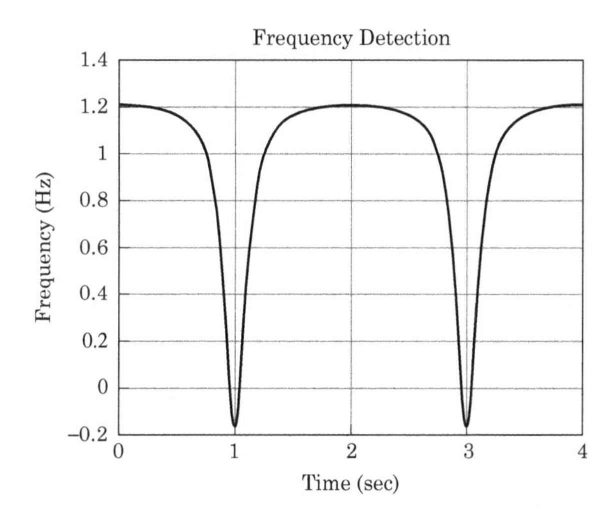

FIGURE 2-44 ■ The instantaneous frequency of the function of Figure 2-41, derived using equation (2.43). Like the amplitude and the phase, the frequency is periodic with a frequency of 0.5 Hz. The mean frequency of this signal is 1 Hz.

Figure 2-44 shows the time versus instantaneous frequency plot of the function of Figure 2-41.

2.4.7 Explaining It All With Phasors

Let us examine the phasor expression of the signal at 0.7 seconds. At that instant, the values of the plots shown in Figure 2-40 through Figure 2-44 are as given in Table 2-5.

Figure 2-45 shows the two phasors that represent the 1 Hz and the 1.5 Hz sine waves of Figure 2-40, plotted at 0.7 seconds.

To build a useful phasor diagram, we select one phasor as a reference. We'll stop the rotation of the reference phasor and place it along the x-axis. Let's use the 1 Hz phasor as the reference. Figure 2-46 shows the results.

The two phasors in Figure 2-46 are identical to the two phasors in Figure 2-43. We have simply rotated the diagram approximately 120° to align the 1 Hz phasor with the x-axis. The complex value of the 1 Hz sine wave in Figure 2-46 is unity. The complex value of the 1.5 Hz sine wave of Figure 2-46 is $-0.4114 + j0.5663 = .0.7\angle 126^\circ$.

Figure 2-47 shows the vector addition or resultant of the two phasors from Figure 2-46.

Figure 2-48 shows the behavior of the phasor plot of Figure 2-46 over two seconds. The 1 Hz phasor is frozen along the x-axis, while the 1.5 Hz phasor rotates CCW about the end of the 1 Hz phasor. The 1.5 Hz phasor rotates once around the tip of the 1 Hz phasor every 2 seconds.

The circle in Figure 2-48 is a map of the possible endpoints of the 1.5 Hz phasors. This circle is the locus. Figure 2-49 shows the resultants of the phasors shown in Figure 2-48.

TABLE 2-5 ■ The complex values of a waveform consisting of two additive sine waves.

Figure 2-40	1 Hz sine wave	$-0.3090 - j0.95111\angle - 108^\circ$
	1.5 Hz sine wave	$0.6657 + j0.21630.7\angle 18^\circ$
Figure 2-41	Sum of 1 Hz and 1.5 Hz sine waves	$0.3567 - j0.73470.82\angle - 64^\circ$
Figure 2-42	Amplitude of sum	0.8168
Figure 2-43	Unwrapped phase of sum	5.1644
Figure 2-44	Instantaneous frequency of sum	1.0584

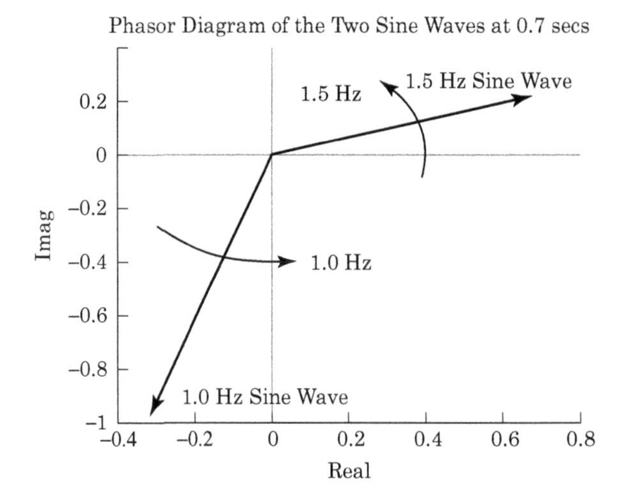

FIGURE 2-45 ■ The two complex sine waves from Figure 2-40, expressed as phasors at $t = 0.7$ seconds. The 1 Hz sine wave has unity amplitude and rotates CCW at 1.0 revolution per second. The 1.5 Hz sine wave has an amplitude of 0.7 and rotates CCW at 1.5 revolutions per second. At $t = 0.7$ seconds, the value of the 1.0 Hz complex sine wave is $-0.3090 - j0.9511 = 1\angle -108^\circ$, whereas the value of the 1.5 Hz complex sine wave is $0.6657 + j0.2163 = 0.7\angle 18^\circ$.

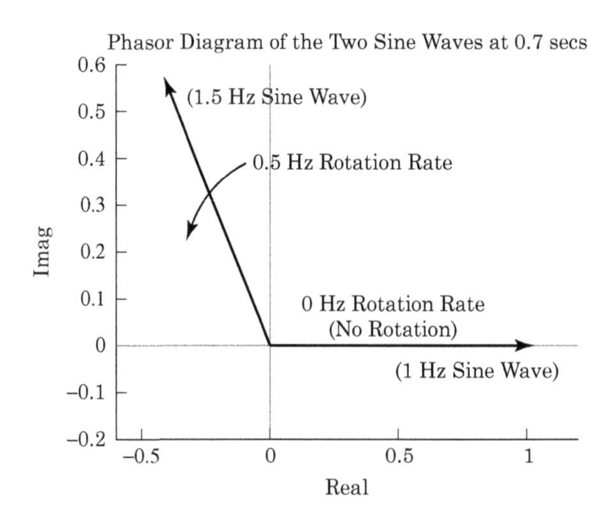

FIGURE 2-46 ■ The two complex sine waves from Figure 2-40, expressed as phasors at $t = 0.7$ seconds. The 1 Hz sine wave is the reference phasor. As such, we stop the rotation of this phasor and set it along the x-axis. The 1.5 Hz phasor continues to rotate around the origin at the difference frequency of $1.5 - 1.0 = 0.5$ Hz. This is the same phasor diagram from Figure 2-43 but rotated approximately 108°.

The geometry of Figure 2-46 through Figure 2-49 will let us find the amplitude, phase, and instantaneous frequency of the signal of Figure 2-41.

2.4.7.1 Amplitude

From Figure 2-49, we can see that the maximum amplitude occurs at $t = 0.0$ seconds, where the two phasors add constructively. The amplitude at this point is $1.0 + 0.7 = 1.7$. Figure 2-49 also reveals that the minimum amplitude occurs at $t = 1.0$ second, where the

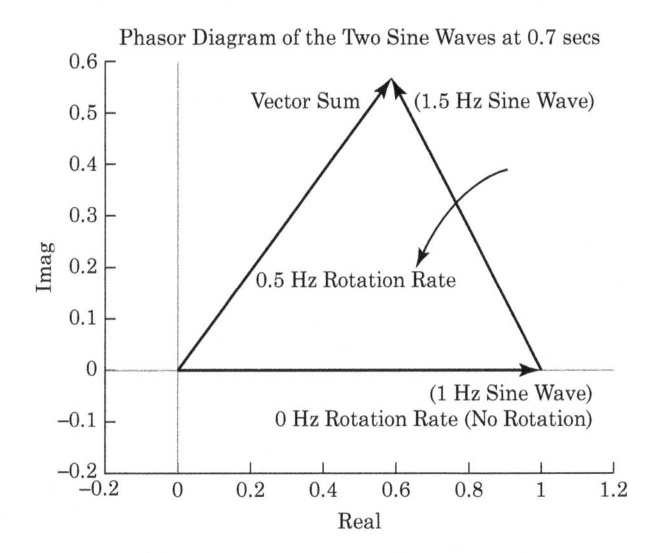

FIGURE 2-47 ■ The vector addition of the two phasors from Figure 2-46. The 1.0 Hz and the 1.5 Hz phasors add vectorially to produce a sum or resultant phasor. The vector sum phasor is $0.5886 + j0.5663$ or $0.8168\angle 43.9°$.

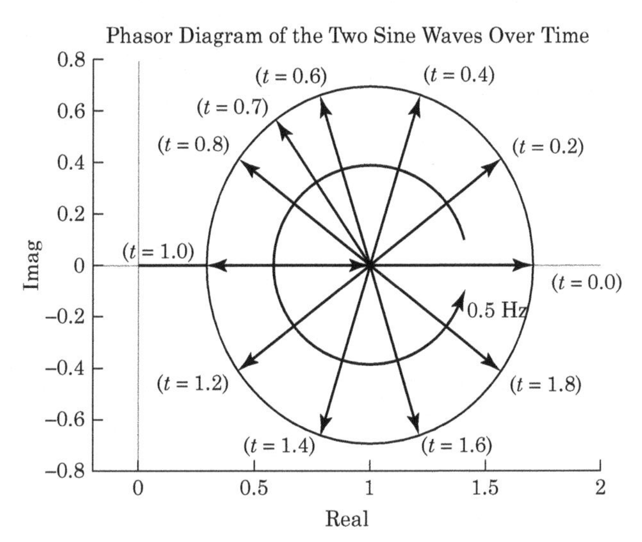

FIGURE 2-48 ■ The behavior of the phasor diagram of Figure 2-46 over a period of 2 seconds. The 1 Hz phasor is frozen along the x-axis, while the 1.5 Hz phasor rotates CCW about the end of the 1 Hz phasor. The circle shows the locus of all the possible endpoints of the 1.5 Hz phasor.

two phasors add destructively. The minimum amplitude is $1.0 - 0.7 = 0.3$. We can also see from the phasor diagram that the amplitude is cyclical with a period of 2 seconds.

The results from the phasor analysis agree with the amplitude waveform of Figure 2-41 (i.e., the maximum amplitude of 1.7, minimum amplitude of 0.3, and a period of 2 seconds).

2.4.7.2 Phase

The phase analysis of the phasor diagram of Figure 2-46 through Figure 2-49 is complicated slightly by the fact that we've "frozen" the 1 Hz phasor to always point along the x-axis. The result of this freezing is that our phase analysis will exhibit a linear bias.

As Figure 2-50 shows, the maximum phase deviation of the resultant occurs when the resultant is just tangent to the circular locus in Figure 2-49 (i.e., when the resultant is tangent to the circle). The tangency occurs when the resultant and the 1.5 Hz phasor form a right angle.

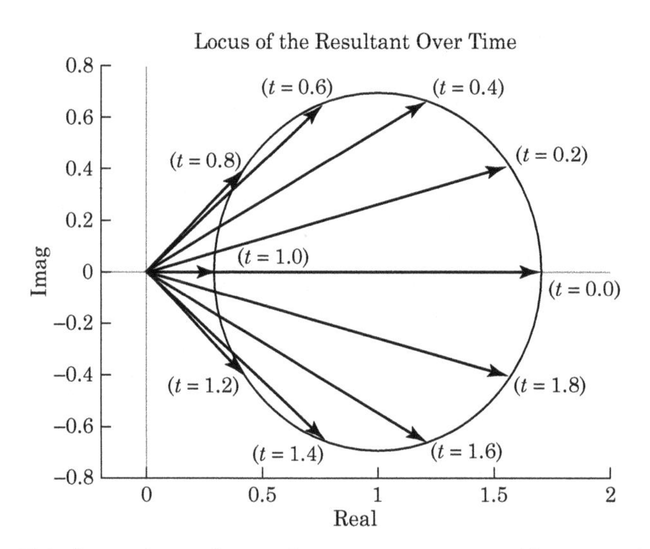

FIGURE 2-49 ■ This figure shows the resultant vectors at several instances in time. Each resultant is the vector sum of 1 Hz phasor and the 1.5 Hz phasor. We can find the amplitude, phase, and instantaneous frequency from this diagram. The geometry of the phasor diagram reveals that the end of the resultant vector will always fall on the circle.

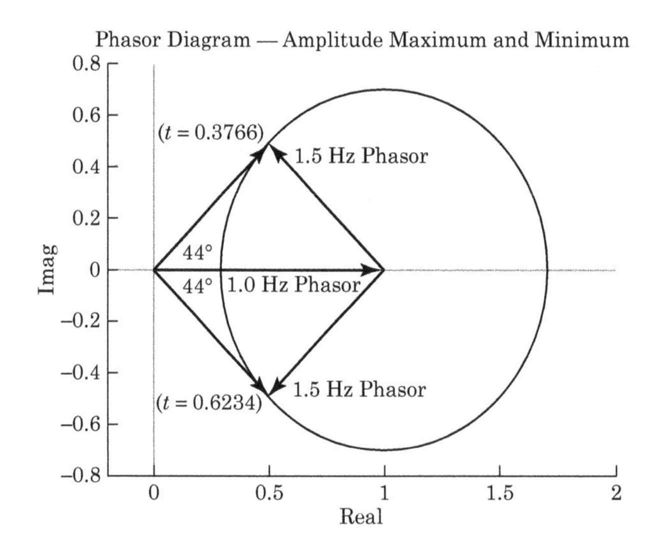

FIGURE 2-50 ■ The maximum phase deviation of the resultant occurs when the resultant is tangent to the locus circle. The two phasors are tangent at $t = 0.3766$ seconds and $t = 0.6234$ seconds, where the phase deviation is 44°.

According to Figure 2-50, the phase of the resultant swings between $+44^\circ$ and -44°. However, the phase plot of Figure 2-43 shows that the phase of the signal is steadily increasing. How do we reconcile the phase plot of Figure 2-43 with the phasor diagram of Figure 2-50?

When we transitioned from the phasor diagram of Figure 2-43 to the phasor diagram of Figure 2-46, we froze the phasor that represents the 1 Hz sinusoid. We arranged the phasor diagram of Figure 2-46 such that the 1 Hz phasor would always point along the x-axis. In effect, we are spinning the phasor diagrams of Figure 2-46 through Figure 2-50 at a 1 Hz rate. The freezing or spinning affects the phase plot.

Figure 2-51 repeats the phase plot of Figure 2-43, but Figure 2-51 also shows the amount of phase for which the 1.0 Hz phasor is responsible.

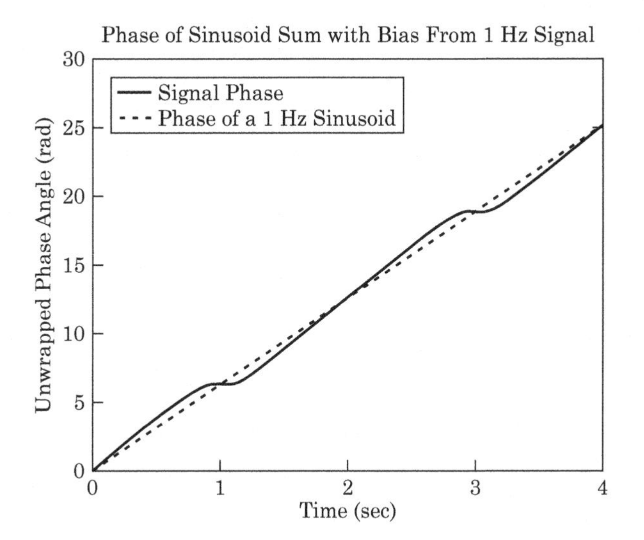

FIGURE 2-51 ■ The solid plot shows the phase response of the sum of two sinusoids. The dashed plot shows the phase of a single 1 Hz sinusoid. The effect of freezing the 1 Hz phasor in the phasor diagrams of Figure 2-46 through Figure 2-50 is to remove the phase introduced by the 1 Hz phasor.

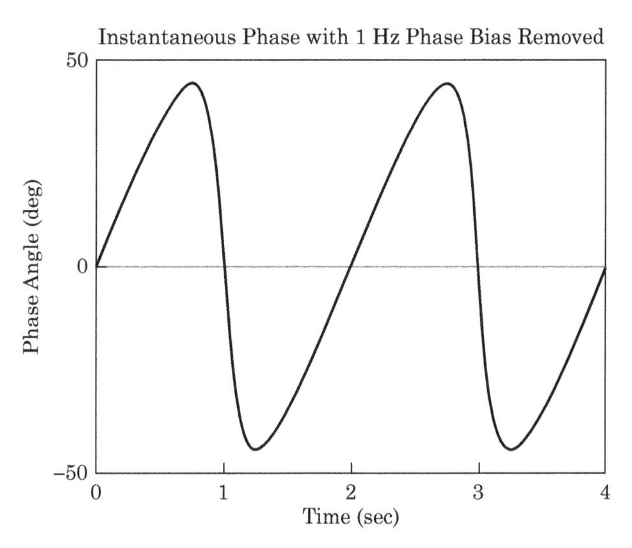

FIGURE 2-52 ■ The phase plot of the sum of two sinusoids. This plot directly describes the phasor diagram of Figure 2-50. As predicted, the maximum phase deviation (i.e., the maximum excursion of the signal phase) is 44°. Note the steep slope of the curve in the neighborhood of 1.0 and 3.0 seconds. Since the frequency is related to the derivative of the phase, we expect the frequency at 1.0 and 3.0 seconds to be relatively high.

Figure 2-52 shows the phase plot of Figure 2-51 when we remove the phase bias caused by the 1 Hz phasor. In other words, Figure 2-52 shows the difference between the solid and dashed plots of Figure 2-51.

2.4.7.3 Frequency

Figure 2-53 shows the instantaneous frequency of the resultant phasor shown in Figure 2-49. Frequency is the derivative of the phase so Figure 2-53 should be recognizable as the derivative of the plot of Figure 2-52.

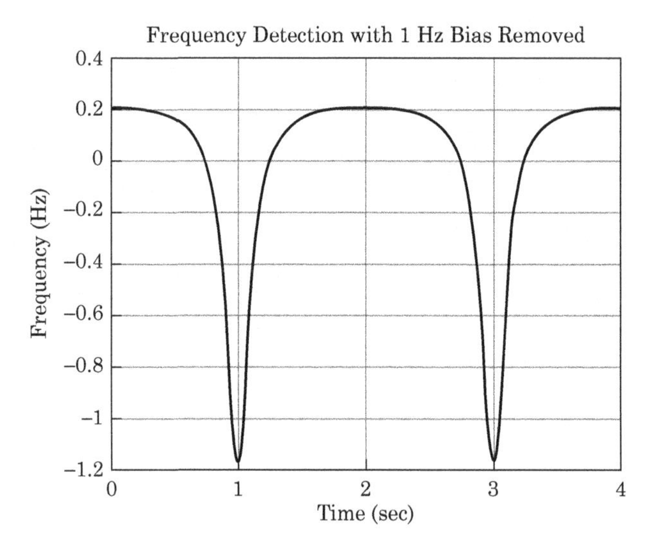

FIGURE 2-53 ■ The frequency plot of the sum of two sinusoids. This plot is related to the derivative of the phase plot of Figure 2-52. The frequency is relatively high when the phase shown in Figure 2-52 exhibits a high rate of change. The shape of this plot is identical to the shape of the frequency plot of Figure 2-44. However, because we've frozen the 1 Hz phasor, the y-axis of this plot is offset by 1 Hz from the frequency plot of Figure 2-44. The mean frequency of this plot is 0 Hz.

2.4.7.4 Frequency Spikes

What causes the frequency spikes in the plot of Figure 2-53? In other words, why does the frequency versus time plot of Figure 2-53 show impulses at $t = 1.0$ and $t = 3.0$ seconds? Can we reconcile those spikes with the phasor diagram of Figure 2-48 and Figure 2-49?

Figure 2-54 shows the resultant phasor as it changes over two 0.1 second intervals. The first interval occurs from $t = 0.0$ to $t = 0.1$ seconds. The second interval occurs between $t = 0.9$ and $t = 1.0$ seconds.

Figure 2-54 shows the source of the frequency spikes present in the plot of Figure 2-53. As time progresses from 0.0 to 0.1 seconds, the phase of the resultant changes from 0° at $t = 0$ seconds to 7° (or 0.122 radian) at $t = 0.1$ seconds. The instantaneous frequency is related to phase change per unit time so the 7° phase change over a 0.1 sec interval implies an approximate frequency of

$$\begin{aligned} \omega &= \frac{\partial\phi}{\partial t} \approx \frac{\Delta\phi}{\Delta t} \\ &= \frac{0.122 - 0.000\,\text{rad}}{0.1\text{ sec}} \\ &= 1.22\frac{\text{rad}}{\text{sec}} \\ &= 0.194\,\text{Hz} \end{aligned} \tag{2.44}$$

which is consistent with Figure 2-53.

The phase change over the 0.1 second interval from 0.9 to 1.0 seconds is 33° (or 0.576 radians). A 33° phase change over 0.1 seconds corresponds to an approximate

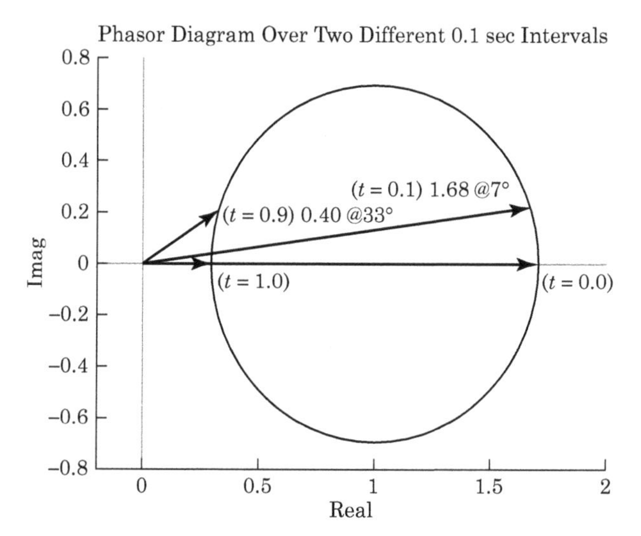

FIGURE 2-54 ■ The phasor diagram of Figure 2-49 over two distinct, 0.1 second intervals. The phase change from $t = 0.0$ to $t = 0.1$ seconds is 7°. The phase change from $t = 0.9$ to $t = 1.0$ seconds is 33°. Since frequency is related to the derivative of phase with respect to time, a larger phase change per unit time results in a higher instantaneous frequency.

frequency of

$$\begin{aligned}\omega &= \frac{\partial \phi}{\partial t} \approx \frac{\Delta \phi}{\Delta t} \\ &= \frac{0.000 - 0.576\,\text{rad}}{0.1\ \text{sec}} \\ &= -5.76\frac{\text{rad}}{\text{sec}} \\ &= -0.916\,\text{Hz}\end{aligned} \tag{2.45}$$

which is also consistent with Figure 2-53.

The origin of the frequency spikes of Figure 2-53 is in the geometry of Figure 2-54. The frequency of the resultant will appear even spikier as the magnitude of the 1.5 Hz signal approaches the magnitude of the 1.0 Hz signal because the phase change of the resultant around $t = 1$ second will be even more radical.

The phasor diagram of Figure 2-49 also explains why the areas of high frequency are in the same neighborhood as the areas of low signal amplitude. Figure 2-55 shows the amplitude and instantaneous frequency plotted together.

2.5 | BAND-LIMITED ADDITIVE WHITE GAUSSIAN NOISE

The type of noise we experience in the receiver design discipline has several names, including Gaussian noise, white noise, and Johnson noise. Each description emphasizes a different characteristic:

- Gaussian noise emphasizes the Gaussian statistical distribution of the time-domain waveform.

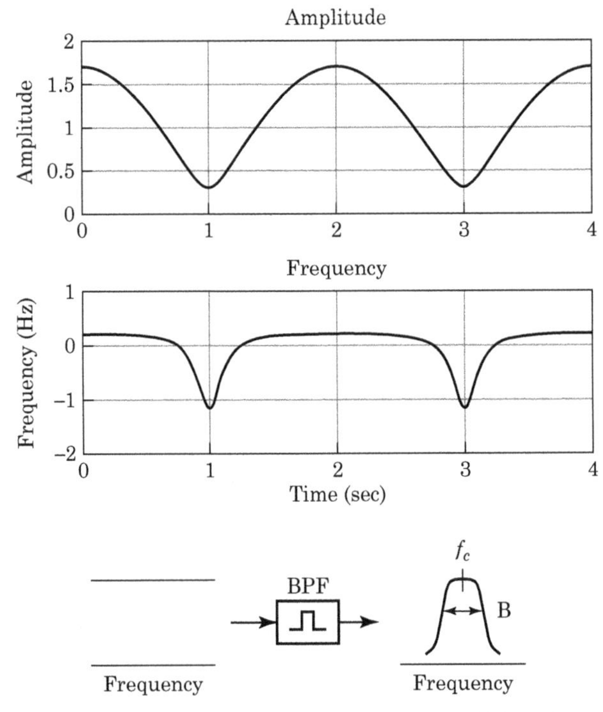

FIGURE 2-55 ■ The amplitude and frequency of the sum of the two sinusoids. The areas of high frequency (at $t = 1.0$ and $t = 3.0$ seconds) also correspond to areas of minimum amplitude. The phasor diagrams of Figure 2-49 and Figure 2-53 explain this effect.

FIGURE 2-56 ■ In nature, we often observe noise with very wideband characteristics, as shown on the left of this figure. Passing spectrally flat noise through a bandpass filter produces the band-limited spectrum shown here.

- White noise emphasizes the flat frequency spectrum of the noise
- Johnson noise and thermal noise both emphasize a particular source of noise common in electronic systems. Johnson and thermal noise are both white and Gaussian.
- Additive noise emphasizes that the noise adds arithmetically to any signal of interest.

We most commonly work with additive white Gaussian noise (AWGN). AWGN is a random, zero-mean, Gaussian statistical process with a flat frequency spectrum. We can't predict the instantaneous value of the waveform at any time in the future, but we can statistically describe the behavior of the noise when averaged over a long period.

Gaussian noise in the wild is very wideband, often extending to several terahertz. We will always view noise through a filter to limit the bandwidth, which is called band-limited additive white Gaussian noise. Figure 2-56 shows the effect of a bandpass filter on the spectrum of white noise.

2.5.1 Time Domain

Describing AWGN as a random, zero-mean, Gaussian process means the values of the time-domain waveform fall into a Gaussian distribution with no 0 Hertz (or DC) component. The standard deviation of the Gaussian process is related to the root mean square (RMS) amplitude of the noise.

When we apply broadband Gaussian noise to a bandpass filter, the filtered noise is centered at approximately frequency f_c (the center frequency of the filter). We can express this filtered noise as a single sinusoid whose amplitude and phase vary randomly over time. The narrower we make the filter's bandwidth, the more slowly the amplitude and phase change.

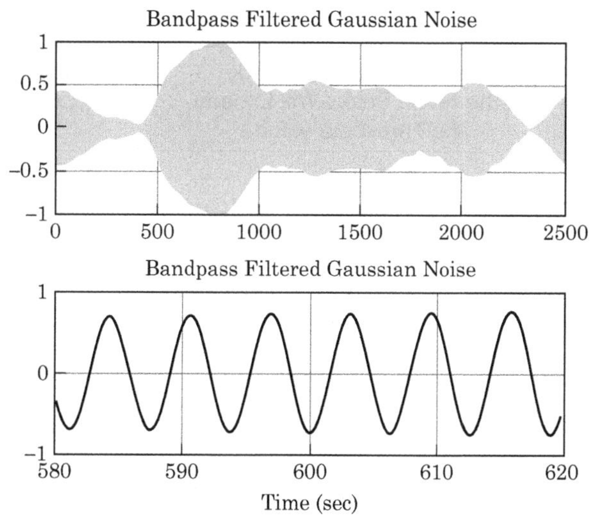

FIGURE 2-57 ■ Bandpass filtered Gaussian noise. The top plot shows the band-limited AWGN over a long time period. The bottom plot shows the AWGN over several cycles of the carrier. The amplitude and phase of the carrier change slowly with respect to the carrier frequency.

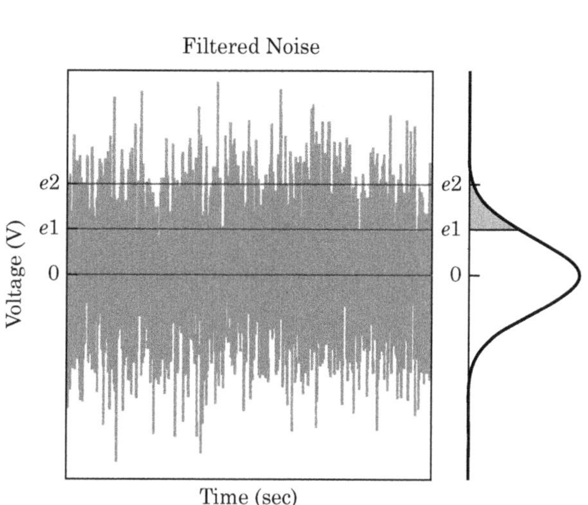

FIGURE 2-58 ■ The time-domain description of Gaussian noise on the left and the Gaussion or normal probability density function (PDF) on the right. The RMS voltage of the noise is the standard deviation of the Gaussian bell curve.

If the filter bandwidth B is a lot smaller than the filter's center frequency fc (i.e., $B \ll f_c$), the amplitude and phase of the filtered noise will exhibit appreciable changes only over many cycles of the carrier. Figure 2-57 shows the effect.

We specify magnitude of the noise voltage in terms of its RMS value, $V_{n,RMS}$. This RMS value is the standard deviation of the noise, and it fits directly into the equations describing the Gaussian distribution.

Figure 2-58 shows an example of Gaussian noise. We've shown the Gaussian bell curve to the right of the time-domain plot in this figure. The shaded area in the Gaussian curve represents the percentage of the time the noise waveform lies between voltage e_1 and voltage e_2.

2.5.1.1 RMS Value

It's meaningful to speak of noise only in statistical terms, such as the RMS or standard deviation of the noise. We can then apply the properties of the Gaussian statistics to predict the behavior of the waveform.

TABLE 2-6 ■ The thresholding characteristics of Gaussian-distributed noise.

Threshold Value	% of Time the Magnitude of the Noise Peaks Will Exceed the Threshold Value	% of Time the Noise Peaks Will Exceed the Threshold Value
$0V_{n,RMS}$	100%	50%
$1V_{n,RMS}$	31.8%	15.9%
$2V_{n,RMS}$	4.56%	2.28%
$3V_{n,RMS}$	0.26%	0.13%
$3.3V_{n,RMS}$	0.097%	0.048%
$4V_{n,RMS}$	63 ppm	32 ppm
$5V_{n,RMS}$	0.57 ppm	0.29 ppm
$6V_{n,RMS}$	$1.98 \cdot 10^{-3}$ ppm	$0.990 \cdot 10^{-3}$ ppm
$7V_{n,RMS}$	$2.6 \cdot 10^{-6}$ ppm	$1.3 \cdot 10^{-6}$ ppm

The following Gaussian noise approximations arise directly from analyzing the Gaussian bell curve. We assume the mean, or DC value, of the noise is 0:

- To a good engineering approximation, the magnitude of the peaks of Gaussian noise will not exceed three times the RMS value of the noise signal (in reality, we'll exceed this value only 0.26% of the time).
- Some specifications speak of a *peak noise voltage*. This specification frequently means 3.3 times the RMS value. Table 2-6 shows that the magnitude of the noise peaks will exceed 3.3 times the RMS value no more than 0.1% of the time.
- Given a noise voltage of $Vn_{,RMS}$, we can construct Table 2-6 using Gaussian statistics.

2.5.2 Zero Crossings

With a Gaussian noise input, the average number of zero crossings per second at the output of a narrow-bandpass filter is

$$N_{Zero\ Crossings} = 2f_0\sqrt{1 + \left(\frac{B_n^2}{12f_0^2}\right)} \tag{2.46}$$

where

f_0 = the center frequency of the filter
B_n = the filter's noise bandwidth

If half of the zero crossings are positive-going and half are negative-going, a frequency counter connected to this system would read a frequency of $N_{Zero\text{-}Crossings}/2$.

We can express equation (2.46) in terms of a filter's lower and upper cutoff frequencies and write

$$\begin{aligned} N_{Zero\ Crossings} &= \sqrt{\frac{f_U^3 - f_L^3}{3(f_U - f_L)}} \\ &= \sqrt{\frac{1}{3}\left(f_U^2 + f_U f_L + f_L^2\right)} \end{aligned} \tag{2.47}$$

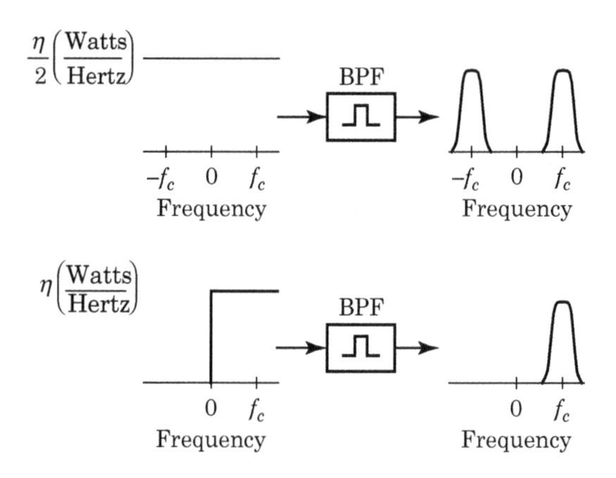

FIGURE 2-59 ■ We can model band-limited AWGN to contain spectral components at positive and negative frequencies (double-sided spectrum) or as containing spectral components only at positive frequencies (single-sided spectrum).

If $f_L = 0$ (i.e., for a lowpass filter)

$$N_{Zero\ Crossings} = 0.577 f_U \tag{2.48}$$

For narrow-bandpass filters (bandwidths of less than 10% of the center frequency), $f_U \cong f_L$, we have

$$\begin{aligned} N_{Zero\ Crossings} &= \frac{f_U + f_L}{2} \\ &= f_0 \end{aligned} \tag{2.49}$$

Passing AWGN through a very narrow-bandpass filter will produce an output very similar to a sine wave of frequency f_0.

2.5.3 Frequency Domain

Like a single sine wave, we can model band-limited AWGN to contain frequency components at both positive and negative frequencies or with only positive frequencies, as shown in figure 2-59. We'll use whatever model is appropriate for our task.

We can use the tools we developed previously to describe AWGN.

The noise power is spectrally flat, and we characterize the PSD of the noise with the quantity η (watts/Hz) for the single-sided spectrum. When modeling AWGN with a double-sided spectrum, the PSD is $\eta/2$ watts/Hz.

In the single-sided spectral case, the total noise power we observe in bandwidth B_n is

$$P_{Noise,Watts} = \eta B_n \tag{2.50}$$

where B_n is the noise bandwidth of the observation.[1]

One interesting consequence of the flat spectral power characteristic is the total noise power (in watts) depends only on the bandwidth through which we view the noise and not on center frequency.

[1] There will be more on noise bandwidth in Chapter 6, but for now assume that *noise bandwidth of the observation* means that we have passed spectrally flat noise through a brick wall bandpass filter of bandwidth B_n.

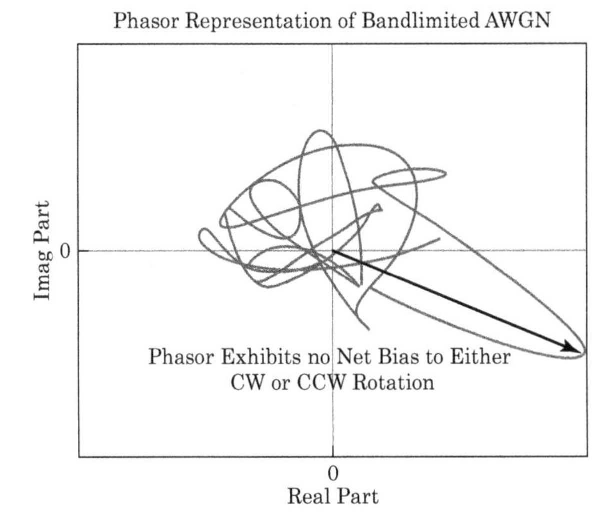

FIGURE 2-60 ■ The phasor representation of band-limited AWGN. We've set the reference frequency to the center frequency of the bandpass filter so there is no net rotation of the phasor around the origin. If the input noise is Gaussian, the amplitude of the phasor is a Rayleigh-distributed random variable. The phase of the vector is a uniformly distributed random variable.

2.5.4 Phasor Domain

Figure 2-60 shows the phasor representation of band-limited AWGN when we've set the reference frequency to the center frequency of the bandpass filter through which we view the noise. The tip of the phasor takes a random path around the origin and there is no bias to either clockwise or counterclockwise rotation. The mean angular velocity of the phasor is zero.

Statistical analysis of the phasor concept reveals that the real and imaginary parts of the phasor are independent Gaussian variables. The real part and imaginary part both follow a Gaussian distribution, and those distributions are independent of each other.

Further analysis reveals that the magnitude of the phasor is a Rayleigh-distributed random variable, while the phase is distributed uniformly over 0 to 2π.

2.6 | AN IDEAL SINE WAVE AND BAND-LIMITED AWGN

2.6.1 Time Domain

Figure 2-61 shows the time-domain waveform of a noisy cosine wave. The equation is

$$f(t) = V_{pk}\cos(\omega_0 t + \theta) + V_{noise}(t) \tag{2.51}$$

where

$V_{noise}(t)$ represents the time-domain noise waveform.

$V_{noise}(t)$ is a random, zero-mean process. The standard deviation of $V_{noise}(t)$ is $V_{noise,RMS}$. The larger $V_{noise,RMS}$, the larger the noise power.

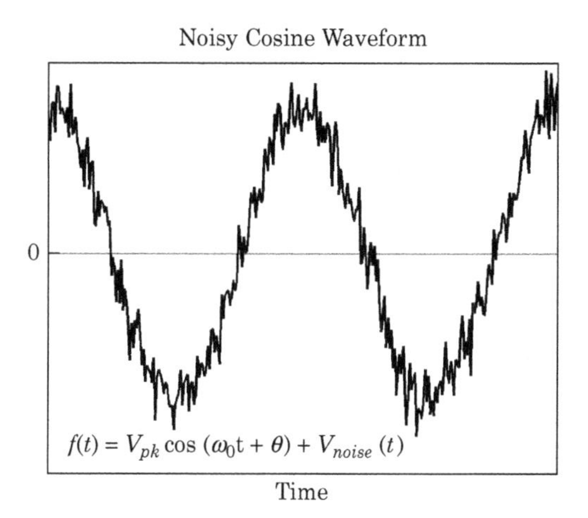

FIGURE 2-61 ■ An ideal cosine waveform corrupted with AWGN.

2.6.2 Zero Crossings

Figure 2-62 shows a close-up view of a zero crossing point from Figure 2-61. For high signal-to-noise ratios (where $V_{pk} \gg V_{noise}$), the zero crossings of the noisy cosine wave and of the ideal cosine wave occur at approximately the same instants. The AWGN creates uncertainty in the zero cross timing. The zero crossings are no longer mathematically deterministic but are now described statistically.

The variable Δt in Figure 2-62 represents the uncertainty in the zero crossings. Δt is a random variable that increases as the signal SNR decreases.

There are several names for the zero crossing uncertainty, such as clock jitter, phase noise, incidental phase modulation, and frequency drift. They all describe the uncertainty in the zero crossings of signal.

The average number of either positive- or negative-going zero crossings per second at the output of a narrow-bandpass filter of rectangular shape when the input is a sine wave

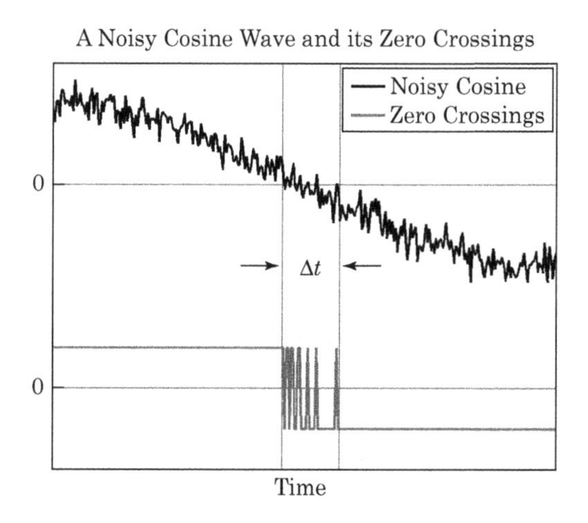

FIGURE 2-62 ■ The zero crossings of a cosine wave corrupted with AWGN.

in Gaussian noise is

$$\frac{\text{Positive Zero Crossings}}{\text{Second}} = \frac{\text{Negative Zero Crossings}}{\text{Second}} = f_0\sqrt{\frac{SNR + 1 + \left(\frac{B^2}{12 f_0^2}\right)}{SNR + 1}} \tag{2.52}$$

where

f_0 = the center frequency of the filter
B = the filter bandwidth
SNR = the signal-to-noise power ratio in linear terms

2.6.3 Frequency Domain

Figure 2-63 shows the frequency-domain representations of an ideal cosine and of band-limited AWGN. We add the two signals together to produce a noisy cosine wave. Before the addition, the cosine had an infinite SNR. Adding the signal to the noise produces a signal with a finite SNR.

Before the addition, the signal power is P_{Signal} and the noise power is ηB_n. After the addition, the SNR of the composite signal is

$$SNR = \frac{P_{Signal}}{\eta B_n} \tag{2.53}$$

The noise power is directly related to the noise bandwidth B_n, while the SNR is inversely proportional to the noise bandwidth.

2.6.4 Phasor

Let the signal vector be **S**, and let the noise vector be **N**. The vector sum of **S** and **N** will be the resultant vector **R**:

$$\overline{R} = \overline{S} + \overline{N} \tag{2.54}$$

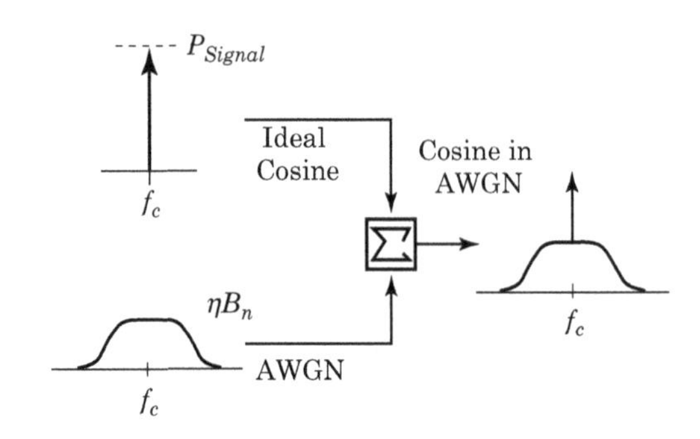

FIGURE 2-63 ■ We generate a noisy cosine wave by arithmetically adding an ideal cosine and band-limited AWGN together.

2.6.4.1 Magnitude Issues

We have to be careful to specify the correct magnitude of the signal and noise phasors. The signal **S** is described by

$$S(t) = V_{pk} \cos(\omega_0 t + \theta) \tag{2.55}$$

which is described with a peak value V_{pk}. The noise **N** is

$$N(t) = V_{noise}(t) \tag{2.56}$$

We can describe the magnitude of the noise phasor only as an RMS value, $V_{noise,RMS}$. In any phasor diagram containing noise, we will represent the magnitude of any signal phasors with their RMS values.

Figure 2-64 shows the phasor diagram of **S** plus **N**. We place the random noise vector at the end of the cosine. We've also drawn standard deviation circles for the noise vector.

The magnitude of the noise vector **N** is Rayleigh distributed. Using the cumulative Rayleigh PDF, we know the noise phasor will reside inside the $\sigma = 1$ circle about 38% of the time. The noise phasor will reside inside the $\sigma = 2$ locus about 86% of the time and it will be inside the $\sigma = 3$ locus 98% of the time.

Using the geometry we previously developed for noiseless sine waves, we can write an expression for the phase and amplitude deviation that the noise impresses on the ideal cosine wave. We know

$$V_{RMS} = \frac{V_{pk}}{\sqrt{2}} \tag{2.57}$$

Figure 2-65 shows the amplitude and phase modulation present on the resultant due to the AWGN.

The amplitude variation of the resultant in Figure 2-65 is

$$\Delta A_{RMS} = 2V_{noise,RMS} \tag{2.58}$$

For the resultant phase variation, referring to Figure 2-65, we can write

$$\sin\left(\frac{\Delta\phi_{RMS}}{2}\right) = \frac{V_{N,RMS}}{V_{S,RMS}} \tag{2.59}$$

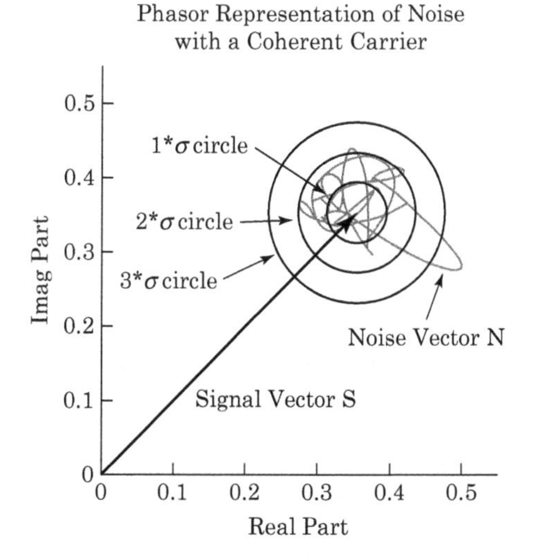

FIGURE 2-64 ■ The phasor representation of a cosine wave combined with band-limited AWGN.

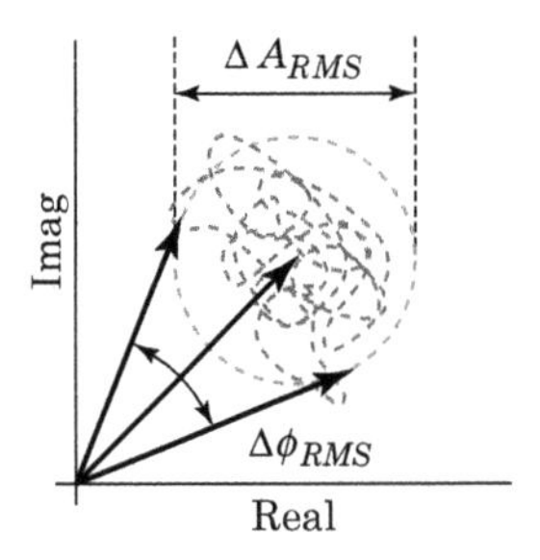

FIGURE 2-65 ■ The phasor representation of a cosine wave combined with band-limited AWGN, emphasizing the amplitude and phase modulation caused by the band-limited AWGN. Note that the values of amplitude and phase deviation are now RMS values. The noise circle represents only one value of standard deviation.

For small values of $\Delta\phi_{RMS}$, we can write

$$\Delta\phi_{RMS} \approx 2\frac{V_{N,RMS}}{V_{S,RMS}} \tag{2.60}$$

Equations (2.58) and (2.59) describe only one value of standard deviation circle.

EXAMPLE

Oscillator Phase Noise

An oscillator delivers 10 dBm of signal power into a 50 Ω load. What is the RMS value of noise voltage that will cause the signal to deviate 1°_{RMS}?

Solution

This is a tricky question, requiring some thought regarding peak and RMS values and involving the quadrature nature of noise. Figure 2-66 shows the geometry.

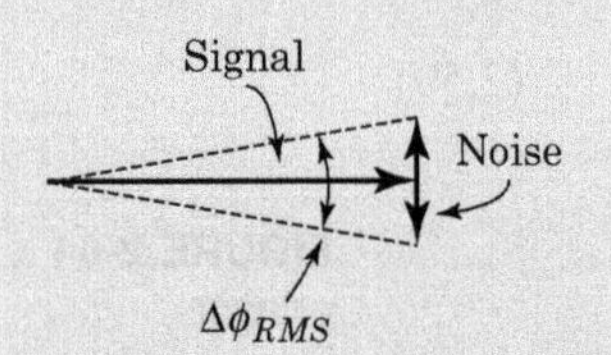

FIGURE 2-66 ■ The geometry of the oscillator phase noise example. What is the magnitude of the noise vector to produce a 1°_{RMS} phase deviation on the composite signal?

A 10 dBm signal corresponds to 0.707 V_{RMS} = 1.000 V_{Pk} in a 50 Ω system. The RMS noise voltage is

$$\begin{aligned} \tan\left(\frac{\Delta\phi_{RMS}}{2}\right) &= \frac{V_{noise,RMS}}{V_{Pk}} \\ \tan(0.5^{\circ}) &= \frac{V_{noise,RMS}}{1.0} \\ \Rightarrow V_{noise,RMS} &= 8.728\,\text{mV}_{RMS} \end{aligned} \tag{2.61}$$

However, the noise voltage we found here accounts only for the noise that is orthogonal to the signal. There is an equal amount of noise present that is collinear with the signal. The effect causes us to multiply the result of equation (2.61) by $\sqrt{2}$; hence, the RMS noise voltage is $V_{noise,RMS} = 12.34\,\text{mV}_{RMS}$.

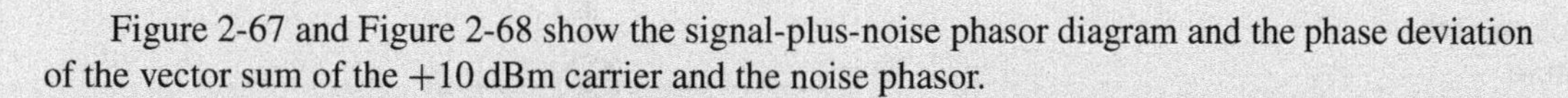

Figure 2-67 and Figure 2-68 show the signal-plus-noise phasor diagram and the phase deviation of the vector sum of the +10 dBm carrier and the noise phasor.

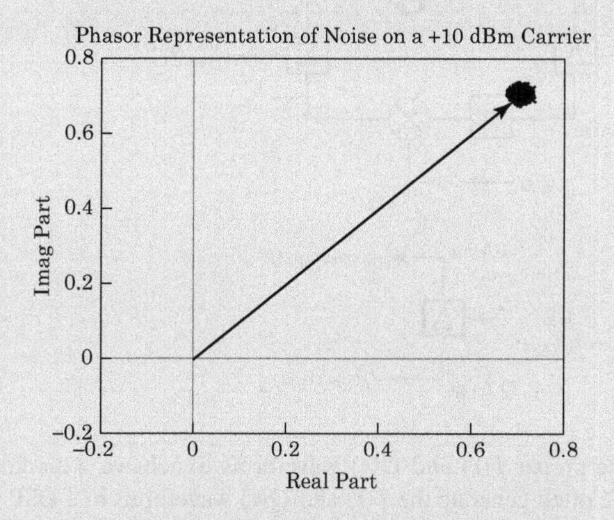

FIGURE 2-67 ■ The signal-plus-noise phasor diagram of a +10 dBm signal combined with sufficient noise to exhibit 1°_{RMS} of phase noise.

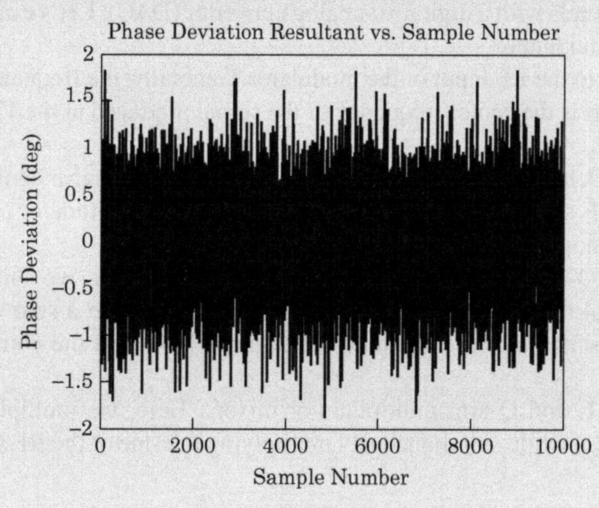

FIGURE 2-68 ■ The angle resulting from the vector addition of the noise phasor of Figure 2-67 and a +10 dBm sine wave. The RMS phase deviation is 1.0°_{RMS}.

2.7 THE QUADRATURE MODULATOR

The quadrature modulator is a general-purpose modulator that allows us to control every aspect of the transmitted waveform. The architecture fits nicely into our discussion of complex signals and, in practice, we can use common, low-frequency DSP techniques to generate any modulated signal.

2.7.1 Quadrature Modulator Architecture

Figure 2-69 shows the basic quadrature modulator. The top signal path is the in-phase arm (I-arm) of the modulator. The bottom signal path (containing the 90° phase shift) is the quadrature arm (Q-arm) of the modulator.

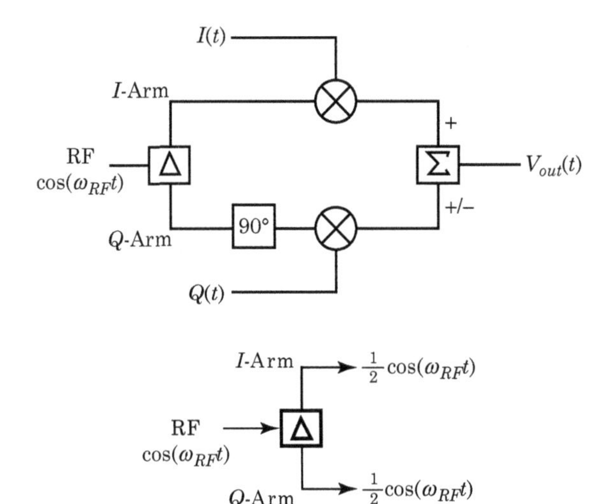

FIGURE 2-69 ■ The architecture of the quadrature modulator.

FIGURE 2-70 ■ The quadrature modulator RF splitter.

Our task is to derive the proper $I(t)$ and $Q(t)$ waveforms to achieve a modulated signal at the $V_{out}(t)$ port. We often generate the $I(t)$ and $Q(t)$ waveforms in a DSP then drive the analog I and Q channels with a digital-to-analog converter (DAC). Let's examine the individual pieces of the modulator.

We apply a quiet carrier to the RF input of the modulator. Generally, the frequency of the signal at the RF input port is the center frequency of the signal produced at the $V_{out}(t)$ port.

Figure 2-70 shows the "Delta-block" or splitter, which generates two equal-magnitude, equal-phase copies of the RF signal. Any deviation from the equal-magnitude or equal-phase criteria results in a nonoptimal modulator.

Figure 2-71 shows the Q-arm in detail. In this arm of the modulator, we shift the phase of the incoming signal by 90°, which changes the cosine wave into a sine wave. The accuracy of the phase shift (and its amplitude) determines how well the complete modulator performs.

Figure 2-72 shows the I- and Q-arm multipliers or mixers. Here, we multiply the cosine of the RF signal by $I(t)$ while simultaneously multiplying the sine of the RF signal by $Q(t)$.

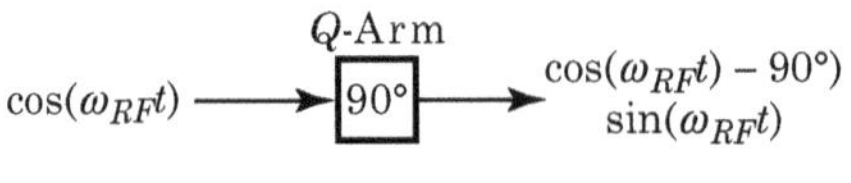

FIGURE 2-71 ■ The Q-arm of the quadrature modulator. The 90° phase shift converts the cosine into a sine.

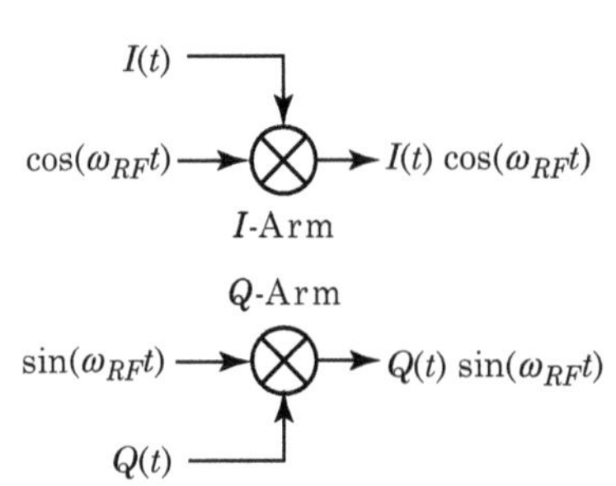

FIGURE 2-72 ■ The I- and Q-arm multipliers or mixers.

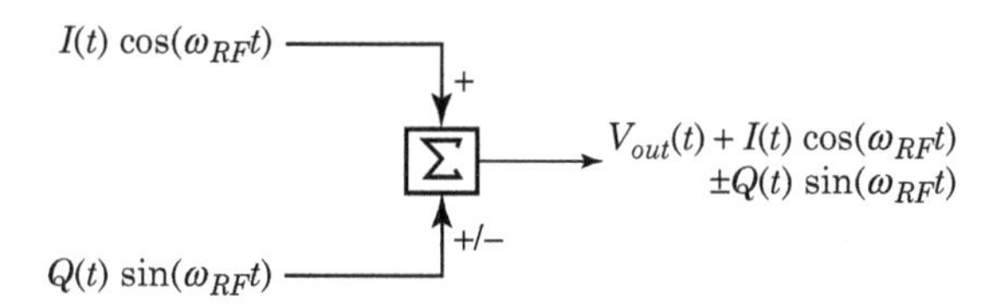

FIGURE 2-73 ■ The final summation of quadrature signals in the quadrature modulator.

The quantity $I(t)$ determines how much $\cos(\omega_{RF}t)$ passes through the I-arm mixer. $Q(t)$ determines how much $\sin(\omega_{RF}t)$ passes through the Q-arm mixer. The ratio of $I(t)$ and $Q(t)$ determines the phase of the RF after we sum.

Finally, we sum the I- and Q-arm signals together in Figure 2-73 to form the output signal $V_{out}(t)$.

The output signal is

$$V_{out}(t) = I(t)\cos(\omega_{RF}t) \pm Q(t)\sin(\omega_{RF}t) \tag{2.62}$$

If we want to account for amplitude and phase imbalance through the system (because the I- and Q-arms aren't exactly matched), we can rewrite (2.62) as

$$V_{out}(t) = I(t)\cos(\omega_{RF}t) \pm Q(t)A_Q\sin(\omega_{RF}t + \phi_Q) \tag{2.63}$$

where A_Q and ϕ_Q represent the differential amplitude and phase imbalances between the I and Q channels.

Often, there's confusion about the $\pm$ sign in equations (2.62) and (2.63). The $\pm$ sign arises from the $\pm$ sign in the Q-arm summer of Figure 2-73. Choosing the plus over the minus sign is a matter of convenience. For example, the minus sign allows us to map the equation nicely to a geometric interpretation. Most signal processing books assume a minus sign, but most RF integrated circuits that perform quadrature modulation use an adder (and hence require a plus sign). We'll use whatever is convenient for our discussion.

2.7.2 Generating a Single Tone at $f = f_{RF}$

Having looked at the quadrature architecture in some detail, let's move on to practical matters. We'll first use the modulator to generate a single tone (or quiet carrier) at the RF input frequency, f_{RF}.

Referring to Figure 2-69 and equation (2.62), setting $I(t) = +1$ and $Q(t) = 0$ produces the results we seek. The I-Arm essentially connects the input signal at f_{RF} to the output of the modulator. The Q-arm is out of the picture because

$$\begin{aligned} Q(t)\sin(\omega_{RF}t) &= 0 \cdot \sin(\omega_{RF}t) \\ &= 0 \end{aligned} \tag{2.64}$$

The output signal is

$$\begin{aligned} V_{out}(t) &= I(t)\cos(\omega_{RF}t) \pm Q(t)\sin(\omega_{RF}t) \\ &= (+1)\cos(\omega_{RF}t) \pm (0)\sin(\omega_{RF}t) \\ &= \cos(\omega_{RF}t) \end{aligned} \tag{2.65}$$

2.7.3 Generating a Single Tone at $f \neq f_{RF}$

Next, we'll generate a single, quiet tone at a frequency different from the input frequency of f_{RF}. Our task is to derive the waveforms $I(t)$ and $Q(t)$, which, when applied to equation (2.62), produce the desired results. We note that applying a cosine at a radian frequency

$\omega_m = 2\pi f_m$ to the I-port of an ideal quadrature modulator and zero to the Q-port produces

$$\begin{aligned} V_{out}(t) &= I(t)\cos(\omega_{RF}t) \pm Q(t)\sin(\omega_{RF}t) \\ &= \cos(\omega_m t)\cos(\omega_{RF}t) \pm 0 \cdot \sin(\omega_{RF}t) \\ &= \frac{1}{2}\cos\left[(\omega_{RF} - \omega_m)t\right] + \frac{1}{2}\cos\left[(\omega_{RF} + \omega_m)t\right] \end{aligned} \tag{2.66}$$

The output of this operation is two cosine waves. The frequency of one cosine wave is $f_{RF} + f_m$ (the sum frequency), whereas the other is $f_{RF} - f_m$ (the difference frequency).

Equation (2.66) suggests that, if we could perform some type of canceling operation at the summer, we might be able to produce a single tone at either the sum or difference tones. Waveforms useful to this concept appear when we apply zero to the $I(t)$ port and a sine wave to $Q(t)$:

$$\begin{aligned} V_{out}(t) &= I(t)\cos(\omega_{RF}t) + Q(t)\sin(\omega_{RF}t) \\ &= 0 \cdot \cos(\omega_{RF}t) + \sin(\omega_m t)\sin(\omega_{RF}t) \\ &= \frac{1}{2}\cos\left[(\omega_{RF} - \omega_m)t\right] - \frac{1}{2}\cos\left[(\omega_{RF} + \omega_m)t\right] \end{aligned} \tag{2.67}$$

We've arbitrarily assumed a plus sign for the $Q(t)$ term. Adding equations (2.66) and (2.67) produces

$$\begin{aligned} V_{sum}(t) &= \left\{\frac{1}{2}\cos\left[(\omega_{RF} - \omega_m)t\right] + \frac{1}{2}\cos\left[(\omega_{RF} + \omega_m)t\right]\right\} \\ &\quad + \left\{\frac{1}{2}\cos\left[(\omega_{RF} - \omega_m)t\right] - \frac{1}{2}\cos\left[(\omega_{RF} + \omega_m)t\right]\right\} \\ &= \cos\left[(\omega_{RF} - \omega_m)t\right] \end{aligned} \tag{2.68}$$

which produces the difference frequency, assuming the cancellation is perfect. Subtracting equation (2.67) from equation (2.66) produces

$$\begin{aligned} V_{sum}(t) &= \left\{\frac{1}{2}\cos\left[(\omega_{RF} - \omega_m)t\right] + \frac{1}{2}\cos\left[(\omega_{RF} + \omega_m)t\right]\right\} \\ &\quad - \left\{\frac{1}{2}\cos\left[(\omega_{RF} - \omega_m)t\right] - \frac{1}{2}\cos\left[(\omega_{RF} + \omega_m)t\right]\right\} \\ &= \cos\left[(\omega_{RF} + \omega_m)t\right] \end{aligned} \tag{2.69}$$

which is a cosine wave at the sum frequency.

With physical hardware, the cancellation is never perfect because of amplitude and phase mismatches through the two arms of the quadrature modulator. However, it is easy to achieve 20 dB suppression of the unwanted sideband.

Figure 2-74 shows the hardware and waveforms used to generate a single tone at the sum frequency. Figure 2-75 shows the waveforms necessary to generate a cosine at the difference frequency. The only difference is the sign of the $Q(t)$ waveform.

2.7.4 Generating a Frequency Shift Keying Signal

We can generate any type of modulated signal we like using a quadrature modulation. Candidates include the following:

- Analog amplitude modulation (AM), analog frequency modulation (FM), or analog phase modulation (PM)
- Digital amplitude modulation (on/off keying [OOK])

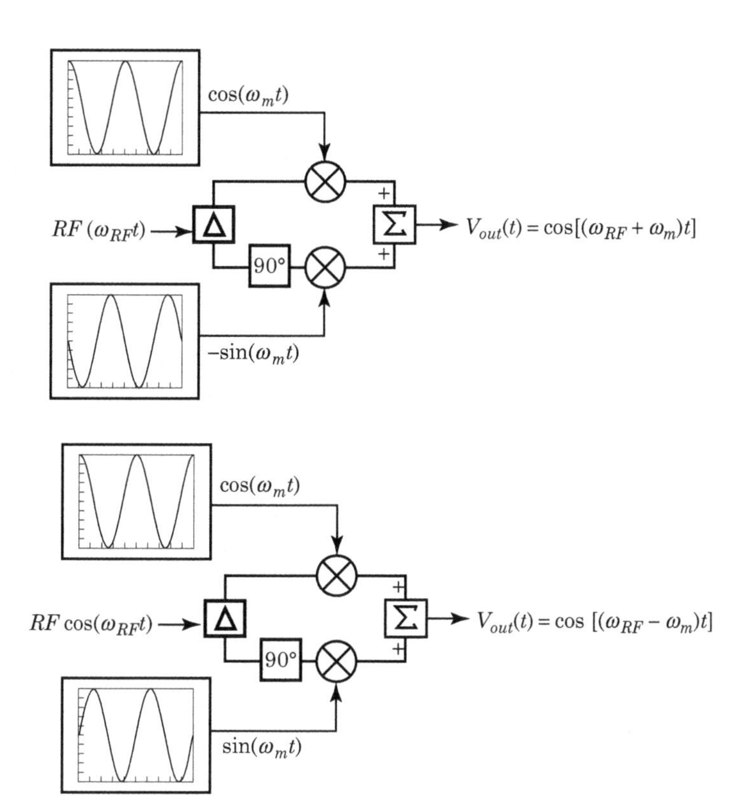

FIGURE 2-74 ■ Producing a single cosine wave at $f_{RF} + f_m$ using a quadrature modulator.

FIGURE 2-75 ■ Producing a single cosine wave at $f_{RF} - f_m$ using a quadrature modulator.

- Two-level frequency shift keying (2LFSK)
- Four-level frequency shift keying (4LFSK)
- Multitone FSK
- Binary phase shift keying (BPSK)
- Quadrature phase shift keying (QPSK)
- M-ary PSK

Given a modulation, our task is to derive the appropriate $I(t)$ and $Q(t)$ waveforms required to express the signal as a modulated wave. Using the concepts we explored earlier, we will generate an FSK waveform at 10k bits per second (10 kbps).

In FSK, we encode a binary "0" as one RF and a binary "1" as a second RF. For this discussion, we'll encode a binary "0" to be the lower frequency $f_0 = f_{RF} - f_m$, and we'll encode the binary "1" to be $f_1 = f_{RF} + f_m$ (see Figure 2-76).

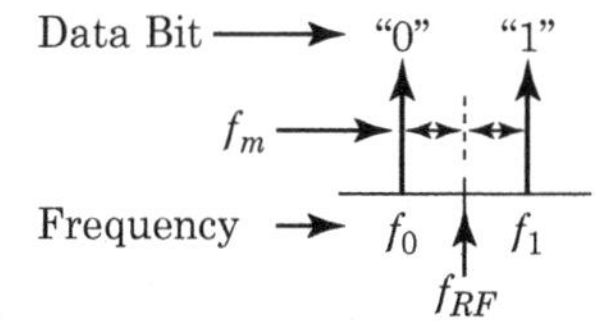

FIGURE 2-76 ■ The frequency plan for a simple two-level FSK modulation scheme. We transmit the 0 or space frequency for 1-bit duration when we want to transmit a logical 0. We transmit the 1 or mark frequency for 1-bit duration when we want to transmit a logical 1. For a 10 kbps data rate, a bit duration would be 0.1 ms or 100 μsec.

The tone spacing relative to the data rate is an important parameter. The value of f_m controls the deviation, or the difference between $f_0$and f_1, of the FSK waveform. Theory tells us that the deviation must be at least half of the bit rate for reliable detection, but we'll deviate by twice the bit rate for clarity in this example. Since we're generting a different tone for every bit, we must change the $I(t)$ and $Q(t)$ signals at each bit transition. For this example, the change occurs every 0.1 ms or at a 10 kHz rate.

In order to generate a tone at $f_0 = f_{RF} - f_m$, we apply a quiet carrier at f_{RF} to the RF port of the modulator. Figure 2-75 reveals that we must apply a cosine wave to the I port of the modulator and a sine wave to the Q port. The frequency of these waves is f_m, or half of the desired FSK deviation. To generate the 1 frequency, we apply a cosine wave to the I port of the modulator and a negative sine wave to the Q port (see Figure 2-74).

Figure 2-77 shows the digital bit stream and the waveforms applied to the I and Q ports. Figure 2-78 shows the data stream and the resulting 2LFSK waveform.

Figure 2-79 shows the spectrum of the 2LFSK signal from Figure 2-78 when the signal is centered at 0 hertz. We mix this signal with 20 kHz in a quadrature modulator to

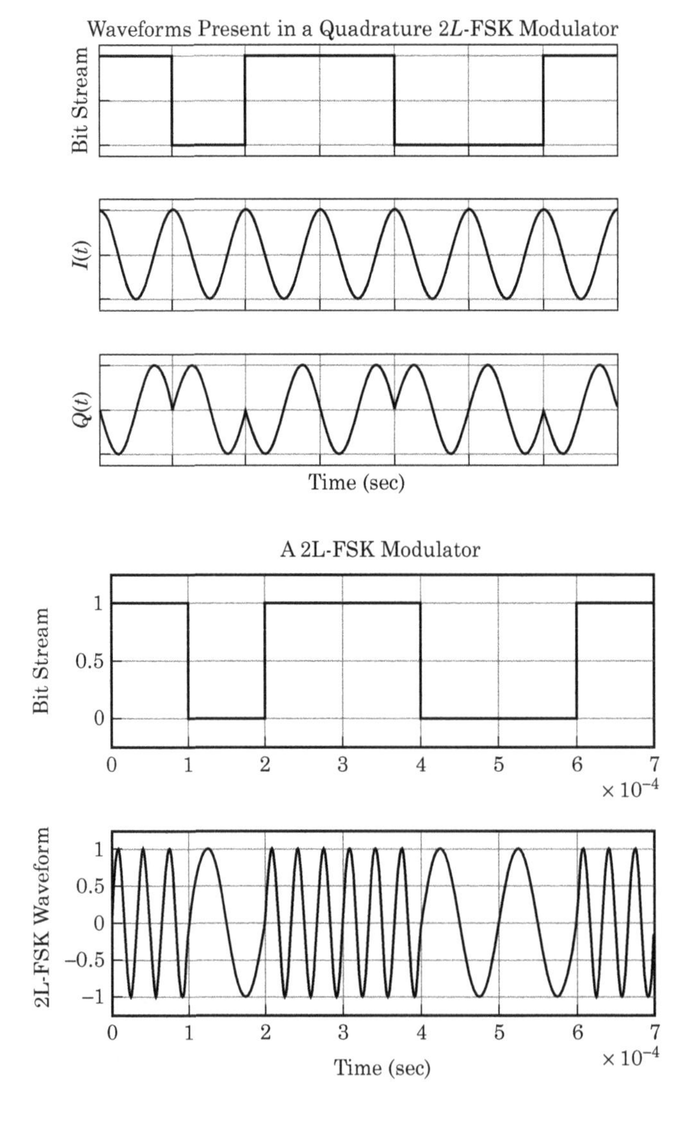

FIGURE 2-77 ■ The bit stream, $I(t)$ and $Q(t)$ waveforms required to generate a 2LFSK in a quadrature modulator. Generating FSK requires changing the $I(t)$ and $Q(t)$ waveforms every bit time.

FIGURE 2-78 ■ The digital data stream and the 2LFSK signal of Figure 2-77 after the complex signal as been converted to a center frequency of 20 kHz. The 0 or space frequency is 10 kHz, while the 1 or mark frequency is 30 kHz. The data rate is 10 kbps.

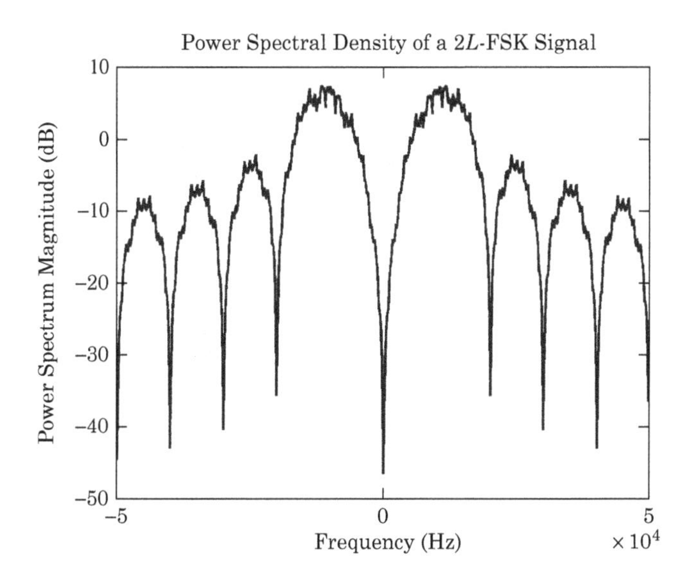

FIGURE 2-79 ■ The baseband spectrum of the 2LFSK signal of figure 2-77. This signal is complex and centered at 0 hertz.

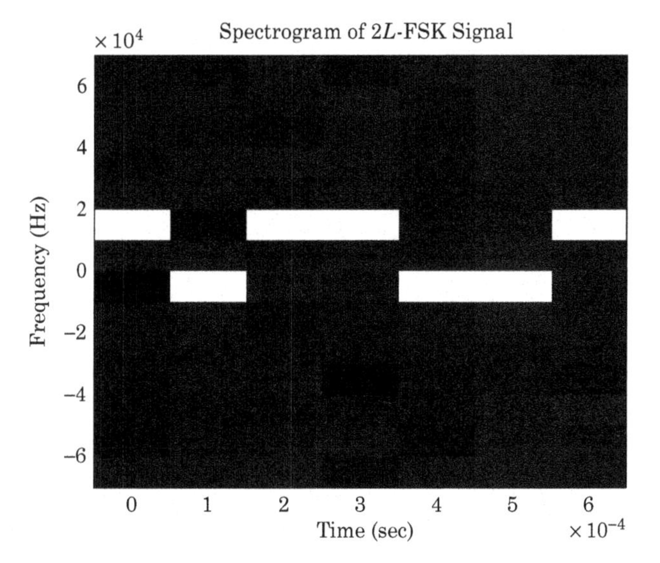

FIGURE 2-80 ■ The time-frequency spectrogram of our 2LFSK signal.

produce the signal of Figure 2-78. Figure 2-80 shows a spectrogram of the FSK signal. The x-axis is time and the y-axis is frequency. The output frequencies at f_0 and f_1 are clearly visible, and the frequency changes at bit interval boundaries based on the bit stream of Figure 2-77.

2.8 ANALOG MODULATION

The modulation and demodulation sections of this book will focus on digital data transmission. We discuss analog modulation formats only because they provide insight into digital modulation schemes.

2.8.1 Double-Sideband Amplitude Modulation (DSBAM)

In DSBAM, we vary the amplitude of a carrier sine wave with the information waveform. The carrier frequency (f_c) is usually much higher than the modulation frequency (f_m).

The distinguishing characteristics of DSBAM are

- Both the upper and lower sidebands are present
- The carrier is present
- The RF bandwidth is twice the bandwidth of the modulating signal

2.8.1.1 Time Domain

The equation describing a DSBAM-modulated signal is

$$V_{DSBAM}(t) = V_{pk}[1 + m_a \cos(\omega_m t)] \cos(\omega_c t) \tag{2.70}$$

where

$\omega_m = 2\pi f_m =$ the modulation frequency (contains the information)
$\omega_c = 2\pi f_c =$ the carrier frequency
$m_a =$ the AM modulation index and $0 <= m_a <= 1.0$ for DSBAM

The modulation frequency f_m is often in the audio range and might equal 5 kHz, while f_c is often in the RF range and it may equal 1 MHz.

We can rewrite equation (2.70) as

$$\begin{aligned} V_{DSBAM}(t) &= V_{pk} \cos(\omega_c t) + V_{pk} m_a \cos(\omega_m t) \cos(\omega_c t) \\ &= V_{pk} \cos(\omega_c t) \\ &\quad + \frac{V_{pk} m_a}{2} \cos[(\omega_c + \omega_m)t] \\ &\quad + \frac{V_{pk} m_a}{2} \cos[(\omega_c - \omega_m)t] \end{aligned} \tag{2.71}$$

where

$\cos(\omega_c t)$ is the carrier.
$\cos[(\omega_c - \omega_m)t]$ is the lower sideband (LSB). The quantity $\omega_c - \omega_m$ is often referred to as the difference frequency or the beat frequency.
$\cos[(\omega_c + \omega_m)t]$ is the upper sideband (USB). The quantity $\omega_c + \omega_m$ is also called sum frequency.

Equation (2.71) tells us that a DSBAM waveform consists of three separate sinusoidal waveforms.

Figure 2-81 shows a time-domain plot of a DSBAM waveform for $m_a = 0.1, 0.5$, and 1.0.

2.8.1.2 Frequency Domain

Figure 2-82 shows the spectrum of DSBAM. The carrier (i.e., the tone at f_c) is always present, and its power doesn't change with modulation index m_a. The tone below the carrier (at $f_c - f_m$) is the LSB. The tone above the carrier (at $f_c + f_m$) is the USB.

As equation (2.71) tells us, the carrier power is constant regardless of the value of the modulation index m_a, but the power in the USB and LSB changes with m_a. As we increase the modulation index, we increase the power in the two sidebands. The power

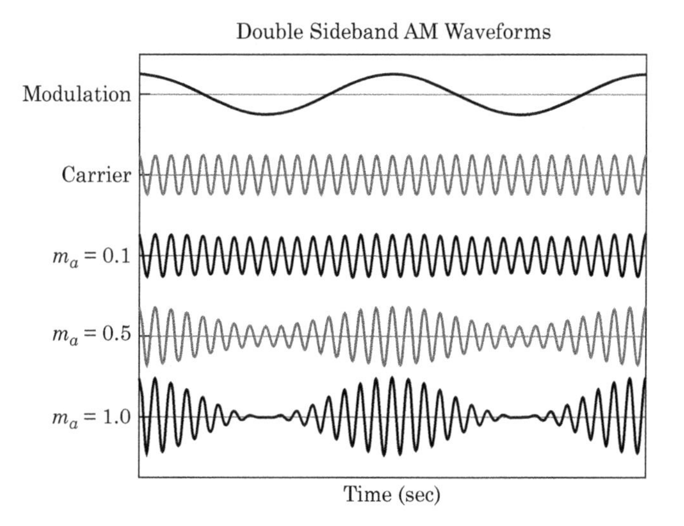

FIGURE 2-81 ■ Time-domain representation of a double-sideband AM signal for $m_a = 0.1, 0.5,$ and 1.0.

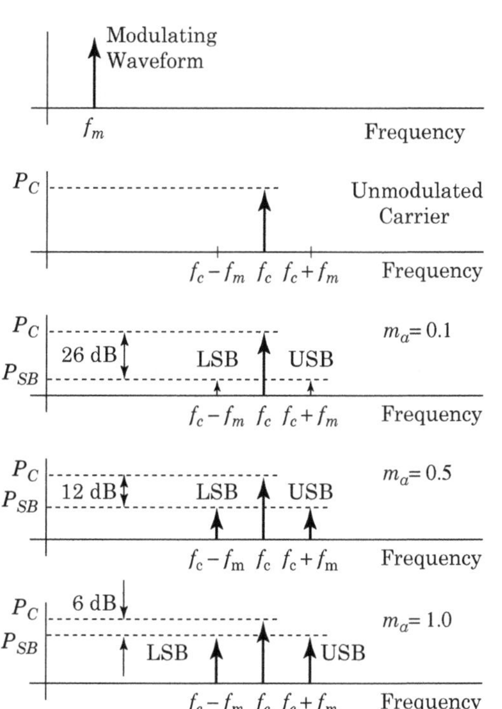

FIGURE 2-82 ■ The frequency-domain representation of a double-sideband AM signal for $m_a = 0.1, 0.5,$ and 1.0.

level of each sideband is

$$P_{Sideband} = P_{Carrier} - 20 \log \left(\frac{m_a}{2} \right) \tag{2.72}$$

where $0 <= m_a <= 1.0$. For 100% modulation (i.e., $m_a = 1$), the sideband levels are

$$20 \log \left(\frac{1}{2} \right) = -6 \text{ dB} \tag{2.73}$$

or 6 dB below the carrier.

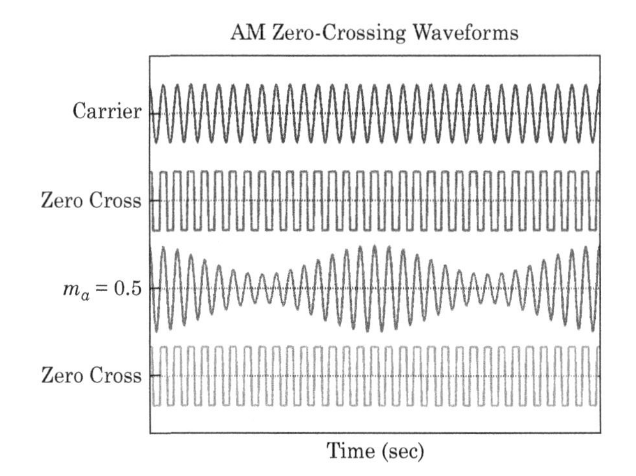

FIGURE 2-83 Zero crossings of a double-sideband AM signal. The zero crossings do not contain information about the modulation applied to the carrier.

2.8.1.3 Zero Crossings

Figure 2-83 shows the zero crossing performance of DSBAM. Applying amplitude modulation to a waveform does not change the zero crossings as long as $0 <= m_a < 1.0$. When $m_a = 1.0$, the waveform goes to zero for a brief period once every $1/f_m$ seconds. This affects the zero crossings and causes some practical implementation problems.

The zero crossings are at the same points as the unmodulated carrier. We often use this effect to remove the amplitude variations from a signal to recover the carrier. This process is called limiting.

2.8.1.4 Phasor

Figure 2-84 shows a phasor diagram of a DSBAM wave modulated with a single sinusoidal carrier. Again, the carrier is frozen in position. The USB appears to rotate counterclockwise at a rate of ω_m. The LSB rotates in a clockwise direction at a rate of ω_m.

There is an interesting the relationship between the phases of the carrier, the LSB and the USB. The angle θ_{LSB} always equals the angle θ_{USB}. In other words, the sidebands are always symmetrical about the carrier. The vector sum of the LSB and USB phasors always

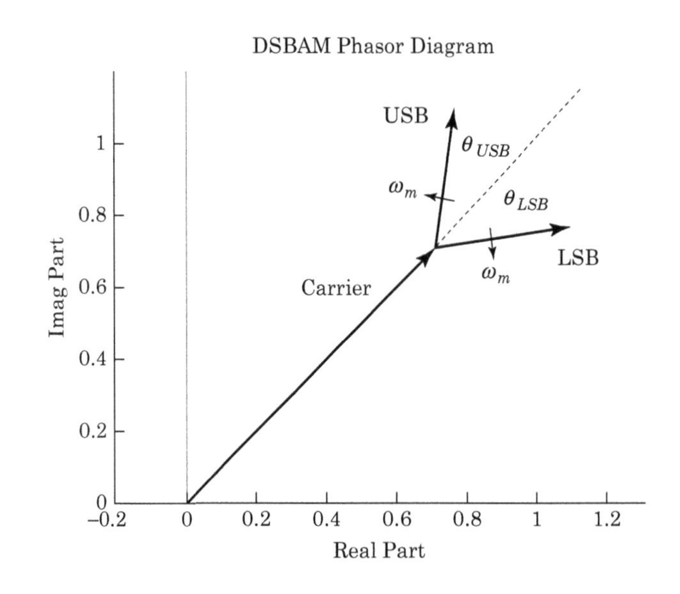

FIGURE 2-84 Phasor representation of a double-sideband AM signal. Note the symmetry of the USB and the LSB with respect to the carrier.

forms a resultant that's directly in line with the carrier. The combination of the carrier and the two sidebands resembles a man doing jumping jacks.

Imagine that the DSBAM signal passes through a medium that delays the lower-frequency components of the signal more than the higher-frequency components. This is unequal group delay. The symmetry of the USB and LSB about the carrier no longer exists, and the vector sum of the USB and LSB is no longer collinear with the carrier vector. This distorts the envelope of the AM waveform. It also causes the entire waveform to exhibit phase modulation. The carrier phase will "wobble" at a rate of f_m.

For DSBAM systems, this conversion of AM to PM usually is not problematic because most DSBAM demodulators are insensitive to phase modulation on the carrier.

2.8.1.5 RF Bandwidth

Figure 2-85 shows a random signal used to amplitude modulate a carrier and the resulting AM waveform.

Figure 2-86 shows the spectrum of the random modulating signal, while Figure 2-87 shows the spectrum of the amplitude modulated signal. The AM signal exhibits two copies of the modulating signal—a USB and an LSB. Both the USB and the LSB are as wide

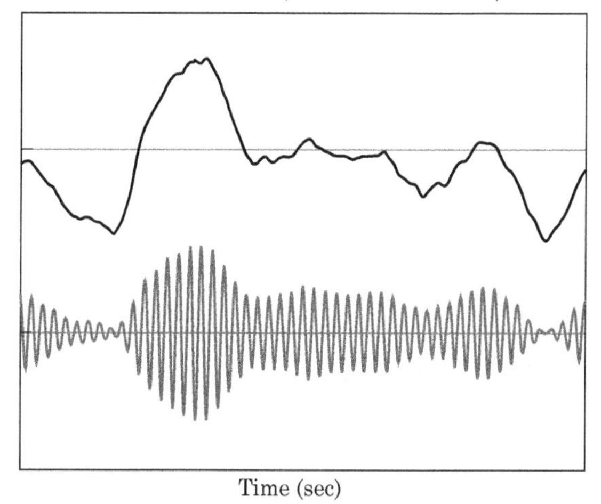

FIGURE 2-85 ■ A signal (top) used to amplitude modulate a carrier (bottom).

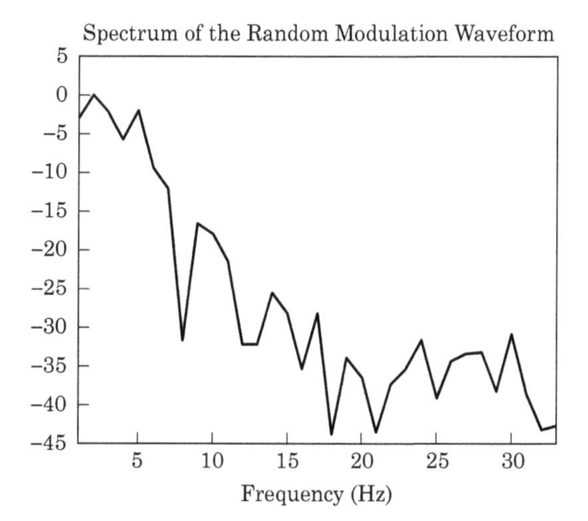

FIGURE 2-86 ■ The spectrum of the modulating signal.

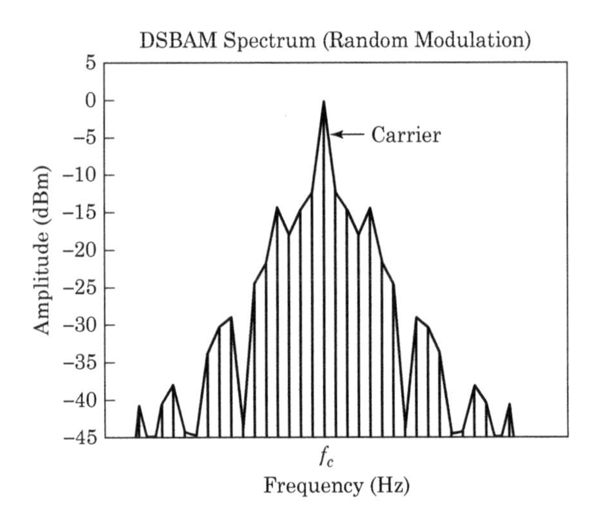

FIGURE 2-87 ■ The spectrum of the AM signal contains two replicas of the modulating signal as the LSB and USB. As a result, the AM spectrum is twice as wide as the spectrum of the modulating signal.

as the original modulating signal. Consequently, the RF spectrum is twice as wide as the baseband signal.

If the highest frequency component in the modulating waveform is $f_{m,max}$, then the RF waveform will extend from $f_c - f_{m,max}$ to $f_c + f_{m,max}$ for a total RF bandwidth of $2f_{m,max}$.

Note that the LSB has experienced a frequency inversion.

2.8.1.6 Generating Analog Amplitude Modulation with a Quadrature Modulator

We can easily generate analog AM with a quadrature modulator. Let V_{AM}(t) be the analog waveform we wish to transmit via AM. We limit the range of V_{AM}(t) to be between −0.5 V and +0.5 V and then add 1 V producing

$$I(t) = [1 + V_{AM}(t)] \tag{2.74}$$

We substitute $I(t)$ from equation (2.74) into the quadrature modulation equation (2.62), and we replace $Q(t)$ with 0 producing

$$\begin{aligned} V_{out}(t) &= I(t)\cos(\omega_{RF}t) \pm Q(t)\sin(\omega_{RF}t) \\ &= [1 + V_{AM}(t)]\cos(\omega_{RF}t) \end{aligned} \tag{2.75}$$

Figure 2-88 shows the AM modulator graphically.

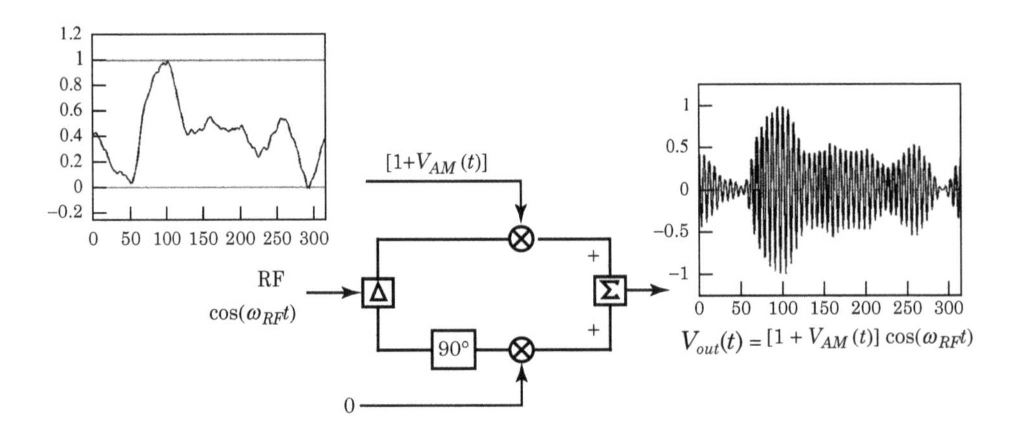

FIGURE 2-88 ■ Generating analog AM using a quadrature modulator. We apply a level-shifted version of the modulating signal to the I port of the modulator and zero to the modulator's Q port.

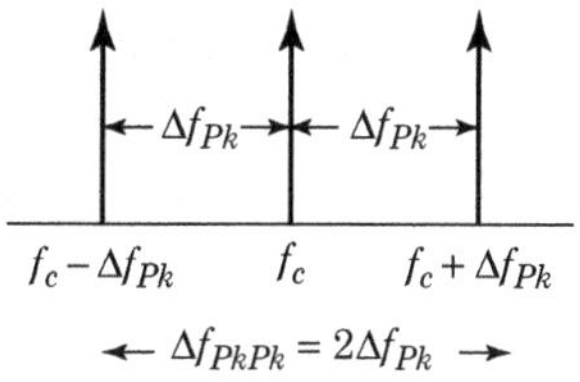

Δf_{Pk} Δf_{Pk}

$f_c - \Delta f_{Pk}$ f_c $f_c + \Delta f_{Pk}$

$\Delta f_{PkPk} = 2\Delta f_{Pk}$

FIGURE 2-89 Definition of the variables used in the frequency modulation equations.

2.8.2 Frequency Modulation

When we frequency modulate a carrier with a modulating waveform, we change the instantaneous frequency of the carrier in step with the instantaneous amplitude of the modulating wave. Figure 2-89 illustrates some of the applicable variables:

$f_c =$ the frequency of the unmodulated carrier

$\Delta f_{pk} =$ the peak frequency deviation

$\Delta f_{pkpk} =$ the peak-to-peak frequency deviation

2.8.2.1 Time Domain

We first examine sinusoidal modulation. For frequency modulation, we force the instantaneous frequency of the carrier to be determined by the modulating waveform. The instantaneous frequency, f_{FM}, of the FM signal is

$$f_{FM}(t) = f_c + \Delta f_{Pk} \cos(\omega_c t) \tag{2.76}$$

The minimum and maximum frequencies are

$$\begin{aligned} f_{FM,min} &= f_c - \Delta f_{Pk} \\ f_{FM,max} &= f_c + \Delta f_{Pk} \end{aligned} \tag{2.77}$$

The instantaneous phase of a signal is the integral of its frequency:

$$\begin{aligned} \phi_{FM}(t) &= 2\pi \int Freq_{FM}(t)\, dt \\ &= 2\pi \int [f_c + \Delta f_{Pk} \cos(\omega_m t)]\, dt \\ &= 2\pi f_c t + 2\pi \Delta f_{Pk} \frac{1}{2\pi f_m} \sin(\omega_m t) \\ &= 2\pi f_c t + \frac{\Delta f_{Pk}}{f_m} \sin(\omega_m t) \end{aligned} \tag{2.78}$$

The time-domain equation for a frequency-modulated waveform is therefore

$$V_{FM}(t) = V_{Pk} \cos\left[\omega_c t + \frac{\Delta f_{Pk}}{f_m} \sin(\omega_m t)\right] \tag{2.79}$$

We often make the substitution

$$\beta = \frac{\Delta f_{Pk}}{f_m} \tag{2.80}$$

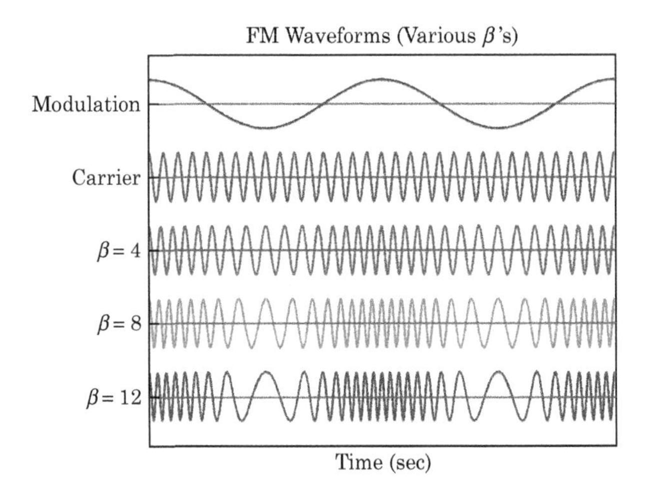

FIGURE 2-90 ■ The time-domain representation of a frequency-modulated waveform for $\beta = 4, 8$, and 12. The amplitude of each FM waveform is constant.

and equation (2.79) becomes

$$V_{FM}(t) = V_{Pk} \cos\left[\omega_c t + \beta \sin(\omega_m t)\right] \tag{2.81}$$

The quantity β is the *modulation index*. The value of β coupled with the value of f_m will lead us into an expression for the RF bandwidth and other FM characteristics.

Figure 2-90 shows the time-domain plots for several values of β.

2.8.2.2 Zero Crossings

The instantaneous frequency of an FM waveform contains all the information of the modulation so we can pass an FM waveform into a limiter and we won't lose any information. Figure 2-91 illustrates this point.

We can use this characteristic to our advantage. We can easily remove any unwanted AM (due to, e.g., multipath, fading, poor filtering) acquired by a signal. We must be wary in our use of limiters, however, because the limiting process can adversely affect a signal's SNR.

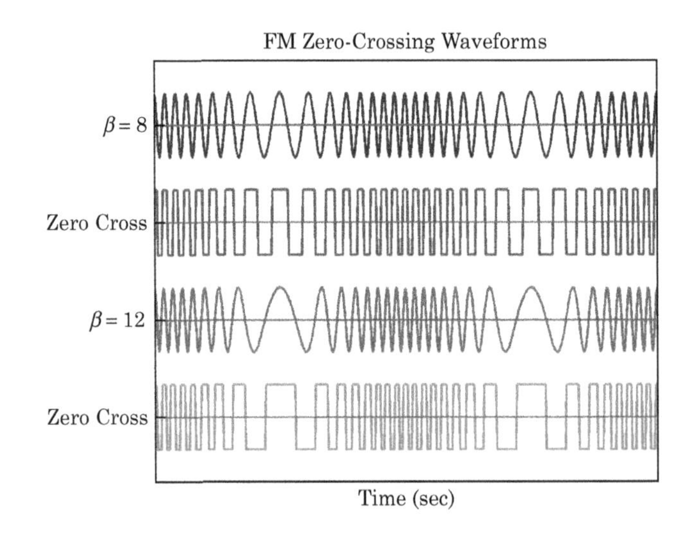

FIGURE 2-91 ■ Zero crossings of a frequency-modulated waveform for $\beta = 8$ and 12. The modulation is preserved in the zero crossings.

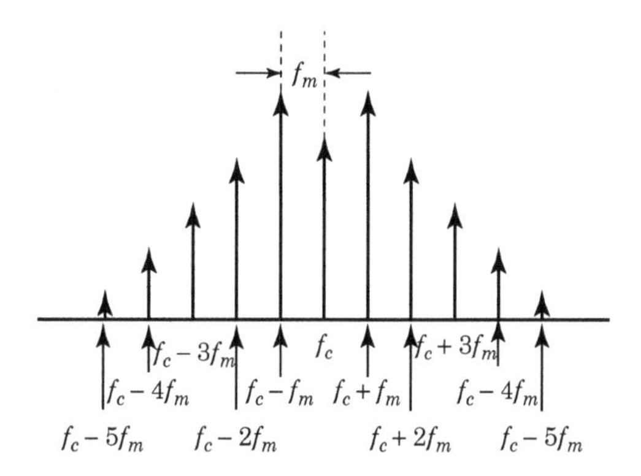

FIGURE 2-92 Frequency-domain representation a frequency-modulated waveform under sinusoidal modulation. Note the symmetry of the sidebands.

2.8.2.3 Frequency Domain

Applying Fourier analysis techniques to equation (2.81) produces

$$\begin{aligned} V_{FM}(t) = {} & V_{Pk} J_0(\beta)\cos(\omega_c t) \\ & + V_{Pk} J_1(\beta)\{\cos[(\omega_c + \omega_m)t] - \cos[(\omega_c - \omega_m)t]\} \\ & + V_{Pk} J_2(\beta)\{\cos[(\omega_c + 2\omega_m)t] + \cos[(\omega_c - 2\omega_m)t]\} \\ & + V_{Pk} J_3(\beta)\{\cos[(\omega_c + 3\omega_m)t] - \cos[(\omega_c - 3\omega_m)t]\} \\ & + V_{Pk} J_4(\beta)\{\cos[(\omega_c + 4\omega_m)t] + \cos[(\omega_c - 4\omega_m)t]\} \\ & + \cdots \\ & + V_{Pk} J_n(\beta)\{\cos[(\omega_c + n\omega_m)t] + \cos[(\omega_c - n\omega_m)t]\} \end{aligned} \tag{2.82}$$

where J_n is the Bessel function of the first kind and order n. Figure 2-92 shows the typical spectrum of an FM waveform.

Figure 2-92 shows several items of interest:

- All of the spectral components are f_m apart.
- There are an infinite number of spectral components, even for the simple case of sinusoidal modulation.
- The spectral components are clustered around the carrier frequency (at f_c).
- The spectral components are symmetrical about the carrier frequency. The power level of the spectral component at $f_c + nf_m$ equals the power level of the spectral component at $f_c - nf_m$ where n is any integer.

2.8.2.4 Phasor

Equation (2.81) and the time-domain plots of Figure 2-90 indicate to us that the magnitude of an FM waveform is constant. The amplitude of the wave never changes as it goes through its modulation cycle. Figure 2-93 shows the vector addition of the components that make up a $\beta = 3$ FM waveform. The phasors always add to place the resultant on the unit circle, regardless of the time of observation.

2.8.2.5 FM Bandwidth (Carson's Rule)

The bandwidth of an FM-modulated signal depends on the maximum frequency with which we modulate the carrier and on Δf_{pk}, the amount of carrier frequency deviation. The generally accepted rule of thumb for the RF bandwidth of an FM signal is given by

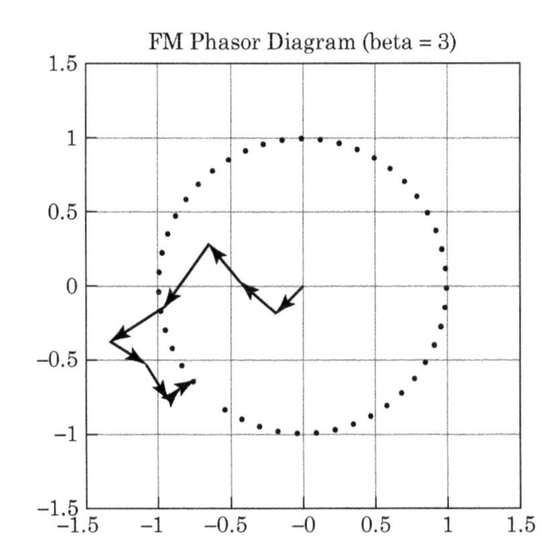

FIGURE 2-93 ■ The phasor diagram of an FM waveform at one moment in time. The phasors always add to place the resultant on the unit circle, regardless of the time of observation. Note the symmetry of the sidebands.

Carson's rule:

$$\begin{aligned} B_{FM} &= 2(\Delta f_{Pk} + f_m) \\ &= 2 f_m\,(\beta + 1) \end{aligned} \tag{2.83}$$

This bandwidth includes approximately 95% of the energy available in the FM waveform. The theoretical 100% bandwidth is infinite.

2.8.3 Phase Modulation (PM)

In phase modulation, the instantaneous phase of the carrier depends on the instantaneous amplitude of the modulating wave. Phase modulation is very similar to frequency modulation, and we'll use many of the results we developed earlier.

2.8.3.1 Time Domain

In a phase-modulated waveform, the instantaneous phase of the carrier is set by the modulating waveform. The instantaneous phase of a sinusoidally modulated PM signal is

$$\phi_{PM}(t) = 2\pi f_c t + \Delta\phi_{Pk}\cos(2\pi f_m t) \tag{2.84}$$

where

$\Delta\phi_{pk}$ = the peak phase shift
f_m = the modulating frequency

The time-domain equation for a phase-modulated wave is

$$\begin{aligned} V_{PM}(t) &= V_{Pk}\cos\left[2\pi f_c t + \Delta\phi_{Pk}\cos(2\pi f_m t)\right] \\ &= V_{Pk}\cos\left[\omega_c t + \Delta\phi_{Pk}\cos(\omega_m t)\right] \end{aligned} \tag{2.85}$$

Note the similarity between equation (2.85) and equation (2.81) (which describes an FM waveform). The equations are identical except for the substitution of $\Delta\phi_{pk}$ for β and a substitution of a cosine for a sine. This similarity allows us to apply many of the FM analysis details to the PM analysis.

Figure 2-94 shows the time-domain plots for several values of $\Delta\phi pk$. As with FM, the amplitudes of all the waveforms are constant.

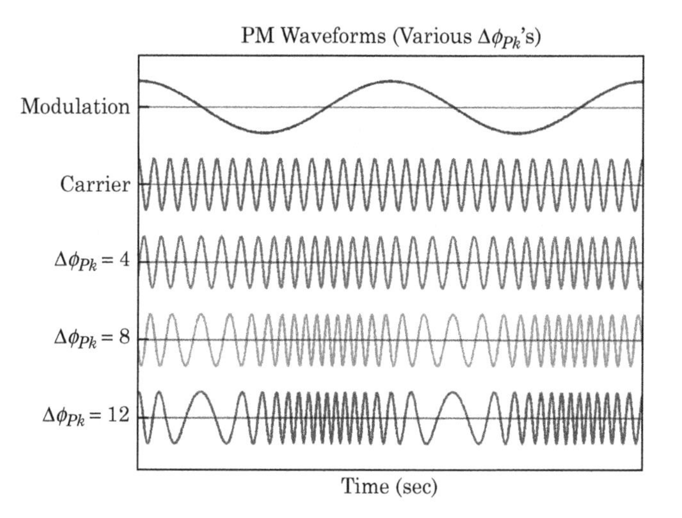

FIGURE 2-94 ■ The time-domain representation of a phase-modulated signal for $\Delta\phi_{Pk}$ of 4, 8, and 12 radians.

2.8.3.2 Comparison of FM and PM Waveforms

Figure 2-95 shows an FM waveform and an equivalent PM waveform on the same graph. Both waveforms are modulated by the same cosine wave and both have a deviation of eight; that is,

$$f_{m,FM} = f_{m,PM} \\ \text{and} \\ \Delta f_{Pk} = \Delta\phi_{Pk} = 8 \tag{2.86}$$

The frequency of the FM waveform is the highest when the modulating waveform has a maximum value. The frequency of the PM waveform is the highest when the derivative of the modulating waveform has a maximum value.

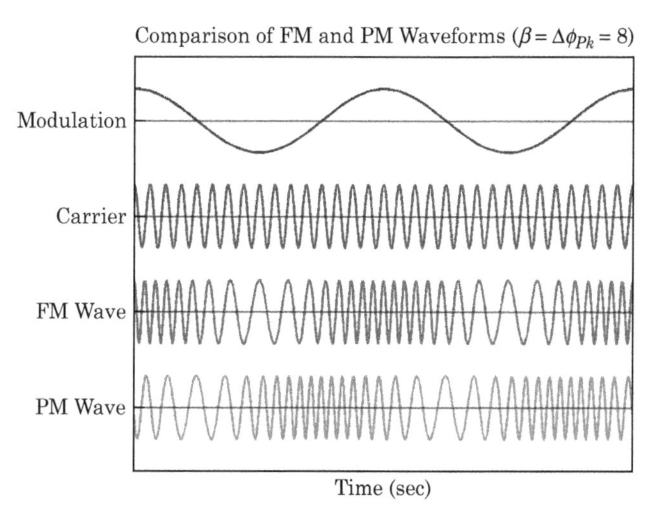

FIGURE 2-95 ■ Comparison of frequency- and phase-modulated waveforms for $\beta = \Delta\phi_{Pk} = 8$. The maximum frequency of the FM signal occurs when the modulating waveform is at its maximum. The maximum frequency of the PM signal occurs when the derivative of the modulating waveform is at its maximum.

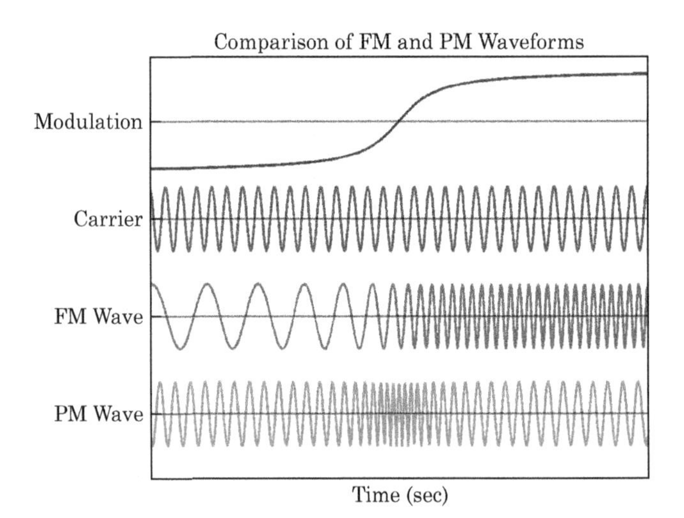

FIGURE 2-96 Comparison of frequency- and phase-modulated waveforms. Maximum frequency of FM signal occurs when modulating waveform is maximum. Maximum frequency of PM signal occurs when the derivative of the modulating waveform is maximum.

Figure 2-96 shows a modulating waveform that emphasizes the difference between the FM and PM signals. The frequency of the FM signal changes directly with the value of the modulating waveform. The frequency of the PM signal stays constant except where the modulating waveform changes value.

2.8.3.3 Zero Crossings

Phase-modulated waveforms behave exactly like frequency modulation waveforms when we pass them through a limiter. As Figure 2-97 shows, we lose no information.

As is the case for FM, we can use a limiter to remove unwanted AM artifacts impressed upon a PM signal.

2.8.3.4 Frequency Domain

The frequency-domain description of a phase-modulated signal is almost identical to the frequency-domain description of a frequency-modulated signal. The major differences involve sign changes and changes of sines and cosines into cosines and sines. We can

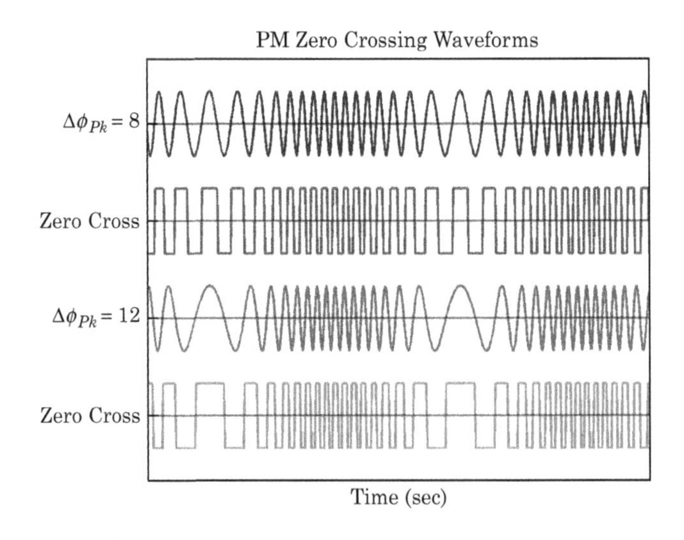

FIGURE 2-97 Zero crossings of a phase-modulated waveform, for $\Delta\phi_{Pk} = 8$ and 12. We lose no information in the limiting process.

rewrite equation (2.85) as

$$\begin{aligned} V_{PM}(t) = {} & V_{Pk} J_0(\Delta\phi_{Pk}) \cos(\omega_c t) \\ & + V_{Pk} J_1(\Delta\phi_{Pk}) \{\sin[(\omega_c + \omega_m)t] + \sin[(\omega_c - \omega_m)t]\} \\ & - V_{Pk} J_2(\Delta\phi_{Pk}) \{\cos[(\omega_c + 2\omega_m)t] + \cos[(\omega_c - 2\omega_m)t]\} \\ & - V_{Pk} J_3(\Delta\phi_{Pk}) \{\sin[(\omega_c + 3\omega_m)t] + \sin[(\omega_c - 3\omega_m)t]\} \\ & + V_{Pk} J_4(\Delta\phi_{Pk}) \{\cos[(\omega_c + 4\omega_m)t] + \cos[(\omega_c - 4\omega_m)t]\} \\ & + \cdots \end{aligned} \tag{2.87}$$

For most purposes, we can consider the spectrum of a PM waveform to be identical to the spectrum for an FM signal. The spectrum characteristics we described for FM signals are valid for PM signals.

2.8.3.5 Phasor

Like the FM waveform we examined earlier, the magnitude of the PM waveform is always constant. The amplitude never changes as it goes through its modulation cycle. Also, the vectorial sum of the vector components always add up to a resultant with a constant magnitude.

2.8.3.6 PM Bandwidth (Carson's Rule Again)

The final RF bandwidth of a PM-modulated wave depends on the maximum frequency we modulate the carrier with and on $\Delta\phi_{pk}$, the maximum phase deviation. The generally accepted rule of thumb for the RF bandwidth of a PM signal is

$$B_{PM} = 2 f_m(\Delta\phi_{Pk} + 1) \tag{2.88}$$

This bandwidth includes approximately 95% of the energy available in the PM waveform. The theoretical 100% bandwidth is infinite.

2.8.3.7 Generating Analog Phase Modulation with a Quadrature Modulator

Generating analog phase modulation with a quadrature modulator is less straightforward than generating AM.

Let's choose an analog waveform that we've forced into the range of −1 V to +1 V. Using a cosine at 0° as the phase reference, we'll transmit an RF signal of $\cos(\omega t + \pi/2)$ for the −1 V level, $\cos(\omega t + 0)$ for the 0 V level and $\cos(\omega t - \pi/2)$ for the +1 V level. We note

$$\cos(\omega t \pm \beta) = \cos(\omega t)\cos(\beta) \mp \sin(\omega t)\sin(\beta) \tag{2.89}$$

implying

$$\begin{aligned} \cos\left(\omega t - \frac{\pi}{2}\right) &= +\sin(\omega t) \\ \cos(\omega t) &= \cos(\omega t) \\ \cos\left(\omega t + \frac{\pi}{2}\right) &= -\sin(\omega t) \end{aligned} \tag{2.90}$$

Our task is to build a map between the modulator input signal to the quadrature I and Q waveforms. Equation (2.62) describes the output of the quadrature modulator:

$$V_{out}(t) = I(t)\cos(\omega_{RF} t) + Q(t)\sin(\omega_{RF} t) \tag{2.91}$$

Inspection of equations (2.90) and (2.91) reveals that, as the modulating signal passes from −1 V through 0 V to +1 V, the quantity I transitions from 0 to +1 then back through 0, whereas Q transitions from −1 to 0 and then to +1.

The instantaneous phase of V_{out} is related to the I and Q waveforms by

$$\tan(\phi) = \frac{Q}{I} \tag{2.92}$$

which we can rewrite as

$$I = \frac{Q}{\tan(\phi)} \tag{2.93}$$

We also have the constraint on V_{out} of constant modulus, which we can express as

$$I^2 + Q^2 = 1 \tag{2.94}$$

Combining equations (2.93) and (2.94) produces

$$\begin{aligned} \frac{Q^2}{\tan^2(\phi)} + Q^2 &= 1 \\ Q^2\left[\frac{1}{\tan^2(\phi)} + 1\right] &= 1 \\ Q^2 &= \frac{\tan^2(\phi)}{1 + \tan^2(\phi)} \end{aligned} \tag{2.95}$$

and

$$\begin{aligned} I^2 &= \frac{Q^2}{\tan^2(\phi)} \\ &= \frac{1}{1 + \tan^2(\phi)} \end{aligned} \tag{2.96}$$

Figure 2-98 shows the plot of the instantaneous I and Q values versus the instantaneous modulating signal.

Let's look at an example. Figure 2-99 shows a signal waveform we would like to impress upon the phase of a quiet carrier producing a phase-modulated signal.

We use the modulating signal of Figure 2-99 and the I and Q mappings of Figure 2-98 to generate the $I(t)$ and $Q(t)$ signals shown in Figure 2-100.

Figure 2-101 shows a section of the modulating signal and a section of the phase-modulated carrier. When the modulating signal is near -1, the phase of the PM signal is

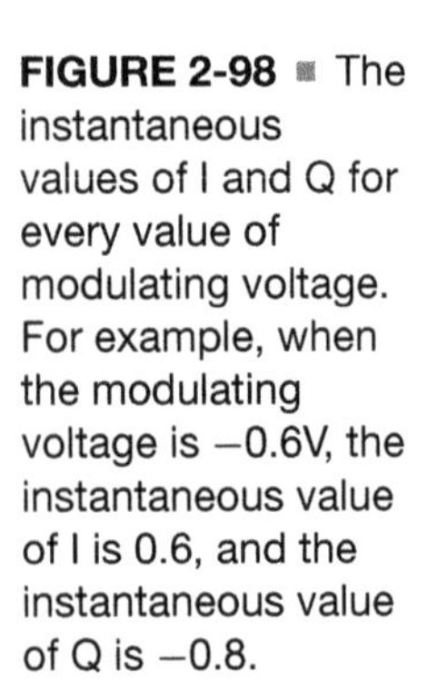
FIGURE 2-98 ■ The instantaneous values of I and Q for every value of modulating voltage. For example, when the modulating voltage is −0.6V, the instantaneous value of I is 0.6, and the instantaneous value of Q is −0.8.

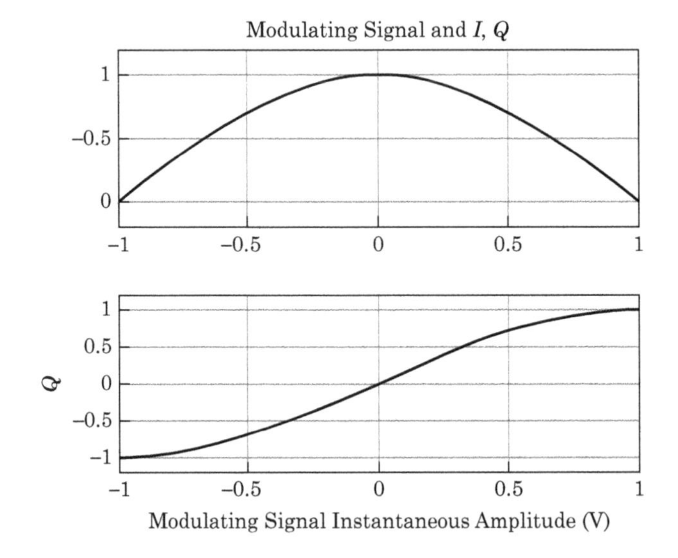

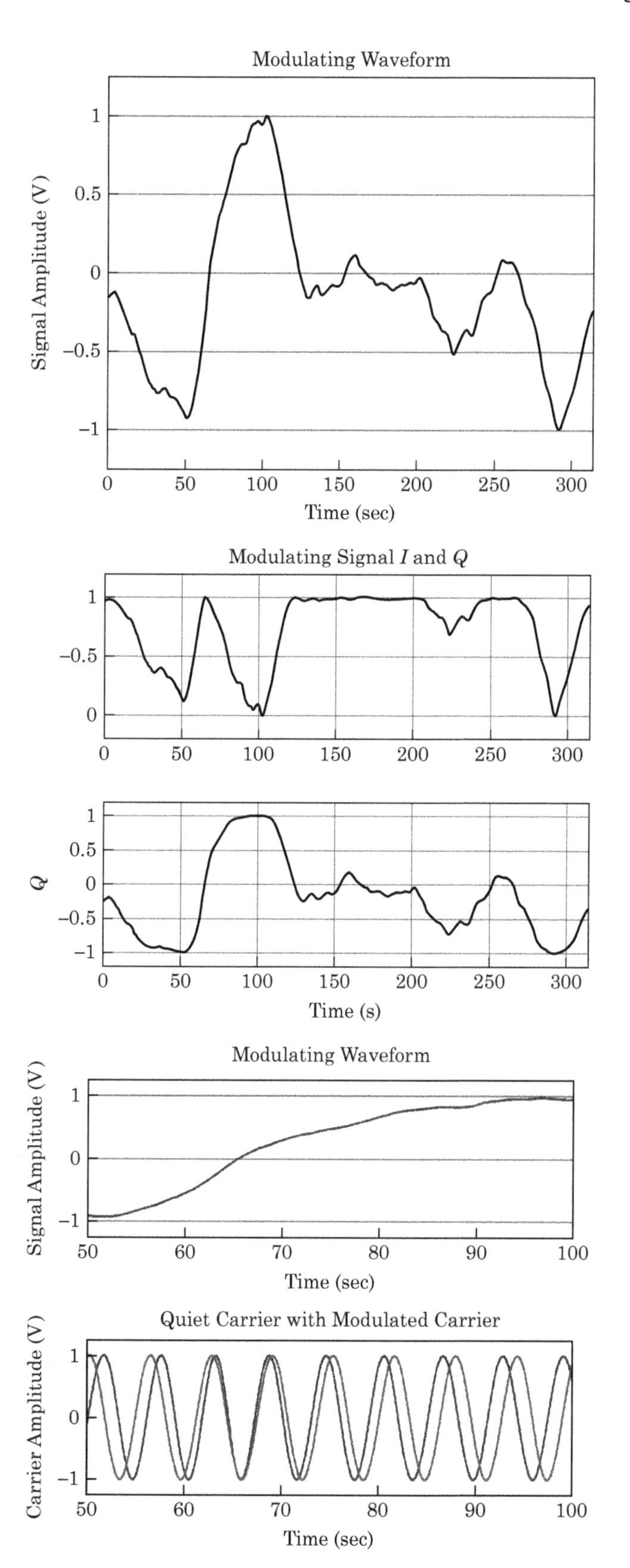

FIGURE 2-99 ■ An information-bearing signal. We'll use this signal to phase modulate a carrier using a quadrature modulator.

FIGURE 2-100 ■ We use the modulating signal of Figure 2-99 combined with the I and Q mapping shown in Figure 2-98 to generate the $I(t)$ and $Q(t)$ signals shown here.

FIGURE 2-101 ■ A portion of the modulating signal of Figure 2-99 and the corresponding portion of the phase-modulated carrier.

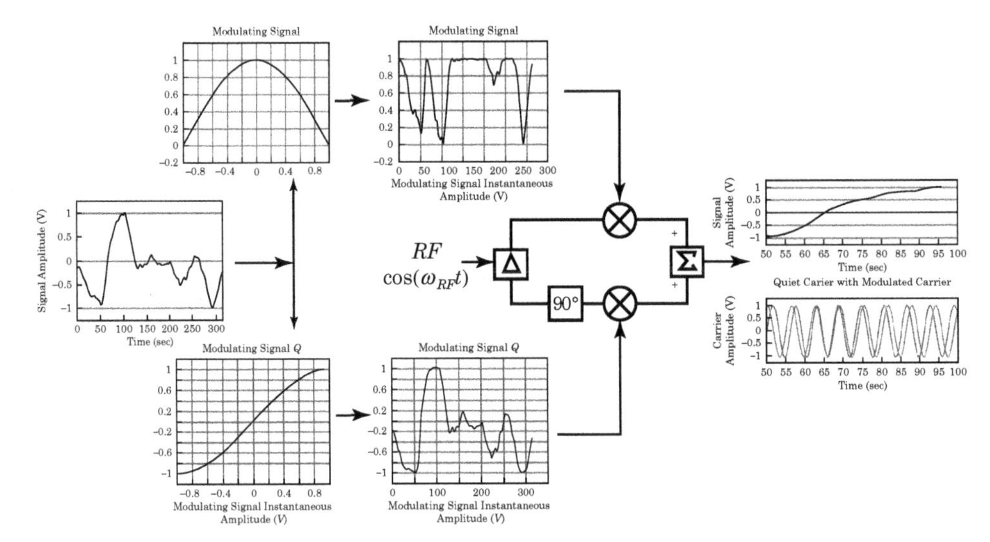

FIGURE 2-102 Generating analog PM using a quadrature modulator. The analog input signal is mapped to its corresponding I (top) and Q (bottom) components. The I and Q components are then applied to the I and Q input ports of the quadrature modulator where they modulate the phase of the RF carrier.

near $-\pi/2$. When the modulating signal is near $+1$, the phase of the PM signal is near $+\pi/2$.

Figure 2-102 shows the quadrature phase modulator graphically.

2.9 | DIGITAL MODULATION

2.9.1 Generating Digitally Modulated Signals

In this section, we discuss using a quadrature modulator to generate signals that carry digital information. Digital signal candidates include the following:

- Amplitude modulation (OOK)
- 2LFSK
- 4LFSK
- Multitone FSK
- BPSK
- QPSK
- M-ary PSK

Given a modulation type and other characteristics, our task distills to deriving the appropriate $I(t)$ and $Q(t)$ waveforms required to express the digital bit stream as a modulated wave.

2.9.2 Bits, Symbols, and Alphabets

Before we discuss digital modulation, we must address the relationship among bits, symbols, and the corresponding quadrature modulator $I(t)$ and $Q(t)$ waveforms.

A bit is the smallest unit of information. One bit contains enough information to discern, for example, a 0 from a 1, an off from an on, or a yes from a no.

The digital data that we wish to transmit come to us as a digital bit stream. We often group or combine several of these bits to produce a symbol. For example, if we grouping together all combinations of two bits form four unique symbols:

- Symbol 0: 00
- Symbol 1: 01
- Symbol 2: 10
- Symbol 3: 11

In this example, the symbol number (0, 1, 2, 3) is equivalent to the binary representation of a 2-bit sequence (i.e., $0 = 00$, $1 = 01$, $2 = 10$ and $3 = 11$). Often, there will be a clear mapping between the symbol name and the bit configuration it represents. At other times, the relationship will be quite convoluted.

The number of bits expressed per symbol is

$$b = \log_2(s) \tag{2.97}$$

where b is the number of bits, and s is the number of symbols. For example, the 26 lowercase letters of the alphabet contain

$$\begin{aligned} b &= \log_2(s) \\ &= \log_2(26) \\ &= 4.7 \, \frac{\text{bits}}{\text{symbol}} \end{aligned} \tag{2.98}$$

assuming that each symbol appears with equal probability.

2.9.3 Bit Rate versus Symbol Rate

The bit rate of a communication system is the number of bits communicated between the transmitter and receiver per second. The symbol rate is the number of symbols transmitted per second. The bit rate and symbol rate are related by

$$\frac{\text{bits}}{\text{sec}} = \frac{\text{bits}}{\text{symbol}} \cdot \frac{\text{symbols}}{\text{sec}} \tag{2.99}$$

For example, if one symbol represents four bits, then equation (2.99) becomes

$$\frac{\text{bits}}{\text{sec}} = 4 \cdot \frac{\text{symbols}}{\text{sec}} \tag{2.100}$$

A system that transmits 1 Mbps at 4 bits/symbol would require us to transmit 250k symbols per second. Similarly, a system that transmits 1 Mbps at 16 bits/symbol requires a 62.5k symbol transmission rate. Figure 2-103 shows an 8-bit per symbol converter using a shift register.

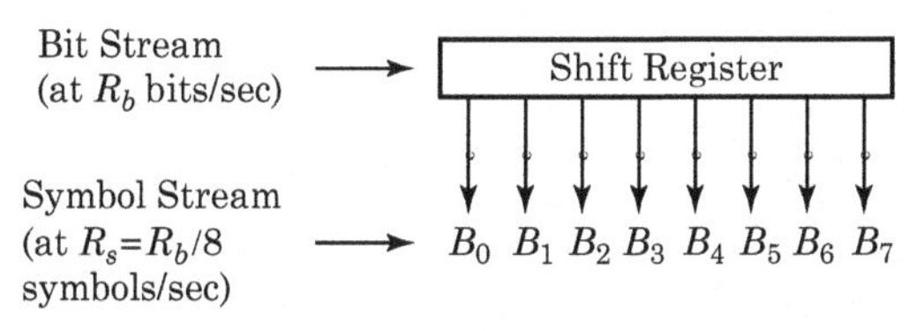

FIGURE 2-103 ■ A shift register used to convert a bit stream into 8-bit symbols. The symbol rate R_S = the bit rate $R_b/8$.

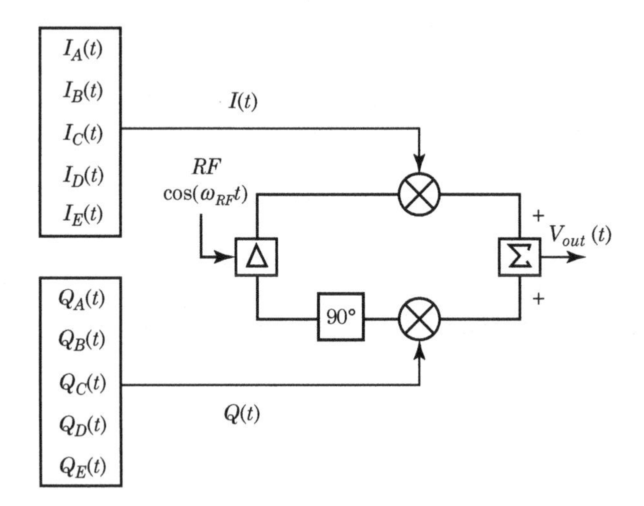

FIGURE 2-104 ■ A quadrature modulator producing five symbols. Each $I(t)$ and $Q(t)$ pair generates a symbol. Each pair is presented to the modulator for one symbol period.

When we build modulated waveforms that represent symbols, our task will be to calculate the proper $I(t)$ and $Q(t)$ waveforms that will accurately represent those symbols. We must develop a unique set of $I(t)$ and $Q(t)$ waveforms for every unique symbol we wish to express. Figure 2-104 shows a quadrature modulator configured to emit a five-symbol waveform consisting of symbols A through E.

2.9.4 Random Data

The requirement of a communications system is to move the user's data from one location to another. We cannot place restrictions on the characteristics of that data. For example, the user may desire to send a long string of binary 0's with an occasional 1 or send a repetitive pattern over and over again. Such nonrandom data play havoc with both our modulators and demodulators. Nonrandom data cause the signals we generate in our transmitter to exhibit narrow spectral lines that waste signal power and confuse demodulators.

There exist common, well-known techniques to randomize user data before they arrive at the modulator and to remove that randomization once the data have left the demodulator. One such randomization technique applies a linear recursive sequence (LRS) to the user's data.

Figure 2-105 shows a BPSK signal modulated by random and nonrandom data. Note the narrow spectral lines in the spectrum generated by nonrandom data.

2.9.5 Amplitude Shift Keying (ASK)

ASK is the digital version of analog amplitude modulation. We change only the amplitude of the carrier in discrete units.

2.9.5.1 Time Domain

Figure 2-106 shows a four-alphabet symbol stream and corresponding ASK RF signal. The symbol rate is 75k symbols/second, and the carrier frequency is 250 kHz. In this example, we're sending two bits per symbol so the bit rate is 150k bits/second.

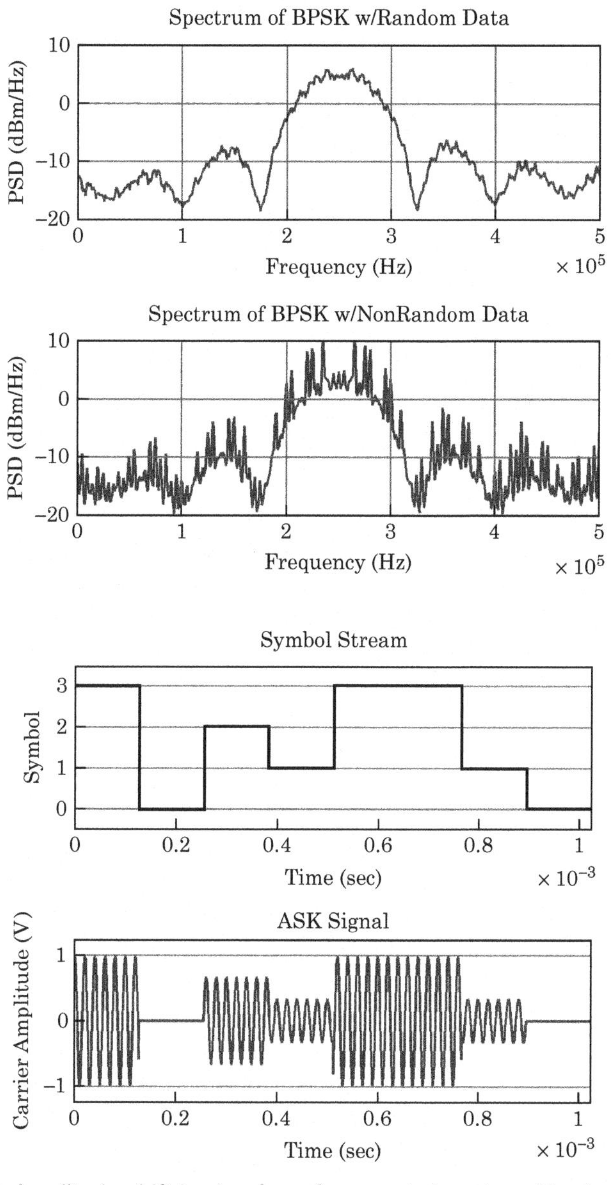

FIGURE 2-105 ■ The spectrum of a signal modulated with nonrandom and random data. A demodulator can lock onto the coherent spikes generated by the nonrandom data and confuse those signals with estimated baud rates or carrier frequencies.

FIGURE 2-106 ■ Amplitude shift keying for a four-symbol system. The top plot shows the symbol stream, and the bottom plot shows the corresponding ASK RF signal. The ASK waveform changes only in amplitude. It does not change phase or frequency. In this example, the symbol rate is 75k symbols/second, and the carrier frequency is 250 kHz. Since we're sending 2 bits/symbol, the bit rate of this signal is 150k bits/second.

2.9.5.2 Frequency Domain

Figure 2-107 shows the spectrum of our example four-level ASK signal. The symbol rate is 75k symbols/second, and the carrier frequency is 250 kHz. The width of the spectral lobes is set by the symbol rate.

The presence of the 250 kHz carrier in Figure 2-107 is explained by examining the time-domain waveform of Figure 2-106. When we transmit symbols 1, 2, or 3, we transmit

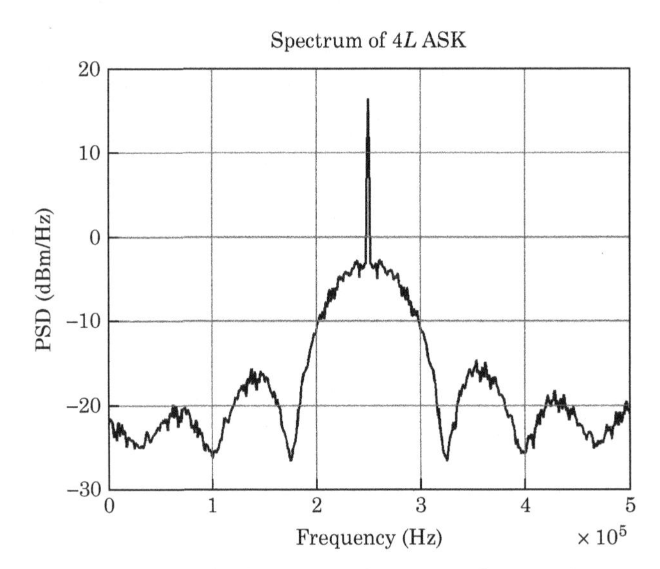

FIGURE 2-107 ■ The spectrum of a four-level ASK signal. Overall, the spectrum shows the $\sin(x)/x$ characteristic associated with a square-shaped symbol stream. Note there is significant carrier present in the spectrum. The carrier contains no information and is a waste of energy from an information transmission perspective. However, we can use the carrier in the demodulator to achieve carrier lock.

a 250 kHz tone for the symbol period. We never change the phase of the carrier, only the amplitude, so the energy present at 250 kHz builds up unencumbered. The result is the narrow spike shown in Figure 2-107.

We can remove the 250 kHz spike if we so desire by changing the phase of the carrier with every symbol. This is not strictly ASK, but we are using the transmitted energy more efficiently. Figure 2-108 shows the spectrum of an ASK signal with random phase applied to each symbol.

2.9.5.3 Phasor Domain

Figure 2-109 shows the phasor diagram of our four-level ASK signal. Each symbol corresponds to a phasor in the positive-x direction. The value of the symbol (0 through 3) determines the magnitude of the phasor. The phase of the phasor is always 0°.

The phasor corresponding to the symbol is valid for one symbol period. At the next symbol period, the diagram changes to the phasor corresponding to the next symbol.

2.9.5.4 Generating ASK with a Quadrature Modulator

We break our serial bit stream into a symbol stream consisting of 2-bit symbols, producing four unique symbols (levels in the case of ASK). We then apply those levels to the I port of the quadrature modulator, as we show in Figure 2-110.

2.9.6 On-Off Keying

OOK is a special case of ASK. We allow the amplitude to take on only two levels, 0 and 1. The effect is to turn the carrier on and off at the bit rate.

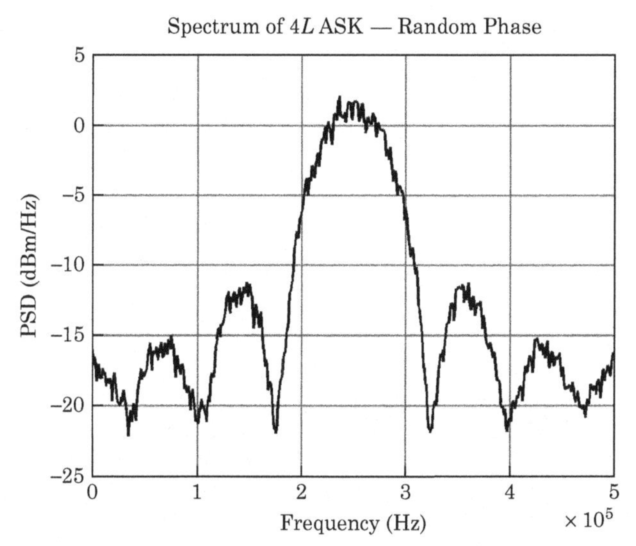

FIGURE 2-108 ■ The spectrum of a four-level ASK signal with a random phase applied to each symbol. The spectrum here is similar to the spectrum of Figure 2-107, but this spectrum contains no carrier. This spectrum more efficiently uses the signal powe—all of the energy goes into the modulation—but this signal will be more difficult to demodulate because achieving carrier lock will be problematic.

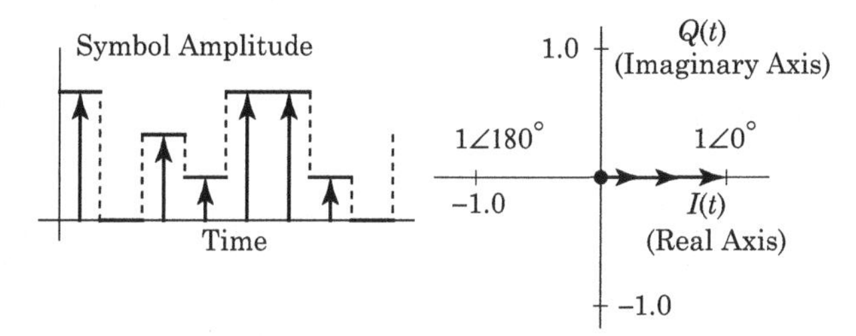

FIGURE 2-109 ■ The phasor diagram of our four-level ASK signal.

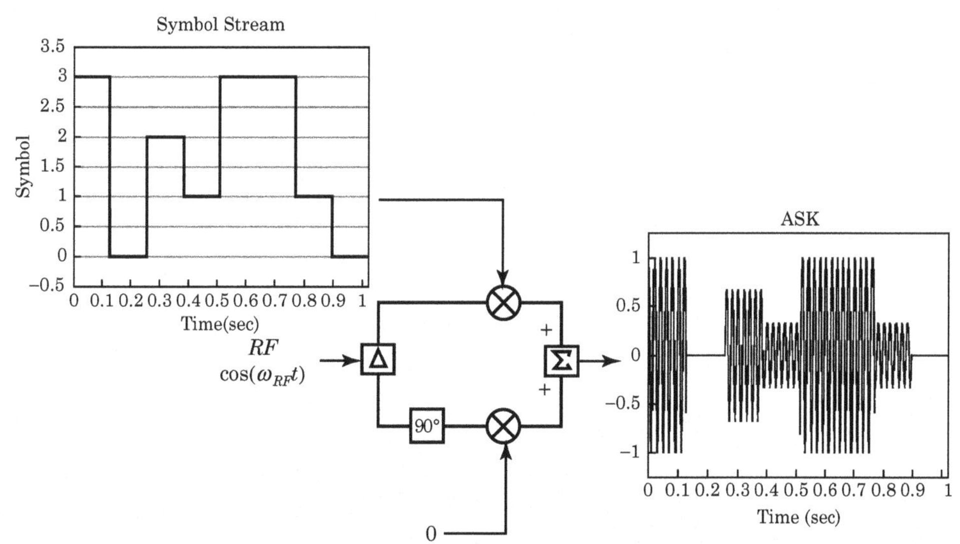

FIGURE 2-110 ■ Generating digital ASK using a quadrature modulator. The symbol stream in the range 0 through 3 is fed into the I port of the quadrature modulator. The output is a cosine whose amplitude is encoded with the symbol stream. Only the amplitude of the RF carrier is affected by the symbol stream. The carrier's phase and frequency are unaffected.

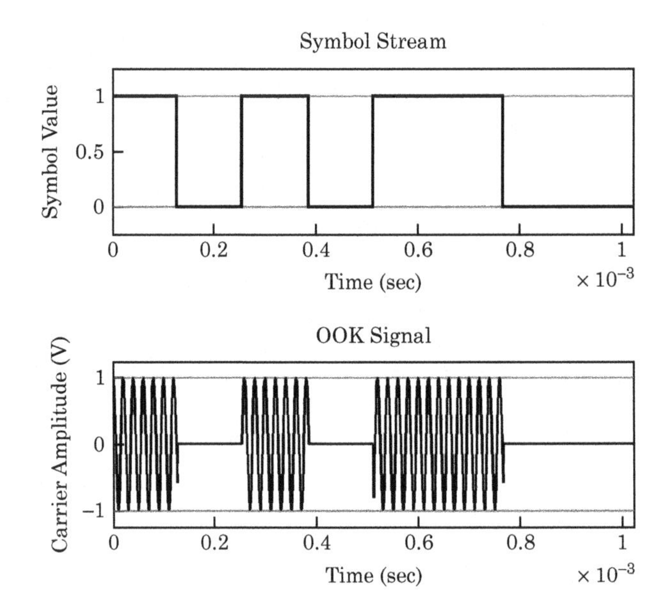

FIGURE 2-111 The time-domain representation of an OOK digital signal. The top plot shows the symbol stream, while the bottom plot shows the corresponding OOK RF signal. The OOK waveform contains only bursts of the carrier and every burst is the same amplitude.

2.9.6.1 Time Domain

Figure 2-111 shows an OOK symbol stream and the corresponding OOK RF signal. The symbol rate is the same as the bit rate in OOK since we're transmitting one bit per symbol. The symbol rate is 75k symbols/second, and the carrier frequency is 250 kHz.

2.9.6.2 Frequency Domain

Figure 2-112 shows the spectrum our OOK signal. The width of the first and largest spectral lobe is set by the symbol rate. Note the similarity between Figure 2-112 and the spectrum of the four-level ASK signal of Figure 2-107, including the large, unmodulated carrier.

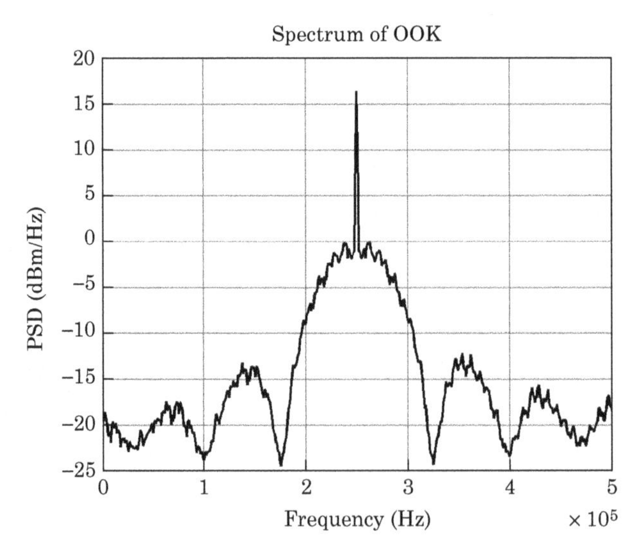

FIGURE 2-112 The spectrum of an OOK signal whose symbol rate is 75k symbols/second and carrier frequency is 250 kHz. Overall, the spectrum shows the $\sin(x)/x$ characteristic associated with a square-shaped symbol stream. Note there is significant carrier present in the spectrum. The carrier contains no information and is a waste of energy from an information transmission perspective.

2.9.6.3 Phasor Domain

Figure 2-113 shows the phasor diagram of our OOK signal. The 1 symbol corresponds to a phasor in the positive-x direction and of unity amplitude. We turn the carrier off during the 0 symbol time. The phase of the phasor is always 0°.

The phasor corresponding to the symbol is valid for one symbol period. At the next symbol period, the diagram changes to the phasor corresponding to the next symbol.

2.9.6.4 Generating OOK with a Quadrature Modulator

Figure 2-114 shows the I and Q waveforms we generate to produce OOK.

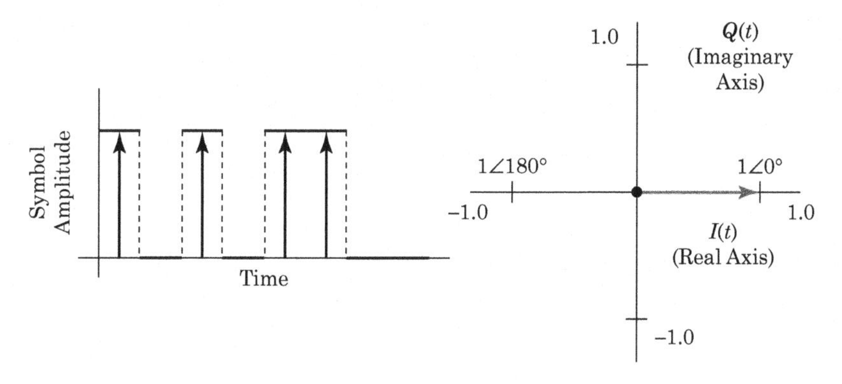

FIGURE 2-113 ■ The phasor diagram of an OOK signal.

2.9.7 Binary Phase Shift Keying

We encode our symbol stream into the phase of the transmitted signal when generating a BPSK waveform. We transmit a 0 by sending a cosine wave at a 0° phase for one symbol duration and transmit a 1 by sending a cosine wave at 180° phase.

$$
\begin{aligned}
V_{BPSK}(t) &= \cos(\omega_{RF}t + n\pi); \quad n = 0, 1 \\
&= \pm\cos(\omega_{RF}t) \\
&= \cos(\omega_{RF}t) \ or \ \cos(\omega_{RF}t + \pi)
\end{aligned}
\tag{2.101}
$$

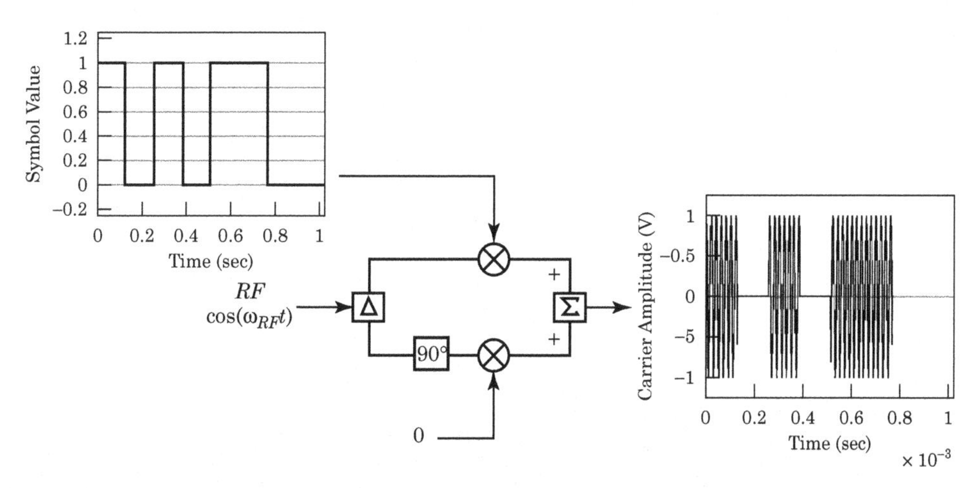

FIGURE 2-114 ■ Generating OOK using a quadrature modulator. The symbol rate is 75k symbols/second and the carrier frequency is 250 kHz. The symbol stream is fed into the I port of the quadrature modulator. The output cosine is turned on when the symbol is a 1 and turned off when the symbol is a 0. Only the amplitude of the RF carrier is affected by the symbol stream. The carrier's phase and frequency are unaffected.

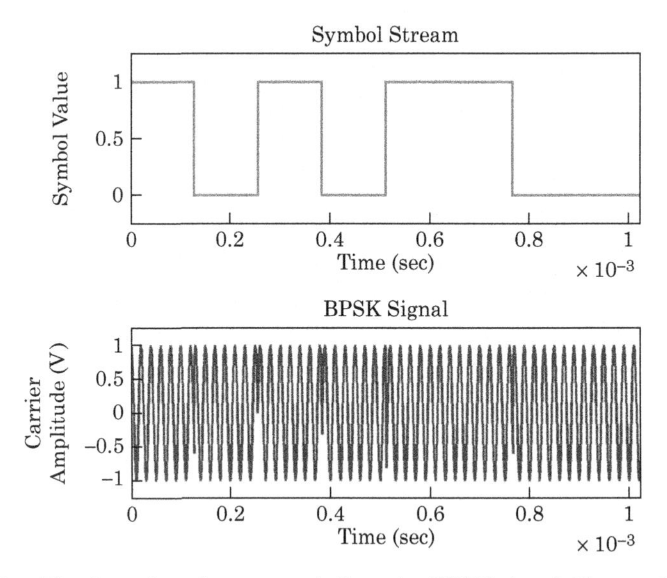

FIGURE 2-115 ■ The time-domain representation of a BPSK signal. The symbol rate is 75k symbols/second, and the carrier frequency is 250 kHz. The top graph is the user's data stream. The bottom graph is the BPSK signal generated by the modulator. Depending on the symbol, we transmit either $\cos(\omega t)$ or $\cos(\omega t + \pi)$. Because this is a 1-bit/symbol waveform, the bit rate equals the symbol rate. Note that the BPSK signal is constant amplitude.

2.9.7.1 Time Domain

Figure 2-115 and equation (2.101) show the time-domain representation of a BPSK signal. When the symbol is a 0, we transmit $\cos(\omega t)$. When the symbol is a 1, we transmit $\cos(\omega t + \pi)$.

2.9.7.2 Frequency Domain

Figure 2-116 shows the frequency domain description of a BPSK signal. The spectrum of BPSK is similar to those of ASK and OOK of the same symbol rate except that the BPSK spectrum shows a notable absence of carrier.

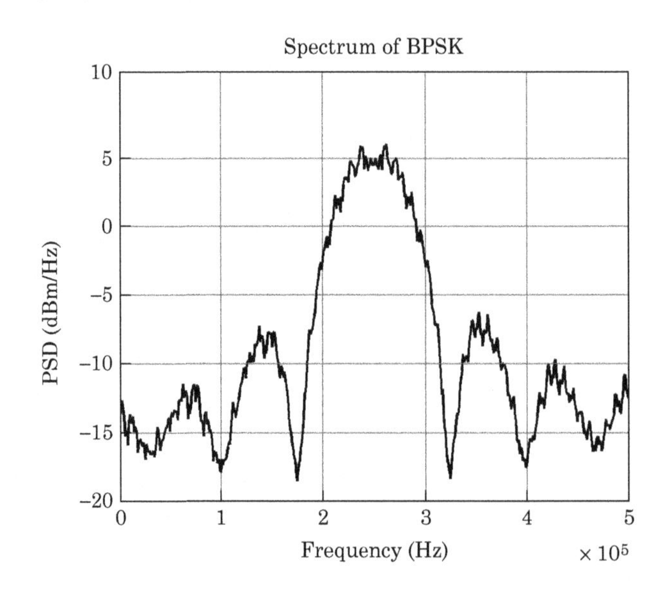

FIGURE 2-116 ■ The spectrum of a BPSK signal. The symbol rate is 75k symbols/second and the carrier frequency is 250 kHz. Note the now familiar $\sin(x)/x$ spectral shape.

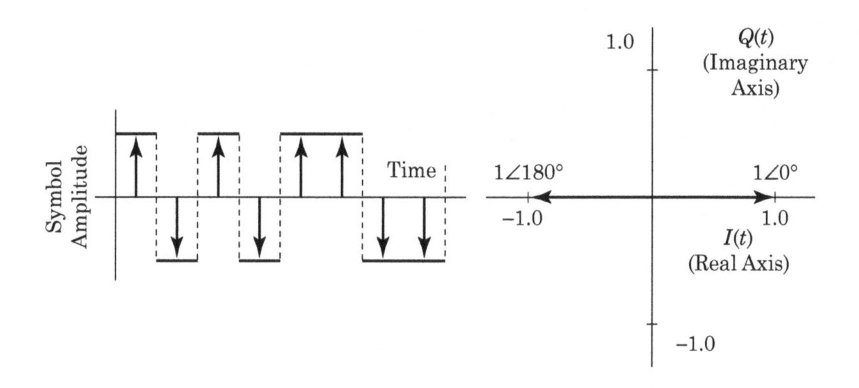

FIGURE 2-117 The phasor diagram of a BPSK signal.

2.9.7.3 Phasor Domain

Figure 2-117 shows the phasor diagram of a BSPK signal. The 1 symbol corresponds to a phasor in the positive-x direction and of unity amplitude. The 0 symbol corresponds to a phasor in the negative-x direction and of unity amplitude.

The phasor corresponding to the symbol is valid for one symbol period. At the next symbol period, the diagram changes to the phasor corresponding to the next symbol.

This phasor diagram is simply a plot of $I(t)$ versus $Q(t)$. We can interpret each $[I(t), Q(t)]$ pair as a vector from the origin.

2.9.7.4 Generating BPSK with a Quadrature Modulator

The quadrature modulator output voltage is

$$V_{out}(t) = I(t)\cos(\omega_{RF}t) \pm Q(t)\sin(\omega_{RF}t) \tag{2.102}$$

One method used to generate BPSK is to set $Q(t)$ to zero and let the user's bit stream drive $I(t)$ by mapping a logical 1 to +1 and a logical 0 to –1.

$$\begin{aligned} V_{BPSK}(t) &= I(t)\cos(\omega_{RF}t) \pm Q(t)\sin(\omega_{RF}t) \\ &= D_I(t)\cos(\omega_{RF}t);\, D_I(t) = -1 \text{ or } +1 \end{aligned} \tag{2.103}$$

Figure 2-118 shows the arrangement of the quadrature modulator.

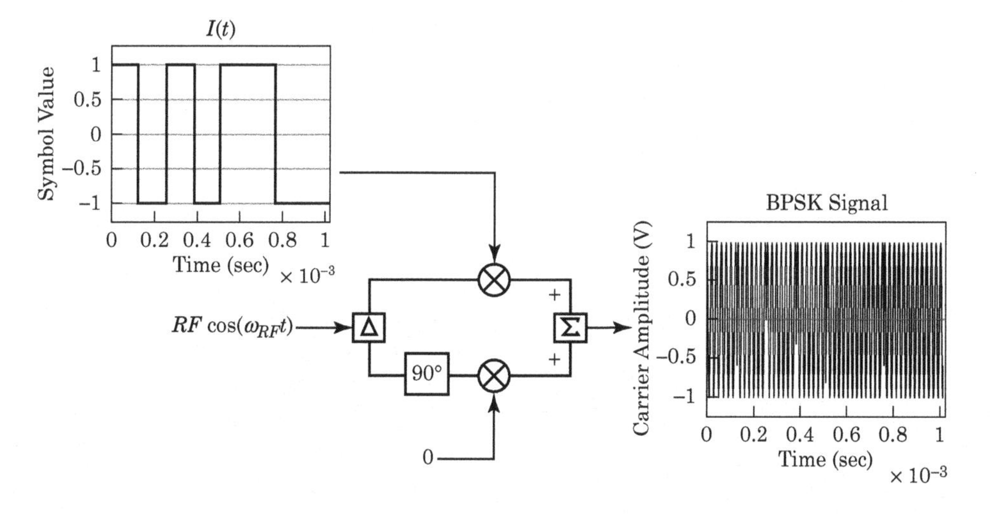

FIGURE 2-118 Generating BPSK with a quadrature modulator. The digital waveform $D_I(t)$ is based on the data to transmit: $D_I(t)$ is -1 or $+1$. We apply $D_I(t)$ to the I port of the quadrature modulator and apply zero to the Q port.

2.9.8 Quadrature Phase Shift Keying

QPSK is a 2 bit/symbol modulation scheme. Like BPSK, we encode information into the phase of the signal, but the scheme allows for the transmission of four possible phases:

Symbol	Transmitted Phase
A	45°
B	135°
C	225°(−135°)
D	315°(−45°)

Having 2 bits/symbol means we must break the user's bit stream into two separate data streams. Thus, the symbol rate is one-half the bit rate, or $R_s = R_b/2$.

2.9.8.1 Time Domain

The time-domain equation that describes QPSK is

$$
\begin{aligned}
V_{QPSK}(t) = \cos\left(\omega_{RF}t + \frac{(2n+1)\,\pi}{4}\right);\quad n = 0, 1, 2, 3 \\
\text{Symbol A} \Rightarrow \cos\left(\omega_{RF}t + \frac{\pi}{4}\right) \\
\text{Symbol B} \Rightarrow \cos\left(\omega_{RF}t + \frac{3\pi}{4}\right) \\
\text{Symbol C} \Rightarrow \cos\left(\omega_{RF}t + \frac{5\pi}{4}\right) \\
\text{Symbol D} \Rightarrow \cos\left(\omega_{RF}t + \frac{7\pi}{4}\right)
\end{aligned}
\qquad (2.104)
$$

For every symbol period, we transmit one of four possible phases of the RF carrier. See Figure 2-119 for an example. Figure 2-120 shows the same system as Figure 1-119 with an emphasis on the 2-bit nature of each symbol.

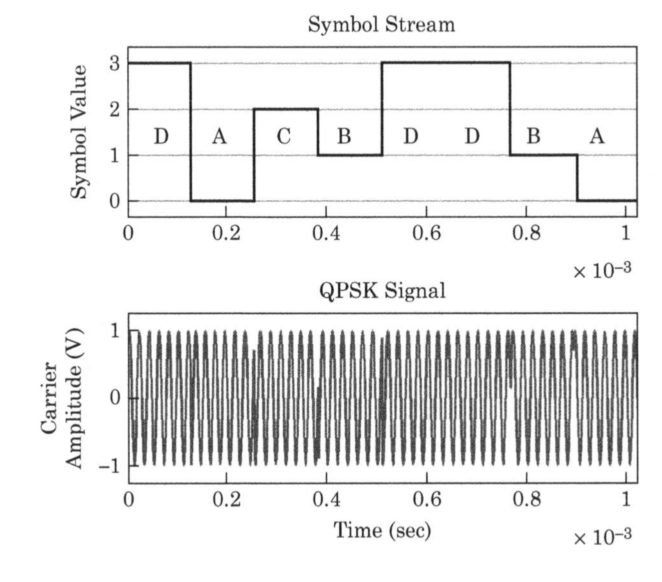

FIGURE 2-119 The time-domain expression of a75k baud QPSK signal. The top plot is the symbol stream, while the bottom plot is the QPSK RF waveform.

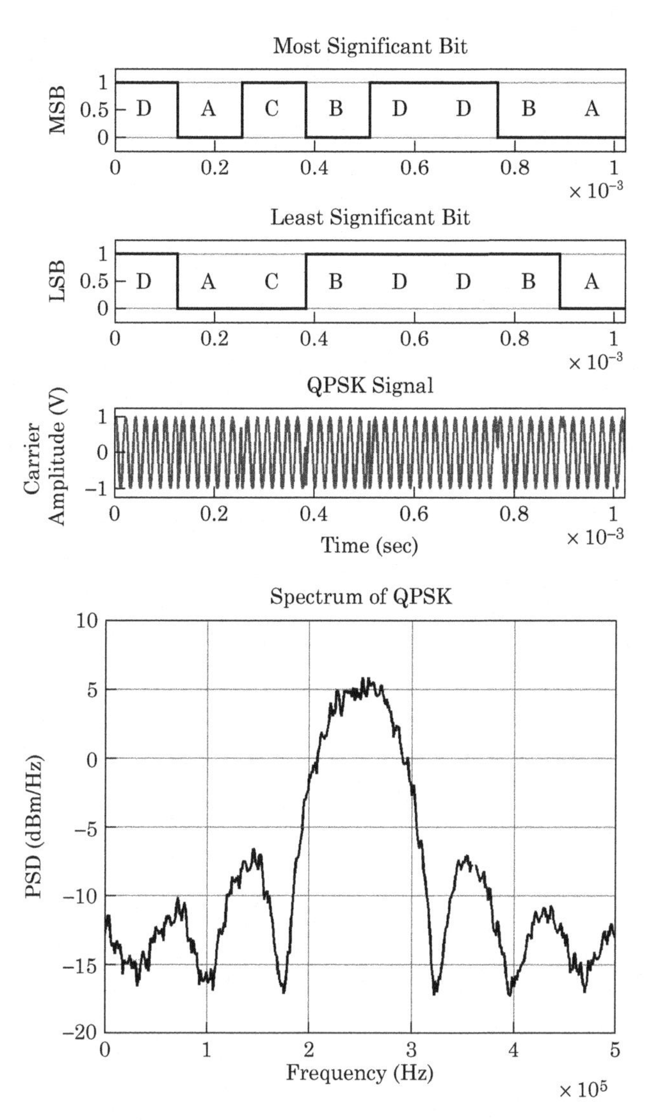

FIGURE 2-120 ■ We've redrawn the symbol stream of Figure 2-119 to emphasize the 2-bit nature of each symbol. The 2 separate bits map to the $I(t)$ and $Q(t)$ waveforms we'll need to produce QPSK with a quadrature modulator.

FIGURE 2-121 ■ The spectrum of a 75k baud QPSK signal centered at 250 kHz.

2.9.8.2 Frequency Domain

Figure 2-121 shows the spectrum of a 75k baud QPSK signal centered at 250 kHz. The $\sin(x)/x$ spectral shape is caused by the rectangular shape of the I- and Q- signals we used to generate the spectrum.

2.9.8.3 Phasor Domain

Figure 2-122 shows the phasor diagram of a QPSK signal.

2.9.8.4 Generating QPSK with a Quadrature Modulator

We now explore the method used to generate the $I(t)$ and $Q(t)$ needed to map our symbol stream into a QPSK RF signal. Equation (2.62) describes the output of a quadrature modulator:

$$V_{out}(t) = I(t)\cos(\omega_{RF}t) \pm Q(t)\sin(\omega_{RF}t) \qquad (2.105)$$

Because $Q(t)$ doesn't equal zero when we generate QPSK, we must choose either the plus or minus sign. The mathematics maps nicely into the phasor diagram when we choose the

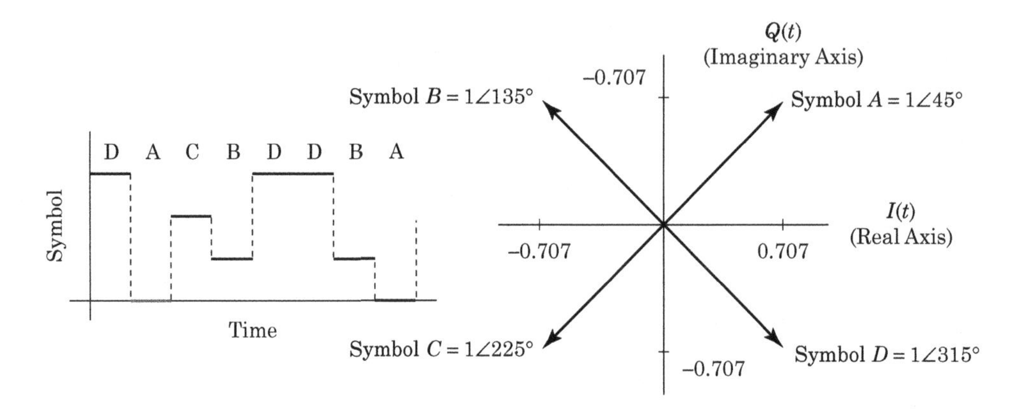

FIGURE 2-122 The phasor diagram of a QPSK signal. In this signal, the symbols A, B, C, and D have been mapped to a cosine at 45°, 135°, 225°, and 315°, respectively.

minus sign, so the modulator equation is

$$V_{QPSK}(t) = I(t)\cos(\omega_{RF}t) - Q(t)\sin(\omega_{RF}t) \tag{2.106}$$

We now derive the $I(t)$ and $Q(t)$ waveforms that generate cosine waves with phase shifts of 45°, 135°, 225° (or −135°), and 315° (or −45°). The following trigonometric relationships will prove helpful:

$$\begin{aligned}
\cos(\alpha \pm \beta) &= \cos(\alpha)\cos(\beta) \mp \sin(\alpha)\sin(\beta) \\
\cos(\omega_{RF}t \pm 45^\circ) &= \cos(\omega_{RF}t)\cos(45^\circ) \mp \sin(\omega_{RF}t)\sin(45^\circ) \\
&= 0.707\cos(\omega_{RF}t) \mp 0.707\sin(\omega_{RF}t) \\
&= I(t)\cos(\omega_{RF}t) - Q(t)\sin(\omega_{RF}t) \\
\cos(\alpha \pm \beta) &= \cos(\alpha)\cos(\beta) \mp \sin(\alpha)\sin(\beta) \\
\cos(\omega_{RF}t \pm 135^\circ) &= \cos(\omega_{RF}t)\cos(135^\circ) \mp \sin(\omega_{RF}t)\sin(135^\circ) \\
&= -0.707\cos(\omega_{RF}t) \mp 0.707\sin(\omega_{RF}t) \\
&= I(t)\cos(\omega_{RF}t) - Q(t)\sin(\omega_{RF}t)
\end{aligned} \tag{2.107}$$

We produce a cosine at 45° if $I(t)$ equals 0.707 while $Q(t)$ equals 0.707. A cosine at −45° results if $I(t)$ equals 0.707 while $Q(t)$ equals −0.707. Similarly, we produce a cosine at 135° if $I(t)$ equals 0.707 while $Q(t)$ equals −0.707. The result is a cosine with a phase shift of −135° if $I(t)$ equals −0.707 while $Q(t)$ equals −0.707 (see Table 2-7).

To generate the QPSK signal, we break the user's bit stream into two separate bit streams. We might label bits 0, 2, 4, 6, ... as stream $D_I(t)$ and then label bits 1, 3, 5, 7, ... as $D_Q(t)$. We feed $D_I(t)$ into the modulator's I port while feeding $D_Q(t)$ into its Q port. Table 2-8 shows the relationship between the user's data and the values of $I(t)$ and $Q(t)$.

TABLE 2-7 Mapping symbols to modulator output phases to the quadrature modulator *I(t)* and *Q(t)* signals.

Symbol	Modulator Output	*I(t)*	*Q(t)*
A	$\cos(\omega_{RF}t + 45^\circ)$	0.707	0.707
B	$\cos(\omega_{RF}t + 135^\circ)$	−0.707	0.707
C	$\cos(\omega_{RF}t - 135^\circ)$	−0.707	−0.707
D	$\cos(\omega_{RF}t - 45^\circ)$	0.707	−0.707

TABLE 2-8 ■ The relationship between the user's data and the values of $I(t)$ and $Q(t)$.

Symbol	Transmitted Phase	User Symbol	$[I(t), Q(t)]$
A	45°	11	[0.707, 0.707]
B	135°	10	[0.707, −0.707]
C	225° (−135°)	00	[−0.707, −0.707]
D	315° (−45°)	01	[−0.707, 0.707]

The quantity $D_I(t)$ controls the amount of $\cos(\omega_{RF}t)$ the system sends to its output while $D_Q(t)$ controls the amount of $\sin(\omega_{RF}t)$ that passes to the output. The modulator output is

$$\begin{aligned} V_{QPSK}(t) &= I(t)\cos(\omega_{RF}t) - Q(t)\sin(\omega_{RF}t) \\ &= D_I(t)\cos(\omega_{RF}t) - D_Q(t)\sin(\omega_{RF}t) \\ D_I(t), D_Q(t) &\in [-0.707, 0.707] \end{aligned} \tag{2.108}$$

Figure 2-123 shows QPSK generation. Like BPSK, QPSK is a constant-amplitude waveform.

2.9.9 Comparison of the Spectrum of BPSK and QPSK

The spectral shapes of BPSK and QPSK are very similar. However, the bandwidth of a digital signal is, in general, a strong function of its symbol or keying rate. For the same bit rate, QPSK changes at half the rate of the BPSK signal (because BPSK is a 1 bit/symbol modulation, whereas QPSK is a 2 bit/symbol modulation). Figure 2-124 compares the BPSK and QPSK spectra.

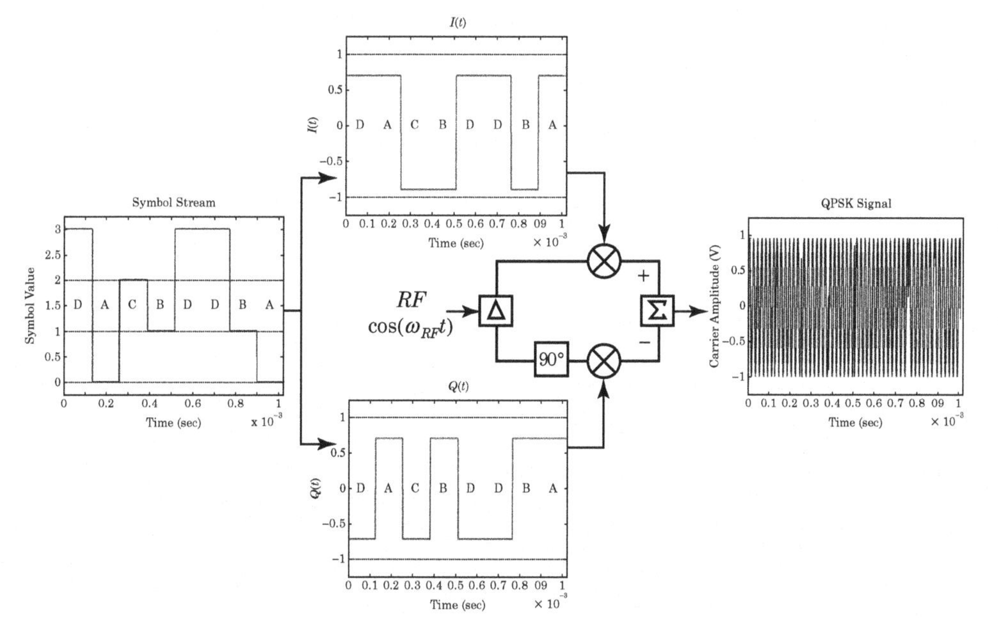

FIGURE 2-123 ■ Generating QPSK with a quadrature modulator. The scheme derives both $D_I(t)$ and $D_Q(t)$ from the user's data stream. Note the minus sign at the input to the summing junction.

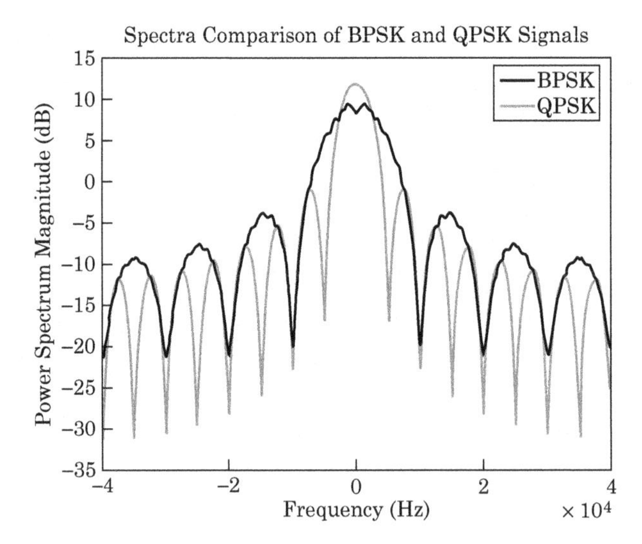

FIGURE 2-124 Comparison of the spectra of BPSK and QPSK signals. Both signals are 10 kbps, but the BPSK signal changes at a 10 kHz rate while the QPSK signal changes at a 5 kHz rate. Hence, the QPSK signal takes less bandwidth than the BPSK signal.

2.9.10 Offset or Staggered QPSK

A common variant of QPSK is known as offset QPSK (OQPSK) or staggered QPSK (SQPSK). The difference between SQPSK and QPSK is the timing of the $I(t)$ and $Q(t)$ waveforms. In QPSK, $I(t)$ and $Q(t)$ change at the same instant. In SQPSK, $I(t)$ and $Q(t)$ still change once every symbol time, but $I(t)$ will change on the baud boundary while $Q(t)$ changes one-half of a symbol time later. Figure 2-125 illustrates the process.

Figure 2-126 shows the constellation diagrams of QPSK and SQPSK with emphasis on the transitions the signal makes between constellation points. The transitions in OQPSK never pass through the origin. This detail has important consequences for a future discussion of spectrum control.

2.9.11 8PSK

Eight-level phase shift keying (8PSK) is the most complex phase modulation we see "in the wild." We encode our symbol stream into one of eight possible phases of the carrier. Thus the symbol rate is one-third of the bit rate, or $R_s = R_b/3$.

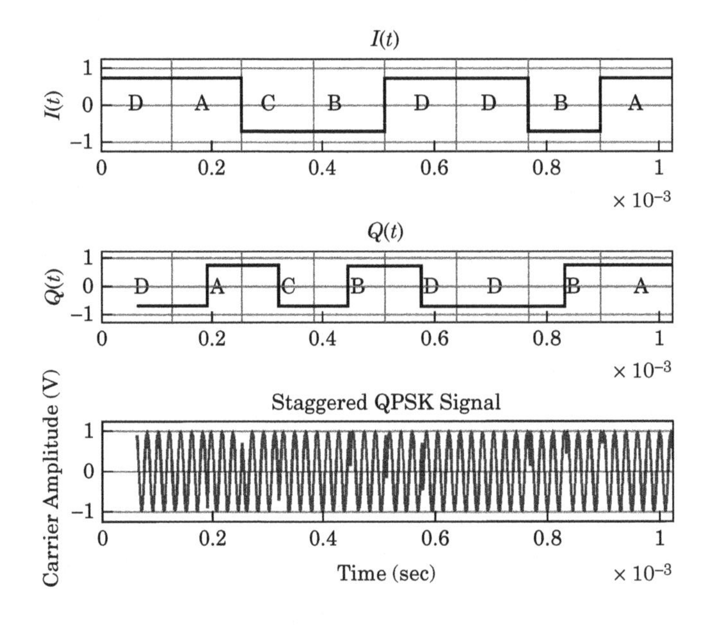

FIGURE 2-125 Staggered QPSK waveforms. Compare this figure to Figure 2-120, which shows the same waveform for QPSK. In staggered QPSK, $I(t)$ and $Q(t)$ never change at the same time. We delay $Q(t)$ by half a symbol duration.

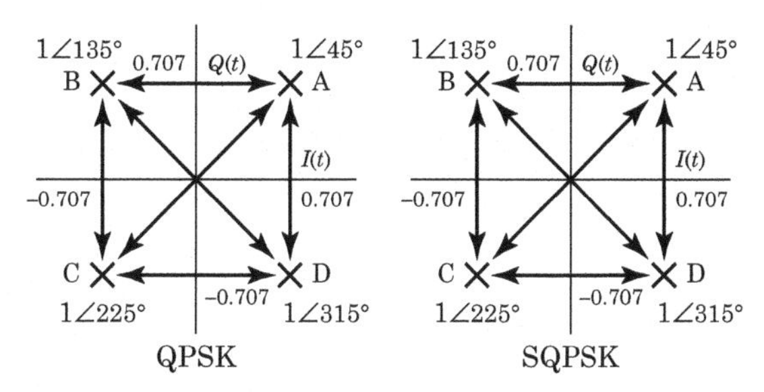

FIGURE 2-126 ■ The constellation diagrams of QPSK and SQPSK, with emphasis on the transitions permitted between symbols. In QPSK, we can make transitions between symbols A and C and between symbols B and D. Both of these transitions cause the carrier to pass through the origin. An OQPSK signal never transitions directly between points A and C or B and D (although it can travel along paths A-B-C or A-D-C). As a result, the OQPSK waveform never passes through the origin.

2.9.11.1 Time Domain

We encode information into the phase of the signal, shown in Table 2-9. 8PSK allows for the transmission of eight possible phases.

The time-domain expression that describes 8PSK is

$$V_{8PSK}(t) = \cos\left(\omega_{RF}t + n\frac{\pi}{4}\right); \quad n = 0, \ldots, 7$$

$$\begin{aligned}
&\text{Symbol A } (n = 0) \Rightarrow \cos(\omega_{RF}t) && \text{Symbol E } (n = 4) \Rightarrow \cos(\omega_{RF}t + \pi) \\
&\text{Symbol B } (n = 1) \Rightarrow \cos\left(\omega_{RF}t + \frac{\pi}{4}\right) && \text{Symbol F } (n = 5) \Rightarrow \cos\left(\omega_{RF}t + \frac{5\pi}{4}\right) \\
&\text{Symbol C } (n = 2) \Rightarrow \cos\left(\omega_{RF}t + \frac{\pi}{2}\right) && \text{Symbol G } (n = 6) \Rightarrow \cos\left(\omega_{RF}t + \frac{3\pi}{2}\right) \\
&\text{Symbol D } (n = 3) \Rightarrow \cos\left(\omega_{RF}t + \frac{3\pi}{4}\right) && \text{Symbol H } (n = 7) \Rightarrow \cos\left(\omega_{RF}t + \frac{7\pi}{4}\right)
\end{aligned} \tag{2.109}$$

Figure 2-127 shows the time domain expression of 75k baud 8PSK signal centered at 250 kHz. The phase of the signal over a given baud time is determined by the 3-bit symbol we transmit during that time period.

TABLE 2-9 ■ 8PSK symbols mapped to the phase of the transmitted waveform.

Symbol	Transmitted Phase
A	0°
B	45°
C	90°
D	135°
E	180°
F	225° (−135°)
G	270° (−90°)
H	315° (−45°)

2.9.11.2 Frequency Domain

Figure 2-128 shows the spectrum of an 8PSK, 75k baud signal centered at 250 kHz. Note the similarity of this spectrum to the QPSK spectrum of Figure 2-121 and to the BPSK spectrum of Figure 2-116.

2.9.11.3 Phasor Domain

Figure 2-129 shows the phasor diagram of an 8PSK signal.

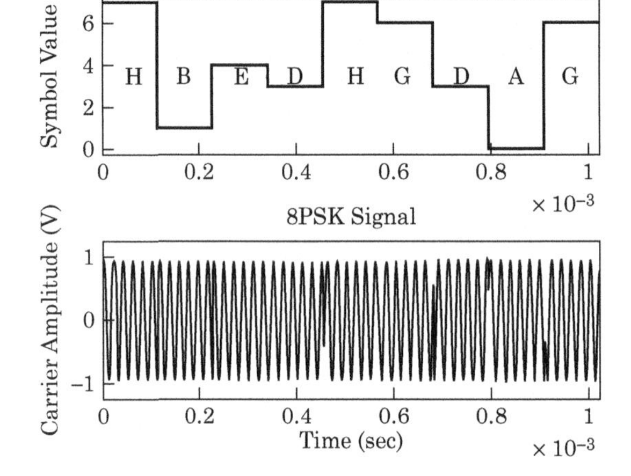

FIGURE 2-127 ■ The time-domain expression of a75k baud 8PSK signal. The top plot is the symbol stream, while the bottom plot is the 8PSK RF waveform.

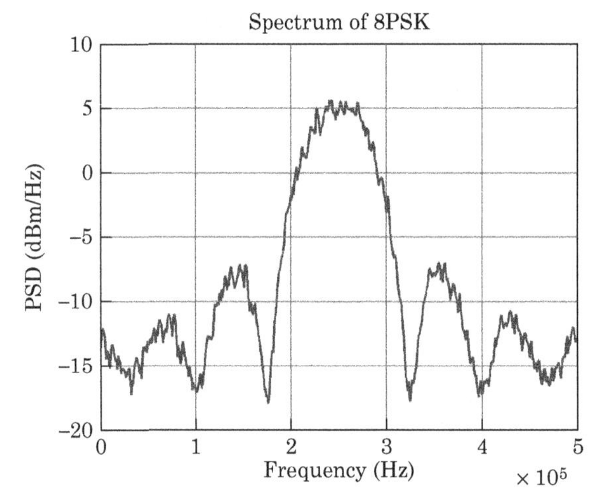

FIGURE 2-128 ■ The spectrum of a75 kbaud 8PSK signal.

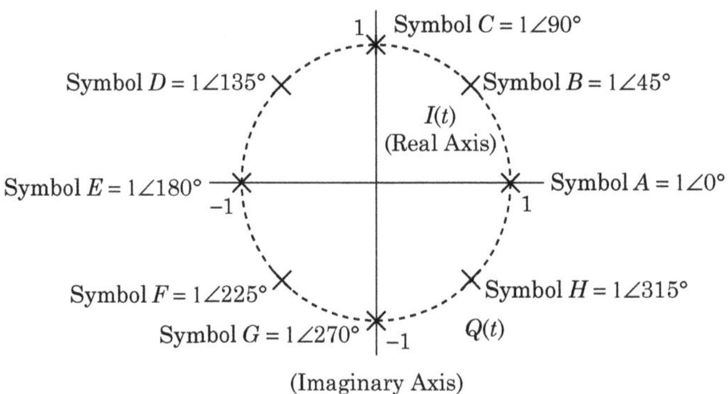

FIGURE 2-129 ■ The phasor diagram of an 8PSK signal. The symbols have been mapped to a phases of a cosine placed at equally spaced intervals around the unit circle.

2.9.11.4 Generating 8SPK with a Complex Modulator

We can calculate the $I(t)$ and $Q(t)$ signals we need to generate 8PSK by inspection of the phasor diagram of Figure 2-129, remembering that our modulator equation will be

$$V_{8PSK}(t) = I(t)\cos(\omega_{RF}t) - Q(t)\sin(\omega_{RF}t) \tag{2.110}$$

The results are given in Table 2-10.

2.9.12 Quadrature Amplitude Modulation (QAM): 16QAM

The PSK signals we have so far discussed were all constant amplitude signals. The time-domain representation changes only in phase, never in amplitude. In the quest to fit more bits into a swatch of spectrum, we will allow the user's symbol stream to be encoded into both the amplitude and phase of the signal. In 16QAM, we encode our symbol stream into one of 16 possible combinations of amplitude and phase of the carrier. Thus, the symbol rate is one-fourth of the bit rate, or $R_s = R_b/4$. Figure 2-130 shows the method used to apply the symbols of Figure 2-129 in a quadrature modulator.

TABLE 2-10 ■ Mapping symbols to modulator output phases to the quadrature modulator *I(t)* and *Q(t)* signals.

Symbol	Modulator Output	*I(t)*	*Q(t)*
A	$\cos(\omega_{RF}t)$	1.000	0.000
B	$\cos(\omega_{RF}t + 45°)$	0.707	0.707
C	$\cos(\omega_{RF}t + 90°)$	0.000	1.000
D	$\cos(\omega_{RF}t + 135°)$	−0.707	0.707
E	$\cos(\omega_{RF}t + 180°)$	−1.000	0.000
F	$\cos(\omega_{RF}t - 135°)$	−0.707	−0.707
G	$\cos(\omega_{RF}t - 90°)$	0.000	−1.000
H	$\cos(\omega_{RF}t - 45°)$	0.707	−0.707

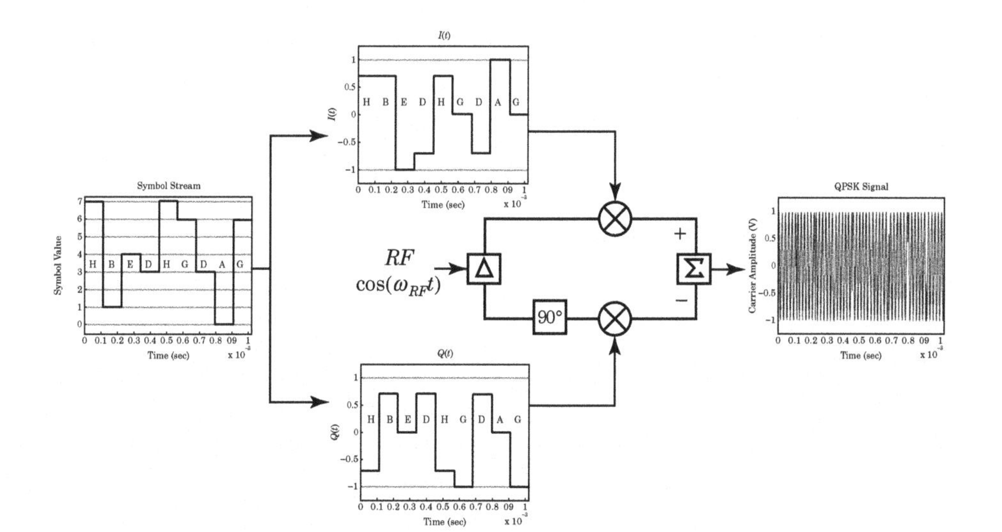

FIGURE 2-130 ■ Generating an 8PSK signal with a quadrature modulator. The scheme derives both $I(t)$ and $Q(t)$ from the user's data stream. Note the minus sign at the input to the summing junction.

2.9.12.1 Time Domain

The goal is to place the constellation points on an equally spaced rectangular grid consisting of 16 points as we show in Figure 2-131.

We set the magnitude of the constellation by forcing the magnitude of the largest points (A, D, M, and P) to be unity. The 16 possible combinations of amplitude and phase are given in Table 2-11.

Figure 2-132 shows the time-domain representation of a 16QAM signal.

2.9.12.2 Frequency Domain

Figure 2-133 shows the spectrum of a 75k baud, 16QAM signal that is centered at 250 kHz. The baseband I- and Q- waveforms are rectangular in shape, causing the $\sin(x)/x$ spectral shape.

2.9.12.3 Phasor Domain

Figure 2-131 shows the phasor diagram of a 16QAM signal.

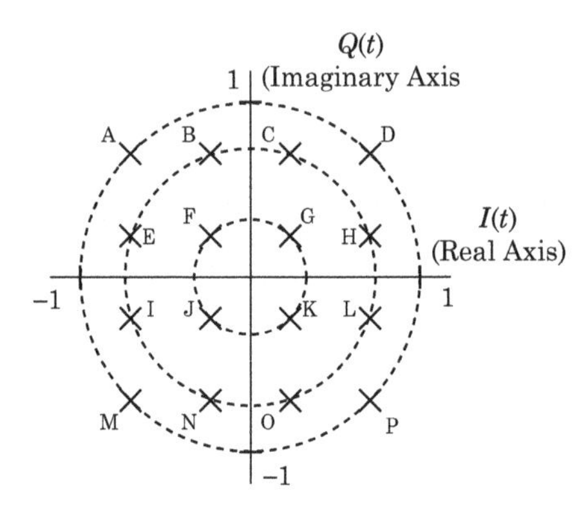

FIGURE 2-131 ■ The constellation of a 16QAM signal. The symbols have been mapped into the amplitude and phase of a cosine. There are three distinct amplitudes and three distinct, rotationally invariant phases.

TABLE 2-11 ■ The positions of the constellation points for a 16QAM signal.

Symbol	Transmitted Amplitude and Phase
A	$1.000\angle 135°$
B	$0.745\angle 108.4°$
C	$0.745\angle 71.6°$
D	$1.000\angle 45°$
E	$0.745\angle 161.6°$
F	$0.333\angle 135°$
G	$0.333\angle 45°$
H	$0.745\angle 18.4°$
I	$0.745\angle 198.4° (0.745\angle -161.6°)$
J	$0.333\angle 225° (0.333\angle -135°)$
K	$0.333\angle 315° (0.333\angle -45°)$
L	$0.745\angle 341.6° (0.745\angle -18.4°)$
M	$1.000\angle 225° (1.000\angle -135°)$
N	$0.745\angle 251.6° (0.745\angle -108.4°)$
O	$0.745\angle 288.4° (0.745\angle -71.6°)$
P	$1.000\angle 315° (1.000\angle -45°)$

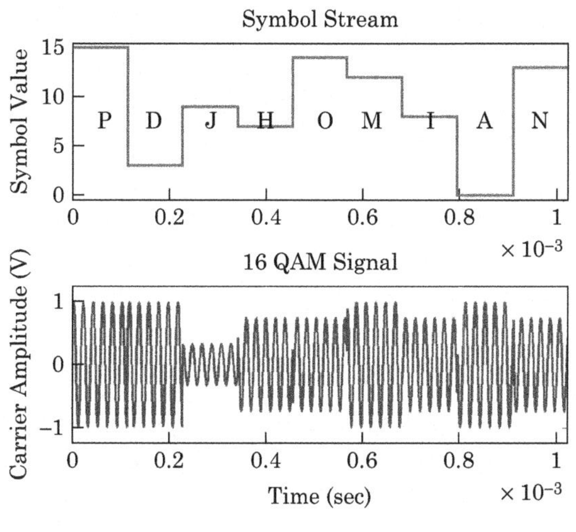

FIGURE 2-132 ■ The time-domain expression of a75k baud 16QAM signal. The amplitude of a 16QAM waveform is not constant.

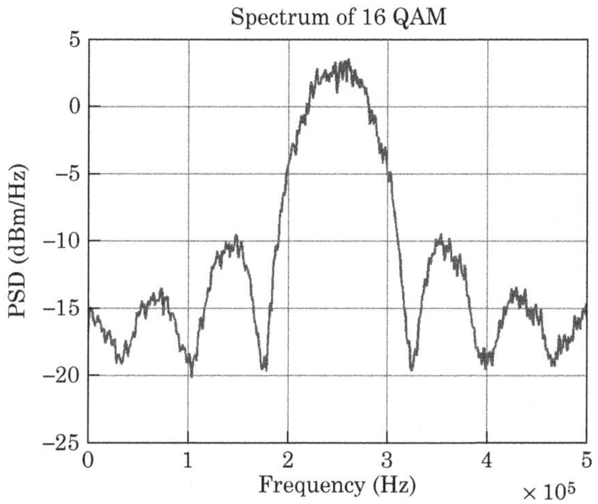

FIGURE 2-133 ■ The spectrum of a75k baud 16QAM signal centered at 250 kHz.

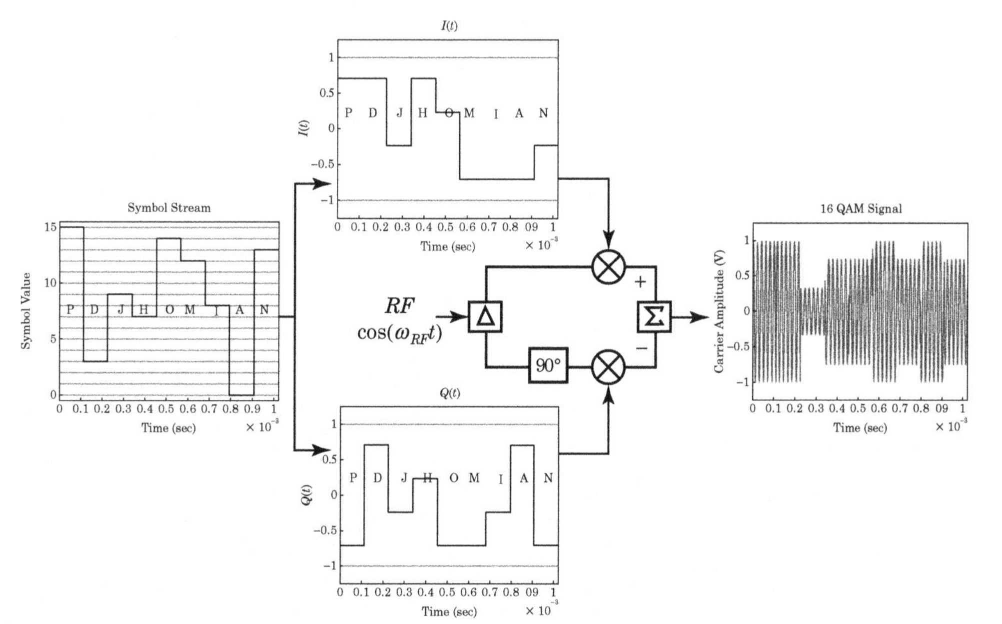

FIGURE 2-134 ■ Generating a 16QAM signal with a quadrature modulator.

2.9.12.4 Generating 16QAM with a Complex Modulator

Figure 2-134 shows the method used to produce a 16QAM signal with a quadrature modulator. The incoming symbols are decomposed into the appropriate I and Q waveforms, then mixed with the quadrature oscillator.

2.9.13 Higher-Order QAM

We can extend the concepts we developed in our discussion of 16QAM signals to higher-order signals. Figure 2-135 shows the constellation diagrams of higher-order QAM signals.

2.9.14 The Noise Performance of PSK and QAM Signals

Adding noise to QAM signals has the effect of turning the small, discrete constellation points into fuzzy balls. As the noise increases, the constellation points become so large that the demodulator begins making mistakes, producing symbol errors in the demodulator. Figure 2-136 shows the effect.

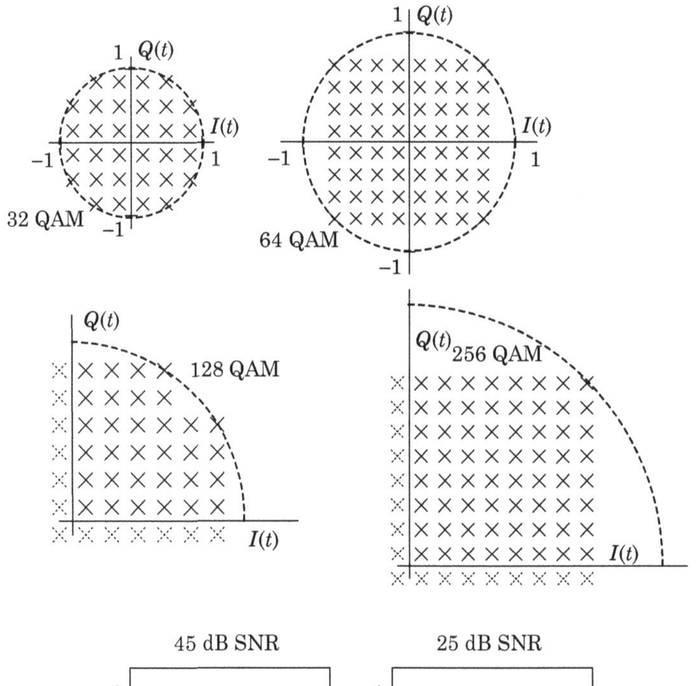

FIGURE 2-135 ■ The constellation diagrams of higher-order QAM. We've shown only one quadrant for the 128QAM and 256QAM cases.

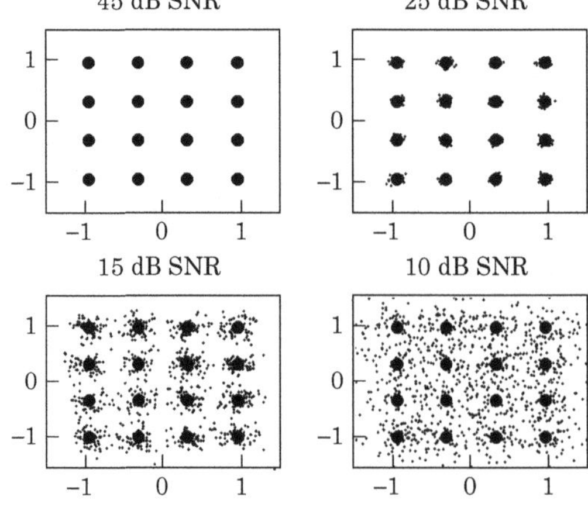

FIGURE 2-136 ■ Constellation diagrams of a 16QAM signal contaminated with various levels of AWGN.

Most complex modulated signals (e.g., 512QAM) require no more than 50 dB of SNR for error-free demodulation.

2.9.15 Frequency Shift Keying Modulation

The major characteristic of FSK is that our symbols consist of one of n possible tones. In 4LFSK, for example, we send one of four possible tones to transmit 2 bits of data. We combine bits from the user's data stream to form separate symbols and then use the symbols to change the characteristics of the transmitted signal via a known methodology.

2.9.16 2LFSK

The characteristics of 2LFSK include the symbol rate and the frequency difference between the tones. There are several common relationships between these two values.

We refer to a 2LFSK signal whose tone spacing equals the data rate as *FSK*. A 2LFSK signal whose tone spacing equals one-half of the data rate is a *minimum shift keyed* (MSK) signal. MSK uses little bandwidth, but it is harder to demodulate.

2.9.16.1 Time Domain

The most general equation for a 2LFSK waveform is

$$V_{2L-FSK}(t) = \cos\left\{\left[\omega_{RF} + D_I(t)\frac{\Delta\omega_{FSK}}{2}\right]t + \phi_I(t)\right\};\quad D_I(t) \in [-1, 1]$$

$$\text{Symbol A}\,[D_I(t) = -1] \Rightarrow V_{2L-FSK,A}(t) = \cos\left\{\left[\omega_{RF} - \frac{\Delta\omega_{FSK}}{2}\right]t + \phi_I(t)\right\} \quad (2.111)$$

$$\text{Symbol B}\,[D_I(t) = +1] \Rightarrow V_{2L-FSK,B}(t) = \cos\left\{\left[\omega_{RF} + \frac{\Delta\omega_{FSK}}{2}\right]t + \phi_I(t)\right\}$$

We deterministically change the value of $D_I(t)$ based upon the user's data. The phase of the cosine $[\phi_I(t)]$ may change with each symbol. This quantity can be a random value, or it can be deterministic depending on the modulator.

Figure 2-137 shows 2LFSK waveforms with various values of $\Delta\omega_{FSK} = 2\pi\Delta F_{FSK}$.

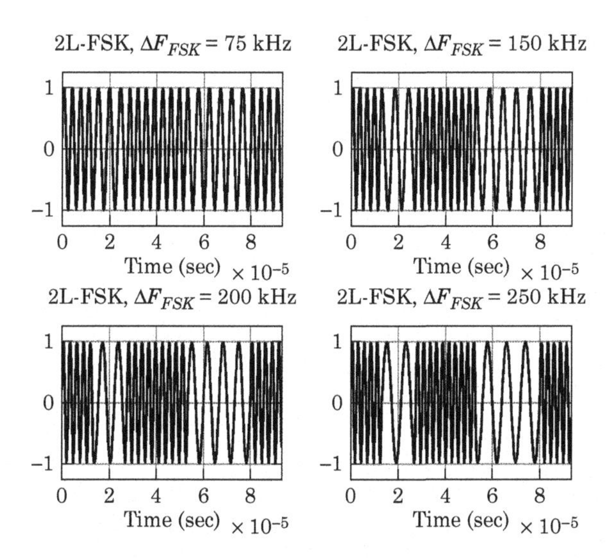

FIGURE 2-137 ■ Time-domain representation of 2LFSK with various values of $\Delta\omega_{FSK} = 2\pi\Delta F_{FSK}$.

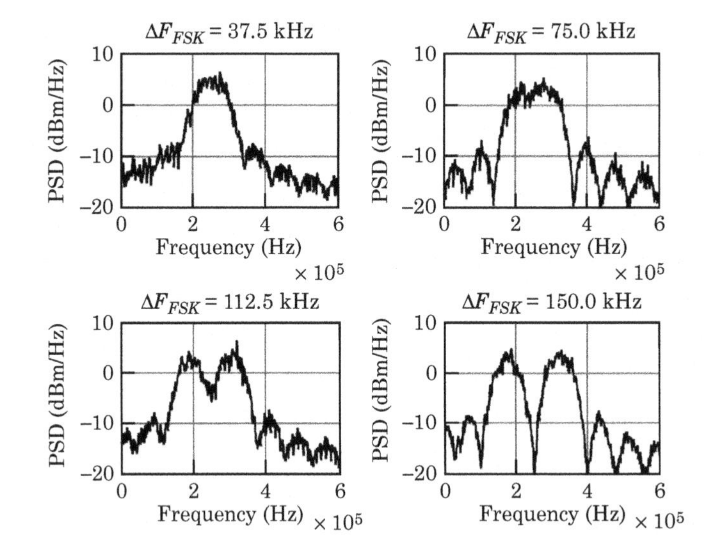

FIGURE 2-138 The power spectral density of several 2LFSK signals. Each signal has a symbol rate of 75 ksps and a ΔF of symbol rate/2 (MSK), symbol rate (conventional FSK), 1.5 * symbol rate, and 2 * symbol rate. Larger values of Δ*F* require more signal spectrum.

2.9.16.2 Frequency Domain

Figure 2-138 shows the power spectral density of several 75 kbps 2LFSK signals centered at 250 kHz. The bandwidth is set by the bit rate and the frequency deviation.

2.9.16.3 Generating 2LFSK with a Complex Modulator

We have shown the methods used to generate a 2LFSK waveform with a quadrature modulator in a prior section.

2.9.17 4LFSK

Four-level FSK is an extension of 2LFSK. In binary or 2LFSK, the system encodes a 0 as one RF (f_0) and a 1 as a second RF (f_1). The symbol rate (R_S) equals the bit rate (R_B). In 4LFSK, we transmit one of four possible tones to send our message. Two data bits generate one of four possible RF tones so the symbol rate is one-half the bit rate. Figure 2-139 shows the bit-to-symbol mapping for generating 2LFSK and 4LFSK.

For this 4LFSK example, we'll use the arbitrary bit-to-symbol map of:

- Lowest frequency: $f_0 = f_{RF} - 1.5\Delta f_{FSK}$ – maps to symbol 00
- $f_1 = f_{RF} - 0.5\Delta f_{FSK}$ – maps to symbol 01
- $f_2 = f_{RF} + 0.5\Delta f_{FSK}$ – maps to symbol 10
- Highest frequency: $f_3 = f_{RF} + 1.5\Delta f_{FSK}$ – maps to symbol 11

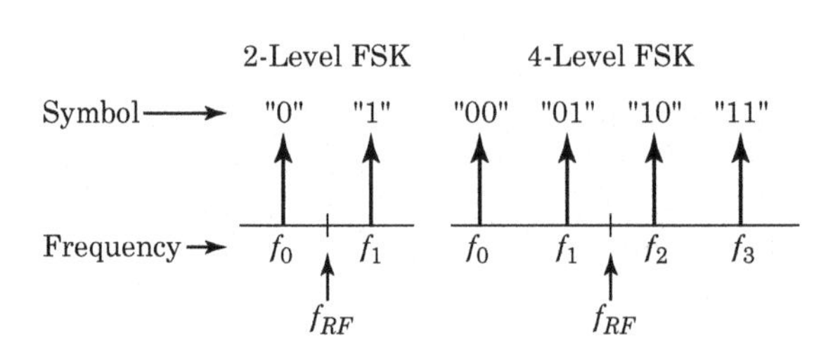

FIGURE 2-139 Symbol (frequency)-to-bit map of 2LFSK and 4LFSK.

The general equation for a 4LFSK waveform is

$$
\begin{aligned}
&V_{4L-FSK}(t) = \cos\{[\omega_{RF} + D_I(t)\Delta\omega_{FSK}]t + \phi_I(t)\};\quad D_I(t) \in [-1.5, -0.5, +0.5, +1.5]\\
&\quad \text{Symbol A}\,[D_I(t) = -1.5] \Rightarrow V_{4L-FSK,A}(t) = \cos\{[\omega_{RF} - 1.5\Delta\omega_{FSK}]t + \phi_I(t)\}\\
&\quad \text{Symbol B}\,[D_I(t) = -0.5] \Rightarrow V_{4L-FSK,B}(t) = \cos\{[\omega_{RF} - 0.5\Delta\omega_{FSK}]t + \phi_I(t)\}\\
&\quad \text{Symbol C}\,[D_I(t) = +0.5] \Rightarrow V_{4L-FSK,B}(t) = \cos\{[\omega_{RF} + 0.5\Delta\omega_{FSK}]t + \phi_I(t)\}\\
&\quad \text{Symbol D}\,[D_I(t) = +1.5] \Rightarrow V_{4L-FSK,A}(t) = \cos\{[\omega_{RF} + 1.5\Delta\omega_{FSK}]t + \phi_I(t)\}
\end{aligned}
\tag{2.112}
$$

It's common to set the frequency between the tones to a ratio of the symbol rate R_S. For the purposes of example, we will separate each tone from its neighbor by twice the symbol rate. That distance is more than required for reliable detection at the receiver, but the resulting graphics appear more intuitive.

Using a quadrature modulator to produce a 4LFSK signal, we must generate various combinations of sine and cosine waves at the symbol rate and at three times the symbol rate to generate the four possible symbols (see Figure 2-140).

When the transmitted symbol is to be 00, we apply $I_{00}(t)$ to the modulator's I port and $Q_{00}(t)$ to its Q port. This combination produces a tone at $f_0 = f_{RF} - 1.5\Delta f_{FSK}$. Similarly, applying $I_{01}(t)$ and $Q_{01}(t)$ produces a tone at $f_1 = f_{RF} - 0.5\Delta f_{FSK}$; $I_{10}(t)$ and $Q_{10}(t)$ produce a tone at $f_2 = f_{RF} + 0.5\Delta f_{FSK}$; and $I_{11}(t)$ and $Q_{11}(t)$ produce a tone at $f_3 = f_{RF} + 1.5\Delta f_{FSK}$. We apply each set of $I(t)$ and $Q(t)$ waveforms for one symbol period and then change according to the next symbol. Figure 2-141 shows a random symbol stream and the I and Q waveforms we must generate to encode that stream as a 4LFSK waveform.

Figure 2-142 shows the spectrum of a 4LFSK signal whose baud rate is 75k symbols/sec and the deviation, Δf_{FSK}, is 150 kHz. The spectrum shows the four symbol frequencies at -225 kHz ($-1.5\Delta f_{FSK}$), -75 kHz($-0.5\Delta f_{FSK}$), $+75$ kHz ($+0.5\Delta f_{FSK}$), and $+225$ kHz ($+1.5\Delta f_{FSK}$). The $\sin(x)/x$ spectral shape of each symbol is due to the suddenness with which the system changes symbols and the high frequencies that arise from a step function.

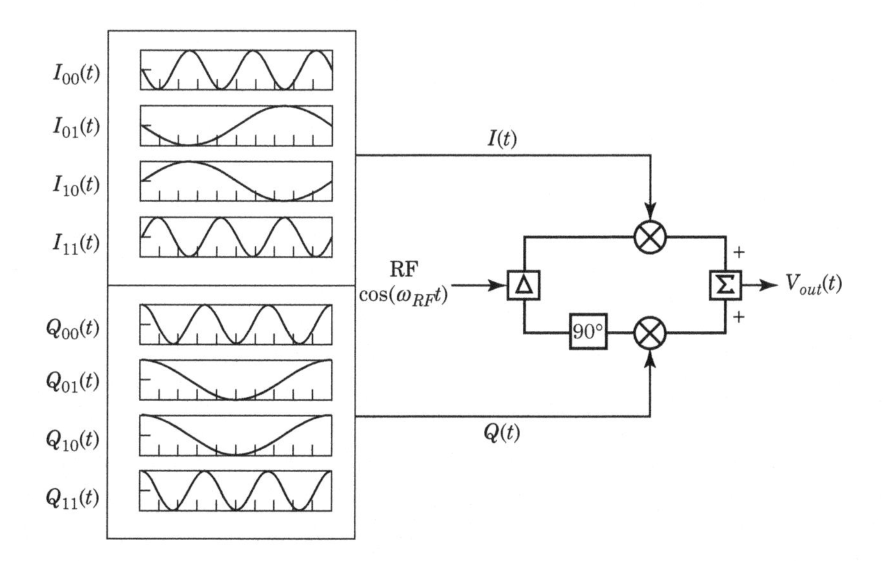

FIGURE 2-140 ■ Generating 4LFSK with a quadrature modulator. The symbol we wish to transmit determines which $I(t)$ and $Q(t)$ to apply at the modulator inputs. The symbol 00 requires $I_{00}(t)$ and $Q_{00}(t)$; a 01 needs $I_{01}(t)$ and $Q_{01}(t)$; a 10 wants $I_{10}(t)$ and $Q_{10}(t)$; while $I_{11}(t)$ and $Q_{11}(t)$ produce symbol 11.

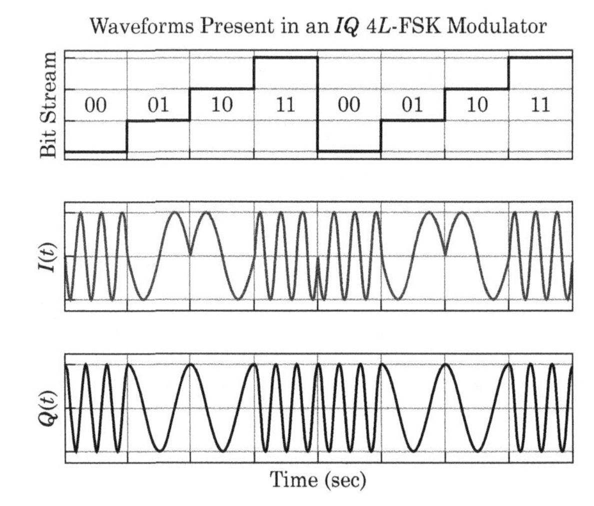

FIGURE 2-141 ■ Waveforms present in a 4LFSK quadrature modulator. The top trace shows the symbol stream to encode. The bottom two traces show the $I(t)$ and $Q(t)$ waveforms we must apply to the modulator. Symbols 00 and 11 both require three cycles of a sine wave, whereas symbols 01 and 01 require only a single sine wave cycle.

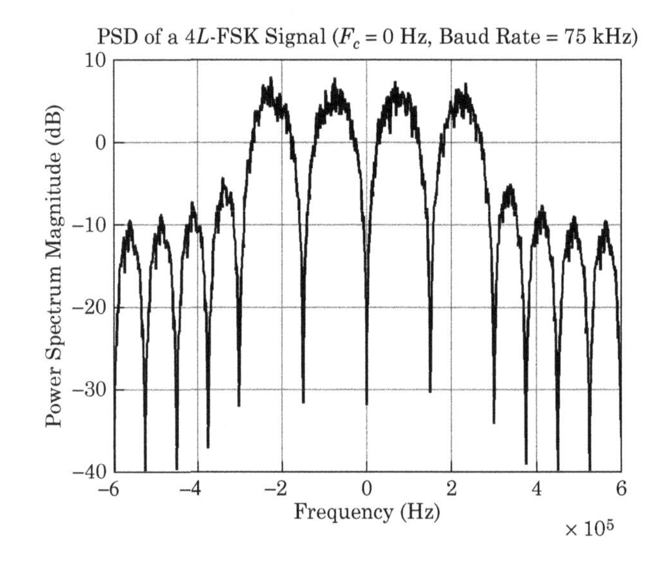

FIGURE 2-142 ■ Spectrum of a 4LFSK signal. The bit rate is 150 kbps, whereas the symbol rate is 75k symbols/second. The four tones are separated by 150 kHz.

Figure 2-143 shows the time-frequency plot (spectrogram) of the 4LFSK signal. The four symbols at frequencies f_0, f_1, f_2, and f_3 are clearly visible. A frequency change occurs every 13.3 μsec, indicating a 75 kHz symbol rate.

2.9.18 Multitone FSK

The concept behind multitone FSK is to break down a high-speed bit stream into N lower-speed bit streams, each of which we encode as its own separate 2LFSK channel. If the bit rate of the high-speed channel is R_b, then the bit rate of each of the N low-speed channels is R_b/N. One primary advantage of this modulation is a decreased symbol rate, which fights multipath propagation effects. Another advantage is that this modulation spreads

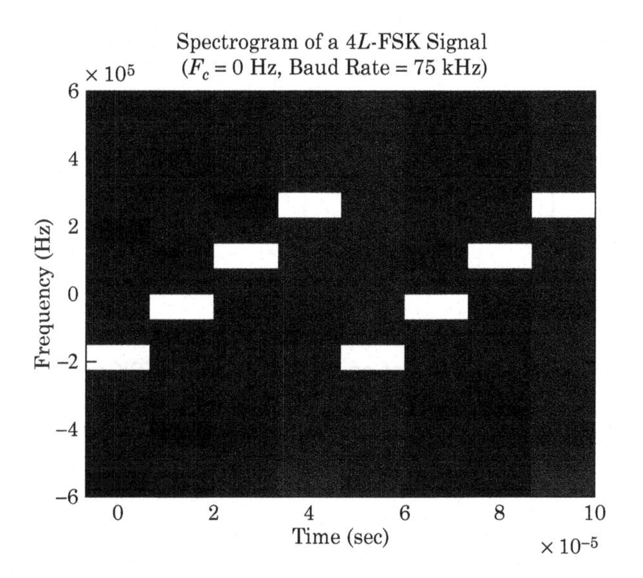

FIGURE 2-143 ■ Spectrogram (time-frequency plot) of a 75k symbols/second 4LFSK signal. The symbol stream used for this plot was (00, 01, 10, 11, 00, 01, 10, 11), which is encoded as (f_0, f_1, f_2, f_3, f_0, f_1, f_1, f_3).

the information over a wide bandwidth. Any propagation nulls or narrowband interference affects only a small number of channels. Multitone FSK is used in adaptive high-frequency (HF) modems, high-defintion TV (HDTV) transmissions, and some power line modems.

For example, we can break our data stream down into 8-bit symbols and then use each 8-bit symbol to generate eight separate FSK channels. Each bit in the symbol determines the state of one FSK channel for the duration of a symbol.

We'll describes this type of modulation primarily by example. Figure 2-144 shows a demultiplexer that absorbs a bit stream of R_B bits/sec and emits 8 bit streams, B_0 to B_7, each running at $R_B/8$ bits/sec.

We use each low-speed $R_B/8$ bit stream to frequency modulate one carrier at a rate of $R_B/8$. Bit stream B_0 modulates a carrier centered at frequency f_{B0}. The modulator will emit frequency $f_{B0,0}$ when $B_0 = 0$ and to emit frequency $f_{B0,1}$ when $B_0 = 1$. Similarly, bit stream B_1 frequency-modulates the carrier at f_{B1}, causing the modulator to emit frequency $f_{B1,0}$ when $B_1 = 0$ and frequency $f_{B1,1}$ when $B_1 = 1$. In general, bit stream B_n causes the modulator to emit frequency $f_{Bn,0}$ when $B_n = 0$ and frequency $f_{Bn,1}$ when $B_n = 1$. Figure 2-145 illustrates this plan.

The modulator always emits eight simultaneous tones during each symbol period. When the system switches over from one symbol to the next, it changes all eight tones at the start of a new symbol. To transmit the symbol $B_7B_6B_5B_4B_3B_2B_1B_0 = 11000100$, for example, we simultaneously produce the following eight tones:

- $f_{B0,0}$ because $B_0 = 0$
- $f_{B1,0}$ because $B_1 = 0$

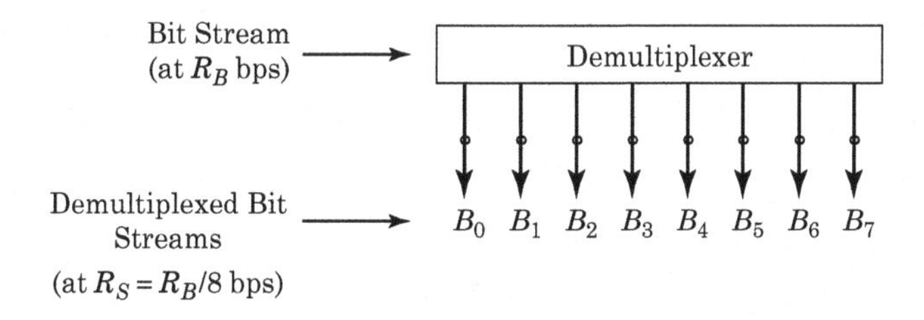

FIGURE 2-144 ■ A demultiplexer takes in a user bit stream at R_B bits/sec and produces 8 low-speed bit streams, each at $R_B/8$ bits/sec.

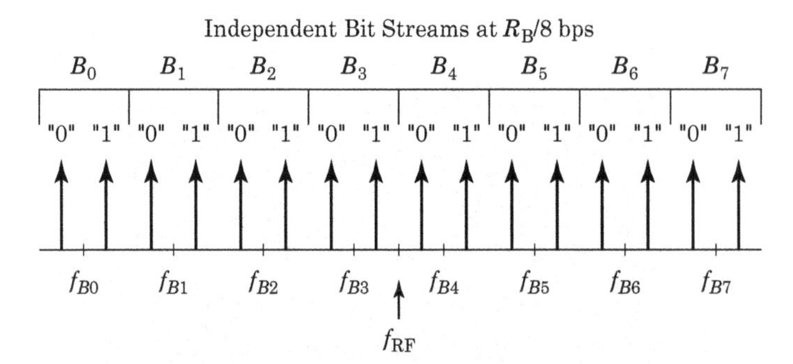

FIGURE 2-145 ■ The frequency plan of an 8-level multitone FSK modulator. Bit stream B_0 controls the frequency of the tone centered about f_{B0}, while bit stream B_1 controls the frequency of the tone at f_{B1}, etc. During one symbol period, the modulator emits either $f_{B0,0}$ or $f_{B0,1}$. At the same time, it emits a tone at either $f_{B1,0}$ or $f_{B1,1}$, and so forth.

- $f_{B2,1}$ because $B_2 = 1$
- $f_{B3,0}$ because $B_3 = 0$
- $f_{B4,0}$ because $B_4 = 0$
- $f_{B5,0}$ because $B_5 = 0$
- $f_{B6,1}$ because $B_6 = 1$
- $f_{B7,1}$ because $B_7 = 1$

Figure 2-146 shows the output spectrum required for this one symbol. This spectrum is valid for only one symbol time. After the modulator has created this 8-bit symbol, it must generate a new symbol based on the next 8 data bits.

There are two general methods for finding $I(t)$ and $Q(t)$: the time-domain method, which is intuitive and straightforward; and the fast Fourier transform (FFT) method, which is computationally efficient and more versatile.

2.9.18.1 The Time-Domain Method

The time-domain method extends the discussion of 4LFSK. To generate a tone at a frequency that is offset Δf_{FSK} from the carrier f_{RF}, we apply sine and cosine waves at a

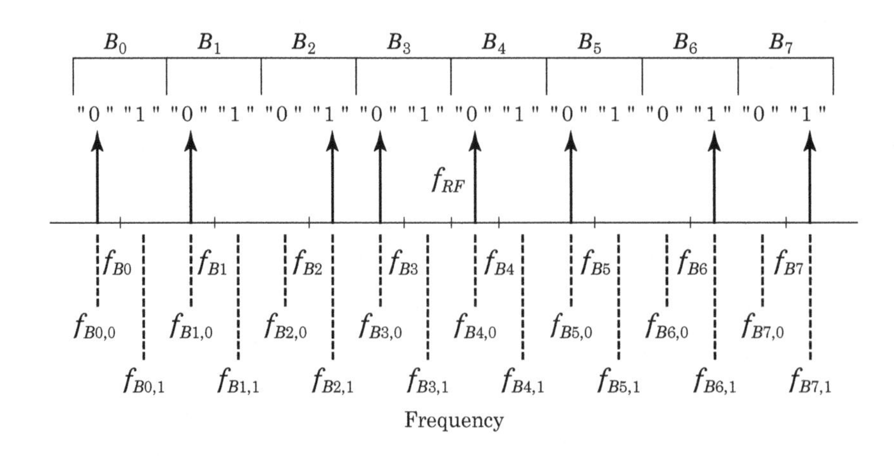

FIGURE 2-146 ■ Multitone FSK spectrum required for symbol $B_7 B_6 B_5 B_4 B_3 B_2 B_1 B_0 =$ 11000100 $= f_{B7,1} f_{B6,1} f_{B5,0} f_{B4,0} f_{B3,0} f_{B2,1} f_{B1,0} f_{B0,0}$. The modulator must produce these eight tones simultaneously.

frequency Δf_{FSK} into the modulator's I and Q ports. The phase relationship between the I and Q signals determines whether the output signal is above or below the RF signal.

We can generate a single sine wave of frequency $f_{RF} + f_{m1}$ by setting $I(t) = \cos(2\pi f_{m1} t)$ and $Q(t) = \sin(2\pi f_{m1} t)$. Superposition applies in a quadrature modulator, so by setting

$$
\begin{gathered}
I(t) = \cos(2\pi f_{m1} t) + \cos(2\pi f_{m2} t) \\
\text{and} \\
Q(t) = \sin(2\pi f_{m1} t) + \sin(2\pi f_{m2} t)
\end{gathered}
\tag{2.113}
$$

we generate two tones at the modulator output at frequencies $f_{RF} + f_{m1}$ and $f_{RF} + f_{m2}$. We can then extend this concept to generate the 8-tone signal of the multitone FSK of Figure 2-146.

Let's consider an example system with a 10 kHz bit rate and a $10\text{k}/8 = 1{,}250$ Hz symbol rate. We'll set the spacing between each tone to twice the symbol rate, or 2,500 Hz. Table 2-12 shows the mapping of each bit to the proper $I(t)$ and $Q(t)$ waveforms.

The symbol $B_7 B_6 B_5 B_4 B_3 B_2 B_1 B_0 = 11000100$ maps to frequencies $f_{B7,1}$, $f_{B6,1}$, $f_{B5,0}$, $f_{B4,0}$, $f_{B3,0}$, $f_{B2,1}$, $f_{B1,0}$ and $f_{B0,0}$ or 18,750 Hz, 13,750 Hz, 6,250 Hz, 1,250 Hz, −3,750 Hz, −6,250 Hz, −13,750 Hz, and −18,750 Hz, respectively. We add all of the $I(t)$ components to produce an aggregate $I_{symbol}(t)$, and add the $Q(t)$ waveforms together

TABLE 2-12 ■ The bit-to-waveform maps for an example multitone FSK system.

Bit Number	0 Frequency	1 Frequency
B_0	$f_{B0,0} = -18{,}750$ Hz $I(t) = \cos[2\pi(-18{,}750)t]$ $Q(t) = \sin[2\pi(-18{,}750)t]$	$f_{B0,1} = -16{,}250$ Hz $I(t) = \cos[2\pi(-16{,}250)t]$ $Q(t) = \sin[2\pi(-16{,}250)t]$
B_1	$f_{B1,0} = -13{,}750$ Hz $I(t) = \cos[2\pi(-13{,}750)t]$ $Q(t) = \sin[2\pi(-13{,}750)t]$	$f_{B1,1} = -11{,}250$ Hz $I(t) = \cos[2\pi(-11{,}250)t]$ $Q(t) = \sin[2\pi(-11{,}250)t]$
B_2	$f_{B2,0} = -8{,}750$ Hz $I(t) = \cos[2\pi(-8{,}750)t]$ $Q(t) = \sin[2\pi(-8{,}750)t]$	$f_{B2,1} = -6{,}250$ Hz $I(t) = \cos[2\pi(-6{,}250)t]$ $Q(t) = \sin[2\pi(-6{,}250)t]$
B_3	$f_{B3,0} = -3{,}750$ Hz $I(t) = \cos[2\pi(-3{,}750)t]$ $Q(t) = \sin[2\pi(-3{,}750)t]$	$f_{B3,1} = -1{,}250$ Hz $I(t) = \cos[2\pi(-1{,}250)t]$ $Q(t) = \sin[2\pi(-1{,}250)t]$
B_4	$f_{B4,0} = 1{,}250$ Hz $I(t) = \cos[2\pi(1{,}250)t]$ $Q(t) = \sin[2\pi(1{,}250)t]$	$f_{B4,1} = 3{,}750$ Hz $I(t) = \cos[2\pi(3{,}750)t]$ $Q(t) = \sin[2\pi(3{,}750)t]$
B_5	$f_{B5,0} = 6{,}250$ Hz $I(t) = \cos[2\pi(6{,}250)t]$ $Q(t) = \sin[2\pi(6{,}250)t]$	$f_{B5,1} = 8{,}750$ Hz $I(t) = \cos[2\pi(8{,}750)t]$ $Q(t) = \sin[2\pi(8{,}750)t]$
B_6	$f_{B6,0} = 11{,}250$ Hz $I(t) = \cos[2\pi(11{,}250)t]$ $Q(t) = \sin[2\pi(11{,}250)t]$	$f_{B6,1} = 13{,}750$ Hz $I(t) = \cos[2\pi(13{,}750)t]$ $Q(t) = \sin[2\pi(13{,}750)t]$
B_7	$f_{B7,0} = 16{,}250$ Hz $I(t) = \cos[2\pi(16{,}250)t]$ $Q(t) = \sin[2\pi(16{,}250)t]$	$f_{B7,1} = 18{,}750$ Hz $I(t) = \cos[2\pi(18{,}750)t]$ $Q(t) = \sin[2\pi(18{,}750)t]$

to generate an aggregate $Q_{symbol}(t)$:

$$
\begin{aligned}
I_{symbol}(t) &= \cos[2\pi f_{B0,0}t] + \cos[2\pi f_{B1,0}t] + \cos[2\pi f_{B2,1}t] + \cos[2\pi f_{B3,0}t] \\
&\quad + \cos[2\pi f_{B4,0}t] + \cos[2\pi f_{B5,0}t] + \cos[2\pi f_{B6,1}t] + \cos[2\pi f_{B7,1}t] \\
&= \cos[2\pi(-18{,}750)t] + \cos[2\pi(-13{,}750)t] + \cos[2\pi(-6{,}250)t] \\
&\quad + \cos[2\pi(-3{,}750)t] + \cos[2\pi(1{,}250)t] + \cos[2\pi(6{,}250)t] \\
&\quad + \cos[2\pi(13{,}750)t] + \cos[2\pi(18{,}750)t] \\
Q_{symbol}(t) &= \sin[2\pi f_{B0,0}t] + \sin[2\pi f_{B1,0}t] + \sin[2\pi f_{B2,1}t] + \sin[2\pi f_{B3,0}t] \\
Q_{symbol}(t) &= \sin[2\pi f_{B0,0}t] + \sin[2\pi f_{B1,0}t] \\
&\quad + \sin[2\pi f_{B2,1}t] + \sin[2\pi f_{B3,0}t] \\
&\quad + \sin[2\pi f_{B4,0}t] + \sin[2\pi f_{B5,0}t] \\
&\quad + \sin[2\pi f_{B6,1}t] + \sin[2\pi f_{B7,1}t] \\
&= \sin[2\pi(-18{,}750)t] + \sin[2\pi(-13{,}750)t] \\
&\quad + \sin[2\pi(-6{,}250)t] + \sin[2\pi(-3{,}750)v] \\
&\quad + \sin[2\pi(1{,}250)t] + \sin[2\pi(6{,}250)t] \\
&\quad + \sin[2\pi(13{,}750)t] + \sin[2\pi(18{,}750)t]
\end{aligned}
\qquad (2.114)
$$

$$
\begin{aligned}
I_{symbol}(t) &= \cos[2\pi f_{B0,0}t] + \cos[2\pi f_{B1,0}t] \\
&\quad + \cos[2\pi f_{B2,1}t] + \cos[2\pi f_{B3,0}t] \\
&\quad + \cos[2\pi f_{B4,0}t] + \cos[2\pi f_{B5,0}t] \\
&\quad + \cos[2\pi f_{B6,1}t] + \cos[2\pi f_{B7,1}t] \\
&= \cos[2\pi(-18{,}750)t] + \cos[2\pi(-13{,}750)t] \\
&\quad + \cos[2\pi(-6{,}250)t] + \cos[2\pi(-3{,}750)t] \\
&\quad + \cos[2\pi(1{,}250)t] + \cos[2\pi(6{,}250)t] \\
&\quad + \cos[2\pi(13{,}750)t] + \cos[2\pi(18{,}750)t]
\end{aligned}
$$

$$
\begin{aligned}
Q_{symbol}(t) &= \sin[2\pi f_{B0,0}t] + \sin[2\pi f_{B1,0}t] \\
&\quad + \sin[2\pi f_{B2,1}t] + \sin[2\pi f_{B3,0}t] \\
&\quad + \sin[2\pi f_{B4,0}t] + \sin[2\pi f_{B5,0}t] \\
&\quad + \sin[2\pi f_{B6,1}t] + \sin[2\pi f_{B7,1}t] \\
&= \sin[2\pi(-18{,}750)t] + \sin[2\pi(-13{,}750)t] \\
&\quad + \sin[2\pi(-6{,}250)t] + \sin[2\pi(-3{,}750)v] \\
&\quad + \sin[2\pi(1{,}250)t] + \sin[2\pi(6{,}250)t] \\
&\quad + \sin[2\pi(13{,}750)t] + \sin[2\pi(18{,}750)t]
\end{aligned}
$$

Figure 2-147 shows the $I_{symbol}(t)$ and $Q_{symbol}(t)$ waveforms for this symbol of equation (2.114).

Figure 2-148 shows the modulator output spectrum for the single symbol [11000100]. The eight output tones and their frequencies indicate the symbol sent.

Repeating this process over many random symbols produces the spectrum shown in Figure 2-149. Figure 2-149 shows 16 tones representing the two possible states of each symbol.

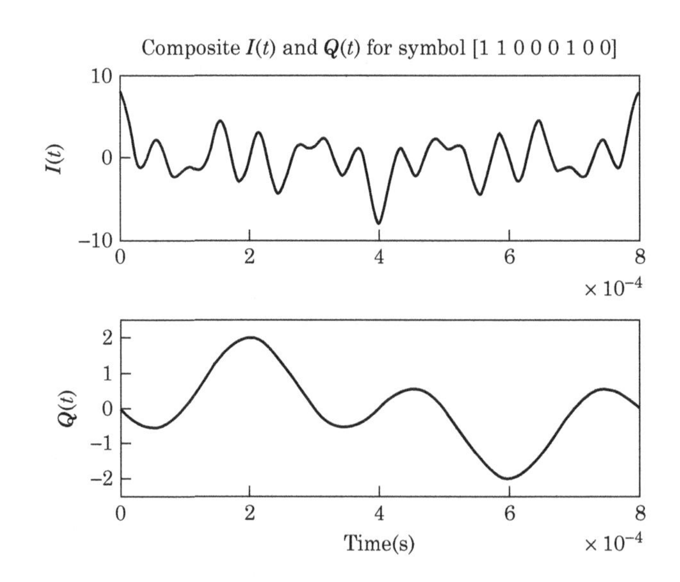

FIGURE 2-147 ■ The $I(t)$ and $Q(t)$ required to produce the 8-Level multitone FSK symbol corresponding to $B_7 B_6 B_5 B_4 B_3 B_2 B_1 B_0 =$ 11000100. The bit rate is 10 kbps, the symbol rate is 10k/8 = 1,250 symbols/second and the frequency deviation is 2.5 kHz.

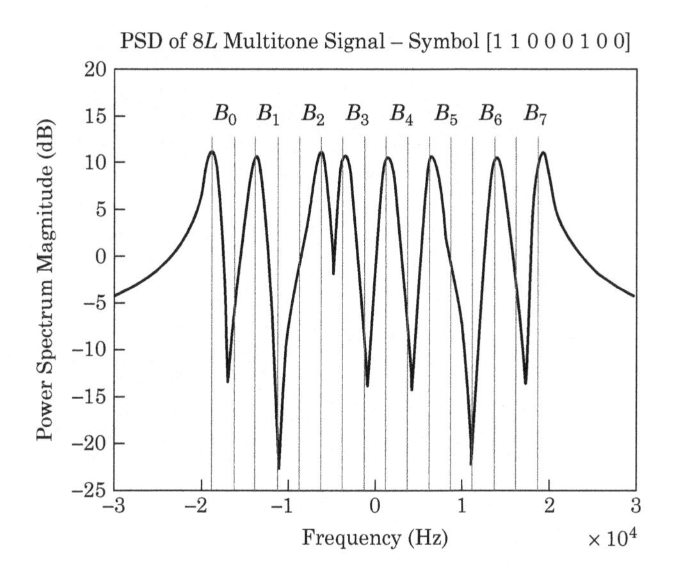

FIGURE 2-148 The frequency spectrum of the 8-level multitone FSK symbol corresponding to $B_7B_6B_5B_4B_3B_2B_1B_0$ = 11000100. The bit rate is 10 kbps, the symbol rate is 10k/8 = 1,250 symbols/second, and the frequency deviation is 2.5 kHz.

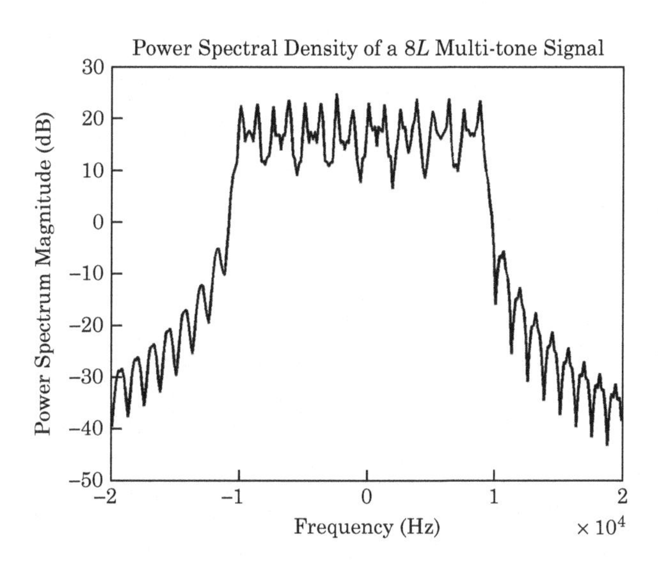

FIGURE 2-149 The frequency spectrum for an 8-level multitone FSK symbol modulated with many random symbols. The bit rate is 10 kHz, the symbol rate is 10k/8 = 1,250 symbols/second, and the frequency deviation is 2.5 kHz.

Figure 2-150 shows the time-frequency spectrogram of an 8-level multitone FSK signal over a five-symbol period. The first symbol is [0 0 0 0 0 0 0 0], which causes each of the eight tones to be their lower frequency. The next symbol is [1 1 1 1 1 1 1 1] and each of the eight tones are at their higher frequency. The symbols used to modulate the signal of Figure 2-150 were as follows:

$[B_7\ B_6\ B_5\ B_4\ B_3\ B_2\ B_1\ B_0]$
[0 0 0 0 0 0 0 0]
[1 1 1 1 1 1 1 1]
[0 1 0 1 0 1 0 1]
[1 0 1 0 1 0 1 0]
[1 1 1 1 1 1 1 1]

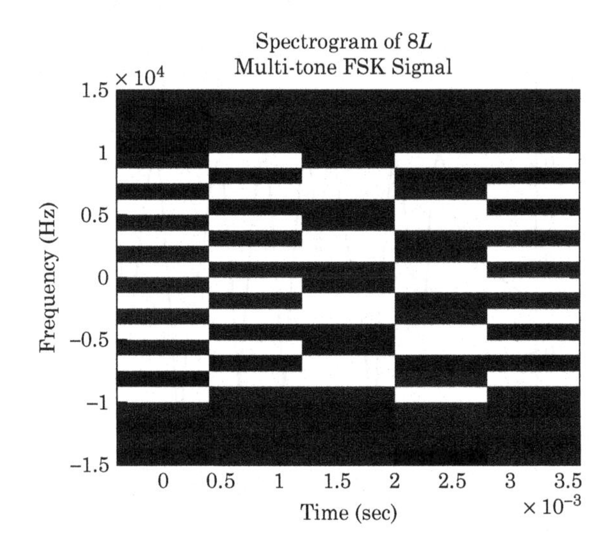

FIGURE 2-150 ■ Spectrogram (time-frequency plot) of an 8-level multitone FSK signal over a five-symbol period. The bit rate is 10 kbps, the symbol rate is 10k/8 = 1,250 symbols/second, and the frequency deviation is 2.5 kHz.

2.9.18.2 Frequency Domain

We can interpret the multitone FSK process in the frequency domain. Each bit in the symbol describes a frequency that we wish to be present at the modulator's output during the symbol's duration:

- If $B_0 = 0$, emit $f_{B0,0}$ otherwise emit $f_{B0,1}$
- If $B_1 = 0$, emit $f_{B1,0}$ otherwise emit $f_{B1,1}$
- If $B_2 = 0$, emit $f_{B2,0}$ otherwise emit $f_{B2,1}$
- If $B_n = 0$, emit $f_{Bn,0}$ otherwise emit $f_{Bn,1}$

Each bit corresponds to one bin of the signal's frequency-domain description. We directly use each bit in the symbol to set a bin in a frequency-domain buffer. Taking the inverse FFT (IFFT) of the frequency-domain buffer will produce a time-domain buffer containing the aggregate $I_{symbol}(t)$ and $Q_{symbol}(t)$ we need. Then, we simply read the time-domain buffer into the I and Q ports of the modulator for the duration of the symbol.

Figure 2-151 shows a 16 byte, frequency-domain buffer. If $B_0 = 0$, we would place a numerical 1 in the bin designated $f_{B0,0}$ and place a numerical 0 in the bin designated $f_{B0,1}$. Similarly, if $B_1 = 1$, we would place a numerical 0 in the bin designated $f_{B1,0}$ and a numerical 1 in the bin designated $f_{B1,1}$. Filling out the buffer for the symbol $[B_7 B_6 B_5 B_4 B_3 B_2 B_1 B_0] = [11000100]$ results in Figure 2-152.

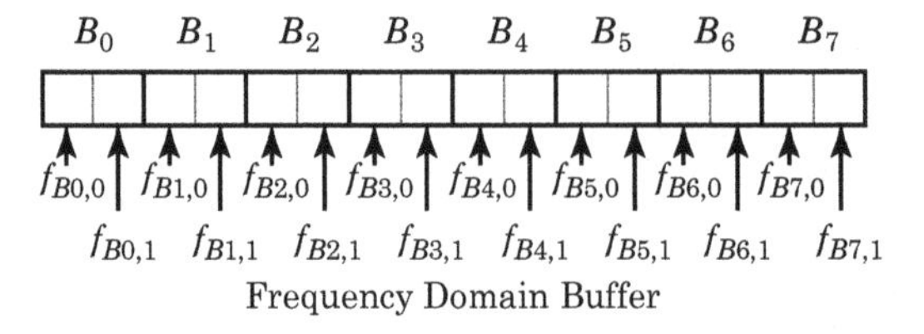

FIGURE 2-151 ■ A simplified frequency-domain buffer used to generate multitone FSK. We place nonzero elements in the positions corresponding to the frequencies we want to generate.

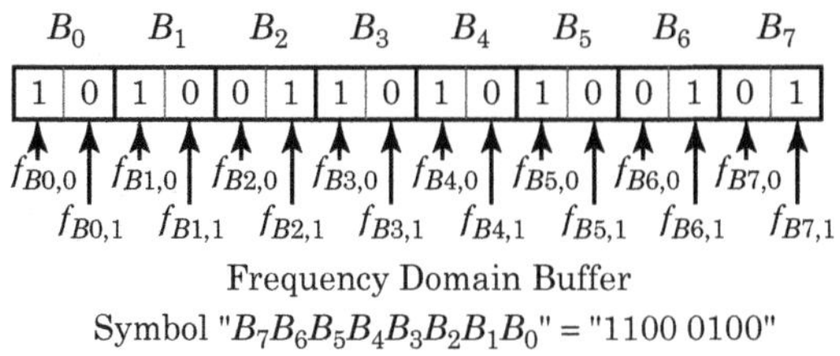

B_0 B_1 B_2 B_3 B_4 B_5 B_6 B_7
1 0 1 0 0 1 1 0 1 0 1 0 0 1 0 1
$f_{B0,0}$ $f_{B1,0}$ $f_{B2,0}$ $f_{B3,0}$ $f_{B4,0}$ $f_{B5,0}$ $f_{B6,0}$ $f_{B7,0}$
$f_{B0,1}$ $f_{B1,1}$ $f_{B2,1}$ $f_{B3,1}$ $f_{B4,1}$ $f_{B5,1}$ $f_{B6,1}$ $f_{B7,1}$
Frequency Domain Buffer
Symbol "$B_7B_6B_5B_4B_3B_2B_1B_0$" = "1100 0100"

FIGURE 2-152 ■ The frequency-domain buffer after filling it in for one particular symbol. A numerical 1 in a particular bin will cause the time-domain buffer to contain the indicated frequency. A numerical 0 in a bin generates nothing at the indicated frequency.

Frequency Domain Buffer
B_0 B_1 B_2 B_3 B_4 B_5 B_6 B_7
1 0 1 0 0 1 1 0 1 0 1 0 0 1 0 1
Inverse FFT
$\int X(f)e^{j2\pi ft}df$
$I_{symbol}(t)$
Time Domain Buffers
$Q_{symbol}(t)$
Time (s)
$\times 10^{-4}$

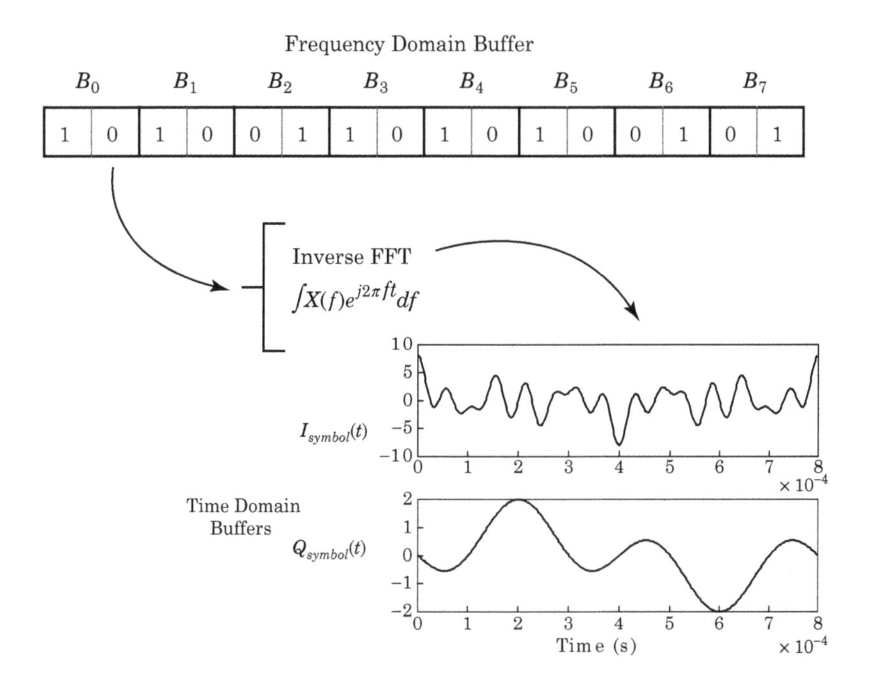

FIGURE 2-153 ■ A top-level overview of the operations we perform to generate multitone FSK using frequency-domain methods. The IFFT of the frequency-domain buffer produces the complex time-domain buffer we require.

We then perform a complex IFFT on the frequency-domain buffer to produce the time-domain buffer.

2.9.19 Implementation Details

Figure 2-153 shows an overview of the operations we perform, omitting many implementation issues. With proper coding, generating the per-symbol $I(t)$ and $Q(t)$ waveforms in a DSP amounts to transferring the user's bits into the frequency-domain buffer and then performing an inverse FFT. The workload per symbol is very low.

2.9.19.1 Going Further

We can easily extend the frequency-domain method to produce more complex forms of modulation. For example, we can independently BPSK modulate individual tones by placing either a $+1$ or -1 into each bin of the frequency-domain buffer. After the IFFT, the $+1$ generates a $0°$ phase tone in the time-domain buffer while the -1 generates a $180°$ phase tone.

If we build a complex frequency buffer (containing both real and imaginary parts), we can QPSK modulate each tone with two symbols per bit:

Symbol to Send	Angle (Degrees)	Real Part of Frequency Buffer	Imaginary Part of Frequency Buffer	Equivalent Complex Number
(00)	+45°	+1	+1	$+1 + j$
(01)	+135°	−1	+1	$-1 + j$
(11)	−135°	−1	−1	$-1 - j$
(10)	−45°	+1	−1	$+1 - j$

For example, the modulator emits a tone at +45° if we place +1 into both the real and imaginary parts of the frequency buffer. Similarly, the modulator emits a tone at +135° when we place −1 into the real part and +1 into the imaginary part of the frequency buffer. The IFFT produces the correct $I(t)$ and $Q(t)$ signals. We can even BPSK-modulate one channel while QPSK modulating a different channel.

2.10 | QUADRATURE MODULATORS, BASEBAND FILTERING, AND SPECTRUM CONTROL

A quadrature modulator can produce many different digital modulation formats: FSK, M-ary FSK, BPSK, QPSK, and OOK. We've been fairly cavalier about the amount of spectrum our signals require.

So far we've concentrated on convincing a quadrature modulator to generate the modulation we desire. We'll now move from those simple, spectrally hungry modulation techniques to more spectrally efficient systems. These techniques are applicable in crowded, channelized systems such as cellular telephones or wireless LANs.

2.10.1 RF Spectrum

All of the systems we've discussed so far exhibit a $\sin(f)/f$-shaped spectrum. This characteristic is a direct result of the square waveform we've been feeding into the I and Q ports of the quadrature modulator.

The $\sin(x)/x$ shape isn't a very desirable spectrum. For instance, in a cellular telephone system we want to pack as many users (signals) as possible into the allocated spectrum, but the spectrum splatter of this signal won't allow us to pack users too closely (see Figure 2-154).

Figure 2-155 shows the root of the problem. Applying a $\sin(f)/f$ spectrum to either the I or Q ports of the complex modulator produces a $\sin(f)/f$ spectrum on the RF port.

We can alter the spectral shape at the RF signal by modifying the spectral shape of $I(t)$ and $Q(t)$. Our task will be to find quadrature signals that exhibit friendly spectral shapes, are easy to generate, and still convey the user's information.

2.10.1.1 Candidate Waveforms

Fourier analysis tells us that there is an inverse relationship between the bandwidth and duration of a signal. A signal that lasts for a long time will exhibit a narrow bandwidth, all other factors being equal. To realize a narrow RF signal bandwidth, the quadrature signals we generate for each symbol must endure over more than one symbol period. The longer the symbol waveform endures, the narrower the final spectrum.

Prior to this discussion, each symbol lasted for only one symbol period and there was no possibility that the information from one symbol could interfere with its neighboring

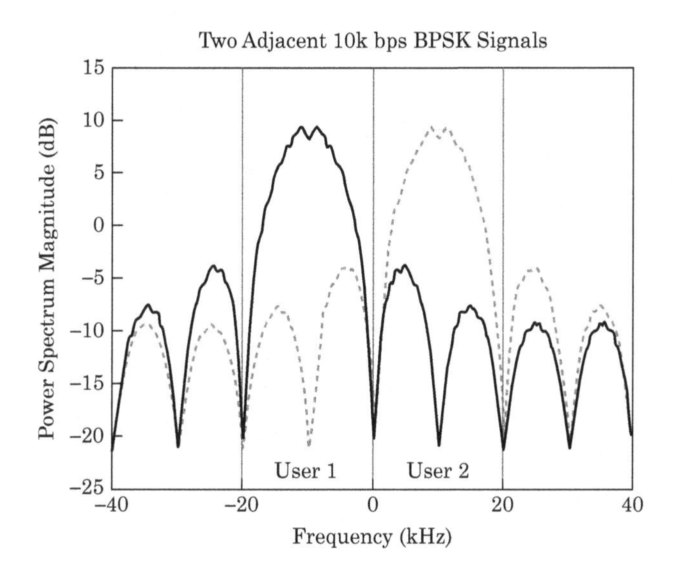

FIGURE 2-154 The spectrum of two adjacent 10 kbps signals. The $\sin(x)/x$ shape of each spectrum causes signal energy from one channel to leak over into the adjacent channel.

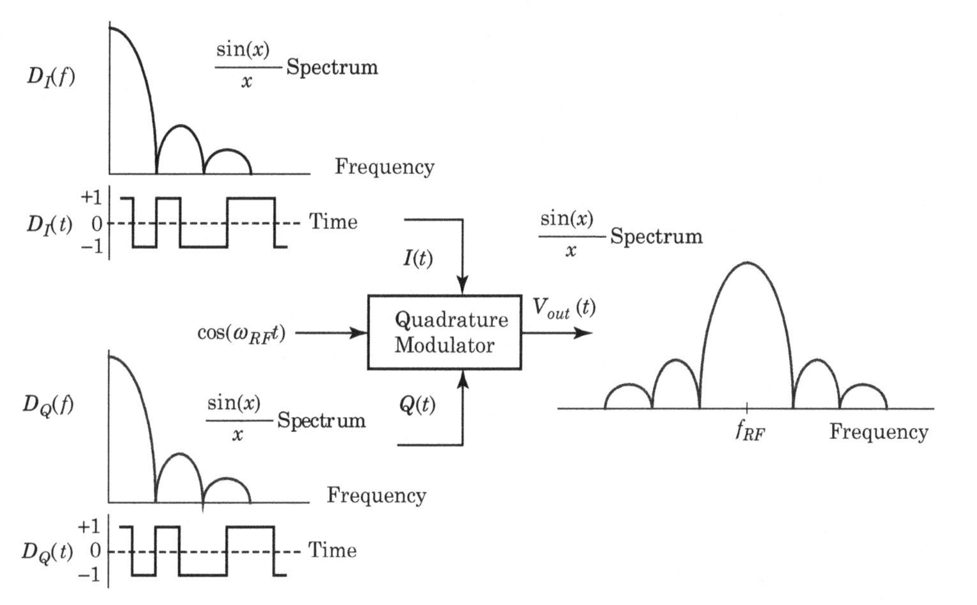

FIGURE 2-155 The bandwidth of the RF signal is a direct function of the bandwidths of the $I(t)$ and $Q(t)$ signals you apply to the quadrature modulator. The $\sin(f)/f$ spectral shape at the RF is a direct result of the $\sin(f)/f$ spectral shape of the square waves you feed into the modulator.

symbols. Allowing the symbols to overlap in time creates the possibility of Intersymbol Interference or ISI. We'll discuss this shortly.

Returning to the topic of spectral shaping, we note in Figure 2-156 that the spectral shape of the RF signal at the output of a quadrature modulator directly corresponds to the spectrum of the I and Q waveforms. We can control the shape of the RF spectrum by controlling the shape of the I and Q spectra.

Figure 2-157 shows that the Fourier transform of a rectangular time-domain waveform is a $\sin(f)/f$ spectrum of infinite duration (i.e. the spectrum theoretically extends over all frequencies). The same figure also shows that an infinite duration, $\sin(t)/t$ waveform transforms into a perfectly square, "brick-wall" spectrum.

A time domain waveform of infinite duration is of little practical use, but we can approximate the infinite waveform by a finite one. Three appropriate spectral shaping waveforms are $\sin(x)/x$ (or $\text{sinc}(x)$), the raised cosine filter and the square root of the raised cosine (or simply the root raised cosine) waveforms.

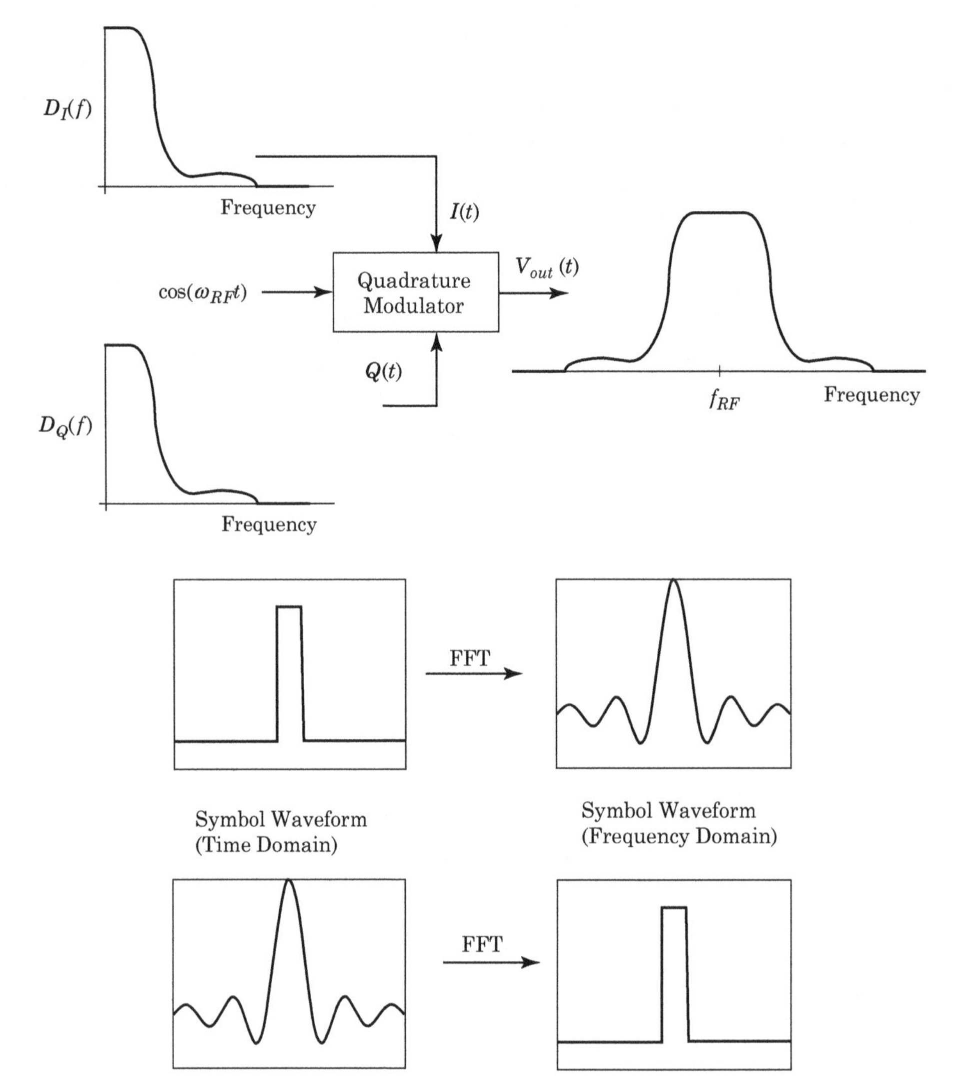

FIGURE 2-156 ■ Changing the $I(t)$ and $Q(t)$ signals to more spectrally efficient waveforms produces a narrower RF spectrum. Any spectral characteristic present on the I or Q signal is impressed on the RF spectrum.

FIGURE 2-157 ■ When developing quadrature waveforms that are spectrally efficient, we can take a brute-force approach and perform an Fourier transform to produce a "brick-wall" spectrum. We'll obtain a $\sin(t)/t$-looking time-domain waveform. The time-domain signal must be of infinite duration to realize the perfectly square spectral shape.

2.10.1.2 Sinc(*x*) Spectral Shaping

The zero-ISI $\sin(x)/x$- or $\text{sinc}(x)$-shaped waveform for a symbol rate R_S is

$$f_{SINC}(t) = \frac{\sin(\pi R_S t)}{(\pi R_S t)} \tag{2.115}$$

This equation produces a $\sin(x)/x$-shaped signal which is unity at time $t = 0$ and zero at times

$$t = \frac{n}{R_S} \quad n = \cdots - 2, -1, 0, 1, 2, \ldots \tag{2.116}$$

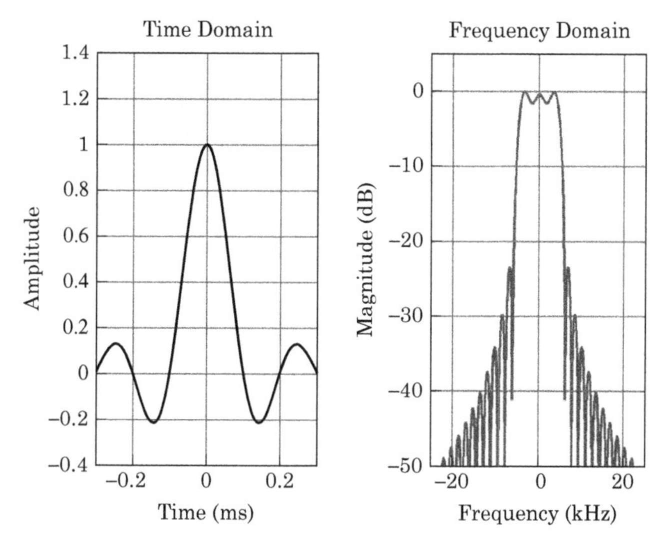

FIGURE 2-158 ■ A truncated sin(x)/x signal and its spectrum; applicable to a 10k symbols/second system. The left graph shows the time-domain sin(x)/x signal, which endures for six symbol periods. The right graph shows the spectrum. This spectrum is much narrower than the spectrum of Figure 2-154.

Figure 2-158 shows a candidate $\sin(x)/x$ signal whose duration is six symbols. The waveform was designed for a 10 kbps system.

The spectrum of the longer-duration $\sin(x)/x$ signal is narrower and squarer shouldered than the unfiltered signal of Figure 2-154.

2.10.1.3 Raised Cosine Spectral Shaping

A more common spectral control filter is the raised cosine filter whose time-domain response is

$$f_{RC}(t) = \frac{\sin(\pi R_S t)}{(\pi R_S t)} \cdot \frac{\cos(R\pi R_S t)}{\left(1 - 4R^2 t^2 R_S^2\right)} \tag{2.117}$$

where

R_s is the symbol rate

R is the rolloff factor ($0 <= R <= 1$). The rolloff factor describes how quickly the filter rolls off in the frequency domain.

Equation (2.117) is a $\sin(x)/x$ function modified by a function involving a cosine (see Figure 2-159).

The relatively high sidelobes in the spctrum of Figure 2-158 is the result of the sudden cessation of the $\sin(x)/x$ waveform. Multiplying the $\sin(x)/x$ by the raised cosine of equation (2.117) results in gentler rolloff and lower sidebands (see Figure 2-160).

2.10.1.4 Rolloff Factor *R*

The rolloff factor, R, of a raised cosine waveform is a measure of the waveform's excess bandwidth. A system that processes R_S symbols/second requires a theorectially minimum bandwidth of R_S hertz. Allowing excess bandwidth enables the receiver to acquire the

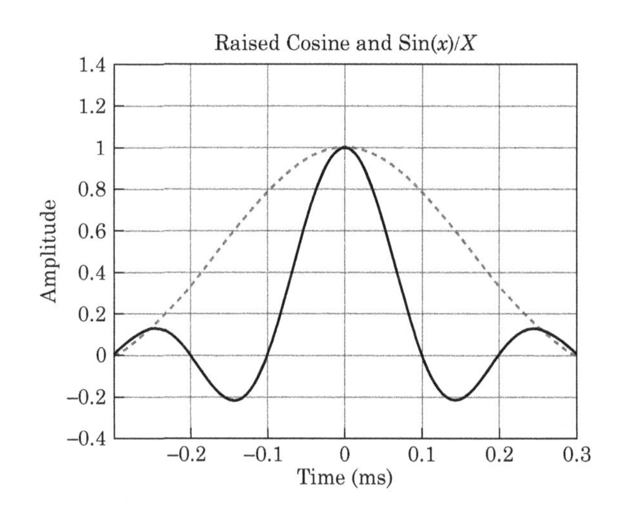

FIGURE 2-159 The raised cosine pulse shaping filter is the product of the $\sin(x)/x$ filter (shown in the solid line) and a raised cosine (shown in the dashed line). The rolloff factor, R, of this waveform is 0.5.

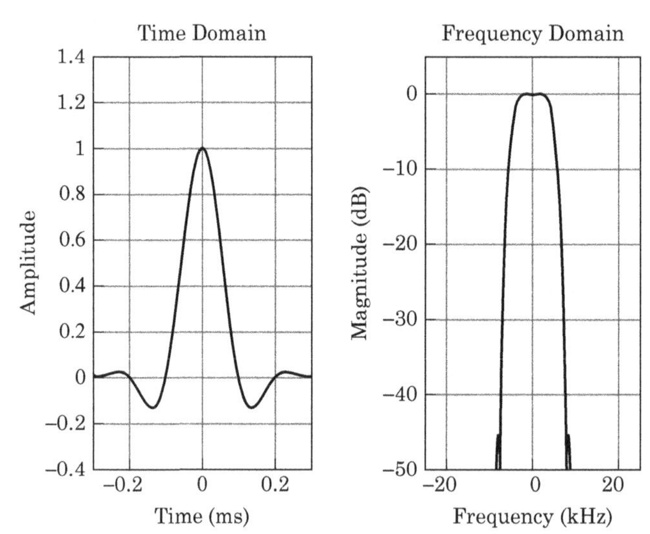

FIGURE 2-160 The time- and frequency-domain response of a raised cosine waveform, applicable to a 10k symbols/second system. The frequency response of this filter has lower sidebands than the simple truncated $\sin(x)/x$ waveform of Figure 2-158.

symbol or baud timing from the transmitted waveform. Values of $R = 0.3$ to $R = 0.5$ are common.

2.10.1.5 Root Raised Cosine Spectral Shaping

The $\sin(x)/x$ and raised cosine filters are inadequate for most practical needs. The sidebands of the truncated $\sin(x)/x$ filter are too high—we can do better. The raised cosine is ideal, but it is not possible to apply a matched filter to such a signal and still maintain zero ISI into the symbol decision engine. The root raised cosine (RRC) waveform solves these problems. The RRC is the square root of the raised cosine waveform, the square root being performed in the frequency domain. The equation describing the impulse response of the RRC filter is a mess:

$$f_{RRC}(t) = 4R\sqrt{R_S}\frac{\cos\left[(1+R)\pi t R_S\right] + \sin\left[(1-R)\pi t R_S\right]/4RR_S t}{\pi\left[1-(4RR_S t)^2\right]} \tag{2.118}$$

Figure 2-161 shows the time- and frequency-domain plots of an RRC filter.

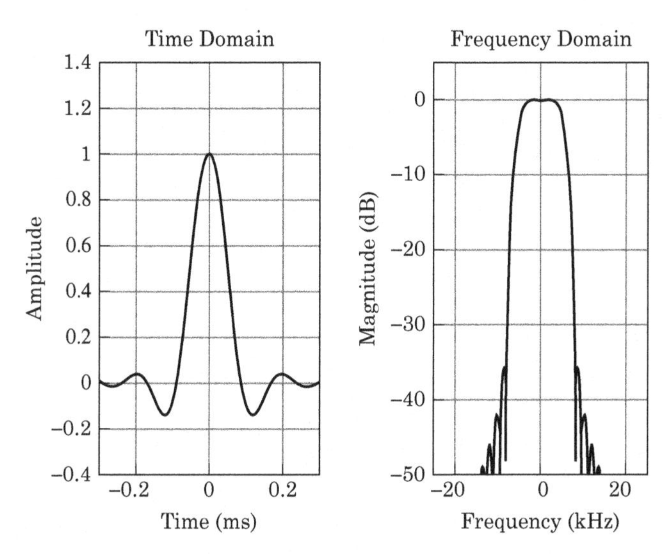

FIGURE 2-161 The time- and frequency-domain response of a root raised cosine waveform, applicable to a 10k symbols/second system. The frequency response of this filter has higher sidebands than the raised cosine waveform shown in Figure 2-160.

2.10.2 Spectrally Shaped BPSK

Let's generate a spectrally shaped BPSK signal that is keyed at 10 kbps and examine some implementation details. In BPSK, we set the Q port of the quadrature modulator to zero and drive the I port with our signaling waveforms. For this example, we will use $\sin(t)/t$ waveforms that are six symbols in duration (the bottom graphs of Figure 2-158).

Previously, we generated BPSK by setting $I(t)$ to one value for the duration of the symbol. The values of $I(t)$ [and $Q(t)$, in more complex systems] depended only on the current symbol. To spectrally shape our signal, we must follow a different process because the waveforms are longer than a single symbol duration. $I(t)$ and $Q(t)$ will now be arbitrary waveforms. We use the user's data stream to select one of the several possible $I(t)$ and $Q(t)$ waveforms:

- If the user symbol is 0: Select waveforms $I_0(t)$ and $Q_0(t)$
- If the user symbol is 1: Select waveforms $I_1(t)$ and $Q_1(t)$
- If the user symbol is N: Select waveforms $I_N(t)$ and $Q_N(t)$

Let us first transmit just one single symbol. There are two symbols in BPSK—0 and 1—so we'll need two symbol waveforms. Symbol 1 can be represented by a $\sin(t)/t$-shaped waveform, appropriately time scaled. The 0 symbol can be represented by a time-scaled $-\sin(t)/t$-shaped waveform.

Figure 2-162 shows the time-domain waveform, which is sent into the modulator's I port to transmit the symbol 1. The symbol starts at time 0 but doesn't reach its full-scale value of unity until 300 μsec (3 symbols) after we've begun sending the symbol. When the $I(t)$ signal is unity, the RF signal at the modulator's output is $1\angle 0^\circ$, which represents the bit to transmit.

Figure 2-163 shows the $I(t)$ signal used when sending the symbol 0. This is the inverse of the symbol one waveform, and the waveform forces the modulator's output to be $1\angle 180^\circ$ for a short period of time at $t = 300\,\mu$sec.

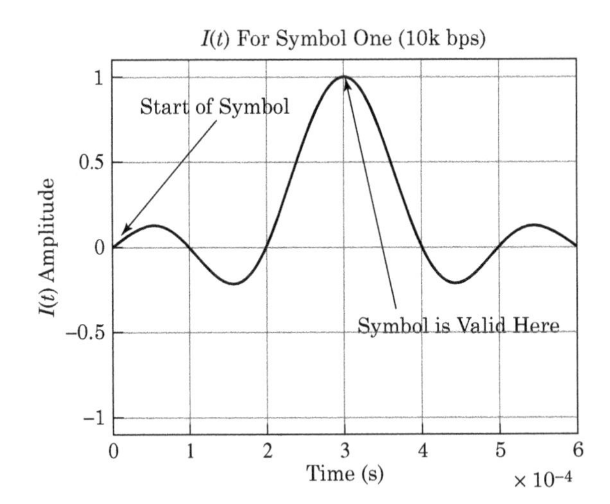

FIGURE 2-162 ■ Time-domain plot of the signal sent to the quadrature modulator's I port whenever the user symbol is a one. This is a truncated $\sin(t)/t$-type signal that is time scaled for at 10k symbols/second rate. At time 0, the start of the symbol, you begin reading this waveform from read-only memory (ROM) and applying it to the modulator's I port. The symbol isn't fully realized or recognizable at the receiver until 300 μsec later, when the I waveform is unity. In this example, the effects of each symbol affect the output of the modulator over six symbol periods.

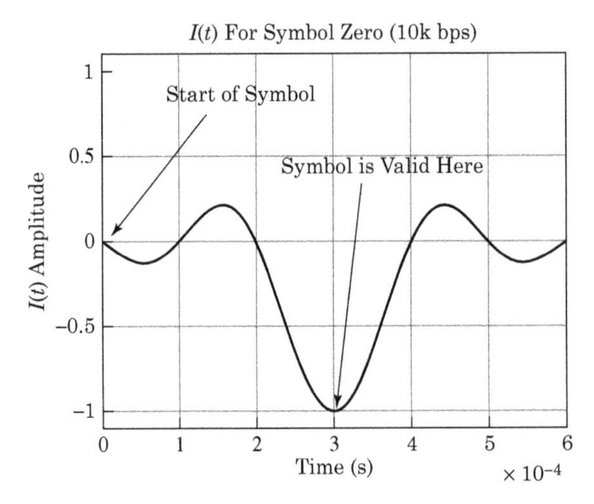

FIGURE 2-163 ■ Time-domain plot of the signal sent to the quadrature modulator's I port whenever symbol 0 is sent. This is the negative of the symbol 1 waveform. This is also a truncated $\sin(t)/t$-type signal, which is time scaled for at 10k symbols/second rate. At time 0, the start of the symbol, we begin reading this waveform from ROM and applying it to the modulator's I port. The symbol isn't fully realized or recognizable at the receiver until 300 μsec later, when the I waveform is −1 or $1\angle 180^\circ$.

2.10.3 Time Scaling

Each user bit in our BPSK signal causes the modulator to emit a $\sin(t)/t$-shaped time-domain waveform to the modulator's I port. Figure 2-162 and Figure 2-163 show the encoded symbols for a single bit. To encode a continuous stream of user bits, we must produce a copy of the $+\sin(t)/t$ (Figure 2-162) waveform every time the user wants to transmit a one and a copy of the $-\sin(t)/t$ waveform (Figure 2-163) every time the user transmits a 0. We must start one of these signals every time a symbol is due from the modulator.

Figure 2-164 shows how to encode the user bit stream 101. At time $T = 0$, we encode the first user bit, a 1, by producing a $+\sin(t)/t$-shaped waveform that starts at time $t = 0$ and continues until time $t = 6T$. We encode the second user bit, a 0, by generating a $-\sin(t)/t$-shaped waveform that starts at time $t = 1T$ and continues until time $t = 7T$. Finally, we encode the last user bit, a 1, by producing a $+\sin(t)/t$-shaped signal that starts at $t = 2T$ and expires at $t = 8T$. Figure 2-165 shows a superposition of these three waveforms.

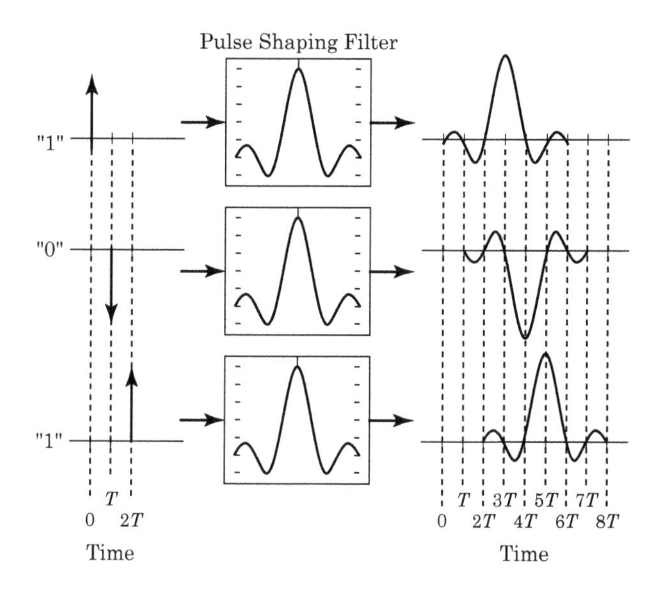

FIGURE 2-164 ■ The modulator produces a $+\sin(t)/t$-shaped waveform every time the user wants to encode a 1. The modulator produces a $-\sin(t)/t$-shaped waveform every time the user wants to encode a 0. The system symbol rate is $1/T$.

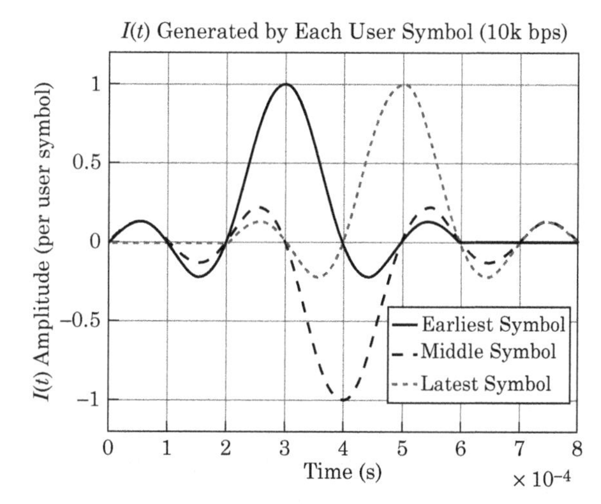

FIGURE 2-165 ■ Encoding the user bit stream [1 0 1]. The waveform labeled "Earliest Symbol" is the $I(t)$ signal you apply to the modulator to encode the first user bit, a 1. The waveform labeled "Middle Symbol" is the $I(t)$ signal required to encode the second user bit, a 0. The waveform labeled "Latest Symbol" is the signal required to encode the final user bit, a 1.

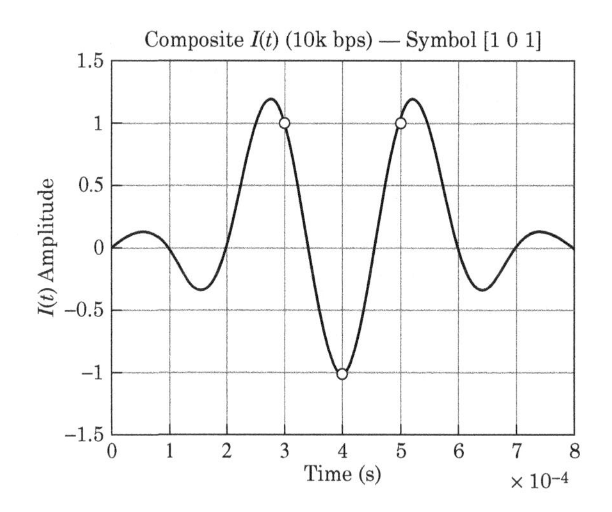

FIGURE 2-166 The aggregate $I(t)$ signal we send to the I port of the quadrature modulator to encode the user bit stream [1 0 1]. This function is the sum of the earliest, middle and latest waveforms shown in Figure 2-165. The circles indicate the zero ISI sampling points.

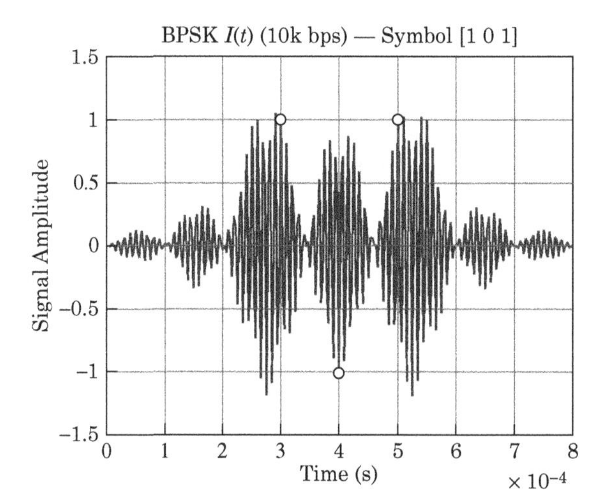

FIGURE 2-167 The RF signal produced by feeding the signal shown in Figure 2-166 into the I port of a quadrature modulator. The circles indicate the zero ISI sampling points.

The $I(t)$ signal we send to the modulator to encode [1 0 1] will be the arithmetic sum of these three $\sin(t)/t$ signals.

To produce the final BPSK signal, the modulator multiplies the aggregate $I(t)$ signal of Figure 2-166 by the RF signal $\cos(\omega_{RF}t)$. Figure 2-167 shows the result of this multiplication.

2.10.4 Time Scaling and Intersymbol Interference

The modulator must emit a symbol once every symbol period, but because we want to control the spectral shape of the RF signal each symbol waveform must last several symbols. Since the symbols overlap in time, there is a clear danger that consecutive symbols will interfere with each other if we are not careful. This is ISI.

Ideally, we would like to arrange the transmission scheme so that we can examine the transmitted signal at some instant in time and derive a single user symbol. When we achieve this goal, our transmitted signal possesses zero ISI.

Our 10k symbol/second system exhibits zero ISI because we examine the transmitted waveform once every 100 μsec and unambiguously derive a single symbol. The circles

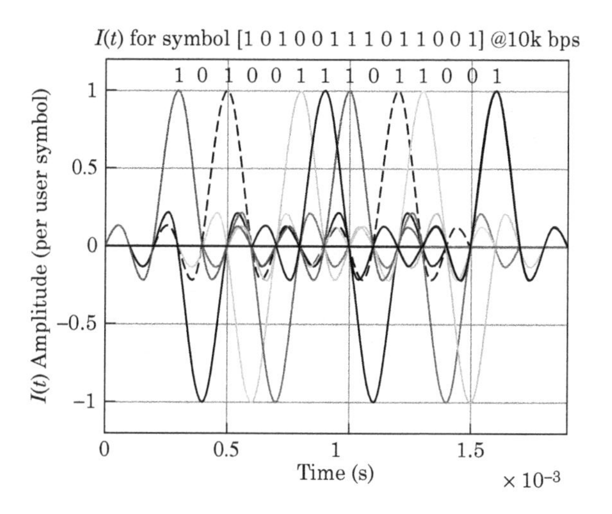

FIGURE 2-168 ■ The individual $\sin(t)/t$-shaped signals produced for an arbitrary user bit stream at 10 kbps. Each $\sin(t)/t$ signal is six bit times long. The positive $\sin(t)/t$ signals represent ones in the user's bit stream, and the negative $\sin(t)/t$ signals represent the zeros in the user's bit stream.

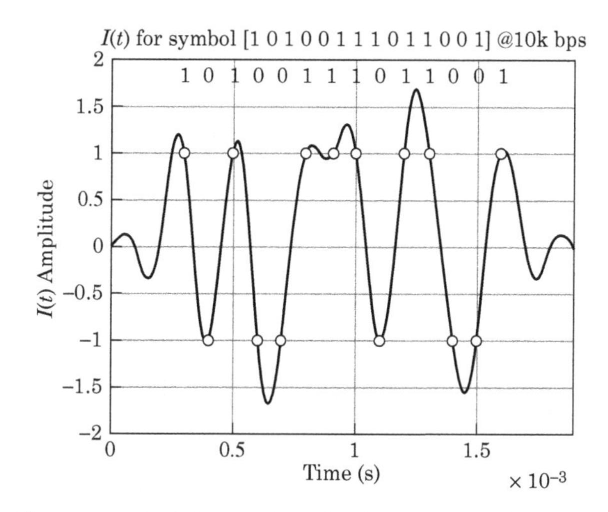

FIGURE 2-169 ■ The aggregate $I(t)$ waveform for the same user bit stream used to generate Figure 2-168. This signal is the sum of all the individual $\sin(t)/t$-shaped waveforms of Figure 2-168. At the zero ISI sampling times (300 through 1600 μsec), the aggregate $I(t)$ signal contains no ISI. The value of $I(t)$ at these instants in time depends only on one user bit.

in Figure 2-166 and Figure 2-167 indicate the zero ISI sampling points for the example system. The receiver can derive the user's [1 0 1] bit stream by measuring the phase of the transmitted signal at 300, 400, and 500 μsec. Any other series of sampling instances (e.g., 325, 425, and 525 μsec) exhibit ISI because the value of the transmitted waveform depends on more than one symbol. The zero ISI $\sin(t)/t$-shaped waveform for a symbol rate R_S is given by equation (2.115).

Figure 2-168 and Figure 2-169 show the relevant waveforms for a long, arbitrary string of user data. Figure 2-168 shows the individual $\sin(t)/t$-shaped signals developed for each user bit. Figure 2-169 shows the aggregate $I(t)$ signal, which is the sum of all the signal in Figure 2-168. Figure 2-170 shows the output of the quadrature modulator.

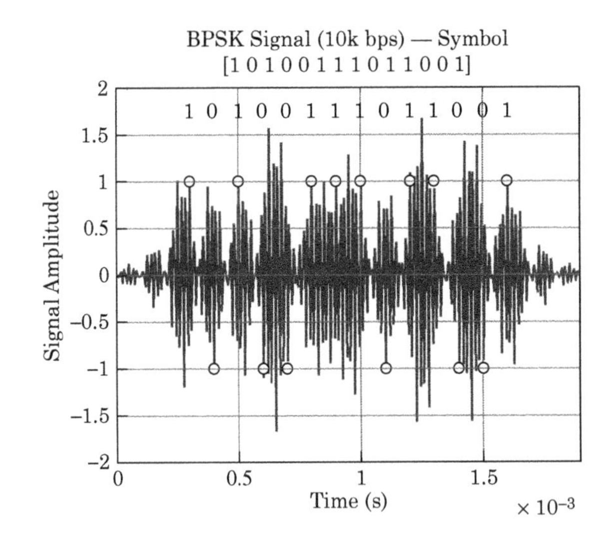

FIGURE 2-170 The quadrature modulator output (at the RF) for the arbitrary user bit stream of Figure 2-168 and Figure 2-169. Circles mark the zero ISI sampling points.

2.11 GENERAL CHARACTERISTICS OF SIGNALS

2.11.1 Average Value

The average value of a signal tells us about the amount of energy present in the signal at 0 Hz or DC. The average value of an analog signal is

$$V_{Avg} = \frac{1}{T_2 - T_1} \int_{T_1}^{T_2} v(t)dt \tag{2.119}$$

The average value of a periodic signal equals the average value of one period of that signal. Similarly, we can approximate the average value of a signal by taking the average values of a series of equally spaced samples.

The average value of a constellation is

$$V_{Avg} = \sum_{i=1}^{N} V_{i,symbol} p_{i,symbol} \tag{2.120}$$

where

$V_{i,symbol}$ is the complex Voltage of the i-th symbol.

$p_{i,symbol}$ is the probability of occurrence of the i-th symbol

If all symbols are equally probable, then equation (2.120) simplifies to

$$V_{Avg} = \frac{1}{N} \sum_{i=1}^{N} V_{i,symbol} \tag{2.121}$$

where

N is the number of symbols in the constellation

2.11.2 RMS Value

The RMS value of a signal describes the power present in the signal. The RMS value of an analog signal is

$$V_{RMS}^2 = \frac{1}{T_2 - T_1} \int_{T_1}^{T_2} v^2(t)dt \tag{2.122}$$

The RMS value of a periodic signal equals the RMS value of one period of that signal. Similarly, we can approximate the RMS value of a signal by taking the RMS values of a series of equally spaced samples.

The RMS value of a constellation is

$$V_{RMS}^2 = \sum_{i=1}^{N} V_{i,symbol}^2 p_{i,symbol} \tag{2.123}$$

where

$V_{i,symbol}^2$ is the square of the magnitude of the complex voltage of the i-th symbol.
$p_{i,symbol}$ is the probability of occurrence of the i-th symbol

If all symbols are equally probable, equation (2.123) simplifies to

$$V_{RMS}^2 = \frac{1}{N} \sum_{i=1}^{N} V_{i,symbol}^2 \tag{2.124}$$

where

N is the number of symbols in the constellation

2.11.3 Peak (Crest) Factor

The peak factor (or crest factor) of a waveform equals the waveform's peak amplitude divided by its RMS value or

$$PF = \frac{|V_{max}|}{V_{RMS}} \tag{2.125}$$

where

V_{max} is the instantaneous Voltage of the symbol with the largest magnitude
V_{RMS} is the RMS value of the signal from equation (2.123)

We often use peak factor to set headroom in RF cascades or when specifying analog-to-digital converters. We find the peak value by inspecting the constellation. The magnitude of the largest symbol is the peak value.

EXAMPLE

Peak Factor of 8QAM

Find the peak factor of the 8QAM signal shown in Figure 2-171 assuming all symbols are equally probable.

Solution

The peak value of the constellation of Figure 2-171 is the magnitude of the largest symbol. The outer four symbols are the largest and are equal in magnitude. The peak value is 2.73.

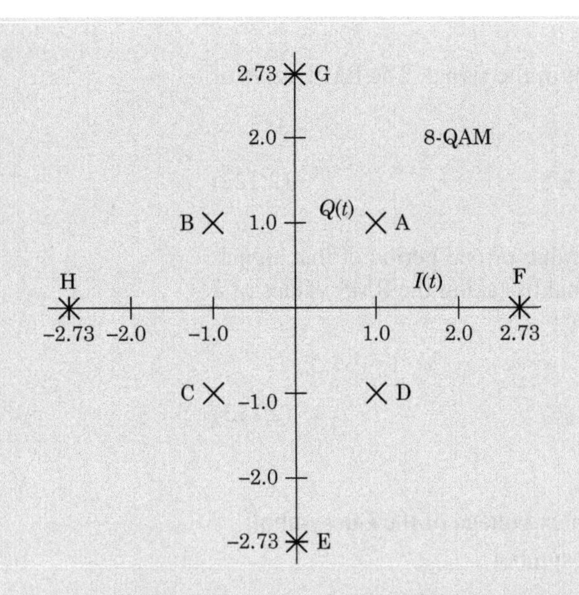

FIGURE 2-171 ■ An 8-symbol QAM constellation.

Equation (2.123) tells us the RMS value. All signals are equally probably so $p_{i,symbol}$ is 1/8 for all i. The RMS value of the constellation is

$$\begin{aligned} V_{RMS}^2 &= \sum_{i=1}^{N} V_{i,symbol}^2 p_{i,symbol} = \sum_{i=1}^{8} V_{i,symbol}^2 \left[\frac{1}{8}\right] \\ &= \frac{|1+j|^2 + |-1+j|^2 + |-1-j|^2 + |1-j|^2 + |-2.73j|^2 + |2.73|^2 + |2.73j|^2 + |-2.73|^2}{8} \\ &= \frac{1}{8}[(1+j)(1-j) + (-1+j)(-1-j) + (-1-j)(-1+j) + (1-j)(1+j) \cdots \\ &\quad +(-2.73j)(2.73j) + (2.73)(2.73) + (2.73j)(-2.73j) + (-2.73)(-2.73)] \\ &= \frac{(2+2+2+2+7.45+7.45+7.45+7.45)}{8} \\ &= 4.725 \\ V_{RMS} &= 2.174 \end{aligned} \tag{2.126}$$

The peak factor is

$$\begin{aligned} PF &= \frac{|V_{max}|}{V_{RMS}} \\ &= \frac{2.73}{2.174} \\ &= 1.256 \end{aligned} \tag{2.127}$$

2.12 | SUMMARY

In this chapter, we've examined signals, noise, and various modulated signals in the time, frequency, zero crossing, and phasor domains. As we move on to new material, we will find all of these expressions useful and necessary to explore fully the topics to come.

2.13 | BIBLIOGRAPHY

Bingham, John A.C., *The Theory and Practice of Modem Design*, John Wiley and Sons, Inc., 1988.

McClaning, Kevin J., "Quadrature Modulator Generates Multitone FSK Using Time-Domain Methods," http://www.chipcenter.com/dsp, 2000.

McClaning, Kevin J., "Digital Modulation Using a Quadrature Modulator for 4-Level and Multitone FSK," http://www.chipcenter.com/dsp, 2000.

McClaning, Kevin J, "Digital Modulation Techniques Using a Quadrature Modulator," http://www.chipcenter.com/dsp, 2000.

Ziemer, R.E., and Tranter, W.H., *Principles of Communications: Systems, Modulation and Noise*, 4th ed., John Wiley and Sons, Inc., 1995.

2.14 | PROBLEMS

1. **Peak Factor of a Signal with Random Data** Calculate the peak factor of the following signals, assuming a random symbol stream feeds the modulator:

   a. QPSK

   b. 64QAM square constellation

2. **Peak Factor of a Signal with Nonrandom Data** Calculate the peak/average ratio of the 8QAM constellation shown in Figure 2-172. Table 2-13 shows the probability of occurance of each symbol.

3. **A Different 16QAM Constellation** Figure 2-173 shows two different 16QAM constellations. The points on the leftmost constellation are equally spaced grid points. The constellation on the right is a V.29 modem constellation. The radii of the three circles in the right-hand are $1/3$, $2/3$, and 1.

   a. For both constellations, calculate the amplitude and phase for each constellation point.

   b. What are the advantages of one constellation over the other? Consider crest factor, performance in Gaussian noise, and immunity to phase noise.

4. **Phasor Diagram and Perfect Cancellation** Figure 2-174 shows a circuit diagam of an antenna array designed to produce a null in the broadside direction to the antenna. Your goal

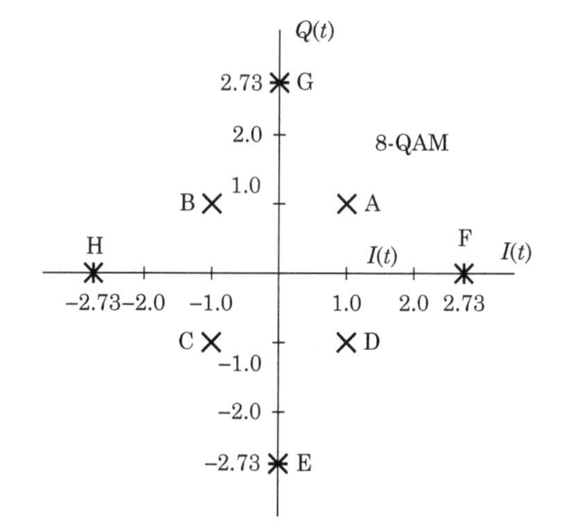

FIGURE 2-172 Constellation of an 8QAM signal.

TABLE 2-13 ■ Symbol probabilities for an 8QAM signal.

Symbol	Prob(Symbol)
A	0.10
B	0.04
C	0.30
D	0.24
E	0.04
F	0.13
G	0.12
H	0.03
Total	1.00

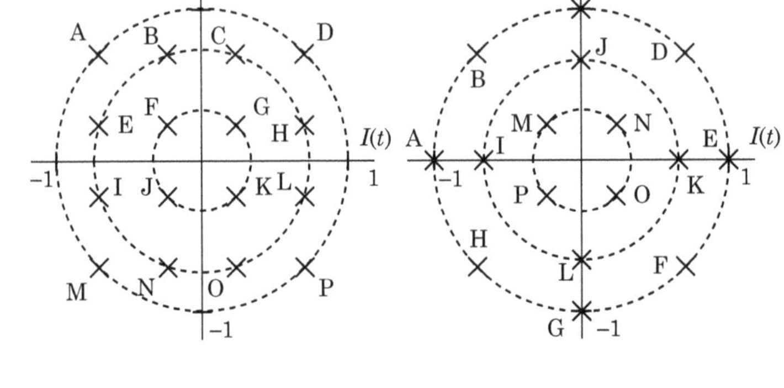

FIGURE 2-173 ■ The constellation diagrams of two different 16QAM signals.

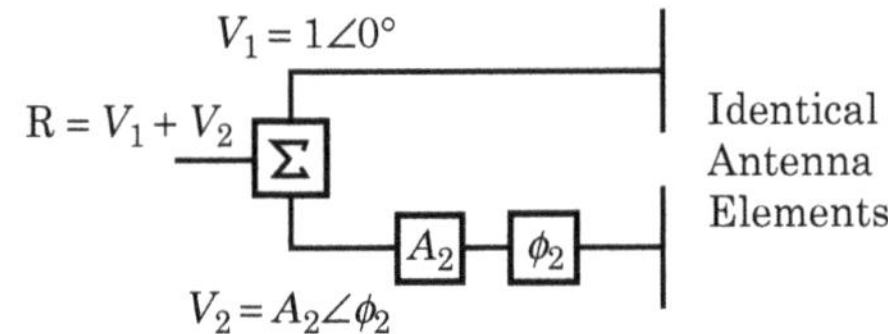

FIGURE 2-174 ■ A phased array antenna system.

is to combine the two signals from each element in such a way as to achieve at least 20 dB of attenuation in received signal power.

Assume $V_1 = 1\angle 0°$ and let $V_2 = A_2\angle\phi_2$. We will vary A_2 and ϕ_2 to achieve cancellation.

a. Draw a phasor diagram showing the conditions for cancellation.

b. If the amplitudes of the two signals are identical (i.e., $A_2 = 1$), how much can the phase vary and still produce the required 20 dB of attenuation?

c. If the phase difference between the two elements is exactly 180° (i.e., $(\phi_2 = 180°)$), what is the allowable variation in the attenuation between the two signals that will still produce the required 20 dB of attenuation?

d. Repeat parts (a) and (b) for a 30 dB attenuation requirement.

5. **Jitter and Phasor Diagrams** We are using a 10 MHz reference signal to clock an analog-to-digital converter (ADC). The 10.000 MHz reference is contaminated with a second, smaller frequency component at 10.001 MHz. The equation of the composite signal $s(t)$ is

$$\begin{aligned} S_C(t) &= S_R(t) + S_{SP}(t) \\ &= \cos(2\pi f_1 t) + A_2 \cos(2\pi f_2 t) \end{aligned} \tag{2.128}$$

where

- $S_C(t)$ is the contaminated reference oscillator.
- $S_R(t)$ is the clean, ideal reference oscillator at 10.000 MHz. The amplitude of the clean oscillator is unity (1) and its frequency is $f_1 = 10.000$ MHz.
- $S_{SP}(t)$ is the spurious component that is polluting our reference oscillator. The frequency of this spurious component is 10.001 MHz. The amplitude of the spurious component is $A_2 = 1/50$.

a. Draw the phasor diagram using $S_R(t)$ as a reference. Show $S_R(t)$, $S_{SP}(t)$ and $S_C(t)$. You may have to draw the diagram not to scale to show all the components.

b. Using $S_R(t)$, the clean, ideal 10.000 MHz oscillator as the reference, find the maximum phase and amplitude deviation of the composite signal $S_C(t)$.

c. Examine the zero crossings of $S_C(t)$ with respect to those of $S_R(t)$. What's the maximum time difference between the zero crossings of the contaminated reference signal $S_C(t)$ and the pure, uncontaminated reference signal $S_R(t)$? In other words, what is the oscillator's timing jitter?

CHAPTER

3

Propagation

Why don't you knock it off with them negative waves?

Sgt. Oddball, Kelly's Heroes

Took a look down a westbound road, Right away I made my choice. Headed out to my big two-wheeler, I was tired of my own voice. Took a bead on the northern plains And just rolled that power on...

Bob Seeger, "Roll Me Away"

Chapter Outline

3.1 INTRODUCTION

The propagation characteristics between two geographically fixed sites will change over time; for no apparent reason. Digital eye patterns will open and close; bit-error rates will vary; frequency nulls will appear and disappear. We can often observe daily, monthly, and yearly variations in the propagation characteristics of a channel. In this chapter, we'll explain some of the causes of these effects and offer some solutions.

3.2 TYPES OF PROPAGATION

We divide propagation up into several types.

3.2.1 Propagation in Free Space

Free-space propagation implies that the dielectric constant (permittivity) and the permeability of the medium are ε_0 and μ_0, respectively. The best example of this type of propagation is a line-of-sight radio link between two spacecraft. There is nothing but vacuum between the transmitting and receiving antennas. This simple channel leads to a simple propagation model.

3.2.2 Propagation through a Homogeneous Medium

The phrase *homogeneous medium* means the dielectric constant, ε, and the permeability, μ, of the medium are both constant and independent of position. Nonturbulent air at a uniform temperature is a good, though impractical, example of a homogeneous medium. The earth's atmosphere is a nonhomogeneous medium.

Often, the only difference between propagation through free space and propagation through a homogeneous medium is that the homogeneous medium exhibits more signal loss than free space. This extra attenuation can be a function of frequency or time.

3.2.3 Propagation through a Nonhomogeneous Medium

The phrase *nonhomogeneous medium* means that the permittivity and permeability of the medium vary with position. In Earth's atmosphere, the μ and ε are functions of humidity, temperature, electrical charge, and pressure. Turbulent air is often nonhomogeneous. Nonhomogeneous media leads to diffraction, reflection, polarization rotation, and other effects.

3.2.4 Multipath Propagation

Multipath propagation is an extreme case of a nonhomogeneous medium, and we'll consider it separately. This type of propagation occurs when two or more signals from the same transmit antenna arrive at the receive antenna at different times. Multipath can occur when multiple unequal length paths exist between the transmitting and receiving antenna. This situation often occurs in cities or indoors where radio waves have many opportunities to reflect off objects in their path.

3.3 | PROPAGATION THROUGH FREE SPACE

The term *free space* implies an environment whose permittivity, ε, and permeability, μ, are both constant. The permittivity of free space is ε_0; the permeability of free space is μ_0 and

$$\begin{aligned}\varepsilon_0 &= 8.85E-12\frac{\text{Farad}}{\text{meter}}\\ \mu_0 &= 1.26E-6\frac{\text{Henry}}{\text{meter}}\end{aligned} \tag{3.1}$$

These values of ε_0 and μ_0 imply that, once radiated, the energy in the radio wave will continue unimpeded. Not even the vacuum of space completely meets this strict definition. However, the concept of free space proves to be an excellent starting point for understanding all types of real world propagation.

3.3.1 Free-Space Path Loss

Free-space path loss is a measurement of the decline in intensity of a transmitted radio wave due to spreading, not the attenuation of a medium. This spreading loss is the most fundamental form of loss incurred by an electromagnetic wave as it propagates. Consider a radio source dumping power into an isotropic antenna in free space. The transmitted energy would radiate out over an ever growing sphere. The surface area of that sphere depends on its distance from the center:

$$A_{Sphere} = 4\pi D^2 \tag{3.2}$$

where

D is the distance from the transmitting antenna

The power density at a distance D from the antenna is the power density. Figure 3-1 shows a radio source dumping power into an isotropic antenna. We will use a second isotropic antenna to receive the transmitting energy.

If the power delivered to the isotropic radiator is P_{TX}, then the power density per unit area at a distance D from the antenna is

$$P_{ISO} = \frac{P_{TX}}{4\pi D^2} \tag{3.3}$$

Figure 3-2 shows the same antenna as Figure 3-1 interpreted as a receiving antenna. Antenna theory tells us that an antenna will present an effective area or effective aperture to incoming energy. The effective aperture for an isotropic antenna is

$$A_{eff,ISO} = \frac{\lambda^2}{4\pi} \tag{3.4}$$

The power intercepted by the receiving antenna will be

$$\begin{aligned} P_{Rx} &= P_{ISO} A_{eff,ISO} \\ &= \left(\frac{P_{Tx}}{4\pi D^2}\right)\left(\frac{\lambda^2}{4\pi}\right) \\ &= P_{Tx}\left(\frac{\lambda}{4\pi D}\right)^2 \end{aligned} \tag{3.5}$$

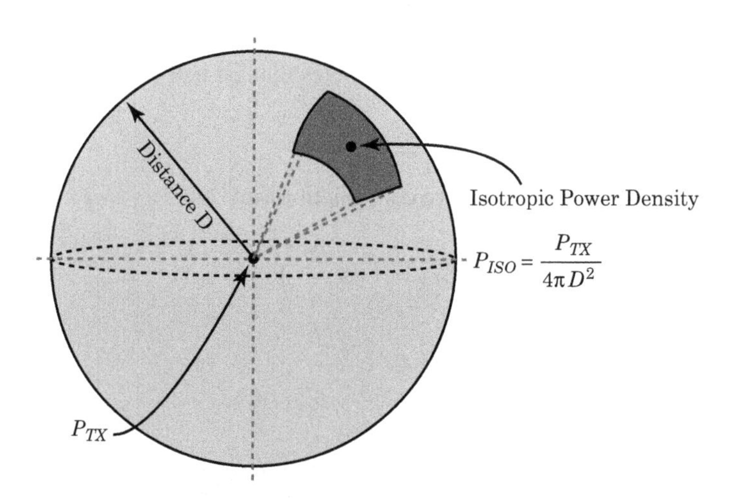

FIGURE 3-1 ■ A transmitting antenna receiving energy from a power source. The power density at a distance D from the antenna is derived using simple geometry.

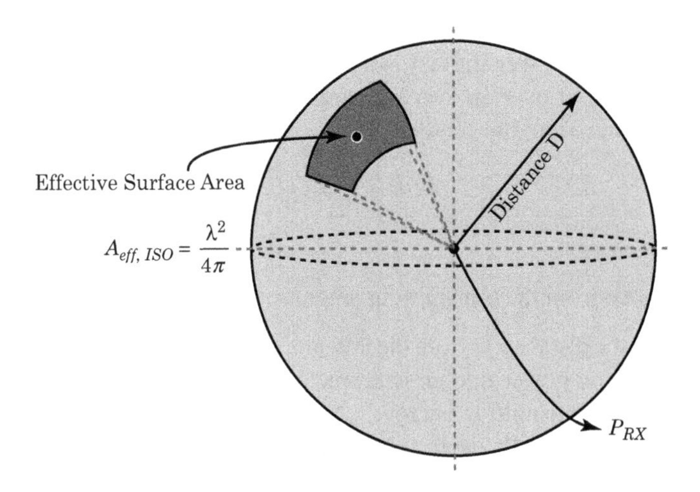

FIGURE 3-2 ■ A receiving antenna intercepting energy from the environment by presenting an effective area to the oncoming wavefront.

Free-space path loss, PL_{FS}, is the ratio of the transmitted power P_{Tx} to the received power P_{Rx} when the system contains only isotropic antennas and

$$\begin{aligned} PL_{FS} &= \frac{P_{Tx}}{P_{Rx}} \\ &= \left(4\pi \frac{D}{\lambda}\right)^2 \end{aligned} \tag{3.6}$$

where

D = the distance between the transmitting and receiving antennas
λ = the wavelength

Free-space path loss is also referred to as spreading loss because the loss mechanism is derived from spreading the energy out over the ever increasing surface area of a sphere.

3.3.2 The Free-Space Path Loss Equation

Equation (3.6) reveals that the free-space path loss is a function of the distance between the two antennas expressed in wavelengths. The path loss increases by a factor of 4 (or 6 dB) every time D/λ doubles. As a space probe travels from Earth to Jupiter, the path loss increases by only 6 dB as the probe completes the last half of its journey. Using

$$c = \lambda f \tag{3.7}$$

we can rewrite equation (3.6) in a more convenient form:

$$PL_{FS} = \left(4\pi \frac{Df}{c}\right)^2 \tag{3.8}$$

where

f = the operating frequency, in hertz
c = the speed of light (3E8 meters/second)
D = the distance between the two antennas, in meters

Converting to dB produces

$$PL_{FS} = 21.98 + 20\log\left(\frac{D}{\lambda}\right) = -147.6 + 20\log(Df) \quad (3.9)$$

EXAMPLE

Path Loss to the Second Sun/Earth Lagrange Point

A geostationary satellite is used to relay signals to Earth from a space observatory at the second sun/Earth LaGrange point. The second Lagrange point is $D_{L2} = 1.5\text{E}6$ km from the center of the earth on the side opposite of the sun. The geostationary satellite orbits directly over the equator at $D_{GEO} = 35{,}790$ km from the surface of the earth. The radius of the earth, R_{Earth}, is 6,375 km, both spacecraft are using isotropic antennas, and the link operates at 1 GHz.

1. A geostationary satellite has an orbit of 1 day. How long will the satellite be out of contact with the observatory per day?
2. Calculate the difference in dB between the minimum and maximum signal received by the satellite from the observatory.

Solution

Figure 3-3 shows the geometry of the satellites and of the earth. The satellite will pass through a shadow once per day, and that shadow is larger than the earth.

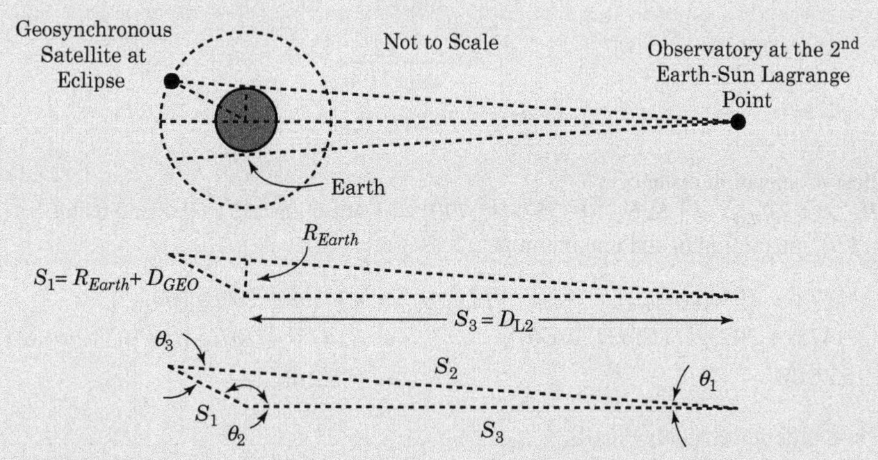

FIGURE 3-3 ■ The earth shadows the geostationary satellite from the observatory. The size of the shadow will be determined by the distance to the observatory, the diameter of the earth, and the orbit of the satellite.

We first find θ_1, the angular extent of the earth as viewed from the Lagrange point as

$$\begin{aligned}\tan(\theta_1) &= \frac{R_{Earth}}{D_{L2}} \\ &= \frac{6375}{1.5E6} \\ &= 4.25E-3 \\ \theta_1 &= 0.243^\circ\end{aligned} \quad (3.10)$$

We use the law of sines to find θ_2, the angle at which the geostationary satellite just goes into eclipse from the observatory. We know

$$\begin{aligned}\frac{S_1}{\sin(\theta_1)} &= \frac{S_2}{\sin(\theta_2)} = \frac{S_3}{\sin(\theta_3)} \\ \frac{D_{GEO}+R_{Earth}}{\sin(\theta_1)} &= \frac{S_2}{\sin(\theta_2)} = \frac{D_{L2}}{\sin(\theta_3)} \\ \frac{35790+6375}{\sin(0.243^\circ)} &= \frac{S_2}{\sin(\theta_2)} = \frac{1.5E6}{\sin(\theta_3)} \\ 9.94E6 &= \frac{S_2}{\sin(\theta_2)} = \frac{1.5E6}{\sin(\theta_3)}\end{aligned} \tag{3.11}$$

Equation (3.11) tells us $\theta_3 = 8.68^\circ$ and $\theta_2 = 180^\circ - (8.68^\circ + 0.243^\circ) = 171.1^\circ$. The angle θ_4 relates to half of the eclipse duration and $\theta_4 = 180^\circ - 171.1^\circ = 8.9^\circ$. The geosynchronous satellite is in eclipse for $2*8.9^\circ = 17.8^\circ$, which corresponds to

$$17.8^\circ \cdot \frac{24\,\text{hours}}{360^\circ} = 1.19\,\text{hours} = 71.2\,\text{minutes} \tag{3.12}$$

The geosynchronous satellite is eclipsed from the observatory for 71.2 minutes per day.

The differential path loss depends on the differential distance of the two satellites. The longest propagation distance is S_2 in Figure 3-3 and

$$\begin{aligned}9.94E6 &= \frac{S_2}{\sin(\theta_2)} \\ &= \frac{S_2}{\sin(171.1^\circ)} \\ S_2 &= 1.54E6\,\text{km}\end{aligned} \tag{3.13}$$

The smallest propagation distance is $D_{L2} - (R_{Earth} + D_{GEO}) = 1.5E6 - (6375 + 35{,}790) = 1.46E6$ km. At 1 GHz and using equation (3.9), the minimum and maximum path losses are

$$\begin{aligned}PL_{FS,Min} &= -147.6 + 20\log\left(fD_{\min}\right) & PL_{FS,Max} &= -147.6 + 20\log\left(fD_{\max}\right) \\ &= -147.6 + 20\log\left[(1E9)\,(1.46E6)\right] & &= -147.6 + 20\log\left[(1E9)\,(1.54E6)\right] \\ &= 155.6\,\text{dB} & &= 156.2\,\text{dB}\end{aligned} \tag{3.14}$$

The path loss difference is only 0.6 dB.

3.4 | PROPAGATION THROUGH A HOMOGENOUS MEDIUM

The free-space path loss calculates the loss of signal power due to geometrical spreading but does not include losses caused by the medium through which the signal is propagating. For example, a 60 GHz signal passing through the atmosphere will experience high attenuation because the oxygen in the atmosphere absorbs energy at that frequency. The signal will experience spreading loss and an excess loss caused by oxygen absorption.

3.4.1 The Atmosphere

Earth's atmosphere is made up mostly of nitrogen, oxygen, and water vapor. Figure 3-4 shows atmospheric absorption split into the two major components associated with oxygen and water vapor. Figure 3-4 shows the excess attenuation due to water and oxygen. The first attenuation peak occurs at 22 GHz and is due to water, while the second attenuation peak is due to oxygen and occurs at 63 GHz. For short-range communication systems, working at 63 GHz prevents the signal from propagating beyond its useful range (and may aid in hiding the communications system).

Figure 3-5 shows the excess attenuation due to precipitation. The raindrops can either absorb or scatter the RF energy.

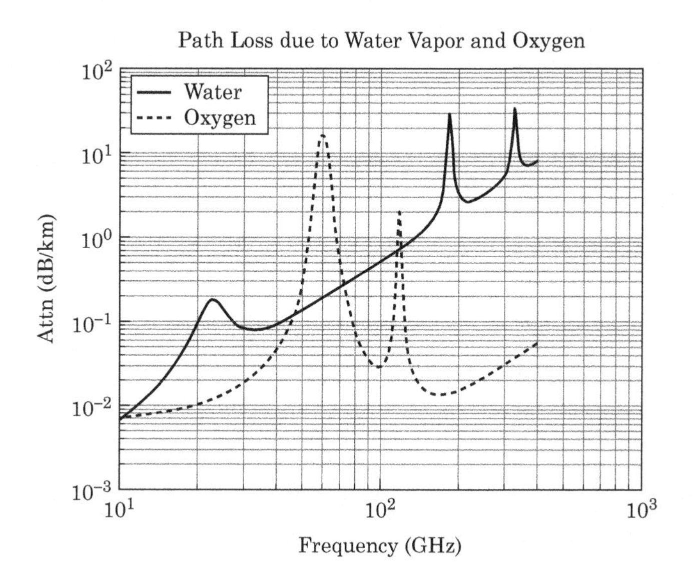

FIGURE 3-4 ■ Radio frequency (RF) signals may be severely attenuated by water vapor and oxygen, depending on their frequency. The attenuation due to water vapor was measured at a standard value of humidity.

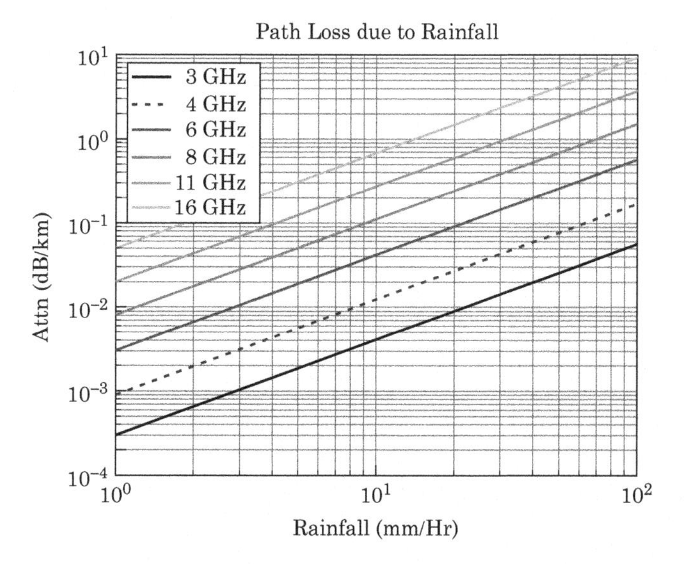

FIGURE 3-5 ■ The effect of rainfall on signal propagation.

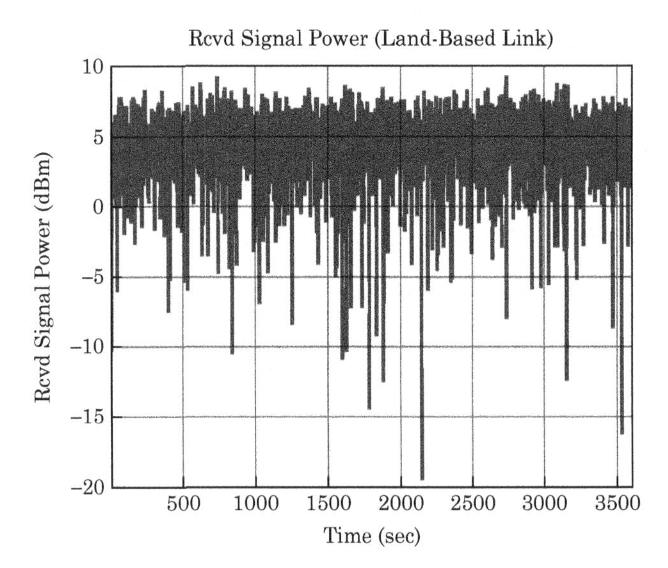

FIGURE 3-6 Signal fading over a land-based radio link observed over an hour. The worst-case excess loss is greater than 30 dB for some periods.

3.4.2 Excess Loss

The total attenuation for a signal is

$$Attn_{Total,dB} = Attn_{FS,dB} + Attn_{H_2O,dB} + Attn_{O_2,dB} + Attn_{Rain,dB} \tag{3.15}$$

3.4.3 Time Varying Excess Loss

Propagation loss is not constant with time. Figure 3-6 shows the received signal strength over time for a system operating over land. The data were taken at 1 second intervals for an hour. The worst-case excess loss is greater than 30 dB.

We often cannot predict the loss of a channel at a particular instant, but we can describe the channel statistically. Our links will suffer high-excess loss for short periods of time. This varying signal power has a large effect on the design of our receiver's gain control stages.

3.4.4 Flat Fading

Excess loss is one type of *flat fading*. Flat fading causes a wideband decrease in received signal power. As Figure 3-7 shows, the entire signal drops uniformly. Rain, snow, dust, or fog (which we model as homogenous but lossy) can cause this. Flat fading can also be

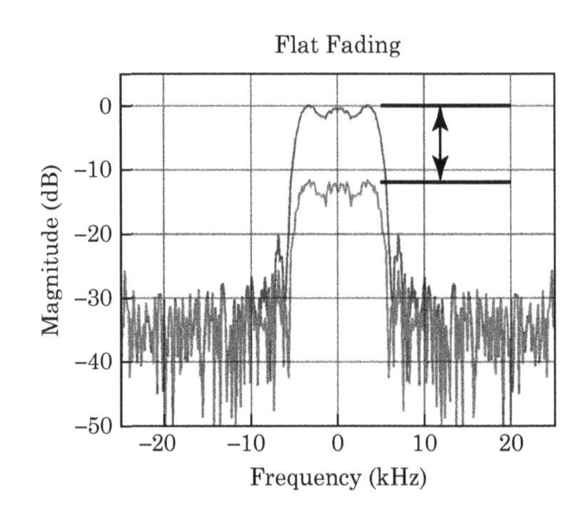

FIGURE 3-7 Flat fading is not frequency selective. The entire signal fades.

caused by refraction through nonhomogeneous media. For example, when the air separates into distinct layers with different temperatures and humidity characteristics, the signal can bend so it no longer is traveling the original path it was on when it left the antenna.

The speed at which the received signal power changes affects the design of the receiver's automatic gain control. Ideally, the receiver should be capable of changing its gain fast enough to keep the signal presented to the demodulator at a constant power.

3.5 | PROPAGATION THROUGH A NONHOMOGENOUS MEDIUM

We have considered the effects of a wave traveling through a homogeneous medium. We modified the exponent of the path loss equation to account for traveling through nonfree space as well as for irregular excess loss that can come from many sources. We found that an easy way to consider the excess loss from nonhomogeneous phenomena is to consider them as a whole and calculate their effect as an additional homogeneous medium loss. This method of dealing with nonhomogeneous mediums works well when considering loss, but loss is not the only effect that a medium can have on a radio wave. Two major effects that we will discuss are *refraction* and *reflection*.

3.5.1 Refraction

Refraction is the bending of an electromagnetic wave due to the changing radio characteristics of the medium, which are determined by the relative permittivity ε_r and relative permeability μ_r. We quantify the propagation characteristic as the index of refraction. The index of refraction of a medium is the ratio of the propagation velocity in free space to the propagation velocity in the medium or

$$\begin{aligned} n &= \frac{c}{v} \\ &= \sqrt{\varepsilon_r \mu_r} \end{aligned} \tag{3.16}$$

where

c = the propagation velocity in free space
v = the propagation velocity in the material
ε_r = the relative permittivity of the material
μ_r = the relative permeability of the material

As a wave passes through a medium whose refractive index changes with position, the wave will bend toward the area of greater refractive index. This effect is responsible for some types of multipath interference and an increase in the distance to the radio horizon.

3.5.2 *K*-Factor

At sea level and at very high frequencies (VHF) and ultrahigh frequencies (UHF), the refractive index of the atmosphere is about 1.0003, and it decreases with height. This decrease in refractive index with height causes a propagating wave to bend away from the earth, as shown in Figure 3-8. The effect is to make the radio horizon farther than the optical horizon, so the signal travels farther than you might think it would (see Figure 3-8).

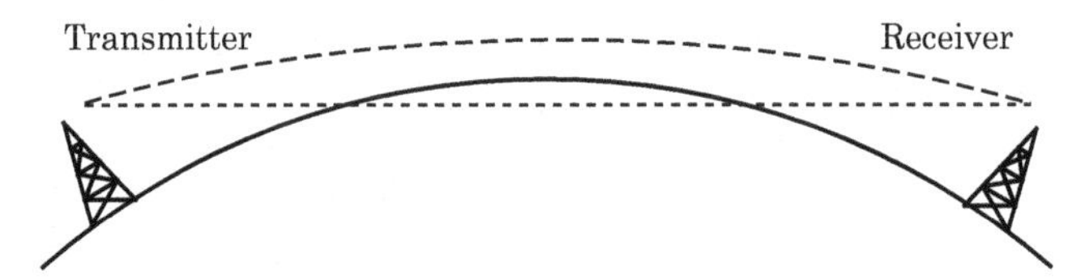

FIGURE 3-8 ■ Changes in the atmosphere's refractive index cause the radio horizon to appear farther than the optical horizon.

For planning purposes, we model this effect by assuming that the ray does travel in a straight line but that the earth is actually a different size. The K-factor defines the relative size of the earth

$$K \approx \frac{r}{r_0} \tag{3.17}$$

where

r = the radius of the earth assuming the ray traveled in a straight line. This is also the radius of curvature of the propagating ray.

r_0 = the physical radius of the earth.

Many designers use a $4/3$ Earth radius propagation model.

3.5.3 Reflection

Reflection occurs at the abrupt junction of two mediums with differing indexes of refraction. Reflections can be either specular or diffuse. A specular reflection is one where the incident wave is reflected intact. A diffuse reflection occurs when the incident wave is broken apart and the resulting wave is highly scattered.

Whether a reflection is specular or diffuse depends greatly on the smoothness of the surface in relation to the wavelength. There is a large gray line between specular and diffuse reflecting surfaces. As a rule of thumb we'll say that a smooth surface is one that has an average departure from flatness of less than $1/4$ of the wavelength. This surface will (in most cases) result in a specular reflection.

3.5.4 Multipath

If the refractive index of the medium is distributed in a complex manner along the propagation path, then two or more waves can make their way from the transmitting antenna to the receiving antenna (see Figure 3-9). As a result, there can be destructive interference if the sum of the phase difference between the paths is an odd multiple of 180° and the magnitudes of the received rays are about the same.

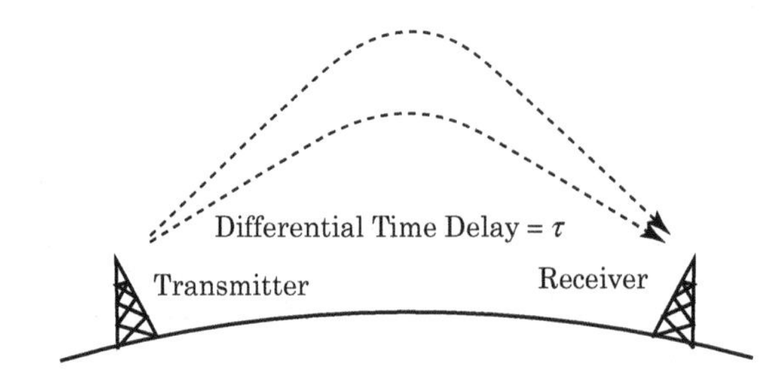

FIGURE 3-9 ■ A nonhomogeneous medium can cause a signal to take two different paths from the transmitting antenna to the receiving antenna.

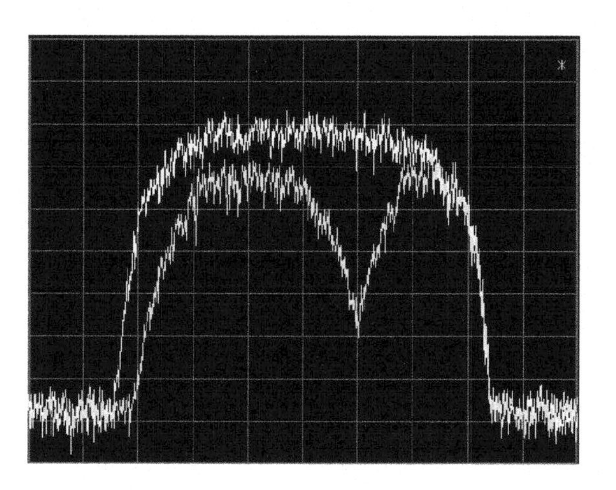

FIGURE 3-10 ■ The spectrum of a signal undergoing frequency-selective multipath. The original signal is spectrally flat with no obvious areas of high attenuation. The multipath signal exhibits areas of high attenuation and low signal power.

Multipath causes *frequency-selective fading* because, since the effect depends on the phase difference between waves, it is a frequency-dependent phenomenon. Figure 3-10 shows a signal undergoing frequency-selective fading.

3.6 | MULTIPATH PROPAGATION

A radio signal spreads out in different directions as it radiates away from the broadcast antenna. Parts of the spreading wave will encounter reflecting surfaces, and the wave will scatter off these objects. In an urban environment, the wave might reflect off buildings, moving trains, or airplanes.

Multipath occurs when a signal takes two or more paths from the transmitting antenna to the receiving antenna. See figure 3-11 for an illustration of the effect. We'll assume that one signal, the *direct ray*, travels directly from the transmitter to the receiver. The direct ray is usually (but not always) the strongest signal present in the receiving antenna.

The other signals (or *rays*) arrive at the receiving antenna via more roundabout paths. These reflected signals eventually find their way to the receiving antenna. In our analysis, we'll assume these indirect rays arrive after the direct ray and that the indirect rays are weaker in power than the direct rays.

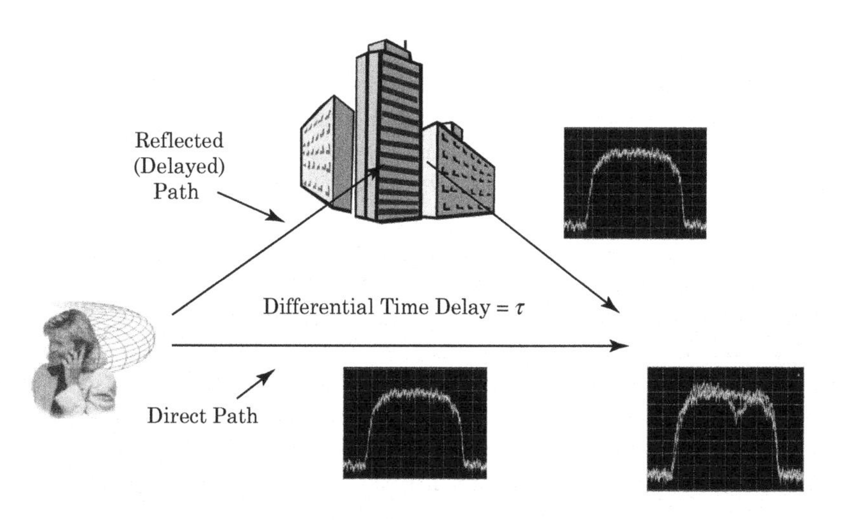

FIGURE 3-11 ■ In a severe multipath environment, a signal might bounce off several different reflectors before it arrives at a receiver. The result can be severe distortion.

The equation for the direct signal is

$$V_D(t) = \cos(\omega_0 t) \tag{3.18}$$

where

$\omega_0 = 2\pi f_0 =$ the angular operating frequency

We'll assume the amplitude of the direct ray is unity and the phase angle is zero. The equations for the reflected rays are

$$\begin{aligned} V_{R,1} &= \rho_1 \cos\left[\omega_0 \left(t - \tau_1\right)\right] \\ V_{R,2} &= \rho_2 \cos\left[\omega_0 \left(t - \tau_2\right)\right] \\ &\vdots \\ V_{R,n} &= \rho_n \cos\left[\omega_0 \left(t - \tau_n\right)\right] \end{aligned} \tag{3.19}$$

where

$\rho_n =$ a real number describing the difference in amplitude between the direct ray and the n-th reflected ray

$\tau_n =$ the time difference of arrival between the direct and the n-th reflected ray. The n-th ray arrives τ_n seconds after the direct ray.

We may find it convenient to rewrite the equations of (3.19) as

$$\begin{aligned} V_{R,1} &= \rho_1 \cos\left(\omega_0 t - \phi_1\right) \\ V_{R,2} &= \rho_2 \cos\left(\omega_0 t - \phi_2\right) \\ &\vdots \\ V_{R,n} &= \rho_n \cos\left(\omega_0 t - \phi_n\right) \end{aligned} \tag{3.20}$$

where

$$\begin{aligned} \phi_1 &= -\omega_0 \tau_1 \\ \phi_2 &= -\omega_0 \tau_2 \\ &\vdots \\ \phi_n &= -\omega_0 \tau_n \end{aligned} \tag{3.21}$$

The scattering mechanism can be more complex than this simple description. Often, a wave will experience a phase shift, polarization change, or some other change when it encounters a scattering surface. For our analysis, we assume that the direct and reflected waves have different amplitudes (described by ρ_n) and that the waves arrive at the receiving antenna at slightly different times (described by τ_n).

The characteristics of multipath channels usually change over time because the geometry of the channel changes. Consider a wireless phone operating in an office environment. The propagation environment contains all sorts of reflectors (e.g., file cabinets, desks, doors, Venetian blinds). As the signal encounters each of these objects, a reflection occurs. The net channel characteristics arise from the sum of all these individual channels. For example, when the user leans back in his chair or begins pacing excitedly during a difficult conversation, the electrical lengths of all the paths change simultaneously. This changes both ρ_n and τ_n for every channel. Even if the user remains perfectly still, the channel will change as people pull out file cabinet drawers or open and close doors. Small differences in the arrival time of the different signal can make a big difference in the received signal quality. Since the propagation characteristics are continually changing, we will eventually describe both ρ_n and τ_n of equation (3.19) in statistical terms.

3.6.1 Two-Ray Analysis

The two-ray multipath analysis assumes that only two rays are present at the receive antenna: the direct ray $V_D(t)$ and a single reflected ray $V_{R,1}(t)$. The two signals present at the receive antenna are

$$V_D(t) = \cos(\omega_0 t) \tag{3.22}$$

and

$$V_{R,1} = \rho_1 \cos\left[\omega_0(t - \tau_1)\right] \tag{3.23}$$

We can rewrite equation (3.23) as

$$\begin{aligned} V_{R,1} &= \rho_1 \cos\left[\omega_0 t - \omega_0 \tau_1\right] \\ &= \rho_1 \cos\left[\omega_0 t + \phi_1\right] \end{aligned} \tag{3.24}$$

where

$$\phi_1 = -\omega_0 \tau_1 \tag{3.25}$$

Our random variables are ρ_1 and ϕ_1. The complete signal present at the receive antenna is

$$\begin{aligned} V_{Rx}(t) &= V_D(t) + V_{R,1}(t) \\ &= \cos(\omega_0 t) + \rho_1 \cos(\omega_0 t + \phi_1) \end{aligned} \tag{3.26}$$

We'll experience a multipath fading event when the received signal power falls below some arbitrary threshold. Analysis of equation (3.26) reveals

$$V_{Rx}(t) = 0 \text{ whenever} \begin{cases} \rho_1 = 1 \\ \quad \text{and} \\ \phi_1 = 180^\circ \end{cases} \tag{3.27}$$

Both of the ρ_1 and ϕ_1 conditions must be true to observe a fading event. Most of our analytical effort will be spent examining equation (3.26) when ρ_1 is in the neighborhood of unity and ϕ_1 is in the neighborhood of 180°.

As Figure 3-12 shows, we can interpret equation (3.26) as the addition of two phasors. The direct ray is represented by a phasor of unity length at 0°, whereas the indirect ray is

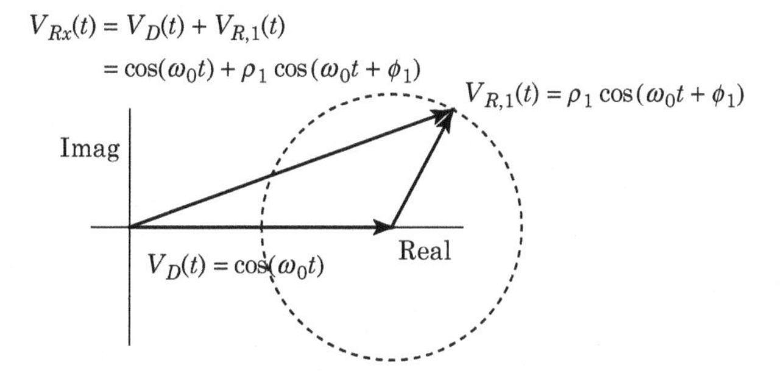

FIGURE 3-12 ▪ The geometric phasor interpretation of equation (3.26). The direct ray has unity magnitude and zero phase. The reflected ray has a magnitude of ρ_1 and a phase angle of ϕ_1.

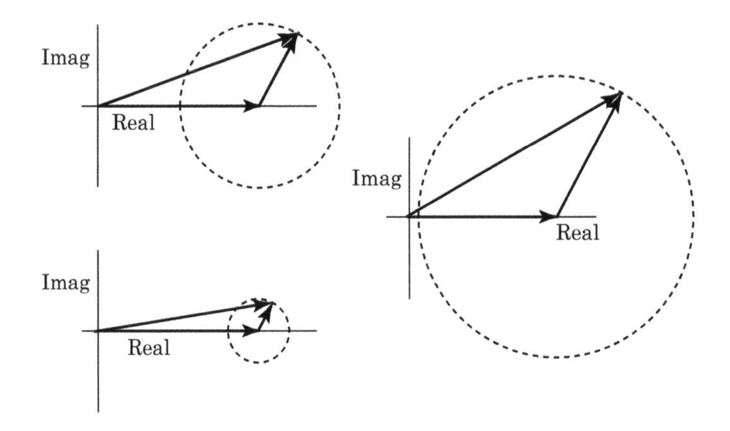

FIGURE 3-13 The phasor interpretation of equation (3.26) for various values of ρ_1.

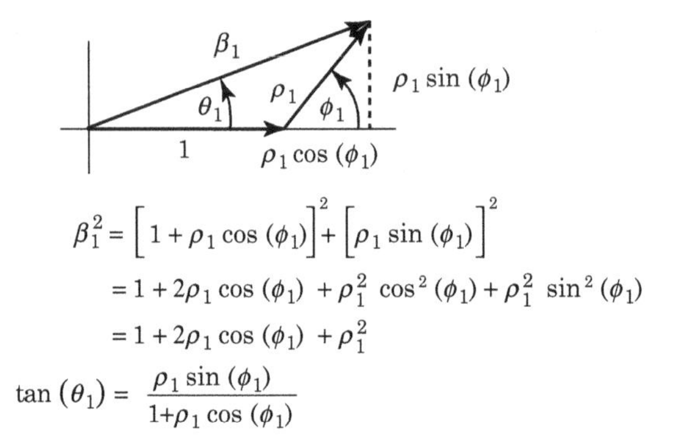

FIGURE 3-14 Phasor analysis of two-ray multipath.

represented by a phasor of length ρ_1 at an angle of ϕ_1. Figure 3-12 also shows the locus of the resultant vector as the phase of the indirect ray varies over 0° to 360°.

Figure 3-13 shows the phasor diagrams for several values of ρ_1. The resultant vector $V_{Rx}(t)$ almost equals the direct ray when ρ_1 is very small. As a result, the receiver sees a nearly undistorted signal. As ρ_1 approaches unity, the phase and amplitude of $V_{Rx}(t)$ change wildly with the phase of the indirect ray. Figure 3-14 formalizes the variables of Figure 3-12 and Figure 3-13.

Vector analysis of Figure 3-14 allows us to rewrite equation (3.26) as a single cosine with an amplitude and phase change

$$\begin{aligned} V_{Rx}(t) &= \cos(\omega_0 t) + \rho_1 \cos(\omega_0 t + \phi_1) \\ &= \beta_1 \cos(\omega_0 t + \theta_1) \end{aligned} \tag{3.28}$$

where

$$\begin{aligned} \beta_1^2 &= 1 + 2\rho_1 \cos(\phi_1) + \rho_1^2 \\ &\text{and} \\ \theta_1 &= \tan^{-1}\left(\frac{\rho_1 \sin(\phi_1)}{1 + \rho_1 \cos(\phi_1)}\right) \end{aligned} \tag{3.29}$$

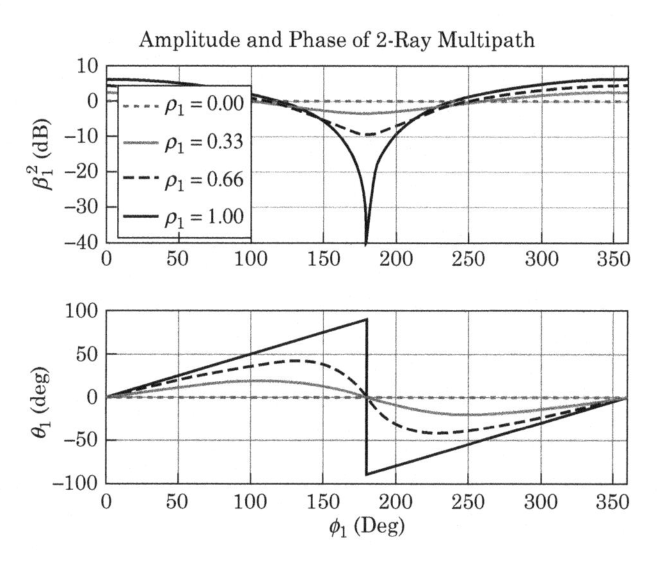

FIGURE 3-15 ■ The phasor interpretation of equation (3.28) for various values of ρ_1.

The ratio of the power in $V_{Rx}(t)$ to the power in the original cosine wave is β_1^2. The change in phase between the original cosine wave and $V_{Rx}(t)$ is θ_1.

Figure 3-15 shows the power and phase shift of the received signal V_{Rx}(t) with respect to the direct ray for various values of ρ_1 and ϕ_1. As expected, the worst signal attenuation occurs when ρ_1 is unity and ϕ_1 is 180°. The received signal is completely canceled under these conditions. The nulls are *deep* and *sharp* when ρ_1 is close to unity. A small change in θ_1 can result in a large change in received signal strength. We see an increase in received signal strength whenever

$$-90^\circ \leq \phi_1 \leq 90^\circ \tag{3.30}$$

The maximum increase is only 6 dB (when ρ_1 is unity and ϕ_1 is 0°). Unfortunately, the same ρ_1 that produces the maximum increase in signal power also produces the worst attenuation. The only difference is the phase ϕ_1.

Multipath affects the phase of the received signal as well as its amplitude. When ρ_1 is in the neighborhood of unity, the received phase changes abruptly when ϕ_1 passes through 180°.

As you might expect, the sudden phase change plays havoc with phase-encoded signals.

EXAMPLE

Multipath Fading

You've probably experienced multipath fading without even knowing it. You pull your car up to a stoplight and the FM radio fades out, but you find you can bring the signal back in by scooting your car forward a few feet.

When the FM signal died, two (or more) signals from the transmitter were arriving at your car exactly out of phase, and they canceled. Moving the car changed the phase difference between the arriving signals by a small amount, and, as Figure 3-15 shows, that can result in a large change in the received signal strength.

EXAMPLE

Multipath Fading

The author once lived along the flight path to an airport. Generally, the television reception was excellent, except when an airplane flew along one particular flight path. The television reception faded in and out as the airplane moved. The rate of change was slow at first, taking perhaps 3 seconds to change from perfect to a completely unwatchable.

As the airplane continued along its approach toward the airport, the rate of change of the signal strength increased until the picture faded in and out several times a second. Finally, as the geometry continued to change, the television reception cycle time gradually decreased back to its 3 second fade-in–fade-out behavior. This entire interference event lasted perhaps 20 seconds from start to finish.

A commercial airliner is an excellent signal reflector, so the signal reflected from the airplane was about as strong as the signal directly from the television transmitter (ρ_1 was about equal to unity). As the airplane moved along its flight path, the changing geometry caused the signals arriving at the receive antenna to add up alternately in phase and out of phase.

3.6.2 Frequency Response

We can describe the effects of multipath in the frequency domain. Equations (3.28) and (3.29) describe the two signals present at our receiving antenna. We're interested in the relationship among $f_0, \tau_1, \theta_1, \beta_1^2$, and ϕ_1. Rewriting equation (3.29) in terms of differential time delay, τ_1, and frequency of operation, ω_0, produces

$$\beta_1^2(\omega_0) = 1 + 2\rho_1 \cos(\omega_0\tau_1) + \rho_1^2$$
$$\text{and} \qquad (3.31)$$
$$\theta_1(\omega_0) = \tan^{-1}\left[\frac{-\rho_1 \sin(\omega_0\tau_1)}{1 + \rho_1 \cos(\omega_0\tau_1)}\right]$$

where τ_1 and ρ_1 are fixed, and ω_0 is the variable. Figure 3-16 shows the amplitude characteristics of multipath in the frequency domain. These graphs show the frequency response of a channel consisting of a direct ray and a reflected ray delayed by 0.2 nsec, 0.5 nsec, 1.0 nsec, and 2.0 nsec. This figure makes it obvious why this effect is sometimes called *frequency-selective multipath.* Figure 3-17 shows the phase response of the multipath channel.

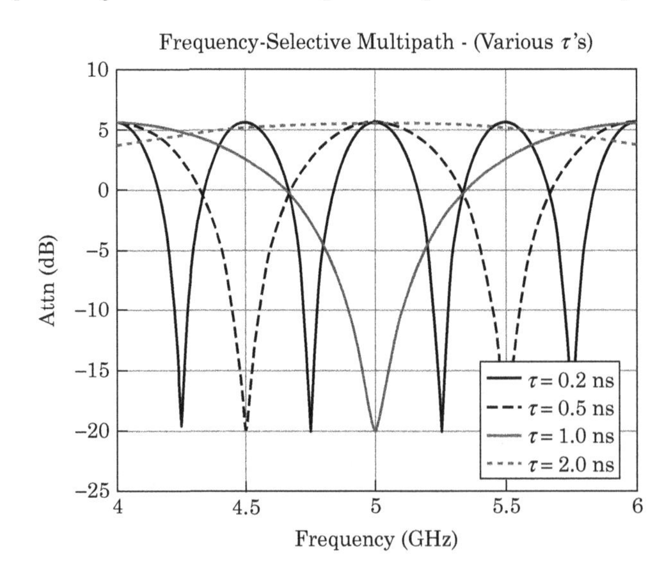

FIGURE 3-16 Two-ray multipath interpreted as a filter. The amplitude response of a multipath channel for various values of differential path delay τ.

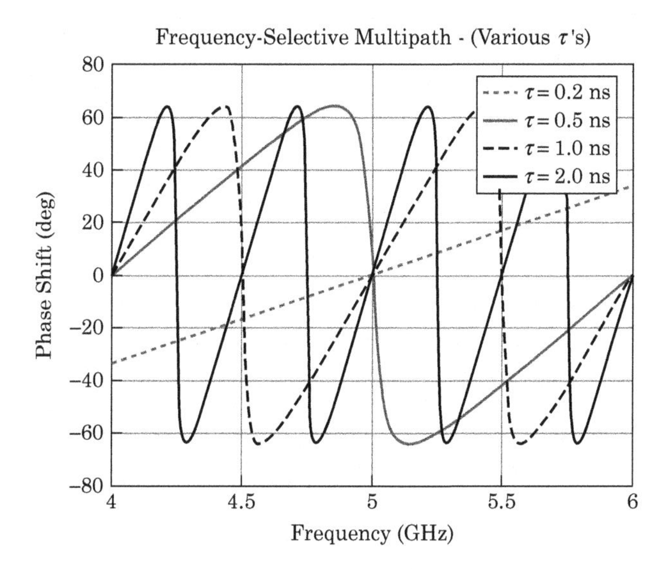

FIGURE 3-17 Two-ray multipath interpreted as a filter. The amplitude response of a multipath channel for various values of differential path delay τ.

Light travels about 1 foot per nsec, so the differential path delays of Figure 3-16 represent differential path lengths of only 2 feet in the worst case.

This simple, two-ray multipath channel acts like a *comb* filter. That is, the channel exhibits regularly spaced intervals of high attenuation, making the transfer function resemble the teeth of a comb. The frequencies of the nulls are

$$f_{null} = \frac{2n+1}{2\tau_1}, \qquad n = 0, 1, 2 \ldots \tag{3.32}$$

Channels with large differential time delays (i.e., large values of τ) produce nulls that are very close together in frequency. Large differential delay produces closely spaced nulls in the frequency response.

In high-frequency (HF) communications, we often bounce the signal off the ionosphere to achieve over-the-horizon communications. The signal can take several paths from the transmitter to the receiver, and the result is that the differential time delay can be on the order of milliseconds, which produces notches separated by only hundreds of hertz. As ρ_1 approaches unity, the more widely the channel characteristics vary with changing time delay.

EXAMPLE

Frequency of the Mulitpath Notches

Derive equation (3.32).

Solution

Equation (3.25) relates the differential time delay τ_1 to the phase between the direct and reflected rays. We will experience a null whenever ϕ_1 is an odd multiple of π so we can write

$$\begin{aligned} \phi_1 &= -\omega_0 \tau_1 \\ (2n+1)\pi &= -2\pi f_{null} \tau_1 \\ f_{null} &= \frac{2n+1}{2\tau_1} \end{aligned} \tag{3.33}$$

3.6.3 Three-Ray Analysis

Let's extend our two-ray analysis of multipath into a three-ray model. The direct ray is given by

$$V_D(t) = \cos(\omega_0 t) \tag{3.34}$$

and the two reflected rays are

$$\begin{gathered} V_{R,1} = \rho_1 \cos\left[\omega_0(t - \tau_1)\right] \\ \text{and} \\ V_{R,2} = \rho_2 \cos\left[\omega_0(t - \tau_2)\right] \end{gathered} \tag{3.35}$$

The complete signal, as seen by the receiving antenna, is

$$\begin{aligned} V_{Rx}(t) &= \cos(\omega_0 t) + \rho_1 \cos\left[\omega_0(t - \tau_1)\right] + \rho_2 \cos\left[\omega_0(t - \tau_2)\right] \\ &= \cos(\omega_0 t) + \rho_1 \cos\left[\omega_0 t + \phi_1\right] + \rho_2 \cos\left[\omega_0 t + \phi_2\right] \end{aligned} \tag{3.36}$$

where

$$\begin{gathered} \phi_1 = -\omega_0 \tau_1 \\ \text{and} \\ \phi_2 = -\omega_0 \tau_2 \end{gathered} \tag{3.37}$$

As in the two-ray analysis, we can rewrite equation (3.36) as

$$V_{Rx}(t) = \beta_2 \cos(\omega_0 t + \theta_2) \tag{3.38}$$

where

$$\begin{aligned} \beta_2^2 = 1 + \rho_1^2 + \rho_2^2 &+ 2\rho_1 \cos(\phi_1) + 2\rho_2 \cos(\phi_2) \\ &+ 2\rho_1\rho_2[\cos(\phi_1)\cos(\phi_2) + \sin(\phi_1)\sin(\phi_2)] \end{aligned} \tag{3.39}$$

and

$$\theta_2 = \tan^{-1}\left[\frac{\rho_1 \sin(\phi_1) + \rho_2 \sin(\phi_2)}{1 + \rho_1 \cos(\phi_1) + \rho_2 \cos(\phi_2)}\right] \tag{3.40}$$

The general analysis of the three-ray problem is not as useful as the two-ray case, but specific examples are helpful. Figure 3-18 shows the amplitude and phase responses for a three-ray channel with the arbitrary values $\rho_1 = 0.85$, $\tau_1 = 0.95$ nsec, $\rho_2 = 0.50$, and $\tau_2 = 2.2$ nsec. Even this simple three-ray model produces many amplitude and phase perturbations, indicating the complexity of a more complicated environment.

3.6.4 *n*-Ray Analysis

We can easily extend the three-ray multipath model to an n-ray model. We assume that n separate signals arrive at the receiving antenna, each with its own ρ and τ. We will experience a multipath fading event whenever the sum of the direct signal and the n reflected signals add to zero.

3.6.4.1 Indoor Multipath

The n-ray model agrees closely with experimentally observed propagation over indoor channels. The receiving antenna collects many different rays from different directions, and most of the received signals exhibit about the same magnitude. These conditions combine to produce a rapidly fluctuating RF environment. Theory and measurements reveal that, under these conditions, the probability density function of the fade statistics is

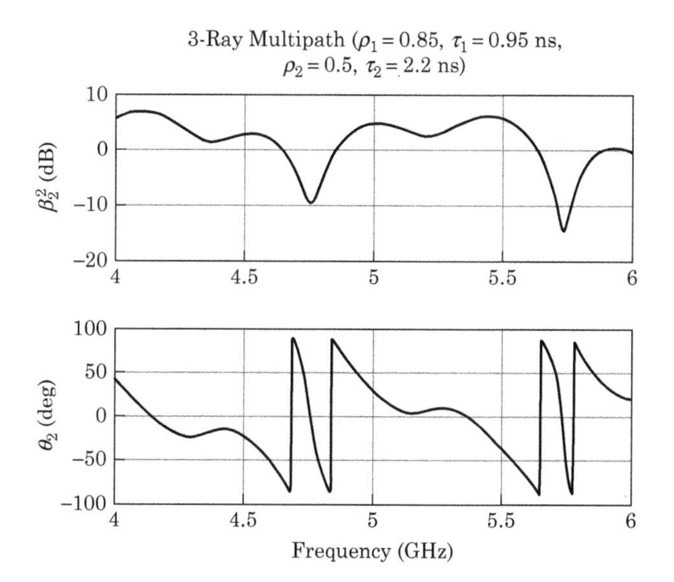

FIGURE 3-18 ■ Three-ray multipath interpreted as a filter. This plot shows the amplitude and phase responses of a three-ray multipath channel for $\rho_1 = 0.85$, $\tau_1 = 0.95$ ns, $\rho_2 = 0.50$, and $\tau_2 = 2.2$ ns.

Ricean. The properties of a Ricean distribution cause the received signal strength to vary widely and rapidly for a large percentage of the time.

3.6.5 Spatial Redistribution of Energy

If there was only one ray present in the environment, the energy available in that ray would be equally distributed over three-dimensional space (three space), and the received signal strength would not depend on the physical location of the antenna. Strong multipath propagation has the effect of redistributing the signal energy over three space. Figure 3-19 through Figure 3-22 show how multipath affects the physical distribution of received energy.

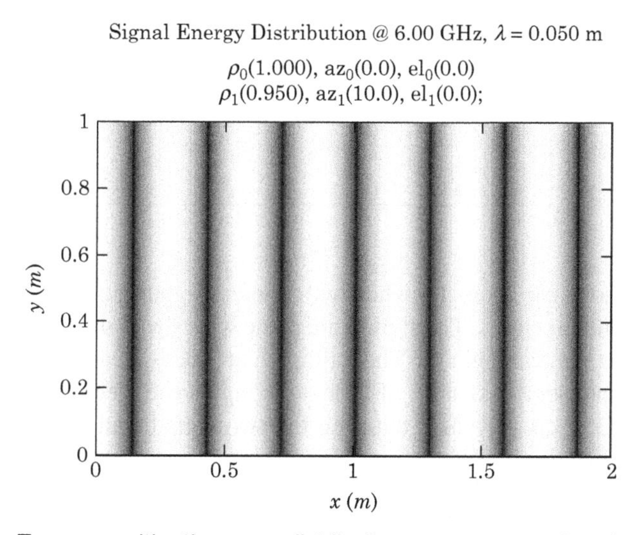

FIGURE 3-19 ■ Two-ray multipath energy distribution across an aperture (or wall). The direct ray ($\rho_0 = 1$) is arriving perpendicular to the wall ($az_0 = 0°$, $el_0 = 0°$), while the single indirect ray ($\rho_1 = 0.95$) is arriving at an azimuth of 10° and an elevation of 0° ($az_1 = 10°$, $el_1 = 0°$). Signal maxima are shown in white, and signal minima are shown as black. The distance between maxima is determined by the frequency (6 GHz here) and the angle of arrival of the indirect ray.

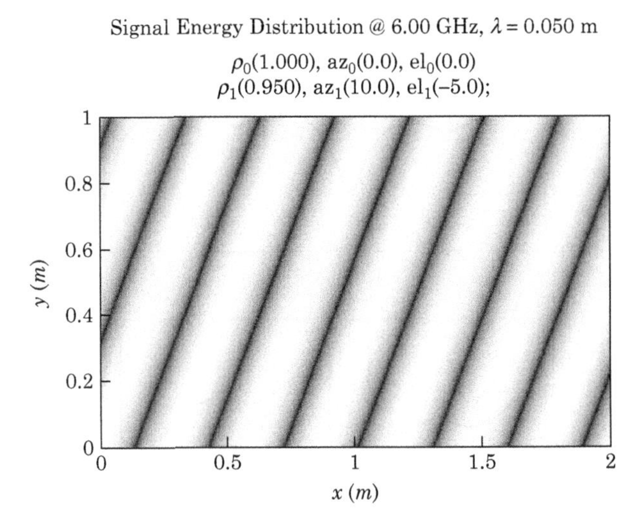

FIGURE 3-20 ■ Two-ray multipath energy distribution across an aperture. This is the same multipath environment as shown in Figure 3-19 except that the indirect ray is arriving from an azimuth of 10° and an elevation of −5° ($az_1 = 10°, el_1 = -5°$). The vertical minima of Figure 3-19 have changed from perfectly vertical to sloped. Bright areas indicate higher signal power densities.

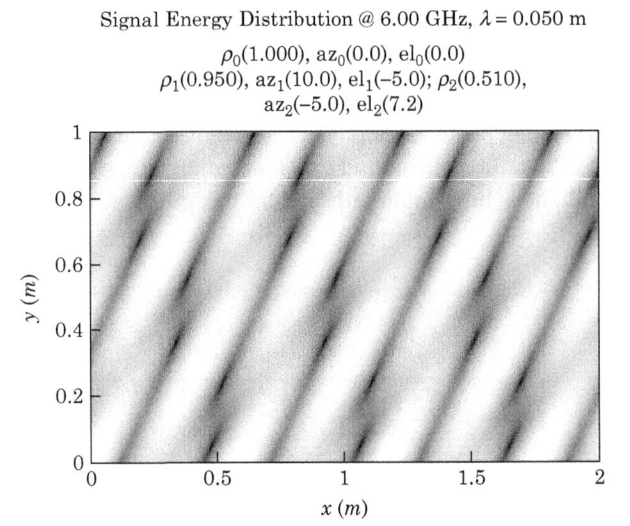

FIGURE 3-21 ■ Three-ray multipath energy distribution across an aperture. This is the same multipath environment as shown in Figure 3-20 with the addition of a second indirect multipath component at ($\rho_2 = 0.51, az_2 = -5°$ and $el_2 = 7.2°$). Bright areas indicate higher signal power densities.

Figure 3-23 shows the geometry when two coherent signals arrive at the same physical space from different directions. The signals are plane waves with different path delays, and they have about the same magnitude. The multipath produces areas of high signal strength (where two lines intersect) and areas of little or no signal (between the signal maxima).

In Figure 3-23, θ_a is the angle between the two signals, and the distance between maxima (and the distance between minima) is

$$\begin{aligned} d_{Max,Max} &= d_{Min,Min} \\ &= \lambda \frac{\cos\left(\dfrac{\theta_a}{2}\right)}{\sin(\theta_a)} \\ &\approx \frac{\lambda}{\theta_a} \text{ for small } \theta_a \end{aligned} \tag{3.41}$$

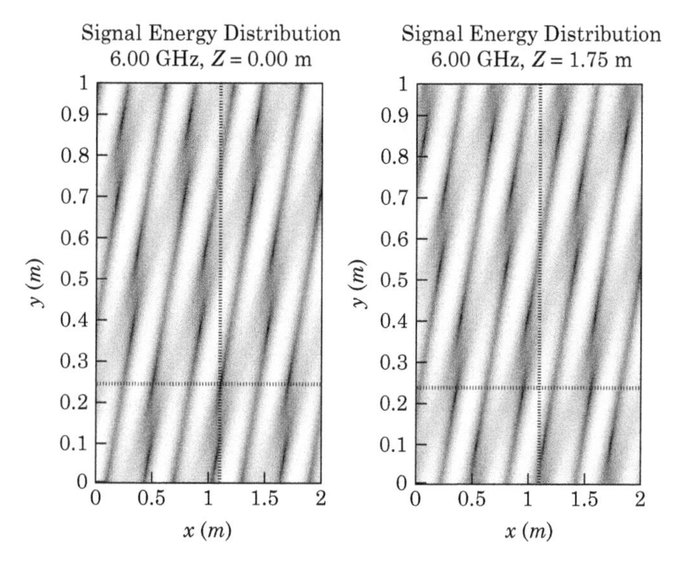

FIGURE 3-22 ■ The three-ray multipath distribution of Figure 3-21 viewed from two different positions. The left figure shows the energy distribution directly at the wall. The right figure shows the energy distribution measured 1.75 m behind the wall. Note that the energy null at (0.25 m, 1.1 m) in the left figure is an energy maximum in the right figure. Bright areas indicate higher signal power densities.

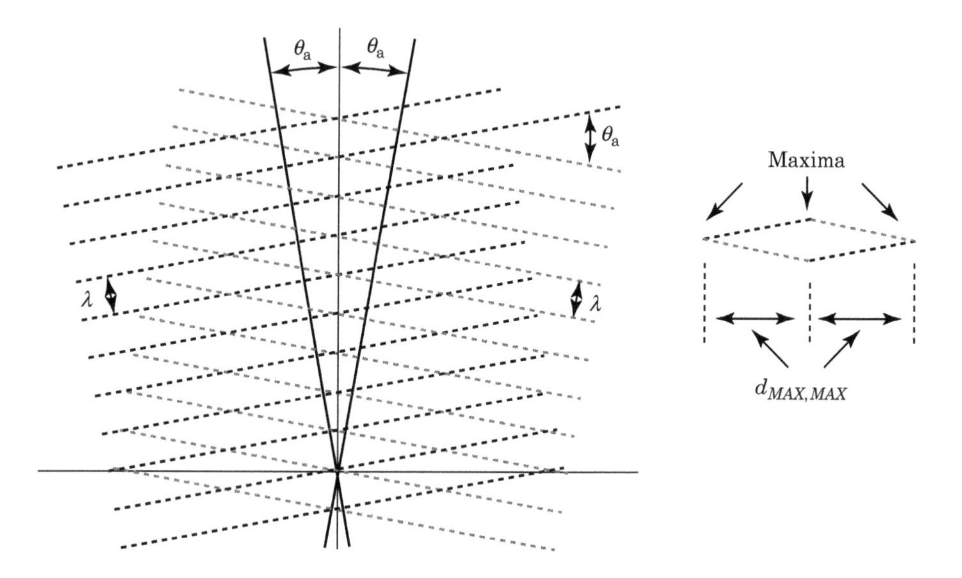

FIGURE 3-23 ■ The geometry of a two-ray multipath environment. The lines represent the signal peaks and are separated by the wavelength λ.

The distance the maxima and minima is half the distance between maxima, so we can write

$$\begin{aligned} d_{Max,Min} &= \frac{d_{Max,Max}}{2} \\ &= \frac{\lambda}{2} \frac{\cos\left(\frac{\theta_a}{2}\right)}{\sin(\theta_a)} \\ &\approx \frac{\lambda}{2\theta_a} \text{ for small } \theta_a \end{aligned} \tag{3.42}$$

EXAMPLE

Multipath Maxima and Minima

Assume that two multipath waves arrive with a θ_a of 3°. If the operating frequency is 4.5 GHz, find the distance between the multipath maxima.

Solution

The wavelength at 4.5 GHz is 0.0667 meters. The angle of arrival is 3° = 0.0524 radians. Using equation (3.41), we can write

$$\begin{aligned} d_{Max,Max} &= \lambda \frac{\cos\left(\frac{\theta_a}{2}\right)}{\sin(\theta_a)} \\ &= (0.0667)\frac{\cos\left(\frac{3^\circ}{2}\right)}{\sin(3^\circ)} \\ &= 1.27 \text{ meters} \end{aligned} \tag{3.43}$$

3.6.6 Multipath Behavior over Time

In the two-ray multipath model, the worst-case situation occurs when ρ_1 is near unity. As the phase varies, the received signal power varies from zero (total cancellation) to four times the power available from a single signal (for in-phase addition).

An environment that reflects signals with little loss is conducive to multipath. The multipath environment over water is normally very bad because the water, being conductive, reflects signals with little loss. Consequently, the multipath signals arriving at the receiving antenna generally exhibit ρ's that are very close to unity. The wave motion of the water causes the phase and magnitude of the reflected signal to vary rapidly. This causes the sum of the direct and reflected rays to vary rapidly.

Urban environments are notorious multipath environments. This is especially true in areas that include tall, mirrored buildings and other metallic structures. The multipath in a city usually isn't as dynamic as it is over water, but it still varies significantly over time.

The situation is very bad inside commercial office buildings. Consider a signal traveling from a transmitting antenna to a receiving antenna—let's use a 900 MHz cordless telephone as an example. The wavelength at 900 MHz is about 33 cm. So, when the difference in path lengths changes by half of a wavelength or 16.5 cm, the received signal strength has the opportunity to vary between +6 dB and –30 dB relative to the strength of the direct signal. If the user of the cordless telephone is agitated and begins to pace about during the conversation, the received signal strength will vary dramatically. The wavelength shortens as the frequency increases and the effect gets worse, and we experience more signal dropouts per unit time.

We must ask questions such as:

- How often will I experience a fade of 20 dB?
- How long will the fade last?
- How many minutes per month will my received signal strength be below a particular value?

3.6.7 Multipath Statistics

We are forced to describe multipath effects statistically, and we must determine the characteristics of the statistics that govern the values of ρ_n and τ_n of equation (3.24). We begin by assuming that both ρ_n and τ_n follow a Gaussian distribution. We can write the equation for the reflected ray in two equivalent forms:

$$V_{R,1} = \rho_1 \cos[\omega_0(t - \tau_1)] \quad \text{and} \quad V_{R,1} = \rho_1 \cos[\omega_0 t + \phi_1] \tag{3.44}$$

where

$$\phi_1 = -\omega_0 \tau_1 \tag{3.45}$$

What are the statistics that describe θ_1 if we assume that τ_1 is a Gaussian-distributed random variable? We find that, if the variation in the time delay, $\Delta\tau_n$, is much greater than the carrier signal period or

$$\Delta\tau_n \ll \frac{1}{f_0} \tag{3.46}$$

the angle θ_1 will be uniformly distributed. All angles between 0.0° and 359.999...° are equally likely. We can show that a Gaussian-distributed ρ_1 and a uniformly distributed θ_1 will combine to produce a resultant whose length is Rayleigh distributed. The Rayleigh distribution tells us that the received signal strength is usually constant but will occasionally suffer a deep fade.

The statistics of ρ_1 and τ_1 also govern the multipath fade event time. Rather loosely, a multipath fade event time is the average amount of time the signal strength falls below a given level. For a system that can withstand fading events of 30 dB, we are interested in the percentage of time our received signal strength will be attenuated by 30 dB or more. We are also interested in the average amount of time that a single fading event will last (i.e., how long will it be from the time the signal drops below the 30 dB level to the time it finally rises above the 30 dB level). This information is will be important when we discuss multipath mitigation.

3.6.8 Multipath Mitigation

Multipath fading does not reduce the average energy of the received signal. The signal energy is redistributed over time and space. If we design our communications links to be insensitive to this energy redistribution, then our link can approach the theoretical performance of an additive white Gaussian noise (AWGN) channel.

Generally, all multipath mitigation techniques use some form of diversity. The transmitter sends sufficient information over two or more statistically separate channels. The receiving antenna collects this energy, and, by careful manipulation, it can recover the transmitter's data. The statistically separate channels can include frequency diversity, time diversity (using data coding and interleaving), and space diversity.

3.6.8.1 Antenna Pattern

If we limit the number of signals our antenna views, we will limit the probability of receiving two separate signals from the same transmitter. We can satisfy this requirement with a narrow-beamwidth, low-sidelobe antenna.

Sometimes we can't control the antenna, however. Consider the case of a cellular telephone. The user may be in an arbitrary location and may move rapidly during a transmission. We must use an omnidirectional antenna to satisfy the user's needs. As we've seen, omnidirectional antennas are not helpful in a multipath environment because they pick up signals from every direction.

3.6.8.2 Frequency Diversity

A signal experiences a fade as it travels from the transmit antenna to the receive antenna because the direct and the reflected signals cancel out. The cancellation occurs when the difference in path lengths between the direct and reflected rays is an odd multiple of 180°. The length of the transmission paths in terms of wavelengths determines the phase of the two arriving signals.

For example, imagine a system that is exhibiting a deep fading event. The magnitude of the received signal is almost zero because the phase between the direct and reflected rays is almost 180°. Now, imagine we change the frequency slightly. Two signals traveling from the same transmit antenna to the same receive antenna will probably travel over the same path if the frequencies of the signals are nearly identical. A signal at frequency f_1 will probably travel over the same path as a signal at f_2 if $f_1 \approx f_2$.

Although the physical paths taken by the signals at f_1 and f_2 are about the same, the path lengths (in wavelength units) are different. At f_1, the path length might be 143,445.278 wavelengths for the direct path and 144,322.778 wavelengths for the reflected path (note that these two path lengths differ by 180° so we are experiencing a fade). Conversely, at f_2, the same physical paths might translate into 142,227.777 wavelengths (direct path) and 142,990.101 wavelengths. Since the phase difference between the direct and reflected signals is no longer 180°, the signal is not in a fade at this frequency.

The instantaneous fading behavior of a channel is a strong function of frequency. If we experience a fade at one frequency, the channel may not be fading at a slightly different frequency. If we transmit the same signal at two frequencies at the same time, we can take advantage of this behavior. When the signal fades on one frequency, there is a good possibility that the signal on the other frequency might not be fading. There is no guarantee, however. It's entirely possible that both channels will be in fade at the same time. All we've really done is to change the fading statistics in our favor.

Studies by the telephone companies indicate that the two frequencies should be separated by at least 2% of the center frequency or by 20 MHz, whichever is greater. Most microwave telephone relay systems that use frequency diversity have separations of 2–5% of the lower frequency. Figure 3-24 shows the received signal power of two channels separated in frequency by 4% of center frequency.

Frequency diversity is effective, but it isn't often used because it wastes spectrum. Fortunately, there are other multipath mitigation methods.

3.6.8.3 Spatial Diversity

We've seen that a multipath fade will occur when the geometry among the transmitting antenna, the receiving antenna, and some reflector produces a 180° phase shift between the direct and delayed signals. If the geometry changes just slightly, the phase relationship between the two received signals changes, and we often see a dramatic increase in received signal strength.

We can use this effect to change the multipath statistics in our favor as Figure 3-25 shows. We can use one transmitting antenna and two, physically separate receiving antennas and allow the receiving equipment to choose the best signal from the two antennas. If the signal from one antenna experiences a fade, the signal from the second antenna might not be in a fade. As with frequency diversity, it's possible that the signals from both receive antennas are in a fade.

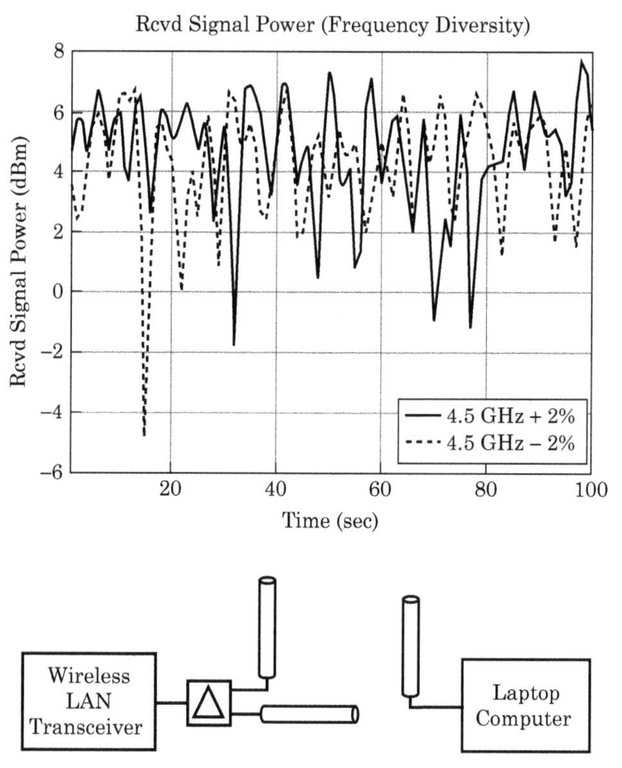

FIGURE 3-24 ■ A recording of frequency diversity transmission. The two signals followed identical paths were spaced 1,080 MHz apart (4% of the center frequency). Note that deep fades do not often occur simultaneously.

FIGURE 3-25 ■ A diagram of a home wireless local area network (WLAN) transceiver. The access point on the left will often use two antennas set at different angles to use path and polarization diversity.

EXAMPLE

Space Diversity

A particular WLAN access point has two antennas physically separated by 4 inches. How many wavelengths separate the two receive antennas?

Solution

Four inches is about 10.1 cm. The wavelength at 2.4 GHz is

$$\begin{aligned}\lambda &= \frac{c}{f}\\ &= \frac{3E8\text{ meters/sec}}{2.4E9/\text{sec}}\\ &= 0.125\text{ meters}\\ &= 12.5\text{ cm}\end{aligned} \tag{3.47}$$

So 4 inches is about 1 wavelength. Experience indicates that you will still achieve significant diversity improvement even when the receive antennas are this close together. The relatively small antenna separation will still improve transmission reliability.

3.6.8.4 Polarization Diversity

Vertically polarized waves tend to reflect off vertically oriented objects, whereas horizontally polarized waves tend to reflect off horizontally oriented objects. Consider a window covered by horizontal Venetian blinds. Assume the blinds are in the open position so only a small amount of metal is presented to the oncoming waves. Vertically polarized waves will pass through the blinds with little attenuation, whereas horizontally polarized waves will be almost completely reflected. From a multipath perspective, the physical paths taken by vertically polarized waves will like be different from the path taken by horizontally polarized waves. If we're experiencing a fade on one polarization, we may not be fading on the other polarization. This is polarization diversity. Studies have suggested that polarization diversity is as effective as space or frequency diversity for multipath protection.

Polarization diversity does present some practical problems. Both the transmitter and receiver require dual-polarization antennas, which can be physically cumbersome.

3.6.8.5 Time Diversity

We can also use time diversity as a weapon against multipath fading. Time diversity is useful mostly when we are transmitting digital data and when we're using forward error correction (FEC).

Once we determine the fading statistics, we can determine how long an average fade will last. For example, if our system can endure a 10 dB reduction in signal strength, we want to know how long the signal will be attenuated by 10 dB or more in a typical fading event. The fade will create a burst of errors, which lasts for the duration of the multipath event. When we know the statistics of a fading event (i.e., we know how long an average fading event will last, and we know the average number of fades per second), we also know how many bits per second we will be unable to decode. We can now design a FEC scheme to correct these data errors.

Finally, we find that we must spread our data out over time to take the best advantage of the FEC and the error statistics. For example, if we experience a fade long enough to obliterate both the data bits and the FEC bits, then that data are lost. We can spread both the data and their associated FEC bits over time so that an average multipath event can't generate unrecoverable errors.

War Story

A form of time diversity is used in audio compact discs. The data are spread out over a large area of the disk. If one section of the disk is damaged or covered by dirt, then the error correction circuitry in the CD player can correct the errors as long as it can also read data from the clean areas of the disk.

3.6.9 Multipath Equalizers

Figure 3-26 shows a signal undergoing multipath distortion. Observed over time, the frequency of the multipath nulls of figure 3-26 will change as the environment changes. We can view multipath as a time-varying filter between the transmitter and receiver that distorts the signal. Multipath equalizers are filters used to mitigate the effects of multipath propagation on the received signal by dynamically adapting to the changing multipath channel.

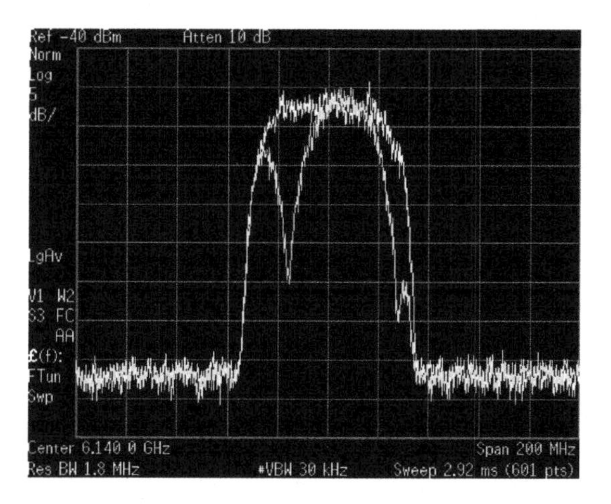

FIGURE 3-26 ■ A signal undergoing severe multipath. We can view the multipath as a time-varying filter and use adaptive equalization techniques to remove it.

One method to measure our channel is to observe the output of the demodulator and adaptively make adjustments to the equalizing filter. This is adaptive, decision-based equalization. A second method is to place a known piece of data in the transmitted signal, that is, place a pre-, mid-, or post-amble in the signal at regular intervals. We examine the *-amble* at the receiver and use that data to configure the equalizing filter.

3.7 | BIBLIOGRAPHY

GTE Lenkhurt Incorporated, "Space Diversity," The GTE Lenkhurt Demodulator, pp. 252–261, 1973.

Lenkhurt Electric Company, "Multipath Fading," The Lenkhurt Demodulator, pp. 842–851, 1967.

Lenkhurt Electric Company, "Microwave Diversity. . . How It Improves Reliability," The Lenkhurt Demodulator, pp. 199–205, March 1961.

Lenkhurt Electric Company, "Antenna Systems for Microwaves—Parts 1 and 2," The Lenkhurt Demodulator, pp. 301–323, May 1963.

Freeman, Roger L., *Telecommunication Transmission Handbook,* Wiley-Interscience, 1991.

Freeman, Roger L., *Radio System Design for Telecommunications (1–100 GHz),* Wiley-Interscience, 1987.

Hata, M., "Empirical Formula for Propagation Loss in Land Mobile Radio Services," *IEEE Transactions on Vehicular Technology,* August 1980.

Feher, Kamilo, *Wireless Digital Communications: Modulation and Spread Spectrum Applications,* Prentice Hall.

Willis, Mike, *Propagation Tutorial*, http://www.mike-willis.com/Tutorial/PF5.htm, May, 2007.

3.8 | PROBLEMS

1. **Cell Coverage** Consider the coverage of a cellular telephone tower under the assumption of propagation through a homogenous medium with $n = 3.4$ (this assumption is wildly inaccurate but for the sake of a homework problem). If the path loss from the cell to any user should not be greater than A_{dB}, find an expression for the radius of the cell in terms of A_{dB} for a frequency of 1,800 MHz.

2. **Free Space Path Loss** The free-space path loss equation describes the loss of signal power as the power emitted by an isotropic radiator spreads out over the surface of a sphere. Free-space path loss is often referred to as spreading loss. The free-space path loss equation is

$$\begin{aligned} PL_{FS} &= \frac{P_{Tx}}{P_{Rx}} \\ &= \left(4\pi\frac{D}{\lambda}\right)^2 \end{aligned} \tag{3.48}$$

where

d = the distance between the transmitting and receiving antennas
λ = the wavelength

Given this geometrical interpretation of free-space path loss or spreading loss, explain why equation (3.51) contains a reference to wavelength.

3. **Three-Ray Multipath** Consider a 980 MHz signal signal experiencing three-ray multipath. The tranmitted signal is $s(t)$. The received signal is

$$r(t) = A_0 s(t) + A_1 s(t - \tau_1) + A_2 s(t - \tau_2) \tag{3.49}$$

where

$A_0 = 0.4$
$A_1 = 1.0,\ \tau_1 = 5.125\ \mu\text{sec}$
$A_2 = 0.7,\ \tau_2 = 8.334\ \mu\text{sec}$

a. What is the distance in meters between the longest and shortest path?
b. If the transmitted signal is a cosine, what are the relative phase angles between the three multipath components?
c. If the transmitted signal is a cosine, are there any values for τ_1 and τ_2 that will cause $r(t)$ to be equal to zero (i.e., infinite attenuation)? If so, give values for τ_1 and τ_2 that will cause $r(t)$ to be zero.

4. **Multipath Statistics** Consider a situation with two-ray multipath. The received signal is

$$\begin{aligned} V_{Rx}(t) &= V_D(t) + V_{R,1}(t) \\ &= \cos(\omega_0 t) + \rho_1 \cos(\omega_0 t + \phi_1) \end{aligned} \tag{3.50}$$

where

$\rho_a = 1.0$
ϕ_1 is a uniformly disturbed random variable whose value lies between 0° and 359.999...°

a. What's the percentage of time that we would expect the received signal $V_{Rx}(t)$ to be less than 20 dB below the direct signal $V_D(t)$? In other words, how often do we expect the received signal to have faded by more than 20 dB?
b. Repeat your analysis for a 30 dB fade.

CHAPTER 4

Antennas

I wanna make contact . . .

Joan Jett

Chapter Outline

4.1 INTRODUCTION

Antennas serve as our interface between guided wave propagation (a transmission line or a waveguide) and unguided wave propagation (free-space propagation). We use a transmitting antenna to launch a signal into free space in the hope that we can receive it at some remote location using a receiving antenna. We impress information on the signal, and, if we work things right, we can retrieve that information at the receiving end of the system.

In this chapter, we will examine the characteristics of antennas as they affect the design of the receiver. We examine metrics describing how efficiently antennas launch and receive signals. We also discuss antenna properties that help mitigate the unkind effects of unguided propagation. Finally, we will develop models that will help us characterize an antenna when we use it in both its transmitting mode and its receiving mode.

4.1.1 Antenna Duality

As we progress through this chapter, we will sometimes refer to an antenna characteristic as though we were using the antenna in a receiving mode. At other times, we will discuss the antenna as though we were using it to transmit. Both configurations are valid, and insight gained from one configuration usually applies to the other configuration.

4.1.2 Maxwell's Equations

Maxwell's equations describe the characteristics of electromagnetic fields under almost all circumstances. They describe the fields as they behave in, for example, conductors, free space, and waveguides. They also describe the fields surrounding antennas. However, Maxwell's equations describe the relationships between the electric and magnetic fields in very concise, compact language. The four equations contain all of the information we require to describe the fields, but, unless you have worked with the equations for some time, they are not at all intuitive. All of the information is there, but it can be hard to derive meaningful answers.

4.1.3 Spatial Preselection in the Receiver

As designers of receiving systems, it is our major goal to limit the number of signals our receiver must process at one time. Ultimate, our receiver is tuned to only one signal at one frequency, so any other signals we allow into our system cannot help us and may degrade the signal of interest (e.g., due to system nonlinearities, poor phase noise).

The receive antenna is our first step in this signal preselection process. If the antenna is more sensitive to signals coming from one direction than another, we can point our antenna in the direction of the desired signal to accentuate the desired signal while attenuating undesired signals that are arriving from different directions.

4.1.4 Spatial Preselection in the Transmitter

Transmitting antennas often exhibit the ability to radiate energy preferentially in a particular direction. This characteristic allows our transmitter to appear that it is broadcasting more power than it would if the antenna radiated energy equally in all directions. Also, by limiting our radiation only to certain directions, we reduce the risk of polluting the electromagnetic environment with unwanted signals, making data transmission easier for us all.

4.2 ANTENNA EQUIVALENT CIRCUITS

We seek to model antennas as simple equivalent circuits, consisting of resistances and reactances. We'll attribute certain antenna characteristics to each component, and then we'll develop relationships between the antenna's behavior and the pieces of the equivalent circuit. However, the components in the equivalent circuits we are about to describe do not really exist as definable lumps in the antenna. They are simply mental models that enable us to make predictions about how an antenna will behave in a given situation.

4.2.1 Transmitting Model

Figure 4-1 shows a transmitting system driving an antenna and the equivalent circuits for the transmitter and antenna. The amplifier model is a voltage source, Vs, connected in series with a resistor, R_S, where R_S represents the output impedance of the amplifier.

Some of the energy fed into the antenna is dissipated uselessly away as heat. We denote this energy as P_{Heat}, and we will model this loss with resistor R_{Loss} of Figure 4-1. Any power dissipated in R_{Loss} is wasted. The remaining energy, P_{Rad}, is radiated into free space. The quantity P_{Rad} is the power dissipated in the radiation resistance, R_{Rad}, and that is the power that the antenna sends into the environment. Our goal is often to maximize the power dissipated in R_{rad}.

In the case of a transmitter, we are not concerned with the noise characteristics of R_{Rad} and R_{Loss} and usually consider them to be noiseless.

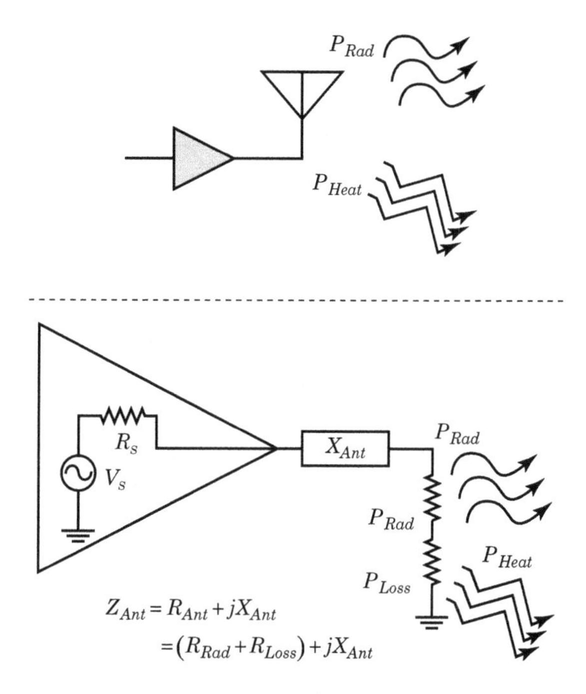

FIGURE 4-1 ■ A transmitting antenna with its equivalent circuit.

For completeness, the transmitting antenna model also shows a reactance X_{Ant} (which can be either capacitive or inductive) in series with loss and radiation resistances. The reactance is undesired and, for most purchased antennas, is insignificant.

Using the notation of Figure 4-1 and maximum power transfer theory, we would like

$$\begin{aligned} R_{Loss} &\gg R_{Rad} \\ R_{Rad} &= R_S \\ X_{Ant} &= 0 \end{aligned} \tag{4.1}$$

The first condition ($R_{Loss} \gg R_{Rad}$) ensures that most of the power from the transmitter is dissipated in the radiation resistance R_{Rad} rather than in the loss resistance R_{Loss}. The second condition ($R_{Rad} = R_S$) ensures that the antenna is matched to the output impedance of the transmitter. The third condition ($X_{Ant} = 0$) also ensures that the antenna is matched to the transmitter's output impedance. Setting X_{Ant} to zero also eases our matching problems if the antenna is to be wideband.

Figure 4-2 shows the power dissipated in the radiation resistance being distributed into free space around the antenna and radiated away. The antenna pattern causes the energy surrounding the antenna to be strong in some directions than in other directions.

FIGURE 4-2 ■ The power dissipated in the radiation resistance of a transmitting antenna is radiated into free space. The antenna pattern causes a nonuniform power distribution surrounding the antenna.

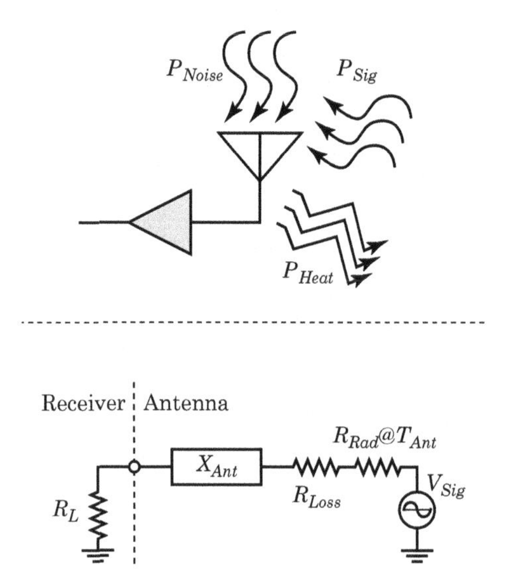

FIGURE 4-3 ■ A receiving antenna with its equivalent circuit.

4.2.2 Receiving Model

Figure 4-3 shows a receiving system and its equivalent circuit. In the receiving case, we are interested in how effectively the antenna captures power present in an electromagnetic wave. If the power in the wave is measured in watts/meter^2, then the antenna presents an *effective area* to the wave with the result of the antenna capturing a given number of watts. This effective area will change with antenna orientation.

The voltage source, V_{Sig}, represents the signal power received by the antenna. The larger the signal the antenna receives, the larger the value of V_{Sig}. The voltage source includes both desired and undesired signals but not the noise received by the antenna. Ultimately, V_{Sig} delivers power to the receiving system, which is modeled by the resistor R_L.

The resistor R_{Loss} represents the electrical losses in the antenna. The reactance X_{Ant} can be either capacitive or inductive. As in the transmitting case, we normally strive for a minimum amount of reactance to make matching and maximum power transfer easier to realize. The resistor R_{Rad} again represents the radiation resistance of the antenna. The noise temperature of R_{Rad} is T_{Ant}. The radiation resistor's noise temperature accounts for the noise present on the output terminals of the antenna.

In the receive mode, we strive for similar relationships among R_{Rad}, R_{Loss}, R_L, and X_{Ant} as we strove for in the transmitting case (see equation (4.1). The first relationship ($R_{Loss} \gg R_{Rad}$) ensures that most of the power developed by V_{Sig} ends up in the receiver and not in R_{Loss}. The second relationship ($R_{Rad} = R_L$) dictates that the antenna and amplifier are matched and ensures maximum power transfer from the antenna to the receiver. The third relationship ($X_{Ant} = 0$) also involves matching and maximum power transfer.

Figure 4-4 shows an antenna with gain operating in its receiving mode. The antenna is more sensitive to signals arriving at the antenna from the direction in which the antenna exhibits high gain.

4.2.3 Received Noise

A receive antenna will capture external noise from its environment at the same time that the antenna is capturing the signal of interest. We generally lump the noise present in the radio frequency (RF) environment into two broad classes.

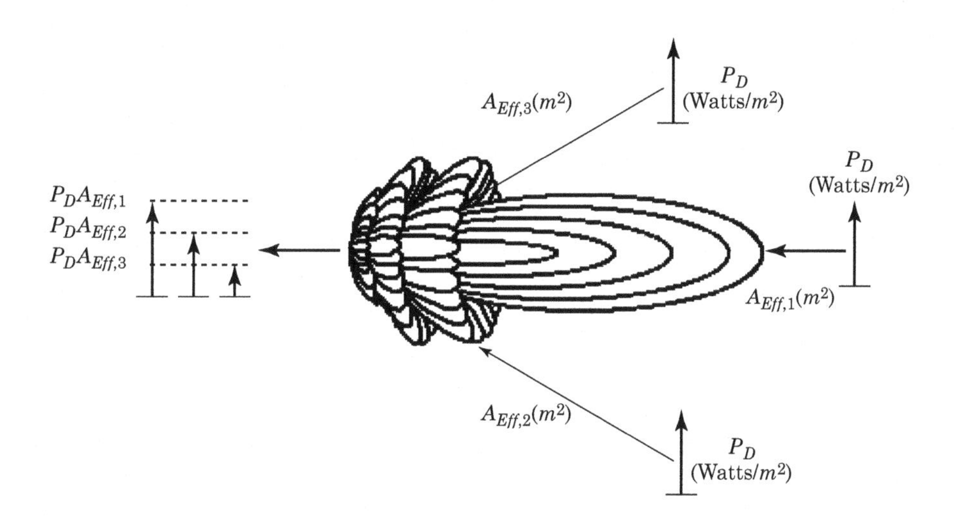

FIGURE 4-4 ■ An antenna used in the receiving mode. The antenna is more sensitive to energy coming from angles at which the antenna exhibits high gain. High gain manifests itself as a large effective area presented to the oncoming wave.

4.2.3.1 Natural Noise

There are electrical noise sources naturally present in nature. For example, one statistic tells us that lightning strikes the earth approximately 100 times a second. These discharges, along with the many more that don't strike the earth, are a significant noise source. Other noise sources include the following:

- The auroras (the Northern and Southern lights)
- Sunspots
- Galactic noise (due to the sun, moon, and any other warm radio source)
- Ground noise (the earth, being a physically warm body, radiates noise. This is a significant noise contribution in ground-station satellite systems)
- Absorption noise (energy passing through the atmosphere is absorbed and then reradiated as noise)
- Atmospheric noise (the continuous, chaotic churning of the atmosphere causes charge separation, and, when those charges recombine, they generate noise)
- Precipitation noise (precipitation generates noise-like changes in signal amplitude as well as charge separation)

Natural noise tends to dominate system performance at low frequencies (below 30 MHz or so).

4.2.3.2 Man-made Noise

There are many man-made sources of electromagnetic energy present on Earth. Uncounted numbers of distant RF transmitters combine to produce a noise-like waveform present in the environment. There are also large numbers of microwave ovens, computers, car ignitions, cellular telephone systems, microwave telephone relay system, wireless baby monitors, and cordless telephones present in the modern world. All of these devices combine to generate RF energy, which has noise-like properties (especially when there are many such signals present in the environment at the same time).

We model received noise with the noise temperature of the radiation resistor of Figure 4-3. We measure the noise power present at the output of the antenna, then set the

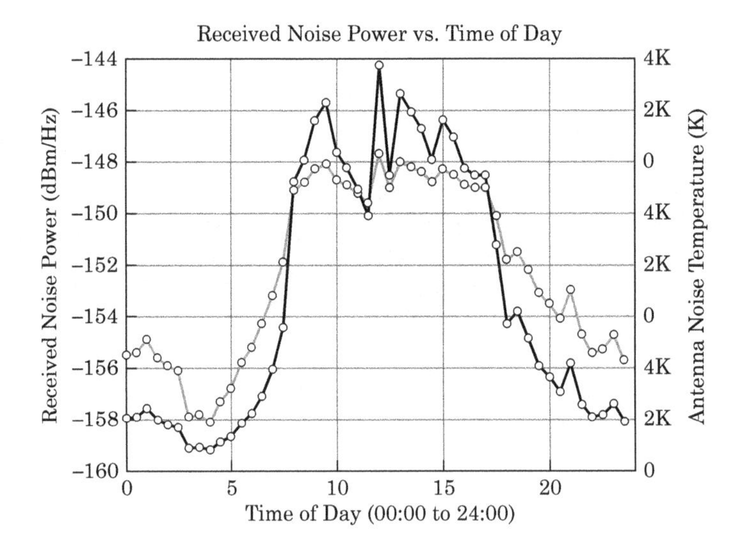

FIGURE 4-5 Received noise power versus time of day of a satellite antenna.

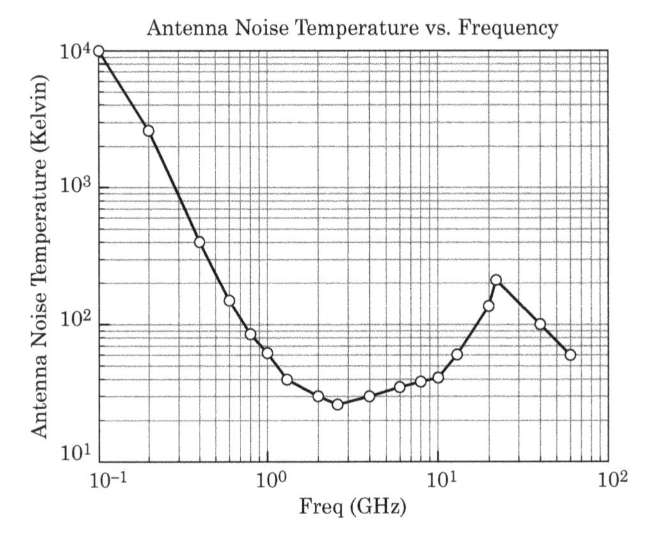

FIGURE 4-6 Received noise power versus frequency of a satellite antenna.

antenna noise temperature (T_{Ant} of R_{Rad} in Figure 4-3) to the appropriate value. The larger the noise power we measure at the output of the antenna, the larger the value of T_{Ant}. Figure 4-5 shows the noise received by a satellite antenna during one 24-hour period.

Figure 4-6 shows the antenna temperature, T_{Ant}, of the radiation resistance of a satellite receiving antenna. The antenna is pointed at an elevation of 5° above the horizon. We see that T_{Ant} is a very strong function of frequency and that it covers almost three orders of magnitude as the frequency changes from 100 MHz to 10 GHz. Note especially the disturbing increase in noise at the low-frequency end of the figure.

The antenna noise increases rapidly with decreasing frequency, and, unfortunately, the trend continues off the graph. Some workers have measured noise temperatures 100,000° K at 1 MHz and 10,000,000° K at 100 kHz. At low frequencies (e.g., in the high-frequency [HF] and lower very high frequency [VHF] bands), we find that the noise present in the external environment is so large that it overwhelms any internally generated noise a receiving system might develop. In the cases where the external noise is a lot larger than the internally generated receiver noise, we are *externally noise limited.*

EXAMPLE

Antenna Noise Temperature

Designers of deep-space probes are very interested in the noise temperature of their Earth station antennas. They have found that the noise temperature of the Earth station antenna changes with the position of the probe. For example, when the probe is between planets, the receive antenna noise temperature is very cold (i.e., the antenna is quiet or low noise) because the only items within the view of the receive antenna are the space probe and the cold space behind it.

However, when the probe is near a planet, the noise temperature of the receive antenna on Earth rises. The temperature of the planet is higher than empty space, so the planet radiates noise. In addition, other geological effects at work in the planet's interior may generate more noise.

Similarly, the noise temperature of the antenna aboard the space probe changes with the position of the earth and the sun. When the sun, the earth, and the space probe all fall along the same line, and then the space probe receives the radio frequency noise from the sun as well as the signal transmitted from the earth.

EXAMPLE

Satellite Antenna Noise Temperature

For about 20 days a year, geosynchronous satellites pass between the earth and sun for several minutes a day (see Figure 4-7). When the geometry is right, the satellite is in front of the sun for 70 minutes or so.

During the time when the satellite is positioned between the earth and sun, the noise temperature of the ground-station receiving antenna increases dramatically. This occurs because the receive antenna sees the electrical noise radiated by the sun as well as the normal background noise.

Just as a matter of information, this geometry causes another problem. For another 20 days a year, the earth is between the geostationary satellite and the sun. While this doesn't cause the ground-station antenna noise temperature to increase, it forces the satellite to run off batteries for the time the sun is eclipsed. Some of the communications satellites require 2,000 watts or so. This is quite a battery problem when you factor in the environmental and reliability problems of space-borne equipment.

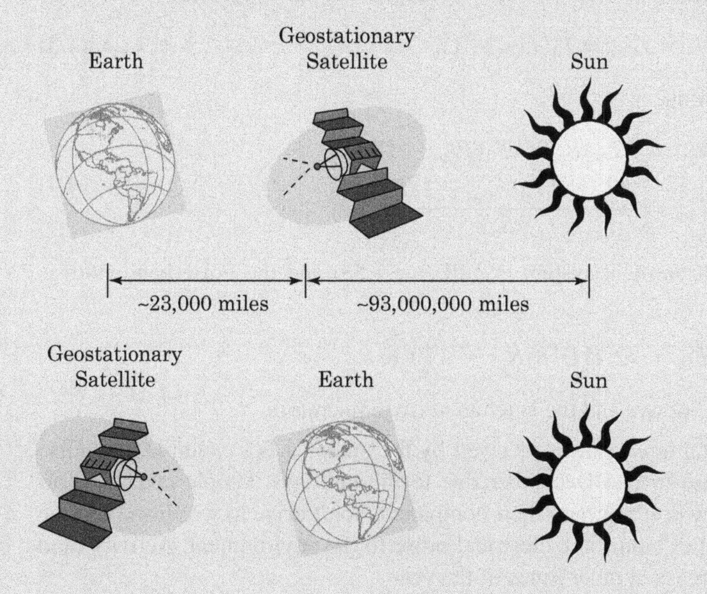

FIGURE 4-7 ■ Geosynchronous satellite geometry.

EXAMPLE

Antenna Noise Temperature

You are hired to measure the noise environment at a particular location. Figure 4-8 shows the system used to measure the noise. The noise figure of the amplifier/spectrum analyzer cascade is $F_{Cas} = 2$ dB, and the measurement bandwidth is 30 kHz. Figure 4-5 shows the data you measured on one particular day.

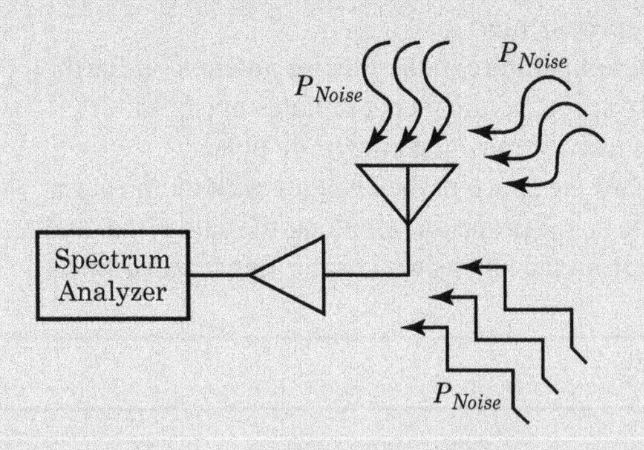

FIGURE 4-8 ■ An antenna connected to an amplifier, followed by a spectrum analyzer.

a. Develop an expression relating the equivalent input noise power dissipated in the spectrum analyzer's input resistor, R_L, to the antenna noise power.

b. Assuming that these data adequately represent the entire population of the noise measurements in which we're interested, comment on the statistics of the antenna noise temperature.

Solution

a. The total noise power dissipated in the spectrum analyzer's load resistor, R_L, is the equivalent input noise power of the system N_{Sys} and

$$N_{Sys} = k(T_{Ant} + T_{Cas})B_n \tag{4.2}$$

Chapter 6 reveals that the relationship between T_{Cas} and the amplifier's noise figure is

$$T_{Cas} = T_0(F_{Cas} - 1) \tag{4.3}$$

Combining these two equations produces

$$\begin{aligned} N_{Sys} &= kT_{Ant}B_n + kT_0(F_{Cas} - 1)B_n \\ \Rightarrow T_{Ant} &= \frac{N_{Sys} - kT_0(F_{Cas} - 1)B_n}{kB_n} \end{aligned} \tag{4.4}$$

When the noise figure of the measurement system is 2 dB (or 1.58) and the noise bandwidth is 30 kHz, equation (4.4) reduces to

$$T_{Ant} = (2.41E18)N_{Sys} - 170^{\circ}\text{K} \tag{4.5}$$

Table 4-1 shows the received noise power and the antenna noise temperature.

b. Many factors contribute to the total noise power received by the antenna. For example, the noise temperature will probably decrease in late December due to the holidays. People take time off from work, and, consequently, they don't turn on their computers, don't drive to work, and use the telephone less. All of these activities contribute electrical noise to the environment. Activity (and hence noise temperature) may increase at other times of the year.

TABLE 4-1 Measured noise power and antenna temperature for the example noise measurement.

Time of Day	Measured Noise Power (dBm)	Calculated Antenna Noise Temperature (° K)	Time of Day	Measured Noise Power (dBm)	Calculated Antenna Noise Temperature (° K)
00:00	−155.5	510	12:00	−147.7	3,930
00:30	−155.4	526	12:30	−149.0	2,870
01:00	−154.9	611	13:00	−148.0	3,657
01:30	−155.6	495	13:30	−148.2	3,484
02:00	−155.9	451	14:00	−148.4	3,320
02:30	−156.1	423	14:30	−148.8	3,013
03:00	−157.9	222	15:00	−148.3	3,401
03:30	−157.8	231	15:30	−148.5	3,240
04:00	−158.1	204	16:00	−148.9	2,940
04:30	−157.3	280	16:30	−149.0	2,870
05:00	−156.8	334	17:00	−149.0	2,870
05:30	−155.8	465	17:30	−150.1	2,190
06:00	−155.2	559	18:00	−151.8	1,425
06:30	−154.3	727	18:30	−151.5	1,539
07:00	−153.2	986	19:00	−152.2	1,285
07:30	−151.9	1,389	19:30	−153.1	1,013
08:00	−149.1	2,800	20:00	−153.5	908
08:30	−148.8	3,013	20:30	−154.1	769
09:00	−148.3	3,401	21:00	−153.0	1,040
09:30	−148.1	3,570	21:30	−154.7	648
10:00	−148.7	3,087	22:00	−155.4	526
10:30	−148.9	2,940	22:30	−155.3	543
11:00	−149.2	2,733	23:00	−154.7	648
11:30	−149.6	2,477	23:30	−155.7	480

4.3 APERTURE

Antenna aperture is most useful when we describe a receiving antenna. An antenna's aperture represents the amount of electrical *area* an antenna presents to an oncoming electromagnetic wave (see Figure 4-9). The power present in the oncoming electromagnetic wave is measured in watts/meter^2, and the effective electrical antenna area is A_{eff}, measured in meter^2. The result is that the antenna *gathers* a given number of watts from the electromagnetic environment. The amount of power gathered by the antenna is directly proportional to the antenna's area or aperture.

Antenna aperture, gain, and directivity are all closely related quantities. Knowing the antenna's gain or directivity (along with a few other quantities such as loss and wavelength) allows the calculation of the antenna's aperture. Similarly, if we know the aperture that the antenna presents to a wave coming from a given direction, we can find the equivalent gain in that direction.

Antenna theory tells us that the relationship between antenna gain G and effective aperture A_{eff} is

$$A_{eff}(\theta, \phi) = G(\theta, \phi)\frac{\lambda^2}{4\pi} \tag{4.6}$$

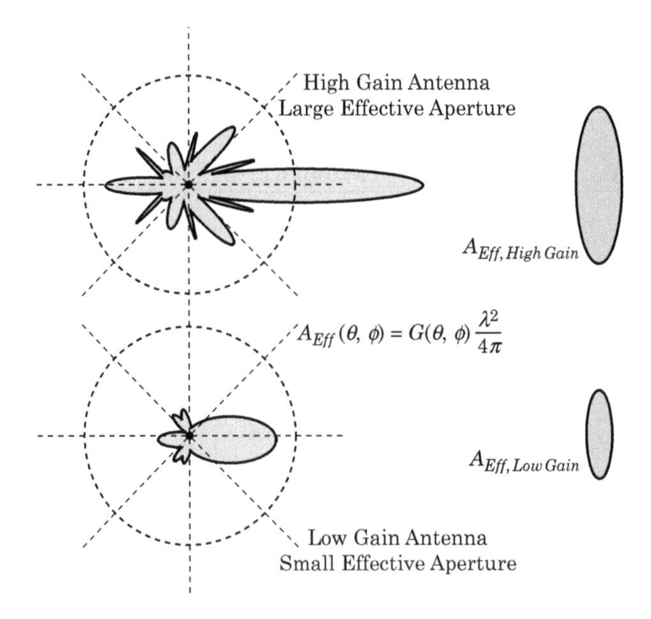

FIGURE 4-9 ■ The relationship between antenna gain and aperture.

Note that A_{eff} is the *effective* area of the antenna, not the physical area. In some cases (e.g., a parabolic dish antenna or a horn antenna), the antenna does actually present some physical area $A_{physical}$ to the oncoming wave. The aperture efficiency, $\eta_{aperture}$, relates the physical area of the antenna to the antenna's effective aperture

$$\eta_{aperature} = \frac{A_{eff}}{A_{physical}} \tag{4.7}$$

4.4 THE ISOTROPIC RADIATOR

The isotropic radiator is the simplest of all antennas if we're considering only the pattern. In the transmitting mode, this antenna accepts power from a source and directs that power equally in all directions (see Figure 4-10). The power density at a distance R from an

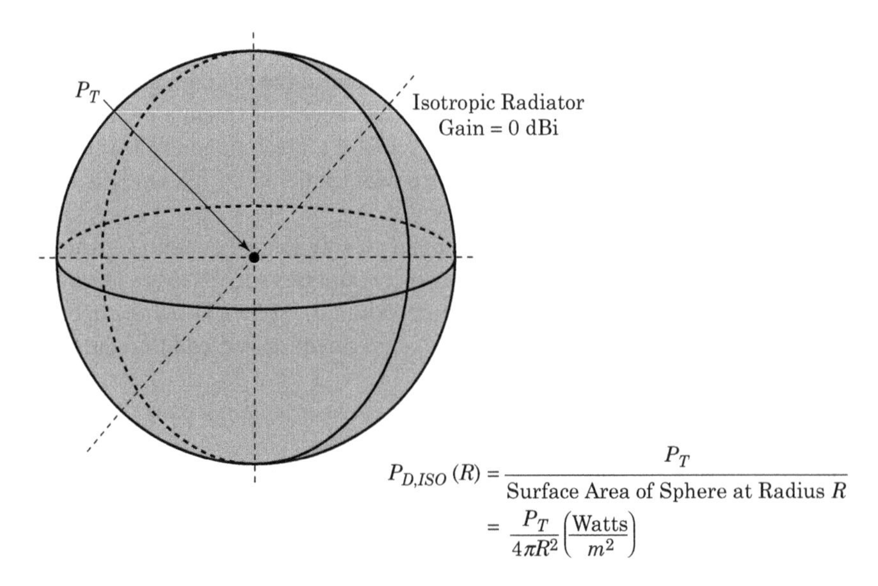

FIGURE 4-10 ■ The isotropic radiator.

isotropic radiator is

$$PD_{D,ISO} = \frac{P_{TX}}{\text{Surface Area of a Sphere}} = \frac{P_{TX}}{4\pi R^2}\left(\frac{\text{Watts}}{\text{m}^2}\right) \tag{4.8}$$

Since the gain of an isotropic radiator is unity in all directions, the aperture is the same in all directions and is equal to

$$A_{Eff,\text{ISO}} = \frac{\lambda^2}{4\pi} \tag{4.9}$$

4.5 | ANTENNA GAIN, BEAMWIDTH, AND APERTURE

The bibliography at the end of this chapter lists several excellent books that describe antennas deeply and with authentic perspective. In this book, we will discuss antennas from a (mostly) black-box perspective, concentrating on antenna aspects that affect receivers and demodulators. Antenna gain, antenna beamwidth, and antenna aperture will be our main topics.

4.5.1 Antenna Gain and Directivity

Antenna gain and directivity both describe the ability of an antenna to direct energy in a given direction. Gain and directivity are equivalent once we account for losses in the antenna. We usually describe antenna gain using a transmitting model, but the concept is also useful in the receive mode. In the receive mode, antenna gain describes the antenna's ability to preferentially receive energy from a given direction. The gain describes how much more signal the antenna will gather above some standard reference antenna, such as an isotropic radiator.

Figure 4-11 shows a garden-variety directional antenna pattern. We generate this pattern by placing the antenna under test (AUT) on a rotating platform. We place a calibrated reference antenna on a fixed platform a known distance from the AUT. We supply a known amount of signal power to the AUT, and then, while we rotate the AUT we measure the amount of power we receive with the reference antenna as a function of angle. We often repeat this measurement over frequency.

4.5.2 Nomenclature

We divide antenna pattern plots such as Figure 4-11 into two discrete regions: (1) regions of intentional radiation; and (2) regions of unintentional radiation. In Figure 4-11, the region of intentional radiation falls over the range of $-4^\circ <= \phi <= +4^\circ$. The rest is unintentional radiation.

The area in the neighborhood of the highest power transmission is the antenna's *main lobe*. The area that is 180° directly behind the main lobe is the antenna's *back lobe*. Any lobes appearing between the main lobe and the back lobe are generally referred to as *sidelobes*.

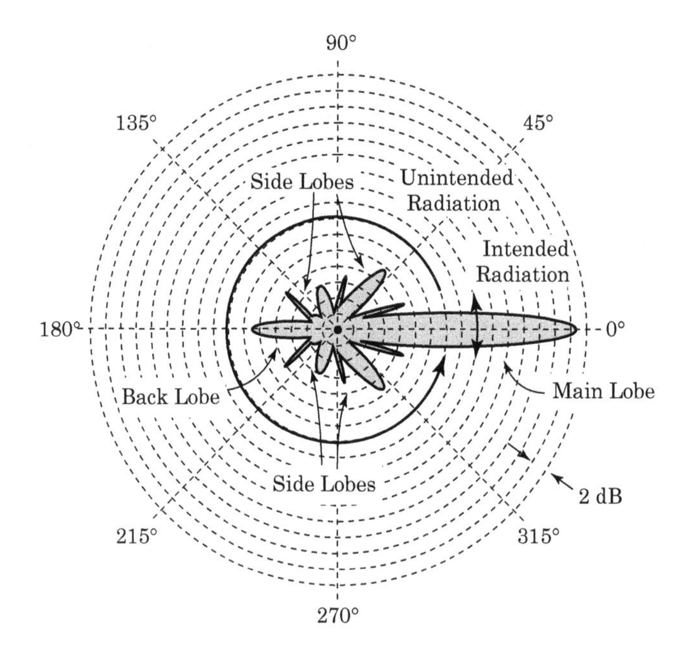

FIGURE 4-11 Generic antenna pattern.

We often speak of sidelobe levels. This refers to the level of a particular sidelobe referenced to the main lobe. In Figure 4-11, the sidelobes at 45° and 315° are both about 23 dB below the main lobe, or simply 23 dB down.

We are also interested in how much energy the antenna radiates in its reverse or backward direction. This is the *fronttoback ratio,* or f/b. The f/b of the antenna pattern of Figure 4-11 is about 20 dB.

Generally, the back lobes and sidelobes of the antenna are not desired, and we try to suppress them with the design of the antenna. In the transmitting mode, energy from the transmitter will spill out uselessly in the direction of the sidelobes and back lobes. In the receive mode, signals and noise from undesired directions can make their way into the antenna through the sidelobes and back lobes. These signals can interfere with the desired signal arriving at the antenna through the main lobe.

Nulls are the positions in the pattern were the gain is zero or very close to zero. These points can be useful to the receiver system designer. If there is an interfering signal coming from a known direction, it might be possible to turn the antenna such that the interferer is placed into a null. Sharp nulls are also be used to determine the direction of a signal.

EXAMPLE

Radar Antennas

In a radar system, we seek to measure the distance and position of remote objects. A radar system performs this measurement using a rotating directional antenna. The antenna rotates while constantly emitting a stream of RF pulses. These pulses bounce off any objects they encounter and return to the radar antenna.

Since the radar system can measure the time it takes for the pulses to travel to the object and back again, it can calculate the distance to the object. The radar system also knows the angular bearing to the object because it can measure the direction of the radar antenna when the system received the strongest return pulses. When we know the radial distance and the angle, we can locate the object in space.

Imagine the effects of antenna sidelobes on this system. When the transmitter emits a pulse, it will travel out into space in the direction of the main lobe and in the direction of the sidelobes. The pulses from the sidelobes will be smaller in magnitude than the pulses in the main lobe, but they are still there.

If any pulse encounters and object, energy will return to the radar antenna and enter the system. The magnitude of the pulse will depend on the radial distance to the object and the antenna gain in the particular direction. For example, if the receiver measures a particular return pulse power, it might be due to an object in the main beam that is very far away or to an object in one of the sidelobes that is very close to the antenna. The antenna sidelobes cause an ambiguity in the radar measurement.

Modern systems use computers to sort through all of the radar returns and categorize them. For example, if the radar returns show several objects, all at the same radial distance but at different angles, the computer will ignore the weakest returns in favor of the strongest. This is especially true if the angles between the phantom objects and the real object are related to the known sidelobe angles of the radar antenna.

Let's say we wanted to fool the radar into thinking a particular target was not at its true position. A jamming aircraft can stand away from the radar and measure the rise and fall in the radar's transmitted pulses as the radar antenna sweeps around. The jammer can calculate the antenna's gain pattern with this information (after all, this situation is equivalent to the antenna gain measurement we described earlier). The jammer can also calculate the antenna's rotational rate.

To fool the radar, the jammer only has to transmit pulses that will arrive at the radar antenna when the radar antenna is pointed in the wrong direction—away from the true position of the target.

Of course, the radar can use techniques to counter this type of deceptive jamming (e.g., it can randomly change the duration between pulses, can emit more complicated, coded pulses, can make better use of computers to keep track of targets and their positions). One of the best countermeasures is to build a very low sidelobe antenna.

EXAMPLE

Satellite Systems

We are concerned about antenna sidelobes in satellite systems for several reasons. A satellite is usually power limited, so it cannot transmit a very large signal consequently we require either a very large Earth station antenna or a very low noise receiving system. The cost of an Earth station receiving antenna is a strong function of its physical size and thus our most cost-effective measure to reduce the ground-station antenna noise temperature.

The main beam of the antenna looks at the satellite and the cold space behind the satellite, so this is not a significant source of noise. However, the back lobe of the antenna often looks at the warm earth. Even though the back lobe of the antenna may be 10 or 20 dB below the main beam, it will still allow a large amount of noise into the antenna. Often, the back lobe and sidelobe noise is the largest source of noise in the receiving system.

A geosynchronous satellite does not change its apparent position in the sky relative to the earth, so we are not required to track the satellite as it moves through the sky—this is a big advantage in terms of cost and simplicity. Physics tells us that all geosynchronous satellites must lie in the same orbital plane, above the equator, and naturally we would like to place as many such satellites in the sky as reliable engineering allows. Consequently, some geostationary satellites are as close as 4° apart. If one satellite lies within the ground-station antenna's main lobe while a second satellite lies within the antenna's sidelobes, the signals from the second satellite could interfere with the signals from the desired satellite.

EXAMPLE

Direction Finding

Antenna patterns can exhibit angles where the gain and aperture of the antenna are nearly zero. The antennas are insensitive in the direction of these antenna nulls. The nulls are often very sharp and deep. Rotating the antenna only a few degrees can cause a signal to change 50 dB or more. We use this characteristic when measuring the direction of arrival of a signal (i.e., we are interested in *direction finding*).

It is common to use a *loop* antenna for direction finding. The antenna consists of a loop of wire, usually less than $\lambda/10$ in circumference. The pattern exhibits two broad, flat areas of gain and two deep nulls. The nulls lie along lines perpendicular to the plane of the loop, and the maximum gain is in the plane of the loop. In other words, if you hold the loop up to your face and look through it (like you might hold a hand mirror), you are looking in the direction of the antenna nulls. One null points through the loop, while the other null is behind your head. If you rotate the loop by 90°, you are viewing the loop from the direction of maximum gain (see Figure 4-12).

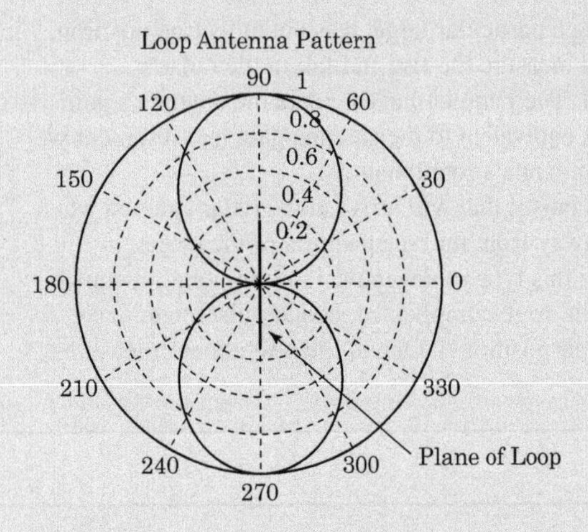

FIGURE 4-12 ■ The pattern of a loop antenna. The pattern is a doughnut-like shape with the null of the pattern perpendicular to the plane of the loop.

When direction finding, the operator finds the signal of interest by rotating the antenna until he gets the maximum signal. However, because the gain characteristic of the antenna is very broad, he doesn't achieve high resolution, but he does know the general direction of the signal (within approximately 30°). To accurately measure the direction, the operator turns the antenna by 90° and adjusts the angle until the signal disappears completely. Since the null is sharp and deep, he can achieve high angular resolution.

This single method gives the direction finder one line of bearing. The operator now has to move the direction finding apparatus and repeat the measurement. The two bearing lines from the two separate measurements should intersect at the location of the transmitter.

4.5.3 Gain Measurement Details

The details of measuring the antenna pattern of Figure 4-11 offer more insight into antenna characteristics. We applied a known amount of power to the AUT and measured the power received by the reference antenna as we rotated the AUT. Unless we carefully quantify the losses present in our test, we've really measured a *relative* antenna pattern. In other words, we know how the AUT performs at, say, $(\phi, \theta) = (20°, 40°)$ compared with how it performs at $(\phi, \theta) = (0°, 0°)$.

If it's worth the expense, we can follow the signal through the test configuration and catalog the losses. The signal we measure from the receive antenna depends on the following:

- The power applied to the antenna under test
- The mismatch efficiency of the test antenna
- The radiation efficiency of the test antenna
- The directivity of the test antenna at the current (ϕ, θ)
- The distance between the AUT and the receive antennas (or, equivalently, the path loss of the signal as it travels from the AUT to the receive antenna)
- The aperture (or, equivalently, the directivity) of the reference antenna
- The radiation efficiency of the reference antenna
- The mismatch efficiency of the reference antenna
- Reflections from the walls of the test chambers

Many of these quantities are difficult to know accurately. However, we can take steps to make sure that none of these quantities changes significantly as we rotate the test antenna.

4.5.4 Gain, Directivity, and Efficiency

If we generate an antenna pattern like Figure 4-11 without carefully examining the antenna losses, we have really measured the directivity of the antenna. We know, in a relative sense, how the antenna distributes energy over space, but we do not know anything about the losses in the antenna. For example, we could place a 10 dB attenuator directly before the AUT and still produce the same antenna pattern. Most of the energy is lost in the attenuator, but that doesn't affect the measurement we're making.

Antenna gain describes the pattern of the AUT while accounting for the antenna's efficiency. The relationship between antenna gain and antenna directivity is

$$G(\theta, \phi) = \eta D(\theta, \phi) \tag{4.10}$$

where

$G(\theta, \phi) =$ the gain of the AUT
$D(\theta, \phi) =$ the directivity of the AUT
$\eta =$ the efficiency of the AUT

The efficiency, η, of the AUT is

$$\eta = \frac{P_{radiated}}{P_{accepted}} \tag{4.11}$$

where

$P_{radiated} =$ the power radiated by the AUT
$P_{accepted} =$ the power accepted by the AUT from its source

As system designers, we are more interested in antenna gain, but antenna directivity is far easier to measure. Usually, most antennas we find in practice have very small losses, so $P_{radiated} \approx P_{accepted}$ and $\eta \approx 1$.

4.5.4.1 Antenna References

As system designers, we would like to compare the gain of a particular antenna with some standard radiator. There are two commonly used standards: (1) the isotropic radiator; and (2) the ideal dipole. The isotropic radiator is the most mathematically convenient antenna but is difficult to realize. The ideal dipole is easier to measure, and simple mathematical tools can be used to convert one reference to the other.

It is conceptually very simple to measure an arbitrary antenna's gain relative to another antenna. After rotating the AUT and taking data, we replace the AUT with our reference antenna, being careful not to change anything else. We then measure the received signal power from the reference antenna, producing our reference power. Finally, we subtract the reference power (in dBm) from each of the AUT data points to produce the relative power gain.

Antenna gain measured with respect to a dipole is measured in dBd (decibels with respect to a dipole). Antenna gain measured with respect to an isotropic radiator is measured in dBi (decibels with respect to an isotropic radiator).

4.5.4.2 Gain and Beamwidth

An antenna produces gain with respect to some reference antenna, typically an isotropic antenna. Figure 4-13 shows the patterns of an isotropic and a nonisotropic antenna. The isotropic pattern is a perfect sphere—the energy is spread out evenly over the surface of the sphere. In some directions, the directive antenna broadcasts more energy than the isotropic

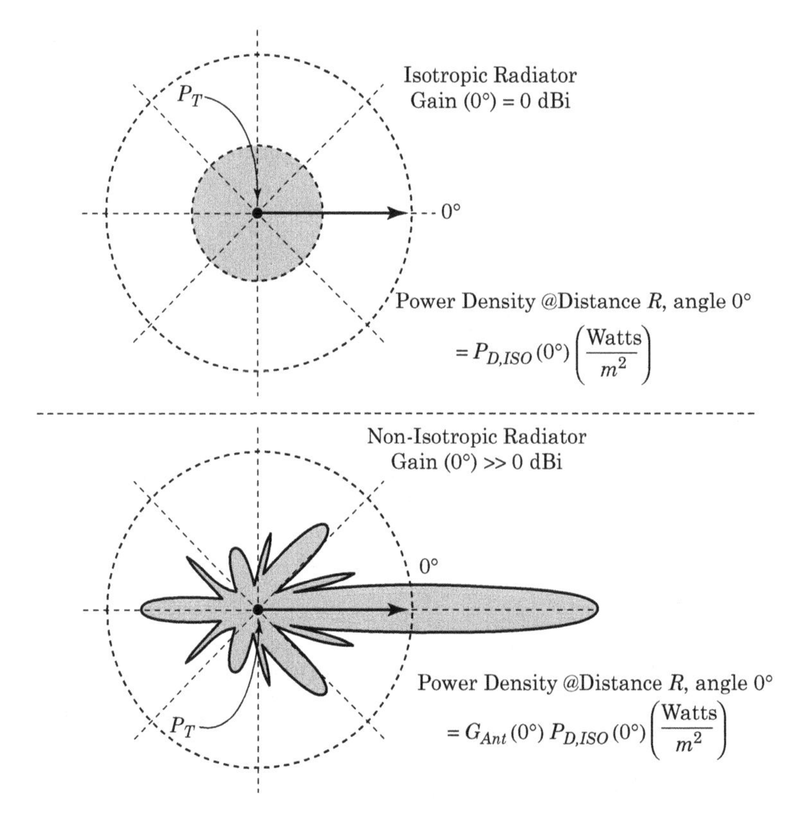

FIGURE 4-13 Power densities resulting from an isotropic antenna and from an antenna exhibiting gain. The power density of the antenna with gain is larger than the power density of the isotropic radiator when we look along the main lobe of the antenna with gain.

antenna. In other directions, the directive antenna transmits significantly less power than the isotropic antenna.

We conclude that an antenna produces gain by radiating energy preferentially in one direction at the expense of radiation in other directions. There is an inverse relationship between gain and the antenna beamwidth. Higher gains imply smaller beamwidths.

EXAMPLE

Gain and Beamwidth

Find an approximate relationship between gain and beamwidth. Find the approximate beamwidth for antennas with the following gain: 10 dBi, 20 dBi, 30 dBi, and 40 dBi.

Solution

Figure 4-14 shows the power density on the surface of a sphere for both an isotropic antenna and for an antenna with gain. The power density, PD, at the surface of the sphere at radius R is

$$\begin{aligned} PD_{A,omni} &= \frac{P_{TX}}{\text{Surface area of radius } R \text{ sphere}} \\ &= \frac{P_{TX}}{4\pi R^2}\left(\frac{\text{Watts}}{\text{m}^2}\right) \end{aligned} \tag{4.12}$$

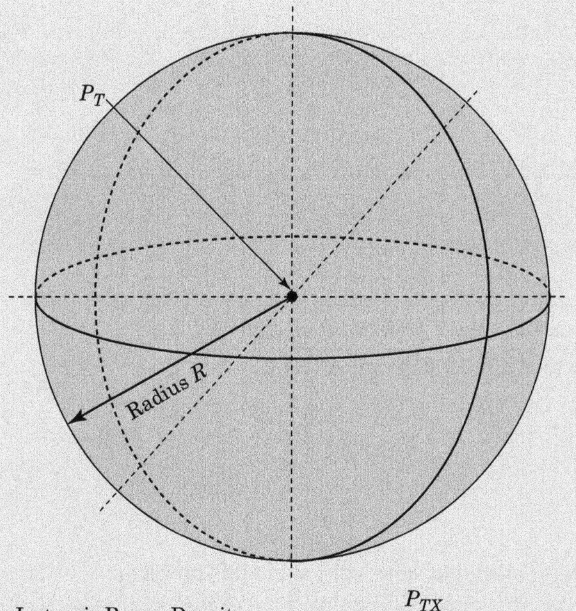

FIGURE 4-14 ■ Power densities on the surface of a sphere resulting from an isotropic antenna and from an antenna exhibiting gain.

$$\begin{aligned} \text{Isotropic Power Density} &= \frac{P_{TX}}{\text{Surface Area of radius } R \text{ Sphere}} \\ &= \frac{P_{TX}}{4\pi R^2}\left(\frac{\text{Watts}}{m^2}\right) \end{aligned}$$

$$\begin{aligned} \text{Power Density } w/\text{Gain} &= G\,\frac{P_{TX}}{\text{Surface Area of radius } R \text{ Sphere}} \\ &= G\,\frac{P_{TX}}{4\pi R^2}\left(\frac{\text{Watts}}{m^2}\right) \end{aligned}$$

For a directional antenna with a gain G, the power density at point A is G times the power density from an omnidirectional antenna or

$$\begin{aligned} PD_{A,Gain} &= G \cdot PD_{A,OMNI} \\ &= G\frac{P_{TX}}{4\pi R^2}\left(\frac{\text{Watts}}{\text{m}^2}\right) \end{aligned} \tag{4.13}$$

An antenna produces gain by selectively radiating power in the desired direction at the expense of radiating power in other directions.

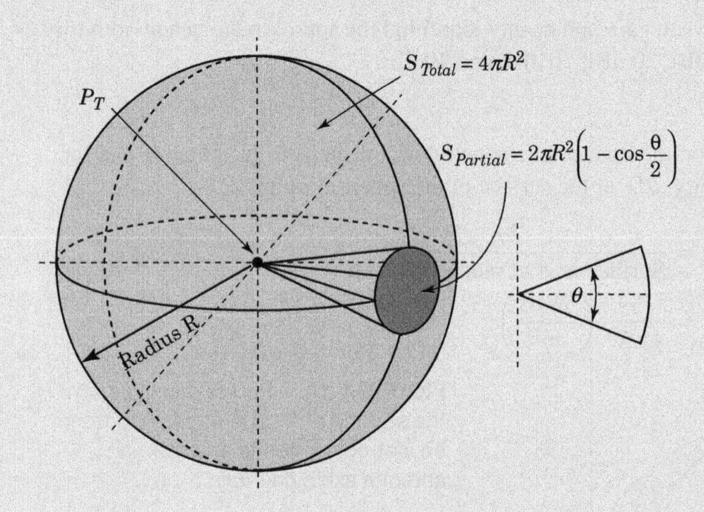

FIGURE 4-15 ■ A geometrical interpretation of antenna gain.

Figure 4-15 shows a geometrical approximation for antenna gain. Combining equations (4.12) and (4.13) produces

$$\begin{aligned} G &= \frac{PD_{A,gain}}{PD_{A,OMNI}} \\ &= \frac{P_{TX}/\text{Surface area of the partial sphere @ Point A (Radius } R)}{P_{TX}/\text{Surface area of the total sphere @ Point A (Radius } R)} \\ &= \frac{4\pi R^2}{2\pi R^2[1-\cos(\theta)]} \\ &= \frac{2}{1-\cos(\theta)} \end{aligned} \tag{4.14}$$

Table 4-2 shows the results of applying equation (4.14) to the values of gain in this problem.

TABLE 4-2 Antenna gain versus beamwidth.

Antenna Gain (dBi)	Beamwidth (°)
10	74
20	23
30	7.2
40	2.3

There are other empirical equations relating gain and beamwidth. Skolnik (1980) gives us

$$G = \frac{20,000}{\theta^2} \quad (\theta \text{ in degrees}) \tag{4.15}$$

while Johnson and Jasik (1984) offer

$$G = \frac{30,000}{\theta^2} \quad (\theta \text{ in degrees}) \tag{4.16}$$

Wolff and Kaul (1988) relate

$$G = \frac{45,000}{\theta^2} \quad (\theta \text{ in degrees}) \tag{4.17}$$

Table 4-3 shows a comparison of these formulae.

TABLE 4-3 Antenna gain versus beamwidth (various authors).

Gain (dBi)	Equation (4.14)	Skolnik (1980) (4.15)	Johnson and Jasik (1984) (4.16)	Wolff and Kaul (1988) (4.17)
10	74°	45°	55°	67°
20	23°	14°	17°	21°
30	7.2°	4.5°	5.4°	6.7°
40	2.3°	1.4°	1.7°	2.1°

Our simple analysis produces a larger beamwidth than the other, more empirical equations.

EXAMPLE

A focused laser diode (e.g., a laser diode in a mount behind a lens) has a gain of about 135 dB. What is the beamwidth of such a radiator? What is the area covered by the beam when it arrives at the earth from geosynchronous orbit (see Figure 4-16)?

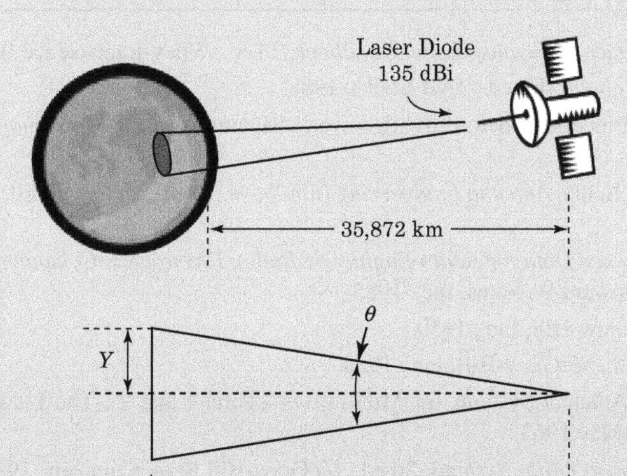

FIGURE 4-16 ■ Laser diode gain and beamwidth example.

Solution

A gain of 135 dB is 31.6E12 in linear terms. Equation (4.16) tells us

$$G = \frac{30,000}{\theta^2} \qquad (\theta \text{ in degrees})$$

$$31.6E12 = \frac{30,000}{\theta^2} \tag{4.18}$$

$$\theta = 30.8E-6^\circ$$

$$= 537.6E-9 \text{ radian}$$

Geosynchronous orbit is about 35,872 km above the surface of the earth. Unfolding the geometry of Figure 4-16 produces

$$\tan\left(\frac{\theta}{2}\right) = \frac{Y}{X}$$

$$\frac{\theta}{2} = \frac{Y}{X} \text{for small } \theta$$

$$Y = (X)\frac{\theta}{2} \tag{4.19}$$

$$= (35{,}872E3)\frac{537.6E-9}{2}$$

$$= 9.6 \text{ meters}$$

The area covered by the beam is a circle whose radius is about 9.6 meters. The area on the ground works out to a little under 300 square meters.

4.6 | BIBLIOGRAPHY

Freeman, Roger L., *Telecommunication Transmission Handbook*, 3d ed., Wiley-Interscience, 1991.

Graf, Rudolf F., *Electronic Databook*, 4th ed., TAB Books, 1988.

Janich, David Z., "RF Signal Processing before the Receiver," *Watkins-Johnson RF Component Catalog,* 1989.

Johnson, Richard C., and Jasik, Henry, *Antenna Engineering Handbook,* 2d ed., McGraw-Hill, Inc., 1984.

Jordan, Edward C. (Ed.), *Reference Data for Radio Engineers: Radio, Electronics, Computer and Communications*, 7th ed., Howard W. Sams, Inc., 1985.

Kraus, John D., *Antennas*, McGraw-Hill, Inc., 1950.

Kraus, John D., *Antennas*, 2d ed., McGraw-Hill, Inc., 1988.

Lenkhurt Electric Company, "Antenna Systems for Microwaves—Parts 1 and 2," The Lenkhurt Demodulator, pp. 301–323, May 1963.

Skolnik, Merrill L., *Introduction to Radar Systems*, 2d ed., McGraw-Hill Book Company, 1980.

Statman, Joseph I., "Optimizing the Galileo Space Communications Link," Jet Propulsion Laboratory, California Institute of Technology, Pasadena, CA. 1996.

Stutzman, Warren L., *Antenna Theory and Design*, John Wiley and Sons, 1981.

Terman, Frederick E., *Electronic and Radio Engineering*, 4th ed., McGraw-Hill, Inc., 1955.

Terman, Frederick E., *Fundamentals of Radio*, McGraw-Hill, Inc., 1938.

Westman, H.P., Karsh, M., Perguini, M.M., and Fujii, W.S. (Eds.), *Reference Data for Radio Engineers*, 5th ed., Howard W. Sams, Inc., 1969.

Wolff, Edward A., and Kaul, Roger. Microwave Engineering and Systems Applications, Wiley-Interscience, Inc., 1988.

4.7 | PROBLEMS

1. **Multiple Choice: Antenna Gain** The gain of an antenna
   a. is proportional to the antenna's aperture in wavelengths.
   b. is proportional to an antenna's physical size in meters.
   c. increases the amount of signal power transmitted in all directions.
   d. is a valid concept only in the transmitting sense.
2. **Multiple choice: Antenna Noise Temperature** The noise temperature of an antenna ...
   a. depends only on the physical temperature of the antenna and its feed lines.
   b. depends largely on the environmental noise present in the antenna's field of view.
   c. depends largely on the antenna's aperture.
   d. depends largely on the antenna's gain.
3. **Geostationary Satellites** Geostationary satellites fly at 35,863 km above the surface of the earth and are typically positioned 2° apart, as viewed from the surface of the earth.
   a. Estimate the minimum antenna gain required to talk with only one satellite.
   b. What other antenna characteristics could you use to separate the signals from adjacent satellites?
4. **Satellite Geometry** Two satellites are 6,370 km above the surface of the earth. The beamwidths of the communications antennas 2.5° and 10°.
   a. Find the gain of each antenna.
   b. Find the footprint (in square km) of the beam on the surface of the earth.
5. **Antenna Patterns and Aperture** Given the antenna pattern shown in Figure 4-17:
   a. What is the aperture of this antenna at 0° and at 45°?
   b. What is the front-to-back ratio of this antenna?
   c. How much attenuation does this antenna present to a signal arriving from 135° with respect to a signal arriving from 0°?

   The frequency of operation is 10 GHz.
6. **Antenna Aperture** A television station transmitting on channel 35 (599 MHz) produces a power density of 1 μwatt/m^2 at the position of a Yagi-Uda antenna whose gain is 8 dBi. Find the power delivered to the receiver by the antenna assuming the antenna is matched to the receiver. Answer in dBm.
7. **Temperature of Mars** At a wavelength of 31.5 mm, the 15 meter radio telescope at the U.S. Naval Research Laboratory has a beamwidth of 0.116°. From this antenna, the planet Mars subtends an angle of 0.005°. When this antenna is focused on cold sky (interstellar space), the antenna noise temperature is measured to be 3°K. When this antenna is focused on Mars, the antenna noise temperature of the antenna increases to 3.24°K.

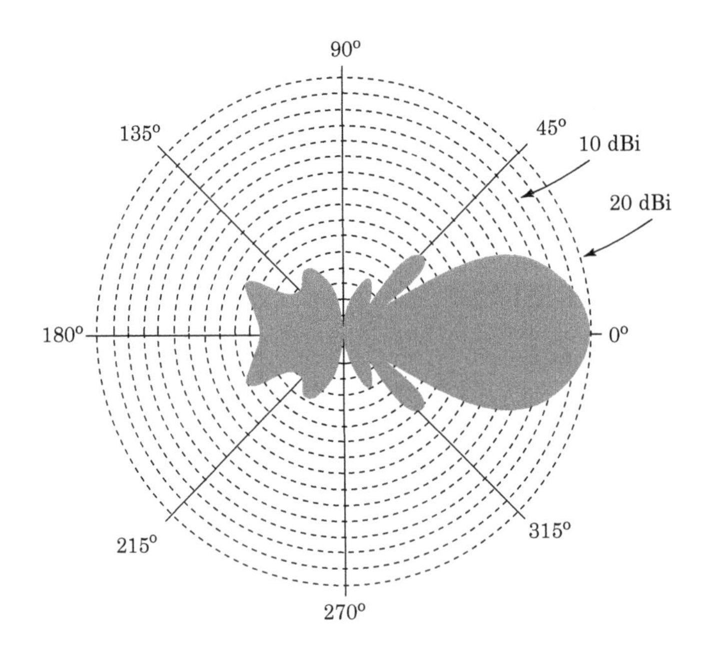

FIGURE 4-17 Antenna pattern.

a. What is the frequency of observation corresponding to the given wavelength?

b. What is the gain of the antenna at the frequency of observation? Be sure to include your units.

c. What would be the gain of an antenna that would view only Mars?

d. Find the average temperature of Mars at a wavelength of 31.5 mm. (Hint: The antenna subtends an area of the sky. Mars subtends a smaller area of the sky. The increase in noise temperature of the observing antenna is caused by a "hot" Mars being present in a small portion of the view of the sky.)

e. Comment on the effect of polarization in this measurement. If the antenna is polarized, does the measured temperature of Mars increase or decrease?

8. **Cell Tower Antenna Gain**

   a. A cell tower antenna has a 360° horizontal (azimuth) coverage and a 10° vertical (elevation) coverage. Using the *surface area of a sphere* geometry concept, estimate the gain of the antenna at 0° elevation.

   b. If we replace the single antenna with 360° of azimuth coverage with three separate antennas, each coverting 120° of azimuth, what is the corresponding increase in antenna gain for each antenna?

9. **Antenna Beamwidths** Two satellites have identical characteristics except that the first satellite orbits the earth in geosynchronous orbit at a height of 22,300 miles and the second satellite orbits nongeosynchronously at 1,200 miles. Both satellites have an antenna gain of 34 dBi. Find the footprint (in square miles) of the beam on the surface of the earth.

CHAPTER 5

Filters

Still a man hears what he wants to hear, and disregards the rest.

Simon and Garfunkel, "*The Boxer*"

Chapter Outline

5.1 INTRODUCTION

A filter's job is to remove the signals we don't want while retaining (and not distorting) the signals we want. The primary discriminator is frequency. However, consider the following:

- How much of the unwanted signals can we tolerate? Our filter won't remove the unwanted signals entirely; the filter will attenuate them by only a finite amount. How much attenuation do we actually need?
- How much of the desired signal we want can we afford to lose in the filtering process? As our filter removes the unwanted signals, it will also attenuate and distort the signals we want to preserve.
- How does the filter alter the wanted signal? The filter may change the magnitude of the wanted signal unevenly over its bandwidth, or it may produce more subtle changes by altering the signal's phase.

5.2 LINEAR SYSTEMS REVIEW

Calling a system *linear* implies several things:

- If we excite a linear system at a particular frequency, we will observe only that frequency in the system. We may observe phase and amplitude changes, but we will never observe frequencies present in the system that we did not apply.
- The transfer function of a linear system does not depend on the magnitude of the input signal. In other words, the ratio of the input to the output will be same at an input level of 1 μV or when the input is 1 MV. If the system performance varies with signal level, the system is not linear.
- Superposition works. We can excite a linear system with a signal S_1 and observe an output O_1. We then excite the system with a signal S_2 and observe an output O_2. When we excite the system with a linear combination of S_1 and S_2, the output will be a linear combination of O_1 and O_2.
- The equations that describe linear systems are well understood and friendly to use. Nonlinear mathematics can be unwieldy.
- We can approximate any device as a linear device when the input signal level is small enough (we'll discuss *small enough* when we discuss linearity). This concept is the basis for s-parameters and small signal models.

5.2.1 Pole/Zero Review—Series RLC

Figure 5-1 shows a series resistor-inductor-capacitor (RLC) circuit and characteristic quantities we will be analyzing.

The voltage $v(t)$ or $V(s)$ will be the input, and we'll consider $i(t)$ or $I(s)$ as the output. The transfer function is

$$\frac{I(s)}{V(s)} = \frac{s/L}{s^2 + s\frac{R}{L} + \frac{1}{LC}} \tag{5.1}$$

The behavior of the circuit in Figure 5-1 is described completely by equation (5.1). Since the roots of the polynomials in the numerator and denominator of equation (5.1) exactly specify the equation, we need only these roots to unambiguously determine the behavior of the circuit in Figure 5-1.

This analysis holds for any transfer function. The roots of the numerator and denominator are all we need to analyze the behavior of any linear system.

We can put equation (5.1) in the form

$$\frac{I(s)}{V(s)} = \frac{s - s_{Z1}}{(s - s_{P1})(s - s_{P2})} \tag{5.2}$$

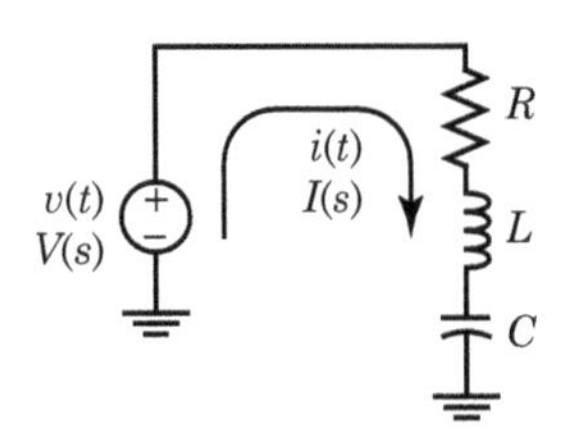

FIGURE 5-1 ■ A simple series RLC circuit.

Combining equations (5.1) and (5.2), we know that the single root of the numerator is

$$s = s_{Z1} \tag{5.3}$$

We'll derive the roots of the denominator from

$$s^2 + s\frac{R}{L} + \frac{1}{LC} = 0 \tag{5.4}$$

which produces

$$s_{P1}, s_{P2} = -\frac{R}{2L} \pm \frac{1}{2}\sqrt{\left(\frac{R}{L}\right)^2 - \frac{4}{LC}} \tag{5.5}$$

When the denominator is nonzero, then the transfer function (5.2) will be zero whenever $s = s_{Z1}$. We call the roots of the numerator the *zeroes of the transfer function.* Similarly, when the numerator is nonzero, the transfer function will be infinite whenever $s = s_{P1}$ or $s = s_{P2}$. We call the roots of the denominator the *poles of the transfer function.*

We can write any transfer function as the ratio of two polynomials

$$H(s) = \frac{(s - s_{Z1})(s - s_{Z2})(s - s_{Z3})\ldots(s - s_{ZN})}{(s - s_{P1})(s - s_{P2})(s - s_{P3})\ldots(s - s_{PM})} \tag{5.6}$$

where

s_{ZN} = the N-th zero of the transfer function
s_{PM} = the M-th pole of the transfer function

Let's assume that the denominator has complex roots, that is,

$$\left(\frac{R}{L}\right)^2 < \frac{4}{LC} \tag{5.7}$$

which forces the quantity under the square root to be negative. When we plot the complex roots on the s-plane, we have a picture that exactly describes the behavior of our series RLC circuit. The roots are plotted in Figure 5-2.

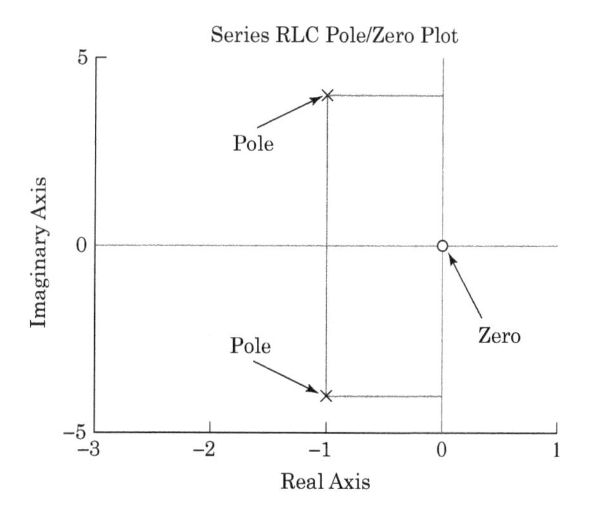

FIGURE 5-2 ■ The pole/zero plot of the simple series RLC circuit of Figure 5-1.

We can use the pole/zero plot of Figure 5-2 to gain insight into the behavior of the transfer function.

5.2.2 Magnitude of $H(j\omega)$

Let us examine the pole/zero plot for clues to the transfer function's magnitude. If we're interested only in the magnitude, we can rewrite equation (5.6) as

$$|H(s)| = \frac{|s - s_{Z1}||s - s_{Z2}||s - s_{Z3}|\ldots(s - s_{ZN})}{|s - s_{P1}||s - s_{P2}||s - s_{P3}|\ldots(s - s_{PM})} \tag{5.8}$$

For a sinusoidal input signal at frequency f, $s = j\omega$ where $\omega = 2\pi f$. The term $|s - s_{ZN}|$ represents the distance between the evaluation frequency s and the N-th zero. Likewise, the term $|s - s_{PM}|$ is the distance between the evaluation frequency s and the M-th pole. In words, equation (5.8) is

$$|H(s)| = \frac{[\text{dist}(s - s_{Z1})][\text{dist}(s - s_{Z2})][\text{dist}(s - s_{Z3})]\ldots[\text{dist}(s - s_{ZN})]}{[\text{dist}(s - s_{P1})][\text{dist}(s - s_{P2})][\text{dist}(s - s_{P3})]\ldots[\text{dist}(s - s_{PM})]} \tag{5.9}$$

where

$\text{dist}(s - s_{ZN})$ = the distance between the evaluation frequency s and the N-th zero
$\text{dist}(s - s_{PM})$ = the distance between the evaluation frequency s and the M-th pole

Figure 5-3 shows the RLC pole/zero plot of Figure 5-2 redrawn to emphasize the magnitude characteristics.

For the RLC circuit of Figure 5-1, equation (5.9) becomes

$$|H(j\omega)| = \frac{\text{dist}(Z_1)}{[\text{dist}(P_1)][\text{dist}(P_2)]} \tag{5.10}$$

where

$\text{dist}(Z_1)$ = the distance from zero #1 to the evaluation frequency
$\text{dist}(P_1)$ = the distance from pole #1 to the evaluation frequency
$\text{dist}(P_2)$ = the distance from pole #2 to the evaluation frequency

The box labeled $s = j\omega$ in Figure 5-3 represents the evaluation frequency.

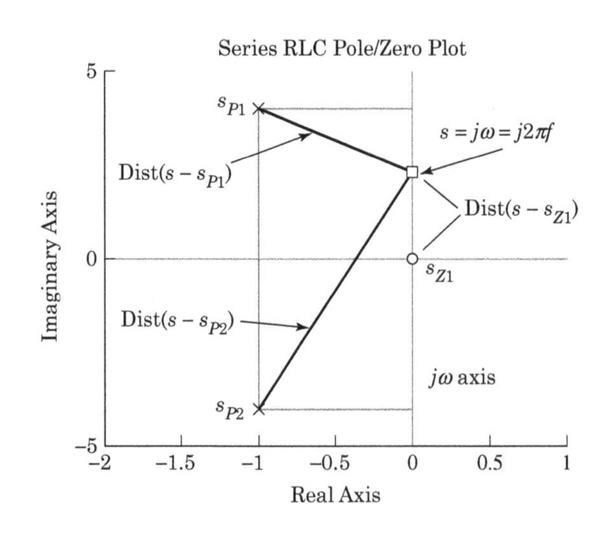

FIGURE 5-3 S-plane description of simple series RLC circuit transfer function emphasizing the magnitude characteristics. The box labeled $s = j\omega$ represents the evaluation frequency.

As the evaluation frequency $s = j\omega$ changes, the distances between s and the various poles and zeroes of the transfer function change. As $j\omega$ approaches a transfer function zero, the distance between $j\omega$ and the zero decreases. This distance will go to zero if the transfer function zero rests directly on the $j\omega$ axis. Since this distance is in the numerator of the transfer function, the magnitude of the transfer function approaches zero whenever $j\omega$ approaches a transfer function zero.

Alternately, as the evaluation frequency $s = j\omega$ approaches either of the two poles, the distance between $j\omega$ and the pole decreases. This distance will also approach zero when the pole is very close to the $j\omega$ axis. Since this distance is in the denominator of the transfer function, the magnitude of the transfer function will get large as $j\omega$ approaches a pole.

We can more concisely say

$$\begin{gathered}\text{As } j\omega \text{ approaches any zero, } |H(j\omega)| \text{ approaches } 0 \\ \text{and} \\ \text{As } j\omega \text{ approaches any pole, } |H(j\omega)| \text{ approaches } \infty\end{gathered} \tag{5.11}$$

5.2.2.1 Large jω

Figure 5-4 shows the RLC pole/zero plot when $j\omega$ is large with respect to the pole/zero constellation.

Examination of Figure 5-4 reveals that, for large $j\omega$, the distance from the evaluation frequency to each pole and zero is about the same, or

$$\begin{aligned}&\text{dist}(s - s_{Z1}) \approx \text{dist}(s - s_{P1}) \approx \text{dist}(s - s_{P2}) \\ &\Rightarrow |H(j\omega)| = \frac{\text{dist}(Z_1)}{[\text{dist}(P_1)][\text{dist}(P_2)]} \approx \frac{1}{[\text{dist}(P_1)]} \approx \frac{1}{[\text{dist}(P_2)]}\end{aligned} \tag{5.12}$$

Although we can't draw any meaningful conclusions about the magnitude response, we can comment on its rate of change. If $|H(j\omega_1)|$ is the magnitude of the transfer function at ω_1 and $|H(j2\omega_1)|$ is the magnitude at $2\omega_1$, then we can write

$$\frac{|H(j\omega_1)|}{|H(j2\omega_1)|} \approx \frac{1/j\omega_1}{1/j2\omega_1} = 2 \tag{5.13}$$

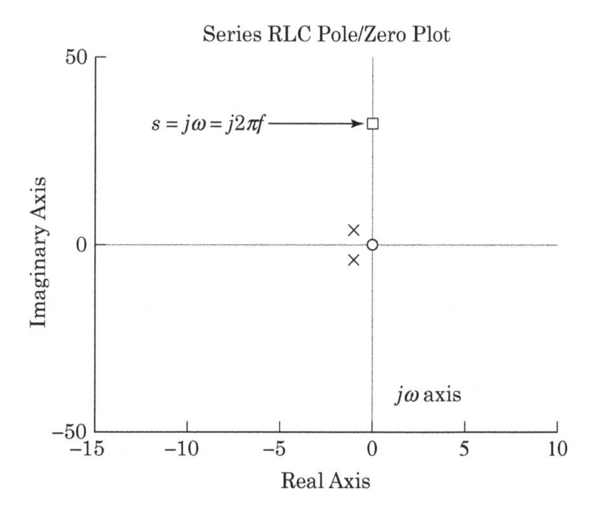

FIGURE 5-4 ■ Simple series RLC circuit transfer function, magnitude characteristics, with large $j\omega$.

In other words, for the simple RLC circuit under large $j\omega$ conditions, if we double the evaluation frequency we halve the magnitude of the transfer function.

A little thought reveals that the magnitude response of any transfer function will drop off at a constant rate under large $j\omega$ conditions. That rate is the *ultimate roll-off* of the transfer function, and it is given by

$$\begin{aligned} \text{Ultimate Rolloff} &= \left(6\frac{\text{dB}}{\text{Octave}}\right)(\#\text{ Poles} - \#\text{ Zeroes}) \\ &= \left(20\frac{\text{dB}}{\text{Decade}}\right)(\#\text{ Poles} - \#\text{ Zeroes}) \end{aligned} \tag{5.14}$$

where an octave refers to a doubling in frequency (i.e., going from f_1 to $2f_1$), and a decade refers to increasing the frequency by an order of magnitude (i.e. going from f_1 to $10f_1$).

EXAMPLE

Transfer Function Ultimate Roll-Off Rate

Given the number of poles and zeroes for the following low-pass filters, predict the ultimate roll-off of each filter:

a. 6 poles, 5 zeroes
b. 7 poles, 6 zeroes
c. 7 poles, 2 zeroes
d. 3 poles, 1 zero

Solution

Using equation (5.14), we find

a. Ultimate roll-off = 6dB/octave (6 poles − 5 zeroes) = 6 dB/octave
b. Ultimate roll-off = 6dB/octave (7 poles − 6 zeroes) = 6 dB/octave
c. Ultimate roll-off = 6dB/octave (7 poles − 2 zeroes) = 30 dB/octave
d. Ultimate roll-off = 6dB/octave (3 poles − 1 zero) = 12 dB/octave

5.2.3 Angle of $H(j\omega)$

We can also use the pole/zero plot to glean information about the transfer function's angle. Figure 5-5 shows the pole/zero plot of our RLC circuit with emphasis on finding the phase of the transfer function.

Emphasizing the transfer function angle, equation (5.6) is

$$\begin{aligned} \angle H(j\omega) = {} & [\angle(s - s_{Z1}) + \angle(s - s_{Z2}) + \angle(s - s_{Z3}) \ldots + \angle(s - s_{ZN})] \\ & - [\angle(s - s_{P1}) + \angle(s - s_{P2}) + \angle(s - s_{P3}) \ldots + \angle(s - s_{PM})] \end{aligned} \tag{5.15}$$

where

$\angle(s - s_{ZN})$ = the angle between the evaluation frequency s and the N-th zero
$\angle(s - s_{PM})$ = the angle between the evaluation frequency s and the M-th pole

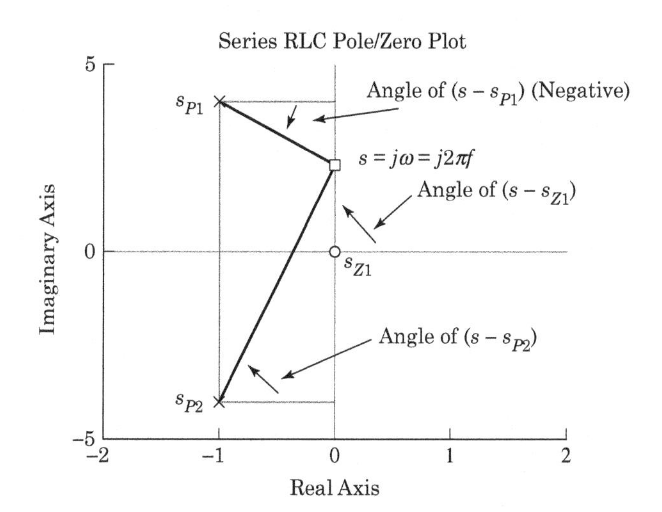

FIGURE 5-5 S-plane description of simple series RLC circuit transfer function emphasizing the phase characteristics.

In words, the angle of $H(s)$, $\angle H(s)$, is the sum of the angles from the zeroes to the evaluation frequency ($s = j\omega$) minus the sum of the angles from the poles to the evaluation frequency.

Note the definition of the angles in Figure 5-5. We measure the angle between a line parallel to the real axis and the line connecting the pole or zero to the evaluation frequency. Counterclockwise represents a positive angle.

From Figure 5-5, the angle of the series RLC transfer function is

$$\angle H(j\omega) = \angle(s - s_{Z1}) - [\angle(s - s_{P1}) + \angle(s - s_{P1})] \tag{5.16}$$

where

$\angle(s - s_{Z1})$ = the angle between the zero and the evaluation frequency
$\angle(s - s_{P1})$ = the angle between the first pole and the evaluation frequency
$\angle(s - s_{P2})$ = the angle between the second pole and the evaluation frequency

Each angle changes with changing frequency. When the evaluation frequency is in the neighborhood of a pole or a zero, the pole or zero will have a significant impact upon the phase response of the transfer function. See the section titled "Dominant Poles and Zeroes" for more information.

5.2.3.1 Large $j\omega$

If $j\omega$ is very large with respect to the pole/zero constellation, we can make some approximations regarding the ultimate phase of the transfer function.

Figure 5-6 shows Figure 5-5 redrawn under large $j\omega$ conditions.

Figure 5-6 suggests that, for large $j\omega$, the angles between s (the frequency of interest) and the poles of the transfer function all approach 90°. The same relationship holds for the angles between s and the zeroes of the transfer function. Equation (5.2), which describes our simple RLC circuit, can be written as

$$\begin{aligned} \angle H\,(j\omega) &= \angle(s - s_{Z1}) - [\angle(s - s_{P1}) + \angle(s - s_{P1})] \\ &\approx 90^\circ - [90^\circ + 90^\circ] \\ &= -90^\circ \end{aligned} \tag{5.17}$$

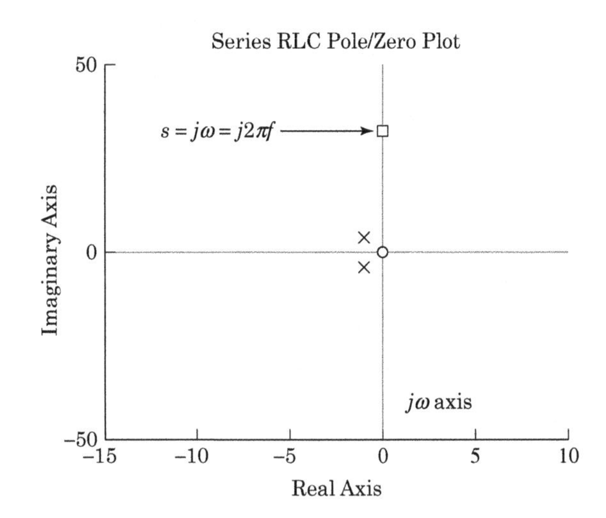

FIGURE 5-6 Simple series RLC circuit transfer function, phase characteristics, with large $j\omega$.

The term *ultimate phase* describes the phase of the transfer function ($\angle H(j\omega)$) attains for large $j\omega$. For large $j\omega$, the ultimate phase of a general transfer function is

$$\text{Ultimate Phase} = -90°[\#\text{Poles} - \#\text{Zeroes}] \tag{5.18}$$

EXAMPLE

Transfer Function Ultimate Phase

Given the number of poles and zeroes for the following low-pass filters, predict the ultimate phase of each filter:

a. 6 poles, 5 zeroes
b. 7 poles, 6 zeroes
c. 7 poles, 2 zeroes
d. 3 poles, 1 zero

Solution

Using equation (5.18), we find

a. Ultimate phase = −90° (6 poles − 5 zeroes) = −90°
b. Ultimate phase = −90° (7 poles − 6 zeroes) = −90°
c. Ultimate phase = −90° (7 poles − 2 zeroes) = −450° = −90°
d. Ultimate phase = −90° (3 poles − 1 zero) = −180°

5.2.4 Dominant Poles and Zeroes

Figure 5-7 shows the pole/zero plot of a transfer function containing five poles.

These poles are very close to the $j\omega$ axis. We know the magnitude response of any transfer function is

$$|H(j\omega)| = \frac{\Pi\ (\text{Distance from } j\omega \text{ to each Zero})}{\Pi\ \text{Distance from } j\omega \text{ to each Pole})} \tag{5.19}$$

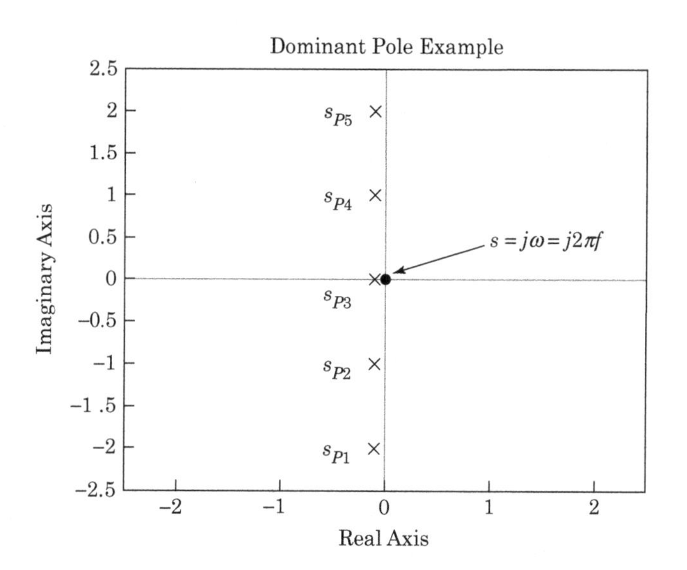

FIGURE 5-7 Pole/zero plot of a transfer function containing five poles.

For Figure 5-7, equation (5.19) is

$$|H(j\omega)| = \frac{1}{[\text{dist}(P_1)][\text{dist}(P_2)][\text{dist}(P_3)][\text{dist}(P_4)][\text{dist}(P_5)]} \tag{5.20}$$

where

dist(P_N) = the distance from pole #N to the evaluation frequency $s = j\omega$

Let's examine the case shown in Figure 5-7, where $j\omega$ is near zero. The transfer function contains a pole s_{P3}, which is very close to $\omega = 0$, so the distance between s_{P3} and $s = j\omega$ will be very small when $\omega = 0$. Equations (5.19) and (5.20) suggest that when the distance between one of the poles and the evaluation frequency is small, then $|H(j\omega)|$ will be large regardless of the distances to the other poles.

In Figure 5-7, the magnitude of the transfer function is controlled almost entirely by s_{P3} when $j\omega$ is in the neighborhood of s_{P3}. We say s_{P3} is the *dominant pole* when $j\omega$ is near zero.

Figure 5-8 shows the case when $j\omega$ is in the neighborhood of s_{P5}.

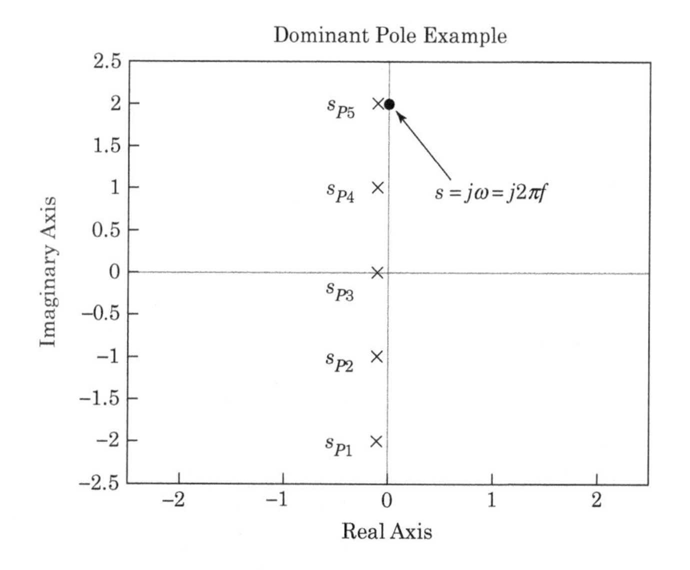

FIGURE 5-8 Pole/zero plot containing five poles. The frequency $j\omega$ is in the neighborhood of pole s_{P5}.

The distance from $s = j\omega$ to s_{P5} is very small, and equation (5.20) tells us that the magnitude of the transfer function will be very large.

The concept of a dominant pole applies whenever the distance between $j\omega$ and any pole is small relative to the other distances on the pole/zero plot. When this condition is true, then the behavior of the transfer function is dominated almost entirely by the closest or dominant pole. Pole s_{P1} is a dominant pole whenever

$$\begin{array}{lcl} d_{P1} \ll d_{P2} & & d_{P1} \ll d_{Z1} \\ d_{P1} \ll d_{P3} & & d_{P1} \ll d_{Z2} \\ d_{P1} \ll d_{P4} & \text{and} & d_{P1} \ll d_{Z3} \\ \vdots & & \vdots \\ d_{P1} \ll d_{Pn} & & d_{P1} \ll d_{Zk} \end{array} \tag{5.21}$$

where

$dP_N =$ the distance from pole #N to the evaluation frequency
$dZ_N =$ the distance from zero #N to the evaluation frequency

Similarly, the concept of a *dominant zero* applies whenever the distance between the evaluation frequency, s, and a zero is much smaller than the other distances on the pole/zero plot. Zero Z_1 is a dominant zero if

$$\begin{array}{lcl} d_{Z1} \ll d_{P1} & & d_{Z1} \ll d_{Z2} \\ d_{Z1} \ll d_{P2} & & d_{Z1} \ll d_{Z3} \\ d_{Z1} \ll d_{P3} & \text{and} & d_{Z1} \ll d_{Z4} \\ \vdots & & \vdots \\ d_{Z1} \ll d_{Pn} & & d_{Z1} \ll d_{Zk} \end{array} \tag{5.22}$$

where

$dP_N =$ the distance from pole #N to the evaluation frequency
$dZ_N =$ the distance from zero #N to the evaluation frequency

5.2.4.1 Magnitude Characteristics

Figure 5-9 shows the pole/zero diagram of a transfer function with several dominant poles and no zeroes.

We've also plotted the magnitude of the transfer function. Note the peakiness of the transfer function as ω increases from s_{P1} (at $\omega = -2$) to s_{P5} (at $\omega = +2$). The transfer function exhibits peakiness when ω is near a dominant pole because the distance between the evaluation frequency and dominant pole changes rapidly over frequency in that neighborhood while the distances between the evaluation frequency and the other poles/zeroes are almost constant.

Figure 5-10 shows an all-zero transfer function that exhibits dominant-zero behavior. Note the *suck-outs,* or areas of high attenuation, in the transfer function magnitude response when ω is in the neighborhood of a dominant zero.

Careful study of Figure 5-9 and Figure 5-10 might give us the idea that we can use the pole/zero concept to construct any arbitrary transfer function by judiciously placing

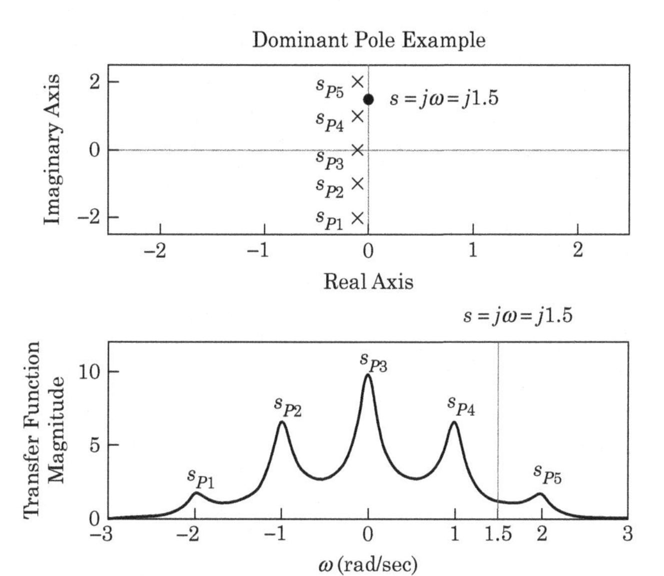

FIGURE 5-9 ■ All pole transfer function with dominant poles.

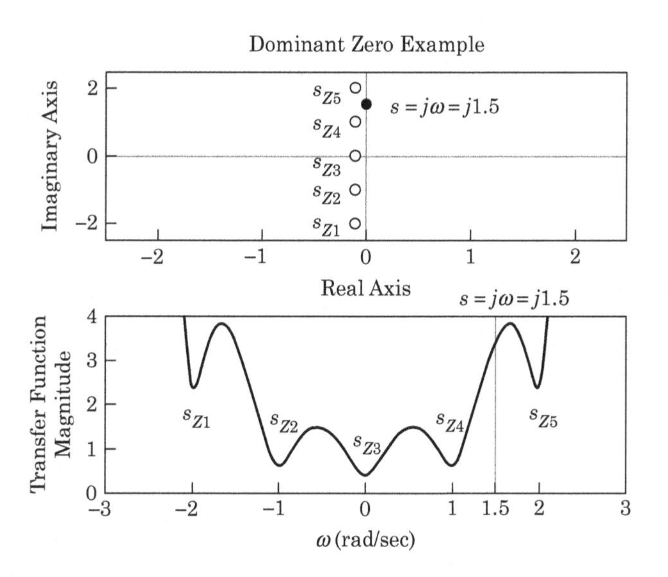

FIGURE 5-10 ■ All-zero transfer function with dominant zeroes.

poles and zeros. This is, in fact, how many filters are realized. Figure 5-11 and Figure 5-12 show the pole/zero and magnitude plots of an elliptic filter.

The positions of the poles and zeroes have been manipulated carefully to produce the desired amplitude response. Sections of the transfer function of Figure 5-12 have been labeled with the pole or zero most responsible for the characteristic.

5.2.4.2 Phase Characteristics

Let's examine the phase response of a transfer function as the input frequency passes by a dominant pole or zero. Figure 5-13 shows the $j\omega$ axis with a dominant zero at f_{Z1}. We've also drawn an expanded version of the $j\omega$ axis around f_{Z1}.

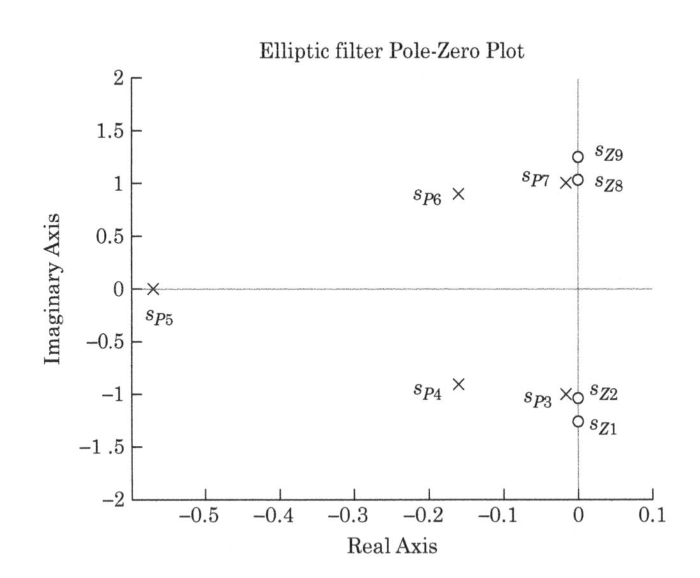

FIGURE 5-11 Pole/zero plot of an elliptical filter. This filter has been designed for a rapid transition from the filter's stopband to its passband.

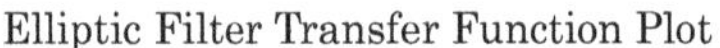
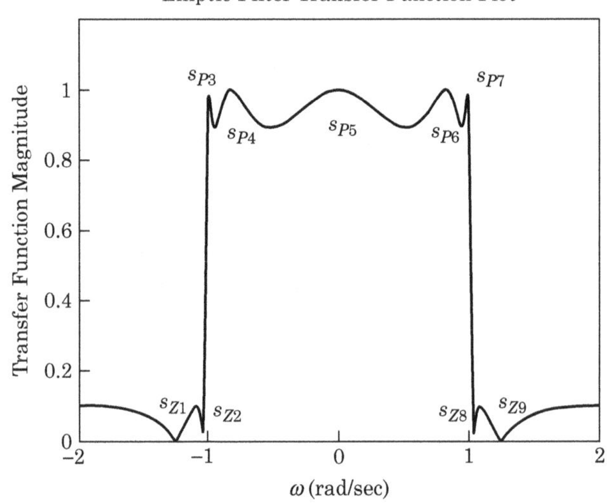

FIGURE 5-12 Magnitude response of an elliptical filter. This filter has been designed for a rapid transition from the filter's stopband to its passband.

Let's set the evaluation frequency at point [1] in Figure 5-13. This is well away from f_{Z1}, and the zero is not dominant at this frequency. The angle θ_Z represents the amount of phase shift contributed by the zero to the overall transfer function. When we're very far away from the zero as we are at point [1], the zero at f_{Z1} contributes almost -90° of phase shift to the transfer function.

As the frequency increases and we move from position [1] to position [2], the phase of the transfer function changes from -90° to -45°. We have to change the frequency quite a bit to get the phase to change by 45°.

Looking at the enlarged portion of Figure 5-13, as we change the frequency from [2] to [3] to [4], we change the phase shift from -45° (at [2]) to 0° (at [3]) to $+45^\circ$ (at [4]). We've changed the total phase shift by 90° in a very small frequency change. We had to move the entire distance from [1] to [2] to produce the same 45° phase shift. In the neighborhood of a dominant pole or zero, the phase of the transfer function changes very rapidly with frequency.

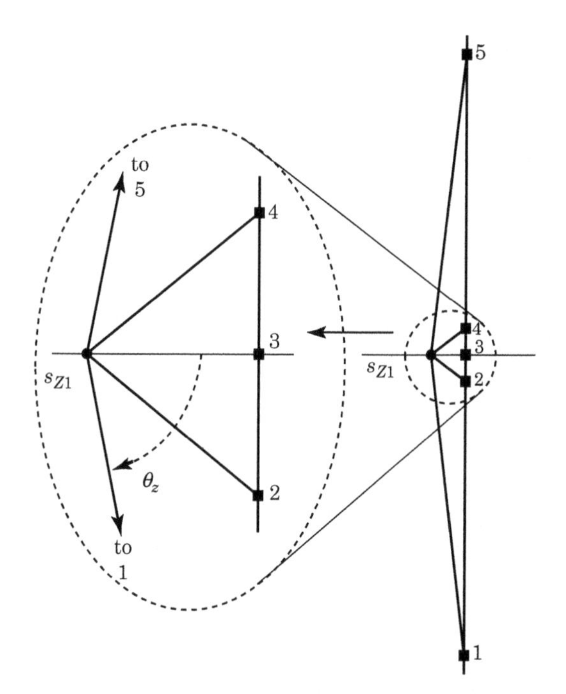

FIGURE 5-13 ■ Transfer function behavior in the neighborhood of a dominant pole.

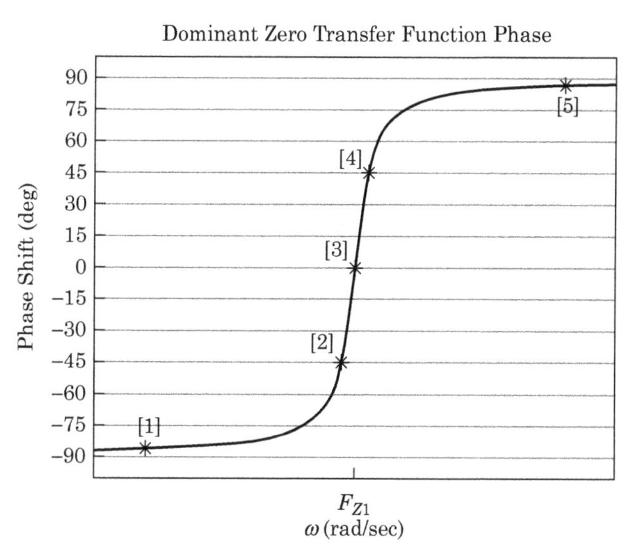

FIGURE 5-14 ■ Transfer function behavior (phase) in the neighborhood of a dominant pole.

Finally, we move from position [4] to position [5] and change another 45°. When we're at position [1] or position [5], we're in the *large* $j\omega$ region of the graph.

Figure 5-14 shows the phase response of the transfer function as we move from position [1] to position [5].

Note how the phase changes rapidly as the frequency changes from [2] to [3] to [4] and how the phase change is fairly mild from [1] to [2] and from [4] to [5]. The phase changes over a 180° range as the frequency moves from [1] to [5].

During this analysis, we've assumed that the zero at f_{Z1} is a dominant zero. In other words, we've assumed that the distances from our evaluation frequency to all of the other poles and that zeroes haven't changed significantly.

The moral here is that we can see very rapid phase changes when we're dealing with poles or zeroes that are close to the $j\omega$ axis (i.e., high-Q poles or zeroes).

EXAMPLE

Dominant Poles and Zeroes in Filter Responses

Figure 5-11 and Figure 5-12 shows the magnitude response of a low-pass filter and its associated pole/zero plot.

a. Reconcile the pole/zero plot with the magnitude response.

b. Are the poles or zeroes responsible for the sharp drops (or suck-outs) in the magnitude response?

Solution

This is an *elliptic filter*. This term is normally used to describe a filter that uses transfer function zeroes to produce a quick transition from the filter's passband to its stopband. They are called elliptic filters because you must solve elliptical integrals to generate the required component values.

a. The filter exhibits severe attenuation when the evaluation frequency is near any of the zeroes (at low and high frequencies). The filter shows only a small amount of attenuation when the frequency is in the neighborhood of the transfer function poles. The pole/zero plot corresponds to the magnitude response.

b. Comparing the pole/zero plot with the filter's magnitude response, we can see that the zeroes are responsible for the magnitude suck-outs.

5.3 | FILTERS AND SYSTEMS

We are concerned with three aspects of a filter's transfer function: (1) the magnitude response; (2) the phase response; and (3) the group delay response. All three metrics are meaningful only when we consider the steady-state performance of a filter.

Figure 5-15 shows the circuit diagram we will be considering.

Figure 5-15 shows a signal source consisting of a voltage source V_S and an internal source resistor labeled R_S. This model is reasonably accurate for signal generators, receiving antennas, and just about any other signal source. We will connect the input port of the filter to the signal source and the output port of the filter to a load resistor R_L. The voltage present at the input of the filter is V_{in}, and the voltage across R_L is V_{out}.

In a radio frequency (RF) environment, we are most often interested in power transfer—not voltage or current. The magnitude response in which we are most interested is

$$\frac{P_{RL}}{P_{Avail}} = \frac{\text{Power Dissipated in } R_L}{\text{Maximum Power Available From the Signal Source}} \tag{5.23}$$

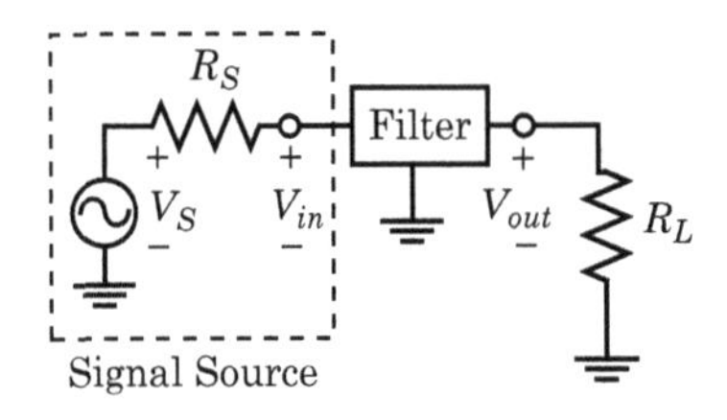

FIGURE 5-15 Circuit model of a filter placed in an radio frequency (RF) system.

where *Maximum Power Available From the Signal Source* is the signal power we would measure in a matched load resistor connected directly to the signal source.

The exact wording of this definition is important. We have previously seen that we will pull the most power from the signal source when the input impedance of the filter is matched to the source resistor, R_S. If the input of the filter is not matched, we will not accept all the available signal power from the source. This definition considers the filter's input impedance.

In a similar sense, if the output impedance of the filter is not matched to R_L, the power delivered to R_L will be less than the maximum power available from the filter.

Finally, this definition considers the filter's internal losses. If both the input and output are properly matched yet the filter dissipates some of the signal power internally (due to component losses), the loss will be expressed in equation (5.23).

5.4 | FILTER TYPES AND TERMINOLOGY

Before proceeding, we must define several filter terms.

5.4.1 Filter Terminology

The following common filter terms refer to a filter's magnitude response in the frequency domain:

Passband: A band of frequencies that a filter will allow to propagate from the filter's input to its output with minimum attenuation. The frequencies within the passband usually contain the signals of interest.

Stopband: A band of frequencies that the filter attenuates severely. Frequencies in the stopband are not signals of interest and can cause problems if they are allowed to propagate further into a system.

Transition band: Lies between a filter's passband and its stopband. Ideally, we would like a filter to transition immediately from its passband (where the filter passes signals without attenuation) to the filter's stopband (where the filter attenuates signals significantly). This characteristic not realizable in the real world.

Figure 5-16 illustrates these terms.

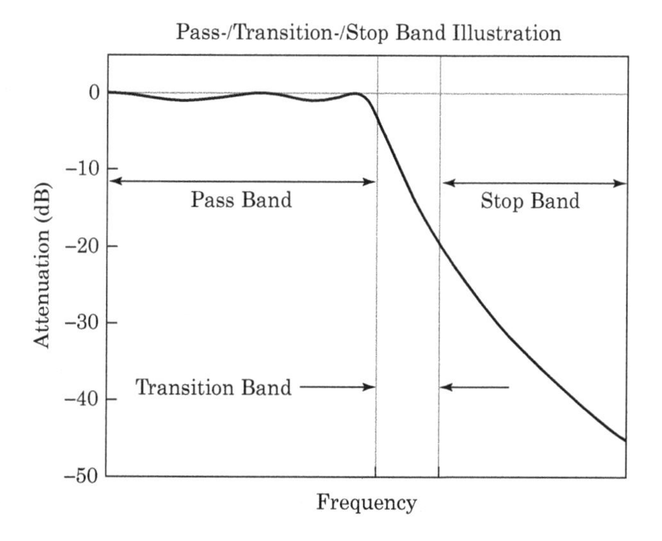

FIGURE 5-16 ■ Illustration of filter passband, transition band, and stopband regions.

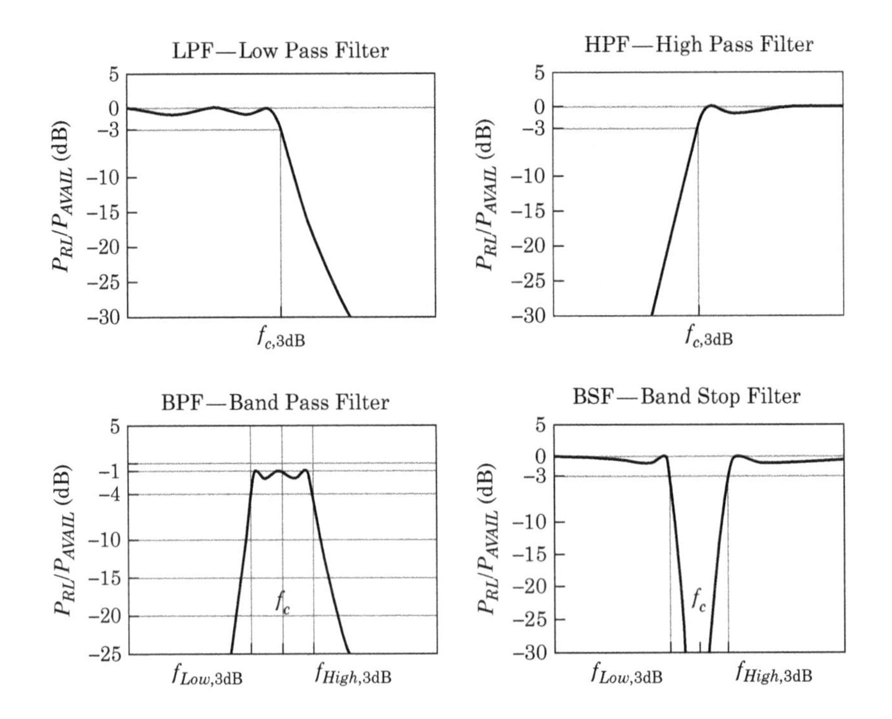

FIGURE 5-17 ■ Ideal low-pass, high-pass filter, band-pass, and band-stop magnitude responses.

The boundaries among the passband, transition band, and the stopband are not always defined clearly. In Figure 5-16, we arbitrarily drew the boundaries at the filter's 3 dB and 20 dB attenuation levels.

5.4.2 Filter Types

We will be discussing four types of filters, labeled according to their magnitude responses. As you read the following paragraphs, refer to Figure 5-17.

5.4.2.1 Low-Pass Filters

These filters pass only the lower frequency components of a signal and attenuate the higher-frequency components. The frequency marking the boundary between the low-frequency passband and high-frequency stopband is the *cutoff frequency*. We normally label the cutoff frequency as f_c (although we'll occasionally use more descriptive labels such as $f_{c,ER}$ or $f_{c,3\,dB}$). We will often abbreviate low-pass filter as LPF.

Due to realization effects, the magnitude response of an LPF often misbehaves at frequencies well above its cutoff frequency. The filter won't provide the attenuation that the mathematics says it will. We must often specify the highest frequency at which we want the LPF to behave.

5.4.2.2 High-Pass Filters

A high-pass filter allows the higher-frequency components of a signal pass while it severely attenuates the lower-frequency components of a signal. The cutoff frequency, f_c, marks the boundary between the high- and low-frequency bands of the filter. We will abbreviate high-pass filter as HPF.

The magnitude response of a HPF does not extend to infinite frequency. Often when building or buying a high-pass filter, we must specify the highest frequency we would like the filter to pass.

5.4.2.3 Band-Pass Filters

Band-pass filters pass only a band of frequencies while attenuating signals both below and above the filter's passband. The center frequency of the filter is f_c, while the lower and upper cutoff frequencies are designated f_{low} and f_{high}, respectively. We will abbreviate band-pass filter as BPF.

Often, a band-pass filter has an intrinsic power loss at the center frequency. This is the *insertion loss* of the filter, which we will designate as *IL*.

5.4.2.4 Band-Stop (or Band-Reject) Filters

Band-stop filters are the exact opposite of band-pass filters. Their job is to pass all frequencies except for a particular band where the filter should provide attenuation. Like the band-pass filter, the center frequency is f_c, while the lower and upper frequencies are f_{low} and f_{high}.

Like the high-pass filter, a band-stop filter's magnitude response will not extend to an arbitrarily high frequency.

5.5 GENERIC FILTER RESPONSES

We will examine the four types of filters with regard to their magnitude, phase, and group delay and how those characteristics relate to the filter's pole/zero plot.

5.5.1 Magnitude Response

Figure 5-18 shows a generalized magnitude response of a low-pass filter.

Areas of the amplitude response of Figure 5-18 are labeled (1) through (4). These numbers refer to particular sections of the filter's response, and we have indicated the same

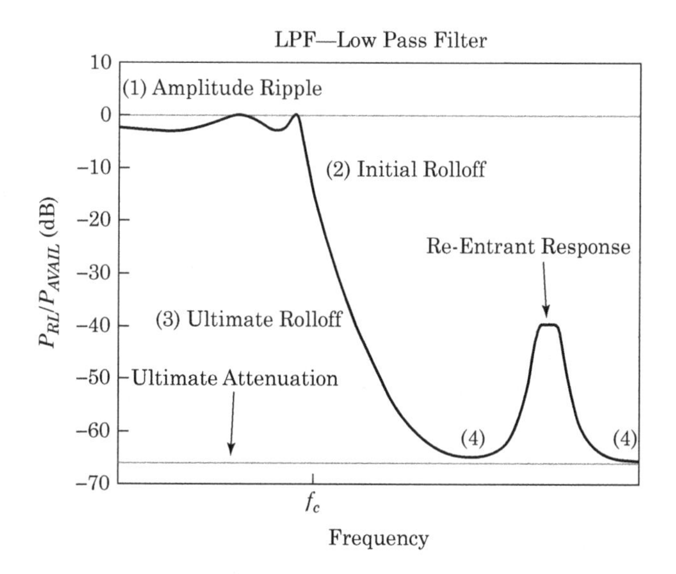

FIGURE 5-18 ■ Low-pass filter magnitude response with important regions labeled. The terms used in this figure also apply to high-pass, band-pass, and band-stop filters with minor modifications.

labels on the plots for the filter's pole/zero plot (Figure 5-19), phase response (Figure 5-21), and the group delay response (Figure 5-23).

Amplitude ripple (Area 1): Variations in the magnitude response of the filter's transfer function, particularly in the passband. Some filters exhibit passband ripple, whereas others show a flat magnitude response. The specifications for amplitude ripple usually read something like "0.5 dB of ripple over the passband."

Initial roll-off (Area 2): Measures how fast the magnitude response of the filter initially drops off just above the filter's cutoff frequency (f_c). The more amplitude ripple we allow in the filter's passband, the steeper the initial roll-off of the filter.

Ultimate roll-off (Area 3): Eventually, the initial roll-off rate gives way to the filter's ultimate roll-off rate. To reach the ultimate roll-off rate, we must be in the large $j\omega$ area of the filter's pole/zero plot. Equation (5.14) gives an expression for the ultimate roll-off of an ideal filter. Due to realization effects, we may never reach the ultimate roll-off rate of equation (5.14) .

Ultimate attenuation (Area 4): Simple filter design equations predict that a filter will roll off forever. That is, at some given frequency far out in the filter's stopband, the mathematics predict huge values of attenuation (e.g., 200 dB). In the real world, the filter eventually reaches a point where it will provide only so much attenuation and no more. This is the ultimate attenuation of the filter. This effect rises from the realization details or from the physical configuration of the filter. Typical values can range from 30 dB in commercial, physically small band-pass filters to 110 dB in large, carefully packaged units.

Reentrant response (Area 5): The behavior of a filter in its stopband. At some frequencies, the filter's magnitude response will creep up from high values of attenuation to much lower values. At some frequency in the stopband, for example, the filter's attenuation will change from 90 dB to 20 dB.

Like ultimate attenuation, reentrant responses are purely a realization problem and caused by, for example, how the filter was built, and what kind of components were used and how they were placed.

5.5.2 Pole/Zero Plot

Figure 5-19 shows the pole/zero plot of the filter from Figure 5-18.

Amplitude ripple (Area 1): Area (1) in Figure 5-19 corresponds to area (1) in Figure 5-18—the filter's passband. The poles of this particular filter are positioned on an ellipse whose long axis runs along the $j\omega$ axis.[1] As $j\omega$ increases from zero radians/sec, it passes by each pole in turn. The closest pole at a particular $j\omega$ becomes slightly dominant, so as $j\omega$ passes each pole we get a bump in the magnitude response of the filter. This is exactly the condition we see in the magnitude response of Figure 5-18.

Initial roll-off (Area 2): Area (2) corresponds to the initial roll-off portion of Figure 5-18. As $j\omega$ begins to pull away from the constellation of poles, the magnitude response drops very quickly because the distances to all of the poles are rapidly increasing. However, $j\omega$ has not increased to the point where the filter exhibits the ultimate roll-off

[1] Equivalently, both foci of the ellipse lie on the $j\omega$ axis.

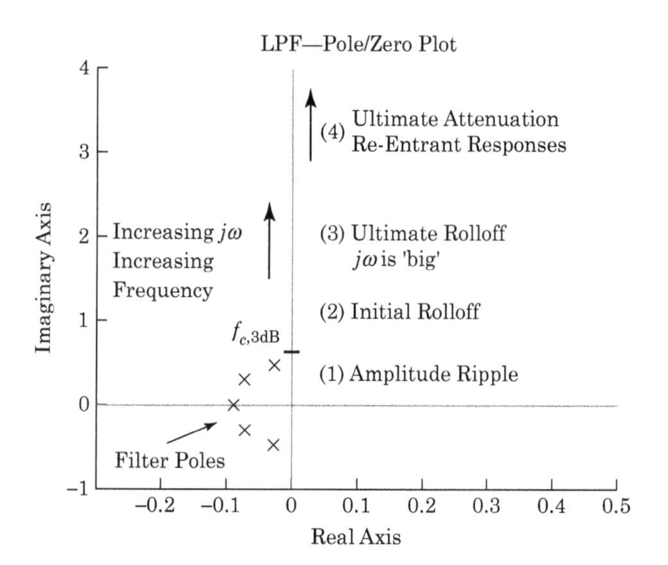

FIGURE 5-19 Low-pass filter pole/zero diagram related to Figure 5-18. The areas numbered (1) through (4) correspond to the same areas of the magnitude plot shown as Figure 5-18.

because we cannot yet make the approximation that the distances to all of the poles and zeroes are equal.

Ultimate roll-off (Area 3): This section of the pole/zero plot corresponds to the ultimate roll-off area of Figure 5-18. We can assume that the distances from $j\omega$ to each pole or zero are equal. In this area, $j\omega$ is *big*.

Ultimate attenuation, reentrant response (Area 4): The simple pole/zero plot of Figure 5-19 does not predict the ultimate attenuation or reentrant responses. According to this figure, the magnitude response should continue to fall off at the ultimate roll-off rate.

Component losses, stray capacitance, stray inductance, and component coupling all play their part in these effects. The component vagaries combine to produce transfer function poles that were not present in the original design. However, these poles are often at relatively high frequencies and do not affect the transfer function until $j\omega$ is large with respect to the filter's cutoff frequency. Figure 5-20 shows the pole/zero plot of a transfer function with these parasitic poles added.

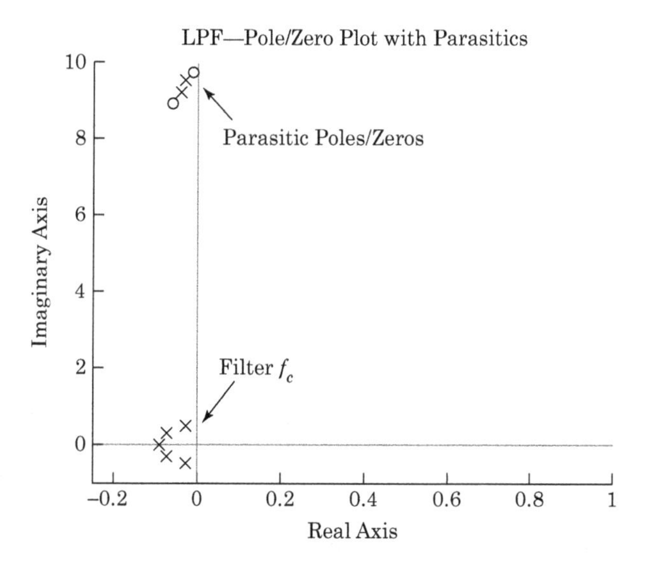

FIGURE 5-20 Pole/zero plot of a low-pass filter containing parasitic poles and zeros. Note the scale change between this figure and Figure 5-19.

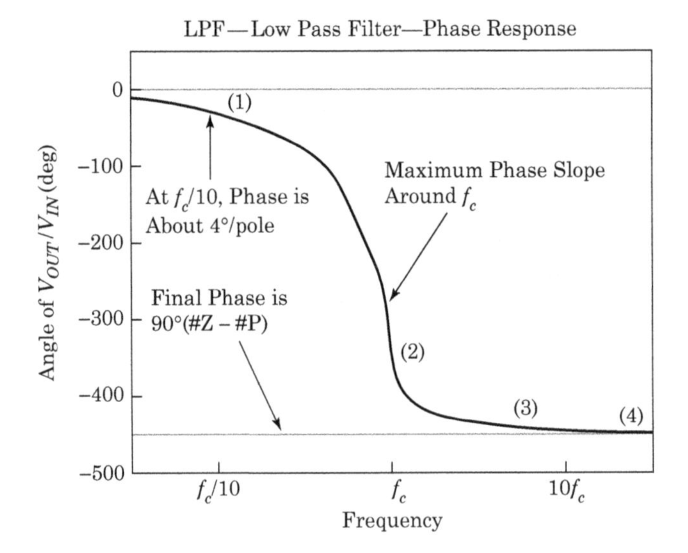

FIGURE 5-21 ■ The phase of the low-pass filter from Figure 5-18. This is the phase of the voltage transfer function of the filter (V_{out}/V_{in}). The frequency scale is logarithmic.

5.5.3 Phase Plot

Figure 5-21 shows the phase plot of the filter from Figure 5-18.

(Area 1) Amplitude ripple: This is the filter's passband. A typical filter will normally exhibit see some phase change beginning at 1/10 of the cutoff frequency. This phase change occurs long before any significant amplitude changes occur. For example, a Butterworth filter exhibits about 4° of phase shift per pole at 1/10 of f_C.

(Area 2) Initial roll-off: As the frequency increases, the slope of the phase transfer function approaches a maximum in the general neighborhood of the cutoff frequency f_C. The phase is approximately 1/2 of the final phase at f_C.

(Area 3) Ultimate roll-off: As the frequency of interest increases beyond f_C, the phase response's slope decreases, and the phase value approaches the ultimate phase value given by equation (5.18).

5.5.4 Group Delay

Let us refer to the circuit diagram of Figure 5-15. At some radian frequency ω_0, the filter produces a phase shift of $-\beta_0$ (the minus sign signifies that V_{out} is delayed in time from V_{in}). Figure 5-22 shows the steady-state time domain plots of V_{out} and V_{in}.

We can interpret the phase shift β_0 as a time delay t_{PD}. We know that one cycle of a sine wave at frequency f_0 requires $1/f_0$ seconds to complete. We also know that there are 2π radians in one cycle. Combining these two facts produces

$$\frac{-\beta_0}{t_{PD}} = \frac{2\pi}{1/f_0} = 2\pi f_0 = \omega_0 \tag{5.24}$$

Rearranging equation (5.24) produces

$$t_{PD} = -\frac{\beta_0}{\omega_0} \tag{5.25}$$

The quantity t_{PD} is *phase delay time* or *carrier delay time*. This is the time required for a sine wave to pass through a filter under steady-state conditions. Figure 5-22 shows that β_0

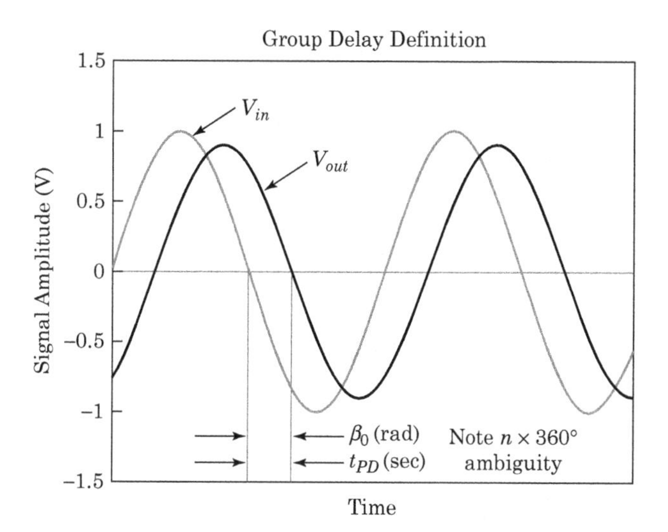

FIGURE 5-22 ■ Group delay in the time domain.

and t_0 represent the same amount of time; the quantities are simply expressed in different formats.

We are not interested in absolute amount of time required for a signal to pass through a filter. We are interested in the differential time delay through the filter. If some frequency components of a signal take longer to pass through a filter than other frequency components, the phase relationship those components will be ruined. This differential delay can cause severe distortion, especially in complex signals.

Group delay is a measure of the differential time delay caused by a filter. The quantity describes how individual frequency components are going to be delayed with respect to other component. We will define group delay as

$$\text{Group Delay} = t_{gd} = -\frac{\partial \beta}{\partial \omega} \tag{5.26}$$

Equation (5.26) represents the slope of the V_{out}/V_{in} phase versus frequency curve. The greater the slope of the V_{out}/V_{in} phase curve, the higher the value of group delay.

A flat group delay curve indicates that all of the frequency components of a signal pass through the filter in the same amount of time. There is no differential delay. A lumpy or uneven group delay curve tells us that some components will take longer to get through the filter than other components. At the output of the filter, the phase relationships among the signal's components will be upset, and the signal will be distorted.

Figure 5-23 shows the phase response of a filter and its corresponding group delay plot.

The group delay plot is the derivative of the phase plot with respect to frequency. If the filter's phase plot exhibits ripple, the group delay plot will be lumpy.

Group delay usually peaks at or near a low-pass filter's cutoff frequency. In a band-pass filter, the group delay usually peaks near the edges of the passband.

Group delay *does not* directly describe the filter's transient response. Group delay is a steady-state measurement while transient measurements speak to non-steady-state conditions.

Often, filter delay and attenuation characteristics are interdependent. The narrower a filter's transition band, the larger the delay peaks. Generally, filters with many poles and

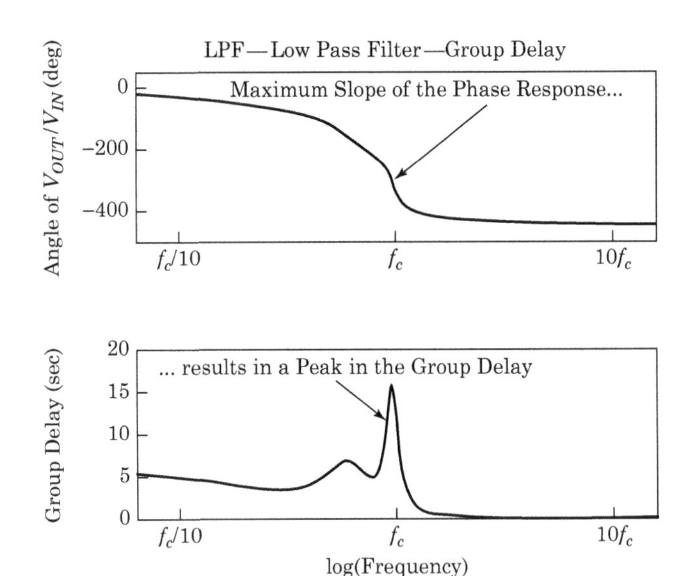

FIGURE 5-23 Group delay related to filter phase response.

filters with close-in stopband zeroes have large delay peaks. Conversely, low selectivity filters with large transition bands tend to exhibit smaller delay peaks.

5.5.4.1 Effects of Group Delay

Inappropriate group delay can unrecoverably distort signals. Amplitude modulation (AM) is immune from the effects of group delay due to the methods we use to demodulate the signal. However, we will use an AM signal to show how group delay can affect modulation.

Using Figure 5-15 as our circuit model, we will assume the signal source consisting of V_S in series with R_S is an AM source and that V_S takes the form of

$$V_s = V_{PK}[1 + m_a \cos(2\pi f_m t)] \cos(2\pi f_c t) \tag{5.27}$$

where

m_a = modulation index ($0 <= m_a <= 1$)
f_c = RF carrier frequency
f_m = modulation frequency

For this example, we will assume a modulation index m_a of 0.5, the carrier frequency f_c is 1 MHz, and the modulating frequency f_m is 1 kHz. Figure 5-24 shows V_S in the time domain, the frequency domain, and as a phasor.

A Fourier analysis of the input waveform V_S shows the signal consists of three separate sine waves:

- The carrier at f_c
- The lower sideband (LSB) is at a frequency of $(f_c - f_m)$
- The upper sideband (USB) is at a frequency of $(f_c + f_m)$

The phasor interpretation of V_s will be the most useful to our present discussion (see Figure 5-25).

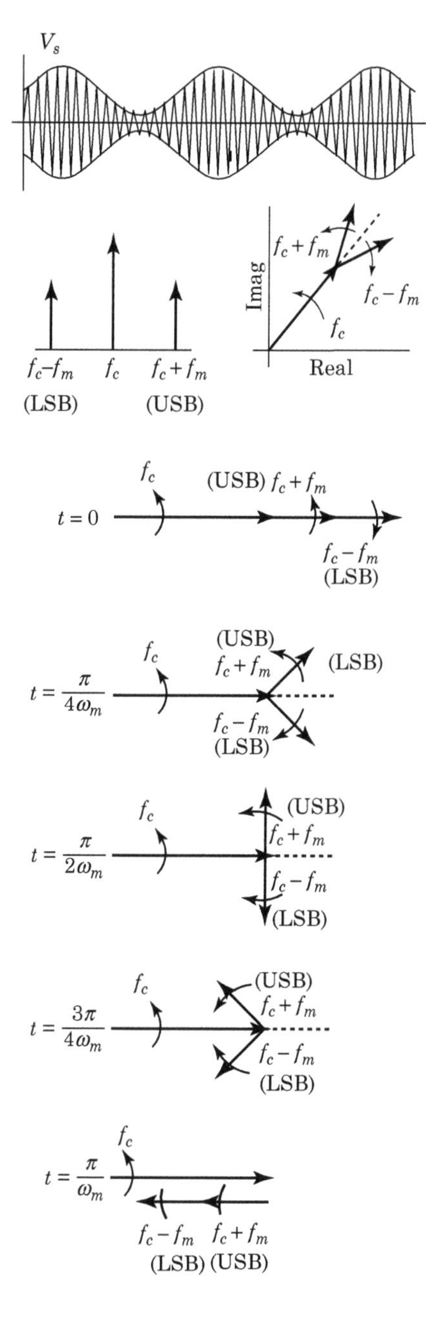

FIGURE 5-24 Amplitude modulation in the time, frequency, and phasor domains.

FIGURE 5-25 Detailed phasor diagram of amplitude modulation. The symmetry of the two sidebands prevents the resultant from exhibiting phase modulation.

The carrier will be our reference and, as such, it will be stationary in the phasor plots of Figure 5-25. We can place the USB at thend of the carrier. When positioned so, the USB rotates counterclockwise (CCW) about the end of the carrier vector. The rotation rate is f_m. Similarly, the LSB appears to rotate clockwise (CW) around the end of the carrier at the same rate.

In an undistorted AM signal, the phases of the LSB and USB are symmetrical about the carrier. The upper and lower sidebands rotate about the end of the carrier at the same rate but in opposite directions. The net effect is the LSB and USB combine to lengthen and shorten the carrier.

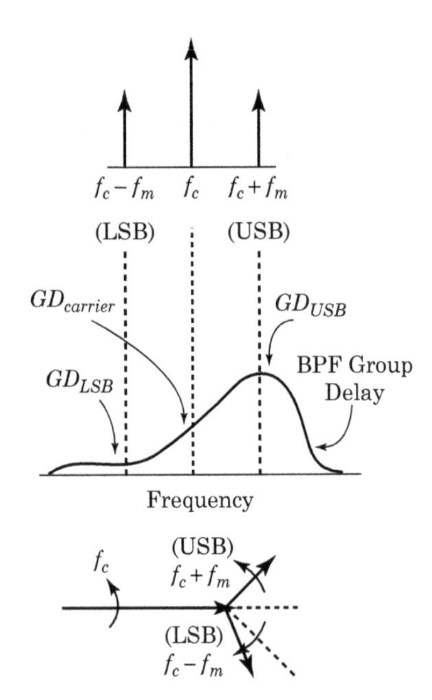

FIGURE 5-26 ■ The effect of filter group delay on amplitude modulation. The unequal group delay of the filter converts the amplitude modulation into phase modulation.

The symmetry of the LSB and USB is critical. If one sideband experiences a time delay (or, equivalently, a phase shift) with respect to the other sideband, the AM waveform will be distorted.

Figure 5-26 shows our AM waveform passing through a filter with a nonsymmetrical group delay.

The filter's group delay has destroyed the symmetry of the USB and LSB with respect to the carrier, so the instantaneous amplitude of the waveform is distorted. Since the angles of the USB and LSB no longer cancel out, the carrier is now shifting in phase at a rate of f_m.

Group delay distortion is easy to visualize for the AM case; however, group delay usually is not an issue in AM systems since the information is carried in the amplitude of the signal. Group delay can be a big problem in frequency- or phase-modulated systems and in digital signals since the information is carried by the phase of the signal. Like AM, the sidebands of FM and PM systems must vectorially add to produce the proper composite waveform. If the sidebands do not combine properly, the distortion produced plays havoc with the demodulator and the demodulated waveforms will suffer badly.

5.6 | CLASSES OF LOW-PASS FILTERS

In this section, we will describe and compare four common types of filters. All of the filters are five-pole low-pass filters, but the low-pass characteristics of each filter will transfer to the equivalent high-pass, band-pass, and band-stop filters.

We'll graph the filter performance on a logarithmic frequency axis. This scale linearizes the plot and brings out aspects of the plots that wouldn't be obvious if we used a linear scale on the x-axis.

In this section, we're going to be very careful to specify filter cutoff frequencies as either $f_{c,3\,\text{dB}}$ (the 3 dB or half-power cutoff frequency, which is the most common specification) or $f_{c,ER}$ (the equal-ripple cutoff frequency normally associated with Chebychev filters, pronounced *Chebi-shev*).

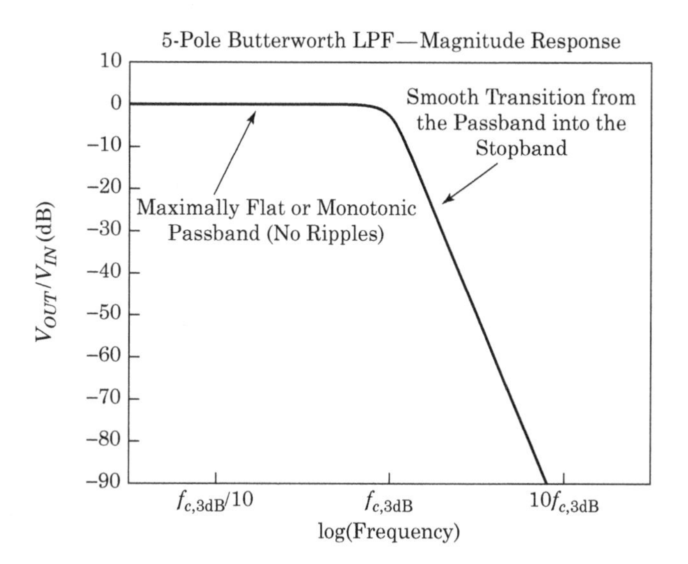

FIGURE 5-27 ■ The magnitude response of a five-pole Butterworth low-pass filter.

5.6.1 Butterworth Low-Pass Filters

The Butterworth filter is the simplest and most common filter. Since its characteristics are simple to describe mathematically, we will use the Butterworth as calculation standard. Whenever we are interested in calculating the attenuation of a filter in a system, we will assume the filter is a Butterworth filter. For more information, see the section "Butterworth Filters in Detail" later in this chapter.

The cutoff frequency of the Butterworth filter is the frequency at which the output power has dropped by 1/2 or 3 dB from its maximum value. This is the 3 dB or half-power frequency, and its symbol is $f_{c,3\,\text{dB}}$.

Figure 5-27 shows the magnitude response of the Butterworth filter.

At low frequencies (well below $f_{c,3\,\text{dB}}$), the filter does not attenuate the signal. As the frequency increases, the attenuation of the filter increases slowly at first and then more rapidly as the frequency increases beyond $f_{c,3\,\text{dB}}$. This type of response is *maximally flat* or *monotonically decreasing*. The response does not exhibit ripples, and the filter exhibits a smooth transition from the passband to the stopband and beyond. The filter reaches its ultimate roll-off rate almost immediately beyond $f_{c,3\,\text{dB}}$.

Figure 5-28 shows the pole/zero plot for a five-pole Butterworth low-pass filter.

Simple Butterworth filters consist only of low-pass poles (no zeroes), which lie on a circle centered about the origin. No one pole is ever dominant, which is why the Butterworth filter exhibits the smooth magnitude response of Figure 5-27.

Figure 5-29 shows the phase and group delay responses for a five-pole Butterworth low-pass filter.

The phase and group delay plots exhibit several interesting features:

- The phase response begins to change long before the magnitude response begins to show any significant attenuation. At $f_{C,3\,\text{dB}}/10$, a Butterworth low-pass filter will exhibit a phase shift of about 4° per pole. Our five-pole Butterworth LPF shows about 20° of phase shift at $f_{C,3\,\text{dB}}/10$. Measurement on a Butterworth filter showed less than 0.1 dB of attenuation and 18° of phase shift at $f_{C,3\,\text{dB}}/10$.
- The phase slope is steepest in the neighborhood of the 3 dB cutoff frequency $f_{C,3\,\text{dB}}$. Since the group delay is the derivative of the phase plot, the group delay has a peak around the same spot.

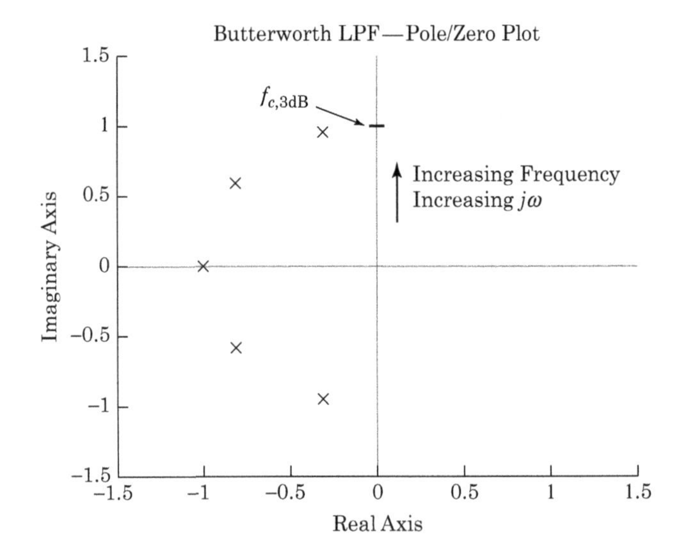

FIGURE 5-28 ■ The pole/zero plot of a five-pole Butterworth low-pass filter.

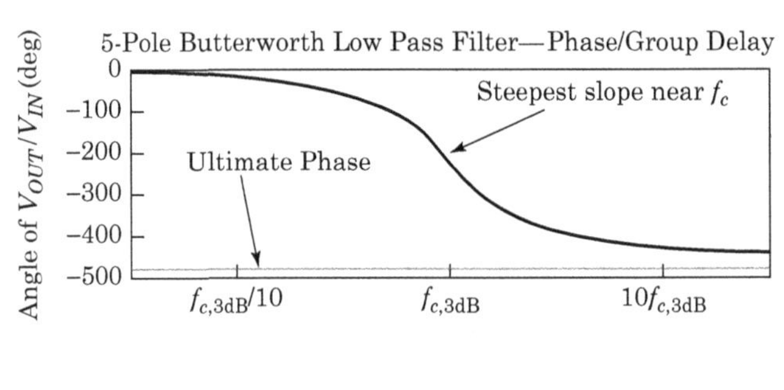

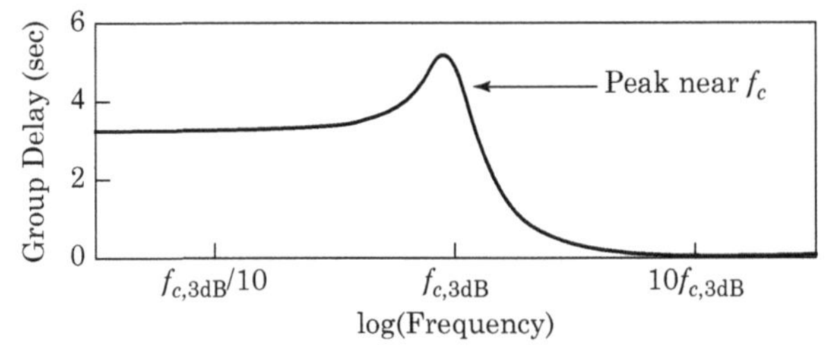

FIGURE 5-29 ■ The phase and group delay responses of a five-pole Butterworth low-pass filter.

- When the frequency increases to well beyond $f_{C,3\,dB}$, the large $j\omega$ approximations are valid, and the filter approaches the ultimate phase given by equation (5.18). A five-pole LPF exhibits a final phase of $-90° \cdot (5-0) = -450°$.
- At very high frequencies (i.e., large $j\omega$), the filter reaches its ultimate phase. However, at $10f_{C,3\,dB}$, we still haven't reached the final phase given by equation (5.18). At $10f_{C,3\,dB}$, we're still about 4° per pole away from the final phase.

5.6.2 Chebychev Low-Pass Filters

The Chebychev filter is the second most common filter in use. As Figure 5-30 shows, the Chebychev's magnitude response displays ripples in the passband and a steeper initial roll-off than the Butterworth. Allowing ripples in the passband (which is undesirable) is the price we have to pay for the steeper initial roll-off (which is desirable).

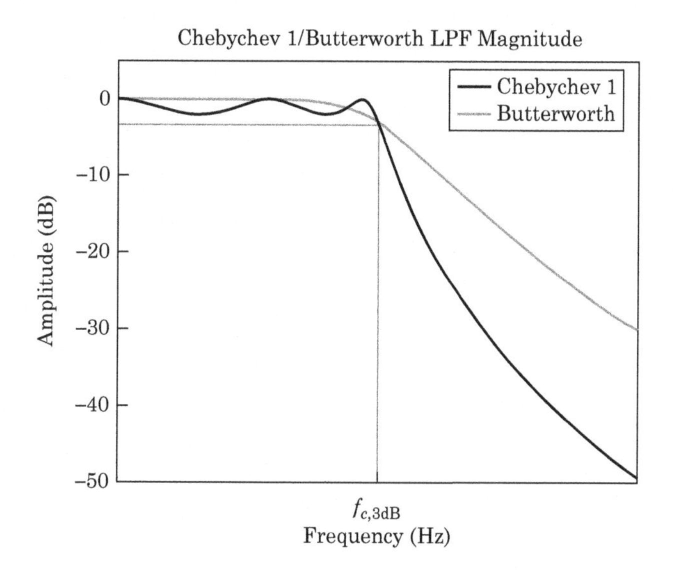

FIGURE 5-30 ■ The magnitude response of a five-pole Chebychev I low-pass filter.

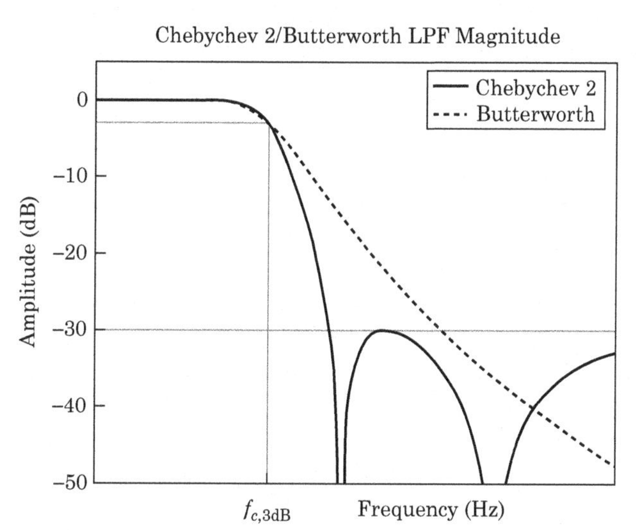

FIGURE 5-31 ■ The magnitude response of a five-pole Chebychev II low-pass filter.

5.6.2.1 Nomenclature

Some authors cite two types of Chebychev filters: Chebychev I and Chebychev II. Figure 5-30 shows a Chebychev I filter. The passband exhibits amplitude ripple, and the theoretical stopband attenuation increases monotonically. Figure 5-31 shows a Chebychev II filter. The passband is flat, but the stopband exhibits amplitude ripple and is not monotonic. In this work, we will refer to the Chebychev I filter simply as a Chebychev filter. We will refer to any filter with a nonmonotonic stopband response as an elliptic filter.

5.6.2.2 Chebychev Characteristics

A Chebychev filter allows us to trade passband ripple for steepness of the initial roll-off. The more passband ripple we allow, the faster the initial roll-off.

There are two cutoff frequencies associated with a Chebychev filter. We'll call them $f_{c,ER}$ and $f_{c,3\,\mathrm{dB}}$. The $f_{c,3\,\mathrm{dB}}$ is the same half-power cutoff frequency we associated with the

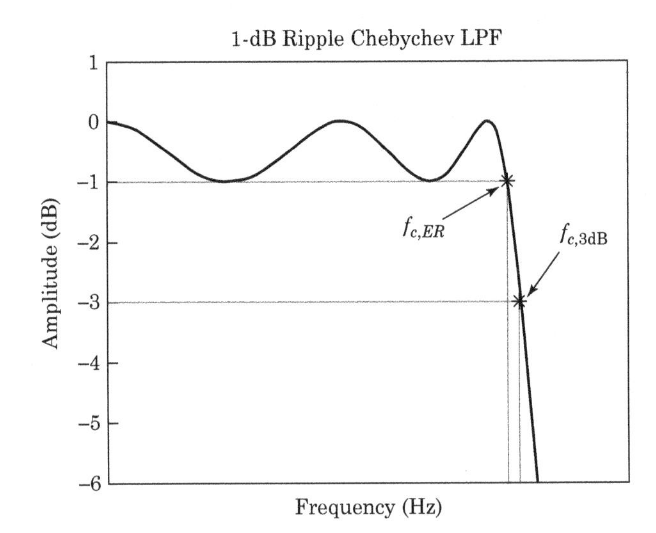

FIGURE 5-32 ▪ The passband magnitude response of a five-pole Chebychev low-pass filter.

Butterworth filter. The equal-ripple cutoff frequency, $f_{c,ER}$, takes a little more explaining. We might specify a Chebychev filter with the following statement:

> A five-pole, 0.1 dB ripple Chebychev low-pass filter with a 25 MHz equal-ripple cutoff frequency.

The 0.1 dB part of the description tells the designer that we will allow the passband to exhibit 0.1 dB of ripple. This passband ripple is the price we will pay for the steeper initial roll-off the Chebychev will provide.

The *25 MHz equal-ripple cutoff frequency* tells us that when the filter heads from the passband into the stopband it will exhibit 0.1 dB of attenuation at 25 MHz. The filter's attenuation will increase as the frequency increases from that point. This is the equal-ripple cutoff frequency or $f_{c,ER}$. The Chebychev filter also has a 3 dB cutoff frequency which we'll call $f_{c,3\,\text{dB}}$ (see Figure 5-32 for details).

When most people speak of the cutoff frequency of a Chebychev low-pass filter or the bandwidth of a Chebychev band-pass filter, they usually mean the equal-ripple cutoff frequency or the equal-ripple bandwidth. In this book, we will specifically state either the equal-ripple cutoff frequency, $f_{c,ER}$, or the 3 dB cutoff frequency, $f_{c,3\,\text{dB}}$. Others may not follow suit.

Since Chebychev filters with more than 3 dB of passband ripple are uncommon, it holds that

$$f_{c,3\,\text{dB}} \geq f_{c,ER} \tag{5.28}$$

The difference between $f_{c,ER}$ and $f_{c,3\,\text{dB}}$ is important when comparing a Chebychev filter with a Butterworth. Figure 5-33 shows a five-pole Chebychev low-pass filter and a five-pole Butterworth low-pass filter. Both have the same 3 dB cutoff frequency, $f_{c,3\,\text{dB}}$. If we were trying to make a comparison between a Butterworth and Chebychev filter for a design, the comparison wouldn't be fair if we used the Butterworth's 3 dB cutoff frequency and the Chebychev's equal-ripple frequency. The test would be biased in favor of the Butterworth.

As Figure 5-33 shows, the initial roll-off of the Chebychev filter is greater than the Butterworth's. However, the ultimate roll-off of both filters is the same because they are

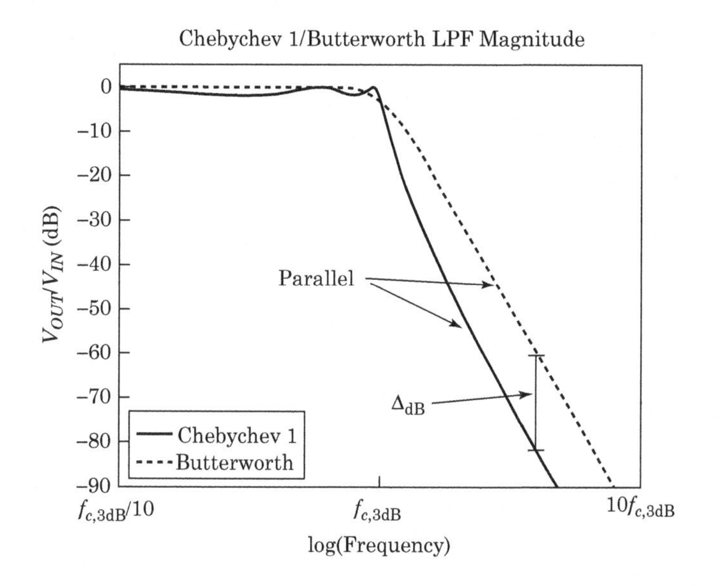

FIGURE 5-33 ■ A comparison of the magnitude responses of a five-pole Butterworth low-pass filter and a five-pole Chebychev I low-pass filter.

both five-pole filters. At any given frequency in the stopband, the Chebychev filter will offer more attenuation than the Butterworth filter. The difference in attenuation between the Butterworth and Chebychev filter (labeled Δ_{dB} in Figure 5-33) increases if we allow more ripple in the passband of the Chebychev.

Figure 5-34 shows the pole/zero plot of a five-pole, Chebychev low-pass filter. Like the Butterworth filter, the Chebychev is an all-pole filter. The poles of a Chebychev low-pass filter lie on an ellipse whose long axis coincides with the $j\omega$ axis. This pole configuration gives the Chebychev its ripply magnitude response. As the frequency rises and we move by each pole in turn, the closest pole is slightly dominant as we pass it. This in turn causes the magnitude ripple. The closer the poles are to the $j\omega$ axis, the more passband ripple the filter will exhibit.

The poles of the Butterworth low-pass filter lie on a circle centered about the origin. This pole placement gave the filter its smooth and uneventful magnitude plot.

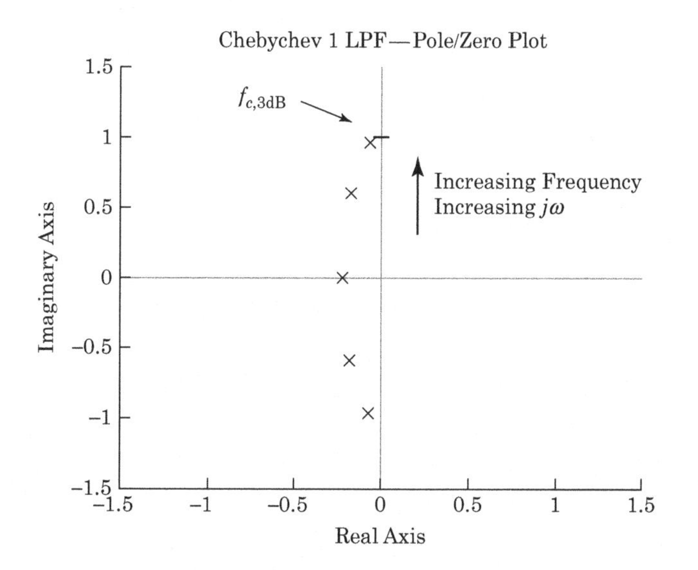

FIGURE 5-34 ■ The pole/zero plot of a five-pole Chebychev I low-pass filter.

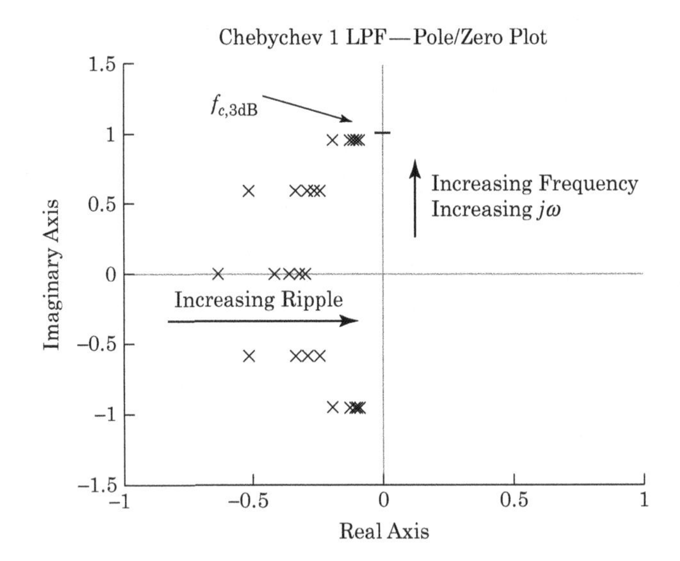

FIGURE 5-35 ■ The relationship between Chebychev pole placement and passband ripple.

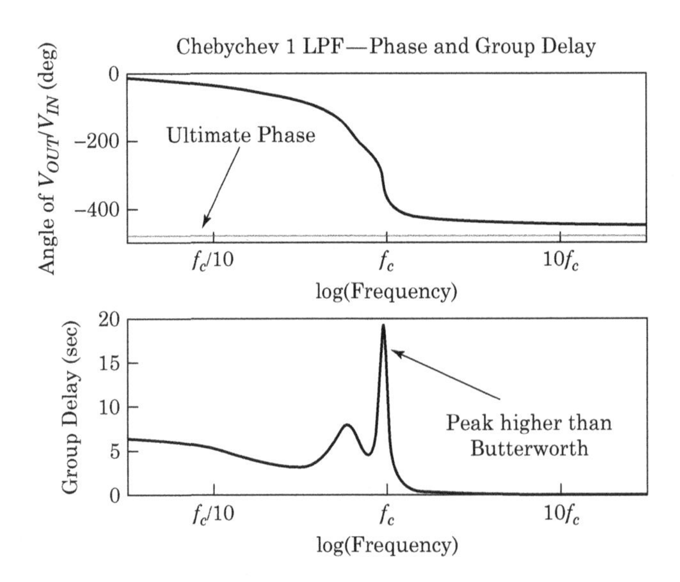

FIGURE 5-36 ■ The phase and group delay responses of a typical five-pole Chebychev.

Figure 5-35 shows the pole placement for several Chebychev filters with varying amounts of passband ripple. The poles of the filters with the largest passband ripple are closest to the $j\omega$ axis.

Figure 5-36 shows the phase and group delay responses of a typical Chebychev low-pass filter.

Overall, these graphs are similar to the phase and group delay plots of the Butterworth filter:

- There is some measurable phase shift at $f_{c,3\,\mathrm{dB}}/10$, long before we see any significant attenuation.
- The phase slope is at its steepest in the neighborhood of the 3 dB cutoff frequency $f_{c,3\,\mathrm{dB}}$. Since the group delay is the derivative of the phase plot, the group delay has a peak around $f_{c,3\,\mathrm{dB}}$.

- Like the Butterworth, when the frequency increases to well beyond $f_{c,3\,\text{dB}}$, the Chebychev filter will eventually reach the ultimate phase given by equation (5.18).
- Unlike the Butterworth, the phase and group delay responses of the Chebychev exhibit ripples due to the pole spacing. The phase ripples cause the group delay of a Chebychev filter to have peaks and valleys. Also, the main group delay peak (in the neighborhood of $f_{c,3\,\text{dB}}$) is higher than in the Butterworth case.
- Like the Butterworth, the Chebychev filter will eventually reach its ultimate phase at very high frequencies (i.e., large $j\omega$). At $10f_{c,3\,\text{dB}}$, we still haven't reached the final phase given by equation (5.18). Like the Butterworth, we'll still be about 4° per pole away from the ultimate phase at $10f_{c,3\,\text{dB}}$.

5.6.3 Filters for the Time Domain

When we compared a Butterworth filter with a Chebychev filter, we performed trade-offs in the frequency domain—passband amplitude ripple versus initial roll-off, for example. We have not discussed how these filters behave when they must process transient events.

The human race has designed many different time-domain filters including Gaussian, Bessel, and equi-ripple group delay filters. We will discuss Bessel filters as an example, but our discussion will apply to the other types as well.

Bessel filters are filters for the time domain. Their magnitude responses are poor, but their phase and group delay characteristics are very good. This means a Bessel filter handles transient phenomena very well. They exhibit little or no ringing or overshoot, and their rise time is near optimal.

5.6.3.1 Bessel Filters

The Bessel filter is designed to have a maximally flat group delay response. The group delay response of a Bessel filter looks similar to the magnitude response of a Butterworth filter. It stays flat throughout the filter's passband and then drops off slowly beyond the filter's cutoff frequency. Figure 5-37 shows the group delay response of a Bessel filter.

The magnitude response of a Bessel filter is gentle and rounded. Figure 5-38 shows the magnitude responses of both the Bessel and Butterworth filters for comparison. The

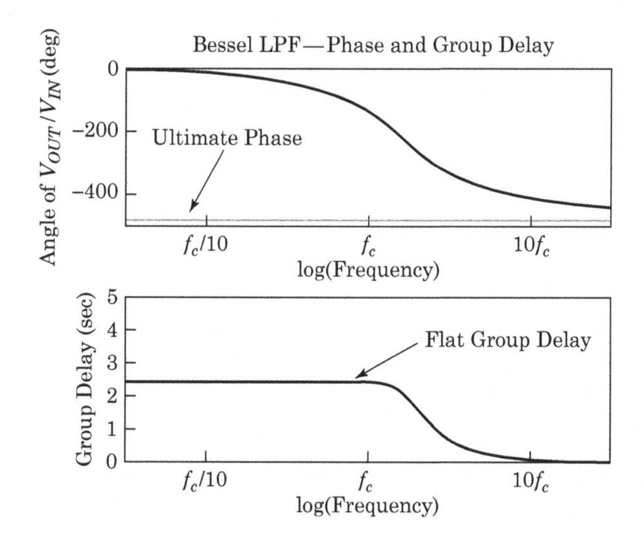

FIGURE 5-37 The phase and group delay responses of a typical five-pole Bessel low-pass filter.

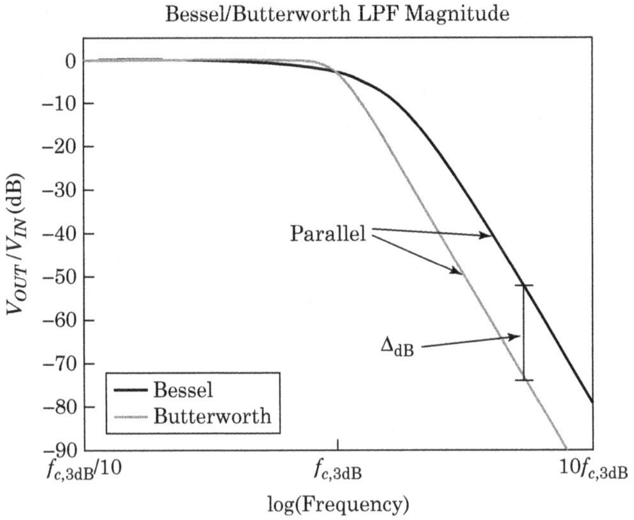

FIGURE 5-38 ■ Comparison of the magnitude responses of a typical Butterworth and Bessel low-pass filter.

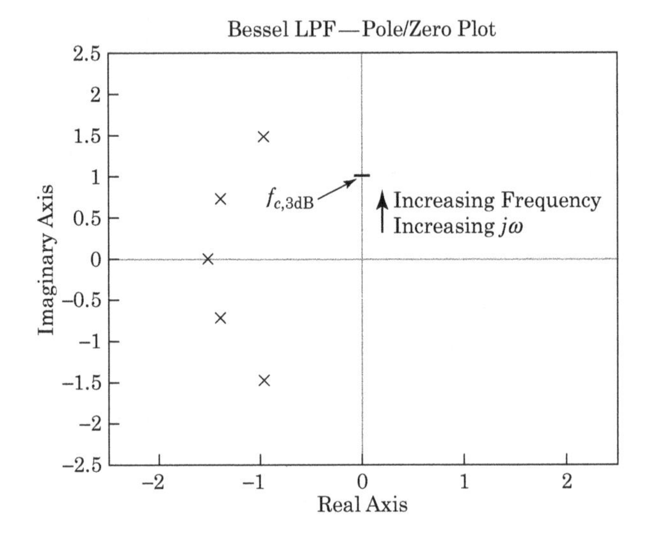

FIGURE 5-39 ■ The pole/zero plot of a five-pole Bessel low-pass filter.

Butterworth filter exhibits less attenuation throughout the passband and more attenuation in the stopband than the Bessel filter. Like the Butterworth, we use the 3 dB cutoff frequency, $f_{c,3\,\mathrm{dB}}$, to specify the filter.

The Bessel filter is another all-pole filter. Figure 5-39 shows the pole constellation for a five-pole Bessel filter.

Although the magnitude response is poor, the phase and group delay performance of the Bessel filter is much better than that of the Butterworth or Chebychev filters. The phase response is linear throughout the passband, and the group delay stays flat until beyond f_c and then drops off gently to zero. This characteristic makes Bessel filters ideal for handling pulses because they exhibit quick rise and fall times.

5.6.3.2 Other Filters for the Time Domain

The Bessel filter is one of many filters designed for its time-domain response. Others include the following:

- *Gaussian magnitude response filters*: Both the magnitude response and impulse response of this filter resemble the familiar Gaussian bell curve of statistics so there isn't any ringing or overshoot for transient inputs. However, the magnitude response doesn't roll off as quickly as even a simple Butterworth filter.
- *Equal-ripple group delay filters*: These filters are designed to have a flat group delay response. In other words, the group delay is nearly constant except for a tightly controlled amount of ripple. The group delay response of this filter looks a little like the magnitude response of a Chebychev filter.

  An equal-ripple group delay filter exhibits some of the time domain properties of a Gaussian or Bessel filter (i.e., good pulse handling capabilities), but the magnitude response is similar to the Butterworth filter. In short, this filter is a compromise between time-domain and frequency-domain characteristics.
- *Transitional filters:* Transitional filters are compromises between flat group delay and a rectangular magnitude response. For example, we might design a low-pass filter with a Gaussian passband response. Then, when the attenuation of the filter reaches, say, 12 dB, we design the filter to have a Chebychev-type roll-off response. This is a Gaussian-Chebychev filter.

  Another type of transitional filter is the Thompson–Butterworth filter. This filter has a Thompson- (or Bessel)- type passband response coupled with a Butterworth stopband roll-off. As you might expect, the performance of these filters lies somewhere between the two filters used to generate the transitional filter.

5.6.4 Elliptic Filters[2]

The Butterworth, Chebychev I, and Gaussian filters we have studied are *all-pole* filters because their transfer functions contain only poles. All-pole filters have a very high theoretical attenuation at high frequencies; however, their transition from passband to stopband is not always as steep as we would like. At times, we are willing to sacrifice attenuation at very high frequencies for a quicker transition from the passband to the stopband. Elliptic filters contain both poles and zeroes in their transfer functions. The zeroes help the filter achieve a sharp transition into the stopband at the expense of reduced stopband attenuation. An elliptic filter is usually more complicated and requires more components than an equivalent all-pole design.

Figure 5-40 is the magnitude response of a typical elliptic filter. The passband can contain ripples like the Chebychev (which again brings up the question of $f_{c,3\,\text{dB}}$ vs. $f_{c,ER}$), or it can be flat like the Butterworth filter. The transfer function zeroes are responsible for the two points of high attenuation in the stopband. By properly placing the poles and zeroes, we can design in almost any amount of initial roll-off we like.

Figure 5-40 also shows a Butterworth filter response. Both are five-pole filters with the same $f_{c,3\,\text{dB}}$, but the elliptical filter contains four zeroes in the stopband. The elliptic filter has a much smaller transition band than the Butterworth. However, the Butterworth's

---

[2]We've lumped all filters with transfer function zeroes into the *elliptic* category. Some authors refer to filters exhibiting passband ripples as *Chebychev I,* filters with stopband ripples as *Chebychev II,* and filters with both passband and stopband ripples as *elliptic* filters. Elliptic filters are often referred to as *Cauer* filters.

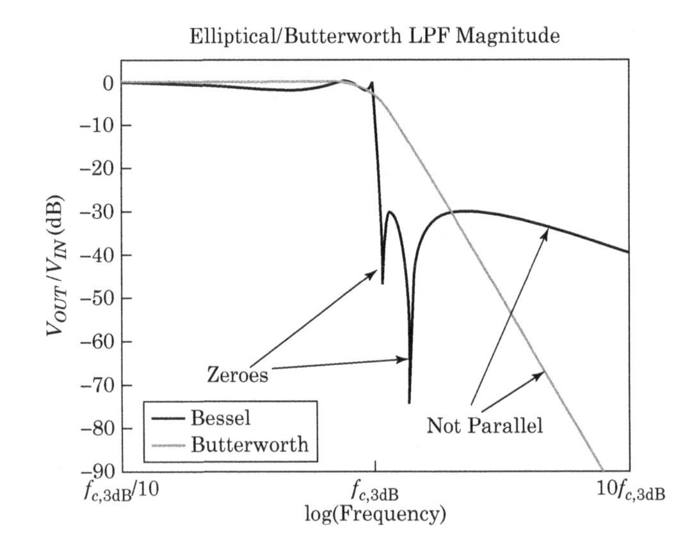

FIGURE 5-40 Comparison of the magnitude response of a typical Butterworth low-pass filter and a typical elliptic low-pass filter.

stopband rejection is better at high frequencies. The elliptic filter is better when we need attenuation very close to the passband.

The ultimate roll-off of an elliptic filter is given by equation (5.14). We used the same equation for the Butterworth, Chebychev, and Bessel filters, but now we have a nonzero number of transfer function zeroes. The elliptic filter we have been using as an example has five poles and four zeroes, so its ultimate roll-off will be only 6 dB/octave. The other five-pole filters produced a 30 dB/octave ultimate roll-off.

Figure 5-41 is the pole/zero plot of our example filter. This is a five-pole, four-zero low-pass filter. The pole positions primarily determine the passband characteristics, whereas the zero positions determine where the points of maximum attenuation occur. We normally position the zeroes on or very near to the $j\omega$ axis. This causes the zero to become dominant whenever $j\omega$ is nearby and makes the dominant zero responsible for the high attenuation at that frequency.

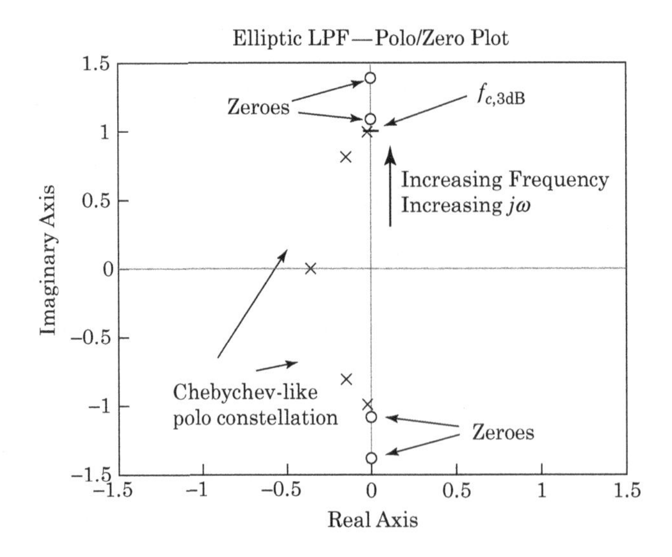

FIGURE 5-41 The pole/zero plot of a five-pole, four-zero elliptic low-pass filter.

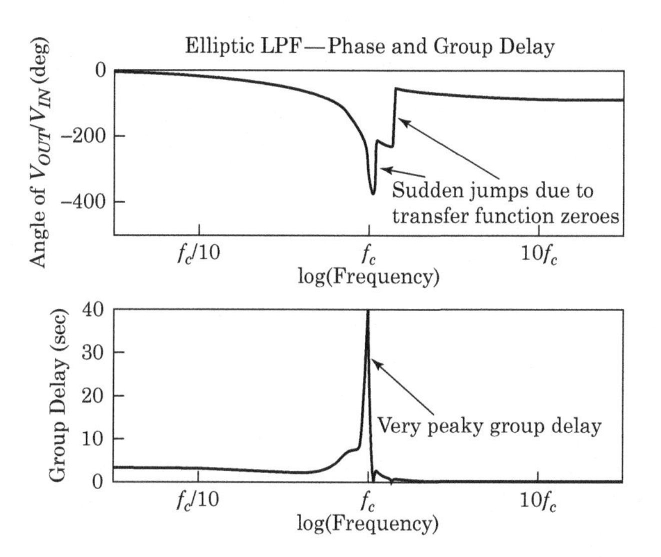

FIGURE 5-42 ▪ The phase and group delay responses of a five-pole, four-zero elliptic low-pass filter.

Figure 5-42 shows the phase and group delay plots of our example elliptic filter. As you might expect of a filter designed solely for its magnitude response, the phase and group delay plots are pretty unsightly.

The phase response shows sudden 180° jumps at the frequencies of the zeroes, f_{Z1} and f_{Z2}. Since the zeroes are dominant when $j\omega$ is in the neighborhood of f_{Z1} and f_{Z2}, they cause a rapid phase change as well as a significant magnitude change.

The sudden jumps in the phase response at f_{Z1} and f_{Z2} produce sudden jumps in the group delay response at the same frequencies. However, since these group delay spikes occur at the frequencies of maximum attenuation, they are rarely problematic.

Depending on the computer program we use to design our filters, we can specify the magnitude response of an elliptic filter in any one of several ways:

- *Specify minimum stopband attenuation*: We can specify that the filter always provides at least A_{min} of rejection throughout the stopband. In Figure 5-40, we chose A_{min} to be 30 dB. The filter algorithm then positions the zeroes appropriately to produce a stopband attenuation always greater than $A_{min} = 30$ dB.
- *Specify initial roll-off:* First, we specify the filter's cutoff frequency, and then we specify a frequency f_{attn}, above the filter's cutoff frequency, where we want to achieve at least A_{attn} dB of rejection. The filter algorithm position the stopband zeroes to achieve the attenuation we desire at the frequency we specified.
- *Specify the frequencies of the zeroes*: It is sometimes convenient to specify the frequencies of the zeroes directly. For example, we may want to place the filter zeroes at the image frequency of a receiver. Other possibilities include using the transfer function zeroes to improve the intermediate frequency (IF) rejection or local oscillator (LO) radiation performance of the same receiver.
- *Complicated*: In the very worst cases, we may have to prune the filter response in a complicated way. The problem may place different rejection requirements on certain frequency bands. Advanced computer programs allow us to directly place the filter poles and zeroes to achieve almost any arbitrary response.

5.7 LOW-PASS FILTER COMPARISON

Figure 5-43, Figure 5-46, Figure 5-44, and Figure 5-45 show the magnitude, pole/zero, phase, and group delay plots of the following filters:

- Butterworth LPF
- Chebychev I LPF with 1 dB of passband ripple
- Bessel LPF
- Elliptic LPF with 1 dB of passband ripple and 30 dB of stopband attenuation

All the filters shown in Figure 5-43 have the same $f_{c,3\,\text{dB}}$. On this scale, it's difficult to make out any significant differences in the passband responses of these filters. In the stopband, we can see that the Butterworth, Chebychev, and Bessel filters all have the same ultimate roll-off (30 dB/octave). We can also see that although the elliptic filter has a smaller ultimate roll-off (6 dB/octave), it provides the most attenuation immediately above the filter's passband. As advertised, the elliptic filter trades attenuation at very high frequencies (well above $f_{c,3\,\text{dB}}$) for attenuation immediately after $f_{c,3\,\text{dB}}$.

Ignoring the elliptic filter for a moment, we can see the Butterworth does perform in a middle-of-the-road fashion. It's not quite as good as the Chebychev but better than the Bessel.

Figure 5-44 shows the phase response of all four filters. At $f_{c,3\,\text{dB}}/10$, the filters exhibit between 15° and 27° of phase shift. The Bessel filter is the smoothest, followed by the Butterworth and then the 1 dB Chebychev. The most abrupt phase plot is the 1.0 dB ripple elliptic filter.

The elliptic filter's phase response is interesting. The phase response of the filter shows a sudden 180° jump as the frequency passes through the frequencies of the two zeroes. This is because the zeroes are very close to the $j\omega$ axis, and they behave as dominant zeroes when the input frequency is close to either of the zero frequencies.

Figure 5-45 shows the passband group delay of our four candidate filters. The Bessel filter has the best group delay response (flatter is better). This filter stays flat well into the

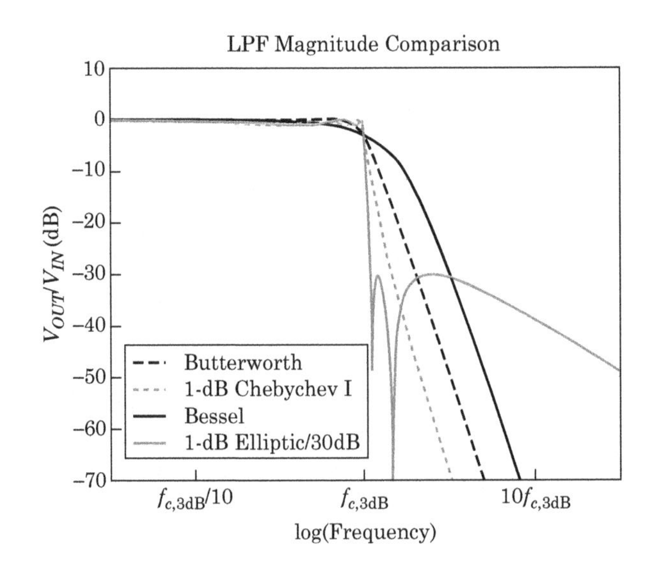

FIGURE 5-43 ■ Comparison of various low-pass filter magnitude responses.

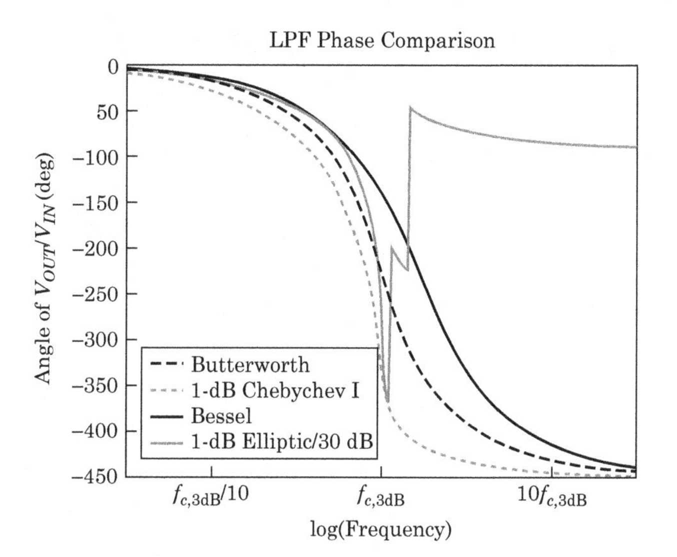

FIGURE 5-44 Comparison of phase responses of four low-pass filters.

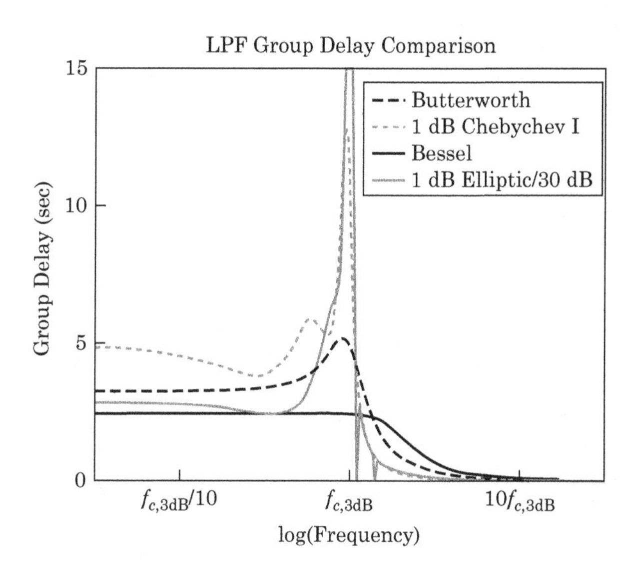

FIGURE 5-45 Comparison of passband group delay of four low-pass filters.

filter's stopband. The Butterworth filter's group delay is the next flattest, followed by the 1.0 dB Chebychev filter and then the 1.0 dB ripple elliptic filter.

Except for the Bessel filter, the group delay tends to get lumpier as we approach the filter's cutoff frequency. In addition, the group delay usually peaks somewhere in the neighborhood of the 3 dB cutoff frequency.

The 1.0 dB ripple Chebychev filter exhibits group delay ripple. The low-ripple elliptic filter, the Butterworth filter, and the Bessel filter rise only once to hit their single peak around $f_{c,3\,dB}$.

Figure 5-46 shows the pole/zero plots of the four filters on the same scale. In both the Butterworth and Bessel filters, a single pole never becomes dominant, so the magnitude responses of these filters are smooth and quiet. The poles of both the Chebychev and elliptic filters are closer to the $j\omega$ axis than in the Butterworth or Bessel case. Single poles do become slightly dominant, and the result is passband ripple.

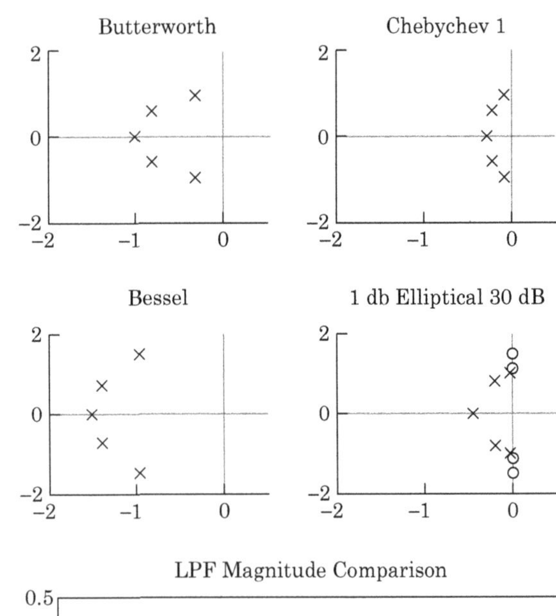

FIGURE 5-46 ■ Comparison of pole/zero constellations of four low-pass filters.

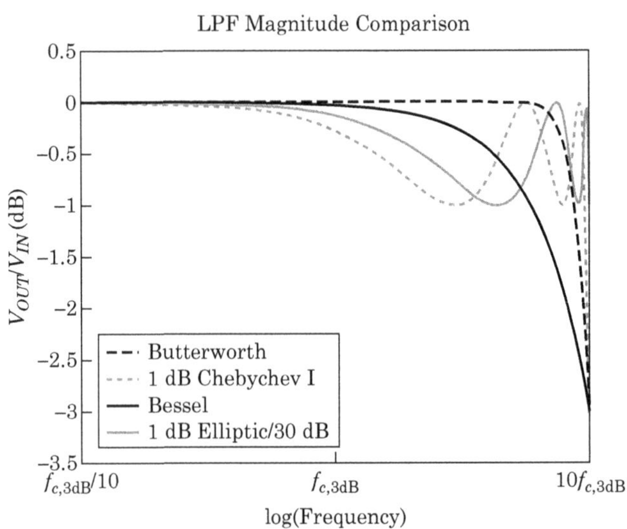

FIGURE 5-47 ■ Comparison of passband magnitude responses of four low-pass filters.

The zeroes in the elliptic filter cause the magnitude response of the filter to transition into its stopband very quickly. The price is increased group delay and a decreased amount of attenuation at high frequencies.

Figure 5-47 shows a close-up view of the passband responses of our four filters. The plot shows clearly that the filters all have the same 3 dB cutoff frequency, $f_{c,3\,\text{dB}}$. The Butterworth filter performs in a middle-of-the-road fashion. It stays reasonably flat throughout most of the passband. The Bessel filter exhibits measurable attenuation even at very low frequencies (below say $f_{c,3\,\text{dB}}/5$).

Figure 5-48 shows the stopband performance of our four filters from $f_{c,3\,\text{dB}}$ to $100 f_{c,3\,\text{dB}}$. The horizontal scale is logarithmic frequency, whereas the vertical scale is attenuation in dB. Almost certainly beyond the ultimate attenuation of any garden-variety filter is 100 dB; 70 to 80 dB is a good rule of thumb.

The Butterworth, Chebychev, and Bessel filters are five-pole, no-zero filters, and they all exhibit an ultimate roll-off of 30 dB/octave. The five-pole, four-zero elliptic filter provides 6 dB/octave of ultimate roll-off. These results are consistent with equation (5.14).

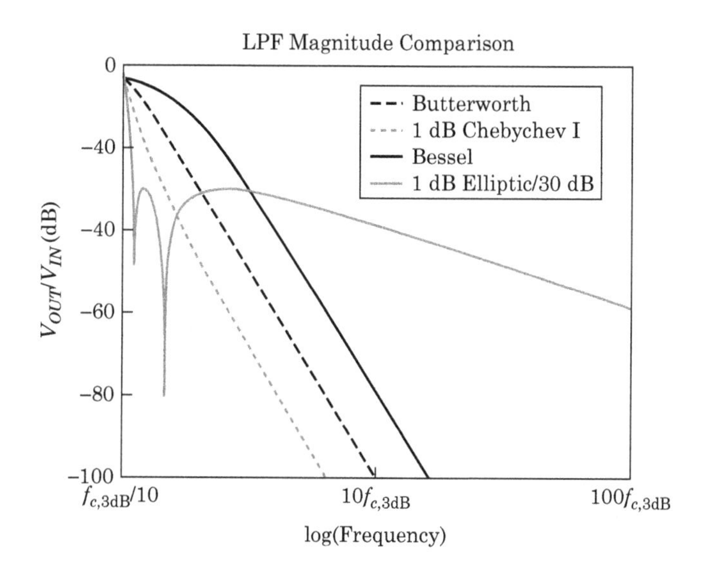

FIGURE 5-48 ■ Comparison of stopband magnitude responses of four low-pass filters.

5.8 | FILTER INPUT/OUTPUT IMPEDANCES

When we ask a computer to design a filter, we are offered several circuit topologies. For example, Figure 5-49 shows two possible realization of a 0.5 dB ripple Chebychev filter with a 3 dB cutoff frequency of 100 MHz. Both filters are designed for a 50 Ω system (i.e., $R_S = R_L = 50\Omega$). We assume that the filters are ideal and the filter components are lossless.

Figure 5-50, Figure 5-51, and Figure 5-52 show the magnitude, phase, and transient responses of these two filters. The two filter topologies perform identical functions.

5.8.1 Filter Element Impedances

Let's examine the impedances of the filter components above and below cutoff. Figure 5-53 shows the series-L filter and the reactance of each element when the frequency is well into the passband ($f_{c,3\,\text{dB}}/100 = 1$ MHz) and when the frequency is well into the stopband ($100f_{c,3\,\text{dB}} = 10$ GHz). Figure 5-54 shows the element impedances of the shunt-C filter.

At 1 MHz, the series elements exhibit low impedances, while the parallel elements show high impedances. As the frequency decreases, the series elements will *short circuit,*

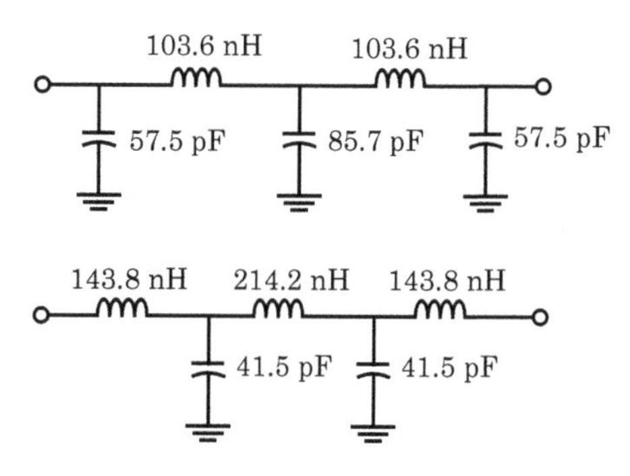

FIGURE 5-49 ■ Two practical realizations of the same five-pole Chebychev low-pass filter.

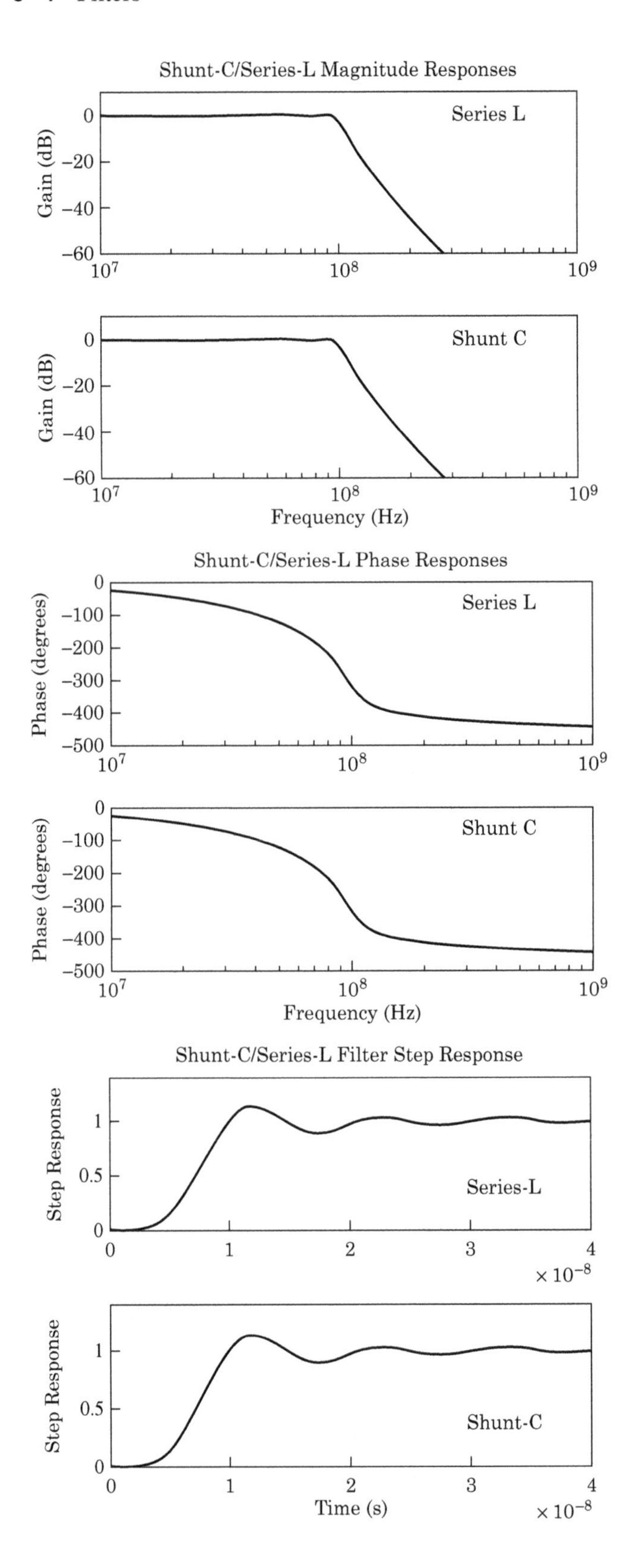

FIGURE 5-50 Magnitude response of the two filters of Figure 5-49.

FIGURE 5-51 Phase response of the two filters of Figure 5-49.

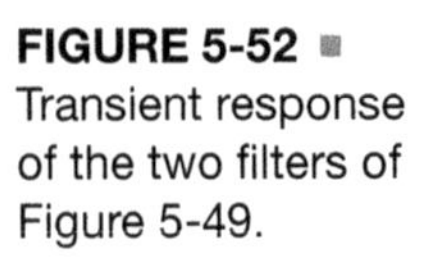

FIGURE 5-52 Transient response of the two filters of Figure 5-49.

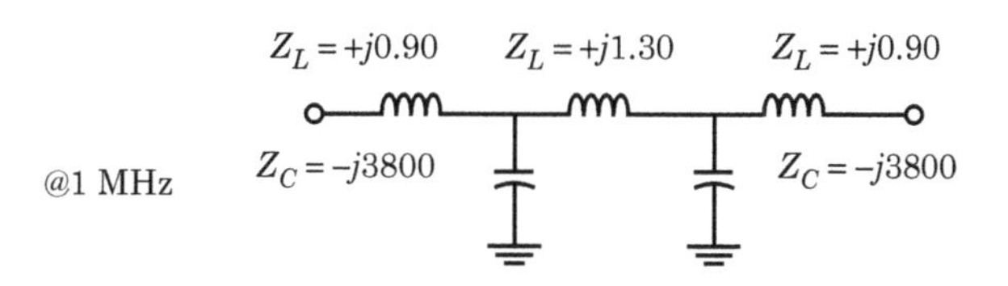

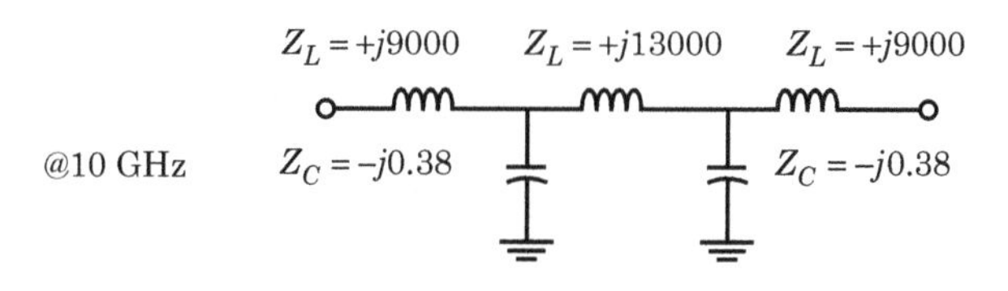

FIGURE 5-53 ■ Element impedances of the series-L filter below and above the filter's cutoff frequency.

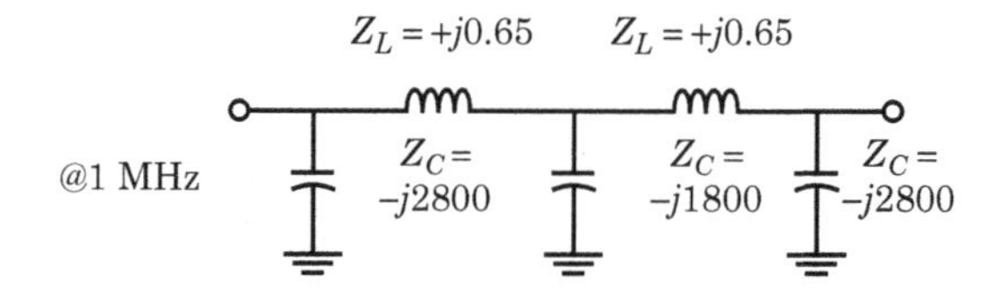

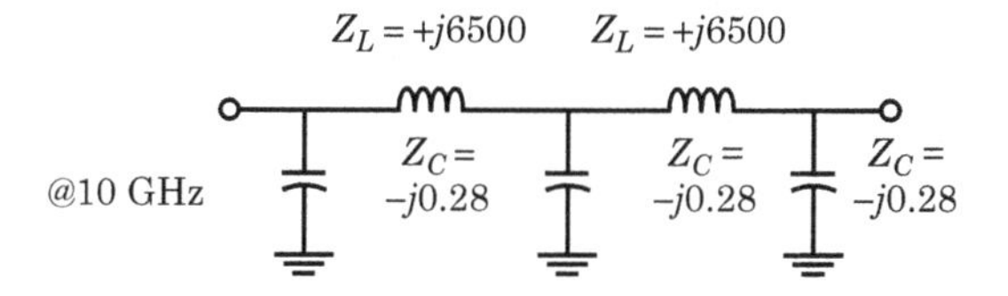

FIGURE 5-54 ■ Element impedances of the shunt-C filter below and above the filter's cutoff frequency.

while the shunt elements will *open circuit*. At low frequencies (well below $f_{c,3\,\text{dB}}$), the source and load resistor are connected directly together. We have a perfect impedance match between the source and load, and we will achieve maximum power transfer.

The series elements become high impedance at 10 GHz (i.e., they open circuit). The shunt elements are low impedances, and they short circuit. At this frequency, the series elements in our filter stop the signal from propagating because they are high impedance. Any signal that does make it through the high-series impedance enters a node connected to ground through a low-impedance component. This process of high-series impedance, low-shunt impedance repeats as the signal continues to move from the source to the load.

This is the fundamental mechanism of filtering, which occurs in all lossless filters, including low-pass, high-pass, band-pass, and band-stop filters. In the passband, the series elements short circuit, while the shunt elements open circuit. The source and load resistors both see 50 Ω when they look into the filter. Everything is matched, and, since the filter is lossless, we experience maximum power transfer from the source to the load resistor.

In a filter's stopband, the series elements open circuit, while the shunt elements short circuit. The source and load both see a high impedance when they look into the filter (due to the first element being a series element). Since the filter is severely mismatched on both the input and output ports, we don't get maximum power transfer. The degree of mismatch determines the amount of power that passes through the filter, so it determines the stopband rejection the filter provides.

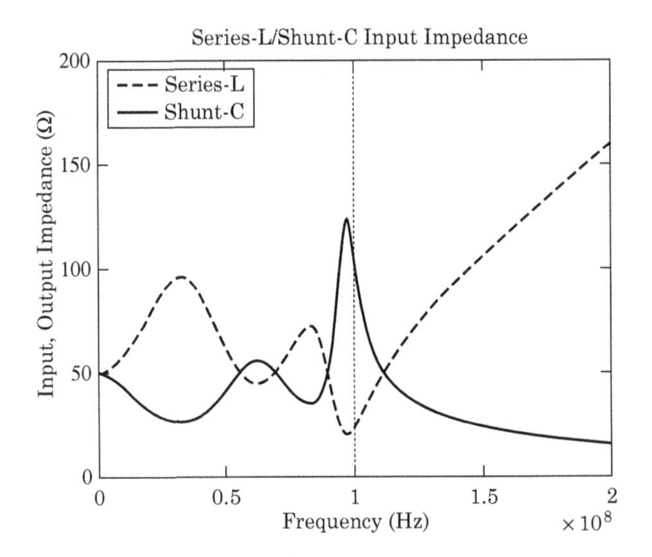

FIGURE 5-55 ■ The input impedances of the two filters of Figure 5-49.

5.8.2 Series Element First, Shunt Element First

When we asked the computer for a 100 MHz, 0.5 dB ripple Chebychev low-pass filter for a 50 Ω system, the computer gave us the two candidates, shown in Figure 5-49. The first filter has a shunt element closest to the source, so this realization will present a low impedance in the stopband. The second filter has a series element closest to the source, so it will present a high impedance to the source in the stopband.

Figure 5-55 shows the input impedance of the two filters plotted over frequency. The series-L filter shows a high impedance in its stopband, while the shunt-C filter presents a low impedance. These mismatches reject power from the source and keep power from the load.

To sum up:

- A filter performs its filtering action by frequency-selective matching. We desire maximum power transfer through the filter in its passband, so the filter opens up and connects the load resistor to the source resistor. The filter is well matched on both its input and output ports.
- In the stopband, a filter presents a severe mismatch to both the source and the load resistors. The filter ports are no longer matched, so the filter does not accept power from the source and will not efficiently deliver power to the load.
- This is a fundamental filtering mechanism. It is present in low-pass, high-pass, band-pass, and band-stop filters.
- The input and output impedances of a filter normally show resistive and reactive components and change with frequency. We express these impedances as voltage standing wave ratio (VSWR), return loss, resistance and reactance, or magnitude and phase angle.
- The most commonly used filter designs are based on reflective rather than absorptive theory. A filter built of lossless components contains no resistive components that can dissipate power, so the filter must suppress stopband energy by reflecting power. At the 3 dB passband edge, half of the incident power is reflected; the return loss has already reduced to 3 dB, and the VSWR is 5.8:1. This effect is important when an out-of-band signal (like a harmonic) is reflected back into a nonlinear device (such as a mixer).

EXAMPLE

Diplexers

Figure 5-56 is a circuit called a diplexer. It's made up of a low-pass filter in parallel with a high-pass filter. It directs signals higher than the cutoff frequency to one port while directing signals lower than the cutoff frequency to the other port.

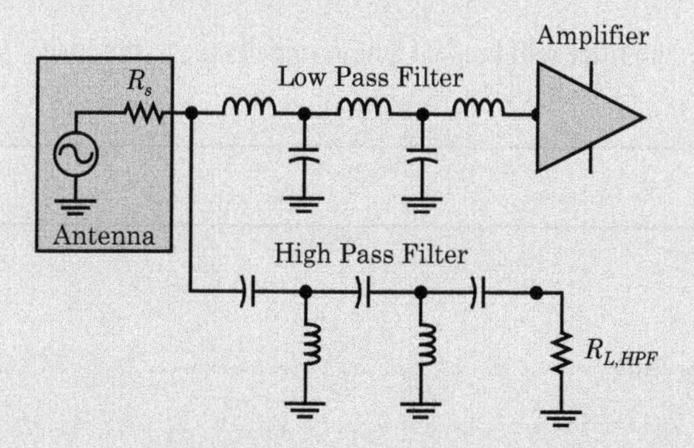

FIGURE 5-56 ■ A diplexer consisting of both a low-pass and a high-pass filter.

Both the high-pass and low-pass filters have the same cutoff frequency, f_C. Below f_C, the high-pass filter presents a high impedance to the source, and the low-pass filter is effectively the only device in the circuit. All the energy below f_C is directed to the amplifier.

The low-pass filter becomes a high impedance above f_C so that section of the circuit is effectively removed. The signals above f_C pass through the high-pass filter and find themselves dissipated in $R_{L,HPF}$.

Due to the high-pass/low-pass arrangement of the filters, the antenna sees a matched load across its entire frequency range.

EXAMPLE

Elliptic Filter I

Figure 5-57 shows the circuit diagram for an elliptic filter:

a. Is this a high-pass filter or a low-pass filter?

b. In the filter's stopband, does this filter present a high- or low-impedance load to the source resistor R_S?

c. In the filter's stopband, does the load resistor see a high or low impedance when it looks into the filter?

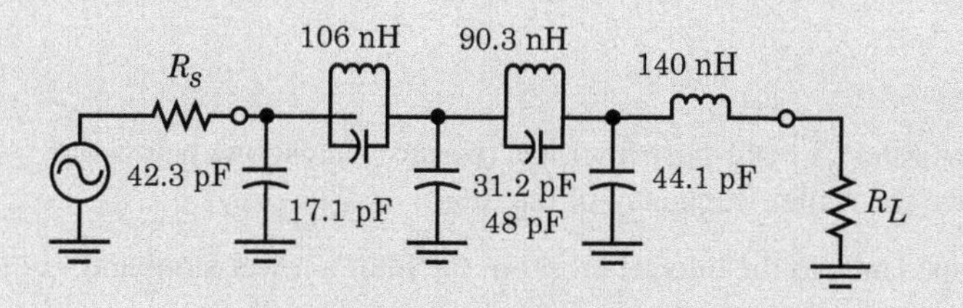

FIGURE 5-57 ■ Elliptic filter circuit diagram.

Solution

Remembering that the impedance of a capacitor approaches zero at high frequencies and the impedance of an inductor approaches infinity at high frequencies:

a. The three shunt capacitors and the series inductor next to R_L tell us that this is a low-pass filter.

b. The shunt capacitor on the filter's source end tells us the filter will present a low impedance to the source in the stopband.

c. The series inductor on the filter's load end tells us the filter will present a high impedance to the load in the stopband.

EXAMPLE

Elliptic Filter II

Figure 5-58 shows the circuit diagram for an elliptic filter:

a. Is this a low-pass filter or a high-pass filter?

b. In the filter's stopband, does this filter present a high or low impedance to the source resistor R_S?

c. In the filter's stopband, does this filter present a high or low impedance to the load resistor R_L?

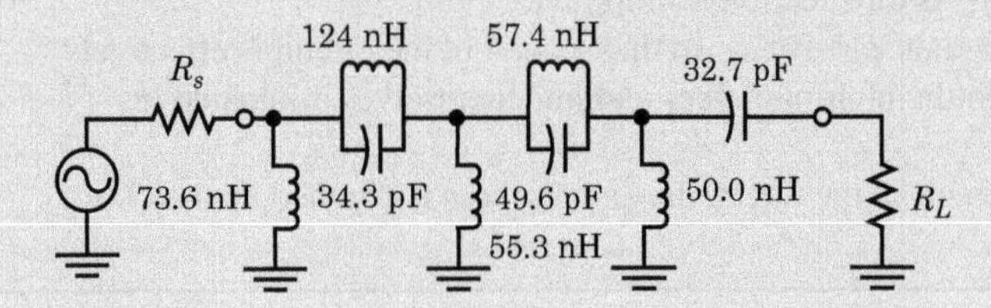

FIGURE 5-58 ■ Elliptic filter circuit diagram.

Solution

Remembering that the impedance of a capacitor approaches zero at high frequencies and the impedance of an inductor approaches infinity at high frequencies:

a. The three shunt inductors and the series capacitor next to R_L indicate that this is a high-pass filter.

b. Since the stopband of a high-pass filter is at low frequencies, we have to examine the behavior of the filter for low frequencies. The element closest to the source is a shunt inductor, so this filter looks like a low impedance in the stopband.

c. At the load end of the first, the first element is a series capacitor. At low frequencies, a capacitor looks like a high impedance, so the filter presents a high impedance to the load in the stopband.

EXAMPLE

Band-Pass Filter

Figure 5-59 shows a Butterworth band-pass filter. A band-pass filter has two stopbands: one below the center frequency of the filter; and one above the center frequency of the filter:

a. Does the filter present a high or low impedance to the outside world in the filter's lower stopband?

b. Does the filter present a high or low impedance to the outside world in the filter's upper stopband?

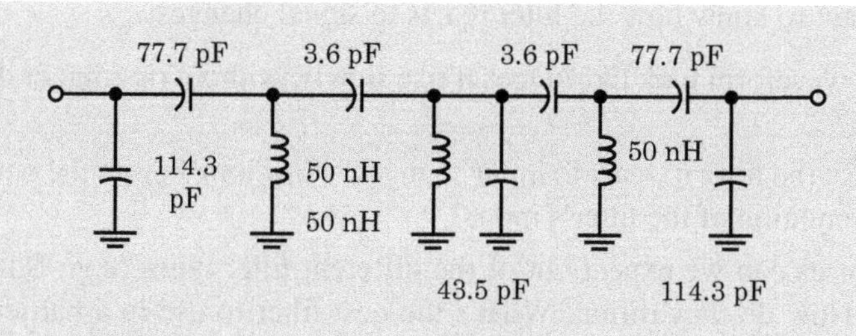

FIGURE 5-59 ■ Butterworth band-pass filter circuit diagram.

Solution

a. Since the input and output ports are both connected to series capacitors, the filter presents a high impedance in its lower stopband.

b. This filter shows a low impedance to the outside world in its upper stopband thanks to the shunt capacitors on its input and output ports.

No matter how complex the filter or how it's realized (e.g., transmission line inductors and cavity resonators), we can often deduce the out-of-band characteristics by examining the elements closest to the source and load.

5.9 | TRANSIENT RESPONSE OF FILTERS

We have so far examined the behavior of filters under steady-state conditions. *Steady-state* implies that nothing in the filter, the load, or the source has changed for a long time and that all transients have had a chance to decay to insignificant levels.

Figure 5-60 shows two transient situations. The left side of the figure shows that the signal source, V_S, has been turned off for a long time. The transients in the filter have settled out. Suddenly, we turn the signal source on and look at the voltage across R_L.

The right side of Figure 5-60 shows a more typical situation. A signal has been processed by the filter for a long time. The old signal's frequency is f_1, its amplitude is A_1, and its phase is θ_1. We suddenly change to a new input signal. The frequency of the new input signal is f_2, the new amplitude is A_2, and the new phase is θ_2.

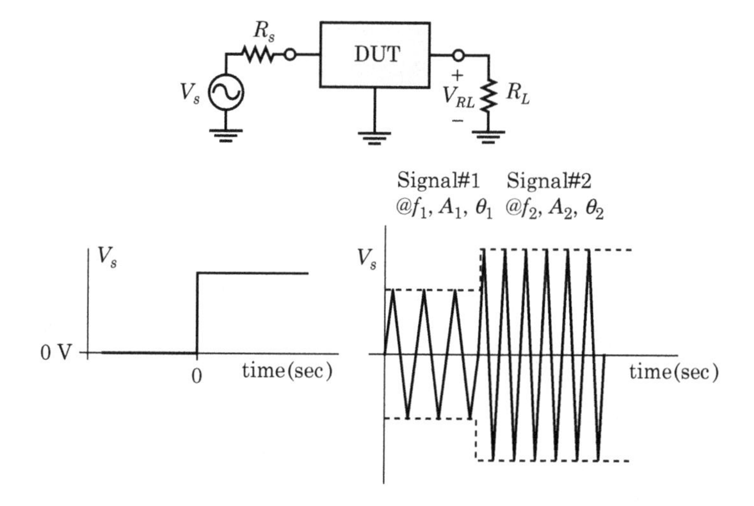

FIGURE 5-60 ■ Two types of transient events involving filters.

In both cases, we want to know how the filter reacts to signal changes:

- What does the output waveform look like? Does it rise slowly, without ringing, or does it ring for a long time?
- How fast can we expect the filter to react to input changes? How long before the output is a reasonable representation of the filter's input?
- What transient responses can we expect out of the different filter types (e.g., Butterworth, Chebychev)? How do they differ? What's the best filter to use in a particular transient situation?

5.9.1 Transient Response of Low-Pass Filters

Figure 5-61 shows the output characteristics of our four low-pass filters: a Butterworth; a 1.0 dB ripple Chebychev I; a 0.1 dB ripple elliptic filter with 55 dB of stopband attenuation; and a Bessel filter. All filters have a 100 MHz 3 dB cutoff frequency.

The Bessel filter is the quickest to rise. It approaches its final value quickly, with very little ringing. However, the Bessel filter has relatively poor magnitude response—it doesn't filter very well. The Bessel filter is a filter chosen for its time-domain properties.

The Butterworth reaches its final value after the Bessel, followed closely by the 0.1 dB passband ripple elliptic filter. The 1.0 dB ripple Chebychev is last. These three filters exhibit significant ringing.

5.9.1.1 Rules of Thumb

- All other filter characteristics being equal, the more ripple a filter exhibits in its passband, the longer the filter will take to respond to a change in its input. In Figure 5-61, the Bessel settles out first, followed by the Butterworth (with no passband ripple). Then the 0.5 dB ripple elliptic filter settles followed by the 1.0 dB ripple Chebychev.
- The sharper the transition band, the longer a filter will take to settle out after its input changes.
- A filter built with dominant poles and zeroes will ring longer than one containing fewer dominant poles and zeros.

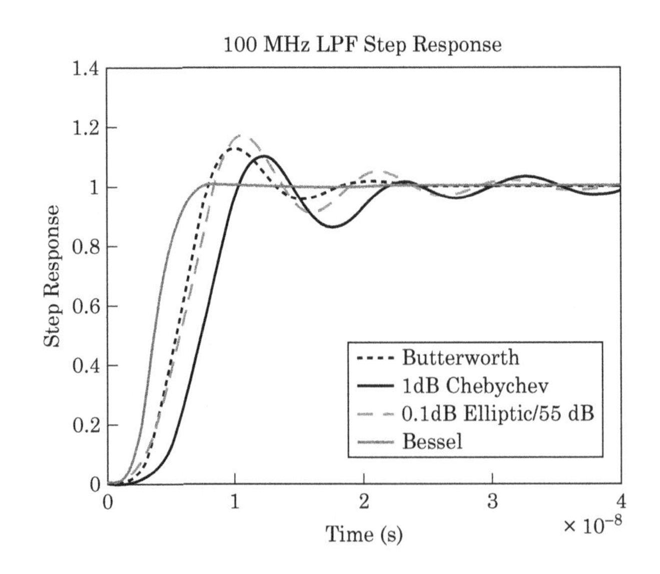

FIGURE 5-61 ■ Comparison of the transient responses of four low-pass filters.

Williams (1981), Zverev (1967), and Blinchikoff and Zverev (1976) provide extensive graphs describing the transient responses of various filters. These books are an excellent source of information on filtering of all types and are highly recommended.

5.10 BAND-PASS FILTERS

Band-pass filters are the most common type of filter used in receivers. As such, they require special attention.

5.10.1 Band-Pass Filter Terminology

Figure 5-62 and Figure 5-63 illustrate various frequency and bandwidth definitions associated with band-pass filters.

We define the following terms:

- IL_{dB}: Insertion loss in dB. This is the minimum amount of attenuation present in the pass band. The minimum value for IL_{dB} usually lies in the neighborhood of the filter's center frequency.
- $f_{U,3\,\text{dB}}, f_{L,3\,\text{dB}}$: Upper and lower 3 dB passband frequencies of the filter. They are the frequencies where the filter provides 3 dB of attenuation below the insertion loss of the filter.

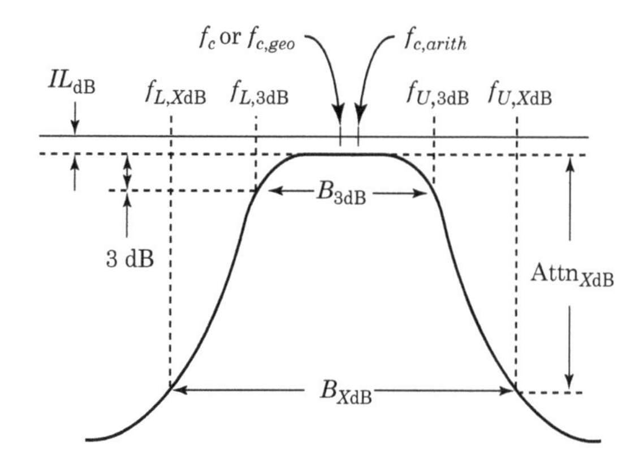

FIGURE 5-62 Frequency and bandwidth definitions for a generic band-pass filter.

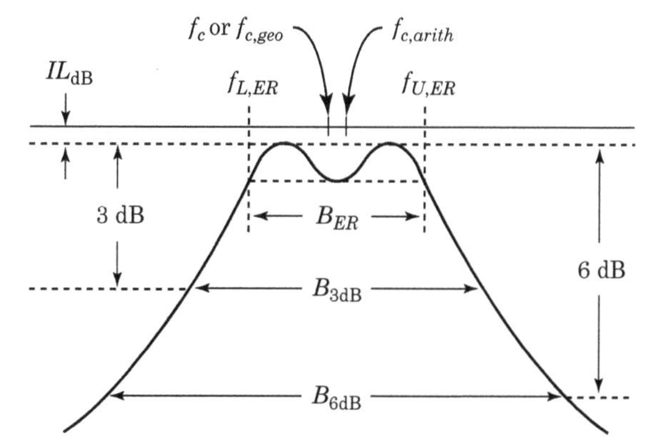

FIGURE 5-63 More frequency and bandwidth definitions for a generic band-pass filter.

- $f_{U,ER}$, $f_{L,ER}$: Upper and lower equal-ripple passband frequencies of a Chebychev band-pass filter. They are equivalent to the equal-ripple cutoff frequency of a Chebychev low-pass filter. Note that the upper and lower equal-ripple frequencies are defined relative to the filter's insertion loss.

5.10.1.1 Band-Pass Filter Center Frequencies

f_c, $f_{c,geo}$: Filter's geometric center frequency. When we encounter f_c in an equation regarding filters, the quantity will be the geometric center frequency of the filter—we will occasionally label it $f_{c,geo}$ when it's necessary. The geometric center frequency of a band-pass filter is not equal to the filter's arithmetic center frequency.

The geometric center frequency of a band-pass filter follows the geometric relationship

$$\frac{f_U}{f_{c,geo}} = \frac{f_{c,geo}}{f_L} \tag{5.29}$$

which simplifies to

$$f_c = f_{c,geo} = \sqrt{f_U f_L} \tag{5.30}$$

The upper and lower frequencies f_U and f_L specified in equations (5.29) and (5.30) could be either the 3 dB frequencies or the equal-ripple frequencies.

$f_{c,arith}$: Filter's arithmetic center frequency. We'll always label the arithmetic center frequency as $f_{c,arith}$, never just f_c. The arithmetic center frequency is not a very useful quantity, and engineers often use the arithmetic center frequency when they should use the geometric center frequency. Most calculations require the geometric center frequency.

The arithmetic center frequency is

$$f_{c,arith} = \frac{f_U + f_L}{2} \tag{5.31}$$

The f_U and f_L values in equation (5.31) can be either the 3 dB values or the equal ripple values. Note that

$$f_{c,arith} > f_{c,geo} \tag{5.32}$$

5.10.1.2 Band-Pass Filter Bandwidths

$B_{3\text{dB}}$: This is the 3 dB bandwidth of the filter. We find the two frequencies where the filter's attenuation is 3 dB below the filter's minimum insertion loss. These two frequencies are $f_{L,3\,\text{dB}}$ and $f_{U,3\,\text{dB}}$, and the 3 dB bandwidth is

$$B_{3\text{dB}} = \left| f_{U,3\,\text{dB}} - f_{L,3\,\text{dB}} \right| \tag{5.33}$$

$B_{6\text{dB}}$: This is the 6 dB bandwidth of the filter. We find the two frequencies where the filter's attenuation is 6 dB below the filter's minimum insertion loss. The 6 dB bandwidth is

$$B_{6\text{dB}} = \left| f_{U,6\text{dB}} - f_{L,6\text{dB}} \right| \tag{5.34}$$

B_{ER}: This is the equal-ripple bandwidth of a Chebychev filter. The equal-ripple bandwidth of the filter is

$$B_{ER} = \left| f_{U,ER} - f_{L,ER} \right| \tag{5.35}$$

where $f_{L,ER}$ and $f_{U,ER}$ are the equal-ripple cutoff frequencies.

Figure 5-63 shows the most general case. At any particular bandwidth defined by points of equal attenuation (x dB in the case of Figure 5-63), we can write

$$\frac{f_{U,X-\mathrm{dB}}}{f_{c,geo}} = \frac{f_{c,geo}}{f_{L,X-\mathrm{dB}}} \tag{5.36}$$

or

$$f_c = f_{c,geo} = \sqrt{f_{U,X-\mathrm{dB}} f_{L,X-\mathrm{dB}}} \tag{5.37}$$

We can also write

$$B_{X-\mathrm{dB}} = \left| f_{U,X-\mathrm{dB}} - f_{L,X-\mathrm{dB}} \right| \tag{5.38}$$

EXAMPLE

Band-Pass Filters

Given a filter with a geometric center frequency f_c of 510 MHz, a 75 MHz 3 dB bandwidth and a 2 dB insertion loss, find:

a. $f_{L,3\,\mathrm{dB}}$ and $f_{U,3\,\mathrm{dB}}$

b. $f_{c,arith}$

Solution

See Figure 5-64.

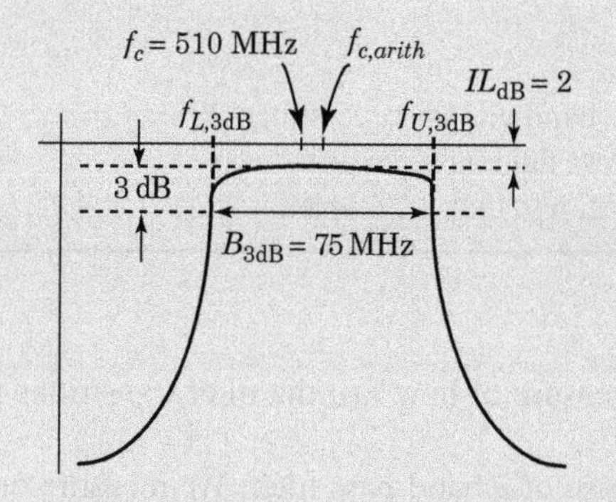

FIGURE 5-64 ■ Band-pass filter example.

a. Substituting $f_c = 510$ MHz and $B_{3\mathrm{dB}} = 75$ MHz into equations (5.30) and (5.31) produces

$$\begin{aligned} &f_{L,3\,\mathrm{dB}}^2 + 75 f_{L,3\,\mathrm{dB}} - 510 = 0 \\ &\Rightarrow f_{L,3\,\mathrm{dB}} = 473.9\,\mathrm{MHz} \end{aligned} \tag{5.39}$$

Substituting $f_{L,3\,\mathrm{dB}} = 473.9$ MHz into equation (5.33) produces $f_{U,3\,\mathrm{dB}} = 548.9$ MHz. Using equation (5.30) as a sanity check reveals

$$\sqrt{(473.9)\,(548.9)} = 510 \text{ MHz} \tag{5.40}$$

b. Using equation (5.31), we know

$$\begin{aligned} f_{c,arith} &= \frac{\left(f_{L,3\,\mathrm{dB}} + f_{U,3\,\mathrm{dB}}\right)}{2} \\ &= \frac{(473.9 + 548.9)}{2} = 511 \text{ MHz} \end{aligned} \tag{5.41}$$

EXAMPLE

Band-Pass Filters

The following band-pass filters all have a geometric center frequency of 700 MHz but different 3 dB bandwidths. Find the difference between $f_{c,geo}$ and $f_{c,arith}$ for the following band-pass filters:

a. $f_{L,3\,\mathrm{dB}} = 682.7$ MHz, $f_{U,3\,\mathrm{dB}} = 717.7$ MHz (this filter's 3 dB bandwidth is 5% of its center frequency)

b. $f_{L,3\,\mathrm{dB}} = 665.9$ MHz, $f_{U,3\,\mathrm{dB}} = 735.9$ MHz (this filter's 3 dB bandwidth is 10% of its center frequency)

c. $f_{L,3\,\mathrm{dB}} = 633.5$ MHz, $f_{U,3\,\mathrm{dB}} = 773.5$ MHz (this filter's 3 dB bandwidth is 20% of its center frequency)

d. $f_{L,3\,\mathrm{dB}} = 432.6$ MHz, $f_{U,3\,\mathrm{dB}} = 1133$ MHz (this filter's 3 dB bandwidth is 100% of its center frequency)

Solution

Using equation (5.31), we can show

a. $f_{c,arith} = 700.2$ MHz. The difference between f_c and $f_{c,arith}$ is 0.2 MHz = 0.029%.

b. $f_{c,arith} = 700.9$ MHz. The difference between f_c and $f_{c,arith}$ is 0.9 MHz = 0.13%.

c. $f_{c,arith} = 703.5$ MHz. The difference between f_c and $f_{c,arith}$ is 3.5 MHz = 0.50%.

d. $f_{c,arith} = 782.8$ MHz. The difference between f_c and $f_{c,arith}$ is 82.8 MHz = 11.8%.

Conclusions

For narrow bandwidth band-pass filters (i.e., those filters whose bandwidths are less than 20% of the filter's center frequency), we can say that $f_{c,geo} = f_{c,arith}$ with less than 0.5% error.

For filters with 100% bandwidth or less, $f_{c,geo} = f_{c,arith}$ with less than 12% error.

5.10.1.3 Shape Factor

The shape factor of a band-pass filter is a measure of how fast the filter transitions from its passband to its stopband.

Figure 5-65 shows the magnitude response of a band-pass filter. We measure or calculate the bandwidth of the filter at two attenuation values: $Attn_{1,\mathrm{dB}}$ and $Attn_{2,\mathrm{dB}}$ (taking the filter's insertion loss into account). The shape factor is

$$\frac{Attn_{1,\mathrm{dB}}}{Attn_{2,\mathrm{dB}}} \quad \text{Shape Factor} = \frac{B_2}{B_1} \tag{5.42}$$

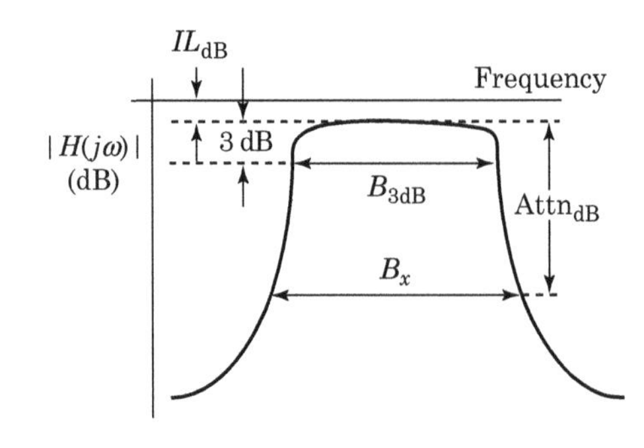

FIGURE 5-65 ■ Band-pass filter attenuation plot illustrating bandwidths for shape factor calculation.

For example, a typical shape factor specification is as follows:

The 60 dB/3 dB shape factor of the filter is 3.3.

The attenuation values are an integral part of the shape factor specification. Simply specifying the shape factor as 4.4 without giving the attenuation values is meaningless.

For all realizable filters

$$\text{Shape Factor} > 1 \tag{5.43}$$

Two common shape factors are the 60 dB/3 dB shape factor and the 60 dB/6 dB shape factor.

EXAMPLE

Shape Factor of a Band-Pass Filter

Figure 5-66 is the magnitude plot of a Chebychev band-pass filter. We designed this filter to pass the 88 to 108 MHz commercial FM broadcast band. It's a seven-pole, 0.5 dB ripple filter with an equal-ripple bandwidth of 20 MHz. The model we used includes the effects of component Q (or internal filter losses).

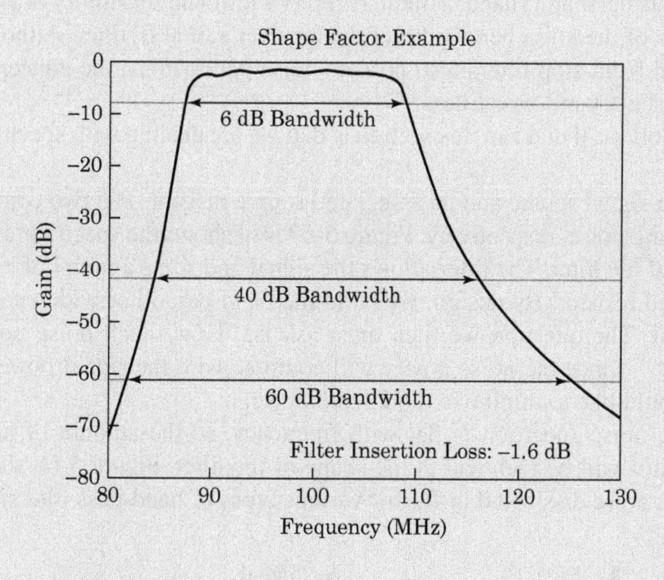

FIGURE 5-66 ■ Band-pass filter shape factor example.

Find

a. the 60 dB/3 dB shape factor

b. the 60 dB/6 dB shape factor

c. the 40 dB/6 dB shape factor

Solution

Measuring from Figure 5-66, we find that the 3 dB bandwidth is about 20.4 MHz, the 6 dB bandwidth is about 21.5 MHz, the 40 dB bandwidth is 31.6 MHz, and the 60 dB bandwidth is 43.6 MHz.

a. 60 dB/3 dB shape factor is

$$\frac{60\,\text{dB}}{3\,\text{dB}}\text{Shape Factor} = \frac{43.6}{20.4} = 2.1 \tag{5.44}$$

b. 60 dB/6 dB shape factor is

$$\frac{60\,\text{dB}}{6\,\text{dB}}\text{Shape Factor} = \frac{43.6}{21.5} = 2.0 \tag{5.45}$$

c. 40 dB/6 dB shape factor is

$$\frac{40\,\text{dB}}{6\,\text{dB}}\text{Shape Factor} = \frac{31.6}{21.5} = 1.5 \tag{5.46}$$

We use shape factor to describe filters that we've actually built. As such, shape factor considers real-world effects such as insertion loss and component tolerance.

5.11 NOISE BANDWIDTH

Noise bandwidth is a filter parameter. It allows us to determine how much noise power a filter will pass based on its passband shape. A radio receiver's ultimate sensitivity or noise floor is a strong function of the noise bandwidth of the receiver's final IF filter. Although low-pass, high-pass, and band-stop filters also possess noise bandwidths, the concept is most useful when we discuss band-pass filters.

The major assumption we'll make in this section is that we are dealing with spectrally flat noise.

Figure 5-67 shows a signal source and its associated source resistor. The two components generate a signal and noise, respectively. Figure 5-67 also shows the spectra present at the input and output of the filter. The filter allows the signal and some amount of noise power to pass to the load resistor. By design, we build the band-pass filter wide enough to pass the entire signal. The question we then must ask is, "How much noise power passes through the filter?" Since the noise power will compete with the signal power in the demodulator, we would like to minimize the filtered noise.

We've assumed the noise spectrum is flat with frequency, so the amount of noise present at the load resistor will be identical to the shape of the filter. Figure 5-68 shows the spectral shape of the noise dissipated in R_L for various types of band-pass filters.

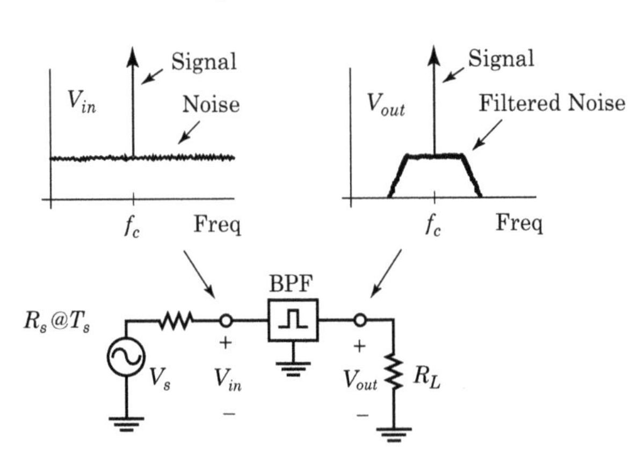

FIGURE 5-67 ■ The effects of a band-pass filter on signal and noise.

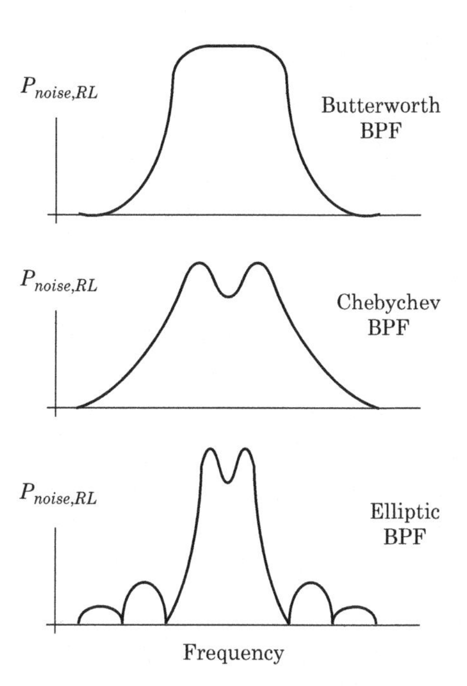

FIGURE 5-68 ■ The noise spectrum present at the load resistor is a direct copy of the final filter's passband shape. Different filters will let different amounts of noise through.

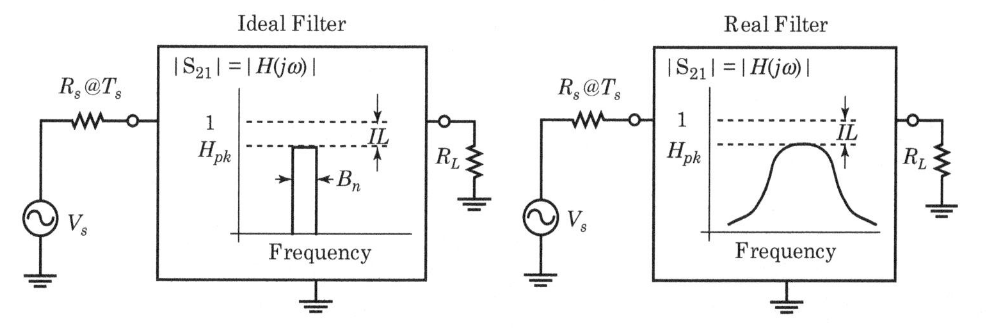

FIGURE 5-69 ■ Illustration of the noise bandwidth of a filter. The two filters let exactly the same amount of noise power through to the load resistor.

For the sake of analysis, we want to replace the arbitrary band-pass filters in Figure 5-68 with a filter that has the same insertion loss but that exhibits a perfectly rectangular passband (see Figure 5-69). If this rectangular filter has the same noise bandwidth, B_n, as the first filter, then the load resistor will dissipate the same amount of noise energy.

The noise bandwidth of any arbitrary filter equals the bandwidth of an ideal rectangular band-pass filter having the same insertion loss. The ideal rectangular filter will allow the same amount of noise power to pass as the original filter allowed. The two load resistors of Figure 5-69 will dissipate the same amount of noise power.

The problem we face is: "Given an arbitrary filter shape, find the noise bandwidth B_n."

5.11.1 Noise Bandwidth Calculation

Given a filter with an arbitrary passband shape, we'd like to find the filter's noise bandwidth B_n. We present both filters of Figure 5-69 with noise whose one-sided power spectral density is η. For the arbitrarily shaped filter, the noise presented to the load resistor is

$$\begin{aligned} N_{arbitrary} &= \int_0^\infty \eta |H\,(j\omega)| \partial f \\ &= \eta \int_0^\infty |H\,(j\omega)| \partial f \end{aligned} \tag{5.47}$$

For the rectangular filter, the quantity $|H(j\omega)|$ is a perfect rectangle, and the noise power in the load resistor is

$$\begin{aligned} N_{\text{rectangular}} &= \eta \int_{f_c - B_n/2}^{f_c + B_n/2} H_{pk} \partial f \\ &= \eta H_{pk} B_n \end{aligned} \tag{5.48}$$

Equating (5.47) with equation (5.48) produces

$$\begin{aligned} B_n \left(\frac{\text{rad}}{\text{sec}} \right) &= \frac{1}{H_{pk}} \int_0^{\infty} |H(j\omega)| \partial \omega \\ B_n(\text{Hz}) &= \frac{1}{H_{pk}} \int_0^{\infty} |H(j\omega)| \partial f \end{aligned} \tag{5.49}$$

Equation (5.49) is valid if the filter's transfer function is presented to you as a power ratio (i.e., $|S_{21}|^2$ or $20\log|S_{21}|$). Most network analyzers present the data to you in this format.

If your data represent a voltage ratio or a voltage transfer function (i.e., $|S_{21}|$), then the noise bandwidth is

$$\begin{aligned} B_n \left(\frac{\text{rad}}{\text{sec}} \right) &= \frac{1}{H_{pk}} \int_0^{\infty} |H(j\omega)|^2 \partial \omega \\ B_n \ (\text{Hz}) &= \frac{1}{H_{pk}} \int_0^{\infty} |H(j\omega)|^2 \partial f \end{aligned} \tag{5.50}$$

Often, the hardest part of evaluating the integral is deciding which type of data you have and then verifying you have the correct units.

EXAMPLE

Noise Bandwidth Calculation

Figure 5-70 shows the linear voltage transfer function of a 10.7 MHz IF band-pass filter with a 3 dB bandwidth of around 350 kHz. Find the noise bandwidth by graphical integration.

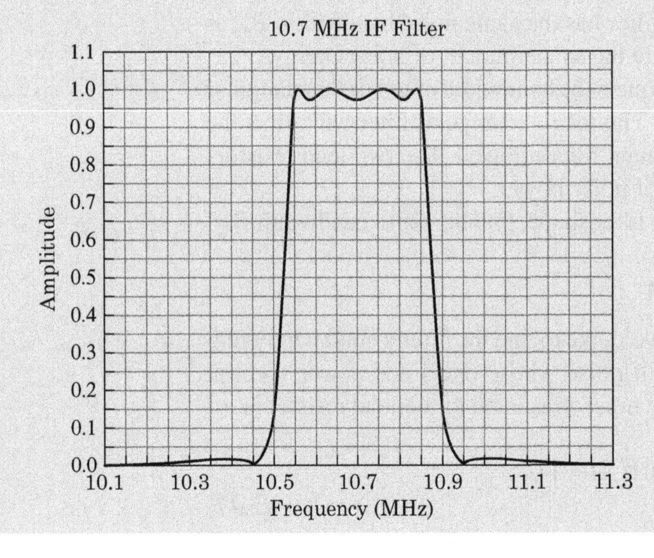

FIGURE 5-70 ■ A 10.7 MHz crystal filter. Find the noise bandwidth of this filter using graphical integration.

Solution

Since we have a voltage transfer function plot, we use equation (5.50). We'll perform the integration by finding the height of each 50 kHz column and squaring the height. Then we'll multiply the height^2 by the width (in hertz) of each column. Finally, we'll add up all these numbers to produce the noise bandwidth.

Figure 5-71 shows some of the definitions we'll use.

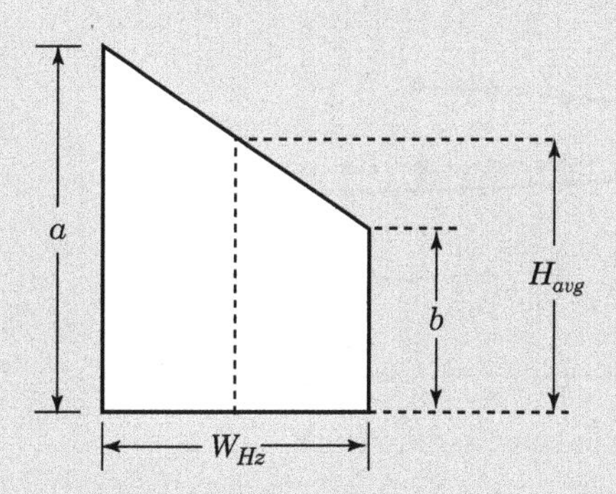

FIGURE 5-71 ■ Definitions for graphical integration in this problem.

We can write

$$H_{avg} = \frac{a+b}{2} \quad and \quad W_{Hz} = 50\text{ kHz} \tag{5.51}$$

Table 5-1 shows the tabulation of the noise bandwidth of the filter.

TABLE 5-1 Noise bandwidth calculation example

	a	b	H_{avg}	$H_{avg}^2 \cdot W_{Hz}$
1	0.0033	0.0008	0.0021	0.2205
2	0.0008	0.0023	0.0016	0.128
3	0.0023	0.0062	0.0043	0.9245
4	0.0062	0.0107	0.0085	3.6125
5	0.0107	0.0155	0.0131	8.5805
6	0.0155	0.0176	0.0166	13.778
7	0.0176	0.0022	0.0099	4.9005
8	0.0022	0.1357	0.069	238.05
9	0.1357	0.9716	0.5537	15329.1845
10	0.9716	0.9779	0.9748	47511.752
11	0.9779	0.9958	0.9869	48698.5805
12	0.9958	0.9716	0.9837	48383.2845
13	0.9716	0.9967	0.9842	48432.482
14	0.9967	0.9770	0.9869	48698.5805
15	0.9770	0.9716	0.9743	47463.0245
16	0.9716	0.1472	0.5594	15646.418
17	0.1472	0.0009	0.0741	274.5405
18	0.0009	0.0173	0.0091	4.1405
19	0.0173	0.0162	0.0168	14.112
20	0.0162	0.0119	0.0141	9.9405
21	0.0119	0.0076	0.0098	4.802
22	0.0076	0.0038	0.0057	1.6245
23	0.0038	0.0007	0.0023	0.2645
24	0.0007	0.0018	0.0013	0.0845
Sum				320743.01

The noise bandwidth of this filter is about 320 kHz.

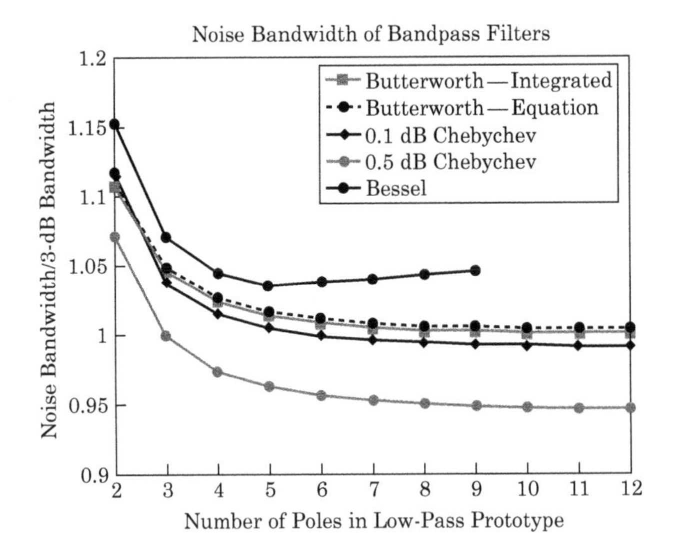

FIGURE 5-72 Noise bandwidth of various band-pass filters.

5.11.2 Noise Bandwidth of Various Band-Pass Filters

Figure 5-72 shows the noise bandwidth of various band-pass filters.

We can derive a rule of thumb from this chart:

> We often approximate the noise bandwidth of any arbitrary band-pass filter by the filter's 3 dB bandwidth. In truth, the noise bandwidth is usually somewhere between the filter's 3 dB and 6 dB points.

5.12 BUTTERWORTH FILTERS IN DETAIL

Butterworth filters are simple and common filters. It's worth taking the time to describe their characteristics in detail for the following reasons:

- The Butterworth equations are straightforward and intuitive. This is a great benefit when things aren't working quite right and you want to know what to do to make them better. We can quickly answer questions like, "How many more poles do I need?" or "What's the attenuation at this frequency?" Simple and intuitive equations are a great help when it comes to telling you what you need to do to make things better.
- Although the numbers we get will not be exactly accurate, they will usually be accurate enough. If you need more accuracy, you can always resort to computer simulation.
- Without going to the trouble of building your filter (or simulating it on a computer), it's nearly impossible to judge what affect realization factors (e.g., component losses or realization approximations) will have on the filter. Even if you used exact equations, they probably wouldn't accurately reflect reality.
- Filter synthesis algorithms may tell the designer he'll need unrealizable components. In response to this problem, a wise filter designer will change the filter topology with various circuit transforms to produce a more realizable design. When you do this,

however, you lose the mathematical purity of the filter. Although the passband hasn't changed much in appearance, the stopband can change dramatically.

For example, the transformed filter might have better rejection at low frequencies than the original design but poorer rejection at higher frequencies.

Since the filter's performance has changed, and since there really isn't any convenient or precise way to mathematically describe the new filter it makes sense to use the Butterworth equations.

If you really need exact numbers, model the filter on a computer, or, better yet, build the filter and measure the performance.

5.12.1 Butterworth Low-Pass Filters

5.12.1.1 Pole Positions

The poles of a Butterworth filter lie on a circle centered about the origin. This pole configuration causes no one pole to be dominant, so the magnitude response of a Butterworth filter is maximally flat or monotonic.

The positions of the poles for a Butterworth low-pass filter with a 3 dB cutoff frequency of f_C are

$$p_k = 2\pi f_{c,3\,\text{dB}}\left[-\sin\left(\frac{(2k-1)\,\pi}{2N}\right) + j\cos\left(\frac{(2k-1)\,\pi}{2N}\right)\right] \qquad (5.52)$$

where

N = the number of poles in the filter
$k = 1, 2, \ldots, N$

The circle containing the Butterworth poles intersects the $j\omega$ axis at the radian 3 dB cutoff frequency $\omega_{c,3\,\text{dB}} = 2\pi f_{c,3\,\text{dB}}$.

EXAMPLE

Butterworth Pole Positions

Find the pole positions for a five-pole Butterworth low-pass filter with a 50 MHz 3 dB cutoff frequency.

Solution

Using $f_{c,3\,\text{dB}} = 50$ MHz, $N = 5$, and $k = 1 \ldots 5$, equation (5.52) produces

$$p_1 = -97.08 \cdot 10^6 + j298.8 \cdot 10^6$$

$$p_2 = -254.2 \cdot 10^6 + j184.7 \cdot 10^6$$

$$p_3 = -314.2 \cdot 10^6 + j0.000$$

$$p_4 = -254.2 \cdot 10^6 - j184.7 \cdot 10^6$$

$$p_5 = -97.08 \cdot 10^6 - j298.8 \cdot 10^6$$

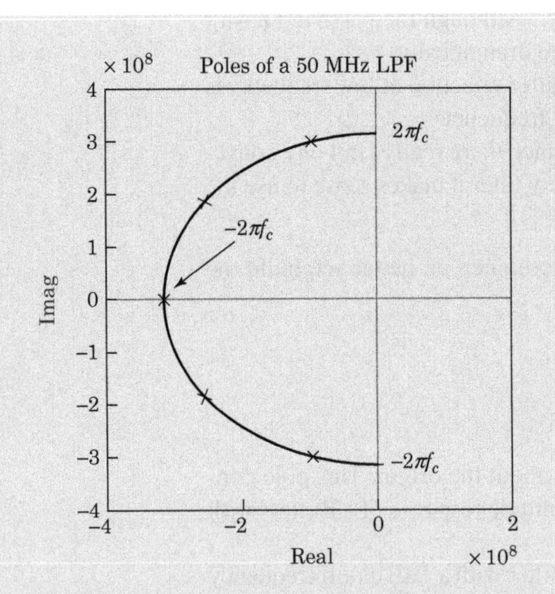

FIGURE 5-73 ■ The pole/zero plot of a 50 MHz LPF. The circle intersects the x- and y-axes at the filter's radian cutoff frequency.

Figure 5-73 shows a plot of these poles. Note that the circle containing the filter poles intersects the $j\omega$ axis at $\omega_{c,3\,\text{dB}} = 314.2 \cdot 10^6$ radian/second $= 2\pi(50 \cdot 10^6)$ hertz.

5.12.1.2 Magnitude Response

Figure 5-74 shows the magnitude response of a Butterworth filter. The response proceeds without ripples, and the attenuation increases steadily as the operating frequency increases beyond the 3 dB cutoff frequency $f_{c,3\,\text{dB}}$.

The attenuation for a lossless low-pass Butterworth filter is

$$Attn_{\text{dB}} = 10 \log \left[1 + \left(\frac{f_x}{f_{c,3\,\text{dB}}} \right)^{2N} \right] \tag{5.53}$$

where

N = the number of poles in the filter
$f_{c,3\,\text{dB}}$ = the filter's 3 dB cutoff frequency
f_x = the frequency of interest

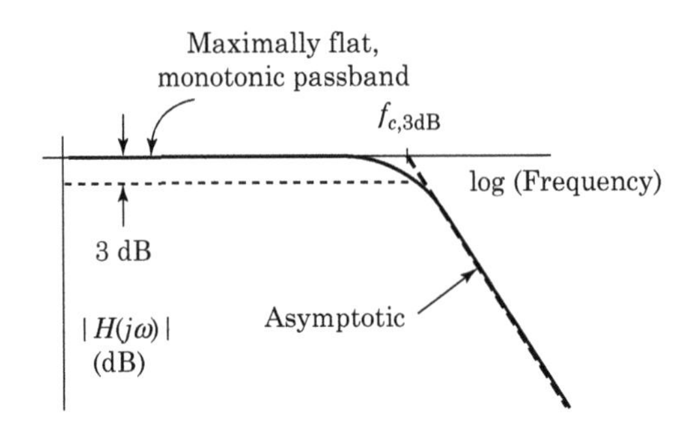

FIGURE 5-74 ■ The magnitude characteristics of a Butterworth low-pass filter.

Equation (5.53) describes the attenuation characteristic for a lossless Butterworth low-pass filter. The lossless approximation is usually good enough for the real world when we're working with low-pass filters. However, we usually have to consider insertion loss when we're dealing with band-pass filters.

EXAMPLE

Butterworth LPF Attenuation

We're given a six-pole Butterworth low-pass filter with 3 dB of attenuation at 120 MHz. Find the attenuation at 650 MHz.

Solution

We know

$N =$ the number of filter poles $= 6$
$f_{c,3\,\text{dB}} =$ the 3 dB cutoff frequency $= 120$ MHz
$f =$ the frequency of interest $= 650$ MHz

Substituting these values into equation (5.53), we find

$$\begin{aligned} Attn_{\text{dB}} &= 10\log\left[1+\left(\frac{f_x}{f_{c,3\,\text{dB}}}\right)^{2N}\right] = 10\log\left[1+\left(\frac{650}{120}\right)^{2(6)}\right] \\ &= 88\text{ dB} \end{aligned} \tag{5.54}$$

EXAMPLE

Butterworth LPF Attenuation

We've found a Butterworth low-pass filter in the lab. We've measured $f_{c,3\,\text{dB}}$ to be 6.6 MHz. We've also measured an attenuation of 30 dB at about 20.9 MHz. How many poles are in the filter?

Solution

We know that the filter's attenuation at 20.9 MHz is 30 dB, so

$f_{c,3\,\text{dB}} =$ the filter's 3 dB cutoff frequency $= 6.6$ MHz
$f_x =$ the evaluation frequency $= 20.9$ MHz
$Attn_{\text{dB}} = 30$ dB at 20.9 MHz

Substituting these values into equation (5.53) and solving for N, we find

$$\begin{aligned} Attn_{\text{dB}} &= 10\log\left[1+\left(\frac{f_x}{f_{c,3\,\text{dB}}}\right)^{2N}\right] \\ 30\text{ dB} &= 10\log\left[1+\left(\frac{20.9}{6.6}\right)^{2N}\right] \\ &\Rightarrow 999 = (3.167)^{2N} \\ &\Rightarrow N = 3\text{ Poles} \end{aligned} \tag{5.55}$$

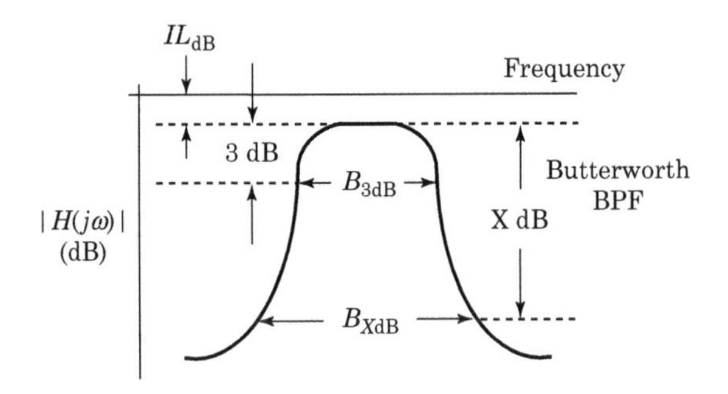

FIGURE 5-75 ■ The magnitude characteristics of a Butterworth band-pass filter with emphasis on bandwidths at various attenuations.

5.12.2 Butterworth Band-Pass Filters

5.12.2.1 Magnitude Response

As defined in Figure 5-75, the absolute attenuation for a lossy Butterworth band-pass filter is

$$Attn_{Abs,dB} = 10\log\left[1 + \left(\frac{B_x}{B_{3dB}}\right)^{2N}\right] + IL_{dB} \tag{5.56}$$

where

$Attn_{Abs,dB}$ = the absolute amount of attenuation the filter provides at B_x
N = the number of poles in the filter
B_{3dB} = the 3 dB bandwidth of the filter
B_x = the bandwidth of interest
IL_{dB} = the insertion loss of the filter in dB. The insertion loss is the minimum amount of signal loss the filter will provide.

Often, though, we're interested in the relative attenuation. The relative attenuation is the difference between the absolute attenuation and the insertion loss or

$$Attn_{Rel,dB} = Attn_{Abs,dB} - IL_{dB}$$
$$\Rightarrow Attn_{Rel,dB} = 10\log\left[1 + \left(\frac{B_x}{B_{3dB}}\right)^{2N}\right] \tag{5.57}$$

EXAMPLE

Butterworth BPF Attenuation

Given a seven-pole band-pass filter with a center frequency of 1.2 GHz, a 3 dB bandwidth of 340 MHz, and an insertion loss of 3.4 dB, find the absolute and relative attenuation at a bandwidth of

a. 500 MHz
b. 780 MHz

Solution

Using equations (5.56) and (5.57) with $N = 7$, $B_{3dB} = 340$ MHz, $IL_{dB} = 3.4$ dB, and

a. $B_X = 500$ MHz. The absolute attenuation is

$$\begin{aligned} Attn_{Rel,dB} &= 10\log\left[1 + \left(\frac{500}{340}\right)^{2(7)}\right] + 3.4 \\ &= 26.9 \text{ dB} \end{aligned} \tag{5.58}$$

We'll use equation (5.57) to find the relative attenuation:

$$\begin{aligned} Attn_{Rel,dB} &= 26.9 - 3.4 \\ &= 23.5 \text{ dB} \end{aligned} \tag{5.59}$$

b. $B_X = 780$ MHz. The absolute attenuation is

$$\begin{aligned} Attn_{Abs,dB} &= 10\log\left[1 + \left(\frac{780}{340}\right)^{2(7)}\right] + 3.4 \\ &= 53.9 \text{ dB} \end{aligned} \tag{5.60}$$

We'll use equation (5.57) to find the relative attenuation:

$$\begin{aligned} Attn_{Rel,dB} &= 53.9 - 3.4 \\ &= 50.5 \text{ dB} \end{aligned} \tag{5.61}$$

EXAMPLE

Shape Factor of a Butterworth Band-Pass filter

Given a six-pole Butterworth band-pass filter with a 275 MHz 3 dB bandwidth:

a. Find the 60 dB/6 dB shape factor

b. Find the 40 dB/6 dB shape factor

Solution

Since this is a relative measurement, we'll use the equation for relative attenuation.

Substituting $N = 6$, $B_{3dB} = 275$ MHz, and $Attn_{dB} = 6$ dB, 40 dB and 60 dB into equation (5.57) and solving for B_{6dB}, B_{40dB} and B_{60dB} produces

$$\begin{aligned} Attn_{Rel,dB} &= 10\log\left[1 + \left(\frac{B_x}{B_{3dB}}\right)^{2N}\right] & Attn_{Rel,dB} &= 10\log\left[1 + \left(\frac{B_x}{B_{3dB}}\right)^{2N}\right] \\ 6 \text{ dB} &= 10\log\left[1 + \left(\frac{B_{6dB}}{275}\right)^{2(6)}\right] & 40 \text{ dB} &= 10\log\left[1 + \left(\frac{B_{40dB}}{275}\right)^{2(6)}\right] \\ &\Rightarrow B_{6dB} = 301 \text{ MHz} & &\Rightarrow B_{6dB} = 592 \text{ MHz} \\ Attn_{Rel,dB} &= 10\log\left[1 + \left(\frac{B_x}{B_{3dB}}\right)^{2N}\right] \\ 60 \text{ dB} &= 10\log\left[1 + \left(\frac{B_{60dB}}{275}\right)^{2(6)}\right] \\ &\Rightarrow B_{6dB} = 870 \text{ MHz} \end{aligned} \tag{5.62}$$

a. The 60 dB/6 dB shape factor is

$$\frac{B_{60dB}}{B_{6dB}} = \frac{870}{301} = 2.89 \tag{5.63}$$

b. The 40 dB/6 dB shape factor is

$$\frac{B_{40dB}}{B_{6dB}} = \frac{592}{301} = 1.97 \tag{5.64}$$

Equations (5.56) and (5.57) are accurate but only marginally useful. Normally, we know the filter's geometric center frequency (f_c or $f_{c,geo}$) and the filter's 3 dB bandwidth B_{3dB}, and we would like to calculate the filter's attenuation at some frequency f_X. Combining equations (5.38), (5.56) and (5.57), we can show that the absolute attenuation of a Butterworth band-pass filter at any frequency f_X is

$$Attn_{Abs,dB} = 10\log\left[1 + \left(\frac{\left|f_x - \dfrac{f_{c,geo}^2}{f_x}\right|}{B_{3dB}}\right)^{2N}\right] + IL_{dB} \tag{5.65}$$

The relative attenuation is

$$Attn_{Rel,dB} = 10\log\left[1 + \left(\frac{\left|f_x - \dfrac{f_{c,geo}^2}{f_x}\right|}{B_{3dB}}\right)^{2N}\right] \tag{5.66}$$

where

$Attn_{Abs,dB}$ = the absolute amount of attenuation the filter provides at f_x
$Attn_{Rel,dB}$ = the relative amount of attenuation the filter provides at f_x
N = the number of poles in the filter
B_{3dB} = the 3 dB bandwidth of the filter
$f_{c,geo}$ = the band-pass filter's center frequency
f_x = the frequency of interest
IL_{dB} = the insertion loss of the filter in dB. The insertion loss is the minimum amount of signal loss the filter will provide.

EXAMPLE

Butterworth Filter Equations

Derive equation (5.65).

Solution

Looking at equations (5.56) and (5.65), we need to express B_x in terms of $f_{c,geo}$, $f_{L,X-dB}$, and $f_{U,X-dB}$ (see Figure 5-76).

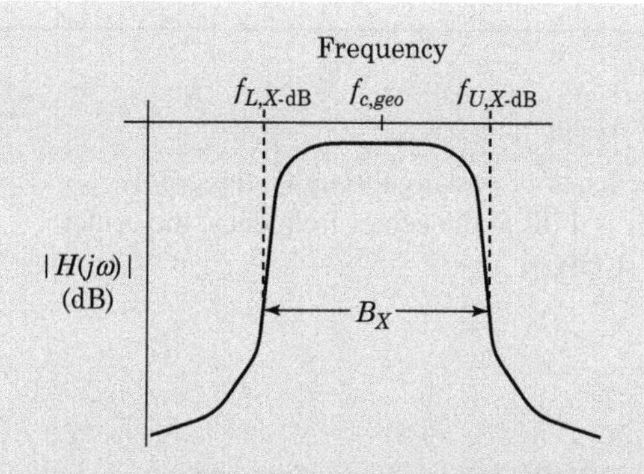

FIGURE 5-76 ■ Bandwidth and frequencies in a Butterworth filter.

Note that two frequencies, $f_{U,X-\text{dB}}$ and $f_{L,X-\text{dB}}$, apply to the bandwidth B_x and

$$B_x = f_{U,X-\text{dB}} - f_{L,X-\text{dB}} \tag{5.67}$$

Let's examine the upper frequency, $f_{U,X-\text{dB}}$. Using equation (5.30) and the previous equation, we find

$$f_{L,X-\text{dB}} = \frac{f_{c,geo}^2}{f_{U,X-\text{dB}}} \tag{5.68}$$

and

$$B_x = f_{U,X-\text{dB}} - \frac{f_{c,geo}^2}{f_{U,X-\text{dB}}} \tag{5.69}$$

Combining equation (5.69) with equation (5.56) produces

$$Attn_{\text{Abs,dB}} = 10\log\left[1 + \left(\frac{\left|f_{U,X-\text{dB}} - \dfrac{f_{c,geo}^2}{f_{U,X-\text{dB}}}\right|}{B_{3\text{dB}}}\right)^{2N}\right] + IL_{\text{dB}} \tag{5.70}$$

Following a similar derivation and expressing equation (5.69) in terms of $f_{L,X-\text{dB}}$ produces

$$Attn_{\text{Abs,dB}} = 10\log\left[1 + \left(\frac{\left|\dfrac{f_{c,geo}^2}{f_{L,X-\text{dB}}} - f_{L,X-\text{dB}}\right|}{B_{3\text{dB}}}\right)^{2N}\right] + IL_{\text{dB}} \tag{5.71}$$

Replacing $f_{L,X-\text{dB}}$ and $f_{U,X-\text{dB}}$ by f_X and taking the absolute value of the appropriate expression produces

$$Attn_{\text{Abs,dB}} = 10\log\left[1 + \left(\frac{\left|f_x - \dfrac{f_{c,geo}^2}{f_x}\right|}{B_{3\text{dB}}}\right)^{2N}\right] + IL_{\text{dB}} \tag{5.72}$$

EXAMPLE

Lossy Butterworth Band-Pass Filter

Find the total attenuation a signal at 1.75 GHz will experience after passing through a three-pole Butterworth band-pass filter. The insertion loss of the filter is 4 dB at the center frequency, the center frequency is 2.7 GHz, and the filter's 3 dB bandwidth is 1.1 GHz.

Solution

We've drawn out the problem in Figure 5-77.

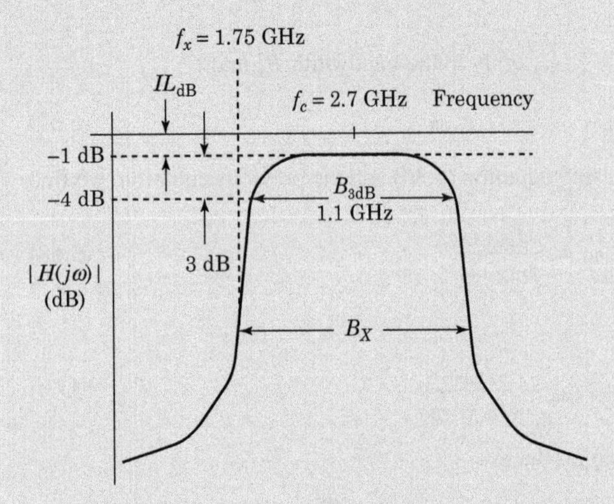

FIGURE 5-77 ■ Bandwidth and frequencies for a Butterworth filter example.

Using equation (5.65) with $N = 3$, $f_{c,geo} = 2.7$ GHz, and $IL_{dB} = 4$ dB, we can write

$$\begin{aligned} Attn_{\text{Abs,dB}} &= 10\log\left[1 + \left(\frac{\left|f_x - \dfrac{f_{c,geo}^2}{f_x}\right|}{B_{3dB}}\right)^{2N}\right] + IL_{dB} \\ &= 10\log\left[1 + \left(\frac{\left|1.75 - \dfrac{2.7^2}{1.75}\right|}{1.1}\right)^{2N}\right] + 4 \\ &= 20.5 + 4 \\ &= 24.5 \text{ dB} \end{aligned} \tag{5.73}$$

5.12.2.2 Noise Bandwidth

The noise bandwidth of a Butterworth band-pass filter is:

$$B_n = B_{3dB}\frac{[\pi/(2N)]}{\sin[(\pi/(2N))} > B_{3dB} \tag{5.74}$$

where

B_n = the noise bandwidth of the Butterworth band-pass filter
N = the number of poles in the filter
B_{3dB} = the 3 dB bandwidth of the filter

The noise bandwidth of a Butterworth band-pass filter is always slightly larger than the filter's 3 dB bandwidth.

EXAMPLE

Butterworth Noise Bandwidth

Using equation (5.74), we built Table FLT-2.

TABLE 5-2 Noise bandwidths of various Butterworth band-pass filters

# of Poles	B_n/B_{3dB}
1 (1-Pole RC Filter)	1.57
2	1.11
3	1.05
4	1.03
5	1.02
6	1.01
7	1.008
8	1.006
9	1.005
10	1.004

5.12.3 Butterworth High-Pass Filters

5.12.3.1 Magnitude Response

Figure 5-78 shows the magnitude response for a Butterworth high-pass filter.

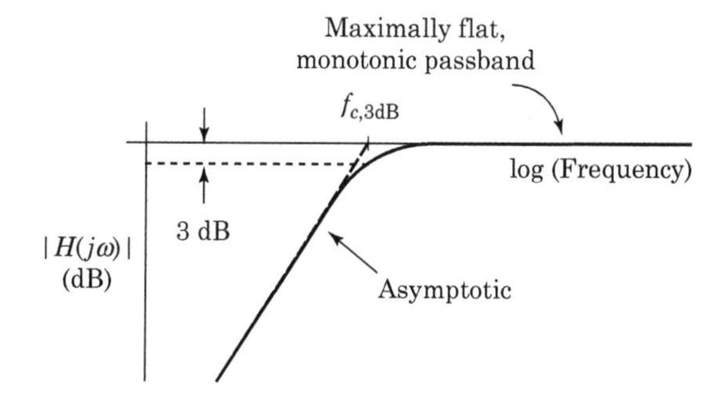

FIGURE 5-78 ■ The magnitude characteristics of a Butterworth high-pass filter.

The equation is

$$Attn_{\text{dB}} = 10\log\left[1 + \left(\frac{f_{c,3\,\text{dB}}}{f_x}\right)^{2N}\right] \tag{5.75}$$

where

N = the number of poles in the filter
$f_{c,3\,\text{dB}}$ = the filter's 3 dB cutoff frequency
f_x = the frequency of interest

Equation (5.75) is very similar to equation (5.53), which describes the attenuation characteristics of a Butterworth low-pass filter.

EXAMPLE

Butterworth High-Pass Filter

Given a three-pole Butterworth high-pass filter with a 3 dB cutoff frequency of 110 MHz, find the attenuation at:

a. the CB band (27 MHz)
b. the FM radio station at 88.3 MHz

Solution

Applying equation (5.75) with $f_c = 110$ MHz, $N = 3$, and

a. $f_x = 27$ MHz produces

$$\begin{aligned} Attn_{\text{dB}} &= 10\log\left[1 + \left(\frac{110}{27}\right)^{2(3)}\right] \\ &= 36.6\text{ dB} \end{aligned} \tag{5.76}$$

b. $f_x = 88.3$ MHz produces

$$\begin{aligned} Attn_{\text{dB}} &= 10\log\left[1 + \left(\frac{110}{88.3}\right)^{2(3)}\right] \\ &= 6.8\text{ dB} \end{aligned} \tag{5.77}$$

5.12.4 Butterworth Band-Stop Filters

5.12.4.1 Magnitude Response

The attenuation of a Butterworth band-stop or band-reject filter is

$$Attn_{\text{dB}} = 10\log\left[1 + \left(\frac{B_{3\text{dB}}}{B_x}\right)^{2N}\right] \tag{5.78}$$

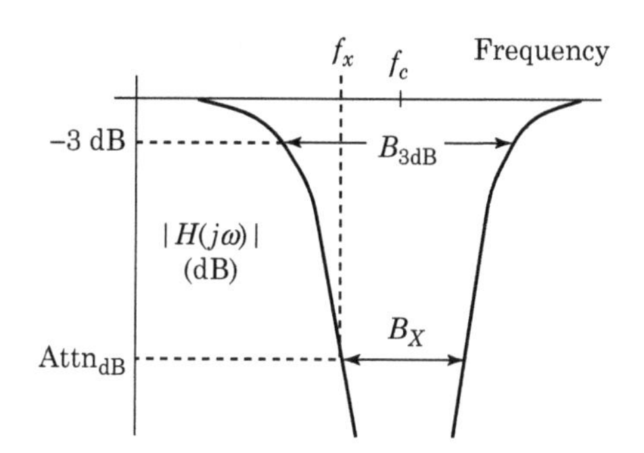

FIGURE 5-79 ■ The magnitude characteristics of a Butterworth band-stop filter showing various bandwidths and frequencies.

or

$$Attn_{dB} = 10\log\left[1+\left(\frac{B_{3dB}}{\left|f_x-\frac{f_{c,geo}^2}{f_x}\right|}\right)^{2N}\right] \tag{5.79}$$

where

$Attn_{dB}$ = the absolute amount of attenuation the filter provides at f_x
N = the number of poles in the filter
B_{3dB} = the 3 dB bandwidth of the filter
$f_{c,geo}$ = the band-pass filter's center frequency
f_x = the frequency of interest

Note the similarity to equations (5.57) and (5.66) for the Butterworth band-pass filter (see Figure 5-79).

EXAMPLE

Butterworth Notch Filter

You need a notch filter to provide at least 10 dB of rejection throughout the FM broadcast band. If you're going to use an eight-pole Butterworth filter, what is the 3 dB bandwidth of the filter?

Solution

The FM broadcast band covers 88 to 108 MHz. We need at least 10 dB of rejection at both 88 and 108 MHz, so the 10 dB bandwidth is

$$\begin{aligned} B_{10dB} &= 108 - 88 \\ &= 20 \text{ MHz} \end{aligned} \tag{5.80}$$

We want to find B_{3dB}. Using equation (5.78), we find

$$\begin{aligned} 10 \text{ dB} &= 10\log\left[1+\left(\frac{B_{3dB}}{20}\right)^{2(8)}\right] \\ \Rightarrow B_{3dB} &= 22.9 \text{ MHz} \end{aligned} \tag{5.81}$$

We usually aren't concerned about insertion loss in low-pass, high-pass, or notch filters. If you are concerned, add the insertion loss to any of the attenuation equations.

5.13 MISCELLANEOUS ITEMS

5.13.1 Isolators

High-performance filters (e.g., those with very critical group delay responses like those used in QPSK demodulators) are often equipped with isolators on both the input and output ports. The isolators always terminate the filter with the proper impedance. The proper termination across the filter's passbands and stopbands assures that the critical performance of the filter won't be spoiled by outside forces.

5.13.2 Nonlinearities

A filter acts as a linear device for low levels of input signal power, but a filter can exhibit nonlinear behavior when the input signals are strong. The threshold can be as low as the unit watt range, depending on the filter.

5.13.3 Power Handling

A filter can pass only so much power. Power handling in a filter is closely related to the factors determining the filter's nonlinearities. The following items can limit the amount of power a filter will pass:

- The filter's insertion loss may cause it to dissipate signal power and heat up. For example, a filter with a 1 dB insertion loss absorbing 100 watts from a source will dissipate about 20 watts internally.
- At high current levels, the magnetic materials used in HF and low VHF filters can saturate. This means that, at high power levels, the inductors behave like resistors and the insertion loss goes up.
- Many filters, particularly narrow band-pass filters, have very high-impedance internal nodes. At even moderate power levels, peak voltages high enough to cause capacitor breakdown or corona may occur. Again, this leads to higher insertion loss and degraded filter performance.
- Crystal filters can exhibit unexpected nonlinearities. At low power levels, the filter may perform beautifully, but for even moderate power changes the filter can exhibit wildly different behavior.

In general, if a filter must pass more than 1 watt of power, you should tell the filter designer to consider the filter's power handling capability.

5.13.4 Vibration Sensitivity

If the components that make up a filter change when subjected to vibration, the filter's performance will change as well. Thus, a filter performance depends on the filter's mechanical stability.

Vibration-induced sidebands may appear on a signal passing through a filter when the filter is subject to acceleration forces due to vibration. Quartz crystal resonators, being piezoelectric devices, convert mechanical to electrical energy. Therefore, the resonant frequency of a crystal is modulated at the frequency of vibration.

The source of vibration can be an internal equipment fan, a transformer, or a speaker. Many manufacturers of precision signal generators shock-mount their crystal filters to avoid this problem.

War Story

One satellite modem manufacturer noticed bursts of bit errors occurring only at certain times of the week. After much research, the company correlated the appearance of the bit error bursts with the routine maintenance performed on some other, unrelated equipment. It turned out that the bit error bursts occurred when the maintenance people slammed the doors to the racks containing the satellite modems. Again, shock mounting the critical filters solved the problem.

5.14 | MATCHED FILTERS

A matched filter is theoretically the best method to discern whether a particular waveform is present in a noisy environment. If we know the statistics of the noise and the characteristics of the particular signal, then we can build a filter that will optimize the output signal-to-noise ratio (SNR) when the signal of interest is present. A matched filter performs this function by coherently adding up each Fourier component in the signal.

We are not interested in preserving the wave shape of the input signal. Instead, we are going to arrange to add up all the signal energy present in the symbol waveform at one particular moment so the output of a matched filter will be a narrow spike when the desired signal is present.

In digital communications systems, we transmit one of N different waveforms to represent one of N different symbols. If we choose our N waveforms correctly, we can build N matched filters; one for each waveform (symbol). We design the m-th filter to detect the m-th waveform while rejecting the other N-1 waveforms.

5.14.1 The Equation

When the input noise is white (i.e., spectrally flat), the equation that describes the response of a matched filter is

$$H_0(f) = S^*(f)e^{-j2\pi Tf} \tag{5.82}$$

where

$H_0(f)$ is the transfer function of the matched filter

$S^*(f)$ is the complex conjugate of the Fourier transform of the known input signal $s(t)$ of duration T sec

T is the duration of the input signal s(t)

Our analysis will be heavily based on Fourier and phasor analysis. Once we cast the problem in the proper domain, the solution and the form of this equation will be obvious.

5.14.2 An Example

Let's look at an example. We'll build a matched filter for a 100 μsec pulse present in a 300 μsec window. Figure 5-80 and Figure 5-81 show the pulse in the time and frequency domains, respectively.

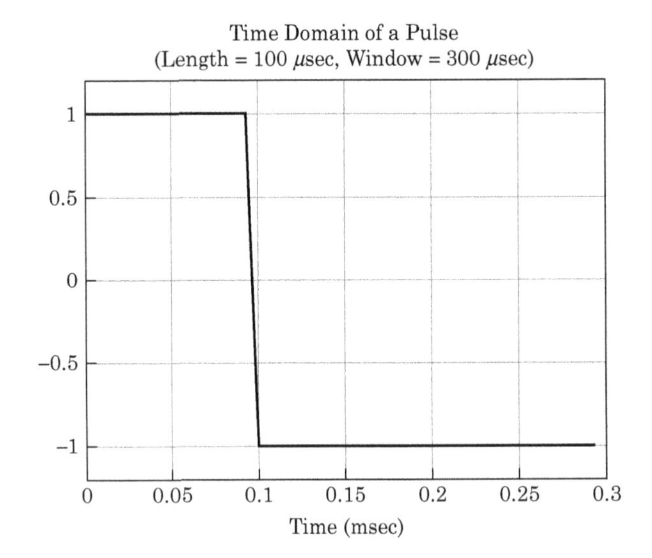

FIGURE 5-80 ■ The waveform for which we will build a matched filter. This waveform consists of a 100 μsec pulse present in a 300 μsec window. We are building a filter to respond to this entire signal—both the 100 μsec one duration and the 200 μsec zero duration.

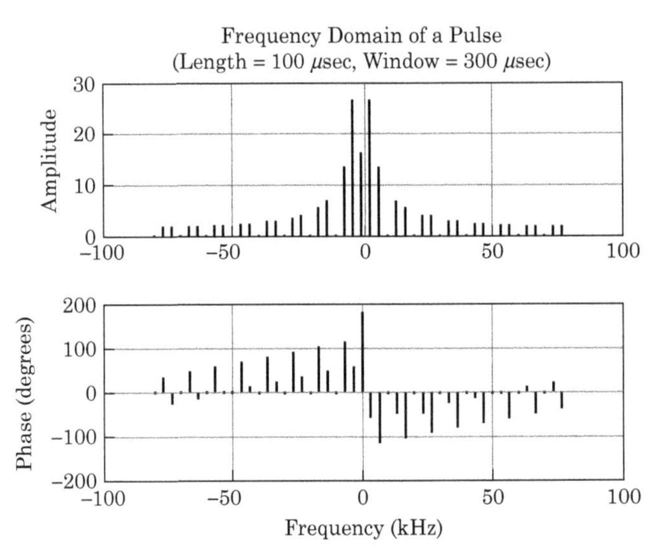

FIGURE 5-81 ■ The frequency-domain version of the signal of Figure 5-80. The frequency- and time-domain versions of the signal are different expressions of the same signal. Each frequency component possesses both a magnitude and a phase component. Both are important.

The discrete frequencies shown in Figure 5-81 are distinct cosine waves with deterministic amplitudes, frequencies and phases. These cosine waves exactly represent the signal of interest. Since the phase and amplitude relationships among the cosine signals of Figure 5-81 are fixed, we can increase the SNR if we can rotate the phases of all the components to line up at one particular instant in time. This process increases the signal strength for a short time period but does not affect additive white Gaussian noise (AWGN).

This coherent addition concept is clearer when we examine at the phasor representation of the signal. To keep things manageable, we'll describe only at the middle five Fourier components of Figure 5-81. Table 5-3 shows the amplitude, phase, and rotation rates of these components.

Figure 5-82 shows the data from Table 5-3 and from Figure 5-81 in phasor format.

The magnitude of the composite signal is the vector sum of all the phasors present in the signal. Figure 5-82 shows the phasor diagram at one particular instant in time. The vector addition of components (1) through (5) determines the amplitude of the signal at that

TABLE 5-3 ■ The rotation frequency, amplitude, and phase of the middle five frequency components of Figure 5-81.

Number	Rotation Frequency[1] (Hz)	Amplitude[2]	Phase (Degrees)[2]
1	−6666	13.270	112.50
2	−3333	26.483	56.25
3	0	16.000	180.00
4	3333	26.483	−56.25
5	6666	13.270	−112.50

[1]The rate at which the phasor rotates about the origin (positive frequency is CCW).
[2]The amplitudes and phases of the cosine wave that makes up the component.

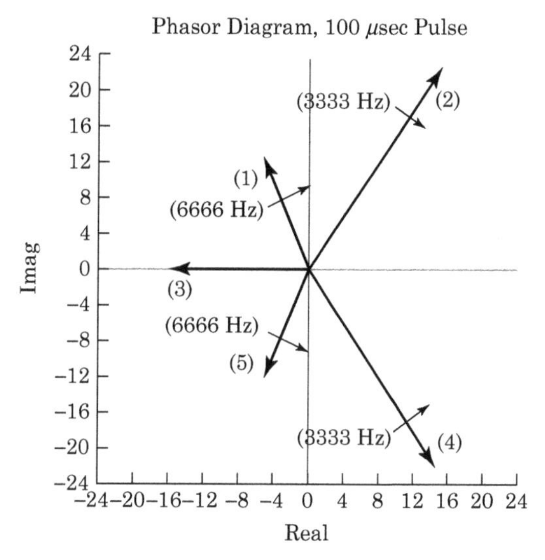

FIGURE 5-82 ■ The phasor diagram for the middle five frequency components of Figure 5-81. Component (1) is $13.27\angle 112.50^\circ$, and it rotates CW about the origin at 6,666 Hz. Component (2) is $26.48\angle 56.25^\circ$, and it rotates CW about the origin at 3333 Hz. Component (3) is $16.00\angle 180.00^\circ$, and it is stationary (i.e., it does not rotate). Component (4) is $26.48\angle -56.25^\circ$, and it rotates CCW about the origin at 3,333 Hz. Component (5) is $13.27\angle -112.5^\circ$, and it rotates CCW about the origin at 6,666 Hz.

instant. We seek to rotate each component of Figure 5-82 so that they are collinear along the real axis. This rotation will maximize the output signal strength at one instant. Figure 5-83 shows the effect. We accomplish this rotation via the $S^*(f)$ portion equation (5.82).

Figure 5-83 shows the increase in detected signal energy after rotating each Fourier component of Figure 5-81 and allow them to add together coherently. Each phasor in Figure 5-83 still rotates about the origin at its original rate. The collinear addition will happen only for a brief instant then collapse.

5.14.3 The Conjugate

The term that performs the rotation in equation (5.82) is the conjugate term

$$S^*(f) \tag{5.83}$$

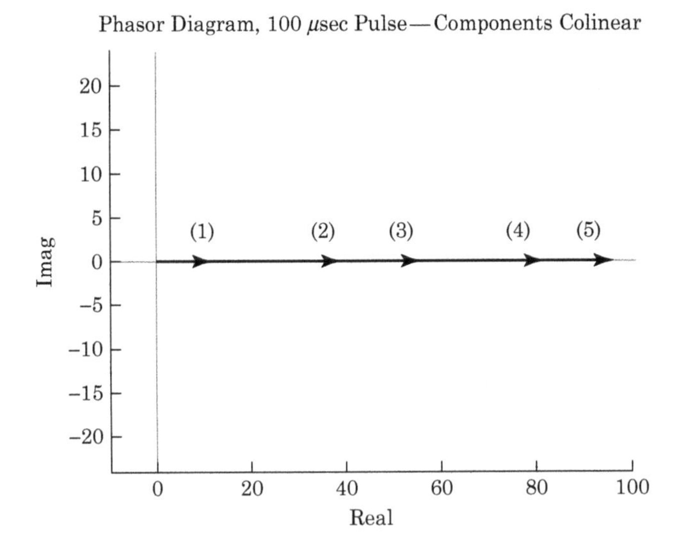

FIGURE 5-83 ■ When we rotate the phase of each component in Figure 5-82, you can achieve the colinear (in phase) addition of all of the Fourier components. This addition increases the signal energy while not increasing the noise energy.

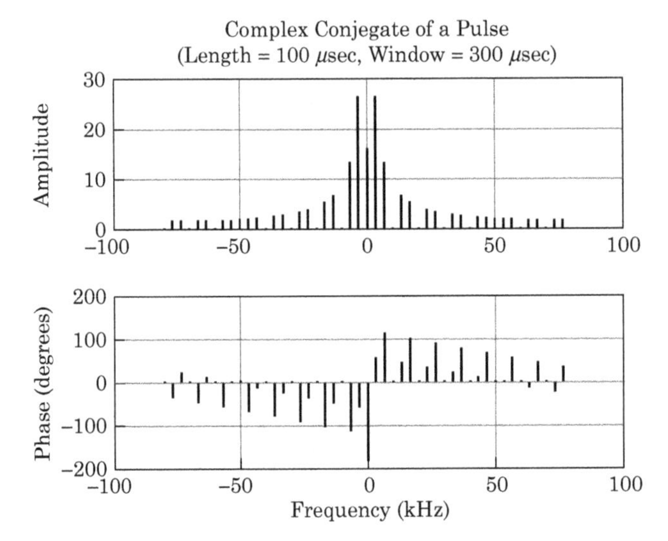

FIGURE 5-84 ■ The complex conjugate of the signal shown in Figure 5-80 and Figure 5-81. The amplitude of each component is unchanged from Figure 5-81, but the phase has been negated. This is the $S^*(f)$ when $S(f)$ is the signal from Figure 5-80 and Figure 5-81.

which is the conjugate of the Fourier transform of the known input signal. Figure 5-84 shows the magnitude and phase of $S^*(f)$.

The phase of $S^*(f)$ rotates each phasor component of each component in Figure 5-82 so that every Fourier component present in the signal $S(f)$ is collinear with the x-axis. The result is that all of the spectral components of the signal to line up on the x-axis, and we achieve the theoretical maximum SNR at the filter's output.

5.14.4 The Magnitude of $S^*(f)$

The top of Figure 5-84 shows the magnitude response of $S^*(f)$. The magnitude of $S^*(f)$ is identical to the magnitude of $S(f)$ because taking the conjugate of a complex function does not change the function's amplitude response. Note the areas of high attenuation at multiples of 10 kHz (e.g., at $+/-$ 10 kHz, $+/-$ 20 kHz, $+/-$ 30 kHz). The signal for which we're building a matched filter (Figure 5-80 and Figure 5-81) contains no energy at multiples of 10 kHz (e.g., at +/ − 10 kHz, +/ − 20 kHz, +/ − 30 kHz), so it would add unnecessary noise to our follow-on processes if we allowed energy to pass through the matched filter at those frequencies. Passing energy in frequency ranges where the signal contains no energy would not help the detection process by adding signal energy and would hinder the detection by allowing noise through unnecessarily.

5.14.5 $e^{-j2\pi fT}$

The last part of the matched filter equation (5.82) is

$$e^{-j2\pi fT} \tag{5.84}$$

which represents a frequency-dependent phase shift. The $S^*(f)$ term of equation (5.82) aligns the phases of the Fourier components of the signal to which we match. This term serves to reverse the phase rotation caused by a simple time delay. In effect, this term delays each Fourier component in the signal to allow the entire signal to contribute to the matched filter output. This term is that it ensures causality. It allows the matched filter to fill with each Fourier component before the final phase alignment.

5.14.6 A Practical Matched Filter

We now design a matched filter to detect the signal shown in Figure 5-80. Figure 5-81 shows the frequency-domain representation of this sampled signal. The signal is sampled, in the time domain, at 160 kHz, making the complete time window 48 samples long. This combination produces a frequency resolution of 3.33 kHz. Applying equation (5.82) produces the matched filter components shown in Figure 5-85.

We normally use a digital signal processing (DSP) finite impulse response (FIR)filter to realize the matched filter. The coefficients of the FIR filter are given by the inverse fast Fourier transform (FFT) of the frequency-domain data shown in Figure 5-85. Figure 5-86 shows the time-domain filter taps.

The FIR filter taps represent the impulse response of the FIR filter. The impulse response of a matched filter is a time-reversed version of signal to which the filter is

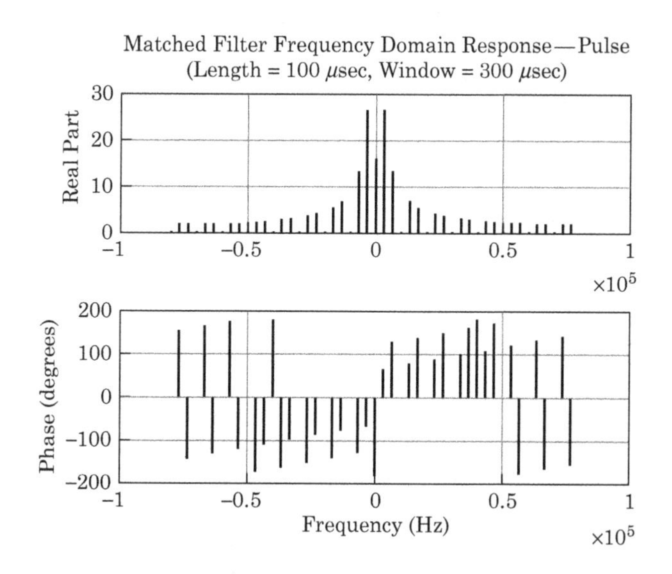

FIGURE 5-85 ■ The frequency-domain representation of the matched filter designed to detect the pulse of Figure 5-80. As equation (5.82) describes, the amplitude response of the filter matches the amplitude of the signal to which we're matching. The phase response is significantly altered from the signal of interest.

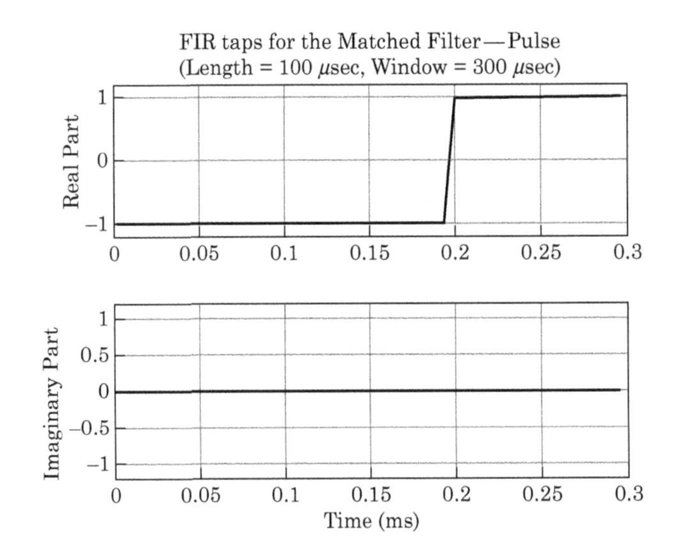

FIGURE 5-86 ■ The FIR filter taps for the matched filter, generated by taking the inverse FFT (IFFT) of the frequency-domain signal of Figure 5-85. Since the real component is so much larger than the imaginary component, we ignore the imaginary component and assume that the filter is real only.

matched. This relationship is clearly apparent by comparing Figure 5-80 and the real part of Figure 5-86.

5.14.7 The Matched Filter in Action

Finally, let us apply the matched filter (the FIR taps of Figure 5-86.) to the matched signal shown in Figure 5-80. Figure 5-87 shows the output.

Figure 5-87 shows the output of the filter matched when we apply the signal of Figure 5-80. We designed the filter to coherently combine all of the Fourier components in

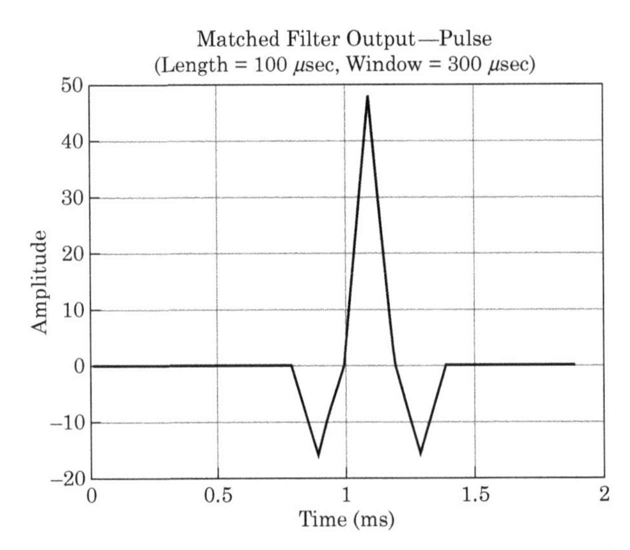

FIGURE 5-87 ■ The output of the FIR filter that is matched to the signal of Figure 5-80. The filter output does not resemble the input signal but was designed to produce a maximum SNR at one instant in time (1.05 msec, in this example). *Note*: Zeroes were added to the beginning and end of the time-domain signal of Figure 5-80 to improve clarity.

the signal together at one instant in time, which produces the peak at 1.05 ms. In spectrally flat, Gaussian noise, the peak at 1.05 ms will exhibit the highest SNR that is theoretically possible.

The demodulator must decide when exactly to sample the output of the matched filter. If the demodulator samples too early or too late, it will be sampling at an instant when all the Fourier components of the signal have not been coherently combined, and we will achieve nonoptimal results.

5.15 | BIBLIOGRAPHY

Williams, Arthur B., *Electronic Filter Design Handbook*, McGraw-Hill Book Company, 1981.

Blinchikoff, Herman J., and Zverev, Anatol I., *Filtering in the Time and Frequency Domains,* John Wiley and Sons, 1976.

Couch II, Leon W., *Digital and Analog Communications Systems,* 4th ed., Macmillan Publishing Company, 1993.

Ferrand, Michael, "Practical Microwave Filter Design," *Applied Microwave Magazine*, p. 120, May 1989.

Freeman, Roger L., *Reference Manual for Telecommunications Engineering*, John Wiley and Sons, Wiley-Interscience, 1985.

Gardner, Floyd M., *Phaselock Techniques*, 2d ed., John Wiley and Sons, 1979.

Orchard, H.J., "The Phase and Envelope Delay of Butterworth and Tchebycheff Filters," *IRE Trans. Circuit Theory,* Vol. CT-7, pp. 180–181, June 1960.

Porter, Jack, "Noise Bandwidth of Chebyshev Filters," *RF Design Magazine*, p. 19, n.d.

White, Donald R.J., *A Handbook on Electrical Filters*, Don White Consultants, Inc., 1980.

Ziemer, R.E., and Tranter, W.E., *Principles of Communications—Systems, Modulation and Noise,* 4th ed., John Wiley and Sons, Inc., 1995.

Zverev, A.I., *Handbook of Filter Synthesis*, John Wiley and Sons, 1967.

5.16 | PROBLEMS

1. **Multiple Choice: Filter Suppression of Signals** A filter suppresses unwanted signals by:
   a. shunting them to ground.
   b. absorbing them internally.
   c. shunting them into a dummy load resistor.
   d. frequency-selective mismatching.

2. **Multiple Choice: Filter Amplitude Response** Which statement is false?
   a. Butterworth filters exhibit a monotonic amplitude response.
   b. Chebychev filters trade passband ripple for a quicker transition from the filter's passband into its stopband.
   c. Chebychev filters contain stopband zeroes.
   d. Gaussian filters are used primarily for their group delay and transient responses.
   e. Stopband zeroes can produce high attenuation of unwanted signals.

3. **Isolators and Circulators** The major filtering mechanism used in RF and microwave filters is selective matching:

   - The filter presents an input and output match in the passband
   - The filter mismatches the input and output in its stopband.

   In general, we don't like mismatched components in systems. Mismatches cause many subtle problems in mixers and oscillators, so we try to avoid them if possible.

   How would you build a filter that presents a broadband match over its passband and its stopband? Hint: Look up two different but related RF components: a *ferrite circulator* and a *ferrite isolator.*

4. **Signal Power through Filters** For the system shown in Figure 5-88, what are the output power levels of the signals at 830, 920, and 1,010 MHz? The component characteristics are

   - BPF1: 910 MHz CF, 30 MHz BW, 2 dB IL
   - BPF2: 930 MHz CF, 20 MHz BW, 5 dB IL
   - All filters are five-pole Butterworth filters
   - Amplifier: Gain = 15 dB

5. **Filters in Cascade** Given the system shown in Figure 5-89, find the gain of the system at the following frequencies:

   a. The aeronautical/navigation band at 4.2–4.4 GHz. This is the frequency range that the system was designed to pass with minimum attenuation.
   b. An interfering signal at 3,140 MHz
   c. The 5.46–5.47 GHz radio-navigation band
   d. The 3.3–3.5 GHz amateur radio band

   When the problem gives a frequency range, find the gain at the band edges (i.e., for part a), find the gain at 4.2 and 4.4 GHz).

   Both filters are Butterworth and exhibit an ultimate attenuation of 85 dB. The center frequency of the BPF is the filter's geometric center frequency.

6. **Amplitude Response for a Chebychev LPF** Find the equation that describes the amplitude response of a Chebychev LPF and a Chebychev BPF.

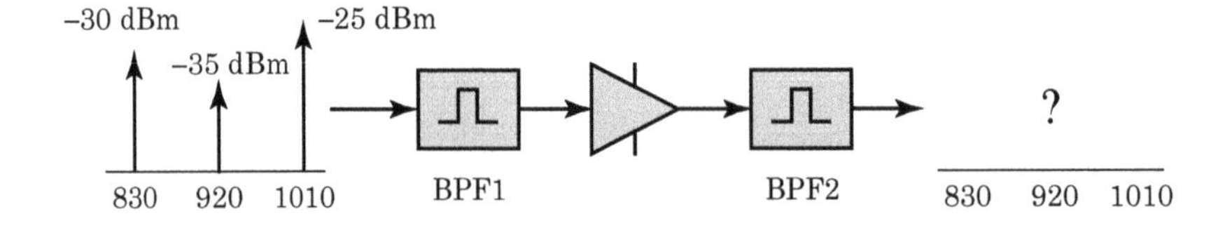

FIGURE 5-88 ■ Attenuation of signals through various filters.

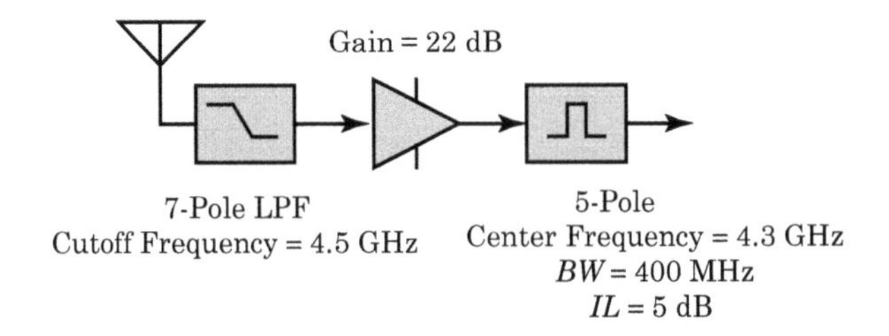

FIGURE 5-89 ■ A radio receiver with filters.

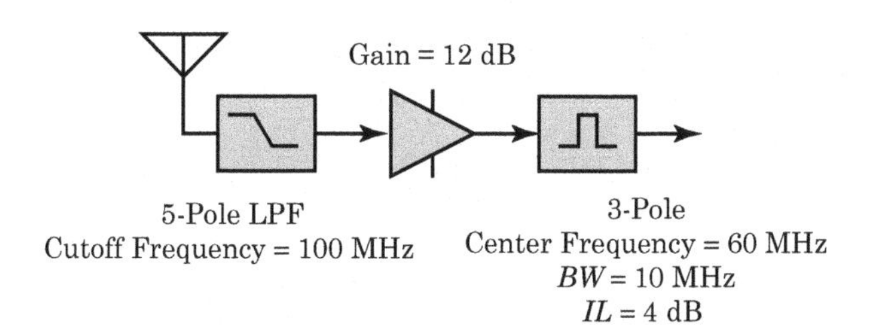

FIGURE 5-90 ■ A radio receiver with filters.

7. **Filter Attenuation** Given the system shown as Figure 5-90, find the signal strength present at the system's output for the following input conditions, assuming both filters exhibit an ultimate attenuation of 65 dB:

   a. 60 MHz at −40 dBm
   b. 175 MHz at −25 dBm

8. **Butterworth BPF Shape factor** Given a three-pole Butterworth band-pass filter with a 200 kHz, 3 dB bandwidth:

   a. Find the 30 dB/3 dB shape factor?
   b. Find the 45 dB/6 dB shape factor?

9. **Ultimate Roll-Off for a Butterworth BPF** Derive the ultimate roll-off rate for a Butterworth BPF.

10. **NF, MDS** Given the system shown as Figure 5-91:

    a. Find the noise figure at the center frequency of the RF BPF.
    b. Find the noise figure at 1,100 MHz.
    c. Find the minimum detectable signal (MDS) at the center frequency of the RF BPF.
    d. How many poles do we need in the RF BPF to bring the signal at 1,100 MHz down to the system's MDS? Use the MDS you calculated for the center frequency of the RF BPF. Let the noise bandwidth be 30 MHz.

11. **Noise BW, MDS** For the cascade of Figure 5-92:

    a. What's the noise band width of this cascade?

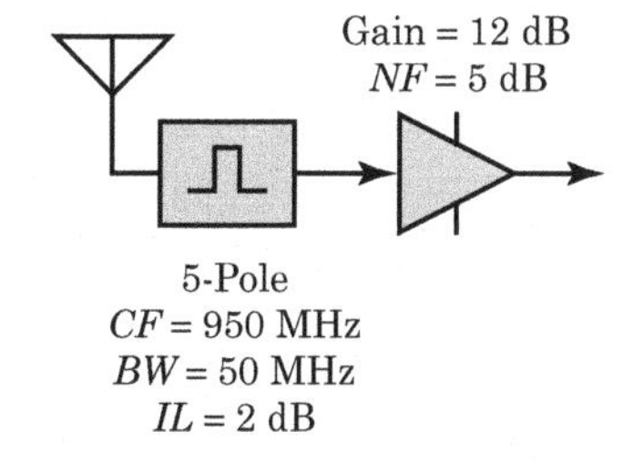

FIGURE 5-91 ■ Simple receiving system.

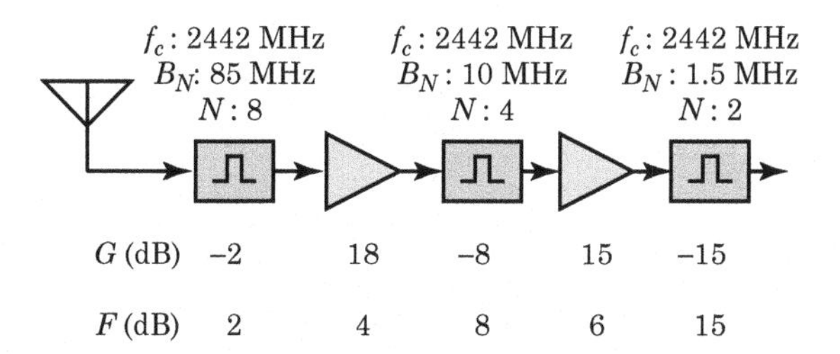

FIGURE 5-92 ■ A simple cascade.

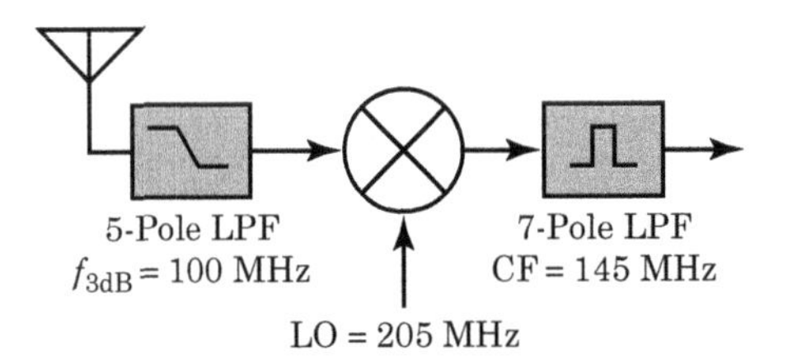

FIGURE 5-93 Simple cascade with a mixer.

b. Find the MDS assuming the antenna noise temperature is 240°K.

c. Find the gain of a signal at 2,465 MHz.

12. **LO, IF, and Image Rejection of an LPF** Given the system shown in Figure 5-93, find the attenuation provided by the front-end low-pass filter at the following frequencies:

 a. the LO frequency of 205 MHz
 b. the IF of 145 MHz
 c. the image frequency of 350 MHz.

13. **Transfer Function Ultimate Roll-Off Rate** Given the number of poles and zeroes for the following low-pass filters described, predict the ultimate roll-off of each filter:

 a. 6 poles, 5 zeroes
 b. 7 poles, 6 zeroes
 c. 7 poles, 2 zeroes
 d. 3 poles, 1 zero

14. **Elliptic Filter Impedances I** Figure 5-94 shows the circuit diagram for an elliptic filter:

 a. Is this a high-pass filter or a low-pass filter?
 b. In the filter's stopband, does this filter present a high- or low-impedance load to the source resistor R_s?
 c. In the filter's stopband, does the load resistor see a high or low impedance when it looks into the filter?

15. **Elliptic Filter Impedances II** Figure FLTPS-95 shows the circuit diagram for an elliptic filter:

 a. Is this a low-pass filter or a high-pass filter?
 b. In the filter's stopband, does this filter present a high or low impedance to the source resistor R_s?
 c. In the filter's stopband, does this filter present a high- or low-impedance to the load resistor R_L?

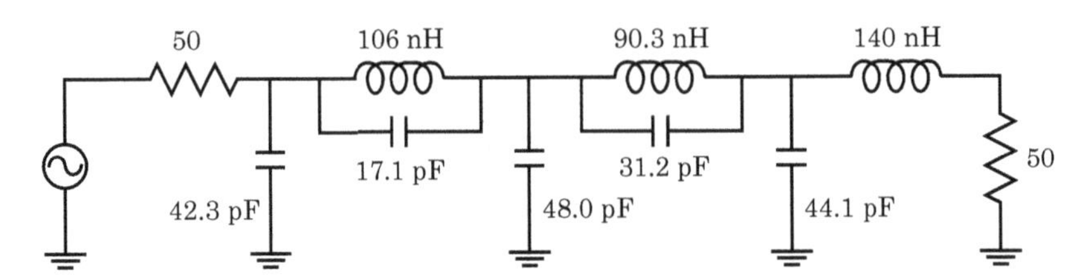

FIGURE 5-94 Filter schematic.

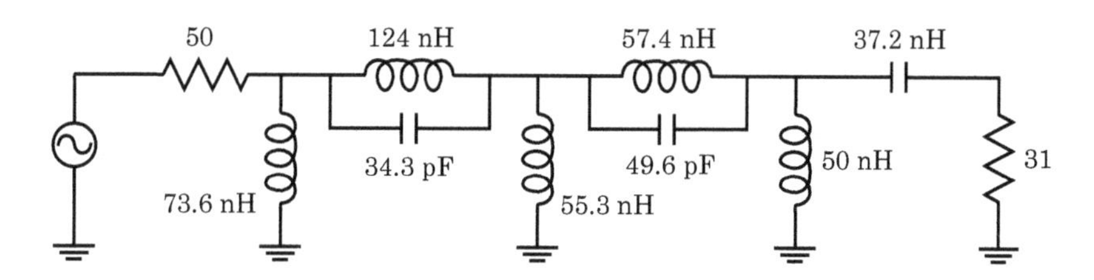

FIGURE 5-95 ■ Filter schematic.

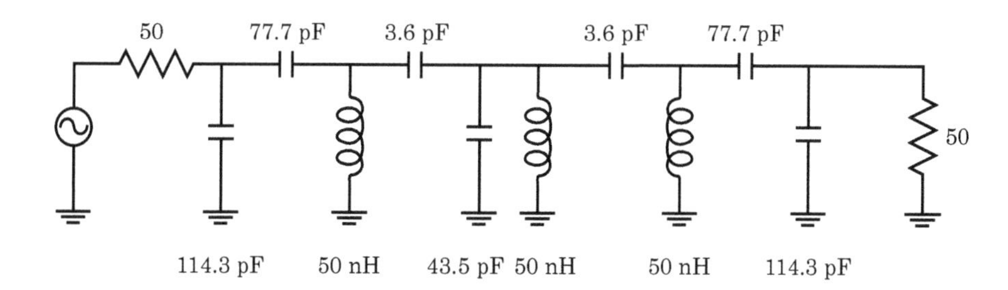

FIGURE 5-96 ■ Filter schematic.

16. **Band-Pass filter Impedances** Figure FLTPS-96 shows a Butterworth band-pass filter. A band-pass filter as two stopbands: one below the center frequency of the filter and one above the center frequency of the filter:

    a. Does the filter present a high or low impedance to the outside world in the filter's lower stopband?

    b. Does the filter present a high or low impedance to the outside world in the filter's upper stopband?

17. **Bandwidths of Band-Pass Filters** The following band-pass filters all have a geometric center frequency of 700 MHz but different 3 dB bandwidths. Find the difference between $f_{c,geo}$ and $f_{c,arith}$ for the following band-pass filters in Mhz and percent:

    a. $f_{L,3\,\text{dB}} = 682.7$ MHz, $f_{U,3\,\text{dB}} = 717.7$ MHz (this filter's 3 dB bandwidth is 5% of its center frequency)

    b. $f_{L,3\,\text{dB}} = 665.9$ MHz, $f_{U,3\,\text{dB}} = 735.9$ MHz (this filter's 3 dB bandwidth is 10% of its center frequency)

    c. $f_{L,3\,\text{dB}} = 633.5$ MHz, $f_{U,3\,\text{dB}} = 773.5$ MHz (this filter's 3 dB bandwidth is 20% of its center frequency)

    d. $f_{L,3\,\text{dB}} = 432.6$ MHz, $f_{U,3\,\text{dB}} = 1133$ MHz (this filter's 3 dB bandwidth is 100% of its center frequency)

18. **Butterworth LPF Magnitude Response**

    a. We're given a six-pole Butterworth low-pass filter with 3 dB of attenuation at 120 MHz. Find the attenuation at 650 MHz.

    b. We've found a Butterworth low-pass filter in the lab. We've measured $f_{c,3\,\text{dB}}$ to be 6.6 MHz. We've also measured an attenuation of 30 dB at about 20.9 MHz. How many poles are in the filter?

19. **Butterworth BPF Shape Factor** Given a six-pole Butterworth band-pass filter with a 275 MHz 3 dB bandwidth:

    a. Find the 60 dB/6 dB shape factor.

    b. Find the 40 dB/6 dB shape factor.

20. **Lossy Butterworth BPF** Find the total attenuation a signal at 1.75 GHz will experience after passing through a three-pole Butterworth band-pass filter. The insertion loss of the filter is 4 dB at the center frequency, the geometric center frequency is 2.7 GHz, and the filter's 3 dB bandwidth is 1.1 GHz.

CHAPTER 6

Noise

Noise is the most impertinent of all forms of interruptions. It is not only an interruption, but also a disruption of thought.

Arthur Schopenhauer

He that loves noise must buy a pig.

John Ray

The noise is so great, one cannot hear God thunder.

R.C. Trench

Chapter Outline

6.1 | INTRODUCTION

Thermal noise places a fundamental limit on the smallest signal a receiving system can process. In this chapter, we will examine noise from a mathematical point of view and then relate our new knowledge to receiving systems.

6.2 | EQUIVALENT MODEL FOR A RADIO FREQUENCY DEVICE

Figure 6-1 shows the general noise model we will use throughout this chapter.

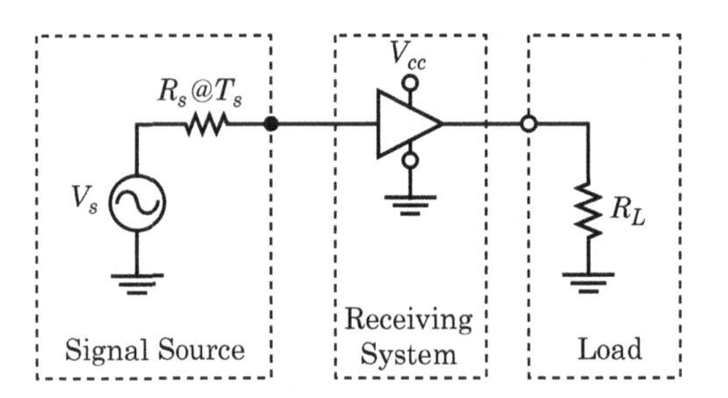

FIGURE 6-1 ■ The noise model of a complete receiving system, including the signal source, the receiving system, and the load resistor.

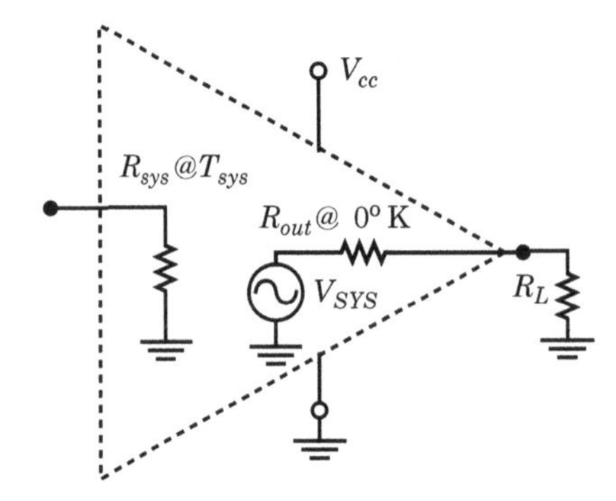

FIGURE 6-2 ■ The noise model of an RF system with emphasis on its noise characteristics.

The *signal source* can be a signal generator, antenna, or another radio frequency (RF) amplifier. We will model any signal source as a voltage source, V_S, in series with a resistor R_S. The resistor R_S represents the output impedance of the signal source.

We will also use R_S to account for the noise present at the output port of the signal source. We'll assume this resistor exists at some noise temperature T_S that is not necessarily the physical temperature of the signal source. The temperature T_S accounts for the noise power present at the output of the signal source.

The receiving system can be a single RF amplifier or a complex system of, for example, amplifiers, mixers, attenuators, and oscillators. We're concerned only with its input and output ports.

Figure 6-2 shows the model we'll use to describe a garden-variety receiving system. The receiver accepts power from the external source. The system measures the power delivered to R_{SYS} and adjusts its internal voltage source V_{SYS} so that a fixed multiple of the input power is delivered to R_L. The waveform developed across the output resistor R_L is a duplicate of the waveform developed across R_{sys}. In equation form

$$P_{RL} = G_P P_{in} \tag{6.1}$$

where

P_{in} = the power delivered to the system's input resistor by some external source (in linear units such as watts or mW)

P_{RL} = the power delivered to the load resistor R_L by the system (in linear units)

G_P = the power gain of the system; usually > 1 (in linear units)

We can write equation (6.1) in a logarithmic format

$$P_{RL,dBm} = G_{P,dB} + P_{in,dBm} \qquad P_{RL,dBW} = G_{P,dB} + P_{in,dBW} \tag{6.2}$$

where

$P_{in,dBm}$ = the power delivered to the receiving system's input resistor by some external source in dBm

$P_{RL,dBm}$ = the power delivered to the load resistor R_L by the receiving system in dBm

$P_{in,dBW}$ = the power delivered to the receiving system's input resistor by some external source in dBW

$P_{RL,dBW}$ = the power delivered to the load resistor R_L by the receiving system in dBW

$G_{P,dB}$ = the power gain of the system; usually > 0 dB; in dB

6.2.1 Noise Temperature

The temperature T_{sys}, of the input resistor R_{sys} is the *noise temperature* of the system. T_{sys} is a direct measure of the noise added by the system. Although there are many sources of noise inside the receiving system, we are going to blame them all on the temperature of the input resistor. We universally assume the noise temperature of the output resistor (R_{out} of Figure 6-2) is 0°K and thus is noiseless.

6.2.2 Power Supply

The receiving system of Figure 6-1 and Figure 6-2 has connections to both a power supply (V_{cc}) and ground. Extraneous noise can leak in through these connections if the power supply isn't adequately filtered.

6.2.3 Matching

Finally, most of our analysis in this chapter assumes that the various components are matched, that is, $R_S = R_{sys} = R_{out} = R_L$. In practice, we often deviate from this ideal.

6.3 | NOISE FUNDAMENTALS

If we measure the alternate current (AC) voltage present across the terminals of a physical resistor, we will observe a noise voltage waveform. This is thermal Gaussian noise.

The physical, fundamental process responsible for the noise involves the physical temperature of the resistor. Since the resistor exists at some temperature above absolute zero, the atoms and electrons inside the material are in constant motion. They *jiggle* in a completely random fashion. The hotter the resistor, the faster the electrons jiggle. This jiggling represents a random current and a random current moving through a resistive material will generate a random voltage across the material. This noise places a limit on the smallest signal we can process.

Other sources of noise have absolutely nothing to do with the thermal effect. However, we often model things as though thermal noise is the cause of all the world's noise problems.

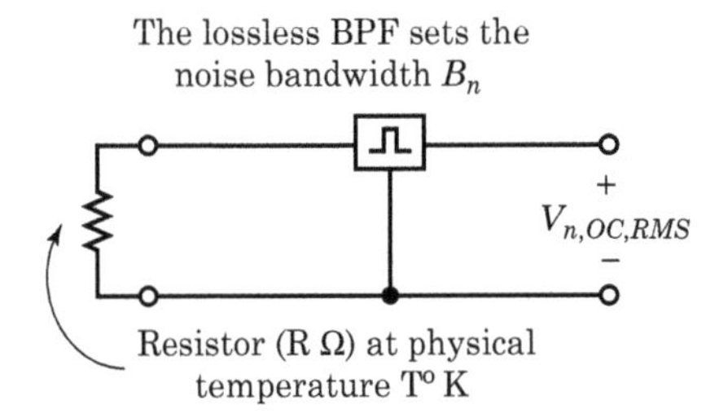

FIGURE 6-3 ■ The fundamental method of measuring thermal noise.

6.3.1 Thermal Noise

Figure 6-3 shows a resistor of resistance $R\,\Omega$ at a physical temperature of $T\,^\circ$K. We connect this resistor to a lossless (and noiseless) bandpass filter whose noise bandwidth is B_n.

6.3.2 Noise Described Statistically

The time-domain representation of thermal noise follows a Gaussian distribution. We require the mean and standard deviation to describe completely this process. The mean value of noise is 0 volts because the RF world is an AC-coupled arena.

The open-circuit, root mean square (RMS) voltage present at the output terminals of Figure 6-3 is

$$V_{n,OC,RMS} = \sqrt{4kTB_nR} \tag{6.3}$$

where

$V_{n,OC,RMS}$ = the open-circuit, RMS noise voltage produced by the resistor. This voltage is the *open-circuit* noise voltage and it is an *RMS* quantity. This noise voltage $V_{n,OC,RMS}$ is the standard deviation value we'll use in our calculations involving Gaussian statistics.

k = Boltzman's constant = $1.381 \cdot 10^{-23}$ Joules/°K (or $1.381 \cdot 10^{-23}$ watt-sec/°K). This number represents the conversion factor between two forms of energy. It gives the average mechanical energy per particle that we can couple out electrically per °K. Another useful expression for Boltzman's constant is $1.381 \cdot 10^{-20}$ milliwatt-sec/°K.

T = temperature in °K. The hotter the resistor, the faster the atoms and electrons jiggle. With increasing temperature, each particle contains more energy on the average, and the noise voltage present across the resistor terminals will rise. We often use *room temperature* (about 290°K) for this number.

B_n = the noise bandwidth of the measurement in hertz. The noise power is spread evenly across the frequency spectrum, so the wider the measurement bandwidth the more noise voltage we will measure. Figure 6-4 shows the effects of changing the noise bandwidth.

R = the value of the resistor in ohms. A noise current forced through a larger resistance will produce a larger noise voltage.

6.3.3 Voltage Source Model

Figure 6-5 shows a model for a noisy resistor. We will replace the noisy resistor by an ideal, noiseless resistor in series with a voltage source. The voltage source's value is $V_{n,OC,RMS}$.

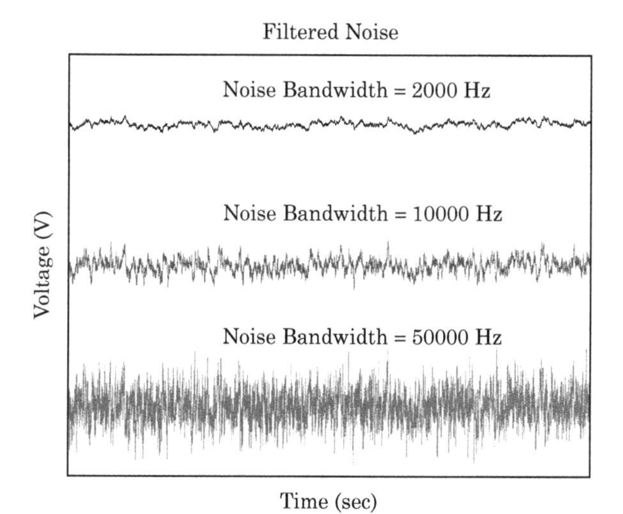

FIGURE 6-4 ■ White or Gaussian noise observed through filters with bandwidths of 2 kHz, 10 kHz, and 50 kHz. Equation (6.3) expresses the relationship between noise bandwidth B_n and $V_{n,OC,RMS}$. The wider the noise bandwidth, the larger the RMS noise voltage.

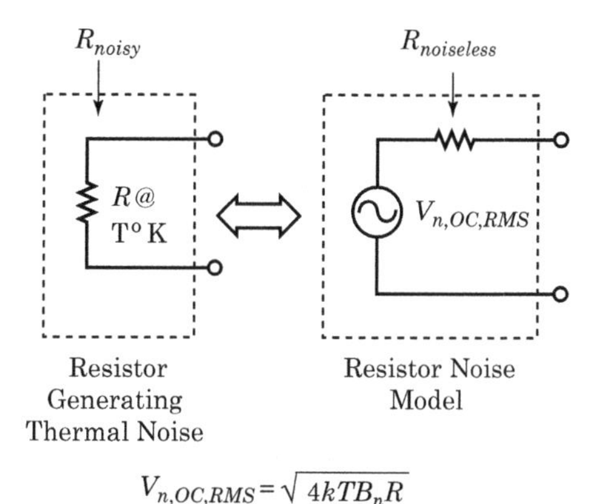

FIGURE 6-5 ■ The voltage source model for a noisy resistor of Figure 6-3. We replace the noisy resistor with a noiseless resistor in series with a Gaussian noise voltage source.

All of the noise phenomena we'll discuss will be based on the voltage source model, equation (6.3), Figure 6-3, and Figure 6-5.

EXAMPLE

Resistor Thermal Noise Voltage

Find the open-circuit, RMS noise voltage present across the following resistors:

a. A 1 MΩ resistor at room temperature (about 290°K) measured in a 200 kHz noise bandwidth

b. A 50 Ω resistor at 1000°K measured in a 10 MHz noise bandwidth

Solution

a. Substituting $T = 290°\text{K}$, $B_n = 200$ kHz, and $R = 1\ \text{M}\Omega$ into equation (6.3), we find $V_{n,OC,RMS} = 56.6\ \mu\text{V}_{\text{RMS}}$.

b. Substituting $T = 1000°\text{K}$, $B_n = 10$ MHz, and $R = 50\ \Omega$ into equation (6.3), we find $V_{n,OC,RMS} = 5.25\ \mu\text{V}_{\text{RMS}}$.

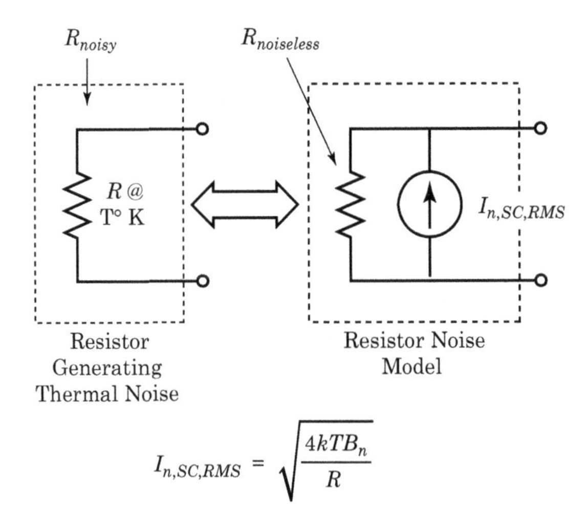

FIGURE 6-6 ■ The current source model for a noisy resistor of Figure 6-3. The noisy resistor is replaced by a noiseless resistor in parallel with a Gaussian noise current source.

6.3.4 Current Source Model

We can also model the noisy resistor of Figure 6-3 as a current source in parallel with an ideal noiseless resistor. Figure 6-6 shows this model.

The value of the current source of Figure 6-6 is

$$I_{n,SC,RMS} = \sqrt{\frac{4kTB_n}{R}} \tag{6.4}$$

where $I_{n,SC,RMS}$ is the the short-circuit, RMS noise current produced by the resistor. This is the RMS *short-circuit* noise current. Like $V_{n,OC,RMS}$, $I_{n,SC,RMS}$ is the standard deviation of the noise current generated by the current source.

EXAMPLE

Resistor Thermal Noise Current

Find the RMS noise current present in the following short-circuited resistors:

a. A 1 MΩ resistor at room temperature (about 290°K) measured in a 200 kHz noise bandwidth

b. A 50 Ω resistor at 1000°K measured in a 10 MHz noise bandwidth

Solution

a. Substituting $T = 290°\text{K}$, $B_n = 200$ kHz, and $R = 1$ MΩ into equation (6.4), we find $I_{n,SC,RMS} = 56.6\ \text{pA}_{\text{RMS}}$.

b. Substituting $T = 1000°\text{K}$, $B_n = 10$ MHz, and $R = 50\,\Omega$ into equation (6.4), we find $I_{n,SC,RMS} = 105\ \text{nA}_{\text{RMS}}$.

EXAMPLE

Equivalency of the Voltage and Current Noise Models

Show that the voltage and current noise models of Figure 6-5 and Figure 6-6 are equivalent models.

Solution

We short the terminals of the voltage model of Figure 6-5 and find the current that flows:

$$
\begin{aligned}
I_{SC,Voltage\,Model} &= \frac{V_{n,OC,RMS}}{R} \\
&= \frac{\sqrt{4kTB_nR}}{R} \\
&= \sqrt{\frac{4kTB_n}{R}} \\
&= I_{n,SC,RMS}
\end{aligned} \tag{6.5}
$$

The open-circuit voltage of the current-source model of Figure 6-6.

$$
\begin{aligned}
V_{OC,Current\,Model} &= I_{n,SC,RMS}R \\
&= \left(\sqrt{\frac{4kTB_n}{R}}\right)R \\
&\sqrt{4kTB_nR} \\
&= V_{n,OC,RMS}
\end{aligned} \tag{6.6}
$$

Since the short-circuit current and the open-circuit voltages are the same, the circuits are equivalent.

EXAMPLE

Expected Values of Resistor Noise

We're given a 50 Ω resistor at room temperature (290°K). We measure the noise voltage through a 100 MHz low-pass filter (assume the filter has a noise bandwidth of 102 MHz).

a. How often will the magnitude of the noise voltage be greater than 16 μV?

b. How often will the magnitude of the noise voltage exceed 16 μV if we heat the resistor to 600°K?

Solution

a. Use equation (6.3) to find $V_{n,OC,RMS}$ for $T = 290°\text{K}$, $B_n = 102$ MHz, and $R = 50\,\Omega$:

$$
\begin{aligned}
V_{n,OC,RMS} &= \sqrt{4kTB_nR} \\
&= \sqrt{4(1.38 \cdot 10^{-23})(290)(102 \cdot 10^6)(50)} \\
&= 9.04\ \mu V_{RMS}
\end{aligned} \tag{6.7}
$$

The standard deviation of the noise, and the standard deviation of the Gaussian bell curve describing the noise, is 9.04 μV. The mean is 0 volts. We normalize this value to a standard Gaussian distribution using

$$z = \frac{x - \mu}{\sigma} \tag{6.8}$$

where

z = the normalized random variable, whose mean is zero and whose standard deviation is unity
μ = the mean of the measured random variable
σ = the standard deviation of the measured random variable
x = the measured random sample we wish to normalize

We can write

$$z_{16\mu V} = \frac{16-0}{9.04} = 1.77 \tag{6.9}$$

16 μV is 1.77 standard deviations out on a Gaussian bell curve whose standard deviation is 9.04 μV.

Standard Gaussian tables tell us the area of the Gaussian curve lying above $z = 1.77$ is 0.0384, meaning that the noise voltage will be greater than 16 μV 3.84% of the time. Using symmetry, we also know that the noise voltage will be less than $-16\ \mu$V for 3.84% of the time. The magnitude of the noise voltage will be greater than 16 μV for 7.68% of the time.

b. As in part **a.**, use equation (6.3) to find $V_{n,OC,RMS}$ with $T = 600°K$, $B_n = 102$ MHz, and $R = 50\ \Omega$

$$\begin{aligned} V_{n,OC,RMS} &= \sqrt{4kTB_nR} \\ &= \sqrt{4(1.38 \cdot 10^{-23})(600)(102 \cdot 10^6)(50)} \\ &= 13.0\ \mu V_{RMS} \end{aligned} \tag{6.10}$$

We use equation (6.8) to normalize 16 μV to a Gaussian curve with a standard deviation of 13 μV

$$z_{16\mu V} = \frac{16-0}{13.0} = 1.23 \tag{6.11}$$

16 μV is 1.23 standard deviations out on the Gaussian curve.

Heating the resistor increased the RMS noise voltage from the resistor. The heating forces the Gaussian curve to flatten out, and since the bell curve is flatter 16 μV doesn't represent as many standard deviations as it did in part **a**.

Standard Gaussian tables tell us that the magnitude of the noise voltage will be greater than 16 μV 10.93% of the time. Symmetry tells us that the instantaneous noise voltage will be less than $-16\ \mu$V for 10.93% of the time; thus, the magnitude of the noise will be greater than 16 μV for 21.9% of the time.

6.3.4.1 Combining Independent Noise Sources

There exist important situation in which two different noise sources are present in a system at the same time. Common examples are transistor amplifiers, operational amplifiers, and receiving systems. Figure 6-7 shows noisy resistors R_1 and R_2 and noiseless R_L connected in a simple circuit. We are interested in the noise voltage present across R_L due to the noise sources in R_1 through R_2.

The noise sources in resistors R_1 and R_2 are independent random processes. Statistical mathematics tells us that we must independently calculate the power delivered to R_L by

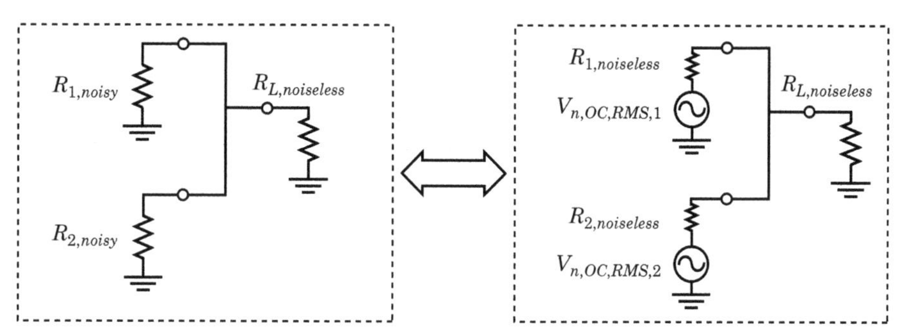

FIGURE 6-7 ■ Combining independent noise sources together. The Gaussian noise generator is not equivalent to the usual sinusoidal voltage sources we often find in circuit theory. Adding the RMS noise voltages generated by each noise source produces an incorrect result.

each noisy resistor and then add the individual powers together to find the total noise power delivered to the load resistor.

We first set all the noise sources to zero, except for one source. We find the power dissipated in R_L due to the active noise source. In Figure 6-7, for example, we first short out $V_{n,OC,RMS,2}$ and leave $V_{n,OC,RMS,1}$ in the circuit. Then, we find the noise power dissipated in R_L due only to $V_{n,OC,RMS,1}$. Next, we set to zero all of the noise sources except $V_{n,OC,RMS,2}$ and find the power dissipated in R_L due to the noise generated by R_2. If there were more noise sources, we proceed onward, examining the power delivered to R_L by each noise source in succession until we find the power delivered to R_L by each noise source in the circuit. Finally, we add all of the powers together to find the total noise power dissipated in R_L.

EXAMPLE

Combining Noise Sources

Figure 6-8 shows two noisy resistors, R_1 and R_2, connected to a noiseless load resistor R_L. Resistor R_1 is a 100 Ω resistor placed in boiling water (100°C = 373°K), R_2 is a 330 Ω resistor at 600°K, and R_L is a 510 Ω noiseless resistor. Find $V_{n,RMS,RL}$, the RMS noise voltage across R_L. Assume the noise bandwidth $B_n = 6$ MHz.

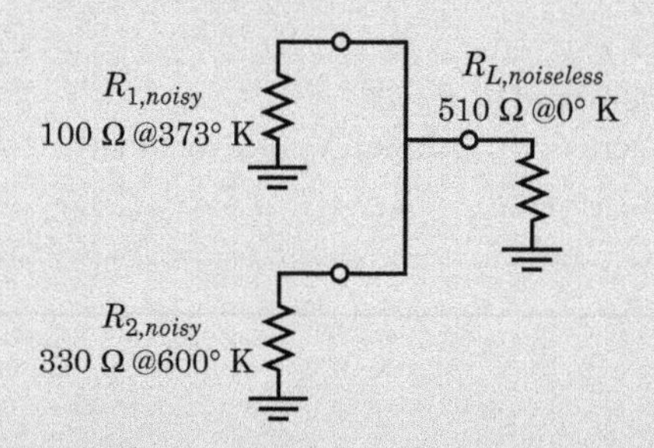

FIGURE 6-8 ■ Two noisy resistors are connected to a noiseless resistor. Find the noise power delivered to the load resistor.

Solution

Figure 6-9 shows the two noisy resistors, R_1 and R_2, replaced by their equivalent noise models of two noiseless resistors $R_{1,noiseless}$ and $R_{2,noiseless}$ in series with their equivalent noise voltage sources.

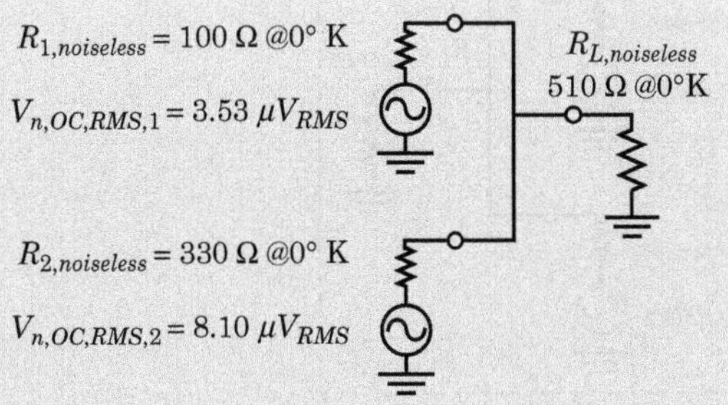

$R_{1,noiseless} = 100\ \Omega$ @$0°$ K

$V_{n,OC,RMS,1} = 3.53\ \mu V_{RMS}$

$R_{L,noiseless}$ $510\ \Omega$ @$0°$K

$R_{2,noiseless} = 330\ \Omega$ @$0°$ K

$V_{n,OC,RMS,2} = 8.10\ \mu V_{RMS}$

FIGURE 6-9 ■ The equivalent circuit of the system shown in Figure 6-8. The two noisy resistors have been replaced by noiseless resistors in series with their equivalent noise voltage generators. The current source equivalents for the noisy resistors would produce the same results.

Equation (6.3) gives us the values of the voltages sources as

$$\begin{aligned} V_{n,OC,RMS,R1} &= \sqrt{4kT_1B_nR_1} \\ &= \sqrt{4\,(1.38E-23)\,(373)\,(6E6)\,(100)} \\ &= 3.53\,\mu\mathrm{V}_{RMS} \end{aligned} \qquad \begin{aligned} V_{n,OC,RMS,R2} &= \sqrt{4kT_2B_nR_2} \\ &= \sqrt{4\,(1.38E-23)\,(600)\,(6E6)\,(330)} \\ &= 8.10\,\mu\mathrm{V}_{RMS} \end{aligned} \tag{6.12}$$

We first short-circuit $V_{n,OC,RMS,R2}$ and find the power delivered to the load resistor by R_1. Using voltage division, we find that the voltage across R_L due to $V_{n,OC,RMS,R1}$ is 2.38 $\mu\mathrm{V_{RMS}}$ and the power delivered to the load is

$$\begin{aligned} P_{noise,RL,R1} &= \frac{V^2_{n,OC,RMS,RL,R1}}{R_L} \\ &= \frac{(2.38E-6)^2}{510} \\ &= 11.1\,\mathrm{fW} \end{aligned} \tag{6.13}$$

Similarly, we find the noise power delivered to R_L by $V_{n,OC,RMS,R2}$ when $V_{n,OC,RMS,R1}$ is shorted. Voltage division reveals the voltage across R_L due to $V_{n,OC,RMS,R2}$ is 1.64 $\mu\mathrm{V_{RMS}}$. The noise power delivered to R_L by $R_{2,noisy}$ is

$$\begin{aligned} P_{noise,RL,R2} &= \frac{V^2_{n,OC,RMS,RL,R2}}{R_L} \\ &= \frac{(1.64E-6)^2}{510} \\ &= 5.26\,\mathrm{fW} \end{aligned} \tag{6.14}$$

We add the noise powers together to find the total noise power delivered to R_L as $P_{noise,RL} = 11.1 + 5.26 = 16.4$ fW, and the RMS noise voltage across R_L is 2.89 $\mu\mathrm{V}_{RMS}$. Note that adding the two noise voltages above produces the *wrong* answer: $V_{n,RMS,RL} \neq 2.38 + 1.64 = 4.02\,\mu\mathrm{V}_{RMS}$.

6.3.5 Noise Power

Receiving systems deal with energy and power. We're now going to examine thermal resistor noise in terms of noise power.

6.4 ONE NOISY RESISTOR

Figure 6-10(A) shows a noisy source resistor $R_{S,noisy}$ connected to a noiseless load resistor $R_{L,noiseless}$. In a receiving system, we almost universally match the value of the load resistor to the source resistor, that is, R_S equals R_L.

Figure 6-10(B) shows the noisy source resistor replaced by its noise voltage source $V_{n,OC,RMS,RS}$ and a noiseless series resistor $R_{S,noiseless}$. The noise power delivered to the load resistor R_L is

$$P_{RL} = \frac{V_{RL}^2}{R_L} \tag{6.15}$$

Since $R_L = R_S$,

$$\begin{aligned} V_{RL} &= \frac{1}{2} V_{n,OC,RMS,RS} \\ &= \frac{1}{2}\sqrt{4kT_S B_n R_S} \end{aligned} \tag{6.16}$$

So

$$\begin{aligned} P_{RL} &= \frac{\left(V_{n,OC,RMS,RS}/2\right)^2}{R_L} \\ &= \frac{\left(\sqrt{4kT_S B_n R_S}/2\right)^2}{R_L} \\ &= \frac{kT_S B_n R_S}{R_L} \end{aligned} \tag{6.17}$$

Since $R_S = R_L$, we can write

$$P_{RL} = kT_s B_n \tag{6.18}$$

In a matched system, the amount of noise that a noisy resistor passes to a noiseless resistor is $kT_S B_n$. This is true regardless of the values of the two resistors. Every resistor will deliver $kT_S B_n$ of noise power to its matched load resistor.

We match the source and load resistances to each other because this allows the load resistor to receive the maximum amount of signal power from the signal source. However, by matching the system, we also force ourselves to receive the maximum amount of noise power as well.

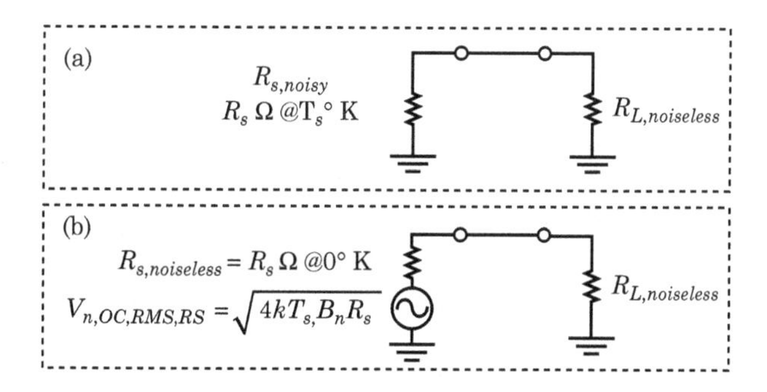

FIGURE 6-10 ■ Calculating the noise delivered by a noisy source resistor to a noiseless load resistor.

EXAMPLE

$kT_S B_n$

How much noise power does a 50 Ω source resistor deliver to a matched load if the source resistor temperature is 100°K? Use a noise bandwidth of 4 kHz.

Solution

Using equation (6.18), we find

$$\begin{aligned} P_{RL} &= kT_s B_n \\ &= (1.38E-23)\,(100)\,(4,000) \\ &= 5.52\,\text{fW} \\ &= -112.6\,\text{dBm} \end{aligned} \tag{6.19}$$

6.4.1 Room Temperature or T_0

As a standard, radio engineers (particularly those dealing with terrestrial radio links) often assume the temperature of the source resistor is room temperature. We use T_0 for room temperature and

$$\begin{aligned} \text{Room Temperature} &= T_0 \\ &= 290°\text{K} \\ &= 17°\text{C} \\ &= 62°\text{F} \end{aligned} \tag{6.20}$$

T_0 is actually a little colder than a comfortable room temperature.

6.4.2 N_0

The noise power delivered to a matched load resistor by a source resistor whose temperature is T_0 is denoted N_0. So

$$N_0 = kT_0 B_n \tag{6.21}$$

Also, we can write

$$\begin{aligned} kT_0 &= \left(1.38 \cdot 10^{-23}\right)(290) \\ &= 4 \cdot 10^{-21}\,\frac{\text{Watts}}{\text{Hz}} \\ &= -174\,\frac{\text{dBm}}{\text{Hz}} \end{aligned} \tag{6.22}$$

Equation (6.22) tells us that in a 1 Hz noise bandwidth a source resistor at room temperature will deliver −174 dBm of power to a matched, noiseless load resistor. We'll use equations (6.21) and (6.22) as starting points for realistic calculations. The noise power delivered by a room temperature source resistor to its matched load resistor in a noise bandwidth of B_n is

$$N_{0,dBm} = -174\,\frac{\text{dBm}}{\text{Hz}} + 10\log\left(B_n\right) \tag{6.23}$$

Equations (6.22) and (6.23) are classic examples of unit misuses. Although logarithmic quantities have no units associated with them, we frequently include bogus units with these equations for clarity.

EXAMPLE

kT_0B_n

How much noise energy does a matched source resistor at room temperature deliver to its load if we observe the noise through a filter with a 200 kHz noise bandwidth? Use equation (6.21) and then resolve with equation (6.23)

Solution

Using equation (6.21)

$$\begin{aligned} N_0 &= kT_0B_n \\ &= (1.38E-23)\,(290)\,(200E3) \\ &= 800\text{E}-18\,\text{W} \\ &= -121.0\,\text{dBm} \end{aligned} \tag{6.24}$$

Using equation (6.23)

$$\begin{aligned} N_{0,dBm} &= -174\,\frac{\text{dBm}}{\text{Hz}} + 10\log(B_n) \\ &= -174 + 10\log(200E3) \\ &= -121.0\,\text{dBm} \end{aligned} \tag{6.25}$$

6.5 | SYSTEM MODEL: TWO NOISY RESISTORS

We can now examine a more realistic model for a receiving system. Figure 6-11 shows an antenna connected to a receiver, which we have drawn as a simple amplifier.

Figure 6-12 shows the model of the receiving antenna. The antenna receives both signal and noise. We model the signal received by the voltage source V_{ant}. We model the

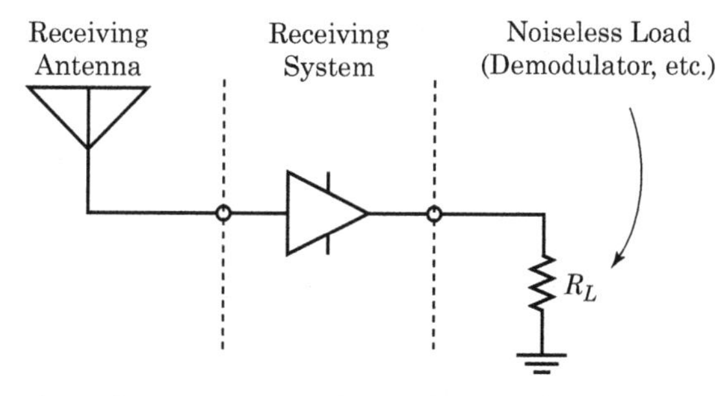

FIGURE 6-11 ■ A noisy antenna connected to a noisy receiver. The total system noise seen at the input is due to noise from the antenna and to noise generated internally in the receiving system.

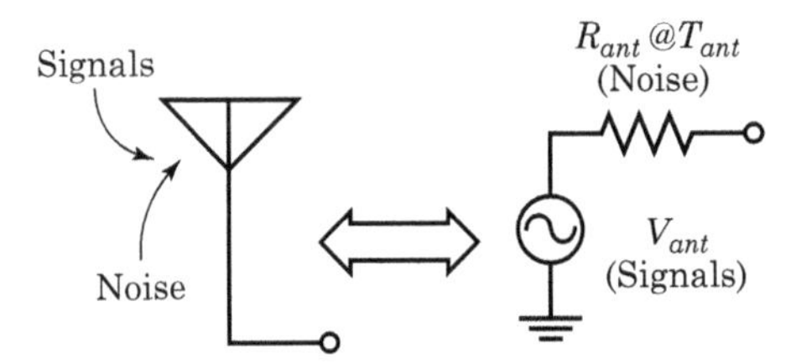

FIGURE 6-12 ■ The model of the receiving antenna of Figure 6-11. The voltage source (labeled V_{ant}) represents the ability of the antenna to receive signals. The larger the signal the antenna receives, the larger the value of V_{ant}. The value of the series resistor R_{ant} represents the output impedance of the antenna (typically 50, 75, or 300 ohms). We take pains to match our antenna to the characteristic impedance of the system. The antenna does not contain a physical resistor; R_{ant} arises from matching the characteristic impedance of free space to our system characteristic impedance.

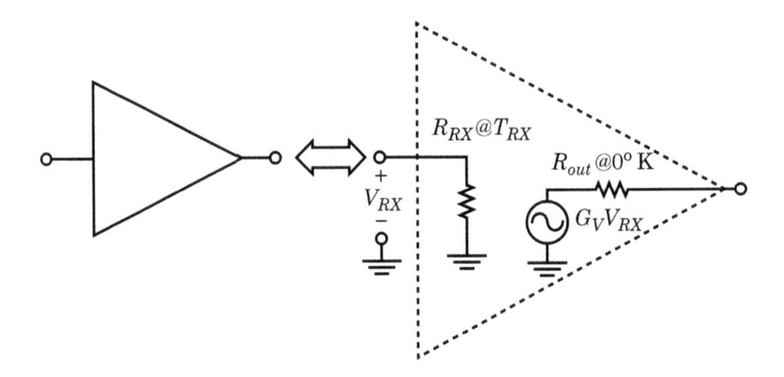

FIGURE 6-13 ■ The model of the receiving system of Figure 6-11, emphasizing the noise performance of the receiver.

noise received by assuming the series resistor, R_{ant}, exists at an *antenna noise temperature* of T_{ant}. The more noise collected by the antenna, the higher T_{ant} becomes.

Figure 6-13 shows the model of the receiver of Figure 6-11, emphasizing the noise performance of the receiver. The receiver accepts both signal and noise from the antenna and adds internal noise to the incoming signal (via R_{RX} @ T_{RX}). The voltage source applies voltage gain G_V to the voltage across the input resistor R_{RX} and sends that voltage to the receiver's output via the noiseless output resistor R_{out}.

The input resistor R_{RX} @ T_{RX} accounts for all of the internally generated noise of the receiving system, regardless of the source.

Figure 6-14 shows the receiving system of Figure 6-11, using the equivalent circuits of the antenna and the receiver. From a noise perspective, the interesting components in Figure 6-14 are the two noisy resistors R_{ant} @ T_{ant} and R_{RX} @ T_{RX}. Every other component in the figure is noiseless. By definition, we will blame all of the noise we see in the system as coming from either R_{ant} or R_{RX}.

Figure 6-15 shows the circuit from Figure 6-14, replacing the two noisy resistors with their noise models. Since the two noise sources are uncorrelated, we must examine the noise performance of the circuit one noise generator at a time.

Case 1: We first derive the noise power delivered to both resistors by the thermal noise generator inside of R_{ant}. We short $V_{n,OC,RMS,RX}$ and leave $V_{n,OC,RMS,ant}$ in the circuit.

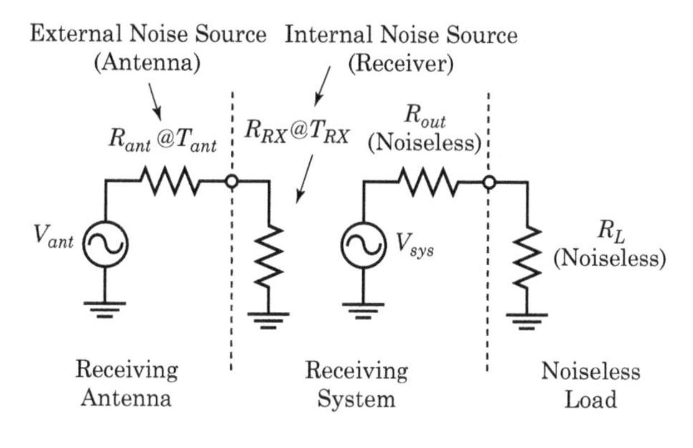

FIGURE 6-14 ■ The receiving system of Figure 6-11, showing the equivalent circuits of the antenna and the receiver.

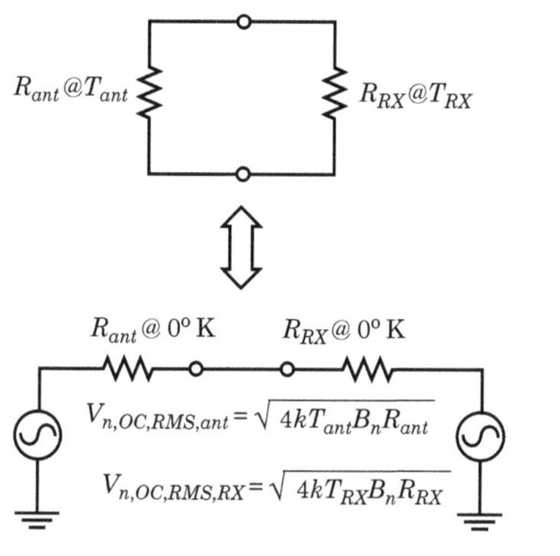

FIGURE 6-15 ■ The receiving system of Figure 6-11 and Figure 6-14, showing only the components contributing to the noise performance of the system.

Using voltage division, we write

$$\begin{aligned} V_{n,RX,Case1} &= V_{n,Rant,Case1} \\ &= \frac{1}{2} V_{n,oc,RMS,ant} \end{aligned} \tag{6.26}$$

where

$V_{n,Rant,Case1}$ = the noise voltage across R_{ant}
$V_{n,RX,Case1}$ = the noise voltage across R_{RX}

Since the $V_{n,ant,Case1} = V_{n,RX,Case1}$, the noise power dissipated by each resistor is identical and equal to

$$\begin{aligned} P_{n,Rant,Case1} &= P_{n,RX,Case2} \\ &= \frac{V^2_{n,OC,RMS,ant}}{R} \\ &= \frac{\left(\sqrt{4kT_{ant}B_nR}/2\right)^2}{R} \\ &= kT_{Ant}B_n \end{aligned} \tag{6.27}$$

The noise power delivered to each resistor is $kT_{ant}B_n$.

Case 2: We next derive the noise power delivered to both resistors by the thermal noise generator inside of R_{RX}. We short $V_{n,OC,RMS,ant}$ and leave $V_{n,OC,RMS,sys}$ active. Using the same techniques, the noise power dissipated in each resistor is the same and equal to

$$\begin{aligned} P_{n,Rant,Case2} &= P_{n,RX,Case2} \\ &= kT_{RX}B_n \end{aligned} \tag{6.28}$$

As with Case 1, the noise power delivered to R_{RX} equals the noise power delivered to R_{ant}. The noise power delivered to each resistor is $kT_{RX}B_n$.

The total noise power dissipated by each resistor is the sum of the noise powers from Case 1 and Case 2:

$$\begin{aligned} P_{n,Rant} &= P_{n,RX} \\ &= k\left(T_{ant} + T_{RX}\right)B_n \end{aligned} \tag{6.29}$$

Equation (6.29) shows us that the entirety of the noise present in our system depends only on the noise temperature of the antenna (T_{ant}) and on the noise temperature of our receiving system (T_{RX}).

EXAMPLE

Power Transfer in Resistors of Unequal Temperatures

Figure 6-16 shows two matched resistors at unequal temperatures. One resistor, R_{Hot}, is submersed in boiling water at 100°C = 373°K. The second resistor, R_{Cold}, is submersed in a dry-ice bath at −79°C = 194°K.

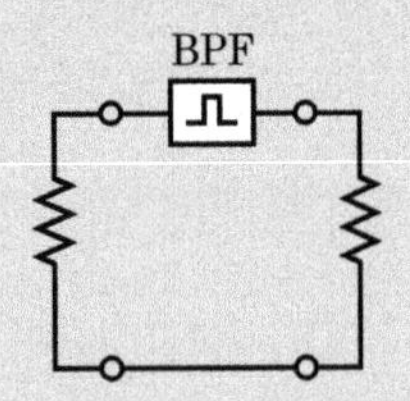

FIGURE 6-16 ■ Power transfer in resistors of unequal temperature.

Find

a. The power delivered to R_{Cold} from R_{Hot}

b. The power delivered to R_{Hot} from R_{Cold}

c. The net power flow and direction

assuming the band-pass filter has a noise bandwidth of 25 MHz and $R_{Hot} = R_{Cold}$.

Solution

We use equations (6.27) and (6.28).

a. The power delivered to R_{Cold} from R_{Hot} is

$$\begin{aligned} P_{\text{RCold from RHot}} &= kT_{Hot}B_n \\ &= \left(1.38 \cdot 10^{-23}\right)(373)\left(25 \cdot 10^6\right) \\ &= 129 \text{ fW} \end{aligned} \tag{6.30}$$

The noise source of R_{Hot} delivers 129 fW to both R_{Hot} and R_{Cold}.

b. The power delivered to R_{Hot} from R_{Cold} is

$$\begin{aligned} P_{\text{RHot from RCold}} &= kT_{Cold}B_n \\ &= \left(1.38 \cdot 10^{-23}\right)(194)\left(25 \cdot 10^6\right) \\ &= 66.9 \text{ fW} \end{aligned} \tag{6.31}$$

The noise source present in R_{Cold} delivers 66.9 fW to both R_{Hot} and R_{Cold}.

c. Since 129 fW flows from R_{Hot} to R_{Cold} and 66.9 fW flows from R_{Cold} to R_{Hot}, there is a net power flow of 62.1 fW from R_{Hot} to R_{Cold}.

In theory, the power flowing from R_{Hot} to R_{Cold} would eventually cause R_{Hot} to cool down and R_{Cold} to heat up until the temperatures of the two resistors were equal (and there would be no net power flow from one resistor to another). In reality, the outside environment supplies the energy necessary to sustain this heat transfer. The boiling water forces R_{Hot} to stay at 100°C, while the dry-ice bath forces the temperature of R_{Cold} to stay at −79°C.

6.5.1 Antenna Noise Models

In a receiving system, we built the antenna and positioned it in the environment to absorb electromagnetic waves. Hopefully some of those waves will be the signals in which we're interested, but some of those waves will be Gaussian noise at the same frequency of the received signals. Since the noise is at the same frequency as the signal, so we can't remove the noise via frequency-selective filtering.

6.5.2 Antenna Noise Temperature

The physical temperature of the antenna has very little to do with the value of T_{ant}. The noise the antenna produces is simply the noise the antenna receives from its environment. Figure 6-17 emphasizes this point.

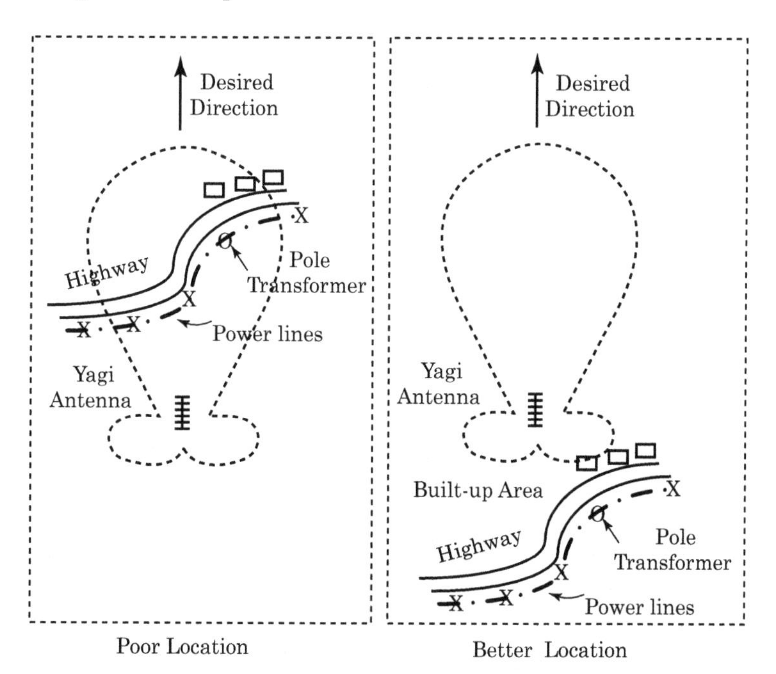

FIGURE 6-17 Since an antenna absorbs noise from the environment, the same antenna can have a widely varying noise temperature, depending on the environment viewed by the antenna. The noise temperature of the leftmost antenna will higher than the noise temperature of the rightmost antenna.

We can reduce the noise temperature of your antenna by repositioning it with respect to sources of external noise.

EXAMPLE

Antenna Noise Temperature

Using Figure 6-12, Figure 6-13, and Figure 6-14 as guides, the signal power we measure from an FM broadcast antenna is −94 dBm. Assume that the characteristic impedance of our system is 50 Ω and that we're operating with a noise bandwidth of 210 kHz:

a. Find V_{ant}.

b. At 2:30 a.m., we measure −125 dBm of noise power from the antenna. Find T_{ant} assuming $T_{RX} \ll T_{ant}$.

c. At 9:00 a.m., the world wakes up. People turn on their computers, start their cars, and otherwise begin generating noise. The noise power measured at the terminals of the antenna increases to −114 dBm. Find T_{ant} assuming $T_{RX} \ll T_{ant}$.

Solution

a. The voltage across the receiver's input resistor $V_{RX} = 1/2V_{ant}$. Using

$$P_{dBm} = 10\log\left(\frac{P_{Watts}}{0.001}\right) \tag{6.32}$$

we know

$$\begin{aligned} -94\text{ dBm} &= 10\log\left(\frac{P_{RX,Watts}}{0.001}\right) \\ &\Rightarrow P_{RX} = 398\text{ fW} \end{aligned} \tag{6.33}$$

and

$$P = \frac{V^2}{R_L} \tag{6.34}$$

with $R_{RX} = 50\,\Omega$ and $P_{RX} = 398$ fW, we find

$$\begin{aligned} 398 \cdot 10^{-15} &= \frac{V_{RX}^2}{50} \\ &\Rightarrow V_{RX} = 4.46\ \mu\text{V}_{RMS} \end{aligned} \tag{6.35}$$

Since $V_{ant} = 2V_{RX}$, we know $V_{ant} = 8.92\ \mu\text{V}_{RMS}$.

b. If $T_{RX} \ll T_{ant}$, the noise present in R_{RX} is due to T_{ant}. Applying equation (6.18) produces $P_{RX} = -125\text{ dBm} = 316 \cdot 10^{-18}$ watts and $B_n = 210$ kHz:

$$\begin{aligned} P_{RX} &= kT_{ant}B_n \\ &\Rightarrow 316 \cdot 10^{-18}\text{ W} = (1.38 \cdot 10^{-23})T_{ant}(210{,}000) \\ &\Rightarrow T_{ant} = 109^\circ\text{K} = -164^\circ\text{C} \end{aligned} \tag{6.36}$$

The noise temperature of the antenna is very cold, reinforcing the concept that the noise temperature of an antenna has little to do with the antenna's physical temperature.

c. Since $T_{RX} \ll T_{ant}$, the noise present in R_{RX} is entirely due to T_{ant}. Following the procedure outlined in part b, we find $P_{RX} = -114\,\text{dBm} = 3.98\,\text{fW}$. With $B_n = 210\,\text{kHz}$, equation (6.18) produces

$$\begin{aligned} P_{RX} &= kT_{ant}B_n \\ 3.98\,\text{fWatts} &= (1.38 \cdot 10^{-23})T_{ant}(210{,}000) \\ &\Rightarrow T_{ant} = 1,374°\text{K} = 1101°\text{C} \end{aligned} \tag{6.37}$$

The noise temperature of the antenna changed even though we haven't changed the system or the physical temperature of antenna.

EXAMPLE

Satellite Antenna Noise Temperature

For about 5 days a year, geosynchronous satellites pass between the earth and sun for several minutes a day (see Figure 6-18). The satellite is typically in front of the sun for 70 minutes or so.

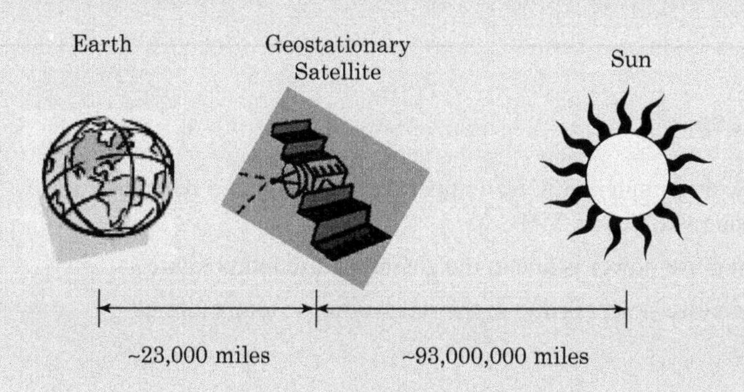

FIGURE 6-18 ■ The geometry of the earth, sun, and satellite causes an increase in the Earth station receiving antenna's noise temperature because the Earth station antenna is forced to look at the sun as well as the satellite.

During the time the satellite is positioned between the earth and sun, the noise temperature of the ground-station receive antenna increases dramatically. This occurs because the receive antenna sees the electrical noise radiated by the sun as well as the normal background noise.

6.6 AMPLIFIER NOISE MODEL

Figure 6-14 and Figure 6-15 show the models that describe the noise performance of a receiving system. We'll use the *two noisy resistor* model we investigated earlier as a mathematical tool.

6.6.1 Equivalent Input Noise Power

Referring to equation (6.29), the total noise power dissipated by R_{RX} of Figure 6-14 is

$$\begin{aligned} N_{in} &= k(T_{ant} + T_{RX})B_n \\ &= kT_{ant}B_n + kT_{RX}B_n \end{aligned} \tag{6.38}$$

where

N_{in} = the total noise power dissipated in R_{sys} due to the sum of the external noise generated by R_{ant} and the internal noise generated by R_{sys}

T_{ant} = the noise temperature of the antenna

T_{RX} = the noise temperature of the receiving system

N_{in} is the total noise power dissipated by R_{sys}, which we refer to as the *equivalent input noise power* or the *system noise floor*. This is the sum of any noise power delivered to the amplifier from external sources plus any noise power generated internally within the amplifier. Converting equation (6.38) into more convenient units produces

$$N_{in,dBW} = 10\log(T_{ant} + T_{RX}) + 10\log(B_n) - 228.6 \tag{6.39}$$

and

$$N_{in,dBm} = 10\log(T_{ant} + T_{RX}) + 10\log(B_n) - 198.6 \tag{6.40}$$

Equations (6.38) through (6.40) describe the amount of noise power present on the input of a receiving system given the temperature of the input source resistor and the noise characteristics of the receiving system.

EXAMPLE

System Noise Floor, General Case

a. Find the noise floor for a system with an antenna noise temperature of 200°K, a receiving noise temperature of 75°K, and a noise bandwidth of 35 MHz.

b. What percentage of the total system noise power is due to the antenna noise temperature?

c. What percentage of the total system noise power is due to the receiver noise temperature?

Solution

We refer to Figure 6-14 and Figure 6-15.

a. Equation (6.38) produces

$$\begin{aligned} N_{in} &= k(T_{ant} + T_{RX})B_n \\ &= (1.38 \cdot 10^{-23})(200 + 75)(35 \cdot 10^6) \\ &= 133 \cdot 10^{-15} \text{ Watts} \\ &= -99 \text{ dBm} \end{aligned} \tag{6.41}$$

or, using equation (6.40),

$$\begin{aligned} N_{in,dBm} &= 10\log(T_{ant} + T_{RX}) + 10\log(B_n) - 198.6 \\ &= 10\log(275) + 10\log(35 \cdot 10^6) - 198.6 \\ &= -99 \text{ dBm} \end{aligned} \tag{6.42}$$

b. The noise power contributed by the antenna is due to the antenna noise temperature, so

$$\begin{aligned} N_{in,ant} &= kT_{ant}B_n \\ &= (1.38 \cdot 10^{-23})(200)(35 \cdot 10^6) \\ &= 96.6 \cdot 10^{-15} \text{ Watts} \end{aligned} \tag{6.43}$$

The percentage of the total input noise power contributed by the antenna is

$$N_{in,\%} = \frac{96.6}{133} = 73\% \tag{6.44}$$

c. The noise power contributed by the receiver is

$$\begin{aligned} N_{in,sys} &= kT_{RX}B_n \\ &= (1.38 \cdot 10^{-23})(75)(35 \cdot 10^6) \\ &= 36.2 \cdot 10^{-15} \text{ Watts} \end{aligned} \tag{6.45}$$

The percentage of the total input noise power contributed by the receiver is

$$N_{in,\%} = \frac{36.2}{133} = 27\% \tag{6.46}$$

6.6.2 Output Noise Power

The amplifier applies its power gain, G_P, to the equivalent input noise power N_{in} and produces

$$\begin{aligned} N_{out} &= G_p k(T_{ant} + T_{RX})B_n \\ &= G_p kT_{ant}B_n + G_p kT_{RX}B_n \end{aligned} \tag{6.47}$$

where N_{out} is the noise power delivered to R_L by the receiver, in linear units.

Converting equation (6.47) into more convenient units produces

$$N_{out,dBW} = G_{p,dB} + 10\log(T_{ant} + T_{RX}) + 10\log(B_n) - 228.6 \tag{6.48}$$

and

$$N_{out,dBm} = G_{p,dB} + 10\log(T_{ant} + T_{RX}) + 10\log(B_n) - 198.6 \tag{6.49}$$

6.7 | SIGNAL-TO-NOISE RATIO

A signal's signal-to-noise ratio (SNR) describes how well we can interpret the received signal. SNR directly affects the signal's *understandability* or *quality*. In the digital domain, SNR relates directly to the expected number of bits we will receive in error, or the bit error rate (BER).

A high SNR means that the waveform contains much more signal power than it does noise power. This implies that the characteristics of the signal coming out of our receiver are almost entirely controlled by the received signal and not by the received noise. A low SNR indicates that the waveform's signal power is not much higher than the waveform's noise power.

Usually, we measure the SNR in the demodulator, but it is equally valid to measure or calculate it at the receiver's input.

6.7.1 Definition

The signal-to-noise ratio is defined as

$$SNR = \frac{\text{Total Signal Power}}{\text{Total Noise Power}} \text{ in a given bandwidth} \tag{6.50}$$

We always imply some observation bandwidth when measuring noise. The filter restricts the noise power present in the signal.

6.7.2 SNR and Signal Quality

For a digital signal, the SNR coupled with information about, for example, the signal's modulation and symbol rate describe to us the expected BER of the received signal. Table 6-1 relates BER in a direct and practical sense to the signal's reception quality.

Problem-free reception of a digital signal requires at least 1E-5 BER, and we'd like to see better than 1E-6. The SNR required for a particular BER depends on the signal's modulation.

Table 6-2 shows BER and SNR, related to modulation type.

SNR doesn't directly describe the demodulated signal quality. If we observe a binary phase shift keying (BPSK) signal with 21 dB of SNR, we can demodulate the signal with plenty of margin. If the signal is 16QAM (quadrature amplitude modulation), 21 dB of SNR is a marginal situation. If the signal is 64QAM, we don't have enough SNR to adequately demodulate the signal.

The numbers shown in Table 6-2 are absolute minimum numbers. These SNRs, taken from the technical literature and verified in testing, are under ideal lab conditions. We often add an implementation margin to cover fading and imperfections in the demodulator and

TABLE 6-1 ■ BERs versus signal quality.

Bit Error Rate (BER)	Signal Quality
>10E-2	The signal is not processable, but some signal characteristics are discernable.
10E-3	A listener can identify the sex and language of a speaker. Complete signal parameters can be measured.
10E-4	The static, pops, and crackles in voice are fatiguing for a listener.
10E-5	The occasional pops in voice are acceptable.
10E-6	There are perfect voice and data.

TABLE 6-2 ■ The SNR required to demodulate particular signal types.

Modulation Type	Required SNR (dB) to Produce 1E-6 BER
BPSK	10.8
QPSK	13.7
8PSK	18.9
16QAM	20.4
32QAM	23.5
64QAM	26.6

front-end equipment and other impairments. In even a very well-designed and -maintained system, these excess losses are at least 3 dB. In most situations, the losses are higher.

6.7.3 Measuring SNR

6.7.3.1 Where to Measure SNR

Ideally, we want to measure SNR at the output of the low-noise amplifier (LNA) that is connected directly to the antenna. This is the first place in the RF system that we can make an accurate measurement, and it is a good reference point. If there's a problem somewhere in the RF path (e.g., bad amplifier, intermittent cable), the signal might suffer excess attenuation or amplifier compression, and the SNR measurement will be inaccurate.

The SNR at the output of the first LNA will be the highest SNR we can measure. The SNR only can degrade as the signal moves through the RF system.

6.7.3.2 Ensuring Noise Lift—Excess Gain

A spectrum analyzer is a complex, expensive, and accurate measurement tool. However, spectrum analyzers are insensitive instruments by design. They are incredibly deaf. Spectrum analyzer noise figures can be as high as 40 dB. By contrast, typical LNAs have noise figures less than 3 dB, and receivers typically exhibit noise figures of 15 dB. When we measure SNR, we want to make sure we're measuring the noise that's coming from the antenna and not the noise that's generated by our test equipment.

The situation is like trying to be heard in a room filled with equipment and cooling fans. You have to shout to be heard over the noise. Similarly, we have to provide gain to a signal before applying it to a noisy device. If we apply the right amount of gain, the device will hear the signal over its own internally generated noise.

A rule of thumb is that, if we're sending a signal into a device whose noise figure is F dB, we have to apply at least $F + 10$ dB of gain before the device to overcome the device's internally generated noise.[1] For example, we have to precede a spectrum analyzer whose noise figure is 40 dB with at least 50 dB of gain. To ensure that we're overcoming the internal noise in a receiver whose noise figure is 15 dB, we have to precede the receiver by at least 25 dB of gain.

The amount of gain we place before a device is the *excess gain*. Sufficient excess gain causes the noise on the spectrum analyzer display to increase or *lift*. Figure 6-19 shows a spectrum analyzer display under three different conditions of excess gain:

- The lowest trace. There is nothing connected to the spectrum analyzer input port. This flat noise trace shows the internally generated noise of the spectrum analyzer.
- The middle trace. There is a signal source directly feeding the spectrum analyzer, but there isn't enough gain before the spectrum analyzer to overcome the analyzer's internally generated noise. The noise level hasn't changed from the value shown in the lowest trace. We're still viewing the analyzer's internal noise.

  Under these conditions, the user might be tempted to report the SNR of the signal as 28 dB (the difference between the 1R and the 1 markers). In reality, we're comparing the

[1] More isn't better here. Too much gain makes our systems susceptible to signal distortion caused by large signals. A gain of F + 10 dB is a minimum amount. A gain of F + 20 dB is the maximum amount. A gain of F + 30 dB is too much.

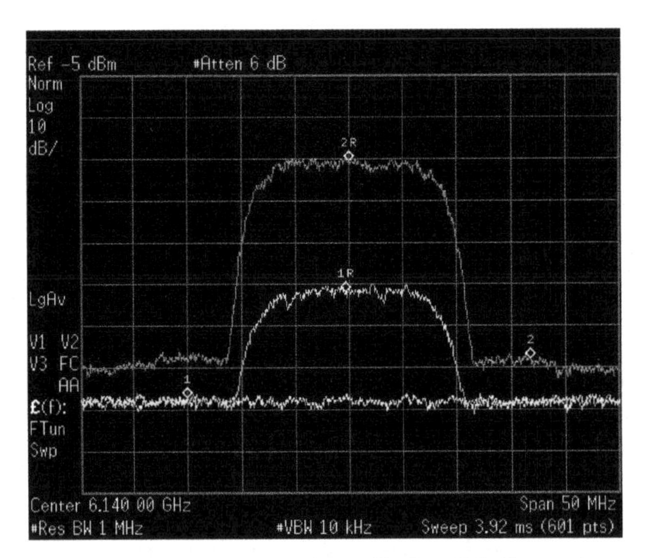

FIGURE 6-19 ■ Noise lift on a spectrum analyzer display. This figure shows a spectrum analyzer trace under three different input conditions. The lowest trace, with no signal, shows the spectrum when there is nothing connected to the spectrum analyzer input port. This flat noise trace shows the internally generated noise of the spectrum analyzer. The middle trace shows a signal with insufficient excess gain. The preceding gain stages are not overcoming the internal noise of the analyzer and the SNR appears to be lower than it truly is. The upper trace shows the same signal with sufficient gain in front of the analyzer to overcome the analyzer's noise floor. The correct SNR of this signal is 47 dB.

signal power from the antenna to the noise power generated internally by the spectrum analyzer. This is not a meaningful measurement.

- The upper trace: The only difference between the middle and upper traces is the addition of a 30 dB amplifier between the signal source and the spectrum analyzer. As expected, the signal has grown by 30, but, more importantly, the noise floor has increased about 10 dB over its previous levels. This is the noise lift we seek.

  Under this condition, the spectrum analyzer is accurately displaying the noise present at the antenna. The user will measure an accurate and useful SNR under these conditions.

The SNR of this signal is the difference between the 2 and 2R markers, or about 47 dB.

As a rule of thumb, you should apply enough gain before the analyzer so that the noise floor shown on the analyzer increases by at least 10 dB. The noise floor can go up as much as 20 dB without causing problems with intermodulation products.

Rules of Thumb: Signal-to-Noise Objectives

Most systems need at least 10 dB of SNR to function effectively. Received signal characteristics such as BER and *understandability* drop off rapidly below 15 dB SNR.

Some of the SNR objectives we strive for include the following:

- 50 dB: telephony
- 45 dB or higher: analog television
- 70 dB: CD quality audio
- $<10^{-6}$: data link bit error rates

Analog television SNR and picture quality:

- 55 dB: very good, noise barely perceptible
- 45 dB: acceptable, some noise visible
- 35 dB: not acceptable, poor picture

For a 10^{-8} bit-error rate, we need:

- 10.6 dB SNR for coherent phase shift keying (PSK)
- 14.5 dB SNR for noncoherent FSK
- 15.0 dB SNR for noncoherent ASK

EXAMPLE

Input Noise Power

Given a system with a 6.2 MHz noise bandwidth, an antenna temperature of 120°K, and a system noise temperature of 220°K, find the output SNR for an input signal power of −77 dBm.

Solution

Using Figure 6-20 and equation (6.49) as a guide, we know that$T_{ant} = 120°$K and $T_{RX} = 220°$K. We also know the signal power delivered to R_{sys} is −77 dBm. The total noise power dissipated in R_{RX} is

$$\begin{aligned} N_{RX,dBm} &= 10\log(T_{ant} + T_{RX}) + 10\log(B_n) - 198.6 \\ &= 10\log(120 + 220) + 10\log(6.2 \cdot 10^6) - 198.6 \\ &= -105.4 \text{ dBm} \end{aligned} \tag{6.51}$$

The noise power dissipated in R_{RX} is −105.3 dBm, while the signal power dissipated in the same resistor is −77 dBm. The signal-to-noise ratio is −77 − (−105.3) = 28.3 dB.

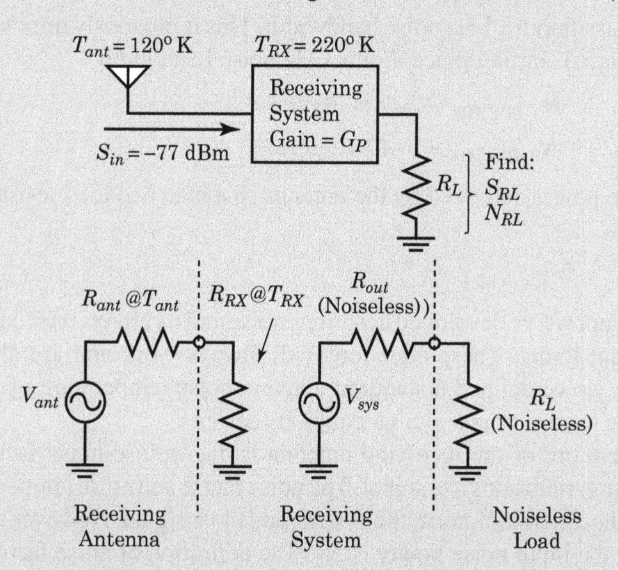

FIGURE 6-20 ■ SNR example. The receiving system and its equivalent noise model.

Since the system applies its gain equally to both the signal and the noise, the output SNR is the same as the input SNR. If the power gain, G_P, of the system was 20 dB, we would measure −57 dBm of signal power and −85.3 dBm of noise power at the output of the amplifier. The output SNR would still be −57 − (−85.3) = 28.3 dB.

6.8 NOISE FACTOR/NOISE FIGURE

In carefully controlled situations, such as microwave telephone links or satellite communications systems, we typically spend a lot of money on the link, and we can afford measure the antenna's noise temperature and take steps to minimize it.

There are also situations in which the antenna noise temperature will be poorly characterized. Such situations include cellular telephone equipment, ham radio gear, commercial broadcasting (AM, FM, and television), and handheld (walkie-talkie) system. In these cases, we simply cannot characterize the noise performance of our antenna. However, we still want answers to questions such as, "How far can I separate the receiver and transmitter and still have a workable system?"

Since we really can't perform an accurate noise analysis of these situations, we do the next best thing. We guess.

6.8.1 N_0

In most communications systems, we simply assume an antenna noise temperature of $T_0 = 290°\text{K}$. A signal source whose noise temperature is T_0 will deliver a noise power of N_0 to a matched load resistor where

$$\begin{aligned} N_{0,\,Watts} &= kT_0B_n \\ &= \left(1.38 \cdot 10^{-23}\right)(290)\,B_n \\ &= \left(4.00 \cdot 10^{-21}\right)B_n \end{aligned} \tag{6.52}$$

Also

$$\begin{aligned} N_{0,dBW} &= 10\log(B_n) - 204 \\ N_{0,dBm} &= 10\log(B_n) - 174 \end{aligned} \tag{6.53}$$

We often start our analyses assuming a 1 Hz noise bandwidth. This is obviously impractical but is useful as a computational starting place. For a 1 Hz noise bandwidth

$$\begin{aligned} N_{0,dBW,1\text{Hz}} &= -204 \text{ dBW} \\ N_{0,dBm,1\text{Hz}} &= -174 \text{ dBm} \end{aligned} \tag{6.54}$$

N_0 describes only the noise power delivered by the antenna to a matched load resistor.

6.8.2 Definitions

The noise temperature concept we've developed describes noise and its effects on receiving systems in the most general terms. These equations will always work and are always applicable. However, once we've defined a standard antenna noise temperature of $T_0 = 290°\text{K}$, we can define some terms to make our calculations easier.

When the noise temperature of the receiving antenna is T_0, we use noise factor to quantify the noise power our system adds to a signal. The noise factor and noise temperature of a system both describe the amount of noise the system adds to a signal. However, noise factor is most useful when the input noise power is N_0. The definition of noise factor is

$$F = \frac{\text{Noise power delivered by a noisy component}}{\text{Noise power delivered by a noiseless component}} \tag{6.55}$$

when the component's input noise power is $N_0 = kT_0B_n$

Occasionally, you'll see other definitions of noise factor but they are all derivations of equation (6.55).

The noise power delivered by a noisy receiver is the equivalent input noise of the receiver times the receiver's power gain or $G_p k(T_0 + T_{RX})B_n$. The noise power delivered by an ideal, noiseless receiver is $G_p k T_0 B_n$. Combining these two expressions with equation (6.55) produces

$$F_{RX} = \frac{G_p k\,(T_0 + T_{RX})\,B_n}{G_p k T_0 B_n} \tag{6.56}$$

Simplifying produces

$$F_{RX} = \frac{T_0 + T_{RX}}{T_0} \qquad F_{RX} \geq 1 \tag{6.57}$$

6.8.3 Noise Factor/Noise Figure

Knowing the difference between noise factor (F – *a* linear term) and noise figure (F_{dB} – *a* term in decibels) is very important for calculation. Many mistakes arise via plugging noise figure into equations that need noise factor. To convert between noise factor and noise figure, use

$$F_{dB} = 10\log(F) \tag{6.58}$$

where

F_{dB} = noise figure
F = noise factor

6.8.4 Noise Factor, Noise Temperature Relationships

Rearranging equation (6.57) produces

$$T_{RX} = T_0(F_{RX} - 1) \tag{6.59}$$

Both T_{RX} and F_{RX} exactly describe the noise performance of a system. They are equivalent terms. However, noise figure is a little handier when the temperature of the input noise is 290°K. Table 6-3 shows several values of noise figure and the equivalent noise temperature.

TABLE 6-3 ■ Noise figure versus noise temperature.

F_{dB}	T_{RX} (°K)	F_{dB}	T_{RX} (°K)
0.0	0.0	6.0	865
0.1	6.8	7.0	1,160
0.2	13.7	8.0	1,540
0.3	20.7	9.0	2,010
0.5	35.4	10.0	2,610
1.0	75.1	15.0	8,880
1.5	120	20.0	28,700
2.0	170	30.0	290,000
3.0	290	40.0	2,900,000
4.0	438	50.0	29,000,000
5.0	627	100.0	$2.9 \cdot 10^{12}$

EXAMPLE

Noise Figure/Noise Factor/Noise Temperature Conversions

a. Find the noise factor and noise temperature of an amplifier whose noise figure is 2 dB.
b. Find the noise figure and noise factor of an amplifier whose noise temperature is 290°K.
c. A third amplifier has a noise factor of 100. Find its noise figure and noise temperature.

Solution

a. Using equation (6.58) with $F_{dB} = 2$ dB implies $F = 1.58$. Equation (6.59) with $F = 1.58$ produces $T_{RX} = 170°$K.
b. $T_{RX} = 290°$K. Using equation (6.59) produces a noise factor of $F = 2$. Using $F = 2$ in equation (6.58) gives a noise figure of $F_{dB} = 3$ dB.
c. Using equations (6.58) and (6.59) with $F = 100$ gives $F_{dB} = 20$ dB and $T_{RX} = 28700°$K.

6.8.5 Equivalent Input Noise Power (Again)

The equivalent input noise power (or noise floor) of a component is the sum of the noise delivered to the component by the external source plus the noise generated internally in the component. The noise floor of a system whose noise figure is F_{RX}, noise temperature is T_{RX} and whose input noise temperature is T_0 is

$$N_{in,equ} = k(T_0 + T_{RX})B_n \tag{6.60}$$

Combining equations (6.57) and (6.60) produces

$$\begin{aligned} N_{in,equ} &= k[T_0 + (F_{RX} - 1)T_0]B_n \\ &= F_{RX} k T_0 B_n \end{aligned} \tag{6.61}$$

Converting equation (6.61) into decibels produces

$$N_{in,equ,dBm} = F_{RX,dB} + 10\log(kT_0) + 10\log(B_n) \tag{6.62}$$

We know

$$10\log(kT_0) = -174\frac{\text{dBm}}{\text{Hz}} \tag{6.63}$$

so equation (6.62) becomes

$$N_{in,equ,dBm} = F_{RX,dB} - 174\frac{\text{dBm}}{\text{Hz}} + 10\log(B_n) \tag{6.64}$$

Equation (6.64) tells us the equivalent input noise power for a system with a noise figure of $F_{RX,dB}$ and a noise bandwidth of B_n when the input noise power (the power delivered by the external source) is kT_0B_n. Let's examine the terms of equation (6.64):

- -174 dBm/Hz represents the T_0 noise temperature of the antenna (or other external source). When we assumed the antenna noise temperature to be 290°K, we effectively set this term to -174 dBm/Hz. Since we've assumed the operating noise bandwidth is

1 Hz, the term $kT_0 = 174$ dBm/Hz is the noise delivered to the system in a 1 Hz noise bandwidth.

- $F_{RX,dB}$ describes the noise generated internally by our receiver. The higher the noise figure, the higher the equivalent input noise power.
- $10\log(B_n)$ represents a bandwidth adjustment to our calculations. The noise is spectrally flat, so the input noise power increases directly the noise bandwidth. To minimize the input noise power, we operate in the smallest bandwidth possible. However, bandwidth is generally not something we have much, if any, control over.

EXAMPLE

System Noise Floor with N_0 Input Noise Power

We've given an antenna whose noise temperature is 290°K. The antenna is connected to a receiver whose noise figure is 7 dB. The noise bandwidth is 6.7 MHz, and the amplifier has a power gain of 16 dB (see Figure 6-21).

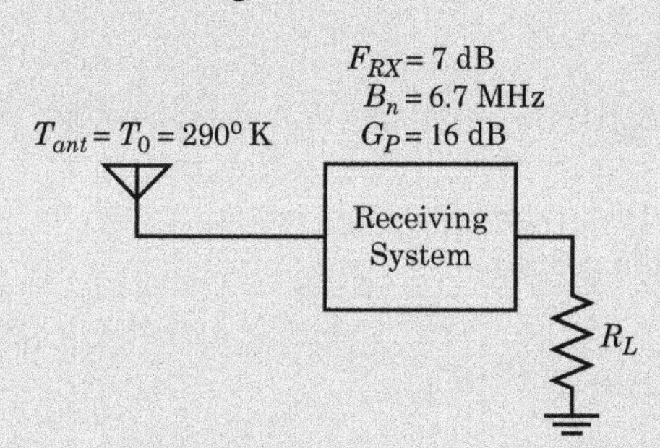

FIGURE 6-21 ■ Input and output noise power example with antenna noise temperature = T_0.

Find the equivalent input noise power (the noise floor) and the output noise power using:

a. Equation (6.64)

b. Equation (6.61)

c. Equation (6.38)

Solution

a. Since the noise temperature of the antenna is T_0, we can use the noise figure of the amplifier directly with equation (6.64) to find the equivalent input noise power:

$$\begin{aligned} N_{in,dBm} &= F_{RX,dB} - 174\frac{\text{dBm}}{\text{Hz}} + 10\log(B_n) \\ &= 7 - 174\frac{\text{dBm}}{\text{Hz}} + 10\log\left(6.7\cdot 10^6\right) \\ &= -99\,\text{dBm} \end{aligned} \tag{6.65}$$

The noise floor is −99 dBm. The amplifier applies its power gain to this noise power, so the output noise power is -99 dBm $+G_{p,dB} = -99 + 16 = -83$ dBm.

b. We first convert the noise figure and gain into linear units:

$$\begin{aligned} 7\text{ dB} &= 10\log(F_{RX}) & 16\text{ dB} &= 10\log(G_p) \\ \Rightarrow F_{RX} &= 5.01 & \Rightarrow G_p &= 39.8 \end{aligned} \tag{6.66}$$

Applying equation (6.61) produces

$$
\begin{aligned}
N_{in,equ} &= F_{RX} k T_0 B_n \\
&= (5.01)\left(1.13 \cdot 10^{-23}\right)(290)\left(6.7 \cdot 10^{6}\right) \\
&= 134.4 \cdot 10^{-15} \text{ Watts} \\
&= -99 \text{ dBm}
\end{aligned}
\tag{6.67}
$$

and

$$
\begin{aligned}
N_{out} &= G_P N_{in} \\
&= (39.8)\left(134.4 \cdot 10^{-15} \text{ Watts}\right) \\
&= 5.35 \cdot 10^{-12} \text{ Watts} \\
&\approx -83 \text{ dBm}
\end{aligned}
\tag{6.68}
$$

These are the same results produced in part a.

c. Equation (6.38) requires noise temperatures. We convert the 7 dB (or 5.01) receiver noise figure into noise temperature using equation (6.59):

$$
\begin{aligned}
T_{RX} &= T_0\left(F_{RX} - 1\right) \\
&= 290\,(5.01 - 1) \\
&= 1160\ ^\circ\text{K}
\end{aligned}
\tag{6.69}
$$

Now we find the equivalent input noise power using equation (6.38)

$$
\begin{aligned}
N_{in} &= k\left(T_{ant} + T_{RX}\right) B_n \\
&= \left(1.38 \cdot 10^{-23}\right)(1160 + 290)\left(6.7 \cdot 10^{6}\right) \\
&= 134.2 \cdot 10^{-15} \text{ Watts} \\
&= -99 \text{ dBm}
\end{aligned}
\tag{6.70}
$$

6.8.6 Signal-to-Noise Ratio and Noise Figure

Let us examine the signal-to-noise ratios at the input and output of a receiver whose noise figure is F_{RX} and whose power gain is G_P. The noise temperature of the external source is $T_0 = 290^\circ$K.

Using Figure 6-22, let S_{in} equal the signal power dissipated in the receiver's input resistor R_{RX}. The noise power delivered to R_{RX} by the external source is $N_0 = kT_0B_n$.

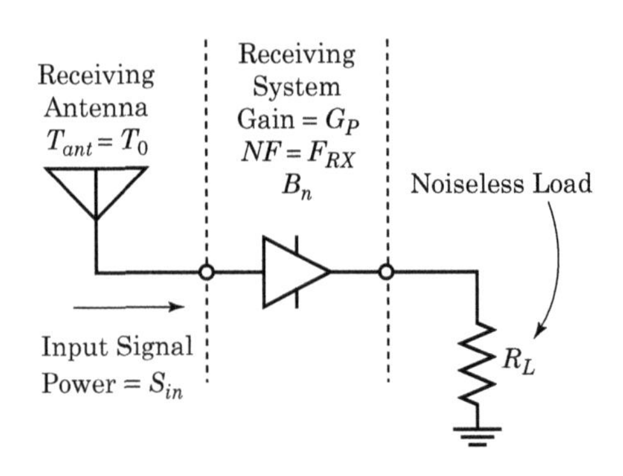

FIGURE 6-22 ■ A receiving system with input power S_{in}, antenna noise temperature T_0, receiver noise figure of F_{RX}, and receiver power gain G_P all observed in noise bandwidth B_n.

Before the receiver adds its internal noise to the signal, the input signal-to-noise ratio is

$$\left(\frac{S}{N}\right)_{in} = \left(\frac{S}{N_0}\right) = \left(\frac{S_{in}}{kT_0 B_n}\right) \tag{6.71}$$

Equation (6.61) tells us the total noise power dissipated in the receiver's input resistor is

$$N_{in} = F_{RX} k T_0 B_n \tag{6.72}$$

and the output signal-to-noise ratio is

$$\left(\frac{S}{N}\right)_{out} = \left(\frac{G_P S_{in}}{G_P F_{RX} k T_0 B_n}\right) \tag{6.73}$$

Combining equations (6.71) and (6.73) produces

$$\frac{(S/N)_{in}}{(S/N)_{out}} = \frac{S_{in}/(kT_0 B_n)}{G_P S_{in}/(G_P F_{RX} k T_0 B_n)} \tag{6.74}$$

so

$$F_{RX} = \frac{(S/N)_{in}}{(S/N)_{out}} \qquad \text{when } \mathrm{N}_{in} = N_0 \tag{6.75}$$

Equation (6.75) tells us the noise factor of a system is a measure of the degradation of the signal-to-noise ratio as the signal passes through the system, when the input noise power is N_0.

EXAMPLE

S/N Degradation

We have a system with an antenna noise temperature of T_0, a gain of 10 dB, a noise figure of 3 dB, and a noise bandwidth of 30 kHz. We apply a signal of –110 dBm to the system:

a. Find the noise power delivered to the system by the antenna.

b. Find the total noise power present at the input of the system. This is the equivalent input noise power.

c. Find the output noise power.

d. Find the output signal power.

e. Verify that the noise figure of the system equals the degradation in the signal-to-noise ratio as the signal passes through the system.

Solution

Since the input noise power is N_0 (i.e., the input noise temperature is 290°K), we can use the equations involving noise figure:

a. The noise delivered to the system by the external amplifier is given by equation (6.23):

$$\begin{aligned} N_{0,dBm} &= -174\,\frac{\text{dBm}}{\text{Hz}} + 10\log(B_n) \\ &= -174 + 10\log(30{,}000) \\ &= -129\text{ dBm} \end{aligned} \tag{6.76}$$

The antenna delivers −129 dBm of noise power to the system. The −129 dBm power level does not include any noise the system adds.

b. Using equation (6.64), we find the total noise dissipated in R_{RX} is

$$\begin{aligned} N_{in,dBm} &= F_{RX,dB} - 174\,\frac{\text{dBm}}{\text{Hz}} + 10\log(B_n) \\ &= 3 - 174 + 10\log(30{,}000) \\ &= -126\text{ dBm} \end{aligned} \tag{6.77}$$

The total noise power dissipated in the system's input is −126 dBm. This answer includes the noise delivered to the system by the antenna and the system's internally generated noise.

c. The output noise power equals the input noise power (in dBm) plus the power gain of the amplifier (in dB):

$$\begin{aligned} N_{out,dBm} &= N_{in,dBm} + G_{p,dB} \\ &= -126 + 10 \\ &= -116\text{ dBm} \end{aligned} \tag{6.78}$$

d. The output signal power is the input signal power (in dBm) plus the amplifier's power gain

$$\begin{aligned} S_{out,dBm} &= S_{in,dBm} + G_{p,dB} \\ &= -110 + 10 \\ &= -100\text{ dBm} \end{aligned} \tag{6.79}$$

e. The signal-to-noise ratio at the system's input is

$$\begin{aligned} \left(\frac{S}{N}\right)_{in} &= -110 - (-129) \\ &= 19\text{ dB} \end{aligned} \tag{6.80}$$

The signal-to-noise ratio at the system's output is

$$\begin{aligned} \left(\frac{S}{N}\right)_{out} &= -110 - (-116) \\ &= 16\text{ dB} \end{aligned} \tag{6.81}$$

The degradation of the signal-to-noise ratio through the system is

$$\begin{aligned} \frac{(S/N)_{in}}{(S/N)_{out}} &= 19 - 16 \\ &= 3\text{ dB} \end{aligned} \tag{6.82}$$

which is the system's noise figure.

6.9 CASCADE PERFORMANCE

The RF system of Figure 6-23 contains several amplifiers in cascade. Given the power gain and the noise characteristics of all the devices in a cascade, we seek the noise performance of the cascade.

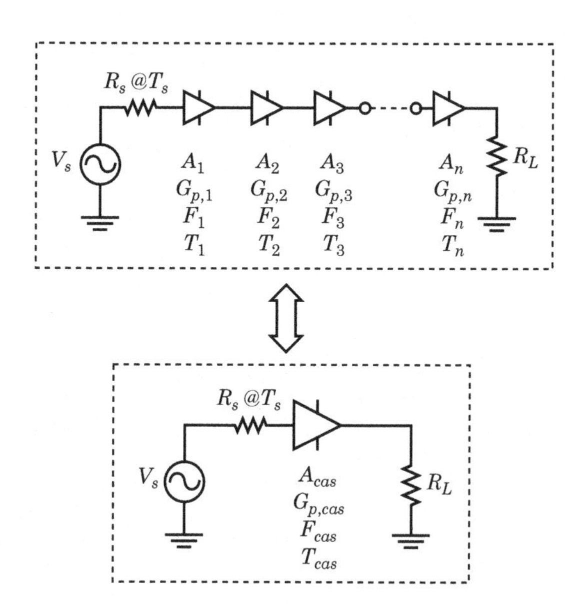

FIGURE 6-23 ■ An n-element cascade and its single-element equivalent.

Let

$G_{p,1}$ = the power gain of the first amplifier in linear terms
$G_{p,2}$ = the power gain of the second amplifier in linear terms
$G_{p,n}$ = the power gain of the n-th amplifier in linear terms
T_1 = the noise temperature of the first amplifier
T_2 = the noise temperature of the second amplifier
T_n = the noise temperature of the n-th amplifier
F_1 = the noise factor of the first amplifier in linear terms
F_2 = the noise factor of the second amplifier in linear terms
F_n = the noise factor of the n-th amplifier in linear terms

We will assume that our entire system is exactly matched to the system's characteristic impedance Z_0. The goal is to describe the cascade as a single amplifier with a power gain of $G_{p,cas}$, a noise temperature of T_{cas}, and a noise factor F_{cas}.

6.9.1 Noise Temperature of a Cascade

Figure 6-24 illustrates the following analysis.

We first apply an input noise power of $kT_{ant}B_n$ to our cascade. Equation (6.38) tells us that the equivalent input noise power to amplifier A_1 is

$$N_{in,A1} = k\left(T_{ant} + T_1\right)B_n \tag{6.83}$$

which includes noise supplied by the external source (T_{ant}) and noise generated internally by $A_1(T_1)$.

Amplifier A_1 applies its power gain to the input noise, so the noise available at the output of A_1 and at the input of A_2 is

$$\begin{aligned} N_{out,A1} &= G_{P1}k\left(T_{ant} + T_1\right)B_n \\ &= k[G_{P1}\left(T_{ant} + T_1\right)]B_n \end{aligned} \tag{6.84}$$

FIGURE 6-24 Graphical description of the noise present in a three-element cascade.

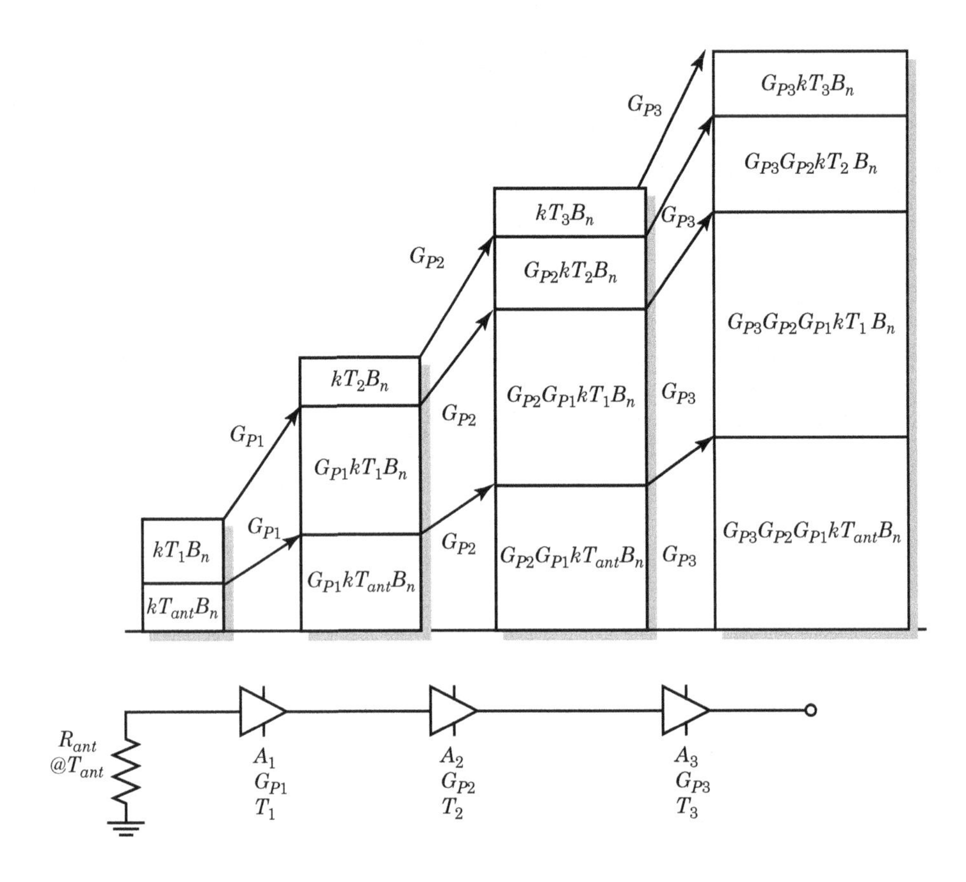

We can think of the term $[G_{P1}(T_{ant} + T_1)]$ as the temperature of the noise applied to A_2's input. Amplifier A_2 accepts this noise power from the external source (i.e., A_1) and adds noise of its own. The noise that amplifier A_2 adds is due to T_2, the noise temperature of A_2's input resistor. The total noise power dissipated in A_2's input resistor is

$$N_{in,A2} = k\left[G_{p,1}\left(T_{ant} + T_1\right) + T_2\right]B_n \tag{6.85}$$

where $N_{in,A2}$ is the sum of the noise delivered by A_1 and to the noise generated internally by A_2.

A_2 applies its power gain, G_{P2}, to the input noise power. The noise power present at the output of A_2 is

$$\begin{aligned} N_{out,A2} &= G_{P2}k[G_{P1}\left(T_{ant} + T_1\right) + T_2]B_n \\ &= k\{G_{P2}[G_{P1}\left(T_{ant} + T_1\right) + T_2]\}B_n \end{aligned} \tag{6.86}$$

The term $\{G_{P2}[G_{P1}(T_{ant} + T_1) + T_2]\}$ is the temperature of the noise available fromA_2. Amplifier A_3 accepts the noise fromA_2 and adds its own noise to the system. The total noise power present at the input of A_3 is

$$N_{in,A3} = k[G_{P2}G_{P1}\left(T_{ant} + T_1\right) + G_{P2}T_2 + T_3]B_n \tag{6.87}$$

The noise power available from the output of amplifier A_3 is

$$N_{out,A3} = k[G_{P3}G_{P2}G_{P1}\left(T_{ant} + T_1\right) + G_{P3}G_{P2}T_2 + G_{P3}T_3]B_n \tag{6.88}$$

We carry this process through to the n-th amplifier and write the noise power available from the output of the n-th amplifier as

$$\begin{aligned} N_{out,An} = k[&G_{Pn} \cdots G_{P3}G_{P2}G_{P1}\,(T_{ant} + T_1) \\ &+ G_{Pn} \cdots G_{P3}G_{P2}T_2 \\ &+ G_{Pn} \cdots G_{P3}T_3 \\ &+ \ldots \\ &+ G_{Pn}T_n]B_n \end{aligned} \tag{6.89}$$

Dividing the output noise power given by equation (6.89) by the power gain of the cascade produces the equivalent input noise power to the cascade:

$$\begin{aligned} N_{in,cas} &= \frac{N_{out,cas}}{G_{cas}} = \frac{N_{out,cas}}{G_{Pn} \cdots G_{P3}G_{P2}G_{P1}} \\ &= k\frac{[G_{Pn} \cdots G_{P3}G_{P2}G_{P1}\,(T_{ant} + T_1) + G_{Pn} \cdots G_{P3}G_{P2}T_2 + G_{Pn} \cdots G_{P3}T_3 + \cdots]}{G_{Pn} \cdots G_{P3}G_{P2}G_{P1}}B_n \\ &= k\left[(T_{ant} + T_1) + \frac{T_2}{G_{P1}} + \frac{T_3}{G_{P1}G_{P2}} + \cdots + \frac{T_n}{G_{P1}G_{P2}G_{P3}\ldots G_{Pn-1}}\right]B_n \end{aligned} \tag{6.90}$$

Looking at the cascade as just a single device with a noise temperature of T_{cas}, equation (6.38) tells us that the equivalent input noise power of the cascade is

$$N_{in,cas} = k\,(T_{ant} + T_{cas})\,B_n \tag{6.91}$$

Equating (6.90) and (6.91) produces

$$k\,(T_{ant} + T_{cas})\,B_n = k\left[(T_{ant} + T_1) + \frac{T_2}{G_{P1}} + \frac{T_3}{G_{P1}G_{P2}} + \cdots + \frac{T_n}{G_{P1}G_{P2}G_{P3}\ldots G_{Pn-1}}\right]B_n \tag{6.92}$$

Canceling out like terms produces an expression for the noise temperature of the cascade

$$T_{cas} = T_1 + \frac{T_2}{G_{P1}} + \frac{T_3}{G_{P1}G_{P2}} + \cdots + \frac{T_n}{G_{P1}G_{P2}G_{P3}\ldots G_{Pn-1}} \tag{6.93}$$

Note that the power gains $G_{P1}, G_{P2}, \ldots, G_{Pn-1}$ are linear, not logarithmic or decibel quantities.

EXAMPLE

Noise Temperature of a Cascade

Find the noise temperature of the cascade shown in Figure 6-25.

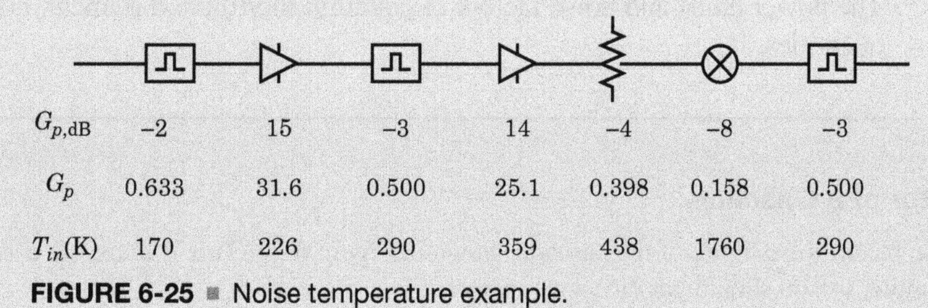

FIGURE 6-25 ■ Noise temperature example.

Solution

Using equation (6.93), we find

$$\begin{aligned}
T_{cas} &= T_1 + \frac{T_2}{G_{P1}} + \frac{T_3}{G_{P1}G_{P2}} + \cdots + \frac{T_n}{G_{P1}G_{P2}G_{P3}\ldots G_{Pn-1}} \\
&= 170 + \frac{226}{(0.633)} + \frac{290}{(0.633)\,(31.6)} + \frac{359}{(0.633)\,(31.6)\,(0.500)} \\
&\quad + \frac{438}{(0.633)\,(31.6)\,(0.500)\,(25.1)} \\
&\quad + \frac{1760}{(0.633)\,(31.6)\,(0.500)\,(25.1)\,(0.398)} \\
&\quad + \frac{290}{(0.633)\,(31.6)\,(0.500)\,(25.1)\,(0.398)\,(0.158)} \\
&= 170 + 357 + 14.5 + 35.9 + 1.8 + 17.6 + 18.3 \\
&= 615\ ^\circ\mathrm{K}
\end{aligned} \tag{6.94}$$

6.9.2 Noise Factor of a Cascade

Combining equation (6.93) with equation (6.59) produces an expression for the noise factor of a cascade given the noise factors of the elements making up the cascade:

$$\begin{aligned}
T_0\,(F_{cas} - 1) &= T_{cas} \\
&= T_1 + \frac{T_2}{G_{P1}} + \frac{T_3}{G_{P1}G_{P2}} + \cdots + \frac{T_n}{G_{P1}G_{P2}G_{P3}\ldots G_{Pn-1}} \\
&= T_0\,(F_1 - 1) + \frac{T_0\,(F_2 - 1)}{G_{P1}} + \frac{T_0\,(F_3 - 1)}{G_{P1}G_{P2}} \\
&\quad + \cdots \\
&\quad + \frac{T_0\,(F_n - 1)}{G_{P1}G_{P2}G_{P3}\ldots G_{Pn-1}}
\end{aligned} \tag{6.95}$$

Simplifying produces

$$F_{cas} = F_1 + \frac{F_2 - 1}{G_{P1}} + \frac{F_3 - 1}{G_{P1}G_{P2}} + \cdots + \frac{F_n - 1}{G_{P1}G_{P2}\ldots G_{Pn-1}} \tag{6.96}$$

The power gains and noise factors of equation (6.96) are the linear, not the logarithmic, quantities.

EXAMPLE

Noise Factor of a Cascade

Find the noise factor/noise figure of the cascade shown in Figure 6-26. This is the same cascade as the previous example, so you should get the same answer.

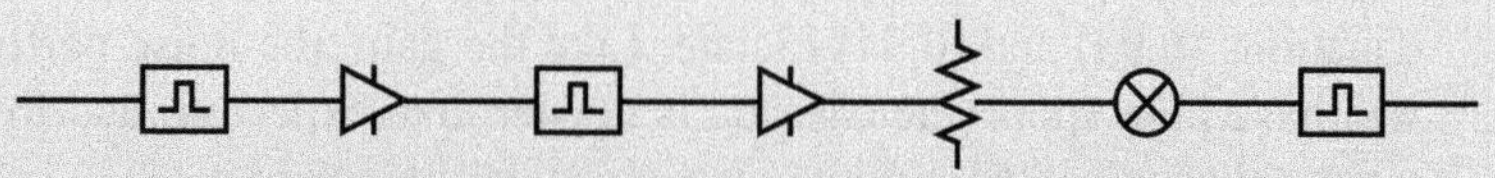

$G_{p,\text{dB}}$	−2	15	−3	14	−4	−8	−3
G_p	0.633	31.6	0.500	25.1	0.398	0.158	0.500
F_{dB}	2	2.5	3	3.5	4	8.5	3
F	1.58	1.78	2.00	2.24	2.51	7.08	2.00

FIGURE 6-26 ■ Noise figure example.

Solution

Using equation (6.96), we find

$$\begin{aligned} F_{cas} &= 1.58 + \frac{1.78-1}{(0.633)} + \frac{2-1}{(0.633)(31.6)} + \frac{2.24-1}{(0.633)(31.6)(0.500)} \\ &\quad + \frac{2.51-1}{(0.633)(31.6)(0.500)(25.1)} \\ &\quad + \frac{7.08-1}{(0.633)(31.6)(0.500)(25.1)(0.398)} \\ &\quad + \frac{2-1}{(0.633)(31.6)(0.500)(25.1)(0.398)(0.158)} \\ &= 1.58 + 1.23 + 0.05 + 0.12 + 0.0060 + 0.071 + 0.63 \\ &= 3.12 \\ &= 4.95\ \text{dB} \end{aligned} \tag{6.97}$$

A noise figure of 4.95 dB is equivalent to a 615°K noise temperature.

6.9.2.1 Real-World Effects

As with gain, the noise temperature and noise figure of a real-world cascade are not always what equations (6.93) and (6.96) indicate. The plot of the noise performance versus frequency of a real-world cascade will exhibit peaks and valleys, similar to the cascade gain for several reasons:

- The terminal impedances and gain ripple of each amplifier in the cascade cause errors when we evaluate the cascade noise equations. Since the noise performance of the cascade is intimately connected to the cascade's gain performance, any error when evaluating the gain will result in error when evaluating the noise performance.
- Like the gain, the noise figure or noise temperature of a component is not constant with time, temperature, and frequency.
- The noise specification associated with a component is valid only when the component has been terminated in its characteristic impedance on both its input and output. If the terminations are wrong, the component can exhibit widely different noise characteristics from what the manufacturer specified. Often the noise performance gets worse as we move away from the characteristic impedance.

These effects combine to produce the noise performance that will, on average, behave as equations (6.93) and (6.94) dictate. Like the gain, the noise performance of the cascade will have high spots (the cascade is noisier than expected) and low spots (the cascade is quieter than expected) on its noise versus frequency plot.

6.10 EXAMINING THE CASCADE EQUATIONS

Equations (6.93) and (6.96) describe to us the noise performance of a cascade given the characteristics of the pieces that make up the cascade.

- It's easy to isolate the contribution each component makes to the noise performance of the cascade. For example, the only place where the noise temperature of amplifier A_3 appears in equation (6.93) is the term

$$\frac{T_3}{G_{p,1}G_{p,2}} \tag{6.98}$$

  This term quantifies the complete contribution amplifier A_3 makes to the cascade's overall noise temperature.
- Since all of the terms in equation (6.93) are positive, the noise temperature of the cascade will always be greater than T_1, the noise temperature of the first amplifier. T_1 is the minimum noise temperature of the cascade.

  So, if we require a low-noise cascade, we'll want to use an amplifier with a low noise temperature for the first amplifier A_1.
- The amount of noise a particular amplifier contributes to the cascade depends on the amount of power gain that precedes the amplifier. Looking at equation (6.93), the amount of noise temperature A_3 adds to the cascade is diminished by the power gain of the first two amplifiers.

  This implies that if we want to build a system with a small noise figure, we should use as much gain as possible as early in the cascade as possible. Later on, however, we'll see that this isn't a very good idea for linearity or signal distortion reasons.

6.11 MINIMUM DETECTABLE SIGNAL

Equation (6.38) tells us that when we connect a receiver whose noise temperature is T_{RX} to an antenna with a noise temperature of T_{ant} the noise power present at the input of the receiver is

$$N_{in} = k\left(T_{ant} + T_{RX}\right)B_n \tag{6.99}$$

The equivalent noise power present on the input to a receiver sets a lower limit on the MDS, the smallest signal our system can detect. We arbitrarily assume that a signal is detectable when the signal power equals the equivalent input noise power or

$$S_{MDS} = k\left(T_{ant} + T_{RX}\right)B_n \tag{6.100}$$

If we apply a signal to the system's input and the signal's power equals the system's MDS, then the signal-to-noise ratio will be unity (or, equivalently, 0 dB).

Although the system may be able to detect a signal whose power is the MDS of the system, the system may not be able to process the signal. The system may not produce a suitable bit error rate or output signal-to-noise ratio.

6.11.1 Input Noise $= N_0$

If we assume that the antenna noise temperature is T_0, then we can derive several more equations that describe the MDS:

$$S_{MDS,T0} = F_{RX} k T_0 B_n \tag{6.101}$$

and

$$S_{MDS,T0,dBm} = F_{Rx,dB} - 174\frac{\text{dBm}}{\text{Hz}} + 10\log(B_n) \tag{6.102}$$

where

F_{RX} = the noise figure of the receiver
T_{RX} = the noise temperature of the receiver
B_n = the noise bandwidth

6.12 NOISE PERFORMANCE OF LOSSY DEVICES

We described resistive attenuators in Chapter 1. We'll now examine the noise performance of these devices.

6.12.1 System without Attenuator

Figure 6-27(a) shows a signal source (V_S in series with R_S) connected directly to a load resistor R_L. The system is matched, so $R_S = R_L$ and the noise temperature of R_S is T_0.

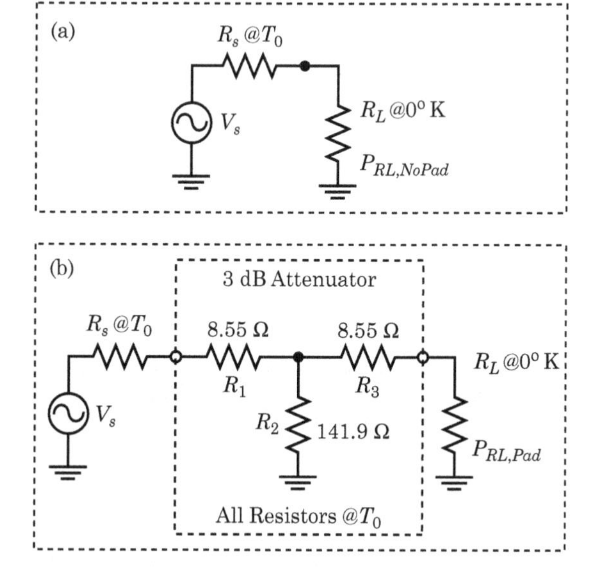

FIGURE 6-27 ■ A system with and without a 3 dB attenuator in the signal path. The signal power dissipated in the load resistor $P_{RL,Pad} = P_{RL,NoPad} - 3$ dB.

We measure a signal power of $P_{RL,no\,pad}$ in R_L. Since R_S is at T_0, R_L will dissipate a noise power of N_0 where

$$\begin{aligned} P_{RL,NoPad} &= N_0 \\ &= kT_0 B_n \end{aligned} \tag{6.103}$$

In a 1 hertz noise bandwidth

$$\begin{aligned} P_{RL,NoPad} &= kT_0 B_n \\ &= (1.38 \cdot 10^{-23})(290)(1) \\ &= 4 \cdot 10^{-21} \text{ Watts} \end{aligned} \tag{6.104}$$

6.12.1.1 System with Attenuator

Figure 6-27(b) shows the same system, but we've placed a resistive attenuator between the signal source and the load resistor. By design, this attenuator has a power loss of 3 dB. We are interested in the total noise power delivered to the load resistor when the attenuator is in place.

Noise is like any other signal, so the noise power generated by the source resistor R_S is attenuated by 3 dB before it reaches the load resistor. However, the resistors that make up the attenuator will generate noise of their own since they are at some nonzero physical temperature. The noise generated by R_1, R_2, and R_3 will contribute to the total noise power dissipated in the load resistor.

We'll assume that the noise temperature of all the resistors except R_L is T_0(290°K). We will also assume that the characteristic impedance of our system is 50 Ω, so $R_S = R_L = 50\ \Omega$. Finally, we'll use a noise bandwidth of 1 hertz.

The first step toward evaluating the effect of the attenuator is to find the values of the noise voltage sources in series with each resistor in the attenuator of Figure 6-27(b). Repeated application of equation (6.3) produces the data shown in Table 6-4.

Figure 6-28 shows the noise-equivalent circuit of Figure 6-27(b) after replacing the noisy resistors with noiseless resistors and noise voltage generators. The noise generators in each resistor are uncorrelated so we must examine the noise power delivered to the load resistor by each noise generator separately and then add up all the noise powers to get the final result.

We first examine the noise generator in R_S. We short all the voltage sources except the 895 pV_{RMS} generator in R_S. Analysis of the loop currents i_1 and i_2 as shown in Figure 6-28 produces

$$\begin{aligned} 895\text{ pV} &= (200.5)\, i_1 + (-141.9)\, i_2 \\ 0 &= (-141.9)\, i_1 + (200.5)\, i_2 \end{aligned} \tag{6.105}$$

TABLE 6-4 ■ The noise voltage sources of the resistors making up the attenuator of Figure 6-27(b).

Resistor Value (Ohms)	$V_{n,OC,RMS}$ (V_{RMS})
$R_S = 50\ \Omega$	895 pV_{RMS}
$R_1 = 8.55\ \Omega$	370 pV_{RMS}
$R_2 = 141.9\ \Omega$	1507 pV_{RMS}
$R_3 = 8.55\ \Omega$	370 pV_{RMS}

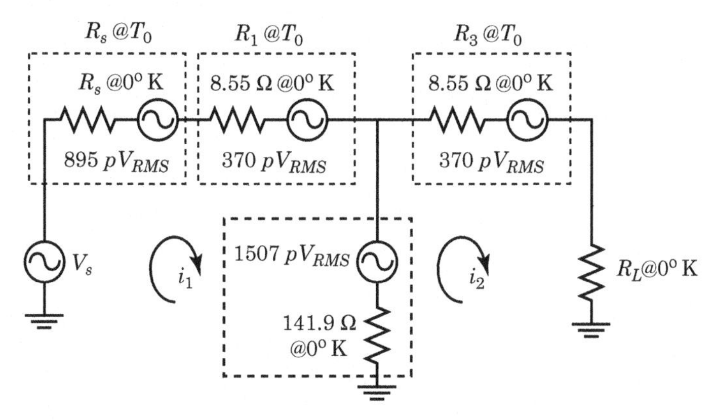

FIGURE 6-28 ■ The system of Figure 6-27(b). The noise-producing resistors have been replaced by their equivalent circuits—a noise generator in series with a noiseless resistor. The currents i_1 and i_2 are shown to facilitate circuit analysis.

We can rewrite these equations in matrix format as

$$\begin{bmatrix} 895\text{p} \\ 0 \end{bmatrix} = \begin{bmatrix} 200.5 & -141.9 \\ -141.9 & 200.5 \end{bmatrix} \begin{bmatrix} i_1 \\ i_2 \end{bmatrix}$$
$$\begin{bmatrix} i_1 \\ i_2 \end{bmatrix} = \begin{bmatrix} 8.943 \text{ pA}_{RMS} \\ 6.33 \text{ pA}_{RMS} \end{bmatrix} \tag{6.106}$$

The noise power delivered to R_L by the noise generator present in R_S is

$$\begin{aligned} P_{Noise\ in\ RL\ due\ to\ RS} &= i^2_{2,RMS} R_L \\ &= \left(6.33 \cdot 10^{-12}\right)^2 (50) \\ &= 2.00 \cdot 10^{-21} \text{ Watts} \end{aligned} \tag{6.107}$$

When the 3 dB attenuator is in the system, the noise power delivered to the load resistor is $2.00 \cdot 10^{-21}$ watts. When the 3 dB attenuator was not in the system, equation (6.104) told us that the noise power delivered to the load resistor was $4 \cdot 10^{-21}$ watts. As advertised, the 3 dB attenuator reduces the power that reaches the load resistor by 1/2 or 3 dB. However, we haven't yet accounted for the noise power generated by R_1, R_2, and R_3.

We analyze the noise effect of R_1 by shorting all the sources except the 370 pV_{RMS} source associated with R_1. The current-loop matrix is

$$\begin{bmatrix} 370\text{p} \\ 0 \end{bmatrix} = \begin{bmatrix} 200.5 & -141.9 \\ -141.9 & 200.5 \end{bmatrix} \begin{bmatrix} i_1 \\ i_2 \end{bmatrix}$$
$$\begin{bmatrix} i_1 \\ i_2 \end{bmatrix} = \begin{bmatrix} 3.697 \text{ pA}_{RMS} \\ 2.62 \text{ pA}_{RMS} \end{bmatrix} \tag{6.108}$$

The noise present in R_L due to the noise generator in R_1 is

$$\begin{aligned} P_{Noise\ in\ RL\ due\ to\ R1} &= i^2_{2,RMS} R_L \\ &= (2.62 \cdot 10^{-12})^2 (50) \\ &= 343 \cdot 10^{-24} \text{ Watts} \end{aligned} \tag{6.109}$$

Applying the same analysis techniques to R_2 and R_3, we find

$$\begin{aligned} P_{Noise\ in\ RL\ due\ to\ R2} &= i^2_{2,RMS} R_L \\ &= \left(4.40 \cdot 10^{-12}\right)^2 (50) \\ &= 969 \cdot 10^{-24}\ \text{Watts} \end{aligned} \tag{6.110}$$

and

$$\begin{aligned} P_{Noise\ in\ RL\ due\ to\ R3} &= i^2_{2,RMS} R_L \\ &= \left(3.70 \cdot 10^{-12}\right)^2 (50) \\ &= 684.5 \cdot 10^{-24}\ \text{Watts} \end{aligned} \tag{6.111}$$

The total noise power delivered to the load resistor is the sum of the noise powers delivered to R_L by R_S, R_1, R_2, and R_3, so we can write

$$\begin{aligned} P_{TotalNoiseinRL} &= P_{Noise\ from\ RS} + P_{Noise\ from\ R1} \\ &\quad + P_{Noise\ from\ R2} + P_{Noise\ from\ R3} \\ &= (200 \cdot 10^{-21}) + (343 \cdot 10^{-24}) + (969 \cdot 10^{-24}) + (684 \cdot 10^{-24}) \\ &= 4.00 \cdot 10^{-21}\ \text{Watts} \end{aligned} \tag{6.112}$$

The total noise power dissipated in R_L due to the noise generated by R_S, R_1, R_2, and R_3 is $4.00 \cdot 10^{-23}$ watts. This is the same noise power delivered to the load resistor of Figure 6-27(a), when the attenuator wasn't in the system.

The 3 dB attenuator of Figure 6-27(b) does attenuate the noise power generated by R_S by 3 dB. However, the resistors that make up the attenuator add just enough noise to raise the noise power dissipated in R_L back to the original value of kT_0B_n. The attenuator does reduce the noise power, but the thermal noise generated by the lossy elements of the attenuator adds the same amount of noise back into the system.

EXAMPLE

Noise Factor of Attenuators at T_0

Figure 6-29 shows a system with a 3 dB attenuator in place between the source and load. The physical temperature of the attenuator is T_0. Find

a. The noise figure of the attenuator

b. The noise temperature of the attenuator

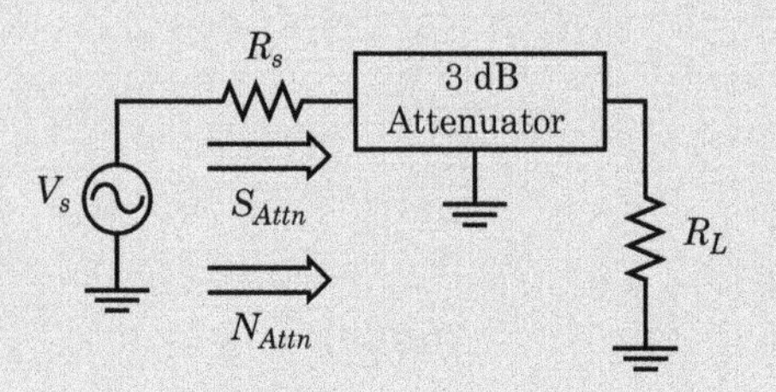

FIGURE 6-29 ■ The noise factor of a resistive attenuator. The physical temperature of the attenuator is T_0.

Solution

a. Let's assume the attenuator absorbs 1 mW of signal power (i.e., $P_{ATTN} = 1\text{ mW} = 0\text{ dBm}$). We've just shown that the source resistor will deliver a noise power of $N_0 = kT_0B_n = 4.00\text{E-}21$ watts $= -174$ dBm to a matched load (in a 1 hertz noise bandwidth). Since the attenuator presents a matched load to the signal source, we know $N_{ATTN} = -174$ dBm. The input signal-to-noise ratio is

$$\left(\frac{S}{N}\right)_{in} = 0 - (-174) \\ = 174\text{ dB} \tag{6.113}$$

On the output side, the attenuator reduces the signal power by 3 dB so the signal power delivered to R_L is -3 dBm. We've just shown that a 3 dB attenuator at room temperature will deliver $N_0 = -174$ dBm of noise power to its load resistor. The output signal-to-noise ratio is

$$\left(\frac{S}{N}\right)_{out} = -3 - (-174) \\ = 171\text{ dB} \tag{6.114}$$

Equation (6.75) tells us that if the input noise to a system isN_0 then the noise figure of a device is the ratio of the input SNR to the output SNR. The noise figure of our attenuator is

$$F_{Attn} = \frac{(S/N)_{in}}{(S/N)_{out}} \\ = 174\text{ dBm} - 171\text{ dBm} \\ = 3\text{ dB} \tag{6.115}$$

The noise figure of our T_0 3 dB attenuator is 3 dB.

b. Equation (6.59) expresses the relationship between noise factor and noise temperature. The noise figure of our attenuator is 3 dB so its noise factor is 2 and its noise temperature is

$$T_{Attn} = T_0\left(F_{Attn} - 1\right) \\ = 290(2 - 1) \\ = 290\ ^\circ\text{K} \tag{6.116}$$

6.12.1.2 Noise Figures of Lossy Devices

The "Noise Factor of Attenuators at T0" example illustrates an important relationship. If the physical temperature of a resistive attenuator is T_0, the noise figure of the attenuator (in decibels) equals the loss of the attenuator (in decibels) or

$$F_{Attenuator,dB} = \text{Attenuator's Loss}_{\text{dB}} \\ \text{When the attenuator's physical temperaure is } T_0 \tag{6.117}$$

assuming that the noise temperature of the attenuator equals the physical temperature.

6.12.1.3 Lossy Devices in Cascade

Figure 6-30 shows a simple two-element cascade consisting of a resistive attenuator, at T_0, followed by an amplifier. Under these conditions, the noise figure of the attenuator equals

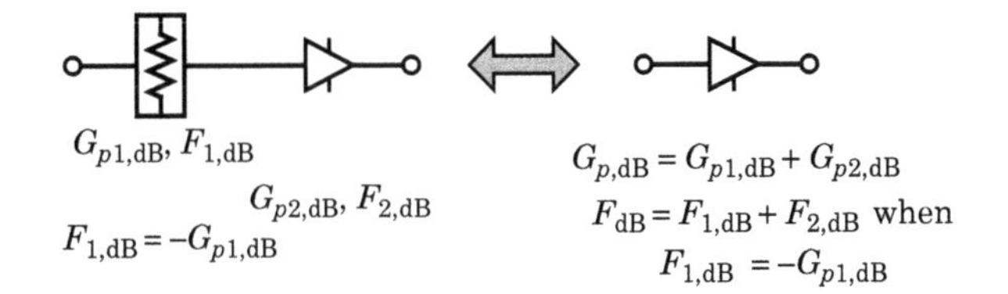

FIGURE 6-30 What is the effect of adding a component to the end of the cascade?

the attenuator value or

$$F_{Attn} = \frac{1}{G_{Attn}}$$
$$\text{or} \tag{6.118}$$
$$F_{Attn,dB} = -G_{Attn,dB}$$

The noise figure of the cascade is

$$F_{cas} = F_1 + \frac{F_2 - 1}{G_1} \tag{6.119}$$

Under the special conditions where the noise figure of the first device is inversely related to its gain, the cascade noise figure becomes

$$\begin{aligned} F_{cas} &= F_1 + \frac{F_2 - 1}{1/F_1} \\ &= F_1 + F_1 (F_2 - 1) \\ &= F_1 F_2 \\ F_{cas,dB} &= F_{1,dB} + F_{2,dB} \end{aligned} \tag{6.120}$$

In words, we can simply add the value of the attenuator to the noise figure of the amplifier to produce the noise figure of the cascade.

EXAMPLE

Lossy Devices in Cascade

Find the noise figure of these two-element cascades.

a. A 10 dB attenuator followed by an amplifier with a 4 dB noise figure
b. A 3 dB attenuator followed by an amplifier with a 6 dB noise figure
c. A 4 dB attenuator, a 6-dB attenuator and an amplifier with a 7 dB noise figure

Solution

Using equation (6.120), we find

a. $F_{cas,dB} = 10 + 4 = 14$ dB
b. $F_{cas,dB} = 3 + 6 = 9$ dB
c. $F_{cas,dB} = 4 + 6 + 7 = 17$ dB

EXAMPLE

Noise Factor of Attenuators at 0°K

Repeat the noise figure analysis of Figure 6-29 assuming that the physical temperature of the attenuator to 0°K. Find

a. The noise figure of the attenuator

b. The noise temperature of the attenuator

Solution

a. Let's assume the signal power delivered to the attenuator (S_{ATTN}) is 1 mW = 0 dBm. Assuming a 1 hertz noise bandwidth, we know the noise power delivered to the attenuator is $N_{ATTN} = N_0 = kT_0 B_n = 4.00 \cdot 10^{-21}$ W $= -174$ dBm. The input signal-to-noise ratio is

$$\begin{aligned} \left(\frac{S}{N}\right)_{in} &= 0 - (-174) \\ &= 174 \text{ dB} \end{aligned} \tag{6.121}$$

At the output, the signal power delivered to the load resistor is still -3 dBm. Since the attenuator is now at 0°K, the attenuator will not generate noise. The noiseless attenuator will decrease the input noise power as much as the input signal power, so the load resistor will dissipate $-174 - 3 = -177$ dBm of noise power. The output signal-to-noise ratio is

$$\begin{aligned} \left(\frac{S}{N}\right)_{out} &= -3 - (-177) \\ &= 174 \text{ dB} \end{aligned} \tag{6.122}$$

Using equation (6.75), the noise figure of our cold attenuator is

$$\begin{aligned} F_{Attn} &= \frac{(S/N)_{in}}{(S/N)_{out}} \\ &= 174 \text{ dBm} - 174 \text{ dBm} \\ &= 0 \text{ dB} \end{aligned} \tag{6.123}$$

The noise figure of our 0°K 3 dB attenuator is 0 dB. When a resistive attenuator is at a physical temperature other than T_0, the noise figure doesn't necessarily equal the attenuator's power loss.

b. Using equation (6.59), we find the noise temperature of our 0°K attenuator is

$$\begin{aligned} T_{Attn} &= T_0 (F_{Attn} - 1) \\ &= 290(1 - 1) \\ &= 0\ ^\circ\text{K} \end{aligned} \tag{6.124}$$

The noise figure of a resistive attenuator depends on its physical temperature. When the attenuator is at room temperature, the noise figure of the attenuator equals the loss of the attenuator. When the attenuator is at absolute zero (0°K), the noise figure of the attenuator is 0 dB.

TABLE 6-5 ■ Physical temperature and noise figure measurements of a 6dB resistive attenuator.

Physical Temperature (°C)	Measured Noise Figure (dB)	Measured Noise Temperature (°K)	Power Gain (dB)
−31.6	5.41	718	−6.05
−21.0	5.52	744	−6.09
+26.4	6.20	919	−6.14
+52.7	6.60	1036	−6.08
+78.0	6.75	1082	−6.10

TABLE 6-6 ■ Physical temperature and noise temperature calculation using the measurements of Table 6-5.

Physical Temperature (°C)	Measured Noise Temperature (°K)	Calculated Noise Temperature Via (6.125) (°K)
−31.6	718	718
−21.0	744	763
+26.4	919	907
+52.7	1036	986
+78.0	1082	1063

The noise temperature of a lossy, resistive attenuator is related to its physical temperature by

$$T_{Attn,Noise} = \frac{T_{Attn,Physical}\left(1 - G_{p,Attn}\right)}{G_{p,Attn}} \tag{6.125}$$

where

$T_{Attn,Noise}$ = the noise temperature of the attenuator in °K
$T_{Attn,Physical}$ = the physical temperature of the attenuator in °K
$G_{p,Attn}$ = the power gain of the attenuator in linear terms

We can verify this relationship experimentally. The author used an HP-8970 noise figure meter to measure the noise figure of a 6 dB attenuator at various temperatures. We generated Table 6-5.

Applying equation (6.125) to Table 6-5 produces Table 6-6.

Measured results agree quite well with theoretical results.

EXAMPLE

Noise Temperature of a Resistive Attenuator

Find the noise temperature and noise figure of a 10 dB attenuator whose physical temperature is

a. 300°C
b. −100°C

Solution

The power gain of a 10 dB attenuator is $-10\,\text{dB} = 1/10 = 0.1$. Using equation (6.125), we find

a. 300°C = 573°K, so the noise temperature of the 10 dB pad is

$$\begin{aligned} T_{Attn,\,Noise} &= \frac{T_{Attn,\,Physical}(1 - G_{p,\,Attn})}{G_{p,Attn}} \\ &= \frac{573(1-0.1)}{0.1} \\ &= 5{,}160°K \end{aligned} \tag{6.126}$$

Equation (6.57) gives us the noise figure

$$\begin{aligned} F_{Attn} &= \frac{T_0 + T_{Attn}}{T_0} \\ &= \frac{290 + 5160}{290} \\ &= 18.8 \\ &= 12.7 \text{ dB} \end{aligned} \tag{6.127}$$

An attenuator that is hotter than room temperature has a noise figure that is higher than its loss.

b. −100°C = 173°K, so the noise temperature is

$$\begin{aligned} T_{Attn,\,Noise} &= \frac{T_{Attn,\,Physical}\left(1 - G_{p,\,Attn}\right)}{G_{p,Attn}} \\ &= \frac{173\,(1-0.1)}{0.1} \\ &= 1{,}560°K \end{aligned} \tag{6.128}$$

Equation (6.57) gives us the noise figure

$$\begin{aligned} F_{Attn} &= \frac{T_0 + T_{Attn}}{T_0} \\ &= \frac{290 + 1560}{290} \\ &= 6.4 \\ &= 8.0 \text{ dB} \end{aligned} \tag{6.129}$$

An attenuator that is cooler than room temperature has a noise figure that is lower than its loss.

There are many topologies for resistive attenuator, including π-, T-, bridged-T, balanced, and distributed attenuators. Although we performed an analysis on only one topology, the analysis holds for any type of resistive attenuator. The analysis is also valid for filters that exhibit insertion loss.

6.12.1.4 Signal Loss without Attenuators

When we want to produce a signal loss in a system, we use attenuators. However, there are other ways of introducing loss into a system without using a resistive attenuator. Such methods include introducing an intentional mismatch into a system, reactive power splitting, and losses via signal spreading (the effect that causes path loss).

Since these methods do not involve resistive losses, they usually don't possess the noise penalties associated with resistive, lossy attenuators.

6.13 BIBLIOGRAPHY

Bryant, James, "Ask the Applications Engineer—8: Op Amp Issues," *Analog Dialogue*, 24-3, 1990.

Freeman, Roger L., *Reference Manual for Telecommunications Engineering*, John Wiley and Sons, Wiley-Interscience, 1985.

Gardner, Floyd M., *Phaselock Techniques*, John Wiley and Sons, 1979.

Ha, Tri T., *Solid-State Microwave Amplifier Design*, John Wiley and Sons, 1981.

Hewlett-Packard Corp., *Fundamentals of RF and Microwave Noise Figure Measurement*, Application Note 57-1, July 1983.

Horowitz, Paul, and Hill, Winfield, *The Art of Electronics*, Cambridge University Press, 1980.

Jorden, Edward C., *Reference Data for Engineers: Radio, Electronics, Computer and Communications*, 7th ed., Howard W. Sams and Co, Inc., 1985.

Motchenbacher, C.D., and Fitchen, F.C., *Low-Noise Electronic Design*, John Wiley and Sons, Wiley-Interscience, 1973.

Perlow, Stewart M., "Basic Facts about Distortion and Gain Saturation," *Applied Microwaves Magazine*, p. 107, May 1989.

Ryan, Al, and Scranton, Tim, "DC Amplifier Noise Revisited," *Analog Dialogue*, 18-1, 1984.

Sklar, Bernard, *Digital Communications—Fundamentals and Applications*, Prentice-Hall, 1988.

Skolnik, Merrill I., *Introduction to Radar Systems*, 2d ed., McGraw-Hill, 1980.

Williams, Richard A., *Communications Systems Analysis and Design—A Systems Approach*, Prentice-Hall, Inc., 1987.

6.14 PROBLEMS

1. **Power from R_{Hot} to R_{Cold}** Figure 6-31 shows a hot resistor connected to a cold resistor through a band-pass filter. Find a general expression for the power flow from R_{HOT} to R_{COLD}.

2. **Resistor Noise Power—Current Source Model** Figure 6-32 shows two noisy resistors, R_1 and R_2, connected to a noiseless load resistor R_L. Resistor R_1 is a 100 Ω resistor placed in boiling water (100°C = 373°K), R_2 is a 330 Ω resistor at 600°K, and R_L is a 510 Ω noiseless resistor. Find $V_{n,RMS,RL}$, the RMS noise voltage across R_L. Assume the noise bandwidth $B_n = 6$ MHz. Use current sources to model the resistors.

3. **Noise Figure Increase Related to Spectrum Analyzer Noise Lift** You connect an amplifier with 22 dB of gain to the input of a spectrum analyzer and observe that the noise on the spectrum analyzer increases by 3 dB. If the noise figure of the spectrum analyzer is 25 dB, what is the noise figure of the amplifier?

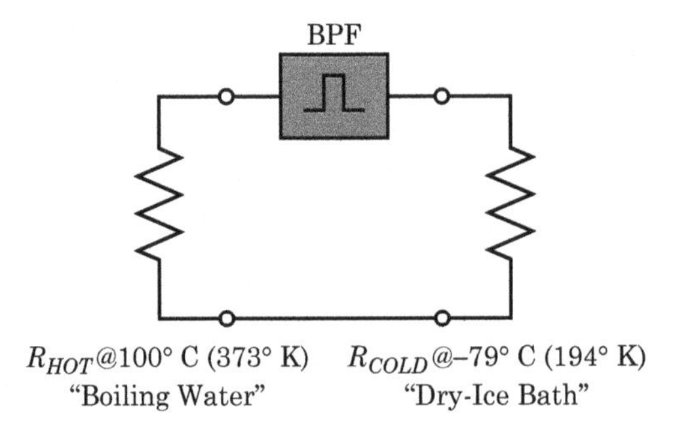

FIGURE 6-31 Power transfer in resistors of unequal temperature.

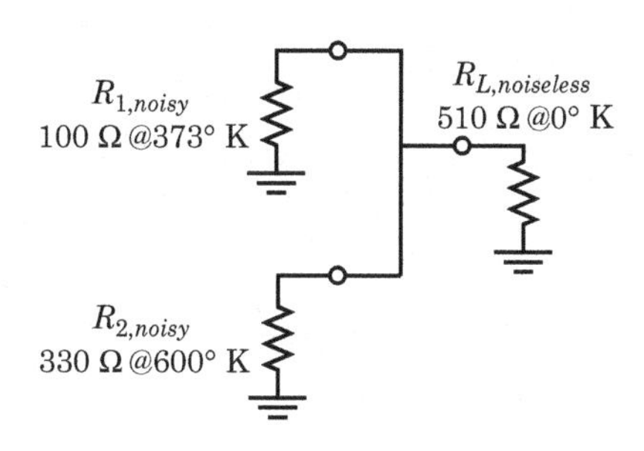

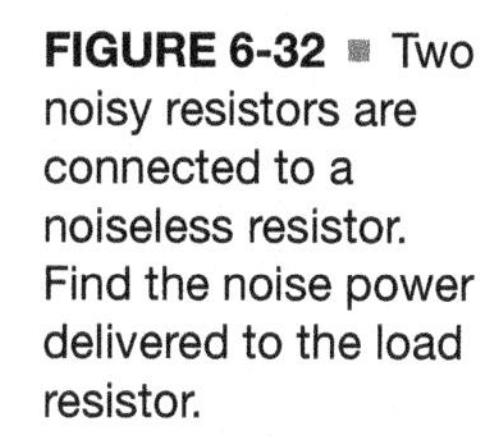

FIGURE 6-32 ■ Two noisy resistors are connected to a noiseless resistor. Find the noise power delivered to the load resistor.

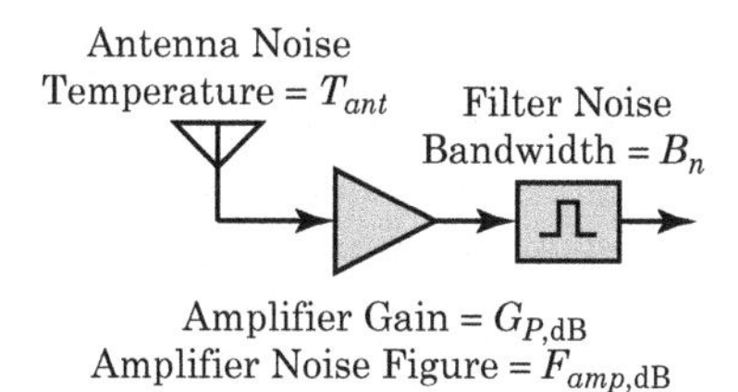

FIGURE 6-33 ■ Simple RF system.

4. **System Noise Floor and SNR** Analyze the system shown in Figure NOIPS-33, assuming that the antenna noise temperature $T_{ant} = 2{,}400°\mathrm{K}$, amplifier noise figure $F_{amp,dB} = 6.7$ dB, amplifier gain $G_{P,dB} = 22$ dB, and filter noise bandwidth $B_n = 50$ kHz.

   The filter is lossless, and the antenna presents a signal whose power is -80 dBm to the input of the amplifier.

   a. Find the total system input noise power. Answer in dBm.
   b. Find the system output noise power. Answer in dBm.
   c. What percentage of the input noise power is the antenna noise temperature responsible for?
   d. What is the SNR of the signal presented by the antenna (i.e., including only antenna noise)? Answer in dB.
   e. What is the SNR of the signal seen by the amplifier (i.e., including both antenna and amplifier noise)? Answer in dB.

5. **System Noise floor and SNR** For the system shown in Figure 6-34:

   a. Find the system noise floor assuming the antenna noise temperature is 290°K.
   b. What percentage of the input noise power is the antenna noise temperature responsible for?
   c. Find the system noise floor assuming the antenna noise temperature is 1,000°K.
   d. What percentage of the input noise power is the antenna noise temperature responsible for?

6. **Noise Figure in Cascade** Find the gain, noise figure and noise temperature of the cascade shown in Figure 6-35. Use the cascaded noise figure equation.

7. **Noise Temperature in Cascade** Find the gain, noise figure, and noise temperature of the cascade in Figure 6-36. Use the cascaded noise temperature equation. Make sure your answer agrees with the answer from Problem 6.

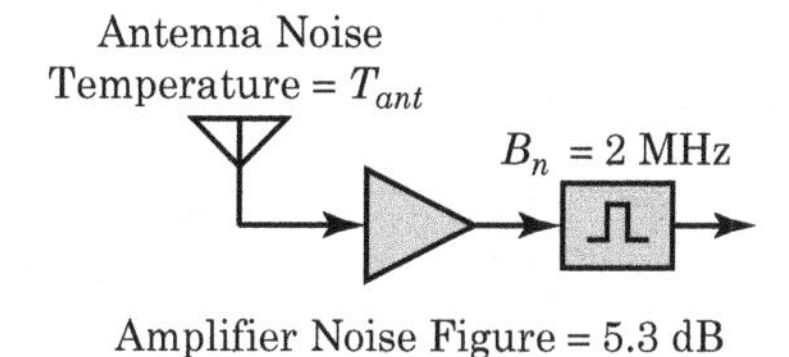

FIGURE 6-34 ■ Simple RF system.

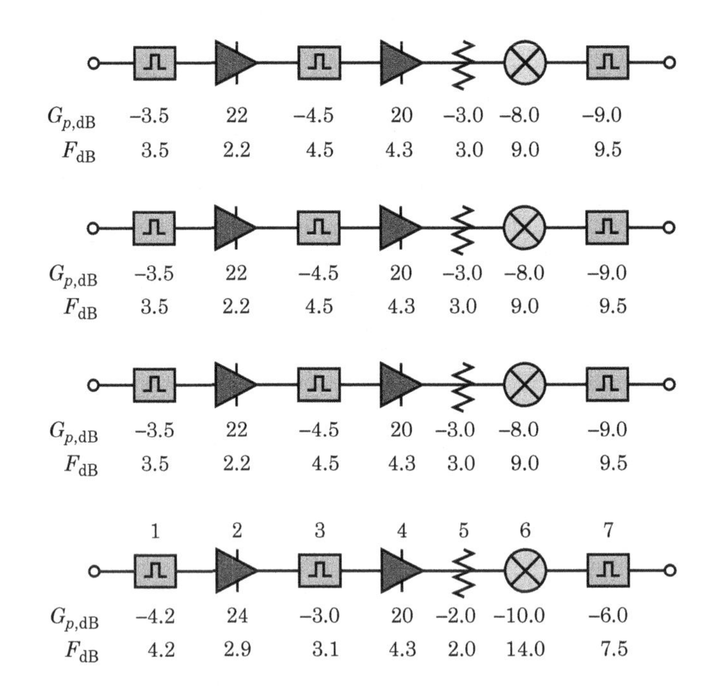

FIGURE 6-35 Noise figures in cascade.

FIGURE 6-36 Noise temperature in cascade.

FIGURE 6-37 Simple RF cascade.

FIGURE 6-38 Noise figure in cascade.

8. **Noise Contributions** Place the components of the cascade in Figure 6-37 in the order of their cascade noise contributions. In other words, which component contributes the most noise to the cascade? Which component contributes the second most noise to the cascade?

9. **Noise Figure in Cascade** Find the gain, noise figure, and noise temperature of the cascade of Figure 6-38. Perform the calculations using the cascaded noise figure equation. Answers in dB, dBm, and °K.

10. **Noise Temperature in Cascade** Convert the noise figures of all the components in the cascade from Problem 9 into noise temperatures. Find the noise temperature of the cascade. Perform the calculations by hand, using the cascaded noise temperature equation. Make sure that your answer agrees with the answer you calculated in Problem 9.

11. **Resistor Noise** Find the open-circuit, RMS noise voltage present across the following resistors:

    a. A 150 kΩ resistor at room temperature (about 290°K) measured in a 300 kHz noise bandwidth.

    b. A 75 Ω resistor at 1500°K measured in a 15 MHz noise bandwidth.

    c. For part a), what percentage of the time can we expect the noise voltage to exceed 60 μV?

    d. For part b), what percentage of the time can we expect the noise voltage to exceed 15 μV?

12. **Resistor Noise Voltage** Find the open-circuit, RMS noise voltage present across the following resistors:

    a. A 1 MΩ resistor at room temperature (about 290°K) measured in a 200 kHz noise bandwidth

    b. A 50 Ω resistor at 1000°K measured in a 10 MHz noise bandwidth

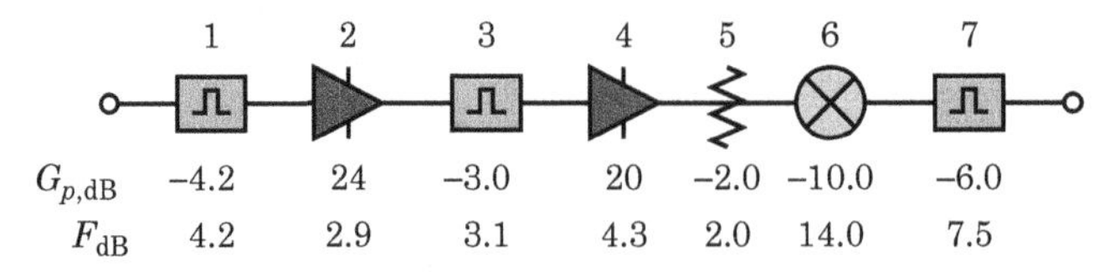

FIGURE 6-39 Noise temperature in cascade.

13. **Multiple Choice: Phasor Representation of Noise** Consider two phasors, P_1 and P_2. P_1 represents a cosine wave at frequency f_1, and P_2 represents a cosine wave at frequency f_2. If we use the phasor P_1 as the system reference (i.e., we freeze P_1 with respect to the complex axes so it does not appear to rotate), then the phasor P_2 rotates about the origin:

    a. Clockwise at a rate of abs$(f_2 - f_1)$
    b. Counterclockwise at a rate of abs$(f_2 - f_1)$
    c. Clockwise at a rate of abs$(f_2 + f_1)$
    d. Counterclockwise at a rate of abs$(f_2 + f_1)$
    e. At a rate of abs$(f_2 - f_1)$ but we don't have enough information to find the direction

14. **Multiple Choice: Setting the System Noise FIgure** In a well-designed radio frequency cascade, the noise figure of the cascade depends largely on:

    a. High gain throughout the system
    b. Low-noise components used throughout the system
    c. The attenuation of any passive devices present in the system prior to the first active device
    d. The noise figure of the first active device in the cascade
    e. Both c) and d)

15. **Noise Figure Comparisons** Arrange these components in order from quietest to noisiest:

    a. An amplifier with a 3.6 dB noise figure
    b. An amplifier with a noise factor of 3.1 (not 3.1 dB but 3.1 in linear terms)
    c. An amplifier with a 10°C noise temperature

16. **More Noise Figure Comparisons** Arrange these components in order from quietest to noisiest:

    a. An amplifier with a 3 dB noise figure
    b. An amplifier with a noise factor of 2.5 (not 2.5 dB but 2.5 in linear terms)
    c. An amplifier with a 0°C noise temperature

17. **Antenna/System Noise Temperature**

    a. Find the noise floor for a system with an antenna noise temperature of 120°K, an amplifier noise temperature of 75°K, and a noise bandwidth of 15 MHz.
    b. What percentage of the input noise power is the antenna noise temperature responsible for?
    c. What percentage of the input noise power is the amplifier noise temperature responsible for?

18. **Minimum Detectable Signal** Find the MDS of the cascade shown in Figure 6-40 assuming the antenna noise temperature is 800°K.

19. **Fixing a Bad Noise Figure** We would like to place an amplifier between the first filter and the mixer to decrease the noise figure of the cascade shown in Figure 6-41. What minimum

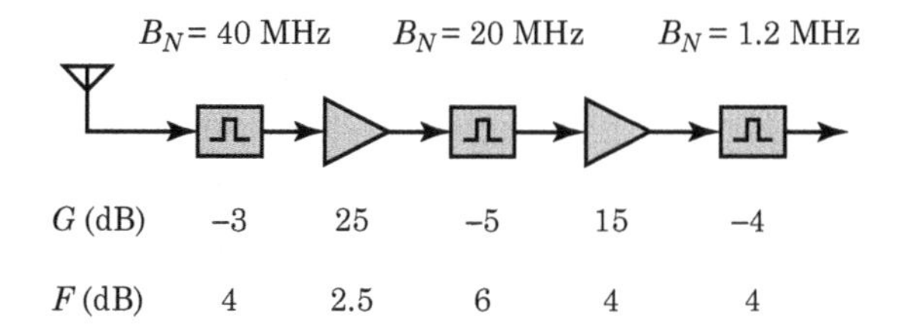

FIGURE 6-40 Minimum detectable signal for a cascade.

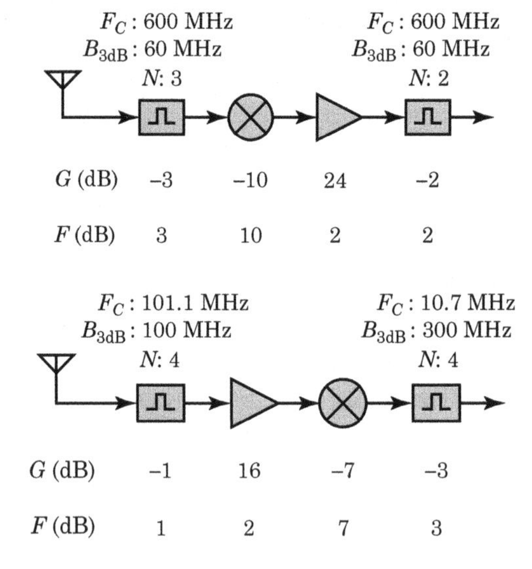

FIGURE 6-41 Fixing a poor noise figure.

FIGURE 6-42 Maximizing SNR.

gain should we use to achieve an overall cascade noise figure of 6 dB? Assume the noise figure of the new amplifer is 2 dB.

20. **SNR and Rearranging Components** Given the system of Figure 6-42, calculate the signal-to-noise ratio at the output of the cascade if we apply a signal power of −90 dBm to the input.

   a. Rearrange the order of the components to provide a better noise figure.
   b. What is the effect of rearranging the components on the linearity of the cascade?

21. **Multiple Choice** If the noise present at the output of our antenna is about equal to our system's internally generated noise floor, we can definitely decrease the amount of noise present at the output of our system by:

   a. Decreasing the noise figure of our system
   b. Pointing the antenna in a different direction
   c. Decreasing the sidelobe levels of the receive antenna
   d. Increasing the gain of our system
   e. Moving the antenna to a different position

22. **Cable Loss and Noise Figure** Given the system shown as Figure 6-43 and the graph of cable attenuation versus frequency shown in Figure 6-44:

   a. Find the noise figure of the system at 2, 4, and 6 GHz.
   b. To what gain should we set the amplifier to overcome the noise figure of the spectrum analyzer at 6 GHz? *Overcoming the noise figure of the spectrum analyzer* means we must have at least $F_{dB,SpectrumAnalyzer} + 10$ dB of gain before the spectrum analyzer.

      Assume the cable length, L, is 50 feet and that the noise figure of the cable equals the loss of the cable (in dB).

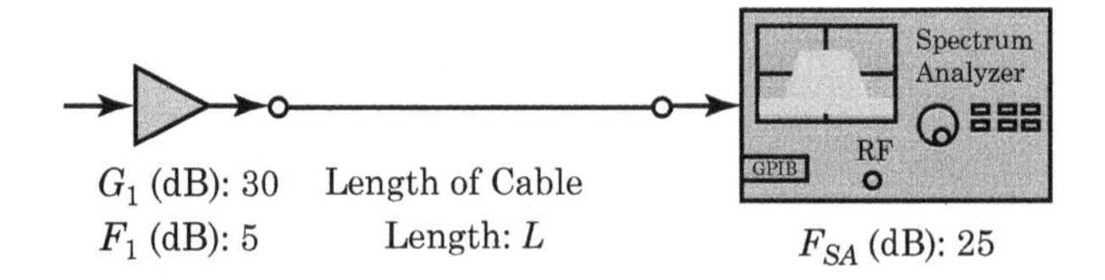

FIGURE 6-43 Improving the noise figure of a spectrum analyzer.

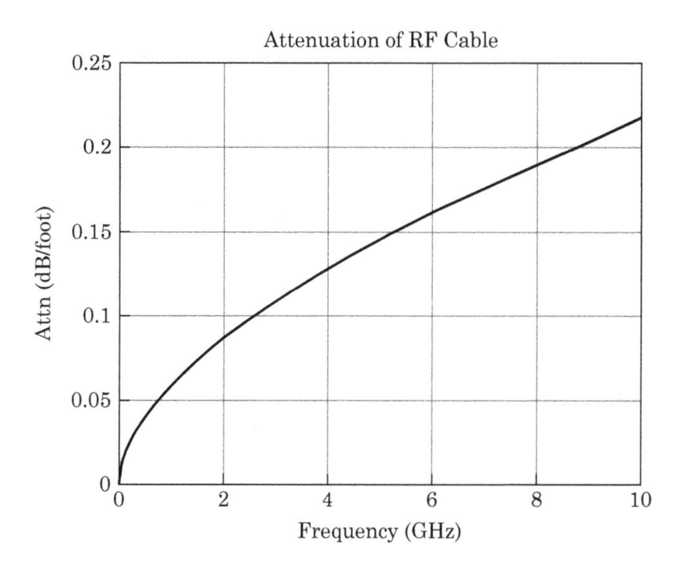

FIGURE 6-44 ■ The attenuation of an RF cable.

TABLE 6-7 ■ A list of available components.

	Attenuator	Amplifier #1	Amplifier #2	Filter
Gain (dB)	−17	12	20	2
Noise Figure (dB)	18	6	3	2.5

23. **MDS** A large, parabolic antenna feeds satellite signals into a receiver with a 12 MHz bandwidth and a 1.5 dB noise figure.

    a. What is the MDS?

    b. Assuming the system needs about 20 dB of SNR to demodulate these signals, what is the minimum signal power the receiver can process?

24. **Order in Cascade** Table 6-7 lists the components available to you to build a receiver:

    a. Which component should go first in the cascade if you want minimum noise figure? Justify your answer.

    b. Which components should go first in the cascade if you maximum noise figure. Justify your answer.

    c. Which arrangement of components will produce the highest gain?

25. **System Noise Floor**

    a. Find the noise floor for a system with an antenna noise temperature of 200°K, an amplifier noise temperature of 75°K, and a noise bandwidth of 35 MHz.

    b. What percentage of the input noise power is the antenna noise temperature responsible for?

    c. What percentage of the input noise power is the amplifier noise temperature responsible for?

26. **Geostationary Satellites** For about 20 days a year, geosynchronous satellites typically experience about 70 minutes of outage. The signal-to-noise ratio of the signal from the satellite decreases so much that your cable TV picture is often unwatchable.

    What might be the cause of this phenomenon?

CHAPTER 7

Linearity

Nothing is more noble, nothing more venerable than fidelity.

Cicero

Chapter Outline

7.1 INTRODUCTION

In Chapter 6, we found that thermal noise set the limit on the smallest signal a receiver can process. Linearity sets the limit on the largest signal we can process. If we apply too large of a signal into a system, the system will alter or distort the signal, rendering it unusable.

7.1.1 Definitions

In the following pages, we'll be discussing signals and their harmonics. We'll use the following nomenclature:

- Fundamental: The frequency is f_0.
- First harmonic: The frequency is also f_0. We'll hardly ever use this term. However, some people define the first harmonic as having a frequency of $2f_0$.
- Second harmonic: The frequency is $2f_0$.
- Third harmonic: The frequency is $3f_0$.
- n-th Harmonic: The frequency is nf_0.

7.2 | LINEAR AND NONLINEAR SYSTEMS

We study linear systems in undergraduate electrical engineering courses. Although every system is nonlinear to some extent, we concentrate on linear systems for several reasons:

- We can model almost every nonlinear system as a linear system over a narrow operating range. For example, radio frequency (RF) amplifiers are nonlinear devices, but if we keep the input signal power small enough we can model the amplifier as a linear device.

  Even simple filters, which consist of common inductors and capacitors, are nonlinear devices. At high power levels, the dielectrics of the capacitors behave nonlinearly, and the inductors saturate. The result is that the value of the component changes with the voltage across it or the current through it. This is a nonlinear process.
- The mathematics describing linear systems is simple, well understood, and intuitive to use. Numerical simulation is quick and uncomplicated. We first model our systems as linear and then describe the system's nonlinearities as deviations from its linear behavior.

7.2.1 Linear Systems

A linear system has the following characteristics:

- The only frequencies present in a linear system are directly traceable to some voltage or current source in the system. If we excite a linear system with a 1 MHz sine wave, we will not discover any voltage or current in the system at another frequency.
- Superposition works. The system behaves identically regardless of the magnitude of the input signal power. The terminal impedances and the various transfer functions all remain the same.
- A linear system does not change its characteristics when we simultaneously apply more than one signal. We can present the system with as many complex waveforms as we like, and the system will behave the same as it did when we applied only one tone.

7.2.2 Weakly Nonlinear Systems

Nonlinear effects are difficult to describe mathematically, and programming the equations into a computer is a difficult task. The numerical simulations tend to be slow, which makes repetitive calculations unwieldy and sluggish. Our philosophy will be to describe systems that are weakly nonlinear. We will put signals into our system that are just strong enough to cause noticeable nonlinear effects but not so strong that the system characteristics change dramatically.

7.2.2.1 Gain Compression

A common nonlinear effect is a change in a system's voltage transfer function with changing power levels. For example, a system may exhibit a voltage gain or power gain when we apply small signals to its input. As we increase the input signal power, we will come to a point where the voltage and power gains change. We commonly refer to this effect as *gain compression*.

7.2.2.2 Harmonic Distortion

Another common nonlinear effect is *harmonic distortion*. When we apply a strong signal to a nonlinear system, we will observe signals at the output of the system that were generated internally inside the system. If we apply a signal whose frequency is f_0 to a system, we'll observe signals at, for example, direct current (DC) (or 0 hertz), f_0, $2f_0$, and $3f_0$, etc. at the output. If we apply two signals, at f_1 and f_2, to the input of a nonlinear system, we will observe signals at $\pm nf_1 \pm mf_2$ at the output of the system. The variables n and m are integers and can include zero.

All systems are nonlinear. Distortion products are *always* present at the output of any nonlinear system. However, if the input signals are weak, the distortion generated by the system may be too small to detect. We often ignore distortion products if their power levels are below the noise.

7.3 | AMPLIFIER TRANSFER CURVE

Figure 7-1 shows a circuit diagram of an RF amplifier and its source and load resistors. The signal voltage at the input of the amplifier is V_{in}, and the signal voltage across the load resistor is V_{out}. We've also shown the power supply terminals (labeled V_{cc} and V_{bb}).

Figure 7-2 shows the large-signal voltage transfer curve of an amplifier. Note the flattening at the top and bottom of the plot and the sharp slope to the curve as the curve passes through (0,0). The flattening is *saturation*, and the slope about (0,0) is related to the voltage gain of the amplifier.

7.3.1 Small Signals

When we drive the amplifier with miniscule signals, we are operating under *small-signal* conditions. The input voltage will be restricted to a very small range around the (0,0) point of Figure 7-2.

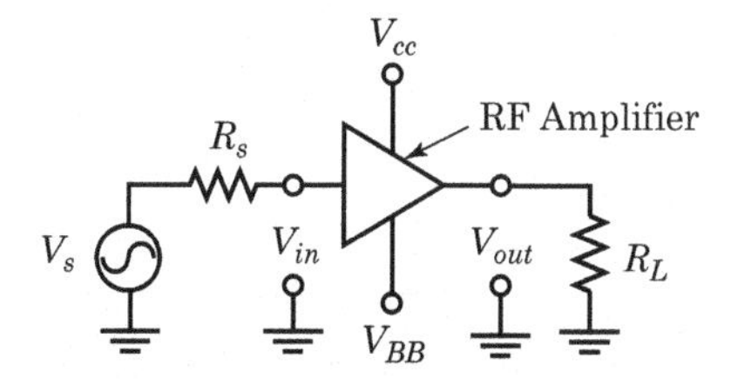

FIGURE 7-1 ■ The circuit diagram of an RF amplifier showing the input and output voltages.

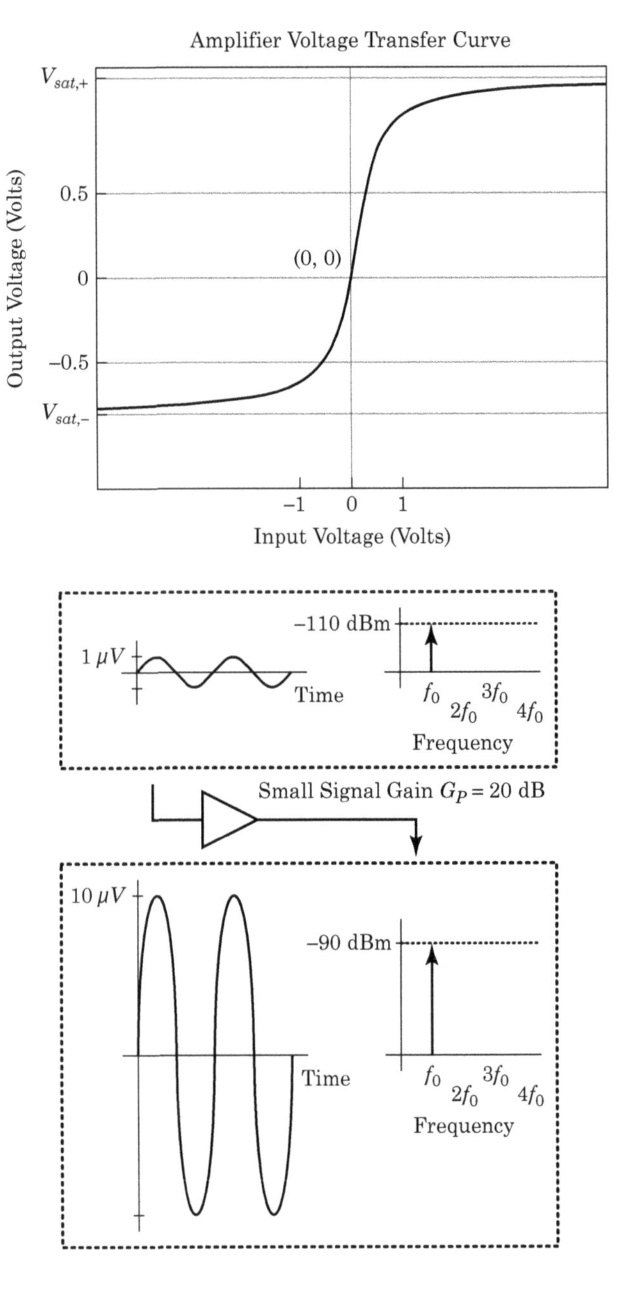

FIGURE 7-2 ■ The large signal voltage transfer curve of the amplifier shown in Figure 7-1.

FIGURE 7-3 ■ A nonlinear amplifier operating under small-signal conditions.

Figure 7-3 shows the amplifier under small signal conditions. The input signal level is 1 μV_{pk}, or −110 dBm in a 50 Ω system. The signal present at the output of the RF amplifier looks very much like the input signal in both the time and frequency domains. The only measurable difference is the increase in voltage and power.

7.3.2 Large Signals

Figure 7-4 shows the input and output signals when we drive the amplifier with a large sine wave. The output is a distorted version of the input. Saturation effects keep the output wave from exceeding limits associated with the power supply voltages and with the bias

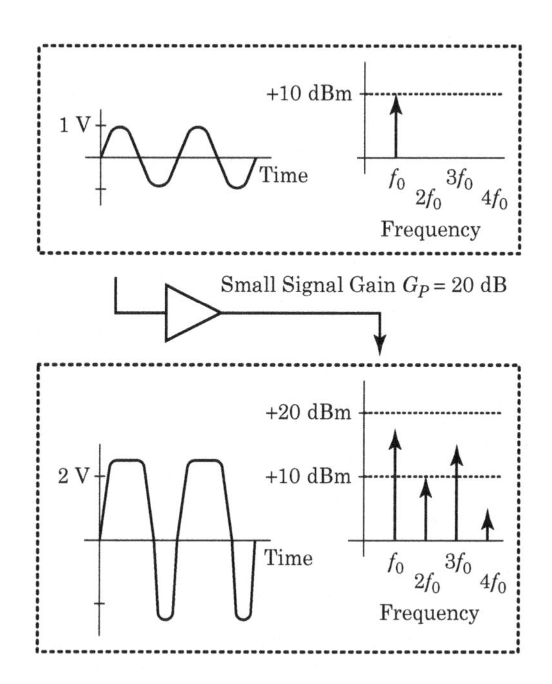

FIGURE 7-4 ■ A nonlinear amplifier operating under large signal conditions.

points of the transistors inside the amplifier. The output waveform is also asymmetrical; that is, the top and bottom halves of the wave are not identical.

Figure 7-4 also shows the spectra of the input and output waveforms. The input is a mathematically pure sine wave, whereas the output shows significant harmonic content. The generation of frequencies not present at the input is a major characteristic of a nonlinear system.

7.3.3 Small Signals

When the amplifier is driven with small signals, it is operating under small signal conditions. Since the input voltage will be restricted to a very small range, we will be operating the amplifier very close to the (0, 0) point shown in Figure 7-2. Figure 7-5 shows that the

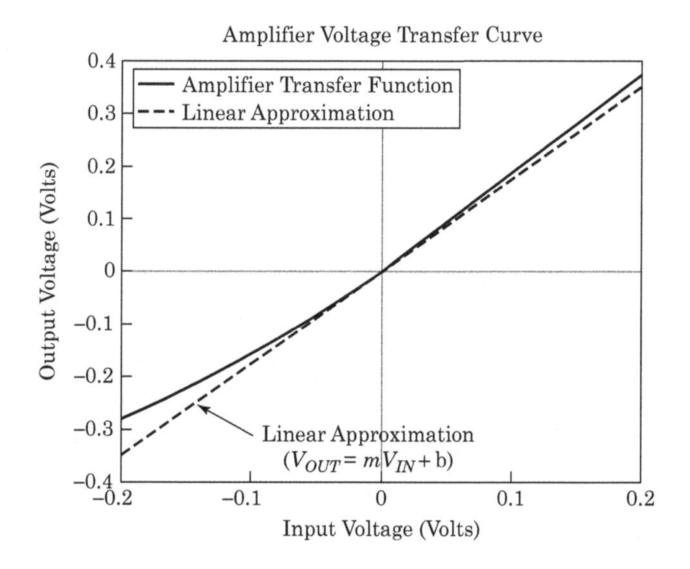

FIGURE 7-5 ■ The voltage transfer curve of Figure 7-2, magnified around (0,0). The linear approximation holds for small input signals.

magnified neighborhood around (0,0) is very nearly a straight line. The slope of the linear approximation is related to the amplifier's power gain.

EXAMPLE

Amplifier Power Gain

Show that the slope of the voltage transfer function of an amplifier is directly related to its power gain. The amplifier operates in a system with characteristic impedance Z_0.

Solution

Use Figure 7-5 as a guide. We've drawn a straight line (i.e., a first-order curve fit) through the (0,0) point, and we've matched the slope of the line to the slope of the amplifier's transfer function at the origin. Let m be the slope of the line. We know

$$V_{out}(t) = mV_{in}(t) + b \tag{7.1}$$

where b is the y-intercept of the line. The y-intercept is zero so

$$\begin{aligned} V_{out}(t) &= mV_{in}(t) \\ \Rightarrow \frac{V_{out}(t)}{V_{in}(t)} &= m \end{aligned} \tag{7.2}$$

Referring to Figure 7-1, the power gain of the amplifier is

$$G_p = \frac{V_{out}^2(t)/R_L}{V_{in}^2(t)/R_{sys}} \tag{7.3}$$

In a system with a Z_0 characteristic impedance, $R_{sys} = R_L = Z_0$ and

$$\begin{aligned} G_p &= \frac{V_{out}^2(t)}{V_{in}^2(t)} \\ &= m^2 \end{aligned} \tag{7.4}$$

The slope of the straight-line approximation of Figure 7-5 is 1.75. The power gain of this amplifier is 1.75^2 or 3.06 = 4.9 dB.

7.3.4 Summary

Figure 7-4 shows amplifier operating under large signal conditions. The input signal voltage is 1 V_{pk}, which is +10 dBm in a 50 Ω system. The amplifier's output signal is severely distorted. In the time domain, the output is obviously different from the input waveform. The frequency domain shows a series of sine waves all harmonically related to the input sine wave. The power levels of the output harmonics are related to the input power level and to the shape of the amplifier's voltage transfer curve. These harmonics are clearly not present on the input, and they were generated inside the amplifier. These signals, generated in nonlinear devices, are often referred to as *intermodulation products* or simply as *intermods*.

7.3.5 Nonlinear Device

This amplifier meets several of the criteria for a nonlinear device:

- Its behavior changes when we apply signals of different powers.
- The amplifier distorts the waveform; producing frequencies not present at the input. This is especially noticeable at high power levels, but the effect occurs at all input power levels.

For single input tones at a frequency of f_0, the frequency of the distortion components take the form of discrete signals at nf_0, where n extends from zero to infinity. For multiple input signals, the expression is more involved.

Some devices are more linear than others. For example, we can pass several thousand watts of RF power through a properly designed transmission line or filter with no visible signs of nonlinear distortion. However, most of the amplifiers and mixers we will be discussing will show significant distortion at input power levels of just 1 mW. From a linearity perspective, the two most troublesome components are amplifiers and mixers. Diode switches and voltage-controlled attenuators can also be troublesome.

7.4 | POLYNOMIAL APPROXIMATIONS

We would like to quantify signals and system from a linearity perspective. How do we compare nonlinear devices? How much distortion will a device generate given a known amount of input power? Polynomial approximations provide the solution.

A polynomial is an equation of the form

$$y = k_0 + k_1 x + k_2 x^2 + k_3 x^3 + \cdots + k_n x^n \tag{7.5}$$

where $k_n x^n$ is the n-th-order term of the polynomial. We can express any arbitrary curve as a polynomial. If we carry enough terms, the error between the curve and its approximation can be arbitrarily small.

7.4.1 RF Amplifier

We now modify equation (7.5) to include variables that are applicable to signals analysis:

$$V_{out}(t) = k_0 + k_1 V_{in}(t) + k_2 V_{in}^2(t) + k_3 V_{in}^3(t) + \cdots + k_n V_{in}^n(t) \tag{7.6}$$

7.4.2 Matching Derivatives

There are several numerical methods available for generating a polynomial to approximate a given curve. The method we choose forces a match between the derivatives of the curve and the derivatives of the polynomial.

Figure 7-6 shows our amplifier transfer function and its first-, second-, and third-order polynomial approximations. The first-order or linear approximation is

$$V_{out}(t) = 1.752 V_{in}(t) \tag{7.7}$$

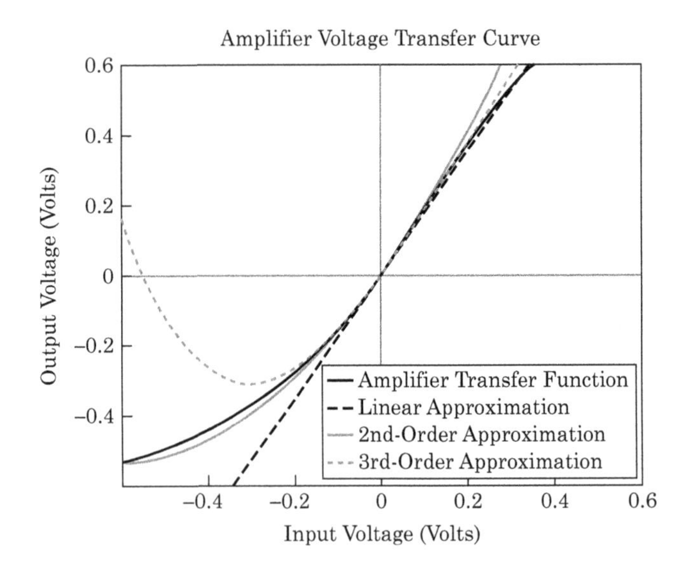

FIGURE 7-6 Amplifier transfer function with its first-, second-, and third-order polynomial approximations.

The second-order approximation is

$$V_{out}(t) = 1.752V_{in}(t) + 1.446V_{in}^2(t) \tag{7.8}$$

and the third-order approximation is

$$V_{out}(t) = 1.752V_{in}(t) + 1.446V_{in}^2(t) - 3.229V_{in}^3(t) \tag{7.9}$$

7.4.2.1 Observations

Close examination of Figure 7-6 reveals that the higher-order polynomials match the original curve better than the lower-order approximations at the (0, 0) point. The only design criteria for the polynomials of Figure 7-6 was that the derivatives of the polynomial matched the derivatives of the original curve at the $V_{in} = 0$, $V_{out} = 0$ point. We also note that the polynomial approximations diverge quickly from the original curve with increasing input signal voltage. The higher-order curves diverge quicker than the lower-order curves.

7.4.3 Weakly Nonlinear

Under small signal conditions, the linear approximation holds. As the input signal increases in power, we must describe the transfer function with a second-order polynomial to account for the distortion we observe. As the input signal level increases further, we must describe the amplifier using the third-order approximation.

7.4.4 Strongly Nonlinear

As we increase the input level, the polynomial approximations no longer describe the amplifier behavior adequately. This characteristic limits our ability to predict the system's performance at very large signal levels. Stated differently, if we do not allow our input signal to become too strong, the following analysis is accurate and useful. If the input signals are sufficiently strong such that the amplifier's transfer curve can't be accurately described by a matching polynomial, our analysis breaks down.

7.5 SINGLE-TONE ANALYSIS

We can approximate the nonlinear voltage transfer function of an RF amplifier with a polynomial of the form of equation (7.6). We now examine the output voltage of a nonlinear device when the input voltage is a single cosine wave described as

$$V_{in}(t) = A\cos(\omega_0 t) \tag{7.10}$$

7.5.1 The k_0 Term

The 0-th term of the polynomial represents the DC present at the output of the amplifier. We are rarely interested in signals whose frequencies are this low so we ignore this term. In most RF devices, this term is usually very small or equal to zero.

7.5.2 The $k_1 V_{in}^1$ Term

The k_1 term represents the power gain of the amplifier. Large k_1's translate into large power gains, while small k_1's translate into small power gains. Equation (7.4) relates the power gain of the amplifier to its k_1 term. The $k_1 V_{in}^1$ term produces an output voltage of

$$V_{out,1}(t) = k_1 A\cos(\omega_0 t) \tag{7.11}$$

This *linear* or *first-order term* is not responsible for any distortion in the amplifier's output. If we apply a voltage V_{in} to a linear device, the output voltage V_{out} will contain only the same frequency and nothing else.

7.5.3 The $k_2 V_{in}^2$ Term

The second-order term of the polynomial is $k_2 V_{in}^2$, and it is responsible for some of the signal distortion present on the amplifier's output. For a sinusoidal input signal of

$$V_{in}(t) = A\cos(\omega_0 t) \tag{7.12}$$

the $k_2 V_{in}^2$ term generates an output voltage of

$$\begin{aligned} V_{out,2}(t) &= k_2 A^2 \cos^2(\omega_0 t) \\ &= \frac{k_2 A^2}{2}[1 + \cos(2\omega_0 t)] \\ &= \frac{k_2 A^2}{2} + \frac{k_2 A^2}{2}\cos(2\omega_0 t) \end{aligned} \tag{7.13}$$

The second-order term produces energy at DC or 0 hertz (the $k_2 A^2/2$ term) and energy at twice the input frequency (the $k_2 A^2/2 \cdot \cos(2\omega_0 t)$ term).

We use the term *second-order distortion* to describe the effects caused by the second-order term of the approximating polynomial.

7.5.4 The $k_3 V_{in}^3$ Term

The third-order term is partially responsible for the amplifier's nonlinear behavior. For a sinusoidal input signal of

$$V_{in}(t) = A\cos(\omega_0 t) \tag{7.14}$$

the third-order term produces

$$\begin{aligned} V_{out,3}(t) &= k_3 A^3 \cos^3(\omega_0 t) \\ &= \frac{k_3 A^3}{4}[3\cos(\omega_0 t) + \cos(3\omega_0 t)] \end{aligned} \tag{7.15}$$

Equation (7.15) exhibits energy at the fundamental frequency (the original input frequency or f_0) and at three times the input frequency (at $3f_0$). This is *third-order distortion.*

7.5.4.1 The $k_4 V_{in}^4$ Term

For a cosinusoidal input, the fourth-order term produces

$$V_{out,4}(t) = \frac{k_4 A^4}{8}[3 + 4\cos(2\omega_0 t) + \cos(4\omega_0 t)] \tag{7.16}$$

The fourth-order term generates energy at DC, at the second harmonic of the input signal, and at the fourth harmonic of the input signal. Fourth-order distortion is similar to second-order distortion except that fourth-order distortion produces energy at the fourth harmonic of the input signal.

7.5.4.2 The $k_5 V_{in}^5$ Term

The fifth-order term produces

$$V_{out,5}(t) = \frac{k_5 A^5}{16}[10\cos(\omega_0 t) + 5\cos(3\omega_0 t) + \cos(5\omega_0 t)] \tag{7.17}$$

which contains energy at the fundamental frequency, the third harmonic, and the fifth harmonic.

The fifth-order term of the polynomial spawns signals at the fundamental, the third, and the fifth harmonics of the input signal, whereas the third-order term creates energy at only the fundamental and the third harmonics.

7.5.4.3 The $k_6 V_{in}^6$ Term

Under cosinusoidal drive, the sixth-order term produces

$$V_{out,6}(t) = \frac{k_6 A^6}{32}[10 + 15\cos(2\omega_0 t) + 6\cos(4\omega_0 t) + \cos(6\omega_0 t)] \tag{7.18}$$

The output has energy at DC, the second, fourth, and sixth harmonics.

7.5.5 Observations

We can draw several conclusions from the data of the previous section.

7.5.5.1 Even n

For even n, the term $\cos^n(\omega_0 t)$ will produce only even harmonics of the fundamental. The harmonics range from DC (which is the 0-th harmonic) up to and including the n-th harmonic. These are the *even-order responses* of the system.

7.5.5.2 Odd n

For odd n, $\cos^n(\omega_0 t)$ will produce only odd harmonics of the fundamental. The harmonics begin with the fundamental or first harmonic up to and including the n-th harmonic. The responses that are odd and greater than one are the *odd-order responses* of the system.

7.5.5.3 *n*-th-Order Polynomial Generates n-th-Order Harmonics

For a given n, $\cos^n(\omega_0 t)$ will produce harmonics up to and including the n-th harmonic, but no higher. For example, $\cos^5(\omega_0 t)$ will generate only odd harmonics up to and including the fifth harmonic.

7.5.5.4 Power of Each Harmonic

The strength of a given harmonic is determined by several terms of the polynomial. For example, if a sixth-order polynomial adequately describes an amplifier's voltage transfer function, the amplitude of the second harmonic will be determined by the second-, fourth-, and sixth-order terms of the polynomial.

7.5.5.5 Power Decreases as *n* Increases

When we operate a system in its weakly nonlinear input range, the amplitude of the n-th harmonic will tend to decrease as n increases. For example, the third harmonic will be smaller than the second harmonic, the fourth harmonic will be smaller than the third harmonic, and so on.

However, if we increase the input signal power to the point where we're operating in the system's *grossly nonlinear* range, then none of these approximations hold. The polynomial approximation is valid only when the input voltage is small.

Some devices, such as mixers and frequency doublers, are built to be purposefully nonlinear. The designers enhance particular nonlinearities and suppress the linear responses. The decreasing amplitude with increasing n rule does not apply to these devices.

7.6 | TWO-TONE ANALYSIS

We have examined the behavior of nonlinear systems when only one sine wave is present at the system's input. This input condition does not reflect real-world conditions. We now analyze nonlinear systems when they are driven by multiple signals.

In practice, our receiver will see multiple signals existing at different frequencies and power levels. Proceeding directly to this general case is difficult, but there is insight available in the analysis of a nonlinear device when we apply two sine waves of equal power levels to its input. We'll use the same analysis technique we used to analyze the amplifier under single-tone input conditions. We'll examine each term of the approximating polynomial [equation (7.6)] under dual-tone input conditions and then draw some general conclusions from the data. The input signal will be

$$\begin{aligned} V_{in}(t) &= A_1 \cos(\omega_1 t) + A_2 \cos(\omega_2 t) \\ &= A_1 \cos(2\pi f_1 t) + A_2 \cos(2\pi f_2 t) \end{aligned} \tag{7.19}$$

where

$\omega_1 = 2\pi f_1$

$\omega_2 = 2\pi f_2$

f_1 = frequency of tone #1

f_2 = frequency of tone #2

A_1, A_2 = the amplitude of each cosine wave

7.6.1 The k_0 Term

This term represents the DC present at the output of the amplifier.

7.6.2 The $k_1 V_{in}^1$ Term

This term again represents the power gain of the amplifier. Large values of k_1 equate to large values of power gain. For the dual-tone input described by equation (7.19), the output of the amplifier due to the $k_1 V_{in}^1$ term is

$$V_{out,1}(t) = k_1[A_1 \cos(\omega_1 t) + A_2 \cos(\omega_2 t)] \tag{7.20}$$

The linear term doesn't change or distort the input signal. It simply applies power gain.

7.6.3 The $k_2 V_{in}^2$ Term

Applying the two-tone input signal to a nonlinear device produces an output voltage of

$$\begin{aligned} V_{out,2}(t) &= k_2[A_1 \cos(\omega_1 t) + A_2 \cos(\omega_2 t)]^2 \\ &= k_2\left\{\frac{A_1^2 + A_2^2}{2} + \frac{A_1^2 \cos(2\omega_1 t)}{2} + \frac{A_2^2 \cos(2\omega_2 t)}{2}\right. \\ &\qquad \left. + A_1 A_2 \cos[(\omega_1 + \omega_2) t] + A_1 A_2 \cos[(\omega_1 - \omega_2) t]\right\} \end{aligned} \tag{7.21}$$

As in the single-tone case, the second-order term produces energy at 0 hertz and at the second harmonic of each input signal (i.e., at $2\omega_1$ and at $2\omega_2$).

Equation (7.21) contains two terms we haven't seen before: $\cos[(\omega_1 + \omega_2)t]$ and $\cos[(\omega_1 - \omega_2)t]$. These two terms are important components so we define:

- The difference frequency as $\Delta\omega = |\omega_1 - \omega_2|$ or $\Delta f = |f_1 - f_2|$
- The sum frequency as $\Sigma\omega = |\omega_1 + \omega_2|$ or $\Sigma f = |f_1 + f_2|$

The spurious signals generated by the second-order term tend be the strongest nonlinear components.

Figure 7-7 shows the portion of the output spectrum of a nonlinear device that is due only to second-order distortion. The output tones are bunched around DC and $f_1 + f_2$. There are no components at either f_1 or f_2.

We often use nonlinear distortion to our advantage. A mixer is a device built with *enhanced* second-order performance. We use the sum and difference products arising from the $k_2 V_{in}^2$ term and other mathematical operations to perform frequency translation, phase detection, and other useful functions. We also use nonlinearities to demodulate signals.

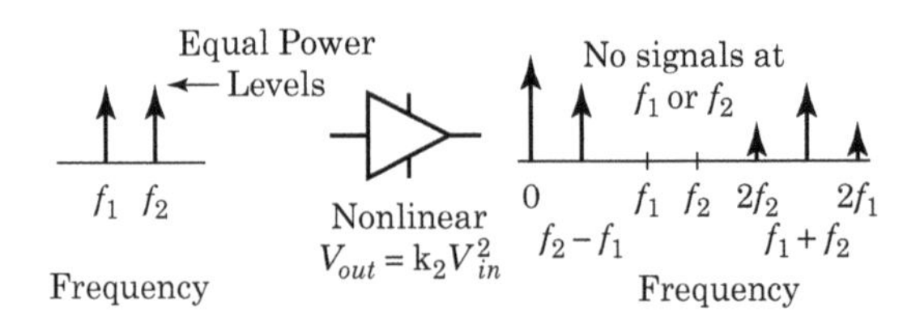

FIGURE 7-7 ■ The output spectrum of a nonlinear device due only to second-order distortion.

EXAMPLE

Second-Order Distortion and the FM Broadcast Band

Find the second-order output frequencies for the following input conditions:

a. $f_1 = 87.9$ MHz, $f_2 = 88.1$ MHz

b. $f_1 = 107.7$ MHz, $f_2 = 107.9$ MHz

c. $f_1 = 87.9$ MHz, $f_2 = 107.9$ MHz

These frequencies represent the center and band edges of the U.S. commercial FM broadcast band.

Solution

Using equation (7.21), we find

a. We have output frequencies at

DC

$2f_1 = 175.8$ MHz

$2f_2 = 176.2$ MHz

$f_2 + f_1 = 88.1 + 87.9 = 176$ MHz

$|f_2 - f_1| = |88.1 - 87.9| = 0.2$ MHz

b. We have output frequencies at

DC

$2f_1 = 215.4$ MHz

$2f_2 = 215.8$ MHz

$f_2 + f_1 = 107.9 + 107.7 = 215.6$ MHz

$|f_2 - f_1| = |107.9 - 107.7| = 0.2$ MHz

c. We have output frequencies at

DC

$2f_1 = 175.8$ MHz

$2f_2 = 215.8$ MHz

$f_2 + f_1 = 107.9 + 87.9 = 195.8$ MHz

$|f_2 - f_1| = |107.9 - 87.9| = 20.0$ MHz

Figure 7-8 shows the output spectrum for these three cases with the appropriate levels marked. For this example, the second-order distortion frequencies possible in this scenario are

DC

0.2 to 20.0 MHz

175.8 to 215.8 MHz

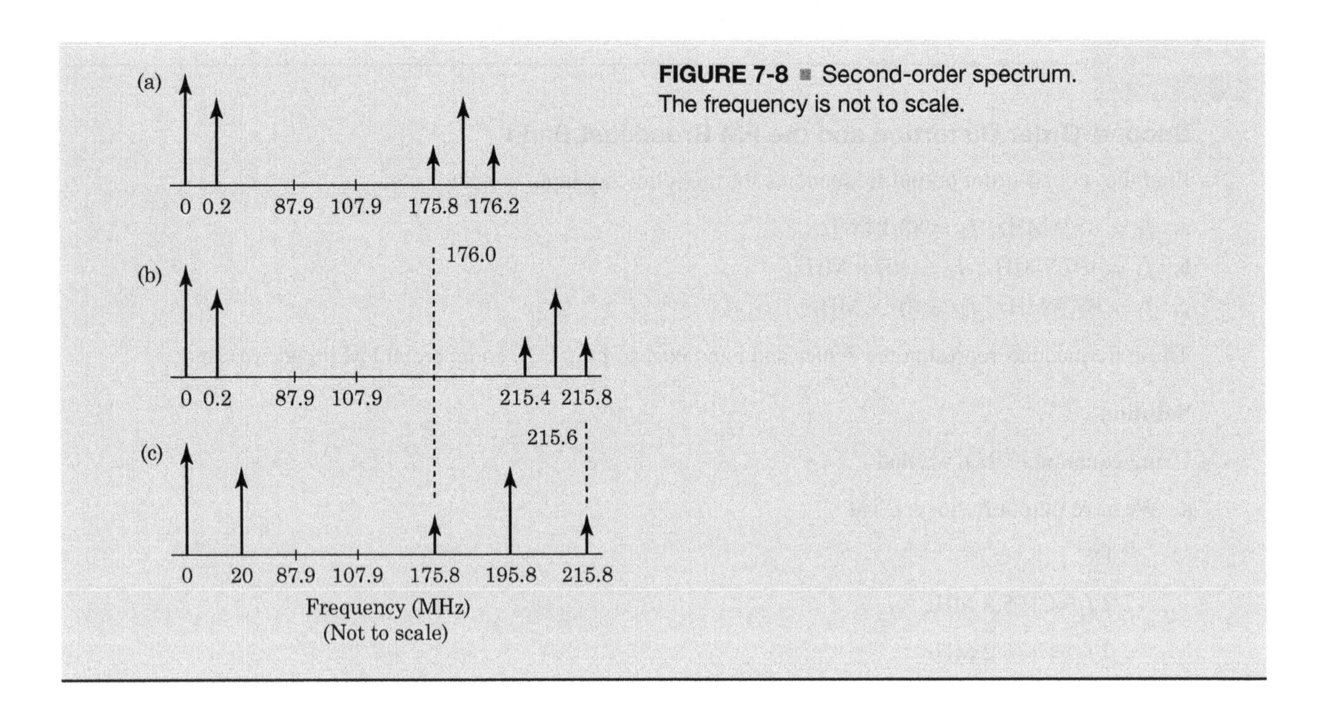

FIGURE 7-8 ■ Second-order spectrum. The frequency is not to scale.

7.6.4 The $k_3 V_{in}^3$ Term

Under two-tone input conditions, the third-order term of the approximating polynomial produces

$$
\begin{aligned}
V_{out,3}(t) &= k_3[A_1\cos(\omega_1 t) + A_2\cos(\omega_2 t)]^3 \\
&= \left(\frac{3A_1^3}{4} + \frac{3A_1A_2^2}{4}\right)\cos(\omega_1 t) + \left(\frac{3A_2^3}{4} + \frac{3A_1^2A_2}{4}\right)\cos(\omega_2 t) \\
&\quad + \frac{A_1^3}{4}\cos(3\omega_1 t) + \frac{A_2^3}{4}\cos(3\omega_2 t) \\
&\quad + \left(\frac{3A_1^2A_2}{4}\right)\cos[(2\omega_1+\omega_2)t] + \left(\frac{3A_1^2A_2}{4}\right)\cos[(2\omega_1-\omega_2)t] \\
&\quad + \left(\frac{3A_1A_2^2}{4}\right)\cos[(2\omega_2+\omega_1)t] + \left(\frac{3A_1A_2^2}{4}\right)\cos[(2\omega_2-\omega_1)t]
\end{aligned}
\tag{7.22}
$$

The third-order term produces energy at the fundamental frequencies, f_1 and f_2, and at three times the input frequencies, $3f_1$ and $3f_2$. We also find several new products at $(2f_1 + f_2)$, $(2f_2 + f_1)$, $(2f_1 - f_2)$ and $(2f_2 - f_1)$. Figure 7-9 shows the portion of the output spectrum of an RF amplifier that is due only to the third-order distortion.

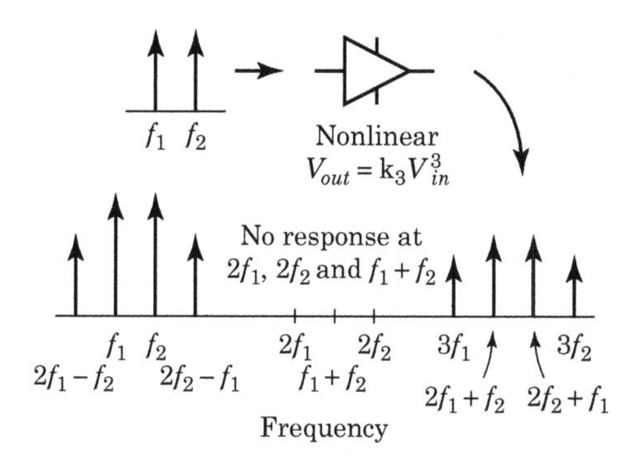

FIGURE 7-9 ■ The output spectrum of a nonlinear device due only to third-order distortion. The signals shown here are due only to the third-order response of the system.

EXAMPLE

Third-Order Distortion and the FM Broadcast Band

Find the third-order output frequencies for the following input conditions:

a. $f_1 = 87.9$ MHz, $f_2 = 88.1$ MHz

b. $f_1 = 107.7$ MHz, $f_2 = 107.9$ MHz

c. $f_1 = 87.9$ MHz, $f_2 = 107.9$ MHz

d. $f_1 = 97.9$ MHz, $f_2 = 98.1$ MHz

Solution

Using equation (7.22), we find

a. We have third-order output frequencies at

$$f_1 = 87.9 \text{ MHz}$$
$$f_2 = 88.1 \text{ MHz}$$
$$3f_1 = 263.7 \text{ MHz}$$
$$3f_2 = 264.3 \text{ MHz}$$
$$2f_2 + f_1 = 2(88.1) + 87.9 = 264.1 \text{ MHz}$$
$$2f_1 + f_2 = 2(87.9) + 88.1 = 263.9 \text{ MHz}$$
$$|2f_2 - f_1| = |2(88.1) - 87.9| = 88.3 \text{ MHz}$$
$$|2f_1 - f_2| = |2(87.9) - 88.1| = 87.7 \text{ MHz}$$

b. We have third-order output frequencies at

$$f_1 = 107.7 \text{ MHz}$$
$$f_2 = 107.9 \text{ MHz}$$
$$3f_1 = 323.1 \text{ MHz}$$
$$3f_2 = 323.7 \text{ MHz}$$
$$2f_2 + f_1 = 2(107.9) + 107.7 = 323.5 \text{ MHz}$$
$$2f_1 + f_2 = 2(107.7) + 107.9 = 323.3 \text{ MHz}$$
$$|2f_2 - f_1| = |2(107.9) - 107.7| = 108.1 \text{ MHz}$$
$$|2f_1 - f_2| = |2(107.7) - 107.9| = 107.5 \text{ MHz}$$

c. We have third-order output frequencies at

$$f_1 = 87.9 \text{ MHz}$$
$$f_2 = 107.9 \text{ MHz}$$
$$3f_1 = 263.7 \text{ MHz}$$
$$3f_2 = 323.7 \text{ MHz}$$
$$2f_2 + f_1 = 2(107.9) + 87.9 = 303.7 \text{ MHz}$$
$$2f_1 + f_2 = 2(87.9) + 107.9 = 283.7 \text{ MHz}$$
$$|2f_2 - f_1| = |2(107.9) - 87.9| = 127.9 \text{ MHz}$$
$$|2f_1 - f_2| = |2(87.9) - 107.9| = 67.9 \text{ MHz}$$

d. We have third-order output frequencies at

$$f_1 = 97.9 \text{ MHz}$$
$$f_2 = 98.1 \text{ MHz}$$
$$3f_1 = 293.7 \text{ MHz}$$
$$3f_2 = 394.3 \text{ MHz}$$
$$2f_2 + f_1 = 2(98.1) + 97.9 = 294.1 \text{ MHz}$$
$$2f_1 + f_2 = 2(97.9) + 98.1 = 293.9 \text{ MHz}$$
$$|2f_2 - f_1| = |2(98.1) - 97.9| = 98.3 \text{ MHz}$$
$$|2f_1 - f_2| = |2(97.9) - 98.1| = 97.7 \text{ MHz}$$

Figure 7-10 shows the output spectrum for these four cases.

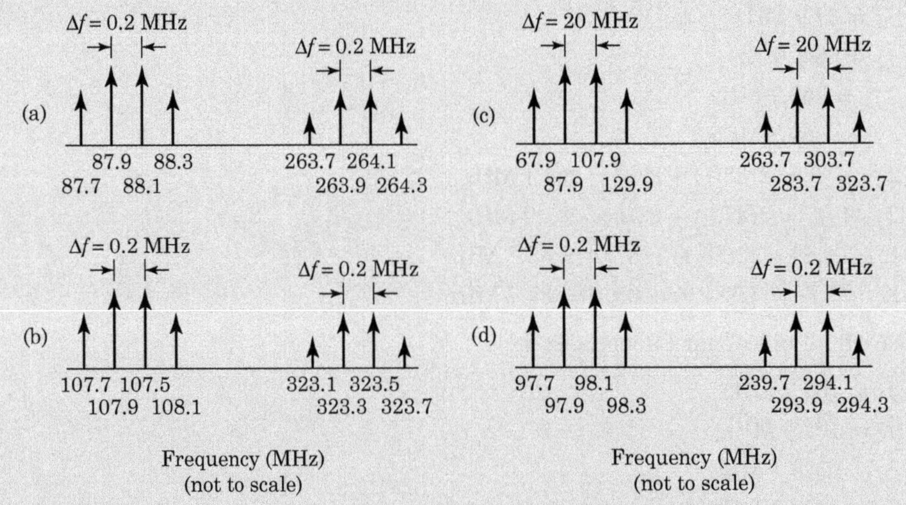

FIGURE 7-10 ■ Third-order spectrum. The frequency is not to scale.

In Figure 7-10(a), Figure 7-10(b) and Figure 7-10(d), the third-order difference products produce signals that lie over other possible FM stations. In Figure 7-10(d), for example, the two third-order products at 97.7 MHz and 98.3 MHz both lie directly on top of valid FM broadcast frequencies.

If we tried to receive a weak station at 97.7 MHz, the two strong signals at 97.9 and 98.1 MHz could combine to mask the 97.7 MHz.

Third-order distortion is a fundamental limitation on the linearity of most systems.

Let us briefly discuss the behavior of the higher-order terms under two-tone sinusoidal drive. We will dispense with giving the exact output equations and satisfy ourselves with the resulting frequencies.

7.6.4.1 The $k_4 V_{in}^4$ Term

Under two-tone sinusoidal drive, the fourth-order term produces signals at

- DC (or 0 Hz)
- $2f_1, 2f_2$
- $4f_1, 4f_2$
- $f_1 \pm f_2$
- $2f_1 \pm 2f_2$
- $3f_1 \pm f_2, 3f_2 \pm f_1$

The fourth-order term produces energy at DC, the second harmonics of f_1 and f_2, and the fourth harmonics of f_1 and f_2. We also observe distortion products at several sum and difference combinations.

7.6.4.2 The $k_5 V_{in}^5$ Term

The fifth-order term of the approximating polynomial produces signals at

- f_1, f_2
- $2f_1 \pm f_2, 2f_2 \pm f_1$
- $3f_1, 3f_2$
- $3f_1 \pm 2f_2, 3f_2 \pm 2f_1$
- $4f_1 \pm f_2, 4f_2 \pm f_1$
- $5f_1, 5f_2$

The fifth-order term produces energy at f_1 and f_2, the third and fifth harmonics of f_1 and f_2, and several sum and difference combinations.

7.6.5 Observations

Figure 7-11 shows the output spectrum of an amplifier when we apply two sine waves at slightly different frequencies, f_1 and f_2.

Analysis of the previous data and Figure 7-11 reveals the following.

7.6.5.1 Neighborhoods

The tones generated by the nonlinear process tend to bunch together in the general neighborhood of the harmonics of the two input signals. When the input signal frequencies are

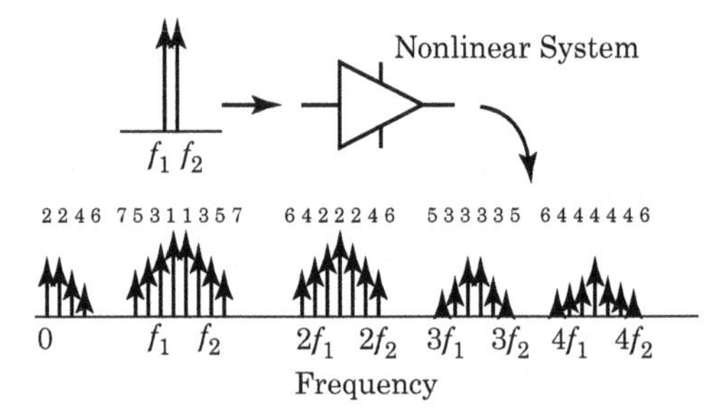

FIGURE 7-11 ■ The output spectrum of a nonlinear device when the input is two equal-amplitude sine waves of slightly different frequencies. The magnitudes and frequencies of the nonlinear output signals are not to scale. The number associated with each spectral element is the order of the lowest term of the approximating polynomial that will generate the element. For example, if a particular spectral element can be generated by the fifth, seventh, ninth, etc. term of the approximating polynomial, we label the element as 5.

f_1 and f_2, and $f_1 \approx f_2$, then the spurious output signals will cluster in the general area of

$$n \cdot \frac{f_1 + f_2}{2} \text{ where } n = 0, 1, 2, 3, \ldots \tag{7.23}$$

7.6.5.2 Separation

The spurious signals tend to cluster in the neighborhoods given by equation (7.23). If we examine only the signals in a particular neighborhood, we find the signals are separated by a frequency Δf where

$$\Delta f = |f_1 - f_2| \tag{7.24}$$

7.6.5.3 Power Decreases as *n* Increases

The power of a particular spurious signal is inversely proportional to its order. For example, a fifth-order spurious signal will likely be stronger than a seventh-order spurious signal.

7.6.5.4 Spurious Frequencies

In equation form, the spurious output signal frequencies are

$$f_{out} = |\pm nf_1 \pm mf_2| \text{ where} \begin{cases} n = 0, 1, 2, 3, \ldots \\ m = 0, 1, 2, 3, \ldots \end{cases} \tag{7.25}$$

where the quantity $(m + n)$ is the *order of the spurious product.*

7.7 DISTORTION SUMMARY

A nonlinear device will generate distortion products whose frequencies are derived from the input frequencies. We modeled the transfer function of a nonlinear device as a polynomial and then analyzed the terms of the polynomial under sinusoidal drive. The approximating polynomial holds only for moderately small power levels. As the input power increases, deviations between the device's actual transfer function and our polynomial approximation increase, and eventually our model falls apart completely.

We limit our analysis to weakly nonlinear systems, that is, those systems driven by just enough power to make the nonlinear aspects of the system noticeable but not so much power that the polynomial model deviates significantly from the true transfer function. Because our model degenerates so quickly with increasing power level, we will make use of two more approximations involving the transfer function of a nonlinear device.

7.7.1 Small Signal Approximation I

For very small input signals (and ignoring the 0 hertz component), we can completely ignore the nonlinear aspects of most devices and can simplify equation (7.6) to

$$V_{out}(t) = k_1 V_{in}(t) \text{ for } V_{in}(t) \approx 0 \tag{7.26}$$

This is our first small signal approximation.

7.7.2 Small Signal Approximation II

When the powers of the input signals are still small yet large enough to generate noticeable distortion, we make a second small signal approximation. We model the transfer function as a third-order polynomial with no DC term (i.e., $k_0 = 0$):

$$V_{out}(t) = k_1 V_{in}(t) + k_2 V_{in}^2(t) + k_3 V_{in}^3(t) \tag{7.27}$$

This polynomial describes a weakly nonlinear device. The input power is still small, and the device does exhibit slight nonlinearity. Equation (7.27) generates only second- and third-order distortion. We ignore higher-order distortion products because if we're applying a large enough input signal to generate higher-order distortion our model is invalid.

7.7.3 Summary

Table 7-1 summarizes the second- and third-order distortion components for single-tone and double-tone input signals.

7.7.4 Two Types of Distortion

From this point, we'll primarily discuss only second-order and third-order distortion. The behavior of the second-order spurious signals is representative of all the even-order spurious signals, while third-order spurious signals are representative of all the odd-order spurious signals. The power of the second- and third-order spurious signals will be the largest of the unwanted products from a nonlinear device.

Our ability to filter away the various nonlinear distortion products has much to do with their frequencies with respect to the desired signals. In some cases, we can design our systems to eliminate second-order distortion. We can then ignore the stronger second-order distortion and concentrate our efforts on reducing the weaker third-order distortion.

First, we'll examine second-order distortion in detail. We'll consider both the frequencies and power levels with respect to the desired signal. We'll also examine ways of characterizing a nonlinear device in the real world. Then we'll repeat this process for the third-order distortion products.

TABLE 7-1 ■ Second- and third-order distortion generated by a nonlinear device under sinusoidal drive conditions.

Signal	2nd-Order Outputs	3rd-Order Outputs
$V_{in}(t) = A\cos(\omega t)$	$\frac{A^2}{2} + \frac{A^2}{2}\cos(2\omega t)$	$\frac{3A^3}{4}\cos(\omega t) + \frac{A^3}{4}\cos(3\omega t)$
$V_{in}(t) = A_1\cos(\omega_1 t) + A_2\cos(\omega_2 t)$	$\frac{A_1^2 + A_2^2}{2}$	$\left(\frac{3A_1^3}{4} + \frac{3A_1A_2^2}{4}\right)\cos(\omega_1 t)$
	$+\frac{A_1^2}{2}\cos(2\omega_1 t)$	$+\left(\frac{3A_2^3}{4} + \frac{3A_1^2A_2}{4}\right)\cos(\omega_2 t)$
	$+\frac{A_2^2}{2}\cos(2\omega_2 t)$	$+\frac{A_1^3}{4}\cos(3\omega_1 t) + \frac{A_2^3}{4}\cos(3\omega_2 t)$
	$+A_1A_2\cos[(\omega_1 + \omega_2)t]$	$+\left(\frac{3A_1^2A_2}{4}\right)\cos[(2\omega_1 + \omega_2)t]$
	$+A_1A_2\cos[(\omega_1 - \omega_2)t]$	$+\left(\frac{3A_1^2A_2}{4}\right)\cos[(2\omega_1 - \omega_2)t]$
		$+\left(\frac{3A_1A_2^2}{4}\right)\cos[(2\omega_2 + \omega_1)t]$
		$+\left(\frac{3A_1A_2^2}{4}\right)\cos[(2\omega_2 - \omega_1)t]$

7.8 | PRESELECTION

Figure 7-12 shows the receiving system model we will analyze. We design receiving systems to process a specific band of frequencies. The band-pass filter (BPF) between the antenna and the first amplifier is wide enough to accept the entire frequency band of interest. Allowing signals that exist outside of the frequency range of interest offers no benefit, and these unwanted signals can seriously degrade the receiver's performance.

For example, suppose you were listening to the FM radio in your car, and you passed another driver who was using his cellular telephone. The car with the cellular phone is probably much closer to you than the FM transmitter, so the undesired cellular telephone signal will be a lot stronger than the desired FM radio station. The band-pass filter between the antenna and first amplifier attenuates the unwanted signal (the cellular telephone frequency) while passing the desired signal (the FM radio station frequency). Without this filter, any large, undesired signal could force our receiver into its nonlinear operating point—producing distortion and perhaps completely masking the signal you want to receive.

We call the initial limiting of frequency *preselection*, and the first filter is a *preselection filter*.

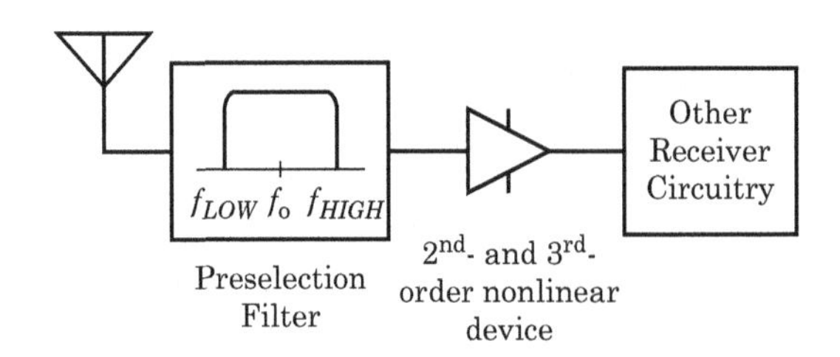

FIGURE 7-12 ■ A simple receiving system with a preselection filter placed between the antenna and the first nonlinear element.

7.9 | SECOND-ORDER DISTORTION

We'll examine second-order effects and how they limit us. We'll start by applying a single tone to the receiver's input and then progress to applying two tones to the system simultaneously.

7.9.1 Preselection and Second-Order Effects

Second-order distortion will generate the largest spurious signals in our system. Anything we can do to minimize the nonlinear second-order effects will go a long way toward helping us achieve a spurious-free system.

7.9.1.1 One Input Signal

For a single-input signal whose frequency is f_0, equation (7.13) tells us that second-order effects will generate distortion only at DC and at $2f_0$. For example, Figure 7-13 shows a wideband antenna connected to an amplifier followed by a receiver. Assume the receiver is tuned to 872.700 MHz.

The antenna will absorb the signal at 872.700 MHz in addition to signals above and below that frequency. The question we ask ourselves is, "Given that I observe a signal at 872.700 MHz at the output of the amplifier, how can I arrange things to be sure that the signal wasn't generated inside the amplifier?"

We will observe a second-order distortion component of frequency f_0 at the output of the amplifier of Figure 7-13 when we apply a signal at $f_0/2$ to the input. In our cellular telephone example, the $f_0/2$ frequency is 436.350 MHz. The single-tone case forces us to place a high-pass filter (HPF) between the antenna and the amplifier. The HPF must supply significant attenuation at $f_0/2$ or 436.350 MHz (see Figure 7-14).

In the more general case, if our receiver tunes from some lower frequency f_{LOW} to some upper frequency f_{HIGH}, then the range of single-tone input signals that will generate in-band distortion due to second-order effects is

$$\frac{f_{LOW}}{2} to \frac{f_{HIGH}}{2} \tag{7.28}$$

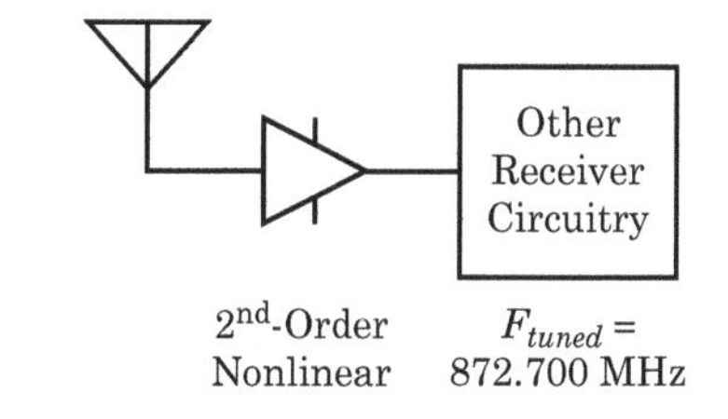

FIGURE 7-13 ■ A wideband antenna connected directly to a nonlinear amplifier without a preselection filter.

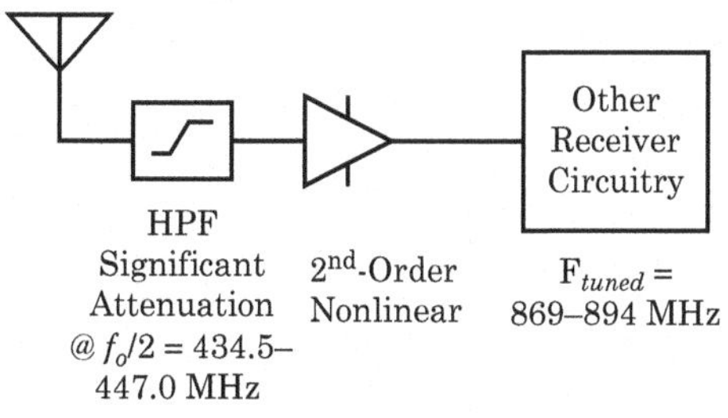

Other Receiver Circuitry
HPF Significant Attenuation @ $f_o/2$ = 434.5–447.0 MHz
2nd-Order Nonlinear
F_{tuned} = 869–894 MHz

FIGURE 7-14 ■ A wideband antenna connected to a nonlinear amplifier through a high-pass preselection filter.

For example, cellular telephone receivers must be sensitive to 869 to 894 MHz. The troublesome $f_0/2$ frequencies are 434.5 to 447.0 MHz. To guard against single-tone, second-order nonlinearities, the HPF of our system in Figure 7-14 must provide high attenuation throughout this range.

7.9.1.2 Two Input Signals

Equation (7.21) describes the second-order outputs for two input signals. If we allow two or more out-of-band signals to enter the receiver, we must consider the sum and difference frequencies in addition to the second harmonic frequencies. Any two signals whose sum or difference frequencies fall within the tuning range of the receiver can cause nonlinear interference. In equation form, any two signals that satisfy

$$f_{tuned} = |f_1 \pm f_2| \tag{7.29}$$

can cause in-band second-order distortion. Both signals must be present in the amplifier at the same time, and they both must be of high power levels to cause distortion.

EXAMPLE

Second-Order Distortion in the Cellular Telephone Band

Our cellular telephone receiver is tuned to 872.700 MHz. What combination of signals below can cause harmful second-order distortion?

a. 861.000 and 11.700 MHz
b. 240.000 MHz and 1113.700 MHz
c. 214.7 and 657.000 MHz

Solution

Using equation (7.21), we find

a. The sum of 861.000 and 11.700 MHz is 872.700 MHz. This is a problematic combination.
b. Subtracting 240.000 from 1113.700 produces 872.7 MHz—problematic.
c. Adding 214.700 and 657.000 produces 871.700 MHz. This combination isn't exactly at the tuned frequency, but it is close.

7.9.1.3 Problematic Frequencies

The system shown in Figure 7-15 consists of a wideband antenna followed by a preselection filter. The output of the filter goes into an amplifier with a second-order characteristic. Finally, we feed the amplifier's output into a receiver with a tuning range of f_{LOW} to f_{HIGH}. The center frequency is f_0.

We want to derive the required filter characteristics to ensure that second-order distortion will not be an issue in this design. The design philosophy is as follows:

- The preselection filter must pass the signals between f_{LOW} and f_{HIGH}.

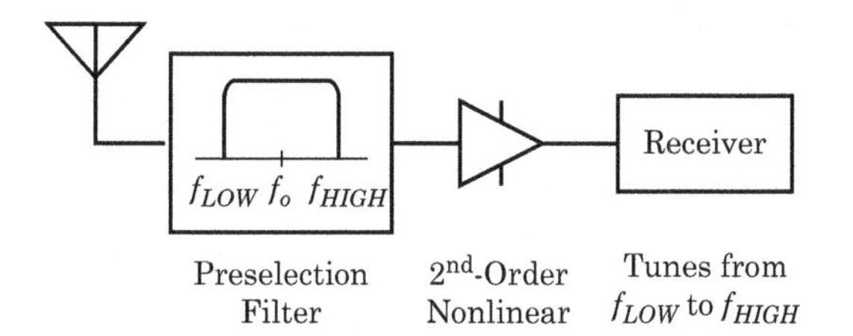

FIGURE 7-15 ■ The analysis of second-order distortion frequencies and the required properties of the preselection filter.

- The filter must suppress signals from $f_{LOW}/2$ to $f_{HIGH}/2$ so their second harmonics won't fall in the receiver's passband. This implies we need at least a high-pass for our preselection filter.
- We must also keep the highest frequency f_{HIGH} less than twice the lowest frequency f_{LOW}. If we don't force this requirement, we could apply a signal at f_{LOW}, and its second harmonic would fall in the receiver passband. The receiver cannot determine if the signal present at $2f_{LOW}$ is from the antenna or if it has risen from second-order distortion in the amplifier.

  A system whose highest frequency less than twice its lowest frequency is a *suboctave* system.
- We must keep signals whose frequency sums lie between f_{LOW} and f_{HIGH} from combining in the amplifier. For example, a signal at $0.32f_0$ will combine with a signal at $0.68f_0$ to produce a signal at f_0.

  Close examination of equation (7.21) reveals that we must suppress only one of the offending carriers. For example, if we apply $f_0/100$ and $99f_0/100$ to the system, we may attenuate only the signal at $f_0/100$ to keep second-order distortion from forming. Again, a high-pass filter will fix this problem.
- We must keep signals whose frequency differences lie between f_{LOW} and f_{HIGH} out of the amplifier. A signal at $3f_0$ will combine with a signal at $2f_0$ to produce a signal at f_0. We need a low-pass filter to fix this problem.

7.9.2 The Second-Order Solution

Since we don't have any control over the signals present at the output of the antenna, we must examine distortion almost from a statistical perspective. What are the most likely events that will cause harmful second-order distortion? We know a single tone at half the tuned frequency will cause problems. This is the most likely event so we must install at least a high-pass filter between the antenna and the first amplifier.

To create in-band interference, two tones must be large in amplitude, and their frequency relationship must allow them to combine through some second-order process to create a signal at the receiver's tuned frequency. Combating this type of distortion requires both a high-pass and a low-pass preselection filter. The sum and difference problem also forces us to keep the bandwidth of our system less than an octave. If we are forced to build a multioctave system, we cannot adequately filter the input to ensure that in-band, nonlinear components are not formed. Figure 7-16 shows a suboctave and a multioctave receiving system.

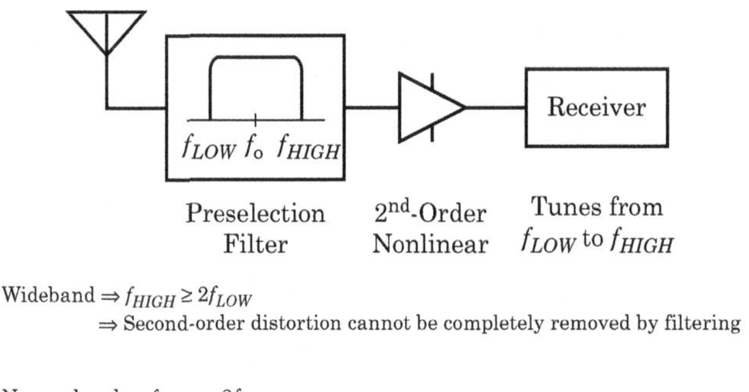

FIGURE 7-16 ■ A suboctave and a multioctave receiving system. Second-order distortion is a problem in systems that must process more than an octave of bandwidth at some point in the system.

7.10 THIRD-ORDER DISTORTION

Second-order distortion produces the strongest intermodulation products. In some situations, we can completely remove second-order effects by the intelligent use of filters. There is no effective way to remove third-order intermodulation products from our systems, making third-order distortion a fundamental limit on the linearity of many common systems.

7.10.1 Preselection and Third-Order Effects

To diminish second-order distortion, we placed a preselection band-pass filter just after our antenna and before the first nonlinear device. We made this filter as narrow as possible to pass only signals we want to process while rejecting all other signals.

7.10.1.1 One Input Signal

Equation (7.15) describes the output of a third-order nonlinearity under sinusoidal drive. If the input frequency is f_1, the third-order output frequencies are f_1 and $3f_1$.

The third-order nonlinearity generates a term at the fundamental frequency f_1. The fundamental component due to the nonlinearity will combine in some way with the fundamental component from the linear term of the nonlinear device to produce the fundamental output signal. Some combination of the undistorted input signal and the distortion makes up the output signal present at f_1.

Real-world effects can change the relative phases between the undistorted signal (at f_1) and the third-order spurious signal (also at f_1). The third-order distortion can increase the apparent power gain of a nonlinear device (if the two signals are in phase) or can decrease the power gain (if the two signals are 180° out of phase).

The spurious signal at $3f_1$ will fall outside of the preselection bandwidth in Figure 7-12 if $f_{HIGH} \leq 3f_{LOW}$. In narrowband systems, when the bandwidth is less than an octave, the $3f_1$ component will always fall outside the preselection bandwidth. However, if the bandwidth so large that we can fit in a 3:1 frequency range, then we can also fit in a 2:1 frequency range and will have to service the more demanding second-order problem.

7.10.1.2 Two Input Signals

Equation (7.22) describes the output voltage of a third-order process when driven by two sinusoids of different amplitudes and frequencies. The third-order term produces energy at the fundamental frequencies (f_1 and f_2) and at three times the input frequencies ($3f_1$ and $3f_2$). We also find signals at $(2f_1 + f_2)$, $(2f_2 + f_1)$, $(2f_1 - f_2)$, and $(2f_2 - f_1)$. We can assume that $f_1 \approx f_2$ since both of these signals must pass through the preselection filter. The third-order components are

- $(2f_1 + f_2)$ and $(2f_2 + f_1)$: When $f_1 \approx f_2$, then

$$2f_1 + f_2 \approx 3f_1 \approx 3f_2 \quad \textit{and} \quad 2f_2 + f_1 \approx 3f_1 \approx 3f_2 \tag{7.30}$$

  These frequencies terms fall somewhere near the third harmonics of the input frequencies. In a sub-octave system, we can filter these products.

- $(2f_1 - f_2)$ and $(2f_2 - f_1)$: When $f_1 \approx f_2$, then

$$2f_1 - f_2 \approx f_1 \approx f_2 \quad \textit{and} \quad 2f_2 - f_1 \approx f_1 \approx f_2 \tag{7.31}$$

  These distortion components will be troublesome. When the input frequencies are approximately equal, these third-order terms are very near the original input frequencies. They are practically impossible to filter away since they fall within the band of operation. We must rely on the raw linearity of the system to avoid this problem.

7.10.1.3 Problematic Frequencies

Figure 7-17 illustrates the problem frequencies. We build a receiving system that is capable of receiving signals of frequencies f_1 through f_4. Our input band-pass filter must be wide enough to pass this frequency range. If we apply two signals at f_2 and f_3, we may observe the output spectrum shown in Figure 7-17. We see the signals at f_2 and f_3 along with the nonlinear distortion at f_1 and f_4.

The follow-on receiving system has no way of knowing whether the signals present at f_1 and f_4 are truly from the antenna or if they are internally generated spurious signals. If there are signals from the antenna at f_1 and f_4, nonlinear distortion could mask them. This effect is a fundamental limitation on linearity.

7.10.2 The Third-Order Solution

The third-order solution is the same solution we developed for the second-order problem—we make the preselection band-pass filter as narrow as possible to limit signals that might combine. However, no preselection filter can prevent the third-order limitation caused by the $2f_1 - f_2$ and $2f_2 - f_1$ components.

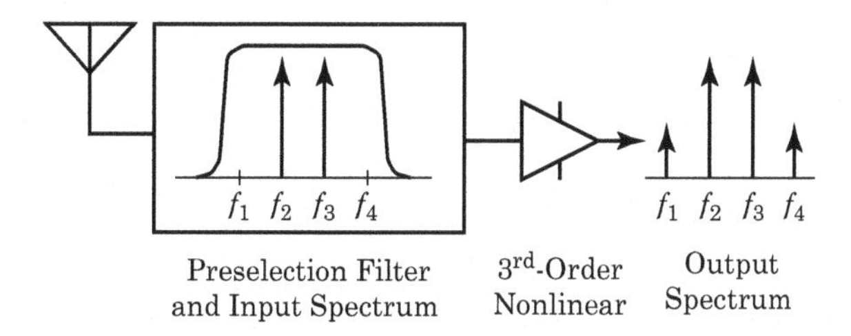

FIGURE 7-17 ■ The third-order problem that limits linearity in most practical receivers. This distortion cannot be removed by filtering.

7.11 NARROWBAND AND WIDEBAND SYSTEMS

We consider a system to be wideband if it contains more than an octave of bandwidth somewhere in its processing chain. A narrowband system never processes more than an octave. We make this discrimination because we do not generate problematic second-order frequencies effects in narrowband systems. Narrowband systems are limited by third-order effects. If possible, we would like to convert wideband problems into narrowband problems.

If we're planning to use a device in narrowband system, we ignore its second-order performance and concentrate on its third-order performance. If we plan to use the same device in a wideband system, we are very interested in its second-order distortion characteristics.

7.11.1 Wideband/Narrowband Examples

A system is narrowband if all of the filters in the processing path are less than an octave wide. A system is wideband if any of the filters in the RF path are more than an octave wide.

Figure 7-18(a) shows an example of a wideband system. This is a 1 to 4 GHz receiver and the preselection filter covers the entire 1–4 GHz range. Since the receiver has a filter whose bandwidth must process more than one octave, this is a wideband system.

Figure 7-18(b) is the same receiver with a slightly different filtering scheme. The receiver tunes over the same frequency range, but it now contains three switched band-pass preselection filters. When the receiver tunes in the 1–1.8 GHz band, we switch in the top filter. When the receiver tunes in the 1.8–2.7 GHz band, we switch in the middle filter.

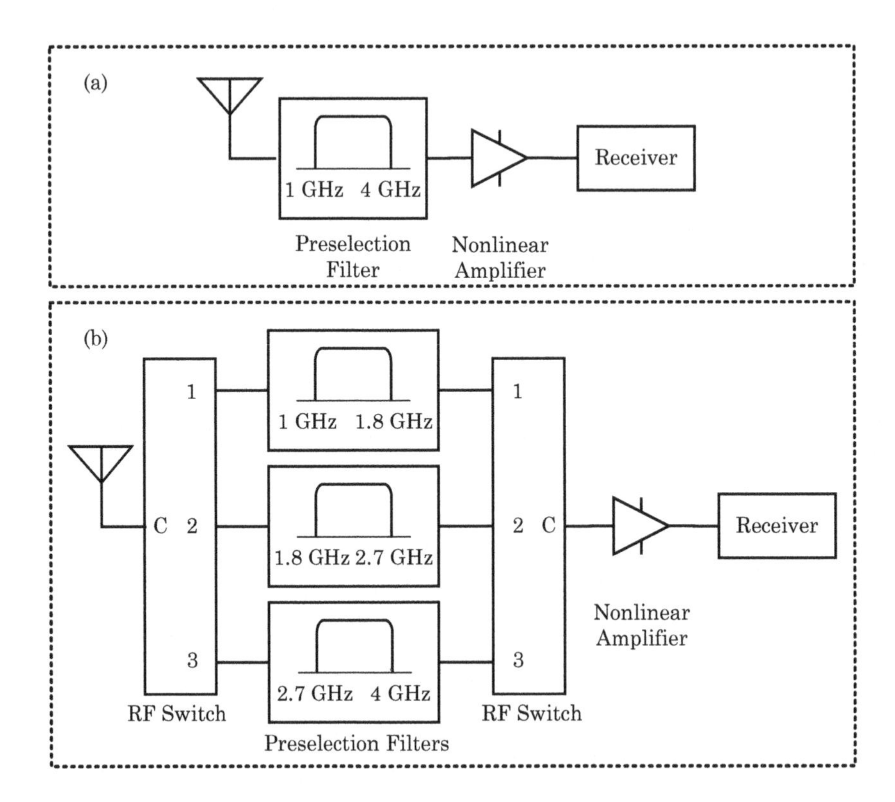

FIGURE 7-18 ■ (a) A wideband receiving system. This system contains a filter that covers more than an octave of bandwidth. (b) The wideband system of (a) has been converted into a narrowband system by the addition of switched filters. None of the filters in the processing chain contains more than an octave of bandwidth.

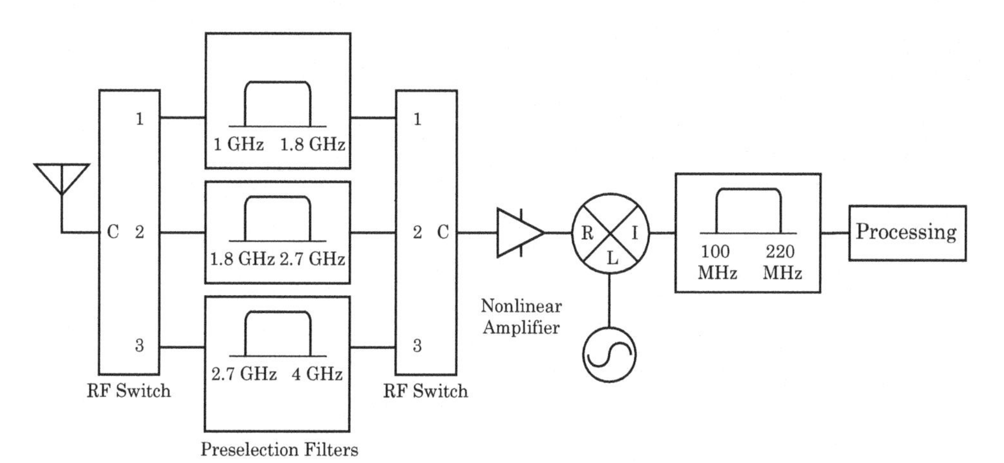

FIGURE 7-19 ■ A wideband receiving system. The preselection structure is narrowband, but the multioctave IF filter defines this system as wideband.

And, finally, when we tune the receiver anywhere in the 2.7–4.0 GHz range, we switch in the bottom filter. Although the receiver still tunes over a 1–4 GHz range, this is now a narrowband system since it never processes more than an octave's worth of bandwidth at one time.

Consider another example. We design a radar-warning receiver to determine if someone is shining a radar on us. The radar signal is 120 MHz wide and might be centered anywhere from 1 to 4 GHz. We will start with the system in Figure 7-18(b), but we'll convert the signal to 160 MHz before processing (Figure 7-19). The intermediate frequency (IF) band-pass filter is centered at 160 MHz, and it has a 120 MHz bandwidth—the filter will pass 100 MHz to 220 MHz.

The 160 MHz IF filter must pass more than an octave of bandwidth, so adding the IF stage converted the receiver back into a wideband system. If we had centered the IF at 700 MHz while keeping the 120 MHz bandwidth, the filter would pass 640 to 760 MHz. No part of this receiver will process more than an octave's worth of frequency, so the system would have remained narrowband.

7.12 | HIGHER-ORDER EFFECTS

We examined second- and third-order distortion in detail because they present fundamental limits on distortion in the cases that interest us. However, real devices exhibit higher-order nonlinearities.

7.12.1 Harmonics

If we apply a signal of frequency f_0 to a nonlinear device, the device will generate harmonics at

$$nf_0 \text{ where } n = 2, 3, 4, \ldots \tag{7.32}$$

Looking at the problem from a receiver perspective, if we are tuned to a signal at frequency f_0, an input signal at

$$f_{out} = \frac{f_0}{n} \text{ where } n = 2, 3, 4, \ldots \tag{7.33}$$

can generate an interfering signal. The high-pass aspect of the preselection filter attenuates these troublesome input signals.

7.12.2 Multiple Input Signals

If we apply multiple input signals to a nonlinear device described by an n-th-order polynomial, the output frequencies are

$$f_{out} = |\pm nf_1 \pm of_2 \pm pf_3 \pm \cdots| \text{ where} \begin{cases} n = 0, 1, 2, \ldots \\ o = 0, 1, 2, \ldots \\ p = 0, 1, 2, \ldots \\ \vdots \end{cases} \tag{7.34}$$

We run into trouble when one of the possible f_{out} frequencies lies on a signal of interest. Clearly, we wish to limit the number of signals a nonlinear device must process because limiting the bandwidth restricts the number of signals that can combine and possibly cause interference.

7.13 SECOND-ORDER INTERCEPT POINT

The intercept concept quantifies the linearity of device in a useful way. We can compare the linearity of different devices and can use the intercept information to calculate the strength of the second-order distortion for a given fundamental input power. Among the questions we answer will be, "Given an input signal power, how much second-order power will a particular nonlinear device generate?" and "What will be the power at each of the second-order frequencies?"

7.13.1 Measuring Nonlinear Devices

To measure the second-order distortion of a nonlinear device, we apply a single tone at frequency f and then measure the output power at f and $2f$. Figure 7-20 shows the input and output spectra.

We connect a signal generator to the device under test (DUT), apply an input, and observe the output on a spectrum analyzer. We must attend to several details:

- The signal generator output is free from second harmonic energy.
- The spectrum analyzer with which we observe the output spectrum does not generate significant second-order distortion.

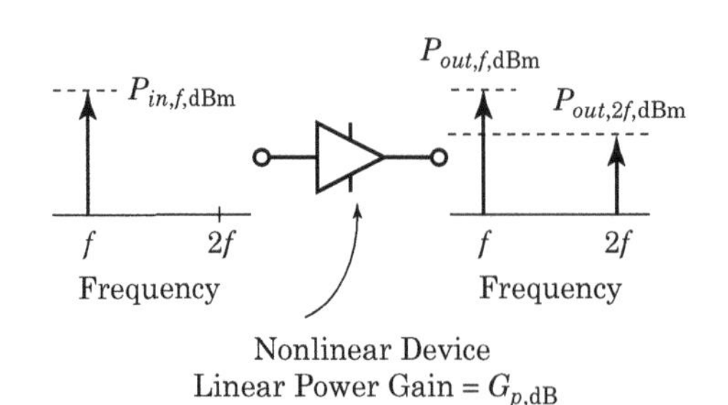

FIGURE 7-20 Measuring the second-order distortion generated by a nonlinear device by observing its input and output spectra.

7.13.2 Definitions

We define the fundamental input power at frequency f to be $P_{in,f,dBm}$, in dBm. The fundamental output power (still at frequency f) is $P_{out,f,dBm}$. We know

$$P_{out,f,dBm} = P_{in,f,dBm} + G_{p,dB} \tag{7.35}$$

where

$G_{p,dB}$ = the power gain of the amplifier in dB.

Let the output second-order distortion power at $2f$ be $P_{out,2f,dBm}$. Although the second-order signal does not exist on the input terminal of the nonlinear device, sometimes it's convenient to describe the equivalent input second-order power as

$$P_{out,2f,dBm} = P_{in,2f,dBm} + G_{p,dB} \tag{7.36}$$

where

$P_{in,2f,dBm}$ is the equivalent second-order input power in dBm.

We measure $P_{out,f,dBm}$ and $P_{out,2f,dBm}$ for different values of $P_{in,f,dBm}$ and graph the measured power in Figure 7-21.

At low power levels, Figure 7-21 shows that the fundamental output power rises with a 1:1 slope with the fundamental input power. The second-order output power rises at a 2:1 slope with increasing input power. We can verify that this behavior fits our second-order model by assuming the device's transfer function is

$$V_{out} = k_2 V_{in}^2 \tag{7.37}$$

and examining the output change for a given input change for small V_{in}.

At high input power, the nonlinear device saturates, and both the fundamental and second-order output powers flatten out because the device simply cannot supply any more signal. Our polynomial approximation is invalid at these high input levels.

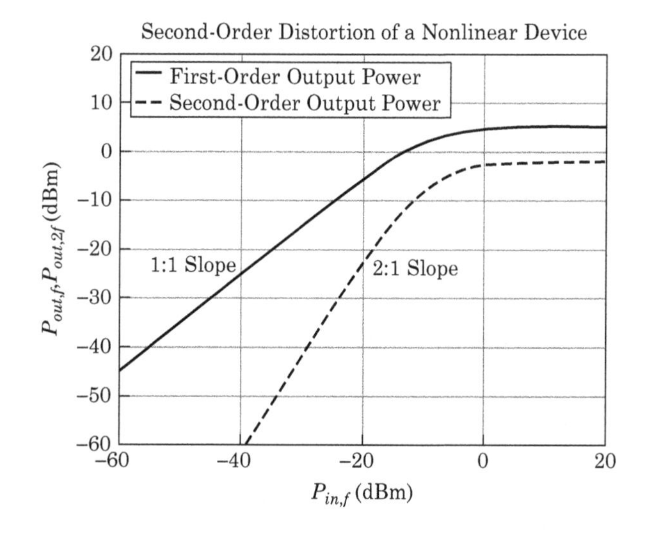

FIGURE 7-21 ■ The measured output power of a nonlinear device, at the fundamental frequency f and at the second-harmonic frequency $2f$.

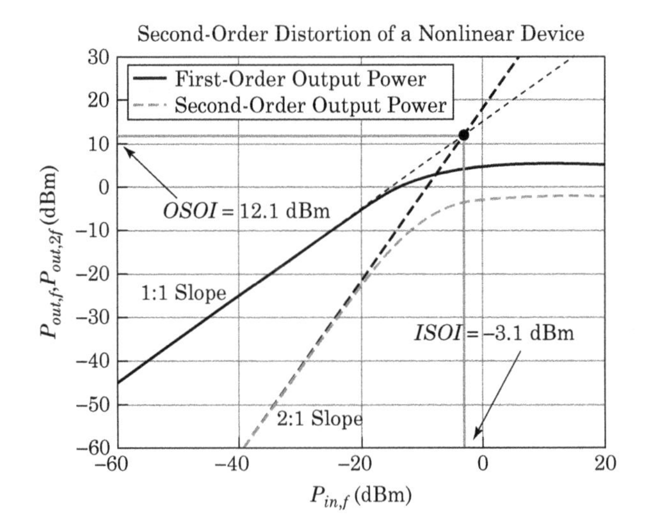

FIGURE 7-22 ▪ The definition of second-order intercept point from the measured output powers at frequencies *f* and 2*f*.

7.13.3 Second-Order Intercept Point

If we extend the fundamental and second-order output power curves as shown in Figure 7-22, the two lines will intersect. The intersection is the second-order intercept (SOI) point of that device. The SOI is a direct measure of the amplifier's second-order performance.

We can read two possible SOI points from Figure 7-22—one for the input and one for the output. The units for input second-order intercept (ISOI) and output second-order intercept (OSOI) are power—we will often use dBm but will occasionally express them in linear terms.

Reading from Figure 7-22, the ISOI of this particular device is –3.1 dBm, whereas the OSOI is +12.1 dBm. We note that

$$OSIO_{dBm} = ISOI_{dBm} + G_{p,dB} \tag{7.38}$$

The SOI point is a mathematical tool that quantifies the nonlinear behavior of a device. Although the SOI point is a power level, we never operate a device at its intercept point. This is a very high power level, and it is well into saturation. For example, the ISOI of the device in Figure 7-22 is −3.1 dBm. If we applied this power level to the input of the amplifier (i.e., $P_{in,f,dBm} = -3.1$ dBm), the fundamental output power level (reading $P_{out,f,dBm}$ from Figure 7-22) is well into the flat part of the curve and is only 4.5 dBm. The output second-order power level ($P_{out,2f,dBm}$) would be −3.4 dBm.

7.13.4 Quantifying Distortion Power

The SOI point relates the fundamental input power level, the fundamental output power level, and the second-order output power level.

The fundamental and second-order lines both pass through the point ($ISOI_{dBm}$, $OSOI_{dBm}$). We know the slope of the fundamental line is unity while the slope of the second-order line is two. The equation for a line with a known slope passing through a known point (x_1, y_1) is

$$y - y_1 = m\,(x - x_1) \tag{7.39}$$

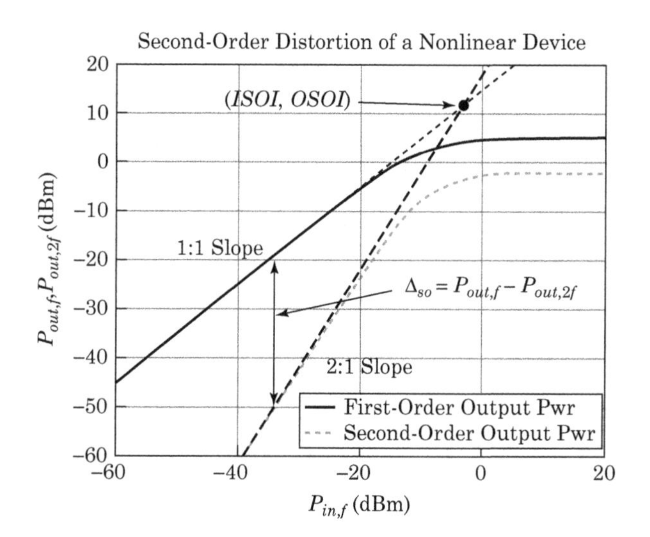

FIGURE 7-23 ■ The geometry describing the relationships between fundamental input power and second-order distortion generated by a nonlinear device.

The equation of the line describing the fundamental input power versus the fundamental output power ($P_{in,f,dBm}$ versus $P_{out,f,dBm}$) is

$$\begin{aligned} P_{out,f,dBm} - OSOI_{dBm} &= 1 \cdot \left(P_{in,f,dBm} - ISOI_{dBm}\right) \\ \Rightarrow P_{out,f,dBm} &= P_{in,f,dBm} - ISOI_{dBm} + OSOI_{dBm} \end{aligned} \tag{7.40}$$

The equation for the line corresponding to the fundamental input power level versus the second-order output power level ($P_{in,f,dBm}$ versus $P_{out,2f,dBm}$) is

$$\begin{aligned} P_{out,2f,dBm} - OSOI_{dBm} &= 2 \cdot \left(P_{in,f,dBm} - ISOI_{dBm}\right) \\ \Rightarrow P_{out,2f,dBm} &= 2P_{in,f,dBm} - 2ISOI_{dBm} + OSOI_{dBm} \end{aligned} \tag{7.41}$$

We define $\Delta_{SO,dB}$ as the difference between the fundamental and second-order output powers (see Figure 7-23) and

$$\begin{aligned} \Delta_{SO,dB} &= P_{out,f,dBm} - P_{out,2f,dBm} \\ &= P_{in,f,dBm} - ISOI_{dBm} + OSOI_{dBm} \\ &\quad - \left(2P_{in,f,dBm} - 2ISOI_{dBm} + OSOI_{dBm}\right) \\ \Delta_{SO,dB} &= ISOI_{dBm} - P_{in,f,dBm} \end{aligned} \tag{7.42}$$

Similarly, we can show

$$\Delta_{SO,dB} = OSIO_{dBm} - P_{out,f,dBm} \tag{7.43}$$

Further manipulation of equations (7.42) and (7.43) reveals

$$P_{in,2f,dBm} = 2P_{in,f,dBm} - ISOI_{dBm} \tag{7.44}$$

and

$$P_{out,2f,dBm} = 2P_{out,f,dBm} - OSOI_{dBm} \tag{7.45}$$

EXAMPLE

SOI and Spurious Power Levels

Figure 7-24 shows an amplifier with a 15 dB power gain, an $ISOI_{dBm}$ of -12 dBm and an input power level of -40 dBm, find

a. The amplifier's $OSOI_{dBm}$

b. The levels of the signals at f and $2f$ at the output of the amplifier

c. The equivalent input second-order power

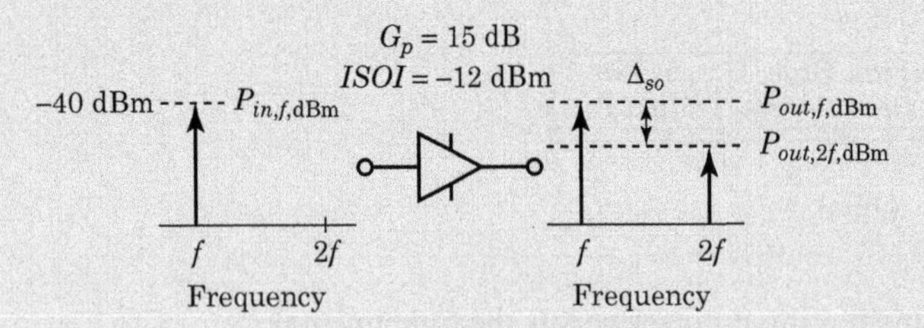

FIGURE 7-24 ■ Find the amplifiers OSOI, the signal powers at the output, and the equivalent second-order input power.

Solution

a. Using equation (7.38), we can write

$$\begin{aligned} OSOI_{dBm} &= ISOI_{dBm} + G_{p,dB} \\ &= -12 + 15 \\ &= +3 \text{ dBm} \end{aligned} \tag{7.46}$$

b. The fundamental output signal power is

$$\begin{aligned} P_{out,f,dBm} &= P_{in,f,dBm} + G_{p,dB} \\ &= -40 + 15 \\ &= -25 \text{ dBm} \end{aligned} \tag{7.47}$$

We use equation (7.42) to find $\Delta_{SO,dB}$:

$$\begin{aligned} \Delta_{SO,dB} &= ISOI_{dBm} - P_{in,f,dBm} \\ &= -12 - (-40) \\ &= 28 \text{ dB} \end{aligned} \tag{7.48}$$

The difference between the fundamental output signal power (at f) and the power of the second harmonic (at $2f$) is $\Delta_{SO} = 28$ dB. The second-order output power is

$$\begin{aligned} P_{out,2f,dBm} &= P_{out,f,dBm} - \Delta_{SO,dB} \\ &= -25 - 28 \\ &= -53 \text{ dBm} \end{aligned} \tag{7.49}$$

c. The equivalent input second-order power is

$$\begin{aligned} P_{in,2f,dBm} &= P_{out,2f,dBm} - G_{p,dB} \\ &= -53 - 15 \\ &= -68 \text{ dBm} \end{aligned} \tag{7.50}$$

The value for $\Delta_{SO,dB}$ is the same on both the input and the output ports of the amplifier.

EXAMPLE

Separating Spurious Signals from Real Signals

Figure 7-25 shows a wideband system covering 4 to 8 GHz and its output spectrum. How can we decide if the signal at 8 GHz is a second-order product or a valid input signal?

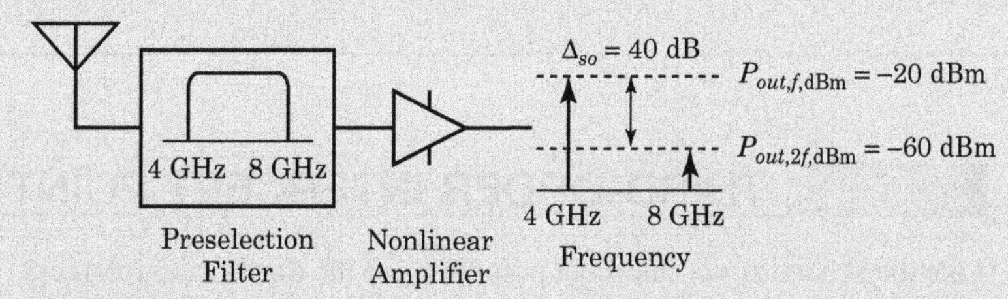

FIGURE 7-25 ■ Identifying signals generated by second-order nonlinear distortion.

Solution

Equation (7.44) and Figure 7-23 both indicate that a 1 dB decrease the fundamental input power will result in a second-order output power drop of 2 dB. Placing a 10 dB attenuator on the input of our system (Figure 7-26) causes any signal presented to the amplifier to drop by 10 dB. If both the 4 and 8 GHz signals are present at the input of the amplifier, then both of these signals will experience a 10 dB decrease in their power. The entire output spectrum would drop by 10 dB if the amplifier is behaving linearly. Part (a) of Figure 7-26 shows the output spectrum if nonlinear processes in the amplifier do not generate the 8 GHz signal. Note that the $\Delta_{SO,dBm}$ has not changed in this case.

If the amplifier is generating the 8 GHz signal, we will observe the output spectrum of Figure 7-26 part (b). The 4 GHz signal will experience a 10 dB power loss before it encounters the amplifier. Equation (7.44) and Figure 7-23 tell us that this 10 dB drop at 4 GHz will cause the 8 GHz signal to drop by 20 dB. Note that the $\Delta_{SO,dBm}$ has changed.

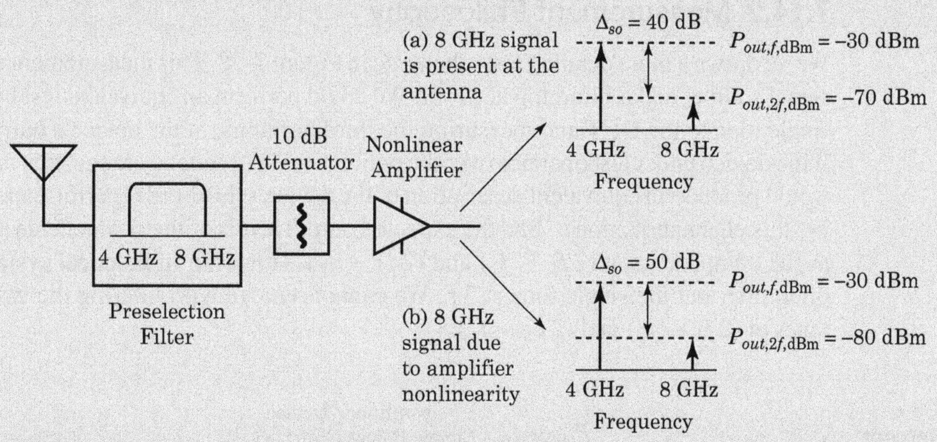

FIGURE 7-26 ■ The effect of placing an attenuator between the antenna and the first nonlinear element of the system shown in Figure 7-25. If the 8 GHz signal is present at the antenna and not due to a second-order device nonlinearity, we obtain the output spectrum shown in (a). Both the 4 and 8 GHz signal powers drop by the value of the attenuator. If the 8 GHz signal is due to a second-order nonlinearity in the amplifier, we will observe the output spectrum shown in part (b). The 4 GHz signal will decrease by the value of the attenuator, whereas the 8 GHz signal power will decrease by twice the value of the attenuator.

This example illustrates a useful method for identifying spurious responses. Attaching an attenuator of n dB on the input of a device will cause a $2n$ dB reduction of an amplifier's second-order distortion components.

There are delinquent cases in which the $2n$ dB rule does not hold. However, anytime you observe a signal drop greater than n dB upon application of n dB attenuator, there is nonlinear behavior occurring in the system.

7.14 THIRD-ORDER INTERCEPT POINT

Like the second-order intercept point, we use the third-order intercept (TOI) point concept to compare the linearity of various devices in a useful way. We also use the concept to calculate the amount of third-order distortion a device is likely to generate under a particular set of input conditions.

7.14.1 Measurement Technique

Figure 7-27 shows the third-order measurement procedure. We radiate the input terminal of a nonlinear device with two tones of equal power. The frequencies are f_1 and f_2. We then measure the output power at f_1 and f_2 and the power of the third-order products at $(2f_1 - f_2)$ and at $(2f_2 - f_1)$. We measure these particular intermodulation products because they are the most troublesome.

Theory tells us that the two third-order tones at $(2f_1 - f_2)$ and $(2f_2 - f_1)$ will exhibit the same power (see equation (7.22) when A_1 equals A_2).

7.14.2 Measurement Philosophy

We've drawn a more detailed test schematic in Figure 7-28. This measurement requires two signal sources and a summing network. We could perform an equivalent test by applying a single tone to the DUT and measuring the third harmonic at the device's output terminal. If the device under test operates over the entire f to $3f$ frequency range, this measurement would produce an equivalent description of the device's third-order performance. We could use this characterization to find the expected output levels of the components at $3f$ as well as the components at $(2f_1 - f_2)$ and $(2f_2 - f_1)$. However, in practical systems, we can often filter out the single tone at $3f$. We cannot remove with filtering the two difference tones at $(2f_1 - f_2)$ and $(2f_2 - f_1)$.

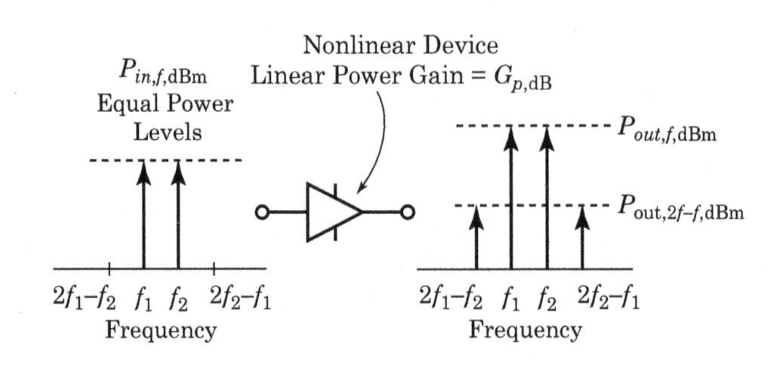

FIGURE 7-27 Third-order intercept point measurement procedure.

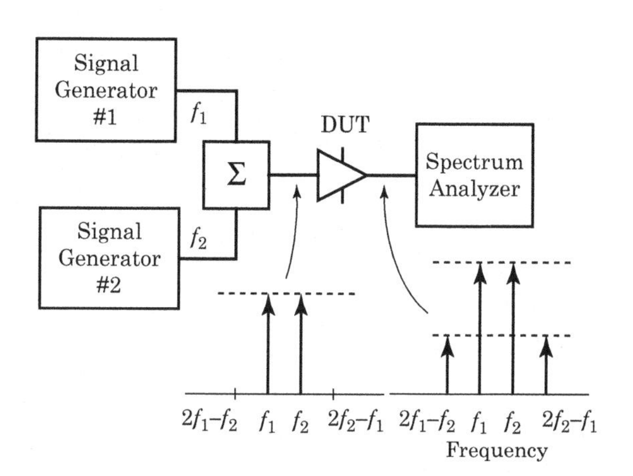

FIGURE 7-28 Detailed test schematic for measuring the third-order intercept point.

7.14.3 Definitions

Referring to Figure 7-27, the fundamental input power of each tone is $P_{in,f,dBm}$. The fundamental output power of each tone is $P_{out,f,dBm}$ and

$$P_{out,f,dBm} = P_{in,f,dBm} + G_{p,dB} \tag{7.51}$$

The amplitudes of each third-order component (at $2f_1 - f_2$ and $2f_2 - f_1$) are equal, so we will label the power of each tone identically as $P_{out,2f-f,dBm}$. As in the second-order case, it is sometimes convenient to reference the third-order tones to the input:

$$P_{out,2f-f,dBm} = P_{in,2f-f,dBm} + G_{p,dB} \tag{7.52}$$

where $P_{in,2f-f,dBm}$ is the equivalent third-order input power. Figure 7-29 shows a graph of $P_{in,f,dBm}$, $P_{out,f,dBm}$, and $P_{out,2f-f,dBm}$ for many different values of $P_{in,f,dBm}$.

Figure 7-29 shows that the fundamental output power, expressed in dBm, rises at a 1:1 slope with increasing input power, also expressed in dBm. The third-order power rises at a 3:1 slope with increasing input power. At high levels of input power, the amplifier saturates. The fundamental and third-order output powers flatten out because the amplifier

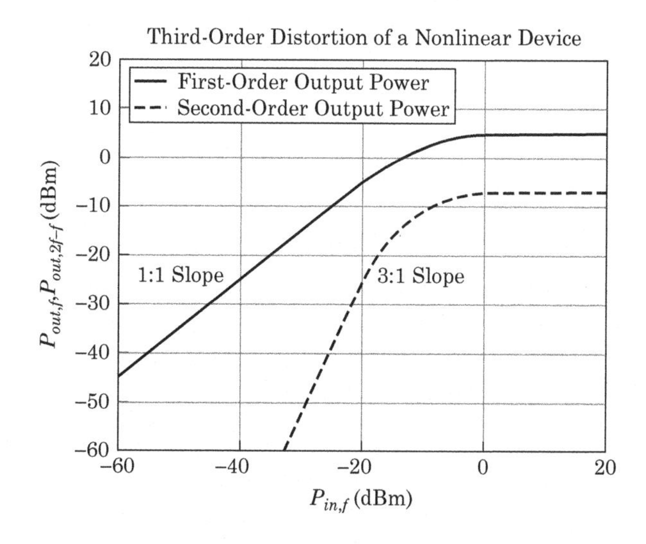

FIGURE 7-29 The measured output power of a nonlinear device at the fundamental frequency f and at the third-order distortion frequencies ($2f_2 - f_1$ or $2f_1 - f_2$).

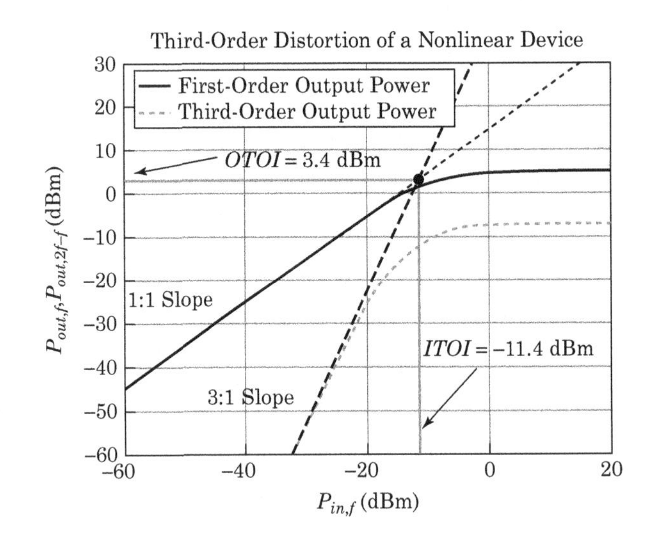

FIGURE 7-30 ■ The geometry used to derive the relationships between fundamental input power and third-order distortion generated by a nonlinear device.

simply cannot supply more output signal. Well before the amplifier saturates, we have moved from the weakly nonlinear region of the amplifier's transfer curve to the strongly nonlinear portion of the curve.

7.14.4 Third-Order Intercept Point

When we extend the fundamental and third-order output power curves of Figure 7-30, the two lines intersect at the TOI point. Like the second-order case, the device possesses both input and output third-order intercept points (denoted *ITOI* and *OTOI*, respectively). The TOI point has units of power.

Figure 7-30 shows the characteristics of a device whose *ITOI* is −11.4 dBm and whose *OTOI* is 3.4 dBm. As in the second-order case, the *ITOI* and *OTOI* of a device are related by

$$OTOI_{dBm} = ITOI_{dBm} + G_{p,dB} \tag{7.53}$$

We never operate a device with an input power of *ITOI* dBm. This is much too high for proper operation. The TOI point is only a mathematical tool that describes the nonlinear behavior of a device. For example, if we apply a fundamental power level of $ITOI = -11.4$ dBm (i.e. $P_{in,f,dBm} = -11.4$ dBm) to the device described by Figure 7-30, the fundamental output power ($P_{out,f,dBm}$) level will be only 1.4 dBm. The output third-order power level ($P_{out,2f-f,dBm}$) would be -11.3 dBm. The device would be in its strongly nonlinear region at this point.

7.14.5 Quantifying Distortion Power

The third-order intercept point sets the relationship between the fundamental input power and the third-order output power. Figure 7-30 shows that the fundamental and third-order lines pass through the point ($ITOI_{dBm}$, $OTOI_{dBm}$). The slope of the fundamental line is unity, while the slope of the third-order line is 3 (recall the slope was 2 for the second-order case). We can write the equation of the fundamental $P_{out,f,dBm}$ line as

$$\begin{aligned} P_{out,f,dBm} - OTOI_{dBm} &= 1 \cdot \left(P_{in,f,dBm} - ITOI_{dBm}\right) \\ &\Rightarrow P_{out,f,dBm} = P_{in,f,dBm} - ITOI_{dBm} + OTOI_{dBm} \end{aligned} \tag{7.54}$$

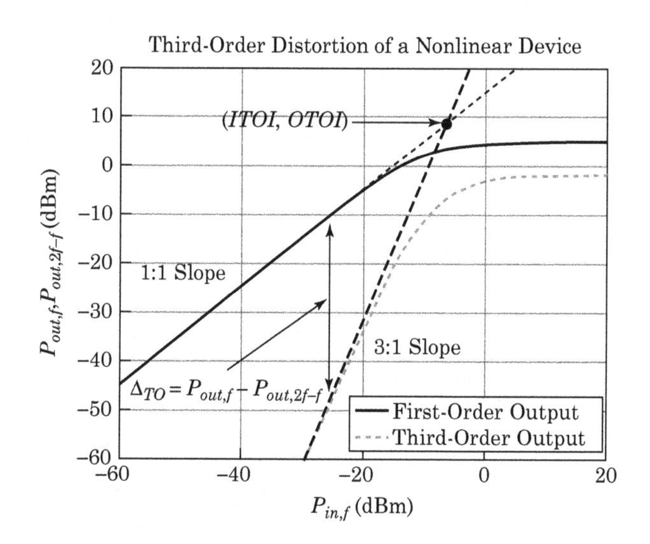

FIGURE 7-31 ■ The definition of third-order intercept point from measured output powers at f and $2f_2 - f_1$ or $2f_1 - f_2$.

The equation for the third-order $P_{out,2f-f,dBm}$ line is

$$P_{out,2f-f,dBm} - OTOI_{dBm} = 3P_{in,f,dBm} - 3ITOI_{dBm}$$
$$\Rightarrow P_{out,2f-f,dBm} = 3P_{in,f,dBm} - 3ITOI_{dBm} + OTOI_{dBm} \quad (7.55)$$

In Figure 7-31, we define $\Delta_{TO,dB}$ as the difference between the fundamental output power and the third-order output power:

$$\begin{aligned}\Delta_{TO,dB} &= P_{out,f,dBm} - P_{out,2f-f,dBm} \\ &= P_{in,f,dBm} - ITOI_{dBm} + OTOI_{dBm} \\ \Delta_{TO,dB} &= 2\left(ITOI_{dBm} - P_{in,f,dBm}\right)\end{aligned} \quad (7.56)$$

We can also show

$$\Delta_{TO,dB} = 2\left(OTOI_{dBm} - P_{out,f,dBm}\right) \quad (7.57)$$

Combining these two equations produces

$$P_{in,2f-f,dBm} = 3P_{in,f,dBm} - 2ITOI_{dBm} \quad (7.58)$$

and

$$P_{out,2f-f,dBm} = 3P_{out,f,dBm} - 2OTOI_{dBm} \quad (7.59)$$

EXAMPLE

Measured Third-Order Intercept

A commercial RF amplifier has 16 dB of power gain, an $OTOI_{dBm}$ of +12 dBm, and an input power level of −30 dBm. Find:

a. The amplifier's *ITOI*

b. The levels of the signals at f_1 and f_2 along with the signal levels at $2f_1 - f_2$ and $2f_2 - f_1$ at the output of the amplifier

c. The equivalent third-order input power

Solution

a. We can rewrite equation (7.53) as

$$\begin{aligned} ITOI_{dBm} &= OTOI_{dBm} - G_{p,dB} \\ &= 12 - 16 \\ &= -4 \text{ dBm} \end{aligned} \tag{7.60}$$

b. The fundamental output signal power is

$$\begin{aligned} P_{out,f,dBm} &= P_{in,f,dBm} + G_{p,dB} \\ &= -30 + 16 \\ &= -14 \text{ dBm} \end{aligned} \tag{7.61}$$

We use equation (7.56) to find $\Delta_{TO,dB}$:

$$\begin{aligned} \Delta_{TO,dB} &= 2\left(ITOI_{dBm} - P_{in,f,dBm}\right) \\ &= 2[-4 - (-30)] \\ &= 52 \text{ dB} \end{aligned} \tag{7.62}$$

The difference between the fundamental output signal power (at f) and the power of the third-order products (at $2f_1 - f_2$ and $2f_{2-f_1}$) is 52 dB. The third-order output power is

$$\begin{aligned} P_{out,2f-f,dBm} &= P_{out,f,dBm} - \Delta_{TO,dB} \\ &= -14 - 52 \\ &= -66 \text{ dBm} \end{aligned} \tag{7.63}$$

c. The equivalent input third-order power is

$$\begin{aligned} P_{in,2f-f,dBm} &= P_{out,2f-f} - G_{p,dB} \\ &= -66 - 16 \\ &= -82 \text{ dBm} \end{aligned} \tag{7.64}$$

Figure 7-32 shows the input and output spectra of the amplifier. Figure 7-33 shows the equivalent input and output spectra. Note that $\Delta_{TO,dB}$ is identical on both the input and the output ports of the amplifier.

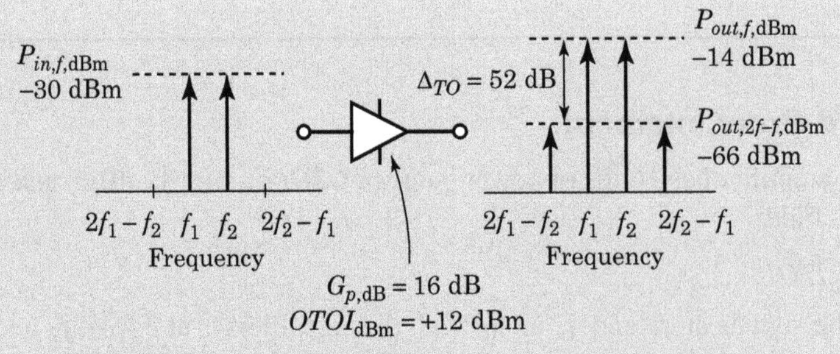

FIGURE 7-32 ■ Input and output spectra of a nonlinear device generating third-order distortion.

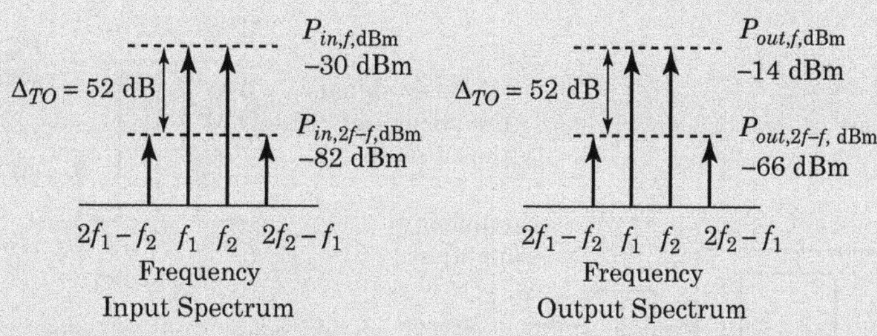

FIGURE 7-33 ■ Output and equivalent input spectra for the example of Figure 7-32.

EXAMPLE

Spurious Signals and TOI

You're building a cellular telephone receiver for the U.S. market where the RF channels are 30 kHz apart. Figure 7-34 shows the spectrum present at the output of an amplifier. The input spectrum contains signals that are spaced 30 kHz apart. We suspect the two outer signals may be due to amplifier nonlinearity. How can we determine this?

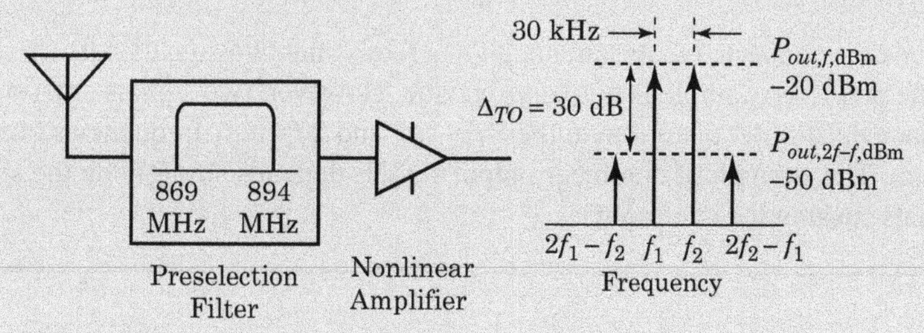

FIGURE 7-34 ■ Identifying signals generated by third-order nonlinear distortion.

Solution

Equation (7.58) tells us that if we decrease the fundamental input power by 1 dB, then the output third-order power will drop by 3 dB. Figure 7-35 shows our system with a 10 dB attenuator placed between the antenna and the amplifier. Figure 7-35(a) shows the output spectrum if two outer tones are due to signals from the antenna. The power in the signals at f_1, f_2, $2f_1 - f_2$, and $2f_2 - f_1$ all drop by the value of the attenuator. Note that $\Delta_{TO,dBm}$ has not changed.

Figure 7-35(b) shows the output spectrum when we place the attenuator in the system if the $2f_1 - f_2$ and $2f_2 - f_1$ tones are due to amplifier nonlinearity. The power in the signals at $f_1, f_2, 2f_1 - f_2$, and $2f_2 - f_1$ have dropped by different amounts, and $\Delta_{TO,dBm}$ has changed.

Like the second-order case, we use this method in practice to identify spurious responses. Putting an n dB attenuator on the input of a device will usually cause a $3n$ dB reduction of the device's third-order distortion.

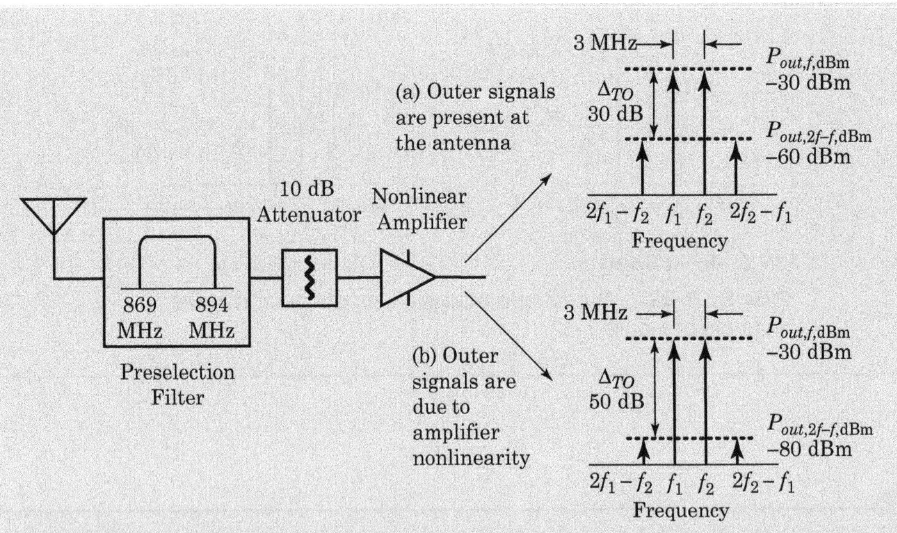

FIGURE 7-35 ■ The effect of placing an attenuator between the antenna and the first nonlinear element of the system shown in Figure 7-34. If the outer signals are present at the antenna and not due to a third-order device nonlinearity, we obtain the output spectrum shown in (a). The outer signal powers drop by the value of the attenuator. If the outer signals are due to a third-order nonlinearity in the amplifier, we will observe the output spectrum shown in part (b). The inner signals will drop by the value of the attenuator, whereas the power in the two outer signals will decrease by three times the value of the attenuator.

When a receiver is connected to an antenna, it's unlikely that two signals will arrive with the same power level so the previous example is not very realistic. However, two signals entering a nonlinear device will produce third-order distortion at the $2f_1 - f_2$ and $2f_2 - f_1$ frequencies even if their power levels are different. The strength of nonlinear output signals depends strongly on the strength of both of the input signals causing the distortion.

7.15 MEASURING AMPLIFIER NONLINEARITY

Equations (7.45) and (7.59) provide an easy way to measure the SOI and the TOI of an unknown device.

7.15.1 Second-Order Measurement

Rewriting equation (7.45) produces

$$OSOI_{dBm} = 2P_{out,f,dBm} - P_{out,2f,dBm} \tag{7.65}$$

We apply a test signal to the nonlinear device and then measure the fundamental output power and the second-order output power. Equation (7.65) provides the output second-order intercept for the amplifier. Figure 7-36 shows the test setup.

We could have also written

$$OSOI_{dBm} = P_{out,f,dBm} + \Delta_{SO,dB}$$
$$\text{and} \tag{7.66}$$
$$ISOI_{dBm} = P_{in,f,dBm} + \Delta_{SO,dB}$$

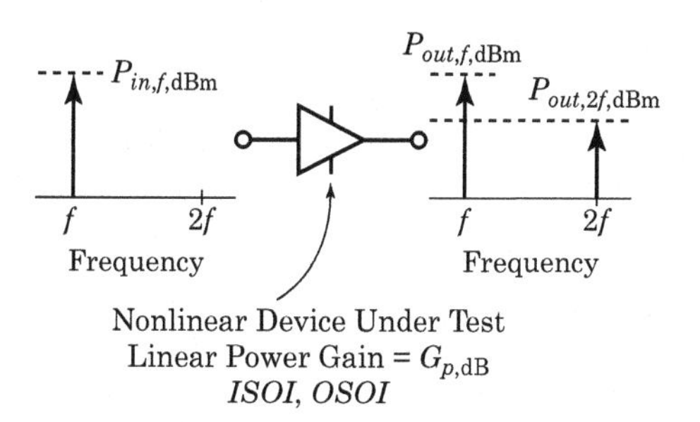

FIGURE 7-36 ■ Measuring the second-order intercept point of a nonlinear device.

We would measure $\Delta_{SO,dB}$ on the output and either $P_{in,f,dBm}$ or $P_{out,f,dBm}$ and use equation (7.66) to find the *SOI* of the DUT.

EXAMPLE

Amplifier Second-Order Measurements

Find the power gain, *ISOI*, and *OSOI* of the DUT shown in Figure 7-37.

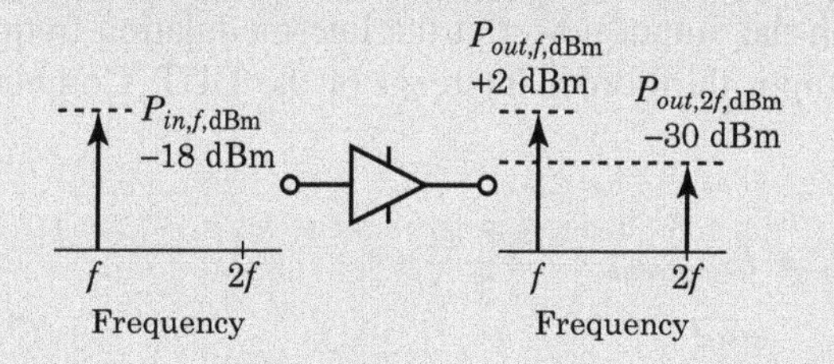

FIGURE 7-37 ■ Measuring the second-order intercept point of a amplifier.

Solution

The power gain is

$$\begin{aligned} G_{p,dB} &= P_{out,f,dBm} - P_{in,f,dBm} \\ &= 2 - (-18) \\ &= 20 \text{ dB} \end{aligned} \tag{7.67}$$

Equation (7.65) gives us the $OSOI_{dBm}$:

$$\begin{aligned} OSOI_{dBm} &= 2P_{out,f,dBm} - P_{out,2f,dBm} \\ &= 2\,(2) - (-30) \\ &= 34 \text{ dBm} \end{aligned} \tag{7.68}$$

The *ISOI* is

$$\begin{aligned} ISOI_{dBm} &= OSOI_{dBm} - G_{p,dB} \\ &= 34 - 20 \\ &= 14 \text{ dBm} \end{aligned} \tag{7.69}$$

We could have used equation (7.66) to find the OSOI:

$$\begin{aligned} OSOI_{dBm} &= P_{out,f,dBm} + \Delta_{SO,dB} \\ &= 2 + 32 \\ &= 34 \text{ dBm} \end{aligned} \tag{7.70}$$

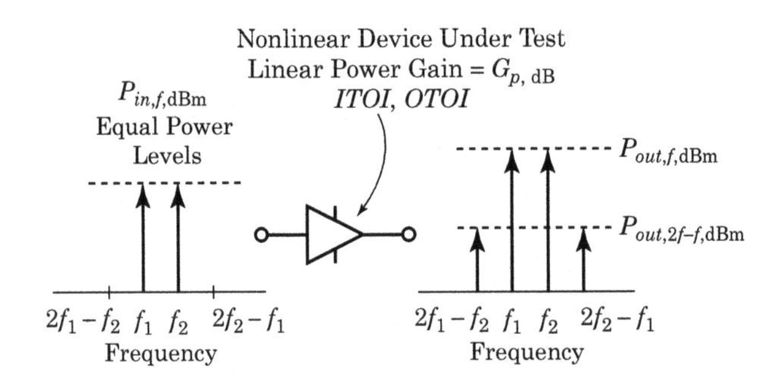

FIGURE 7-38 Measuring the third-order intercept point of a amplifier.

7.15.2 Third-Order Measurements

We measure the *TOI* of a device using equation (7.59). Rewriting this equation produces

$$OTOI_{dBm} = \frac{3}{2}P_{out,f,dBm} - \frac{1}{2}P_{out,2f-f,dBm} \tag{7.71}$$

Figure 7-38 shows the test setup. We apply two signals to the DUT and measure the fundamental output power along with the output power at the intermodulation frequencies. Equation (7.71) gives us the output third-order intercept of the DUT. Combining equations (7.71) and (7.57) produces

$$\begin{aligned} OTOI_{dBm} &= P_{out,f,dBm} + \frac{\Delta_{TO,dB}}{2} \\ &\textit{and} \\ ITOI_{dBm} &= P_{in,f,dBm} + \frac{\Delta_{TO,dB}}{2} \end{aligned} \tag{7.72}$$

These equations allow us to measure the fundamental power (either input or output) and the $\Delta_{TO,dB}$ to find the TOI of the DUT.

EXAMPLE

Amplifier Third-Order Measurements

Find the power gain, *ITOI*, and *OTOI* of the DUT shown in Figure 7-39.

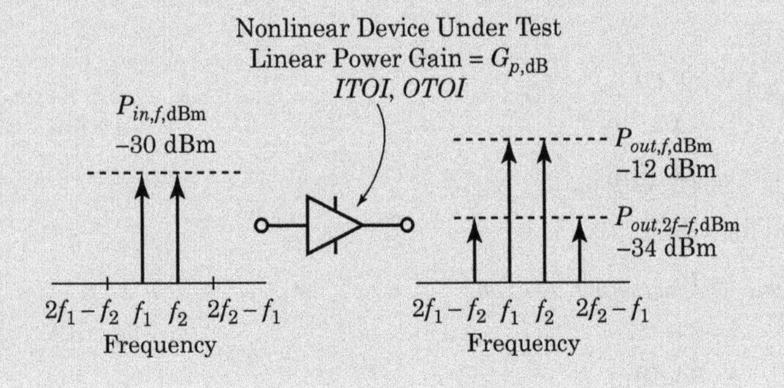

FIGURE 7-39 Find the amplifiers *TOI* and linear power gain.

Solution

The power gain is

$$\begin{aligned} G_{p,dB} &= P_{out,f,dBm} - P_{in,f,dBm} \\ &= -12 - (-30) \\ &= 18 \text{ dB} \end{aligned} \tag{7.73}$$

Equation (7.71) gives us the *OTOI*:

$$\begin{aligned} OTOI_{dBm} &= \frac{3}{2} P_{out,f,dBm} - \frac{1}{2} P_{out,2f-f,dBm} \\ &= \frac{3}{2}(-12) - \frac{1}{2}(-34) \\ &= -1 \text{ dBm} \end{aligned} \tag{7.74}$$

The *ITOI* is

$$\begin{aligned} ITOI_{dBm} &= OTOI_{dBm} - G_{p,dB} \\ &= -1 - 18 \\ &= -19 \text{ dBm} \end{aligned} \tag{7.75}$$

We could have used equation (2.29) to find the *OTOI*:

$$\begin{aligned} OTOI_{dBm} &= P_{out,f,dBm} + \frac{\Delta_{TO,dB}}{2} \\ &= -12 + \frac{22}{2} \\ &= -1 \text{ dBm} \end{aligned} \tag{7.76}$$

7.16 GAIN COMPRESSION/OUTPUT SATURATION

Two other linearity measures are the 1 dB gain compression point and the saturated output power. Figure 7-40 shows output power versus input power for a nonlinear device. The curve is linear at low input power levels. An n dB increase on the input port results in an n dB increase in output power.

As the input power increases, the amplifier is unable to supply the necessary output power to the load. This is a gradual process; the amplifier's gain usually drops off gently with increasing input power.

7.16.1 1 dB Compression Point

Figure 7-40 shows a linear approximation to the $P_{out,dBm}$ versus $P_{in,dBm}$ curve. At low power levels, the linear approximation follows the exact curve very well. However, the two curves depart at higher power levels.

Figure 7-41 shows the same two curves blown up in the area where the two curves just begin to differ. The point where the two curves deviate by 1 dB is the amplifier's 1 dB compression point or simply the compression point.

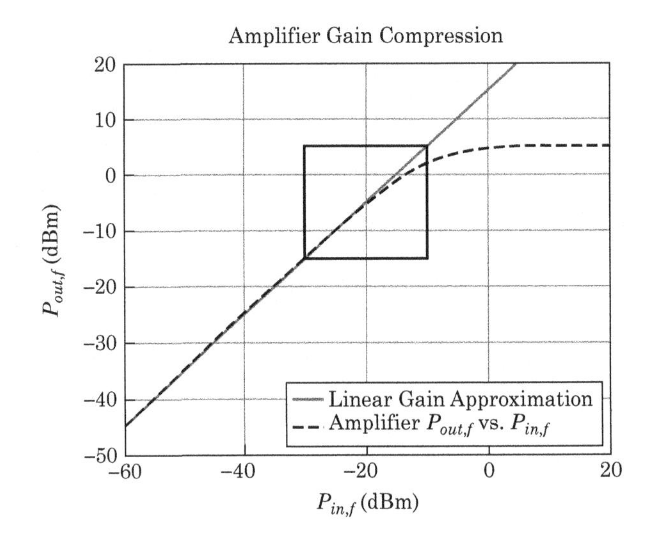

FIGURE 7-40 ■ The input power versus the output power for a nonlinear device. Note the output saturation.

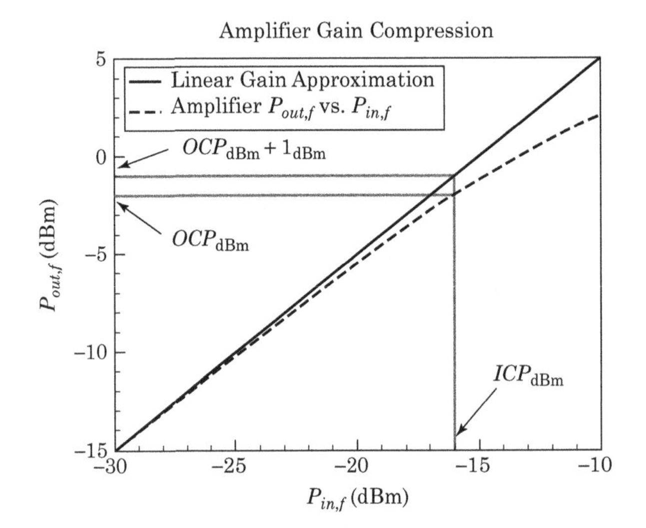

FIGURE 7-41 ■ Figure 7-40 enlarged around the device's 1 dB compression point.

The device has both an output compression point (OCP_{dBm}) and an input compression point (ICP_{dBm}). The input compression point is the input power at which the amplifier's gain has dropped by 1 dB. The output compression point is the amplifier's output power when the input power is *ICP* dBm. The *ICP* and the *OCP* are related:

$$OCP_{dBm} = ICP_{dBm} + \left(G_{p,dB} - 1\right) \tag{7.77}$$

This equation is a little different from the other equations that relate the input and output parameters of a device. Usually the equation takes the form of

$$Output\,Spec_{dBm} = Input\,Spec_{dBm} + G_{p,dB} \tag{7.78}$$

The *ICP* and the *OCP* of a device are related by the power gain minus one. Some literature defines the OCP_{dBm} as $ICP_{dBm} + G_{P,dBm}$. This is the point labeled $OCP_{dBm} + 1$ on Figure 7-41. We'll define it as equation (7.77) states, but be careful when communicating with others.

The 1 dB compression point is an arbitrarily boundary between the small signal operating range of a device and its large-signal operating range. Before the concepts of SOI and TOI, the 1 dB compression point was the primary linearity specification. However, the intercept concept allows us to calculate useful data given the input power levels. The 1 dB compression point provides no such information.

7.16.2 Saturated Output Power

Saturated output power is the maximum fundamental power a device will produce under any input conditions. Figure 7-40 shows the P_{out} versus P_{in} curve for a nonlinear device. This device saturates at about 5.2 dBm of output power.

The saturated output power of an amplifier refers to the maximum amount of power the amplifier will produce at its output port. For example, a 10 watt amplifier can produce one sine wave at 10 watts, 2 sine waves at 5 watts each, or 100 sine waves at 0.1 watts each. The amplifier can also produce a signal whose bandwidth is 20 MHz and whose power spectral density is 0.25 watts/MHz.

There is a caveat to this simple-minded example related to crest factor. When we add two or more signals together, there is an opportunity for constructive interference, and the instantaneous voltage of the combined signal can be much larger than the root mean square (RMS) value would indicate. Under these conditions, the peaks of the signal will saturate even though the power level of the signal is well below the saturation point of the amplifier. This effect can cause the received bit error rate to be higher than the signal's SNR would indicate. This is especially true when dealing with multicarrier modulation (e.g., orthogonal frequency-division multiplexing [OFDM]).

EXAMPLE

Satellite Transponders and Linearity

Figure 7-42 shows the simplified RF chain of a typical satellite transponder. The received signal, broadcast from the earth, passes through the receive antenna, a band-pass filter, and into an amplifier. The mixer and oscillator combination then converts the received signal to a new frequency. After band-pass filtering again, we apply the signal to a power amplifier, then to another band-pass filter, and finally into the transmit antenna.

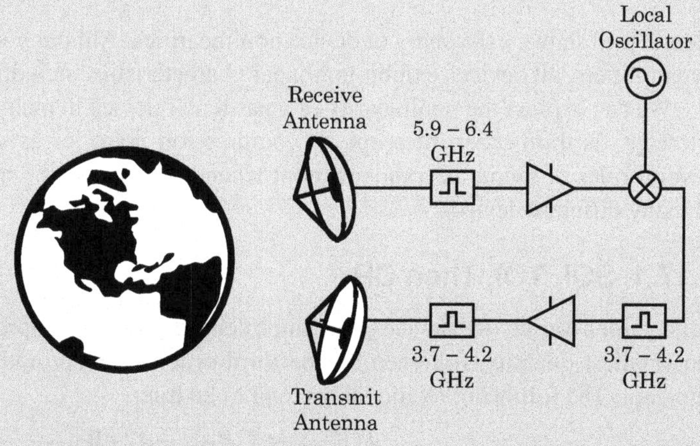

FIGURE 7-42 ■ The RF chain of a typical "bent-pipe satellite transponder.

The common term for this architecture is the bent pipe. To the user, the satellite seems to receive signals that are heading away from the earth and then sends them into a bent pipe for rebroadcast back to Earth. The power gains of all the components are fixed. There is a one-for-one correspondence between the amount of power we transmit to the satellite and the strength of the signal we receive when the signal finally returns from the satellite.

Many cable television channels pass through these satellite systems. The signals are broadcast from the earth to the satellite at 5.925 GHz to 6.425 GHz. The signals are converted to 3.700 to 4.200 GHz aboard the satellite and are rebroadcast back to the subscribers on the earth.

These satellite systems service different users who are physically separated on the earth. A user pays a fee based on the bandwidth (or *transponder space*) he requires. The more spectrum the user requires, the larger his bill.

There is another facet to this problem. If the total transponder bandwidth is 500 MHz, for example, and the power amplifier produces 1,000 watts, then each user is limited to

$$\frac{1000 \text{ Watts}}{500 \text{ MHz}} = 2\frac{\text{Watts}}{\text{MHz}} \tag{7.79}$$

unless he's willing to pay a premium.

Whenever we have several signals present in a nonlinear device at the same time, the device will generate spurious signals. This applies directly to the satellite transponder. When the input power from each user is within the limits specified by the satellite operator, the spurious signals generated by the satellite are not strong enough to cause harm. When one user increases his transmitted power beyond the agreed on limits, he causes distortion in *all* the signals passing through the amplifier, not just his own signal. In addition, due to other nonlinear effects in the receiver, the strongest signal will steal power from the other signals.

All of the users must cooperate for the satellite transponder to work properly. The nonlinear distortion in the power amplifier limits the total amount of power that the satellite transponder can generate. Higher power translates directly into more users for the satellite and, hence, into more money for the satellite owners.

7.17 | COMPARISON OF NONLINEAR SPECIFICATIONS

Figure 7-43 shows a summary of device nonlinearities. Although we've discussed primarily amplifiers, all devices exhibit nonlinear characteristics including mixers and filters.

We can express the nonlinearity of a particular device in many ways: as second-order intercept, as third-order intercept, as compression point, or as saturated output power. Several rules of thumb become apparent when we examine the nonlinear characteristics of many different devices:

7.17.1 SOI, TOI, Then CP

For a garden-variety RF device (e.g., amplifier, mixer), the compression point is generally the smallest quantity, followed by the third-order intercept and then the second-order intercept. The following relationships tend to be true:

$$\begin{aligned} TOI_{dBm} &\approx CP_{dBm} + 12 \text{ dB} \\ SOI_{dBm} &\approx CP_{dBm} + 27 \text{ dB} \end{aligned} \tag{7.80}$$

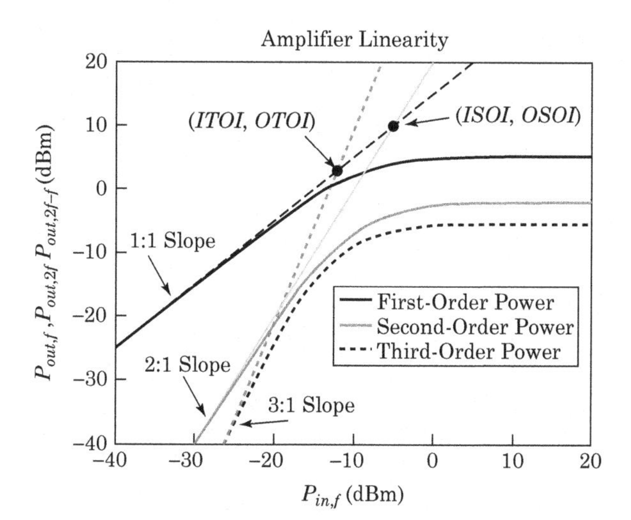

FIGURE 7-43 ■ Summary of device nonlinearities.

For a third-order polynomial approximation, theory says that the output compression point of a nonlinear device is 10.6 dB below the output third-order intercept or

$$OCP_{dBm} \approx OTOI_{dBm} - 10.6 \tag{7.81}$$

which implies

$$ICP_{dBm} \approx ITOI_{dBm} - 9.6 \tag{7.82}$$

7.17.2 Balanced Devices

Balanced devices are designed with symmetrical output structure to reduce second-order distortion and to increase the output power of the device. Unless the word *balanced* appears in the specifications or data sheet, the device probably isn't balanced. If a device is balanced, then the following relationships usually hold:

$$\begin{aligned} SOI_{dBm} &\approx TOI_{dBm} + (30 \text{ or } 40 \text{ dB}) \\ TOI_{dBm} &\approx CP_{dBm} + (10 \text{ or } 15 \text{ dB}) \end{aligned} \tag{7.83}$$

7.17.3 Slopes

If we express the power levels in dB, we've seen that the second-order output power of a weakly nonlinear drops off at a 2:1 slope with decreasing input power. Third-order power drops off at a 3:1 slope. This relationship holds if the input power is at least 20 dB below the ICP_{dBm}.

7.17.4 Worst-Case Scenario

From a linearity perspective, the worst-case scenario is having to process lots of large, undesired signals in a nonlinear device at the same time as a small, desired signal. The large, undesired signals may cause nonlinear distortion to distort or bury the small desired signal. Unfortunately, the worst-case scenario represents the real world. The antenna will supply many signals at a variety of frequencies and amplitudes. Since the possible intermodulation products fall at

$$f_{out} = |\pm n f_1 \pm o f_2 \pm p f_3 \pm \cdots| \text{ where} \begin{cases} n = 0, 1, 2, \ldots \\ o = 0, 1, 2, \ldots \\ p = 0, 1, 2, \ldots \\ \vdots \end{cases} \tag{7.84}$$

we can see why linearity is critical.

7.18 | NONLINEARITIES IN CASCADE

We now explore the effects of cascading nonlinear devices together. How does the linearity of a cascade of components compare with the linearity of the pieces we used to build the cascade?

We will apply a signal to the cascade and then calculate the distortion power generated by each individual device. We will mathematically transfer the distortion power of each device to the cascade's output and then sum the individual powers together. Finally, we will compare the total distortion power with the fundamental output power and calculate the output linearity specification.

7.18.1 Nomenclature Refresher

Figure 7-44 shows the three-element cascade that we'll analyze. We know the gain, *SOI*, and *TOI* of each element and want to find the cascade *SOI* and *TOI*. We define:

G_{Pn} = the power gain of the n-th element in the cascade

G_{Pcas} = the power gain of the cascade

$ITOI_n$ = the input third-order intercept of the n-th element in the cascade

$OTOI_n$ = the output third-order intercept of the n-th element in the cascade

$ISOI_n$ = the input second-order intercept of the n-th element in the cascade

$OSOI_n$ = the output second-order intercept of the n-th element in the cascade

P_f = signal power at the fundamental frequency

P_{2f-f} = the third-order power at $2f_1 - f_2$ or at $2f_2 - f_1$

Also, let

$P_{in,f,n}$ = the fundamental input power of each tone applied to element #n of the cascade. The two tones are at frequencies f_1 and f_2.

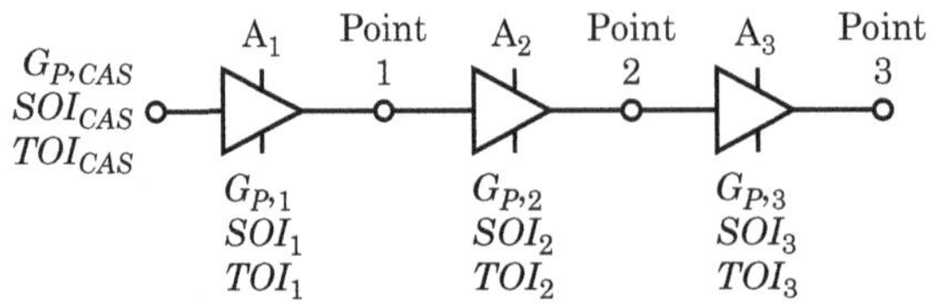

FIGURE 7-44 ■ A simple three-element cascade. We know the gain, SOI, and TOI of each element and want to find the cascade SOI and TOI.

$P_{out,2f-f,n}$ = the third-order output power of one tone generated by cascade element #n. If the two input signals are at frequencies f_1 and f_2, then the third-order signals will be at $2f_2 - f_1$ and $2f_1 - f_2$.

$P_{out,2f-f,n,pt3}$ = the third-order output power generated by cascade element #n after it's been mathematically transferred to the end of the cascade (Point 3 in Figure 7-44)

All terms are assumed linear if they lack a *dB* or *dBm* subscript. An *out* or *in* subscript ties a parameter to the output port or input port, respectively.

7.18.2 Third-Order Intercept

Equation (7.59) expresses the relationship among the fundamental output power, the third-order output power, and the *OTOI* of the nonlinear device:

$$P_{out,2f-f,dBm} = 3P_{out,f,dBm} - 2OTOI_{dBm} \tag{7.85}$$

We also know

$$P_{out,f,dBm} = P_{in,f,dBm} + G_{p,dB} \tag{7.86}$$

The linear equivalents for these two equations are

$$\begin{aligned} P_{out,2f-f} &= \frac{P^3_{out,f}}{OTOI^2} \\ &\textit{and} \\ P_{out,f} &= P_{in,f}G_p \end{aligned} \tag{7.87}$$

Equation (7.59) and its linear equivalent, equation (7.87), tell us the third-order distortion power is proportional to the fundamental input power cubed and inversely proportional to the amplifier's OTOI squared.

7.18.2.1 Amplifier #1

Each tone we apply to the input of amplifier #1 has a power of $P_{in,1}$. The fundamental and third-order output powers from amplifier #1 are

$$P_{out,f,1} = G_{p,1}P_{in,f,1} \tag{7.88}$$

and

$$\begin{aligned} P_{out,2f-f,1} &= \frac{P^3_{out,f,1}}{OTOI_1^2} \\ &= P^3_{in,f,1}\frac{G^3_{p,1}}{OTOI_1^2} \end{aligned} \tag{7.89}$$

At the end of the cascade (Point 3 in Figure 7-44), the third-order output power due to amplifier #1 is

$$\begin{aligned} P_{out,2f-f,1,pt3} &= P_{out,2f-f,1}G_{p,2}G_{p,3} \\ &= P^3_{in,f,1}\frac{G^3_{p,1}G_{p,2}G_{p,3}}{OTOI_1^2} \end{aligned} \tag{7.90}$$

7.18.2.2 Amplifier #2

The fundamental signal power present at the input of amplifier #2 is

$$P_{in,\,f,2} = P_{out,f,1} = P_{in,\,f,1}G_{p,1} \tag{7.91}$$

The fundamental and third-order powers at the output of amplifier #2 are

$$P_{out,f,2} = G_{p,1}G_{p,2}P_{in,\,f,1} \tag{7.92}$$

and

$$\begin{aligned} P_{out,2f-f,2} &= \frac{P^3_{out,f,2}}{OTOI^2_2} \\ &= P^3_{in,\,f,1}\frac{G^3_{p,1}G^3_{p,2}}{OTOI^2_2} \end{aligned} \tag{7.93}$$

The third-order power present at the output of the cascade due to amplifier #2 is

$$\begin{aligned} P_{out,2f-f,2,pt3} &= P_{out,2f-f,2}G_{p,3} \\ &= P^3_{in,\,f,1}\frac{G^3_{p,1}G^3_{p,2}G_{p,3}}{OTOI^2_2} \end{aligned} \tag{7.94}$$

7.18.2.3 Amplifier #3

The fundamental power applied to amplifier #3 is

$$P_{in,\,f,3} = G_{p,1}G_{p,2}P_{in,\,f,1} \tag{7.95}$$

The fundamental output power of amplifier #3 is

$$P_{out,f,3} = G_{p,1}G_{p,2}G_{p,3}P_{in,\,f,1} \tag{7.96}$$

The third-order power present at the output port of the cascade due to amplifier #3 is

$$\begin{aligned} P_{out,2f-f,3} &= P_{out,2f-f,3,pt3} \\ &= \frac{P^3_{out,f,3}}{OTOI^2_3} \\ &= P^3_{in,\,f,1}\frac{G^3_{p,1}G^3_{p,2}G^3_{p,3}}{OTOI^2_3} \end{aligned} \tag{7.97}$$

7.18.2.4 Third-Order Power Summation

We now know the third-order distortion power generated by the three separate devices that is present at the output of the last amplifier in the cascade:

- $P_{out,2f-f,1,pt3}$ is the third-order power generated by amplifier #1. We multiplied this power by the gain of the last two stages to bring it to Point 3, the output of the cascade.
- $P_{out,2f-f,2,pt3}$ is the distortion power generated by amplifier #2. We have mathematically moved this power to the end of the cascade.
- $P_{out,2f-f,3,pt3}$ describes the third-order distortion power generated by amplifier #3 present at the end of the cascade.

7.18.2.5 Coherent versus Noncoherent Summation

The third-order distortion products present at Point 3 of Figure 7-44 are all at the same frequency. To accurately combine the three signals, we must know the relative phases of the signals as they exist at the end of the cascade. If all of the signals are in phase, the vector sum of these signals will be one quantity. If some of the signals are in phase while others have different phases, the vector sum of the signals is a different quantity. Unfortunately, we do not know this information (and it's not easily available).

The worst-case result (i.e., the most nonlinear cascade) results when we assume that the distortion voltages are all exactly in phase and add directly together. This is *coherent summation*. Coherent summation produces the most conservative answer, but this answer is usually too conservative. The term *noncoherent summation* means that we just add the signal powers together. This doesn't produce a worst-case result but usually models the behavior of most systems.

In the real world, we usually observe results that suggest that signals are adding noncoherently. However, at some frequency, the distortion will add together in a coherent fashion so we'll see the coherent numbers. This is especially true when dealing with multicarrier modulation (e.g., OFDM).

We'll derive both equations. Use the noncoherent summation results as a mostly right value, whereas the coherent summation data represents a worst-case result.

7.18.2.6 Coherent Summation TOI Equation

We first convert the nonlinear power components present at the output of the cascade into voltages, perform the addition, and then convert the sum back into power. We'll leave the details to the interested reader. The coherent sum of several signals is

$$P_{total,coherent} = \left[P_1^{1/2} + P_2^{1/2} + P_3^{1/2} + \cdots + P_n^{1/2}\right]^2 \tag{7.98}$$

The total third-order power available at the output of the cascade is

$$P_{out,2f-f,cas} = P_{in,f,cas}^3 \left[\sqrt{\frac{G_{p,1}^3 G_{p,2} G_{p,3}}{OTOI_1^2}} + \sqrt{\frac{G_{p,1}^3 G_{p,2}^3 G_{p,3}}{OTOI_2^2}} + \sqrt{\frac{G_{p,1}^3 G_{p,2}^3 G_{p,3}^3}{OTOI_3^2}}\right]^2 \tag{7.99}$$

The cascade gain is

$$G_{p,cas} = G_{p,1} G_{p,2} G_{p,3} \tag{7.100}$$

Combining these two equations produces

$$\begin{aligned} P_{out,2f-f,cas} &= P_{in,f,cas}^3 G_{p,cas}^3 \left[\sqrt{\frac{1}{G_{p,2}^2 G_{p,3}^2 OTOI_1^2}} + \sqrt{\frac{1}{G_{p,3}^2 OTOI_2^2}} + \sqrt{\frac{1}{OTOI_3^2}}\right]^2 \\ &= P_{out,f,cas}^3 G_{p,cas}^3 \left[\frac{1}{G_{p,2} G_{p,3} OTOI_1} + \frac{1}{G_{p,3} OTOI_2} + \frac{1}{OTOI_3}\right]^2 \end{aligned} \tag{7.101}$$

The third-order output power for the cascade is

$$P_{out,2f-f} = \frac{P_{out,f}^3}{OTOI^2} \tag{7.102}$$

These equations produce an expression for the cascade third-order intercept

$$\frac{P^3_{out,f,cas}}{OTOI^2_{cas}} = P^3_{out,f,cas}\left[\frac{1}{G_{p,2}G_{p,3}OTOI_1} + \frac{1}{G_{p,3}OTOI_2} + \frac{1}{OTOI_3}\right]^2 \tag{7.103}$$

which simplifies to

$$\frac{1}{OTOI_{cas}} = \frac{1}{G_{p,2}G_{p,3}OTOI_1} + \frac{1}{G_{p,3}OTOI_2} + \frac{1}{OTOI_3} \tag{7.104}$$

or, in terms of *ITOI*,

$$\frac{1}{ITOI_{cas}} = \frac{1}{ITOI_1} + \frac{1}{ITOI_2/G_{p,1}} + \frac{1}{ITOI_3/G_{p,1}G_{p,2}} \tag{7.105}$$

More algebra for an n-element cascade reveals

$$\begin{aligned}\frac{1}{OTOI_{cas}} = &\frac{1}{G_{p,2}G_{p,3}G_{p,4}\cdots G_{p,n-1}OTOI_1}\\ &+\frac{1}{G_{p,3}G_{p,4}\cdots G_{p,n-1}OTOI_2}\\ &+\cdots\\ &+\frac{1}{OTOI_n}\end{aligned} \tag{7.106}$$

and

$$\begin{aligned}\frac{1}{ITOI_{cas}} = &\frac{1}{ITOI_1}\\ &+\frac{1}{ITOI_2/G_{p,1}}\\ &+\cdots\\ &+\frac{1}{ITOI_n/G_{p,1}G_{p,2}\cdots G_{p,n-1}}\end{aligned} \tag{7.107}$$

7.18.2.7 Noncoherent Summation TOI Equation

Noncoherent summation assumes the phases of the third-order distortion from each cascade element are random. We simply add the distortion powers together:

$$P_{total,noncoherent} = P_1 + P_2 + P_3 + \cdots + P_n \tag{7.108}$$

The total third-order power available at the output of the cascade is

$$\begin{aligned}P_{out,2f-f,cas} &= P^3_{in,f,cas}\left[\frac{G^3_{p,1}G_{p,2}G_{p,3}}{OTOI_1^2} + \frac{G^3_{p,1}G^3_{p,2}G_{p,3}}{OTOI_2^2} + \frac{G^3_{p,1}G^3_{p,2}G^3_{p,3}}{OTOI_3^2}\right]\\ &= P^3_{out,f,cas}\left[\frac{1}{G^2_{p,2}G^2_{p,3}OTOI_1^2} + \frac{1}{G^2_{p,3}OTOI_2^2} + \frac{1}{OTOI_3^2}\right]\end{aligned} \tag{7.109}$$

Again,

$$P_{out,2f-f} = \frac{P^3_{out,f}}{OTOI^2} \tag{7.110}$$

which combines to produce

$$\frac{P^3_{out,f,cas}}{OTOI^2_{cas}} = P^3_{out,f,cas}\left[\frac{1}{G^2_{p,2}G^2_{p,3}OTOI^2_1} + \frac{1}{G^2_{p,3}OTOI^2_2} + \frac{1}{OTOI^2_3}\right] \tag{7.111}$$

Simplifying reveals

$$\frac{1}{(OTOI_{cas})^2} = \frac{1}{(G_{p,2}G_{p,3}OTOI_1)^2} + \frac{1}{(G_{p,3}OTOI_2)^2} + \frac{1}{(OTOI_3)^2} \tag{7.112}$$

and

$$\frac{1}{(ITOI_{cas})^2} = \frac{1}{(ITOI_1)^2} + \frac{1}{(ITOI_2/G_{p,1})^2} + \frac{1}{(ITOI_3/G_{p,1}G_{p,2})^2} \tag{7.113}$$

The equations describing an n-element cascade are

$$\begin{aligned}\frac{1}{(OTOI_{cas})^2} = &\frac{1}{(G_{p,2}G_{p,3}G_{p,4}\ldots G_{p,n}OTOI_1)^2}\\ &+\frac{1}{(G_{p,3}G_{p,4}\ldots G_{p,n}OTOI_2)^2}\\ &+\cdots\\ &+\frac{1}{(OTOI_n)^2}\end{aligned} \tag{7.114}$$

and

$$\begin{aligned}\frac{1}{(ITOI_{cas})^2} = &\frac{1}{(ITOI_1)^2}\\ &+\frac{1}{(ITOI_2/G_{p,1})^2}\\ &+\cdots\\ &+\frac{1}{(ITOI_n/G_{p,1}G_{p,2}G_{p,3}\ldots G_{p,n-1})^2}\end{aligned} \tag{7.115}$$

7.18.2.8 Examining the TO Cascade Equation

The cascade equations (7.104) through (7.107) and (7.112) through (7.115) provide a great deal of insight into system design. To keep the conversation simple, we'll discuss only equations (7.106) and (7.107).

Component Contribution We first note that the linearity contribution of each component in the cascade is neatly tied up in a single term of the cascade equations. In a three-element cascade, for example, the third component's contribution to the cascade's linearity is found only in the term

$$\frac{1}{ITOI_3/G_{p,1}G_{p,2}} \tag{7.116}$$

This is the only place in the cascade equation that contains the *ITOI* of the third element. This characteristic will make cascade analysis much easier.

Imaging When we ask about the linearity contribution of the third element in a cascade, the only term we have to analyze is

$$\frac{1}{ITOI_3/G_{p,1}G_{p,2}} \tag{7.117}$$

If we think of the $ITOI_3$ as power, for example, then the term

$$ITOI_3/G_{p,1}G_{p,2} \tag{7.118}$$

looks like we're mathematically moving that power level to the input of the cascade. We can think of every term in the cascade equation in this manner. When we perform cascade analysis in later chapters, we'll translate the TOI of each device in the cascade to a common port to look for the weak link in the cascade's TOI chain.

The linearity of a cascade depends on the linearity of each component and on the power gain surrounding the component. Components preceded by a lot of gain will experience higher signal levels, so they will generate higher levels of distortion.

As an example, if we apply a −80 dBm signal to the cascade of Figure 7-45, then amplifier #1 will see a −80 dBm signal. Amplifier #2 will see a −60 dBm signal because of the 20 dB power gain of amplifier #1. Although amplifiers #1 and #2 have the same TOI, amplifier #2 will generate higher levels of distortion because it sees a larger signal. Amplifier #3 will see a −40 dBm signal and will generate the most distortion power of all the devices in the cascade. Amplifier #3 will dominate the TOI of the cascade.

In a well-designed system, the translated TOI of each component should be about equal. If all the translated TOIs are equal, then all of the elements in the cascade go nonlinear at the same input power level. If the translated TOI of one element is too small, the weak component will dominate the TOI of the entire cascade.

Resistors in Parallel Equations (7.104) through (7.107) resemble the equation that describes resistors in parallel:

$$\frac{1}{R_p} = \frac{1}{R_1} + \frac{1}{R_2} + \frac{1}{R_3} \tag{7.119}$$

Figure 7-46 shows the similarities among equations (7.104), (7.105), and (7.119). The value of each resistor represents the TOI of each component when it is referenced to a common port. When we add a new device to a cascade, we add another parallel resistor to the equivalent circuits and we lower the TOI (resistance) of the cascade. We can never

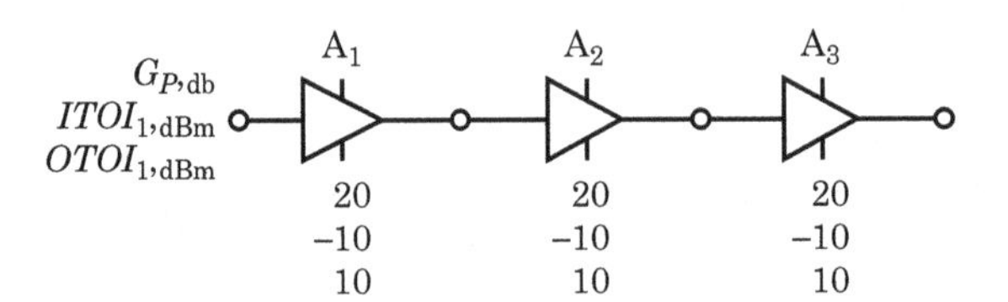

FIGURE 7-45 ■ A simple three-element cascade with known numerical values for the gain and TOI of each element.

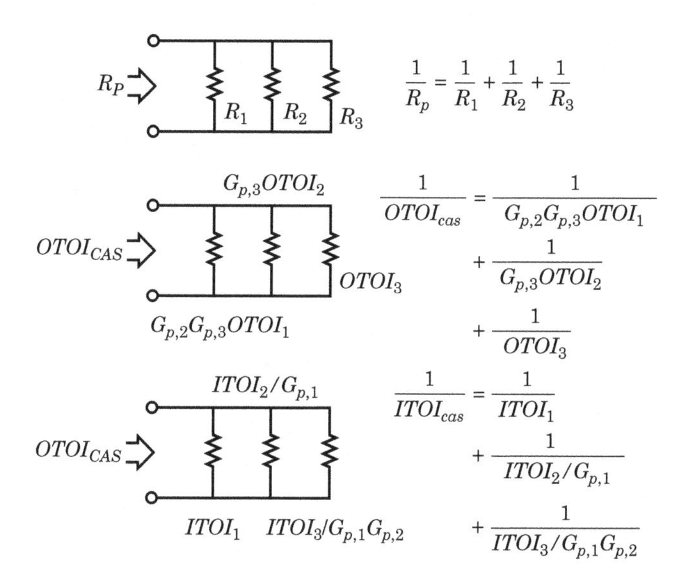

FIGURE 7-46 ■ The cascade TOI equation is similar to the the equation describing resistors in parallel.

improve the linearity of the cascade by adding another component; we can only worsen the linearity.

7.18.3 Second-Order Intercept

We will analyze the system of Figure 7-44 for second-order effects. We know the gain and SOI of each component and seek the SOI of the cascade. Equation (7.45), repeated here, relates the fundamental and second-order power levels present in a nonlinear system:

$$P_{out,2f,dBm} = 2P_{out,f,dBm} - OSOI_{dBm} \tag{7.120}$$

We also know

$$P_{out,f,dBm} = P_{in,f,dBm} + G_{p,dB} \tag{7.121}$$

Converting these equations to linear expressions produces

$$P_{out,2f} = \frac{P^2_{out,f}}{OSOI} \tag{7.122}$$

and

$$P_{out,f} = P_{in,f}G_p \tag{7.123}$$

7.18.3.1 More Definitions

Let

$P_{in,f,n}$ = the fundamental input power applied to amplifier #n

$P_{out,2f,n}$ = the second-order output power generated by amplifier #n. If the input signal is at a frequency of f, then the second-order output signal will be at $2f$.

$P_{out,2f,n,pt3}$ = the second-order output power generated by amplifier #n present at the end of the cascade (Point 3 in Figure 7-44). The second-order signals generated by an amplifier will experience the power gain of the succeeding stages.

7.18.3.2 Amplifier #1

We apply an input power of $P_{in,f,1}$ to amplifier #1 and calculate the fundamental and second-order output powers:

$$P_{out,f,1} = P_{in,\,f,1} G_{p,1} \tag{7.124}$$

and

$$\begin{aligned} P_{out,2f,1} &= \frac{P_{out,f,1}^2}{OSOI_1} \\ &= P_{in,\,f,1}^2 \frac{G_{p,1}^2}{OSOI_1} \end{aligned} \tag{7.125}$$

At point 3 in Figure 7-44 (the end of the cascade), the second-order output power due to amplifier #1 is

$$\begin{aligned} P_{out,2f,1,pt3} &= P_{out,2f,1} G_{p,2} G_{p,3} \\ &= P_{in,\,f,1}^2 \frac{G_{p,1}^2 G_{p,2} G_{p,3}}{OSOI_1} \end{aligned} \tag{7.126}$$

7.18.3.3 Amplifier #2

The input signal we applied to the input of amplifier #1 experiences the power gain of the first amplifier then arrives at the input to amplifier #2. The fundamental input power to amplifier #2 is

$$P_{in,\,f,2} = P_{out,f,1} = P_{in,\,f,1} G_{p,1} \tag{7.127}$$

The fundamental and second-order powers at the output of amplifier #2 are

$$P_{out,f,2} = P_{in,\,f,1} G_{p,1} G_{p,2} \tag{7.128}$$

and

$$\begin{aligned} P_{out,2f,2} &= \frac{P_{out,f,2}^2}{OSOI_2} \\ &= P_{in,\,f,1}^2 \frac{G_{p,1}^2 G_{p,2}^2}{OSOI_2} \end{aligned} \tag{7.129}$$

The second-order output power present at the end of the cascade due to amplifier #2 is

$$\begin{aligned} P_{out,2f,2,pt3} &= P_{out,2f,2} G_{p,3} \\ &= P_{in,\,f,1}^2 \frac{G_{p,1}^2 G_{p,2}^2 G_{p,3}}{OSOI_2} \end{aligned} \tag{7.130}$$

7.18.3.4 Amplifier #3

The fundamental input power to amplifier #3 is

$$P_{in,f,3} = P_{in,f,1} G_{p,1} G_{p,2} \tag{7.131}$$

The fundamental output power of amplifier #3 is

$$P_{out,f,3} = P_{in,f,1} G_{p,1} G_{p,2} G_{p,3} \tag{7.132}$$

The second-order power at the output port of the cascade generated in amplifier #3 is

$$\begin{aligned}P_{out,2f,3} &= P_{out,2f,3,pt3} \\ &= \frac{P_{out,f,3}^2}{OSOI_3} \\ &= P_{in,f,1}^2 \frac{G_{p,1}^2 G_{p,2}^2 G_{p,3}^2}{OSOI_3}\end{aligned} \tag{7.133}$$

7.18.3.5 Second-Order Power Summation

We now have three expressions that describe the second-order distortion power present at the output of the last amplifier in the cascade:

- $P_{out,2f,1,pt3}$ describes the distortion power generated by amplifier #1. We multiplied this power by the gain of the last two stages to bring it to Point 3, the output of the cascade.
- $P_{out,2f,2,pt3}$ describes the distortion power generated by amplifier #2. We've mathematically moved this power to the end of the cascade.
- $P_{out,2f,3,pt3}$ describes the second-order distortion power generated by amplifier #3 present at the end of the cascade.

7.18.3.6 Coherent Summation SOI Equation

The coherent sum of several signals is

$$P_{total,coherent} = \left[P_1^{1/2} + P_2^{1/2} + P_3^{1/2} + \cdots + P_n^{1/2}\right]^2 \tag{7.134}$$

The total second-order power available at the output of the cascade is

$$P_{out,2f,cas} = P_{in,f,1}^2 \left[\sqrt{\frac{G_{p,1}^2 G_{p,2} G_{p,3}}{OSOI_1}} + \sqrt{\frac{G_{p,1}^2 G_{p,2}^2 G_{p,3}}{OSOI_2}} + \sqrt{\frac{G_{p,1}^2 G_{p,2}^2 G_{p,3}^2}{OSOI_3}}\right]^2 \tag{7.135}$$

The cascade gain is

$$G_{p,cas} = G_{p,1} G_{p,2} G_{p,3} \tag{7.136}$$

Combining the cascade gain with equation (7.135) produces

$$\begin{aligned}P_{out,2f,cas} &= P_{in,f,cas}^2 G_{p,cas}^2 \left[\sqrt{\frac{1}{G_{p,2} G_{p,3} OSOI_1}} + \sqrt{\frac{1}{G_{p,3} OSOI_2}} + \sqrt{\frac{1}{OSOI_3}}\right]^2 \\ &= P_{out,f,cas}^2 \left[\sqrt{\frac{1}{G_{p,2} G_{p,3} OSOI_1}} + \sqrt{\frac{1}{G_{p,3} OSOI_2}} + \sqrt{\frac{1}{OSOI_3}}\right]^2\end{aligned} \tag{7.137}$$

Looking at the entire cascade as a single unit, the second-order output power of the cascade is

$$P_{out,2f,cas} = \frac{P_{out,f,cas}^2}{OSOI_{cas}} \tag{7.138}$$

Combining these two equations produces

$$\frac{P^2_{out,f,cas}}{OSOI_{cas}} = P^2_{out,f,cas}\left[\sqrt{\frac{1}{G_{p,2}G_{p,3}OSOI_1}} + \sqrt{\frac{1}{G_{p,3}OSOI_2}} + \sqrt{\frac{1}{OSOI_3}}\right]^2 \tag{7.139}$$

which simplifies to

$$\frac{1}{\sqrt{OSOI_{cas}}} = \frac{1}{\sqrt{G_{p,2}G_{p,3}OSOI_1}} + \frac{1}{\sqrt{G_{p,3}OSOI_2}} + \frac{1}{\sqrt{OSOI_3}} \tag{7.140}$$

The general expression (for n devices in cascade) is

$$\begin{aligned}\frac{1}{\sqrt{OSOI_{cas}}} = &\frac{1}{\sqrt{G_{p,2}G_{p,3}G_{p,4}\ldots G_{p,n-1}OSOI_1}}\\ &+\frac{1}{\sqrt{G_{p,3}G_{p,4}\ldots G_{p,n-1}OSOI_2}}\\ &+\cdots\\ &+\frac{1}{\sqrt{OSOI_n}}\end{aligned} \tag{7.141}$$

The input second-order intercept point for the three-element cascade of Figure 7-44 is

$$\frac{1}{\sqrt{ISOI_{cas}}} = \frac{1}{\sqrt{ISOI_1}} + \frac{1}{\sqrt{ISOI_2/G_{p,1}}} + \frac{1}{\sqrt{ISOI_3/G_{p,1}G_{p,2}}} \tag{7.142}$$

For n devices in cascade, the expression is

$$\begin{aligned}\frac{1}{\sqrt{ISOI_{cas}}} = &\frac{1}{\sqrt{ISOI_1}}\\ &+\frac{1}{\sqrt{ISOI_2/G_{p,1}}}\\ &+\cdots\\ &+\frac{1}{\sqrt{ISOI_n/G_{p,1}G_{p,2}\ldots G_{p,n-1}}}\end{aligned} \tag{7.143}$$

EXAMPLE

SOI Cascade (Coherent Addition)

Figure 7-47 shows three devices in cascade. Find the $ISOI$ of the cascade.

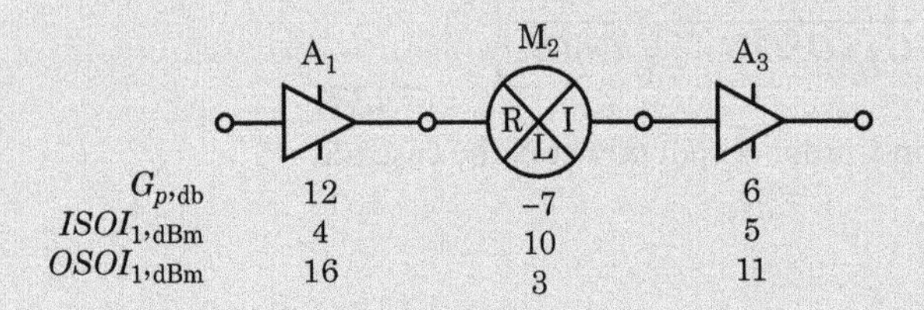

FIGURE 7-47 ■ A simple three-element cascade with known numerical values for the gain and SOI of each element.

Solution

The problem contains a mixture of *ISOI*s and *OSOI*s. We first move the SOI of each device to the input of the cascade and then apply equation (7.143) to the result.

Amplifier #1: We know the $ISOI_{1,dBm}$ at the input of the cascade so

$$ISOI_{amp1,input} = 4 \text{ dBm} = 2.51 \text{ mW} \tag{7.144}$$

Mixer #2: Moving the *ISOI* of the mixer to the input of the cascade produces

$$\begin{aligned} ISOI_{mix2,input} &= ISOI_{dBm,2} - G_{P,dB,1} \\ &= 10 - 12 \\ &= -2 \text{ dBm} \\ &= 0.63 \text{ mW} \end{aligned} \tag{7.145}$$

Amplifier #3: The equivalent input *SOI* of amplifier #3 is

$$\begin{aligned} ISOI_{amp3,input} &= ISOI_{dBm,3} - G_{P,dB,2} - G_{P,dB,1} \\ &= 5 - (-7) - 12 \\ &= 0 \text{ dBm} \\ &= 1.00 \text{ mW} \end{aligned} \tag{7.146}$$

Plugging the equivalent input *SOI* into equation (7.143) produces

$$\begin{aligned} \frac{1}{\sqrt{ISOI_{cas}}} &= \frac{1}{\sqrt{2.51}} \quad \text{(Amp1 contribution)} \\ &+ \frac{1}{\sqrt{0.63}} \quad \text{(Mix1 contribution)} \\ &+ \frac{1}{\sqrt{1.00}} \quad \text{(Amp3 contribution)} \\ ISOI_{cas} &= 0.120 \text{ mW} = -9.2 \text{ dBm} \end{aligned} \tag{7.147}$$

7.18.3.7 Noncoherent Summation SOI Equation

We calculated the equivalent SOI of a cascade by mathematically moving the distortion generated by the components in the cascade to a single point. We then summed up the distortion components assuming that they were all in phase (i.e., coherent addition). This produces the worst-case results described by equations (7.141) and (7.143).

However, in the real world, the distortion components are not always in phase, and we find that the results of equations (7.141) and (7.143) are often too conservative. If we assume the phases of the distortion components are random by the time they arrive at a signal point, we can simply add the powers of all the distortion components together.

For a simple, three-element cascade, noncoherent summation produces the following equations:

$$\frac{1}{OSOI_{cas}} = \frac{1}{G_{p,2}G_{p,3}OSOI_1} + \frac{1}{G_{p,3}OSOI_2} + \frac{1}{OSOI_3} \tag{7.148}$$

and

$$\frac{1}{ISOI_{cas}} = \frac{1}{ISOI_1} + \frac{1}{ISOI_2/G_{p,1}} + \frac{1}{ISOI_3/G_{p,1}G_{p,2}} \tag{7.149}$$

For the general n-element cascade, noncoherent summation produces

$$\begin{aligned}\frac{1}{OSOI_{cas}} = &\frac{1}{G_{p,2}G_{p,3}G_{p,4}\ldots G_{p,n-1}OSOI_1}\\ &+\frac{1}{G_{p,3}G_{p,4}\ldots G_{p,n-1}OSOI_2}\\ &+\cdots\\ &+\frac{1}{OSOI_n}\end{aligned} \tag{7.150}$$

and

$$\begin{aligned}\frac{1}{ISOI_{cas}} = &\frac{1}{ISOI_1}\\ &+\frac{1}{ISOI_2/G_{p,1}}\\ &+\cdots\\ &+\frac{1}{ISOI_n/G_{p,1}G_{p,2}\ldots G_{p,n-1}}\end{aligned} \tag{7.151}$$

EXAMPLE

SOI Cascade (Noncoherent Addition)

Find the *ISOI* of the cascade of Figure 7-47 assuming noncoherent addition.

Solution

Applying equation (7.151) produces

$$\begin{aligned}\frac{1}{ISOI_{cas}} = &\frac{1}{2.51}\text{ (Amp1 contribution)}\\ &+\frac{1}{0.63}\text{ (Mix1 contribution)}\\ &+\frac{1}{1.00}\text{ (Amp3 contribution)}\\ ISOI_{cas} = &\ 0.335\text{mW} = -4.8\text{dBm}\end{aligned} \tag{7.152}$$

The noncoherent assumption implies that our system's *ISOI* is −4.8 dBm. The coherent assumption produces a cascade *ISOI* of −9.2 dBm. Noncoherent addition predicts that the cascade will have a higher SOI (i.e., it will be a more linear system) than coherent addition.

7.18.3.8 Examining the SOI Cascade Equations

Equations (7.141) and (7.143) relate the cascade's second-order performance to the second-order performance of the components in the cascade. The *component contributions, imaging,* and *resistors in parallel* observations we made for the third-order cascade equations apply equally well to the second-order equations.

7.19 | COMPRESSION POINT

The problem of compression points in cascade does not easily lend itself to analysis. Our approach will be to use the rule of thumb from equation (7.80) and state

$$CP_{dBm} \approx TOI_{dBm} - (10 \text{ or } 15 \text{ dB}) \tag{7.153}$$

We calculate the cascade *TOI* then infer the cascade *CP* from the *TOI*.

7.20 | DISTORTION NOTES

In this section, we will discuss a number of idiosyncrasies regarding distortion and nonlinearities.

7.20.1 Third-Order Measurement Difficulties

Upon performing a third-order intercept test, we will occasionally observe the output spectrum shown in Figure 7-48, in which the two nonlinear products are not present at the same power level. We will discuss the causes of this effect in the chapter on Cascades but we must ask, "What is the third-order intercept of this device?" If we use the power of the larger signal, the calculation results in a smaller device *TOI* (i.e. the device will look less linear). Measuring the smaller power produces a larger device *TOI* and the device will appear to be more linear. We might also average the powers of the two components.

The way ahead is not clear. The users should pick a method that is suitable for the particular application.

7.20.2 Input Specs vs. Output Specs

Note the difference between input and output specifications:

$$\begin{aligned} OTOI_{dBm} &= ITOI_{dBm} + G_{p,dB} \\ OSOI_{dBm} &= ISOI_{dBm} + G_{p,dB} \\ OCP_{dBm} &= ICP_{dBm} + G_{p,dB} - 1 \end{aligned} \tag{7.154}$$

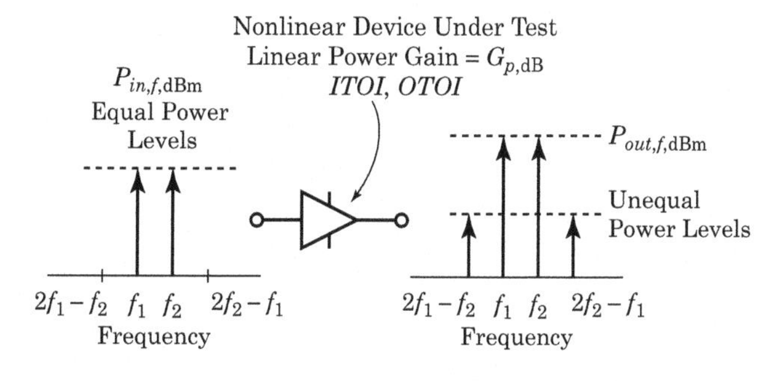

FIGURE 7-48 Third order intercept measurement difficulties. The distortion power levels are not identical.

Engineers are occasionally imprecise and may occasionally not accurately express which specification they are using. If the reference port isn't given, how can we tell whether we're using the input specification or the output specification?

Usually, the vendor will specify the number on the port where it looks the best. Since a larger intercept point is better than smaller one, a vendor will usually specify the linearity of a device on the output port if the device has gain. If the device is lossy (i.e. a mixer or filter), the vendor will usually specify the input numbers.

For example, an amplifier with a 10-dB power gain has an *OSOI* of +7 dBm and an *ISOI* of −3 dBm. The vendor will usually write the specification as "a +7 dBm third-order intercept." However, a mixer might have a gain of -10 dB (a loss of 10 dB), an *OTOI* of +7 dBm and an *ITOI* of +17 dBm. The vendor will usually specify "a +17 dBm third-order intercept."

7.20.3 Model Inadequacy

Occasionally, the results we measure in practice will deviate from the theory presented here. The theory is still sound but we have made several simplifying assumptions during our analysis of nonlinear systems. In practice, it's easy to operate our systems in the regions where these assumptions aren't true. The analysis falls apart and we get bad results.

We've all ready discussed the "weakly nonlinear" approximation and what they mean. Our analysis assumed a small input power level. If we apply too much power to any device, it will no longer behave in a "weakly nonlinear" fashion and this analysis is no longer valid.

Another simplifying assumption was that the transfer function of any nonlinear device can be modeled as a cubic polynomial with a zero DC level or

$$V_{out}(t) = k_1 V_{in}(t) + k_2 V_{in}^2(t) + k_3 V_{in}^3(t) \tag{7.155}$$

If this assumption was always true, then the output spectrum of any third-order nonlinear device will always show a specific symmetry with regard to power levels of the individual components. Figure 7-49(a) shows such a symmetrical spectrum. However, we often observe spectra like Figure 7-49(b). The amplitude of the nonlinear products have lost their symmetry.

Several other mechanisms cause our model to break down and produce nonsymmetrical spectra:

- We have assumed that the gain and port impedances of the device stay constant over frequency and input power. This is not always the case.

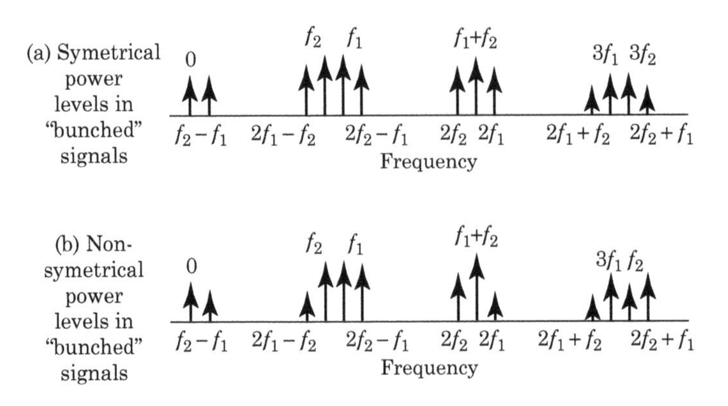

FIGURE 7-49 Symmetrical and nonsymmetrical output spectra. A simpleanalysis predicts a symmetrical output spectrum (a). In practice, we often observe a nonsymmetrical output spectrum (b).

- The third-order relationship between V_{in} and V_{out} describes a memoryless system. The output depends only upon the instantaneous input voltage and nothing else. There is no sense of history. In reality, the devices we build contain inductors and capacitors-these devices have memory. The present state depends on their instantaneous inputs and upon the past inputs.

However, the approximations inherent in the simple third-order equation hold under the most common conditions.

7.20.4 Linearity and Power Consumption

The reader may ask, "If linearity is such a problem, why not just build our systems out of very linear components and be done with it? Then, we don't have to worry about *SOI*, *TOI* or any other nonlinear specification."

Linearity in a particular device always costs us something. There is a strong correlation between the linearity of a device and its DC power consumption. An amplifier exhibiting a high *TOI* will tend to use more DC power than a device with a low *TOI*. The same relationship holds for *SOI* and for *CP*.

A mixer with a high *TOI* will usually require more local oscillator (LO) drive power than a mixer with a low *TOI*.

However, filters and simple resistive attenuators are usually very linear devices. The *TOI* of a bandpass filter, for example, can be several thousand Watts. At high power levels, however, the inductors can saturate and the dielectrics of the capacitors can exhibit nonlinear properties.

7.20.5 Coherent vs. Noncoherent Addition

Coherent addition produces a more conservative result than does noncoherent addition. The coherent assumption will indicate that a cascade is less linear than the same cascade analyzed using noncoherent summation. When evaluating a system, we perform the calculations for both the coherent and noncoherent cases.

In our experience designing low-noise receiving systems, we have found that well-designed cascades usually behave as though the distortion products are adding up noncoherently. For the most part, these systems have achieved the equivalent of noncoherent summation plus one or two dB.

With wideband systems, the cascade *SOI* or *TOI* will stay at noncoherent levels over most of the frequency range of the system. However, over narrow frequency ranges, the *SOI* and *TOI* will increase to coherent summation levels.

In a well-designed system (where the equivalent intercept points of all the devices are equal), the difference between coherent and noncoherent summation is 4 to 5 dB.

7.20.6 Second-Order Distortion and Mixers

A mixer is a device built with **enhanced** second-order performance. We use the sum and difference products arising from the $k_2 V_{in}^2$ term to perform frequency translation, phase detection and other functions. We'll discuss mixers in Chapter 8.

7.21 NONLINEARITIES AND MODULATED SIGNALS

Nonlinearities affect modulated signals in useful and harmful ways. In this section, we'll briefly discuss how nonlinear processes affect modulated signals.

7.21.1 One Modulated Signal

We first explore the effects on a single, modulated signal as it passes through a nonlinear device. Our input waveform is

$$V_{in}(t) = [1 + A(t)] \cos [\omega t + \phi(t)] \tag{7.156}$$

where

$\phi(t)$ represents the PM or FM modulation present on the signal.
$A(t)$ represents the AM modulation present on the signal and

$$0 \leq |A(t)| \leq 1 \tag{7.157}$$

7.21.1.1 Second-Order Output

When the waveform of equation (7.156) undergoes a second-order process, the output waveform is

$$\begin{aligned} V_{out,2}(t) &= V_{in}^2(t) \\ &= [1 + A(t)]^2 \cos^2 [\omega t + \phi(t)] \\ &= \frac{[1 + A(t)]^2}{2} + \frac{[1 + A(t)]^2}{2} \cos [2\omega t + 2\phi(t)] \\ &= \frac{1}{2} + A(t) + \frac{A^2(t)}{2} + \left[\frac{1}{2} + A(t) + \frac{A^2(t)}{2}\right] \cos [2\omega t + 2\phi(t)] \end{aligned} \tag{7.158}$$

The second term on the last line of equation (7.158) [the $A(t)$] is the amplitude modulation present on the input signal [equation (7.156)]. We can low-pass equation (7.158) to remove the high-frequency components:

$$V_{out,2,LowPassed}(t) = \frac{1}{2} + A(t) + \frac{A^2(t)}{2} \tag{7.159}$$

The low-passed, second-order process returns DC, the AM signal and a distortion term involving $A^2(t)$. Thus, we can use a device with second-order characteristics to demodulate an AM waveform. We can recover the AM present on $V_{in}(t)$ even though the input waveform exhibits angle modulation [i.e. the FM or PM represented by $\phi(t)$].

The output waveform also contains the term $A^2(t)/2$ which is the AM signal squared. If the magnitude of $A(t)$ is much smaller than unity, then the second-order distortion will be very small and we can often ignore it.

If the input signal was a quiet carrier (i.e. both $A(t)$ and $\phi(t)$ were both zero), then the output waveform would be

$$V_{out,2}^2(t) = \frac{1}{2} + \frac{1}{2} \cos(2\omega t) \tag{7.160}$$

The output frequency is twice the input frequency. We can use a squaring device to double the frequency of an oscillator.

If the input signal contained only phase modulation (i.e. $A(t) = 0$ and $\phi(t) \neq 0$), then the output waveform would be

$$V_{out,2}(t) = \frac{1}{2} + \frac{1}{2}\cos\left[2\omega t + 2\phi(t)\right] \tag{7.161}$$

The deviation at the second harmonic has doubled via the $2\phi(t)$ term of equation (7.158). This effect accounts for the increase in the phase noise of an oscillator when we double the frequency of an oscillator. We discuss this concept more deeply in Chapter 9 on Oscillators.

7.21.1.2 Third-Order Outputs

If we apply our input waveform [equation (7.156)] to a third-order nonlinear device, the output is

$$\begin{aligned} V_{out,3}(t) &= V_{in}^3(t) \\ &= [1 + A(t)]^3 \cos^3\left[\omega t + \phi(t)\right] \\ &= \left[1 + 3A(t) + 3A^2(t) + A^3(t)\right] \\ &\quad \cdot \left\{\frac{3}{4}\cos\left[\omega t + \phi(t)\right] + \frac{1}{4}\cos\left[3\omega t + 3\phi(t)\right]\right\} \end{aligned} \tag{7.162}$$

The signal at the fundamental frequency and at the third harmonic of the fundamental frequency exhibit several AM distortion components. The deviation of the signal present at the third harmonic, expressed as $3\phi(t)$, has increased three-fold.

7.21.1.3 Frequency Deviation

The deviation of a signal will increase directly with the nonlinear order n. If we pass a signal with a 100 kHz bandwidth through a fourth-order device, the output signal will exhibit a 400 kHz deviation. This is one way to determine the order of a signal generated by a nonlinear effect. For example, an FM radio station has an FCC-defined bandwidth of about 150 kHz. If you find an FM radio station with a bandwidth of 3(150) = 450 kHz, it is likely due to third-order distortion somewhere in the system.

EXAMPLE

Nonlinearities and Deviation

While tuning through the spectrum, you find a signal that sounds a lot like a commercial cellular telephone channel. The signal is FM and sounds the best when the receiver's IF bandwidth is 100 kHz. What is the order of the distortion term causing this signal?

Solution

Since a normal, nondistorted cellular telephone signal has an RF bandwidth of about 20 kHz, we judge that this signal was probably generated by a fifth-order nonlinearity.

7.21.2 One Modulated Signal, One Quiet Carrier

We now apply two signals simultaneously to our nonlinear device-one signal is modulated while the other signal contains no modulation. The modulated signal is:

$$V_{mod}(t) = [1 + A(t)]\cos[\omega_1 t + \phi(t)] \tag{7.163}$$

We can describe the quiet carrier as

$$V_{quiet}(t) = \cos(\omega_2 t) \tag{7.164}$$

The input signal is the sum of these two waveforms or

$$\begin{aligned} V_{in}(t) &= V_{mod}(t) + V_{quiet}(t) \\ &= [1 + A(t)]\cos[\omega_1 t + \phi(t)] + \cos(\omega_2 t) \end{aligned} \tag{7.165}$$

7.21.3 Second-Order Outputs

Applying a second-order process to the input waveform of equation (7.165) produces

$$\begin{aligned} V_{out,2}(t) &= V_{in}^2(t) \\ &= \{[1 + A(t)]\cos[\omega_1 t + \phi(t)] + \cos(\omega_2 t)\}^2 \\ &= \frac{[1 + A(t)]^2}{2} + \frac{1}{2} + \frac{[1 + A(t)]^2}{2}\cos[2\omega_1 t + 2\phi(t)] \\ &\quad + \frac{1}{2}\cos(2\omega_2 t) \\ &\quad + [1 + A(t)]\cos[(\omega_1 + \omega_2)t + \phi(t)] \\ &\quad + [1 + A(t)]\cos[(\omega_1 - \omega_2)t + \phi(t)] \end{aligned} \tag{7.166}$$

We again see the AM demodulation (with the associated distortion) and the signals generated at the second harmonics of the two input signals. The interesting components of the output waveform are centered about $\omega_1 \pm \omega_2$:

$$\begin{gathered} [1 + A(t)]\cos[(\omega_1 + \omega_2)t + \phi(t)] \\ \textit{and} \\ [1 + A(t)]\cos[(\omega_1 - \omega_2)t + \phi(t)] \end{gathered} \tag{7.167}$$

If these two components do not overlap in the frequency domain, we've moved our modulated signal to two new frequencies *without distortion.* Both the AM and FM are intact and undistorted. This effect is invaluable because it is the mathematical basis for mixing or frequency translation.

7.21.4 Two Modulated Signals

We next examine the effects of a nonlinear device on two modulated signals. Each signal will have both AM and FM (or PM). The signals are

$$V_{mod,1}(t) = [1 + A_1(t)]\cos[\omega_1 t + \phi_1(t)] \tag{7.168}$$

and

$$V_{mod,2}(t) = [1 + A_2(t)]\cos[\omega_2 t + \phi_2(t)] \tag{7.169}$$

The input signal is the sum of these two waveforms or

$$\begin{aligned} V_{in}(t) &= V_{mod,1}(t) + V_{mod,2}(t) \\ &= [1 + A_1(t)]\cos[\omega_1 t + \phi_1(t)] + [1 + A_2(t)]\cos[\omega_2 t + \phi_2(t)] \end{aligned} \tag{7.170}$$

7.21.4.1 Second-Order Outputs

Applying a of a second-order process to equation (7.170) produces

$$\begin{aligned} V_{out,2}(t) &= V_{in}^2(t) \\ &= \{[1 + A_1(t)]\cos[\omega_1 t + \phi_1(t)] + [1 + A_2(t)]\cos[\omega_2 t + \phi_2(t)]\}^2 \\ &= [1 + A_1(t)]^2 \cos^2[\omega_1 t + \phi_1(t)] \\ &\quad + 2[1 + A_1(t)][1 + A_2(t)]\cos[\omega_1 t + \phi_1(t)]\cos[\omega_2 t + \phi_2(t)] \\ &\quad + [1 + A_2(t)]^2 \cos^2[\omega_2 t + \phi_2(t)] \end{aligned} \tag{7.171}$$

so

$$\begin{aligned} V_{out,2}(t) &= \frac{[1 + A_1(t)]^2}{2} + \frac{[1 + A_2(t)]^2}{2} \\ &\quad + \frac{[1 + A_1(t)]^2}{2}\cos[2\omega_1 t + 2\phi_1(t)] \\ &\quad + \frac{[1 + A_2(t)]^2}{2}\cos[2\omega_2 t + 2\phi_2(t)] \\ &\quad + [1 + A_1(t) + A_2(t) + A_1(t)A_2(t)]\cos[(\omega_1 + \omega_2)t + \phi_1(t) + \phi_2(t)] \\ &\quad + [1 + A_1(t) + A_2(t) + A_1(t)A_2(t)]\cos[(\omega_1 - \omega_2)t + \phi_1(t) - \phi_2(t)] \end{aligned} \tag{7.172}$$

The modulation of the input signals is distributed to the second-order output terms in various ways. In the chapter on Mixers, we will examine the use of second-order nonlinearities to translate a signal from one center frequency to another. Referring to equation (1.53), the two terms we'll use to perform this translation are

$$\begin{aligned} &[1 + A_1(t) + A_2(t) + A_1(t)A_2(t)]\cos[(\omega_1 + \omega_2)t + \phi_1(t) + \phi_2(t)] \\ &\qquad\qquad \textit{and} \\ &[1 + A_1(t) + A_2(t) + A_1(t)A_2(t)]\cos[(\omega_1 - \omega_2)t + \phi_1(t) - \phi_2(t)] \end{aligned} \tag{7.173}$$

We strive make one signal a quiet carrier i.e. it has no amplitude or phase modulation. This is our Local Oscillator or LO. However, the LO will always posses some irreducible amplitude or phase modulation. Equation (7.172) states that any modulation present on the LO will be transferred to both of the output signals.

We can often ignore the LO amplitude modulation (i.e. we can assume that $A_2(t) = 0$) in realistic situations. When the local oscillator exhibits only phase noise, then the last two terms of equation (7.172) become

$$\begin{aligned} &[1 + A_1(t)]\cos[(\omega_1 + \omega_2)t + \phi_1(t) + \phi_2(t)] \\ &\qquad\qquad \textit{and} \\ &[1 + A_1(t)]\cos[(\omega_1 - \omega_2)t + \phi_1(t) - \phi_2(t)] \end{aligned} \tag{7.174}$$

The oscillator phase noise is transferred directly to the two output signals. This effect often limits the processing of radar signals and it can limit the ultimate bit-error rate of digital phase-modulated signals.

7.21.5 Cross Modulation

If we apply two modulated signals [equation (1.51)] to a nonlinear device whose transfer characteristic is

$$V_{out}(t) = k_1 V_{in}(t) + k_2 V_{in}^2(t) + k_3 V_{in}^3(t) \tag{7.175}$$

the output signal will be a combination of the linear, second-order and third-order components of the device.

Let's examine all of the contributions to the output signal power present at one particular frequency, say f_1. Examination of equations (7.170), (7.172) and other analyses reveals that the total signal voltage present at f_1 will be

$$\left\{ k_1[1 + A_1(t)] + k_3 \frac{3[1 + A_1(t)]^3}{4} + k_3 \frac{3[1 + A_1(t)][1 + A_2(t)]^2}{2} \right\} \cos[\omega_1 t + \phi_1(t)] \tag{7.176}$$

where k_1 and k_3 refer to the amplifier's transfer polynomial.

The signal present at f_1 contains the AM from the input signal at f_2. This is due to the third-order distortion of the amplifier. In general, if we examine the output present at one frequency, we'll often find that the distortion of the amplifier has impressed the modulation of one signal onto a second signal. This is called "cross modulation."

War Story—Radio Controlled Pterodactyls

Sometime in the early 80s, there was some controversy about whether pterodactyls could fly, how far they could fly, how much they could lift, etc. A paleobiologist built a radio-controlled pterodactyl and flew it about a dozen times near the college where he worked. It worked wonderfully.

The professor began performing demonstrations around the country and eventually took his machine to Andrews Air Force Base. For a while, the pterodactyl flew flawlessly. It banked and dived until the professor decided to take the bird up a little higher. Suddenly, the pterodactyl went haywire—it froze up, shuddered in the air, then dropped suddenly and shattered on the ground.

Follow-on forensics reveals that when the pterodactyl flew high it got within sight of a radar transmitter at the air force base. The radar jammed the remote-control receivers in the pterodactyl, which put the beast into a deadly spin. Any strong signal can jam a receiver—not just a strong signal that lies within your band of interest. As receiver designers, one of our main jobs is to limit bandwidth as quickly as we can.

7.22 | BIBLIOGRAPHY

Blattenberger, Kirt, "Cascaded 1 dB Compression Point (P_{1dB})," http://www.rfcafe.com/references/electrical/p1db.htm, 2010.

Gross, Brian P., "Calculating the Cascade Intercept Point of Communications Receivers," *Ham Radio*, p. 50, August 1980.

Ha, Tri T., *Solid-State Microwave Amplifier Design*, John Wiley and Sons; 1981.

Hewlett-Packard Corp., *Fundamentals of RF and Microwave Noise Figure Measurement*, Application Note 57-1, July 1983.

Horowitz, Paul, and Hill, Winfield, *The Art of Electronics*, Cambridge University Press, 1980.

Perlow, Stewart M., "Basic Facts about Distortion and Gain Saturation," *Applied Microwaves*, p. 107, May 1989.

Terman, Frederick Emmons, *Electronic and Radio Engineering*, McGraw-Hill, 1955.

Williams, Richard A., *Communications Systems Analysis and Design—A Systems Approach*, Prentice-Hall, Inc., 1987.

7.23 | PROBLEMS

1. **Filter Calculations/Spurious Frequencies** Answer the following questions for the system and input spectrum shown in Figure 7-50. The component characteristics are

   - BPF_1: 910 MHz CF, 30 MHz BW, 2 dB IL
   - BPF_2: 930 MHz CF, 20 MHz BW, 5 dB IL
   - All filters are five-pole Butterworth filters
   - Amplifier gain = 15 dB

   a. What are the output power levels of the signals at 830, 920, and 1,010 MHz? Assume that the amplifier is perfectly linear for part A.

   b. If the amplifier is nonlinear, what are the possible frequencies you might see at the amplifier's output? Perform the analysis up to and including fifth-order nonlinearities.

2. **Spurious Signal Power** Given the nonlinear system shown in Figure 7-51, if the power of input signal A increases by 1 dB, how much would the power of output signal B increase if

   a. Signal B is generated by a second-order system nonlinearity?

   b. Signal B is generated by a third-order system nonlinearity?

   c. If the power of signal A increases by 1 dB and power of output signal B increases by 2.5 dB, what can you conclude about the origin of signal B?

3. **Spurious Responses Essay** Under what conditions is second-order distortion more of a concern than third-order distortion? Why?

4. **Spurious Responses Essay** Why is third-order distortion a fundamental limit to the nonlinear performance of a system? In other words, we worry about second-order distortion under some conditions. We worry about third-order distortion under other conditions, but we usually stop our analysis at third-order distortion. Why?

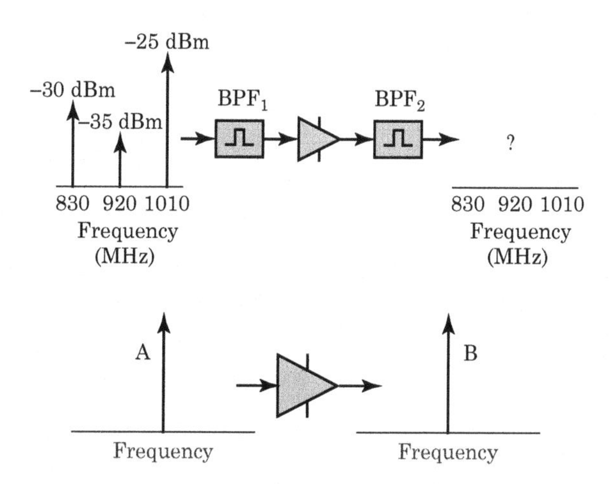

FIGURE 7-50 ■ Filter calculations and spurious signals.

FIGURE 7-51 ■ A signal passing through a nonlinear amplifier.

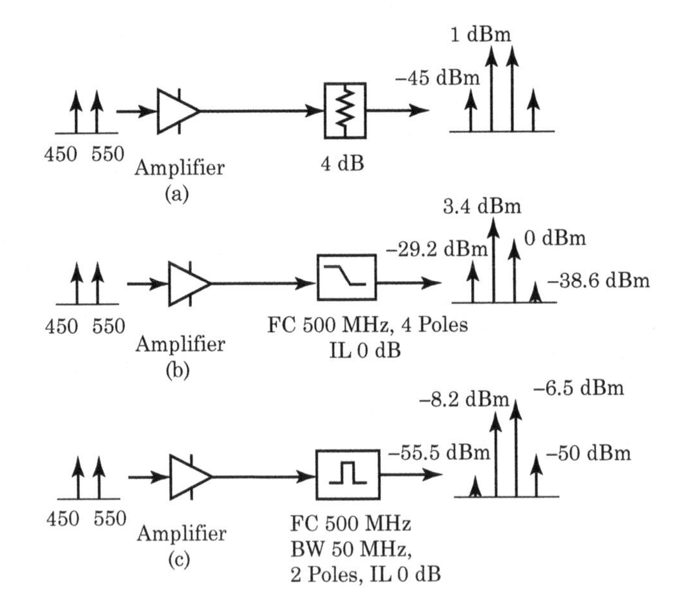

FIGURE 7-52 Three amplifiers with different *TOI* specifications.

5. **Linearity Ranking** Arrange the amplifiers of Figure 7-52 in order from least linear to most linear. Each input tone is −25 dBm. The lower frequency tone is 450 MHz; the upper frequency tone is 550 MHz.

   Assume all filters are Butterworth. Also, look at both the input and output *TOI* to make your linearity judgments.

6. **Noise Figure versus Linearity** Answer the following questions with either *linearity* or *noise figure*, indicating whether linearity or noise figure is the more important consideration. Explain your answer.

   a. An urban environment
   b. A desert environment
   c. A cable TV repeater amplifier
   d. A TV station power amplifier
   e. A television amplifier placed directly after the TV antenna
   f. A probe of the planet Saturn

7. **Multiple Choice: *TOI*** An amplifier has an *ITOI* of 0 dBm. If you apply a 0 dBm signal to this amplifier, it is:

   a. Operating in its linear region but just barely. Any increase in signal power will push the amplifier into distortion.
   b. Operating well into saturation.
   c. Operating well below saturation.
   d. Operating just at its 1 dB input compression point (ICP).

8. **Multiple Choice: Two Signals, One in Saturation** A system has an *ITOI* of −20 dBm. Two complex modulated signals are applied to the system. One signal is at −20 dBm, and the other signal is at −80 dBm. Which statements are true?

   a. The signal at −20 dBm will be heavily distorted, while the signal at −80 dBm will pass through the system unscathed.
   b. Both signals will be heavily distorted.

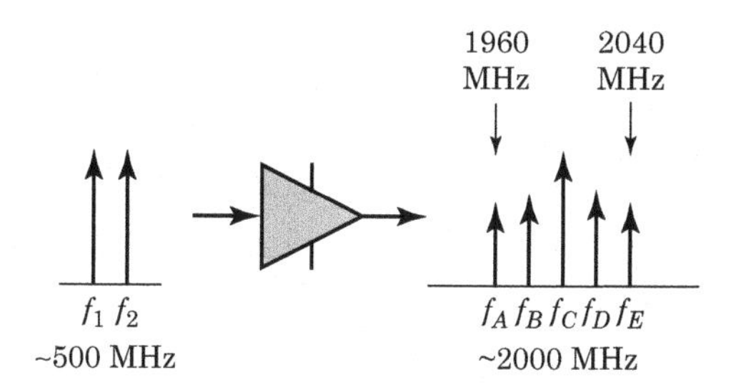

FIGURE 7-53 ■ A nonlinear amplifier generating spurious signals.

c. Both signals will pass through the amplifier without significant distortion.

d. The amplifier will draw significantly more current from its power supply to process the larger signal.

9. Multiple Choice: Linearity Specifications Which characteristics are, at least partially, measurements of a system's linearity:

a. Spurious-free dynamic range
b. Noise figure
c. Third-order intercept
d. Minimum detectable signal

10. Multiple Choice: Wideband Designers True or False: Wideband receiver designers are more interested in third-order intercept linearity specifications than they are in second-order intercepts.

11. More Spurious Signals Running a test, a design engineer applies two signals to an amplifier. The frequencies of the two signals are in the neighborhood of 500 MHz. The engineer observes five equally spaced signals at the output of the amplifier. The highest frequency observed is 2,040 MHz. The lowest frequency observed is 1,960 MHz. Figure 7-53 shows the input and output spectra.

a. What are the two input and five output frequencies?
b. What is the equation that generates each spurious signal (e.g., $4f_1, 3f_1 - f_2$)?
c. What is the highest possible order of the equation that produced this result?

12. Filtering Third-Order Distortion

a. Given the system shown in Figure 7-54, find the gain, noise figure, coherent *ITOI*, and noncoherent *ITOI* of the three-element amplifier cascade. Ignore the filter in these calculations.

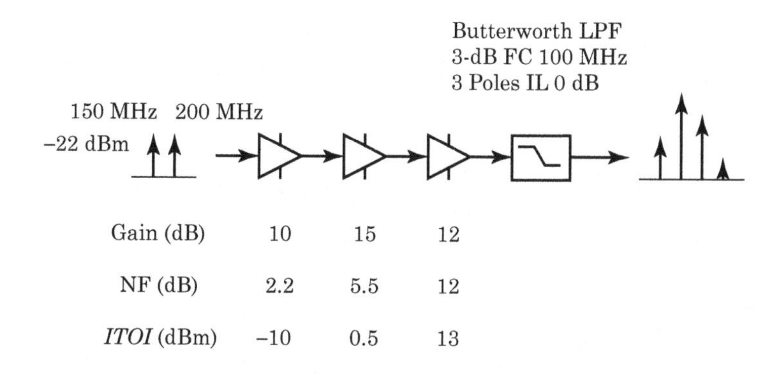

FIGURE 7-54 ■ A system exhibiting nonlinear behavior.

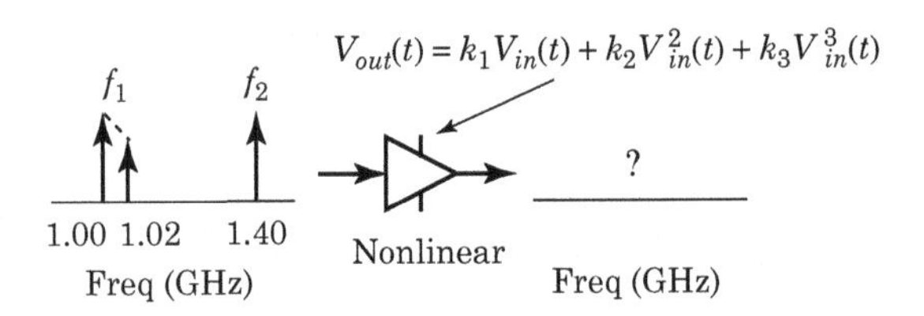

FIGURE 7-55 ■ Modulated and unmodulated signals applied to a nonlinear device.

b. We apply two input tones to the cascade. Both tones are at –22 dBm. Tone #1 is at 150 MHz, and tone #2 is at 200 MHz. Find the power and frequency of each of the four tones shown on the output of the LPF.

13. **Differences between Linear and Nonlinear Systems** Describe three ways linear systems differ from nonlinear systems.

14. **Second-Order Distortion versus Third-Order Distortion** Describe the situations in which second-order distortion, versus third-order distortion, is the limiting factor to a system's linearity.

15. **Nonlinear Processes and Modulation** Figure 7-55 shows a modulated signal and an unmodulated signal being fed into a nonlinear amplifier. One signal, labeled f_1, is frequency modulated with a bandwidth of 20 MHz. The other signal, labeled f_2, is an unmodulated carrier. The amplifier's voltage transfer function is a third-order polynomial.

    Find the frequencies of the signals present on the output of the amplifier. Describe the modulation present on each of the output signals. What is the bandwidth of each output signal?

    Hint: A frequency-modulated (FM) signal is a single sine wave whose frequency changes over time. In this example, the single sine wave moves from 1.00 to 1.02 GHz.

16. **Nonlinearity in Cascade**

    a. Find the gain, *NF*, *ITOI*, *OTOI*, *ISOI*, and *OSOI* of the cascade shown as Figure 7-56. Use the coherent equations.

    b. What are the *TOI* and *SOI* of each component when referenced to the input of the cascade?

17. **Linearity of Components** Arrange these components in order from least linear to most linear:

    a. An amplifier of 20 dB of gain and a –4 dBW *OTOI*

    b. An amplifier whose gain is 14 dB and whose *ITOI* is 8 mW

    c. An amplifier with 10 dB of power gain and a –3 dBm *ITOI*

18. **Second-Order Distortion** A commercial FM radio receiver covers 88.1 to 108.1 MHz. The channel spacing is 200 kHz. In other words, the valid stations fall at

$$88.1, 88.3, 88.5, \ldots, 107.5, 107.9, 108.1 \text{ MHz}$$

FIGURE 7-56 ■ An RF cascade.

	(1)	(2)	(3)	(4)	(5)
$G_{P,\text{dB}}$	18	–7	–4	19	–9
F_{dB}	3	14	4	9	8
$ISOI_{\text{dBm}}$	–3	–8	40	–7	10
$ITOI_{\text{dBm}}$	2	1	55	2	20

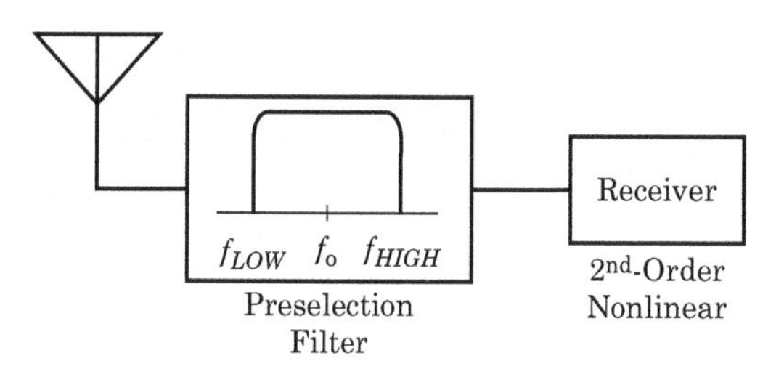

FIGURE 7-57 ▪ A receiver proceeded by a preselection filter.

Which combination of the following input signals, if applied to the receiver, could cause harmful second-order distortion? What is the nature of the problem?

a. 88.10 and 11.60 MHz

b. 108.10 and 11.60 MHz

c. 55.70 and 43.50 MHz

d. 49.55 and 57.55 MHz

e. 40.10 and 33.4 MHz

19. **Filtering and Second-Order distortion** As a general rule, we place a band-pass filter between the antenna and the first active element of a receiver to restrict the number of signals applied to the first active element (see Figure 7-57).

By restricting the number of signals applied to the receiver, we reduce the chance that strong signals can combine in a nonlinear fashion to produce unwanted distortion products that will affect signals in the receiver's passband.

For a commercial FM broadcast receiver that covers 88.1 to 108.1 MHz, what combination of input signals sets the upper and lower limits (f_{LOW} and f_{HIGH}) of the preselection filter?

In other words, to find f_{LOW}, examine the spectrum below the preselection filter's passband. Find the two highest-frequency signals that could combine via a second-order process to produce a signal that falls in the receiver's passband.

To find f_{HIGH}, examine the spectrum above the preselection filter's passband. Find the two lowest-frequency signals that could combine via a second-order process to produce a signal that falls in the receiver's passband.

20. **Spurious Signal Identification** Figure 7-58 and Figure 7-59 show an ISM-band receiver. Figure 7-58 shows the receiver tuned to 920 MHz. Figure 7-59 shows the receiver tuned to 920.5 MHz. The input spectra are identical, but the output spectra are different.

a. Describe the source of each signal (A through J). Some signals present in one spectrum may be obscured by stronger signal in the other spectrum. Signals with identical letters in each plot are from identical sources.

b. Give one reason signal B is at –15 dBm instead of –10 dBm.

c. What is the *OTOI* of this receiver?

21. **Cable TV Amplifier Specification** Cable television amplifiers often list a linearity specification called *composite triple beat* (CTB). Look up composite triple beat on the Web.

Why do you think the linearity of cable television amplifiers is specified this way (with CTB) rather than some other linearity specification (like third-order intercept or compression point)? What effect are they trying to measure?

22. **Second-Order Distortion** A commercial FM radio receiver is tuned to 103.1 MHz. What combination of signals below can cause harmful *second-order* distortion? In other words, which combination of two signals can combine in such a way as to generate a 103.1 MHz signal? Why?

a. 114.800 and 11.700 MHz

b. 102.7 and 102.9 MHz

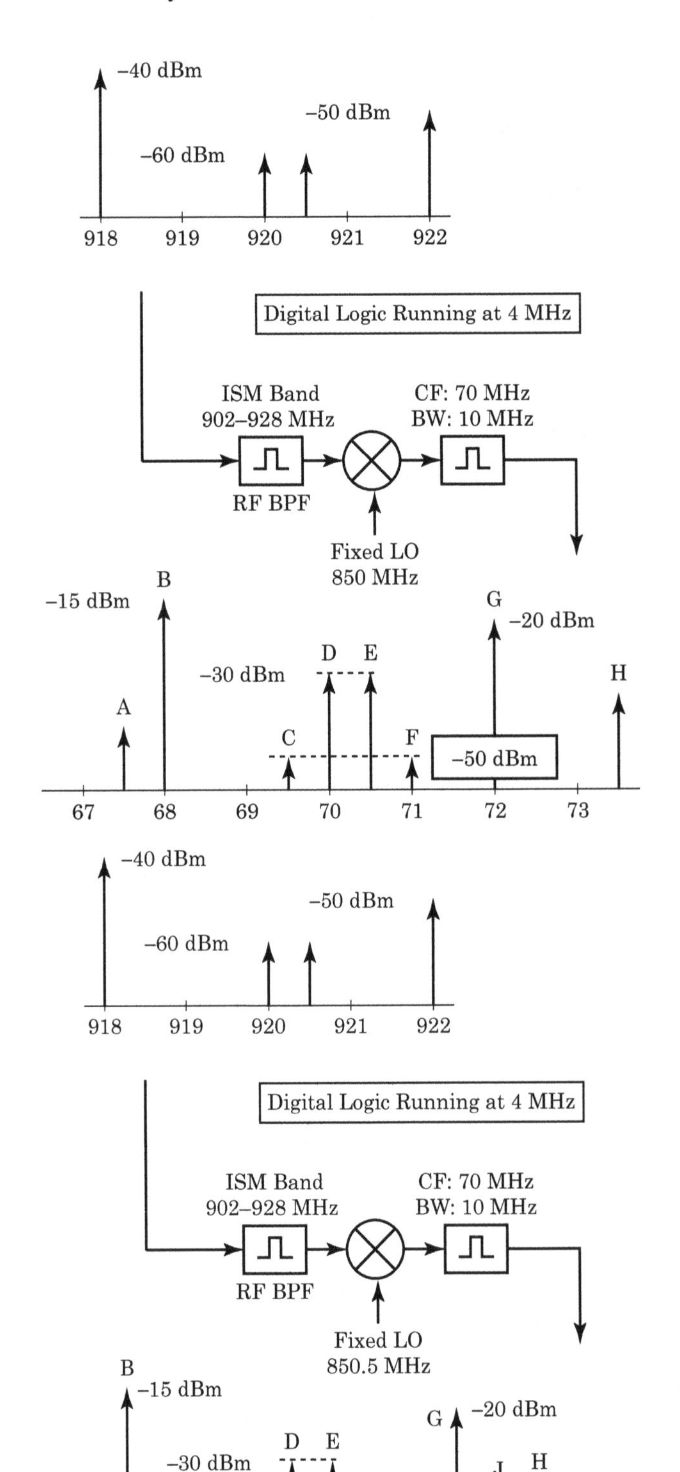

FIGURE 7-58 ■ A receiver processing many signals. The receiver is tuned to 920 MHz.

FIGURE 7-59 ■ The receiver is tuned to 920.5 MHz.

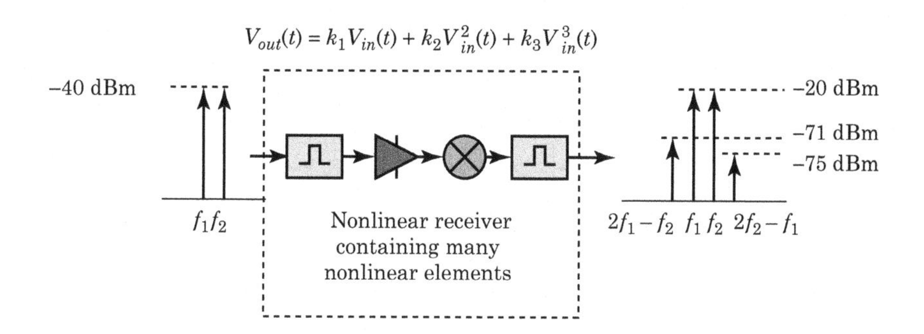

FIGURE 7-60 ■ A third-order test on a complex receiver.

c. 152.000 MHz and 255.100 MHz

d. 214.7 and 657.000 MHz

e. 51.55 MHz

23. Measurement of Third-Order Intercept Figure 7-60 shows a *TOI* measurement of a complex receiving system. The two output tones were measured to have different powers.

a. What would you specify the *ITOI* of this receiver to be and why?

b. What might have caused this effect?

24. Tone Spacing Figure 7-61 shows a *TOI* test on a cascade. The receiver is tuned to 750 MHz, and the two test tones are placed below the tuned frequency of the receiver.

- In one test, we measure the *TOI* of the cascade by placing the two test tones at f_1 = 749.9 MHz and f_2 = 749.8 MHz.
- In a second test, we measure the *TOI* of the cascade by placing the two test tones at f_1 = 749.0 MHz and f_2 = 748.0 MHz.

Which test will show a higher (i.e., better) *TOI*? Why?

25. Tone Spacing Given the amplifier and input spectrum shown in Figure 7-62, draw the output spectrum assuming that the amplifier exhibits only a second-order response.

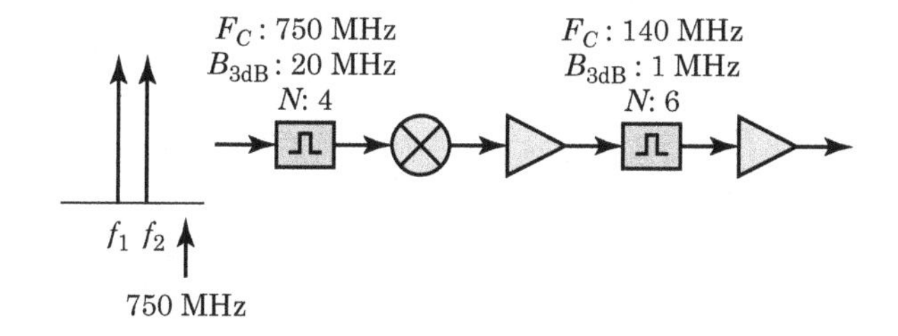

FIGURE 7-61 ■ Running a *TOI* test on an RF cascade.

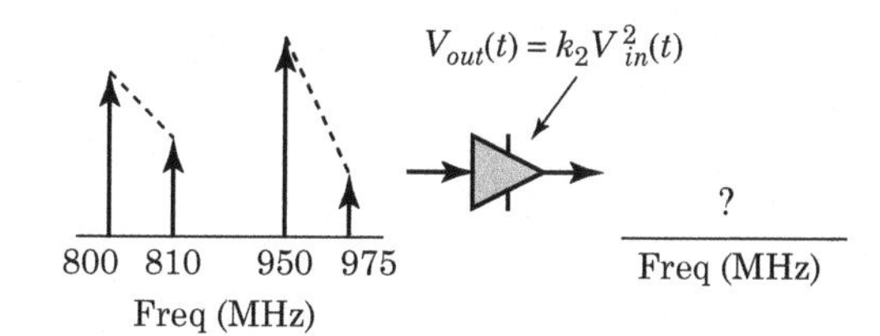

FIGURE 7-62 ■ Spectrum applied to a second-order nonlinear device.

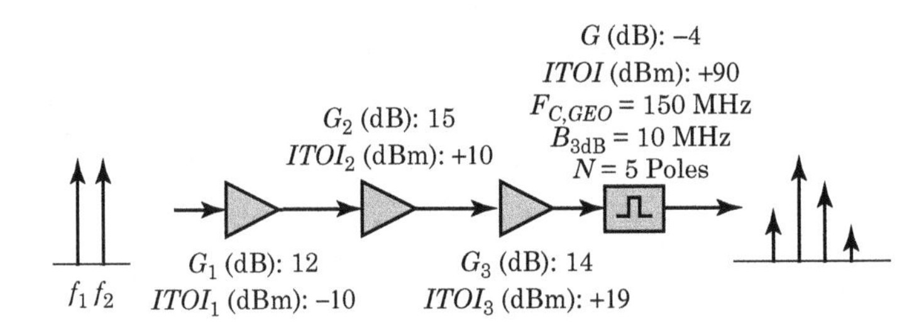

FIGURE 7-63 ■ A nonlinear RF system processing two input signals.

26. Signal Power Given the system shown in Figure 7-63 and the following data:

- Input signal #1: $f_1 = 148$ MHz, $P_{dBW} = -61$ dBW
- Input signal #2: $f_2 = 153$ MHz, $P_{dBW} = -61$ dBW

Find

a. The frequency of each of the four output signals in MHz, if the additional two tones are generated by third-order distortion within the cascade

b. The power level of each of the four output frequencies in dBm

Assume coherent addition of the third-order components. The band-pass filter is a Butterworth filter. Note that the output power levels shown in the figure may not be indicative of the actual power levels calculated.

27. Noise Figure/Linearity Optimization Figure 7-64 shows a cascade. Assume coherent addition of the nonlinear components.

a. Which component is contributing the most noise to this cascade and why?

b. Which component is limiting the second-order intercept of this cascade and why?

28. Filtering Third-Order Distortion Given the system shown in Figure 7-65.

a. Find the gain, noise figure, coherent *ITOI*, and noncoherent *ITOI* of the three-element amplifier cascade. Ignore the filter in these calculations.

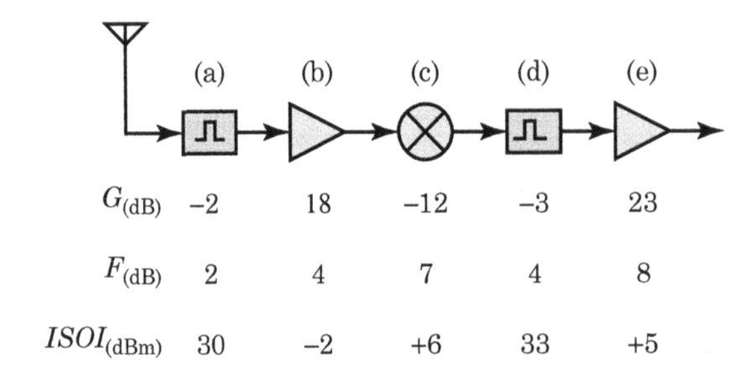

FIGURE 7-64 ■ Nonlinear cascade.

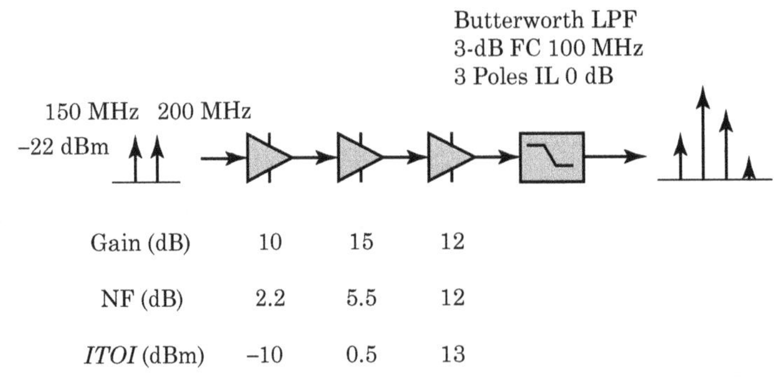

FIGURE 7-65 ■ Filtering and third-order intercept.

	#1	#2	#3	#4	#5
G_{dB}	−3	15	−7	−3	15
F_{dB}	3	2.5	7	3	4
$OTOI_{dBm}$	40	5	5	40	6

FIGURE 7-66 ■ A five-element cascade.

b. We apply two input tones to the cascade. Both tones are at −22 dBm. Tone #1 is at 150 MHz, and tone #2 is at 200 MHz. Find the power and frequency of each of the four tones shown on the output of the LPF.

29. Differences between Linear and Nonlinear Systems Describe three ways linear systems differ from nonlinear systems.

30. *n*-th-Order Intercept

a. Derive the formula for the n-th-order signal levels in terms of the fundamental input signal level and input n-th-order intercept (*INOI*).

b. Derive the spur-free dynamic range (SFDR) in terms of N, the *MDS*, and the device's *INOI*.

31. Optimizing a Cascade For the receiver chain shown in Figure 7-66:

a. Arrange the components in order of the amount of noise they contribute to the cascade.

b. Arrange the components in order of the amount of distortion they contribute to the cascade.

c. Find the cascade *MDS* and *SFDR* for a 1 MHz noise bandwidth.

d. Attempt to improve the *SFDR* by adjusting the gain and placement of the components while maintaining the overall cascade gain. Find the new values for F, *ITOI*, *OTOI*, and *SFDR* for each trial.

32. Spurious Signals Running a test, a design engineer applies two signals to an amplifier. The frequencies of the two signals are in the neighborhood of 900 MHz. The engineer observes eight equally spaced signals at the output of the amplifier. The highest frequency observed is 2,830 MHz. The lowest frequency observed is 2,690 MHz.

a. What are the two input and eight output frequencies?

b. What is the equation that generates each spurious signal (e.g., $4f_1$, $3f_1 - f_2$).

c. What is the highest possible order of the equation that produced this result?

33. Second-Order Distortion A cellular telephone receiver is tuned to 872.700 MHz. What combination of signals below can cause harmful second-order distortion?

a. 861.000 and 11.700 MHz

b. 240.000 MHz and 1113.700 MHz

c. 214.7 and 657.000 MHz

d. 436.35 MHz

34. Second-Order Distortion and Modulation Describe the effects of passing the following signals through a second-order process. What happens to the modulation? Can we still retrieve the transmitted data stream $D(t)$ from the signal?

a. BPSK: A BPSK signal is described by

$$\begin{aligned} V_{BPSK}(t) &= \cos\left(\omega_{RF}t + D(t)\pi\right); \quad D(t) \in [0, 1] \\ &= \pm\cos\left(\omega_{RF}t\right) \\ &= \cos\left(\omega_{RF}t\right) \ or\ \cos\left(\omega_{RF}t + \pi\right) \end{aligned} \tag{7.177}$$

b. QPSK: A QPSK signal is described by

$$\begin{aligned} V_{QPSK}(t) &= \cos\left(\omega_{RF}t + \frac{(2D(t)+1)\pi}{4}\right); \quad D(t) \in= [0, 1, 2, 3] \\ &= \cos\left(\omega_{RF}t + \frac{\pi}{4}\right), \cos\left(\omega_{RF}t + \frac{3\pi}{4}\right), \\ &\cos\left(\omega_{RF}t + \frac{5\pi}{4}\right) \ or\ \cos\left(\omega_{RF}t + \frac{7\pi}{4}\right) \end{aligned} \tag{7.178}$$

35. Third-Order Distortion and Modulation Describe the effects of passing the following signals through a third-order process. What happens to the modulation? Can we still retrieve the transmitted data stream $D(t)$ from the signal?

a. BPSK: A BPSK signal is described by

$$\begin{aligned} V_{BPSK}(t) &= \cos\left(\omega_{RF}t + D(t)\pi\right); \quad D(t) \in [0, 1] \\ &= \pm\cos\left(\omega_{RF}t\right) \\ &= \cos\left(\omega_{RF}t\right) \ or\ \cos\left(\omega_{RF}t + \pi\right) \end{aligned} \tag{7.179}$$

b. FSK: An FSK signal is described by

$$\begin{aligned} V_{FSK}(t) &= \cos\{[\omega_{RF} + \omega_{\Delta}D(t)]t\}; \quad D(t) \in= [-1, 1] \\ &= \cos\{[\omega_{RF} + \omega_{\Delta}]t\} \ or\ \cos\{[\omega_{RF} - \omega_{\Delta}]t\} \end{aligned} \tag{7.180}$$

36. Specifying and Measuring *TOI* Figure 7-67 shows the real-life TOI measurement of a complex receiving system. The two output tones were measured to have different powers.

a. What would you specify the *ITOI* of this receiver to be and why?

b. What might have caused this effect?

37. Third-Order Response of a Quiet Carrier Summed with a Modulated Signal Given a signal that is the sum of a quiet carrier and a generic modulated signal,

$$\begin{aligned} V_{in}(t) &= V_{mod}(t) + V_{quiet}(t) \\ &= [1 + A(t)]\cos\left[\omega_1 t + \phi(t)\right] + \cos\left(\omega_2 t\right) \end{aligned} \tag{7.181}$$

find the third-order output expression and comment on the output signals.

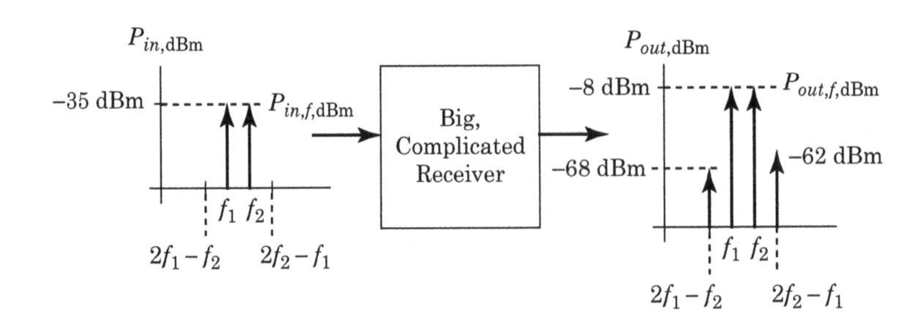

FIGURE 7-67 ■ A real life *TOI* measurement.

38. Third-Order Response of Two Summed Modulated Signals Given a signal that is the sum of two generic modulated signals. The signals are

$$V_{mod,1}(t) = [1 + A_1(t)]\cos[\omega_1 t + \phi_1(t)] \tag{7.182}$$

and

$$V_{mod,2}(t) = [1 + A_2(t)]\cos[\omega_2 t + \phi_2(t)] \tag{7.183}$$

The input signal is the sum of these two waveforms or

$$\begin{aligned} V_{in}(t) &= V_{mod,1}(t) + V_{mod,2}(t) \\ &= [1 + A_1(t)]\cos[\omega_1 t + \phi_1(t)] + [1 + A_2(t)]\cos[\omega_2 t + \phi_2(t)] \end{aligned} \tag{7.184}$$

39. Second-Order Distortion Power Fill in the blanks on Figure 7-68. Show your work.

40. Third-Order Distortion Power Fill in the blanks on Figure 7-69. Show your work.

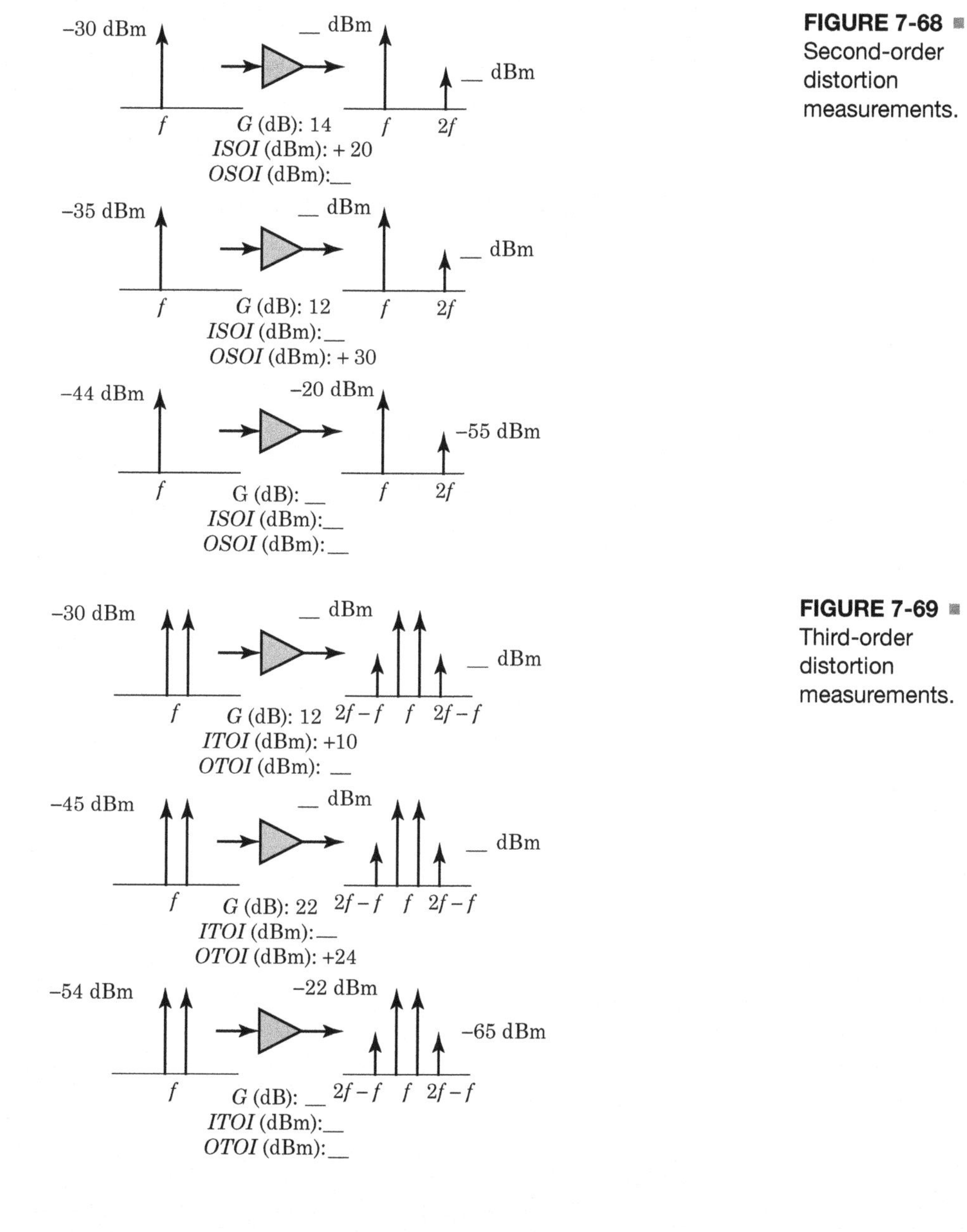

FIGURE 7-68 ■ Second-order distortion measurements.

FIGURE 7-69 ■ Third-order distortion measurements.

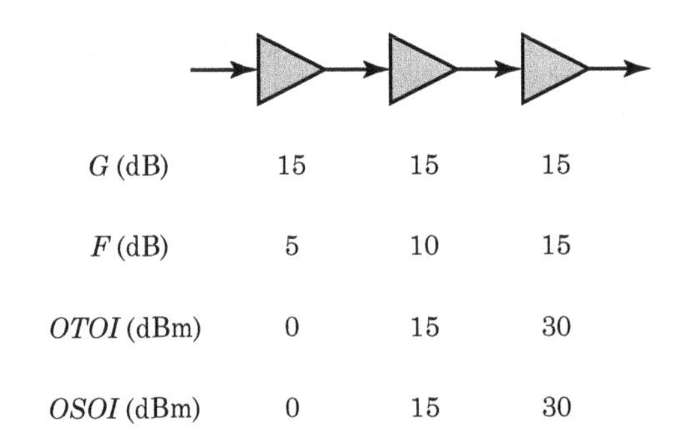

FIGURE 7-70 ■ A three-element nonlinear cascade.

41. **Distortion and Cascades** For the cascade shown in Figure 7-70, find:

a. The cascade's *MDS* assuming a noise bandwidth of 20 MHz and an antenna noise temperature of T_0.

b. The cascade's *ITOI* assuming coherent addition and the cascade's *ITOI* assuming noncoherent addition.

c. The cascade's *ISOI* assuming coherent addition and the cascade's *ISOI* assuming noncoherent addition.

d. The cascade's second-order and third-order limited spur-free dynamic ranges, assuming coherent summation.

CHAPTER

8

Mixers

Translation increases the faults of a work and spoils its beauties.

Voltaire

Chapter Outline

8.1 INTRODUCTION

A mixer is a device used to translate signals from one center frequency to another while keeping the modulation intact. We perform the translation for several reasons.

8.1.1 Filtering

When building a receiver, we must filter the signal of interest before we apply the signal to an analog-to-digital converter for follow on demodulation. We filter to reduce noise and to attenuate unwanted signals.

We move signals to an intermediate frequency (or IF) based almost entirely on our filtering needs. We move the signal of interest from its original center frequency to some IF and perform the filtering at a frequency of our choosing. For example, it's often easier to realize narrowband filters at low center frequencies, but it's better to handle large bandwidth signals at high center frequencies. The realizable insertion loss, percent bandwidth, and physical size of the filter are all strong functions of the filter center frequency.

8.1.2 Frequency Assignments

Our on-the-air frequencies are usually assigned to us by an external organization (e.g., the Federal Communication Commission [FCC]). These frequencies are often chosen based on external requirements that have nothing to do with our filtering or demodulator requirements. We prefer to perform all the complex receiver functions at a frequency we choose and then to translate the signal of interest to the center frequency to which we have been assigned.

Consider a satellite system. The transmitter's power, center frequency, and bandwidth are tightly controlled. To move the most data through this channel, satellite transmitters use complex modulation schemes to limit the transmitted bandwidth while keeping the data rate as high as possible. The transmitter performs the modulation at one frequency (usually 70 MHz), and then a separate piece of equipment translates the modulated waveform to the desired center frequency.

This method produces a system with carefully controlled characteristics because the modulator has to perform at only one frequency.

8.1.3 Antenna Size

The physical parameters of an antenna, expressed in wavelengths, determine the electrical properties of the structure. Antennas at low frequencies must be physically large. Alternatively, the mechanical tolerances required for very high-frequency antennas may be difficult or expensive to realize. If we design a device that must be physically small, we will be forced to use a small antenna, so it would be more efficient to operate at a high frequency.

8.1.4 Propagation

The attenuation, multipath, and reflection characteristics of a propagation channel are strong functions of frequency. The atmosphere exhibits points of high attenuation due to water, oxygen, and other atmospheric components. For example, there is a point of high attenuation at 60 GHz due to energy absorption by oxygen (the *oxygen line*). Normally, we want to avoid transmitting at absorption frequencies because of the high attenuation. However, if we don't want anyone to listen to our communications and we have to transmit only a short distance, these frequencies are ideal. Some military communications links operate at 60 GHz for just this purpose.

Some satellite-to-satellite links operate at 60 GHz to avoid interference from earth-bound communications. The signal passes easily through the vacuum of space but will not pass through the atmosphere.

8.1.5 Component Availability

As designers, we are often in search of less expensive and better-performing components. Once a standard develops, economic factors allow designers to invest effort in designing components that work within the standard. For example, most commercial FM radios use a 10.7 MHz IF, so we find many cheap electronic components designed to work at that frequency. Similarly, the television industry has brought us amplifiers, filters, and other useful components that perform at 45 MHz. The satellite and military industries favor 70 and 140 MHz IFs, so there is a plethora of components available at these frequencies.

8.2 FREQUENCY TRANSLATION MECHANISMS

We will examine the theoretical mechanisms available to move signals from one frequency to another. We want to avoid distortion and keep the modulation of the signal intact.

8.2.1 Amplifier Distortion

A nonlinear device will produce output signals that are not present at the input. Figure 8-1 shows the input and output spectrum of a garden-variety nonlinear device. The spectrum of Figure 8-1 has the following characteristics:

- The output frequencies are given by

$$f_{out} = |nf_1 \pm mf_2| \tag{8.1}$$

  where

$$n = 0, 1, 2, \ldots$$
$$m = 0, 1, 2, \ldots$$
$$n + m = \text{the order of the response}$$

- If we space the two input signals so that $f_1 \approx f_2$, then the output signals tend to bunch together about the harmonics of f_1 and f_2. The bunched signals are Δf apart where $\Delta f = |f_1 - f_2|$.
- The higher the order, the smaller the output power of the signal. The *conversion loss* increases with increasing order.

8.2.1.1 Second-Order Response

The second-order nonlinear response produces the strongest output signal. The second-order operation is

$$V_{out,2}(t) = k_2 V_{in}^2(t) \tag{8.2}$$

We'll assume V_{in} is the sum of two signals: a modulated signal V_{mod} and an unmodulated cosine wave V_{unmod}. We can write

$$\begin{aligned} V_{unmod}(t) &= \cos(\omega_2 t) \\ V_{mod}(t) &= A(t)\cos\left[\omega_1 t + \phi(t)\right] \end{aligned} \tag{8.3}$$

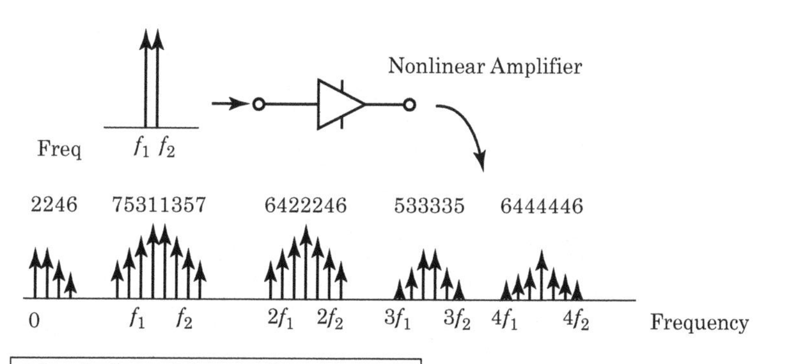

FIGURE 8-1 ■ The input and output spectra of a nonlinear device.

and

$$V_{in}(t) = V_{mod}(t) + V_{unmod}(t) \\ = A(t)\cos[\omega_1 t + \phi(t)] + \cos(\omega_2 t) \tag{8.4}$$

where

$A(t)$ = AM information waveform
$\phi(t)$ = PM or FM information waveform
$\omega_1 = 2\pi f_1$ = carrier frequency of the modulated wave
$\omega_2 = 2\pi f_2$ = carrier frequency of the unmodulated wave

The second-order output is

$$\begin{aligned} V_{out,2}(t) &= k_2[V_{mod}(t) + V_{unmod}(t)]^2 \\ &= k_2\{A(t)\cos[\omega_1 t + \phi(t)] + \cos(\omega_2 t)\}^2 \\ &= \frac{k_2[1 + A^2(t)]}{2} + \frac{k_2 A^2(t)}{2}\cos[2\omega_1 t + 2\phi(t)] \\ &\quad + \frac{k_2}{2}\cos(2\omega_2 t) \\ &\quad + k_2 A(t)\cos[(\omega_1 + \omega_2)t + \phi(t)] \\ &\quad + k_2 A(t)\cos[(\omega_1 - \omega_2)t + \phi(t)] \end{aligned} \tag{8.5}$$

The last two terms of equation (8.5) at $\omega_1 \pm \omega_2$ are the signals we desire. These signals contain the undistorted amplitude and phase modulation of the modulated signal but are at different center frequencies. The signal at $\omega_1 + \omega_2$ is the *sum* product, whereas the signal at $\omega_1 - \omega_2$ is the *difference* product.

We filter out the unwanted second-order products and pass the desired signals. We design the conversion scheme to ensure that the unwanted products fall far away from the desired products and thus are easy to filter.

8.2.2 Amplifier Difficulties

Figure 8-2 shows how we might use a nonlinear amplifier as a mixer. We connect the antenna to a band-pass filter (BPF_1) to limit the number of signals that pass into the amplifier. The output of BPF_1 is connected to the input port of the amplifier. We also connect the amplifier's input port to a locally generated sine wave—a local oscillator (LO). The frequency of the LO is f_2.

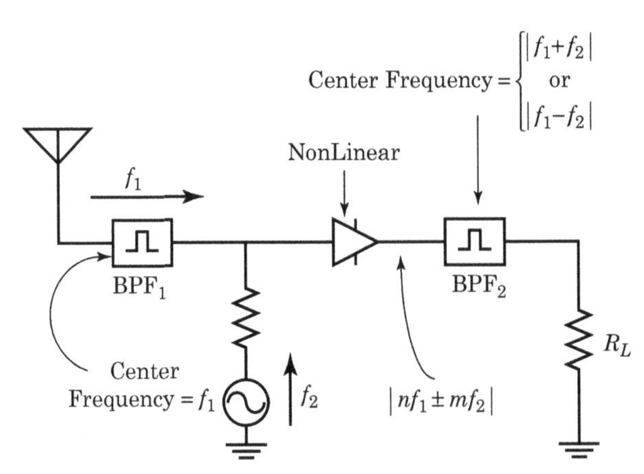

FIGURE 8-2 ■ Using a nonlinear device as a mixer.

The local oscillator signal must be strong enough to force the amplifier to generate nonlinear output products. The signals present on the output of the amplifier will be f_1 and its harmonics, f_2 and its harmonics, and the nonlinear intermodulation products described by equation (8.1). In Figure 8-2, we use BPF_2 to select the signal existing at either the sum $(f_1 + f_2)$ or the difference $(f_1 - f_2)$ frequency.

This architecture is widely used in commercial equipment but exhibits several problems:

- The output of the amplifier contains many signals we don't want including f_1, f_2, and all their sum and difference products.
- The LO at f_2 must be very large to ensure that we're operating the amplifier in its nonlinear region. Consequently, we'll see high-power signals at the amplifier's output at the LO frequency and its harmonics. A garden-variety LO can be 90 dB higher than the signals at the sum or difference frequencies.
- The signals at f_1 and f_2 experience the amplifier's first-order power gain so they will be larger than the desired signals at $f_1 + f_2$ and $f_1 - f_2$.
- A large portion of the local oscillator will travel out of the antenna. This raises interference questions as well as wastes power.

In summary, we can use the second-order characteristics of an amplifier to generate the signals in which we are interested, but this is not a high-performance solution.

8.2.3 Time-Domain Multiply

The trigonometric identity involving the product of cosines provides a second solution:

$$\cos(A)\cos(B) = \frac{1}{2}\cos(A+B) + \frac{1}{2}\cos(A-B) \tag{8.6}$$

This equation expresses exactly the operation we want to perform. We multiply two signals together in the time domain to produce only two output signals: one at the sum frequency and one at the difference frequency. No other frequency components are present.

Combining the input signal described in equation (8.4) with equation (8.6) produces

$$\begin{aligned} V_{mult}(t) &= V_{mod}(t)V_{unmod}(t) \\ &= A(t)\cos\left[\omega_1 t + \phi(t)\right]\cos(\omega_2 t) \\ &= \frac{A(t)}{2}\cos\left[(\omega_1 + \omega_2)t + \phi(t)\right] \\ &\quad + \frac{A(t)}{2}\cos\left[(\omega_1 - \omega_2)t + \phi(t)\right] \end{aligned} \tag{8.7}$$

Both the amplitude and phase modulation of $V_{mod}(t)$ pass through the multiply unscathed.

At high frequencies, it is difficult to perform the multiplication in a mathematically precise way, but we can approximate the operation. These implementation details result in higher levels of spurious output signals than equation (8.7) indicates.

Equation (8.7) illustrates a fundamental property of mixing. The $A(t)/2$ terms guarantee that we will lose at least 6 dB in the conversion process.

8.3 | NOMENCLATURE

Figure 8-3 shows the block diagram of a frequency converter or mixer.

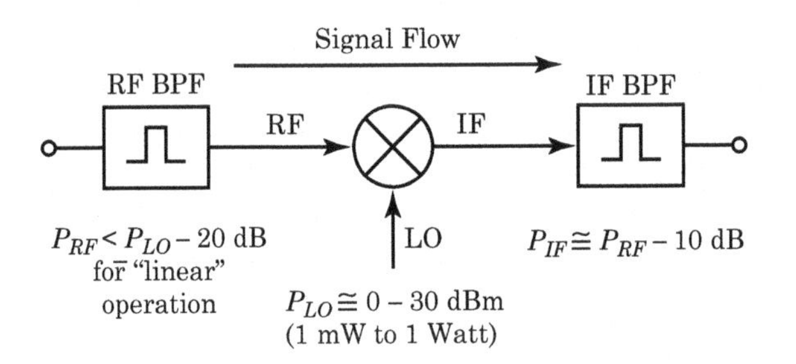

FIGURE 8-3 ■ A typical mixer used to convert signals at the radio frequency (RF) to some intermediate frequency (IF) using a local oscillator (LO).

8.3.1 Ports

A mixer has three ports.

8.3.1.1 Radio Frequency Port

The radio frequency (RF) port is the mixer's input port. We apply the signal that is carrying the modulation to this port. The mixer's purpose is to change the center frequency of the signal present at the RF port to the sum and difference frequencies and place those signals on the IF port. The IF BPF then selects either the sum or difference signal and passes it further downstream.

In most systems, the signal power entering the mixer's RF port should be at least 20 dB below the LO power. This is a realization issue whose causes we'll describe later. If the RF power exceeds this number, the mixer will still operate, but it will generate more unwanted signals than necessary.

8.3.1.2 Intermediate Frequency Port

The mixer's IF port is usually the output port. The sum and difference frequencies appear at this port along with the unwanted spurious signals. The IF port is almost universally connected to a band-pass filter, which removes the unwanted signals and passes the desired signals.

The sum and difference products present at IF port are approximately 10 dB below the level of the RF signal that we applied to the RF port. Common passive mixers have about a 10 dB conversion loss.

8.3.1.3 Local Oscillator Port

We generate the LO internally in the receiver. The frequency of the LO (along with the RF and IF band-pass filters) determine which signals present at the input of the RF BPF pass through to the output of the IF BPF.

The LO power often falls between 0 dBm to 30 dBm. The LO power should be the strongest signal present in the mixer by at least 20 dB. The LO should be free of amplitude and phase noise.

8.3.1.4 Port Interchangeability

Most garden-variety, passive mixers have ports labeled RF, IF, and LO. Despite the labels, these ports are often interchangeable. We can apply the RF signal to the IF port, apply the LO to the IF port, and remove the signal from the RF port. These different configurations are sometimes convenient to accommodate the frequency ranges of the three ports. However, the mixer's performance will likely be different from the data sheet's description.

Some specialized mixers contain active devices or special circuitry to cancel undesired signals, so their ports aren't interchangeable.

When we use the term RF, for example, we'll be referring to the signal present at the RF frequency, not to a specific mixer port. The same concept holds true of the terms IF and LO.

8.3.2 Frequency Translation Equations

Regardless of the method we use perform the translation, the equation that quantifies the relationship between the signals present at a mixer's RF, IF, and LO ports is

$$f_{IF} = f_{LO} \pm f_{RF} \tag{8.8}$$

Figure 8-4 illustrates the conversion process graphically. We apply a single tone at f_{RF} into the mixer's RF port. We place a second tone at f_{LO} into the mixer's LO port, and then we examine the spectrum present on the output port.

Initially, we will ignore everything coming out of the mixer's IF port except the two signals at the sum ($f_{IF,SUM} = f_{LO} + f_{RF}$) and difference ($f_{IF,DIFF} = f_{LO} - f_{RF}$) frequencies. We will also refer to the signal at the sum frequency as the *upper sideband* and the signal at the difference frequency as the *lower sideband.*

Figure 8-4 shows the graphical symmetry between the two IFs and the LO frequency. If we start at the LO frequency and move down by f_{RF} hertz, we find the signal at the difference frequency. If we start at the LO frequency and move up by f_{RF} hertz, we find the signal at the sum frequency.

8.3.2.1 0 Hertz

Although we did not expressly show its position, the location of 0 hertz on Figure 8-4 has a strong impact on the interpretation of that figure. Figure 8-5 shows the output spectrum when both IF components leaving the mixer fall above 0 hertz. Figure 8-6 shows the other possible outcome. The difference frequency leaving the mixer, as described by equation (8.8), falls below 0 hertz. What is the most useful interpretation of negative frequency?

8.3.2.2 Absolute Value

We know

$$\cos(-\omega t) = \cos(\omega t) \tag{8.9}$$

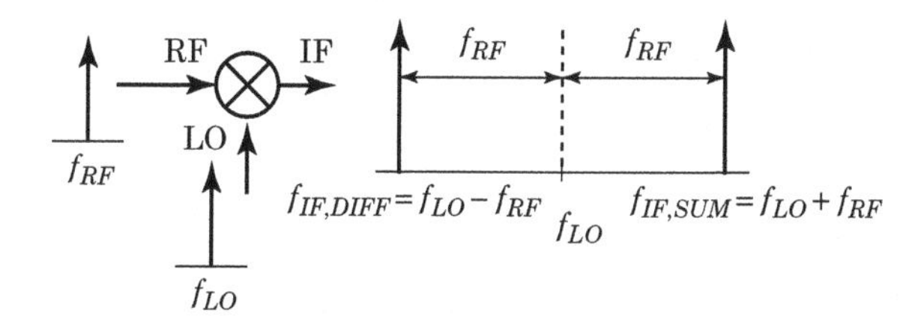

FIGURE 8-4 ■ A frequency conversion and the resulting spectrum.

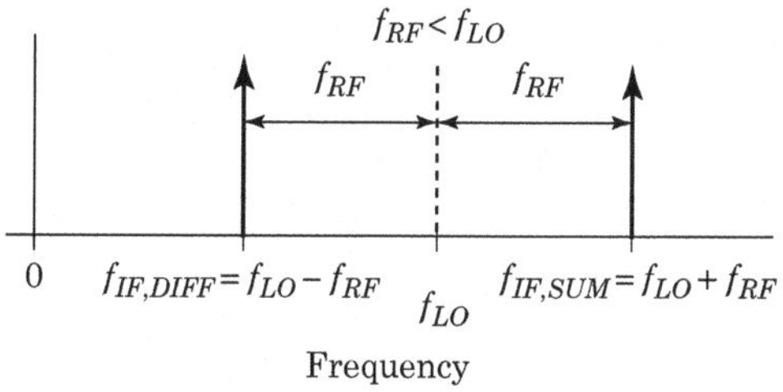

FIGURE 8-5 ■ Conversion spectra when $f_{RF} < f_{LO}$. The mixer output components all lie above 0 hertz.

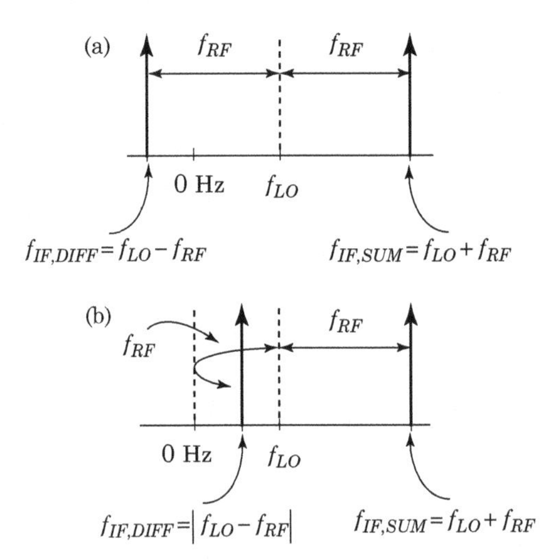

FIGURE 8-6 ■ Conversion spectra when $f_{RF} > f_{LO}$. The difference component from the mixer lies below 0 hertz (or DC) and requires interpretation.

which indicates that we may simply take the absolute value of any frequency produced by equation (8.8). We can rewrite equation (8.8) as

$$f_{IF} = |f_{LO} \pm f_{RF}| \tag{8.10}$$

Equation (8.10) indicates that the spectrum of Figure 8-6(a) is equivalent to the spectrum of Figure 8-6(b). The lower sideband, which was formerly below zero, mirrors off the y-axis and falls into the positive frequency domain.

We will consider the sign of the frequency whenever it's useful to solve the problem. Similarly, we will ignore the sign when it's not necessary.

8.3.3 Three Forms

We can express equation (8.10) as three separate equations when we fully expand the "±" signs. The practical result "±" sign in equation (8.10) is that there are always two possible solutions anytime we solve for one of the variables.

If we know the LO frequency and RF, then the RF signal will be simultaneously converted to two different IFs. These two IFs are

$$f_{IF} = f_{LO} \pm f_{RF} \tag{8.11}$$

If we know the LO frequency and IF, we can find the two possible RFs that will be converted to the IF by the mixer:

$$f_{RF} = f_{LO} \pm f_{IF} \tag{8.12}$$

The third equation tells us the two possible LO frequencies that will convert some RF signal to a given IF:

$$f_{LO} = f_{RF} \pm f_{IF} \tag{8.13}$$

Figure 8-7 shows these equations graphically. A useful way to interpret these graphs is in terms of starting points and distances traveled. For example, the top of Figure 8-7 illustrates equation (8.11). If we begin at f_{LO} and move a distance f_{RF} to either side, we find the two possible solutions of the equation.

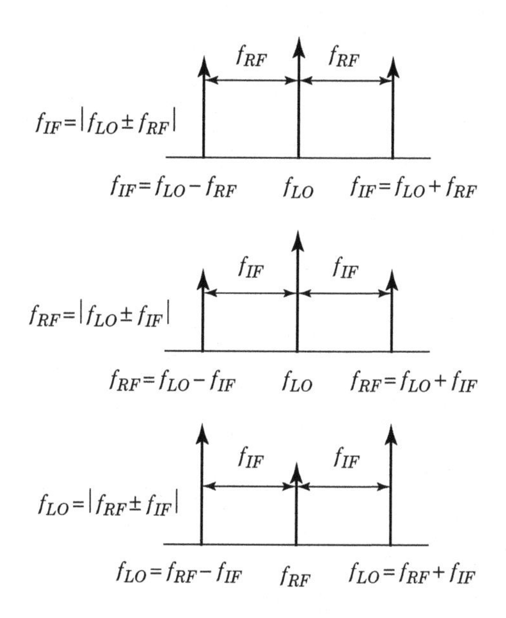

FIGURE 8-7 ■ A graphical illustration of equations (8.11) through (8.13). When one sideband falls below 0 hertz, we use the absolute value of the frequency.

EXAMPLE

Conversion Equations

a. What two possible LO frequencies can we use to convert a commercial FM radio station broadcasting at 106.9 MHz down to a 10.7 MHz IF?

b. What are the center frequencies of the two possible signals that can be converted to an IF of 70 MHz using an LO frequency of 935 MHz?

c. If we mix a signal centered at 215.75 MHz with 256.25 MHz local oscillator, what are the center frequencies of the two resulting signals?

Solution

a. Equation (8.13) produces

$$\begin{aligned} f_{LO} &= f_{RF} \pm f_{IF} \\ &= 106.9 \pm 10.7 \\ &= 117.6 \quad \text{or} \quad 96.2 \text{ MHz} \end{aligned} \tag{8.14}$$

b. This calls for equation (8.12):

$$\begin{aligned} f_{RF} &= f_{LO} \pm f_{IF} \\ &= 935 \pm 70 \\ &= 1{,}005 \quad \text{or} \quad 865 \text{ MHz} \end{aligned} \tag{8.15}$$

c. We'll use equation (8.11):

$$\begin{aligned} f_{IF} &= f_{LO} \pm f_{RF} \\ &= 256.25 \pm 215.75 \\ &= 472.0 \quad \text{or} \quad 40.5 \text{ MHz} \end{aligned} \tag{8.16}$$

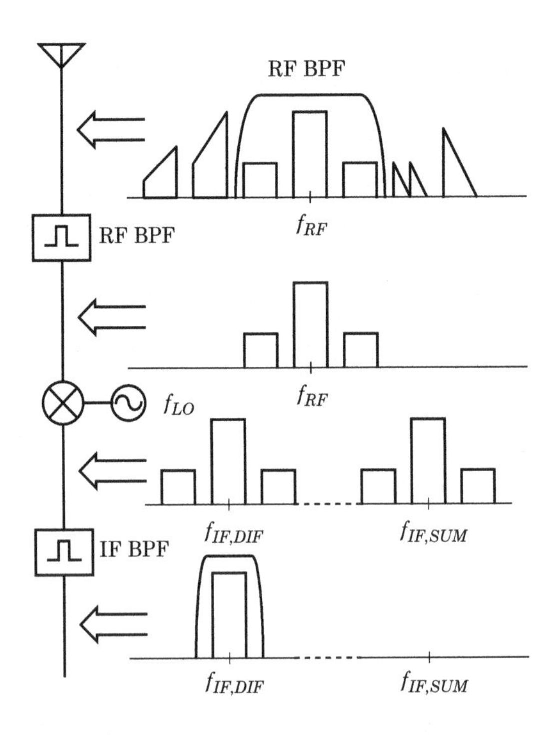

FIGURE 8-8 ■ The conversion and filtering processes in a single-conversion receiver.

8.3.4 Frequency Translation and Filters

Figure 8-8 illustrates the conversion process in practice. Our antenna receives many signals at different frequencies, but we are interested in only one signal. Ideally, we would like to remove all the unwanted signals from the antenna using the RF BPF, but because of practical concerns we usually can't. Instead, the RF filter will remove only signals that are grossly different in frequency from the desired signal.

The filtered RF spectrum then travels to the mixer's RF port, and all of the signals undergo the frequency conversion. We get two copies of the RF spectrum: one centered about $f_{IF,DIFF}$ and one centered about $f_{IF,SUM}$. Finally, the IF filter passes the one signal we want to process.

8.3.4.1 Practical Considerations

Why do we go through the trouble of converting the signal of interest to the IF before filtering it? Why don't we just filter the signal at the RF? The answer is that we are usually unable to filter the signal adequately at its RF. Let's examine the FM broadcast band as an example.

We desire that our commercial FM broadcast receiver to be able to process all of the 101 possible FM stations. Our design choices are:

- Build 101 band-pass filters just wide enough to pass one particular FM station, and then switch to a new filter whenever the user changes stations. This would be bulky and expensive. It would also be difficult and expensive to build filters of such small percentage bandwidths, especially on a production basis.
- Build a narrow, tunable RF band-pass filter, and change its center frequency as the user changes stations. Again, this is a difficult to realize and expensive solution because of the small percentage bandwidths and the extra circuitry involved.

- Let the RF band-pass filter pass all 101 stations, and then filter out the particular station we want at a later stage. This is exactly the solution used in practice.

Receiver design often boils down to solving a filtering problem. If we can't filter a signal at a particular frequency, we move the signal to a frequency where we can filter it. In a commercial FM broadcast receiver, we convert the signal of interest to 10.7 MHz, no matter what the RF was originally. When we want to change the station, we have to change only the LO frequency, and the mixer will convert a different station down to the 10.7 MHz IF. This technique has the tremendous advantage that everything following the IF filter has to work at only the IF.

8.3.5 Conversion Loss

When a mixer converts a signal from its RF to some IF, the signal at the IF has usually lost some of its power. This is the mixer's conversion loss. Figure 8-9 illustrates the concept. If, after the conversion, we will be using the lower sideband IF signal (at $f_{IF,DIFF}$), the mixer's conversion loss is

$$\begin{aligned} CL_{LSB} &= \frac{\text{Signal Power at} f_{RF} - f_{LO}}{\text{Signal Power at} f_{RF}} \\ &= \frac{P_{DIFF}}{P_{RF}} \end{aligned} \tag{8.17}$$

If we will be using the IF signal at the upper sideband frequency ($f_{IF,SUM}$), then the mixer's conversion loss is

$$\begin{aligned} CL_{USB} &= \frac{\text{Signal Power at} f_{RF} + f_{LO}}{\text{Signal Power at} f_{RF}} \\ &= \frac{P_{SUM}}{P_{RF}} \end{aligned} \tag{8.18}$$

Note that P_{SUM} does not necessarily equal P_{DIFF} (although they are often very close). Mixer data sheets usually give only one conversion loss specification, implying that $CL_{SUM} = CL_{DIFF}$ and that $P_{SUM} = P_{DIFF}$.

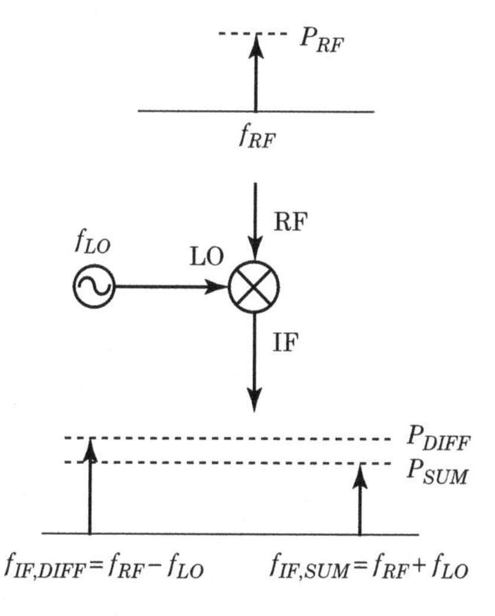

FIGURE 8-9 ■ The definition of mixer conversion loss.

EXAMPLE

Mixer Conversion Loss

The power of the RF signal applied to the mixer's input port is –45 dBm. The power measured at the mixer's IF port at the sum frequency is –60 dBm, while the power measured at the difference frequency is –63 dBm. What is the mixer's conversion loss?

Solution

The upper sideband conversion loss is

$$\begin{aligned} CL_{USB} &= \frac{P_{SUM}}{P_{RF}} \\ &= P_{SUM,dB} - P_{RF,dB} \\ &= -45 - (-60) \\ &= 15 \text{ dB} \end{aligned} \tag{8.19}$$

The lower sideband conversion loss is

$$\begin{aligned} CL_{LSB} &= \frac{P_{DIFF}}{P_{RF}} \\ &= P_{DIFF,dB} - P_{RF,dB} \\ &= -45 - (-63) \\ &= 18 \text{ dB} \end{aligned} \tag{8.20}$$

8.3.6 Port-to-Port Isolation

Due to realization effects, we find that signals applied to one port of a mixer will leak through the mixer to the other two ports. This leakage occurs without frequency translation.

In other words, if we apply a signal at f_{LO} to the LO port of a mixer, we will see a signal at f_{LO} on both the IF and RF ports of the mixer. Identical effects occur when we apply signal to both the RF and IF ports. The amount of isolation (or the amount of attenuation) among the three mixer ports are important and useful numbers.

Referring to Figure 8-10, the three mixer isolation specifications are as follows.

8.3.6.1 LO:RF Isolation

This is the amount of signal attenuation between a mixer's LO and RF ports. The isolation is usually bilateral; that is, it applies for signals traveling from the LO port to the RF port and for signals traveling from the RF port to the LO port.

As receiver designers, we are more interested in the amount of LO power that leaks onto the mixer's RF port than how much of the RF signal leaks to the LO port. We define

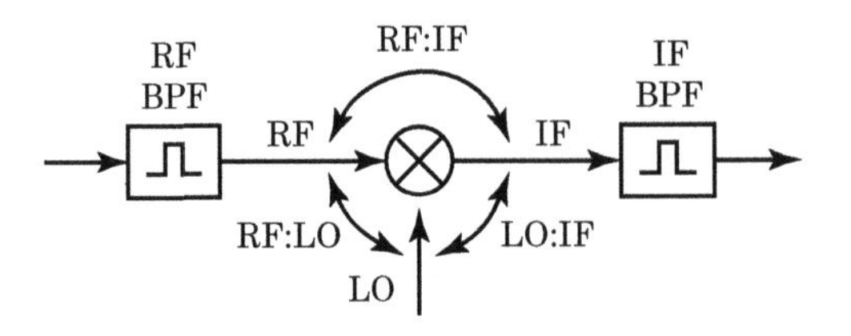

FIGURE 8-10 Mixer port-to-port isolation.

the LO:RF isolation as

$$Isol_{LO:RF} = \frac{\text{LO power present at the mixer's RF port (at } f_{LO})}{\text{LO power entering the mixer's LO port (also at } f_{LO})} \tag{8.21}$$

Since the LO is a high-power signal (>7 dBm), even a mixer with a large LO:RF isolation can exhibit a very large LO signal at the RF port. One of the jobs of the RF filter in Figure 8-10 is to attenuate the LO leaving the RF port.

8.3.6.2 LO:IF Isolation

This is the amount of attenuation between the LO and IF ports of the mixer. Like the LO:RF isolation, this attenuation is often bilateral. As designers of receiving systems, we're interested the amount of LO power that appears on the IF port of the mixer. We define the LO:IF isolation as

$$Isol_{LO:IF} = \frac{\text{LO power present at the IF port (at } f_{LO})}{\text{LO power entering the LO port (also at } f_{LO})} \tag{8.22}$$

Since the LO is a high-power signal, even slight LO:IF leakage places a large LO signal at the mixer's IF port. One task of the IF band-pass filter in Figure 8-10 is to attenuate this LO leakage.

8.3.6.3 RF:IF Isolation

The RF:IF isolation may seem like another name for the conversion loss. Not true. To measure the RF:IF isolation, we apply a signal at f_{RF} to the RF port of the mixer and then measure how much RF signal is present on the IF port of the mixer (still at f_{RF}). The leakage has not undergone the frequency conversion process.

The RF:IF isolation applies to signals present at the same frequency on the RF and IF ports, while conversion loss applies to signals at different frequencies (f_{RF} on the RF port and $f_{RF} \pm f_{LO}$ on the IF port). The RF:IF isolation is

$$Isol_{LO:IF} = \frac{\text{RF power present at the IF port (at } f_{RF})}{\text{RF power entering the RF port (still at } f_{RF})} \tag{8.23}$$

EXAMPLE

Mixer Conversion Loss and Isolation

Given the system shown as Figure 8-11 and the following mixer specifications:

$CL_{dB} = 8$ dB

RF:IF Isolation = 30 dB (about average)

LO:IF Isolation = 45 dB (unusually high)

LO:RF Isolation = 25 dB (about average)

find

a. The strength of the signals at 407 MHz, 243 MHz, 82 MHz, and 325 MHz on the IF port

b. The strength of the signal at 325 MHz on the RF port

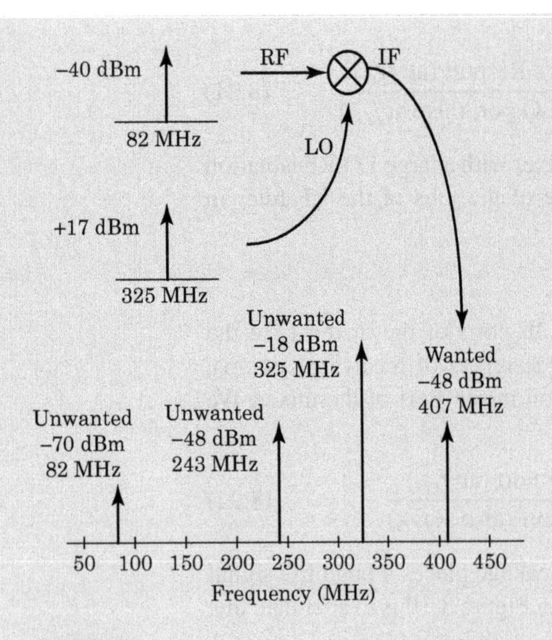

FIGURE 8-11 ■ Wanted and unwanted signal present at a mixer's IF port.

Solution

Using equation (8.11), the sum and difference frequencies present on the IF port are

$$\begin{aligned} f_{IF} &= f_{LO} \pm f_{RF} \\ &= 325 \pm 82 \\ &= 243 \text{ MHz and } 407 \text{ MHz} \end{aligned} \tag{8.24}$$

a. The conversion loss applies to the signals at 243 and 407 MHz, so the level of these signals is

$$\begin{aligned} P_{IF,243} &= P_{IF,407} \\ &= P_{RF,82,dBm} - CL_{dB} \\ &= -40 - 8 \\ &= -48 \text{ dBm} \end{aligned} \tag{8.25}$$

The RF:IF isolation applies to signals at the RF port leaking through the mixer to the IF port, so the level of the signal at 82 MHz is

$$\begin{aligned} P_{IF,82} &= P_{RF,82,dBm} - Isol_{RF:IF,dB} \\ &= -40 - 30 \\ &= -70 \text{ dBm} \end{aligned} \tag{8.26}$$

The LO:IF isolation applies to leakage from the LO port to the IF port. The level of the signal at the IF port at 325 MHz is

$$\begin{aligned} P_{IF,325,dBm} &= P_{LO,325,dBm} - Isol_{LO:IF,dB} \\ &= 7 - 45 \\ &= -38 \text{ dBm} \end{aligned} \tag{8.27}$$

Figure 8-11 shows the spectrum of the signals present at the IF port. The signal due to LO leakage (at 325 MHz and −38 dBm) is stronger than the desired signals at 243 and 407 MHz and −48 dBm. This is a common situation.

b. The LO:RF isolation applies to LO leakage to the RF port. At the RF port, the strength of the signal at 325 MHz is

$$\begin{aligned} P_{RF,325,dBm} &= P_{LO,325,dBm} - Isol_{LO:RF,dB} \\ &= 7 - 25 \\ &= -18 \text{ dBm} \end{aligned} \tag{8.28}$$

8.3.7 Mixer Isolation and Its Problems

Interport isolation causes problems in several flavors.

8.3.7.1 Antenna Radiation

Figure 8-12 shows one problem. The LO:RF isolation of the mixer can allow the LO to radiate out of the antenna port. The LO is a high-power signal that can interfere with other receivers.

For example, the LO of a commercial U.S. FM radio receiver runs from about 99 MHz to about 119 MHz. The commercial aircraft band lies from 108 MHz to 132 MHz. The two bands overlap, and there exists a possibility that the LO of an FM receiver could interfere with aircraft communications equipment. This is the reason that commercial airlines forbid the use of an FM radio or other electronic equipment during take-off and landings. If your equipment radiates a signal that is collocated with one of the important aircraft frequencies, then your receiver could interfere with the pilot's communications.

Assuming that a mixer in a particular receiver requires an LO power of +7 dBm and the mixer provides an LO:RF isolation of 40 dB (this is an unusually large number), there will be a –33 dBm signal present at the antenna. This is a large signal.

8.3.7.2 Intermixer Isolation

Very often, we build receivers with two separate mixers and two separate local oscillators. These are *dual conversion* receivers.

Figure 8-13 shows a simple block diagram of a dual conversion receiver. In a poorly designed receiver, significant amounts of LO_1 will make it out of mixer #1 (via its LO:IF isolation) and into the RF port of mixer #2. Likewise, LO_2 can make it out of mixer #2 (via its LO:IF isolation) and make its way into mixer #1.

Any signal that enters a mixer participates in the conversion process. In mixer #1, LO_2 will mix with LO_1 as well as any RF signals present in the mixer. In mixer #2, LO_1 will mix with LO_2 as well as any other signal present in the mixer.

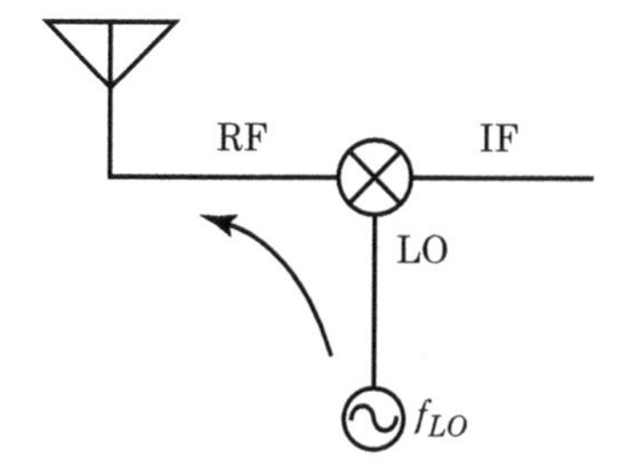

FIGURE 8-12 ■ LO:RF isolation causes a strong LO signal to radiate from a receiver's antenna.

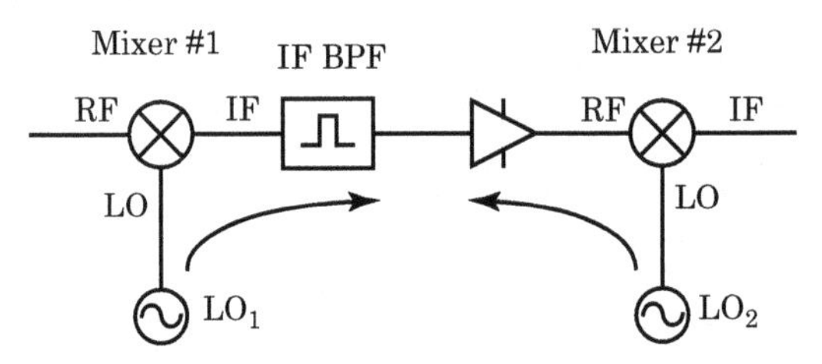

FIGURE 8-13 ■ Mixer interport isolation problem in a dual conversion receiver. Good design keeps LO_1 out of mixer #2 and keeps LO_2 out of mixer #1.

The result of all this unplanned-for mixing action is that various signals will be generated at many different frequencies. This is one cause of internally generated spurious responses or spurs. Spurs are internally generated signals that, to the casual observer, appear to originate at the receiver's antenna port.

We can minimize internally generated spurious signals in our designs by proper conversion scheme design and carefully selecting filters and other components.

The receiver local oscillators are not the only source of unwanted signals that cause internally generated spurious signals. Any other oscillator or signal source present in the receiver can contribute signals. The receiver's microprocessor and its power switching supply have historically been major contributors to this problem.

8.3.8 Mixers in a Cascade

When we use a mixer in a cascade, we treat it just like any other component. It has a loss (or a gain), a noise figure, a TOI, and a SOI. For the purposes of the cascade gain and linearity calculations, we ignore the frequency translation aspects of the device.

EXAMPLE

Mixers in a Cascade

Find the gain, noise figure, and ITOI of the cascade shown in Figure 8-14.

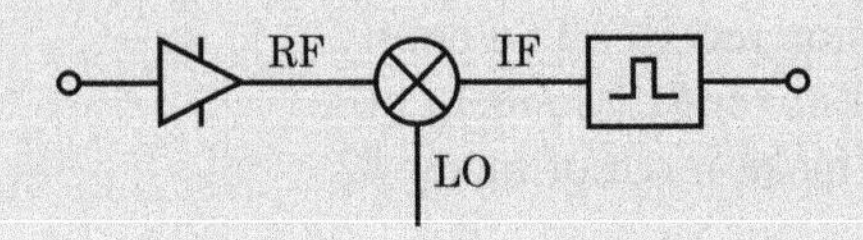

FIGURE 8-14 ■ Mixers in cascade.

G_p	15 dB	−6.0 dB	−2.0 dB
NF	2.5 dB	7.0 dB	2.0 dB
ITOI	4.0 dBm	18 dBm	30 dBm

Solution

Using the mixer's conversion loss as its gain and the equations developed in Chapters 6 and 7, the results are

$$G_{p,cas} = 7 \text{ dB}$$
$$NF_{dB} = 3.0 \text{ dB}$$
$$ITOI_{dBm} = 0.4 \text{ dBm}$$

8.4 | BLOCK VERSUS CHANNELIZED SYSTEMS

We divide our frequency translation tasks into two different types: block conversions and channelized conversions.

8.4.1 Block Conversions

In a block conversion, we move an entire band of signals from one center frequency to another using a fixed, single-frequency local oscillator.

Figure 8-15 shows the conversion scheme of a 1990s era commercial satellite television receive only (TVRO) system. The TVRO receiver collects television signals from geostationary satellites. The satellite's C-band transponders transmit 24 television channels at 3,700 MHz to 4,200 MHz (a 500 MHz bandwidth centered at 3,950 MHz).

After the outdoor antenna captures the TVRO signal, we must move the signal to the house, which can be several hundred feet away. This is a commercial application, and we are worried about cost so we'd like to use cheap, lossy cable between the antenna and the house. We place a block converter at the receive antenna's feed point to convert the 3,950 MHz signal to a 1,200 MHz IF. We then move the 1,200 MHz IF signal to the house, taking advantage of the lower cable losses at 1,200 MHz.

We have not yet selected a channel from the 24 possible channels. We have only translated a swatch of spectrum from one center frequency to another using a single-frequency local oscillator. Since we have moved the entire spectrum to the house, several users can access the same antenna at once.

8.4.2 Channelized Conversions

A channelized system uses a tuning local oscillator to select only one signal out of a crowded spectrum. Figure 8-16 shows the architecture of a commercial FM receiver. The

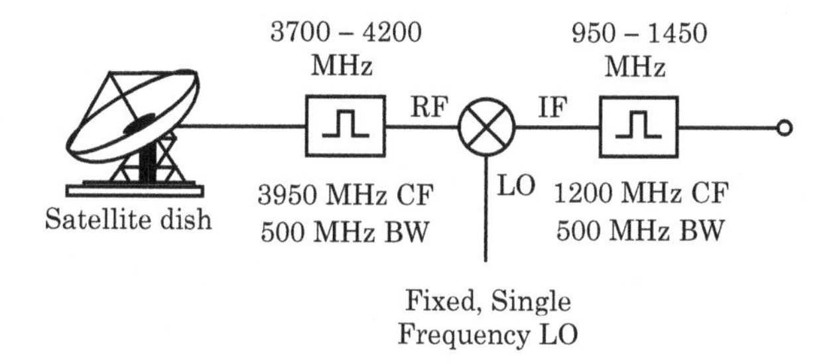

FIGURE 8-15 ■ A TVRO frequency conversion scheme. This is a block conversion scheme.

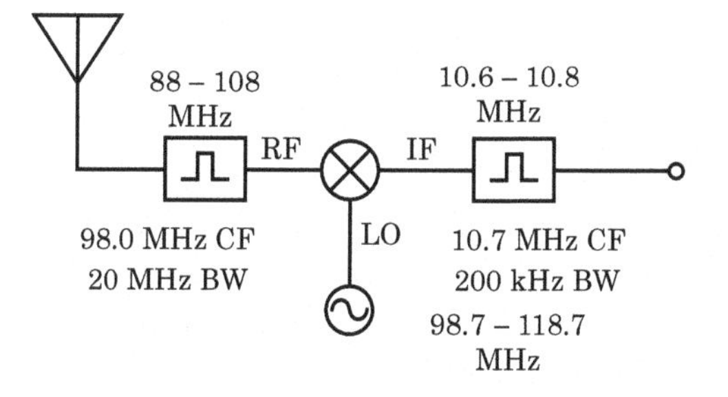

FIGURE 8-16 ■ A commercial FM receiver frequency conversion scheme. This is a channelized conversion scheme.

FM broadcast band in the United States runs from about 88 MHz to 108 MHz. In an FM receiver, we apply the entire 20 MHz FM spectrum into the mixer and tune the LO. We select the frequency of the LO so that the IF filter passes only the signal we desire and severely attenuates the others.

For example, we want to listen to the station broadcasting at 99.1 MHz. We set the LO frequency of Figure 8-16 to 88.4 MHz, which translates the desired signal from 99.1 MHz to 10.7 MHz (the center frequency of the IF filter). The desired signal, now at 10.7 MHz, passes through the IF filter without attenuation, while the stations below 98.9 MHz and above 99.3 MHz are considerably reduced in power. The FM demodulator then processes the signal and we hear music.

8.5 | CONVERSION SCHEME DESIGN

Conversion scheme design is a highly iterative process. We start with an initial guess, examine the consequences and problems, change things a little bit, and examine the problems. We continue this process until we arrive at a workable system.

We usually start with several questions or specifications:

- What is the RF of the spectrum we want to convert? This specification will contribute to the choice of the first IF and the selection of the first LO frequencies.
- Do we want to perform a block conversion or a channelized conversion? We fix the LO to a single frequency in a block conversion, while a channelized system demands a tunable LO.
- In a channelized conversion, what is the bandwidth of the RF signals? The bandwidth affects the final IF selection as well as the accuracy of the LOs.
- What kind of modulation is present on the RF signals? This places demand on the phase noise of the LO, which, in turn, can limit the selection of the IF (high IFs tend to require high-frequency Los, which can have poor phase noise).
- Are there cost, manufacturability, physical size, power, or other special features? This category tends to override many of the other specifications of the radio. Often, these miscellaneous, nontechnical considerations force us into using techniques or components that we would rather avoid.

  In a battery-powered receiver, for example, we spend a good deal of time designing a power supply to convert the changing battery voltage into a stable, well-regulated supply for the receiver. Switching power supplies will generate large, low-frequency current spikes around the system, which will take effort to properly filter.

  Small physical size forces many packaging issues such as heat removal and the physical size of your components. The component size issue forces us into common IFs because of the variety of physically small components available at these frequencies.

8.5.1 TVRO Example

Figure 8-17 shows a block diagram of a satellite TVRO receiver. The signal from the satellite covers 3,700 to 4,200 MHz (a 3,950 MHz center frequency), and we desire to convert this spectral band to 1,200 MHz. These details determine the bandwidths and center frequencies of both the RF and IF band-pass filters.

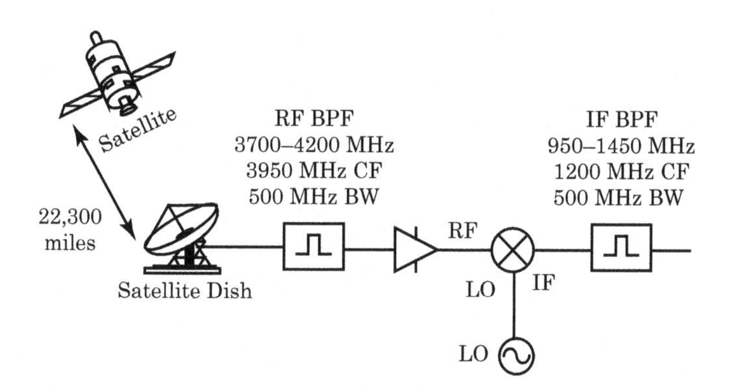

FIGURE 8-17 ■ A TVRO receiving system.

Equation (8.13) provides two possible LO frequencies that will convert a signal from 3,950 MHz to 1,200 MHz. Applying equation (8.13) produces

$$\begin{aligned} f_{LO} &= f_{RF} \pm f_{IF} \\ &= 3{,}950 \pm 1{,}200 \text{ MHz} \\ &= 2{,}750 \text{ MHz} \quad \text{or} \quad 5{,}150 \text{ MHz} \end{aligned} \tag{8.29}$$

We can select either 2,750 MHz or 5,150 MHz as our LO frequency. Both will convert a signal from 3,950 MHz down to 1,200 MHz with equal efficiency. Is there a difference?

8.5.1.1 Band Edges

We will examine the band edges of the RF (3,700 MHz and 4,200 MHz). Using $f_{LO} =$ 2,750 MHz and Figure 8-18(a) as a guide, equation (8.11) produces

$$\begin{aligned} \text{Lower Band Edge} &\Rightarrow 3{,}700 - 2{,}750 = 950 \text{ MHz} \\ \text{Upper Band Edge} &\Rightarrow 4{,}200 - 2{,}750 = 1{,}450 \text{ MHz} \end{aligned} \tag{8.30}$$

A signal at 3,700 MHz converts to 950 MHz, whereas a signal at 4,200 MHz converts to 1,450 MHz. We haven't changed anything in the conversion process when $f_{LO} =$ 2,750 MHz.

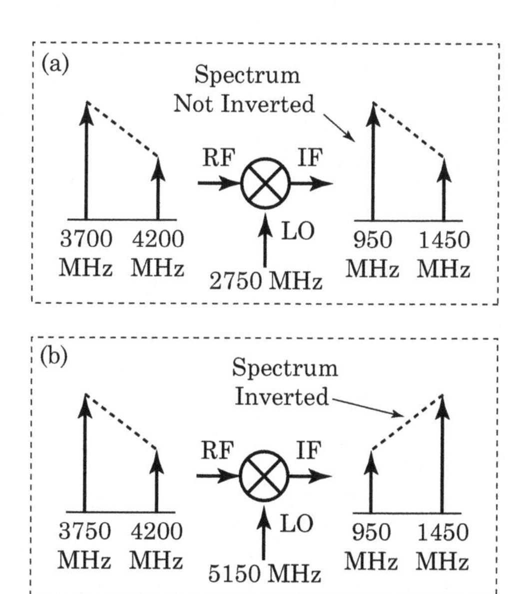

FIGURE 8-18 ■ The TVRO RF and IF spectra for the two possible LO frequencies. The IF spectrum in (a) has not experienced a frequency inversion. The IF spectrum in (b) has experienced a frequency inversion.

At the other LO frequency, $f_{LO} = 5{,}150$ MHz, equation (8.11) and Figure 8-18(b) produce

$$\begin{aligned} \text{Lower Band Edge} &\Rightarrow 5{,}150 - 3{,}700 = 1{,}450\,\text{MHz} \\ \text{Upper Band Edge} &\Rightarrow 5{,}150 - 4{,}200 = 950\,\text{MHz} \end{aligned} \tag{8.31}$$

This LO selection produces a frequency inversion. Signals that were above the center frequency at the RF port of the mixer are now below the center frequency at the IF port. Even for signals that exhibit complex modulation, this frequency inversion is not problematic. We have not altered the information content of the signal, but we must alter the demodulation process to accommodate for the inversion.

8.5.2 High-Side and Low-Side LO

We have a choice of two LO frequencies to convert a signal at the RF to a signal at the IF. We define the following terms:

- Low-side LO (LSLO; also known as low-side injection): The LO frequency is less than the RF.
- High-side LO (HSLO; also known as high-side injection): The LO frequency is greater than the RF.

In equation form, these definitions correspond to

$$\begin{aligned} \text{Low-Side LO} &\Rightarrow f_{LO} < f_{RF} \\ \text{High-Side LO} &\Rightarrow f_{LO} > f_{RF} \end{aligned} \tag{8.32}$$

8.5.3 LO Frequency Calculation

In a channelized system, we have several similar signals (or channels) spread over some frequency range. We desire to convert each individual signal to some common IF where the signal will be processed, digitized, or demodulated. We select the desired signal by changing the frequency of the LO. Given the channelized conversion scheme of Figure 8-19, we must find the HSLO and LSLO frequency ranges. We first define:

- $f_{RF,L}$ = the lowest frequency at the RF port to be converted to the IF center frequency
- $f_{RF,H}$ = the highest frequency at the RF port to be converted to the IF center frequency
- f_{IFCF} = the center frequency of the IF BPF

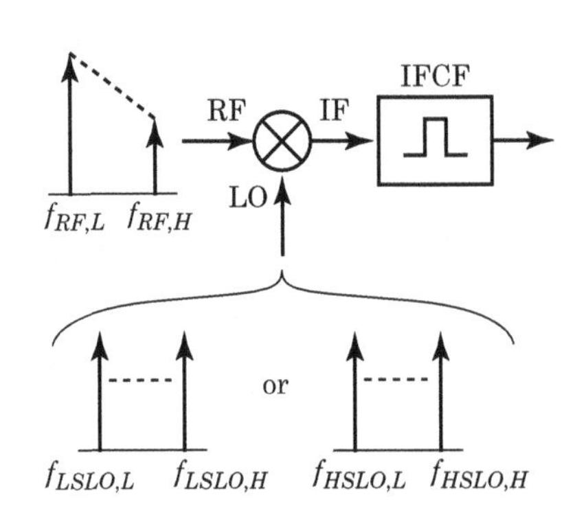

FIGURE 8-19 ■ There exists two possible choices of LO frequencies that will convert a given band of RF to some IF center frequency.

For the low-side LO:

- $f_{LSLO,L}$ = the lowest LO frequency required to perform the conversion.
- $f_{LSLO,H}$ = the highest LO frequency required to perform the conversion.

For the high-side LO:

- $f_{HSLO,L}$ = the lowest LO frequency required to perform the conversion.
- $f_{HSLO,H}$ = the highest LO frequency required to perform the conversion.

Figure 8-20 clarifies the definitions of $f_{RF,L}$ and $f_{RF,H}$. We will often draw only the lower and upper RF to simplify our diagrams.

Equation (8.13) and Figure 8-21 provide some insight into the LO selection process. Figure 8-21 shows that the low-side LO frequency, f_{LSLO}, is a distance of f_{IF} below the RF. The high-side LO frequency, f_{HSLO}, is a distance f_{IF} above the RF.

8.5.3.1 Low-Side LO

Combining equation (8.13) with Figure 8-20 and Figure 8-21 results in expressions for the low-side LO frequencies:

$$
\begin{aligned}
f_{LSLO,L} &= f_{RF,L} - f_{IFCF} \\
f_{LSLO,H} &= f_{RF,H} - f_{IFCF}
\end{aligned}
\tag{8.33}
$$

See Figure 8-22(a).

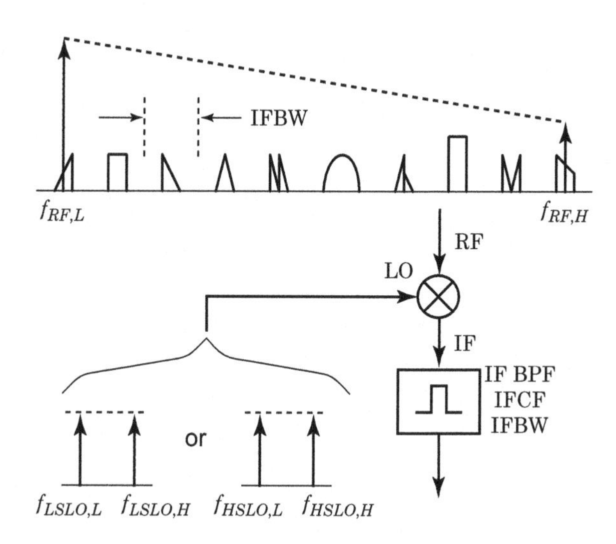

FIGURE 8-20 ■ The RF spectrum and the two possible LO choices for a channelized system.

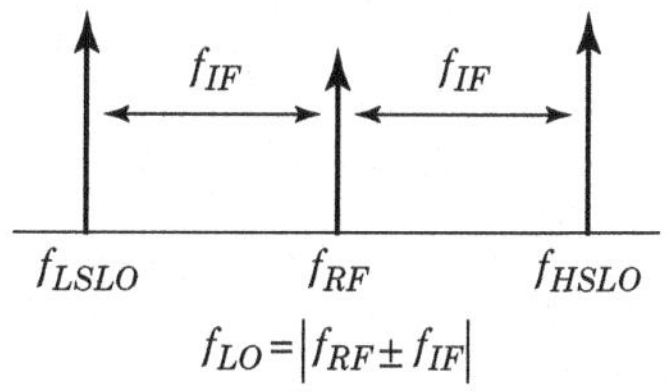

f_{IF} f_{IF}
f_{LSLO} f_{RF} f_{HSLO}
$f_{LO} = |f_{RF} \pm f_{IF}|$

FIGURE 8-21 ■ A graphical interpretation of Equation (8.13).

FIGURE 8-22 ■ Graphical representations of equations (8.33) and (8.34). Given the required RF passband range, find the two possible ranges of LO frequencies.

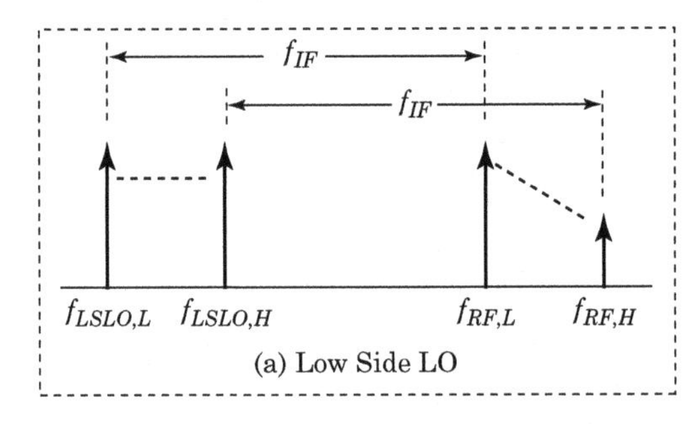

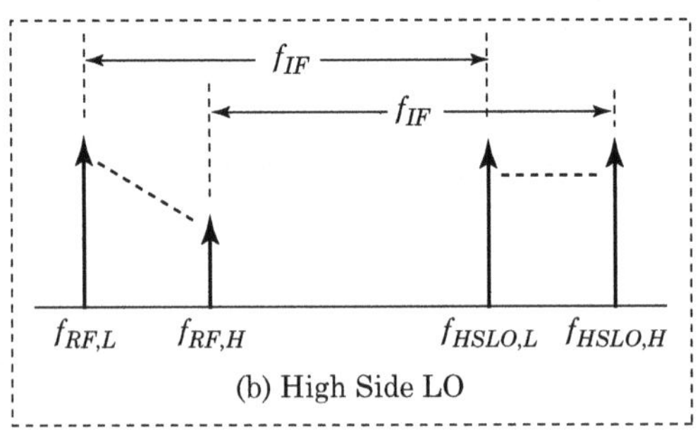

8.5.3.2 High-Side LO

Similarly, the expressions for the high-side LO frequencies are

$$\begin{aligned} f_{HSLO,L} &= f_{RF,L} + f_{IFCF} \\ f_{HSLO,H} &= f_{RF,H} + f_{IFCF} \end{aligned} \tag{8.34}$$

See Figure 8-22(b).

EXAMPLE

High-Side/Low-Side LO

One older cellular telephone system operates at 825 to 890 MHz. Each channel is 30 kHz apart and about 20 kHz wide. If we use a 45 MHz IF, find the HSLO and LSLO frequency ranges for this channelized system.

Solution

Figure 8-23(a) shows the problem. Examining the band edges of the RF ($f_{RF,L} = 825$ and $f_{RF,H} = 890$ MHz), we find the LSLO using equation (8.33):

$$\begin{aligned} f_{LSLO,L} &= f_{RF,L} - f_{IFCF} \\ &= 825 - 45 = 780 \text{ MHz} \\ f_{LSLO,H} &= f_{RF,H} - f_{IFCF} \\ &= 890 - 45 = 845 \text{ MHz} \end{aligned} \tag{8.35}$$

The HSLO comes from equation (8.34):

$$\begin{aligned} f_{HSLO,L} &= f_{RF,L} + f_{IFCF} \\ &= 825 + 45 = 870 \text{ MHz} \\ f_{HSLO,H} &= f_{RF,H} + f_{IFCF} \\ &= 890 + 45 = 935 \text{ MHz} \end{aligned} \tag{8.36}$$

Figure 8-23(b) shows the two LO ranges.

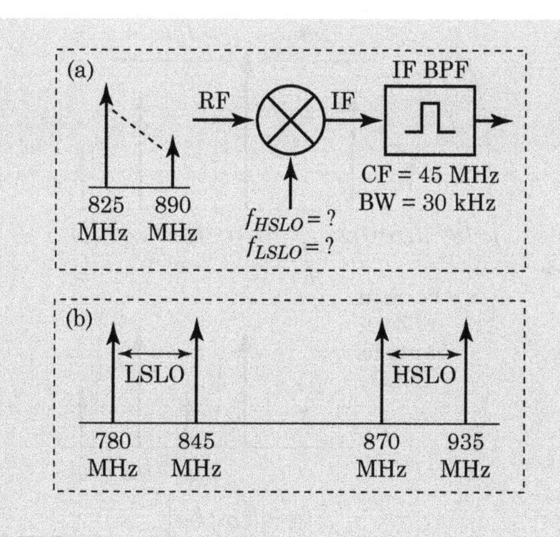

FIGURE 8-23 ■ Given the RF range, find the two possible LO frequency ranges.

8.6 | FREQUENCY INVERSION

Some conversion schemes produce a spectral or frequency inversion. The inversion does not distort or damage the signal, but it does change how we interpret the signal. For example, in a digital frequency shift keying (FSK) system, transmitting a tone below center frequency may correspond to a binary zero, while transmitting a tone above center frequency may represent a binary one. A system with frequency inversion will invert all the bits. The data are still present, but we must make adjustments to produce the correct data. We place an inverter somewhere in the data path to reinvert the data.

Which conversion schemes produce a frequency inversion and why?

We are given an LO frequency and an RF and we want to examine the IF spectrum, so we use equation (8.11). Figure 8-24 shows the graphical interpretation of equation (8.11). We start at f_{LO} and move up in frequency a distance f_{RF} to find one output sideband. We move f_{RF} down in frequency from f_{LO} to find the other sideband.

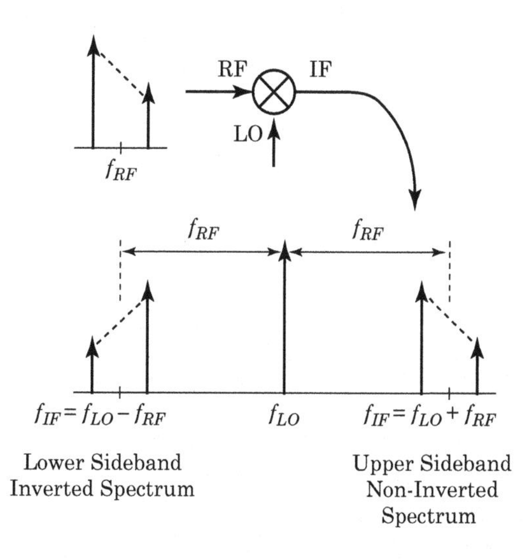

FIGURE 8-24 ■ A single conversion. If we ignore the position of 0 hertz, the lower sideband is always spectrally inverted.

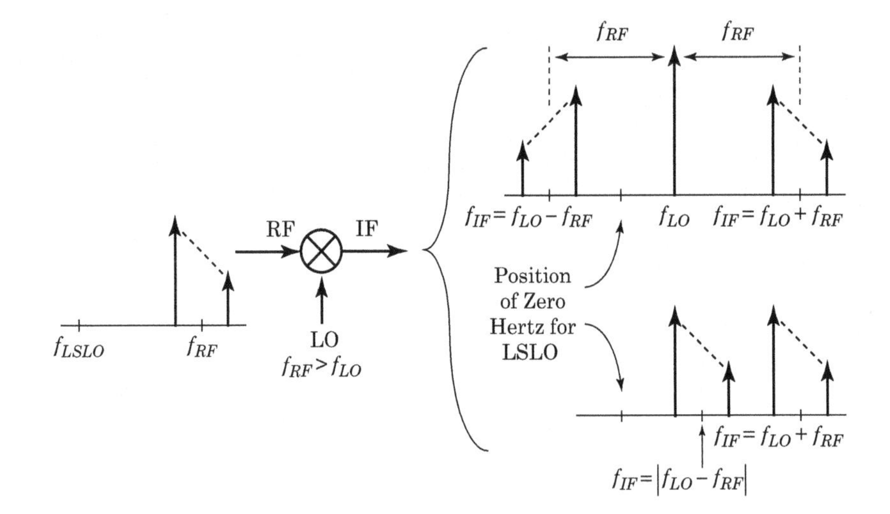

FIGURE 8-25 ■ Spectral reinversion of the lower sideband when the IF spectrum falls below 0 hertz. This conversion scheme uses an LSLO.

Ignoring the position of 0 hertz, Figure 8-24 indicates that we will always experience frequency inversion if we select the lower sideband. However, if the lower sideband falls below 0 hertz, the spectrum will be reinverted.

8.6.1 LSLO

We first examine the LSLO (Figure 8-25). The LO frequency is always less than the RF, and 0 hertz lies between f_{LO} and $f_{IF} = f_{LO} - f_{RF}$. The position of 0 hertz indicates that the lower sideband is made up entirely of negative frequencies. When we take the absolute values of the lower and upper band edges, we restore the noninverted spectrum shown at the bottom of Figure 8-25. Since the lower sideband always lies below 0 hertz when we use an LSLO, we will never experience a frequency inversion with an LSLO.

8.6.2 HSLO

Figure 8-26 shows the situation with an HSLO. Figure 8-26 is almost the same as Figure 8-25 except that the position of 0 hertz has changed. With an HSLO, f_{LO} is greater than f_{RF}, so the position of 0 hertz is always below the lower sideband. Since the lower sideband spectrum is always above 0 hertz, the spectrum stays inverted.

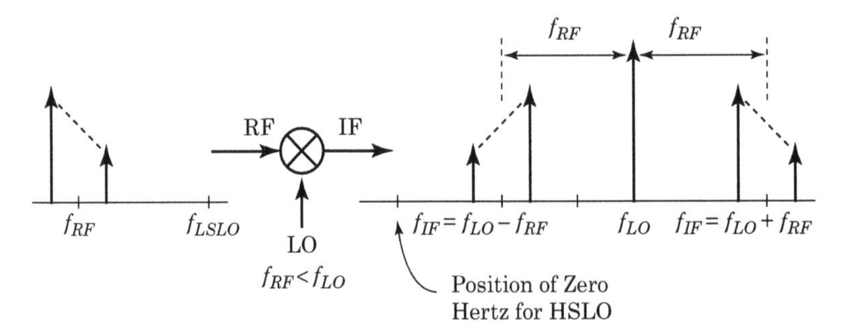

FIGURE 8-26 ■ The lower sideband of an HSLO never falls below 0 hertz, so the lower sideband does not experience a frequency reinversion. Hence, the lower sideband of an HSLO conversion is always inverted.

8.6.3 The Bottom Line

We experience frequency inversion only when we use an HSLO, and we're going to select the lower sideband (at $f_{IF} = f_{LO} - f_{RF}$). In equation form

$$\text{Spectrum Inversion when} \begin{cases} \text{Using HSLO} \\ \text{and} \\ \text{Selecting the Lower Sideband} \end{cases} \tag{8.37}$$

EXAMPLE

Cellular Telephone and Spectrum Inversion

In a previous example, we examined an older cellular telephone system with an RF of 825 to 890 MHz. The IF is 45 MHz. We found two possible LO ranges to perform this conversion: the LSLO of 780 to 845 MHz and the HSLO of 870 to 935 MHz. Does either of these conversions produce a spectrum inversion?

Solution

Equation (8.37) tells us that a frequency inversion occurs only with the HSLO and only when we use the lower sideband.

The midband HSLO frequency is 902.5 MHz, and the midband RF is 857.5 MHz. The sum and difference products of this conversion are 45 MHz and 1,760 MHz. Since the desired IF is the lower sideband, we conclude that we will see a frequency inversion.

Let's check ourselves. Figure 8-27 shows one channel centered at 829.230 MHz. This channel is about 20 kHz wide, and the band edges of this channel are 829.220 and 829.240 MHz. We'll mix this channel with an HSLO at

$$\begin{aligned} f_{HSLO} &= f_{RF} + f_{IF} \\ &= 829.230 + 45 \\ &= 874.230 \text{ MHz} \end{aligned} \tag{8.38}$$

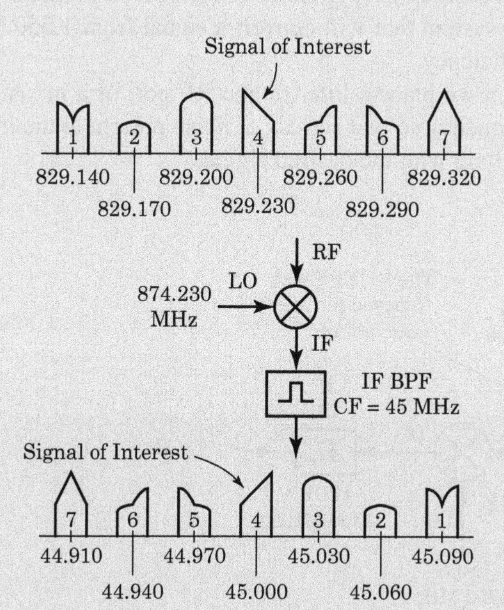

FIGURE 8-27 ■ Cellular telephone conversion scheme showing a frequency inversion. The scheme uses an HSLO, and the IF BPF selects the lower sideband.

The lower RF band edge (at $f_{RF} = 823.220$ MHz) of this one channel emerges from the mixer at

$$\begin{aligned} f_{IF} &= f_{HSLO} - f_{RF} \\ &= 874.230 - 829.220 \\ &= 45.010 \text{ MHz} \end{aligned} \tag{8.39}$$

Using a similar procedure, we find that the signal at the upper band edge of the channel (at $f_{RF} = 823.220$ MHz) emerges from the mixer at $f_{IF} = 44.990$ MHz.

The signal does experience frequency inversion.

8.7 IMAGE FREQUENCIES

Let us examine the TVRO example of Figure 8-28. We view a signal traveling from the satellite at a center frequency of 3,950 MHz. Our block converter is using an LSLO of 2,750 MHz to convert the incoming TVRO signal to a 1,200 MHz center frequency. For the purposes of example, we've left the RF filter out of Figure 8-28.

We know the LO frequency and the IF. Since we have removed the RF band-pass filter, we'll treat that quantity as an unknown and use equation (8.12) to analyze the system. Equation (8.12) produces

$$\begin{aligned} f_{RF} &= f_{LO} \pm f_{IF} \\ &= 2{,}750 \pm 1{,}200 \\ &= 1{,}550 \text{ MHz} \quad \text{or} \quad 3,950\text{MHz} \end{aligned} \tag{8.40}$$

This analysis shows two frequencies that, when placed into the RF port of this mixer, will convert to 1,200 MHz. One is the desired frequency of 3,950 MHz; the other is the *image frequency* at 1,550 MHz. While designing a system to convert 3,950 MHz down to 1,200 MHz, we have also designed a system that will convert a signal from 1,550 MHz down to 1,200 MHz with the same efficiency.

This image response is one reason we place a filter on the RF port of a mixer. The RF filter must attenuate the image frequency so that signals or noise present at the image frequency aren't converted to the IF along with the desired signals.

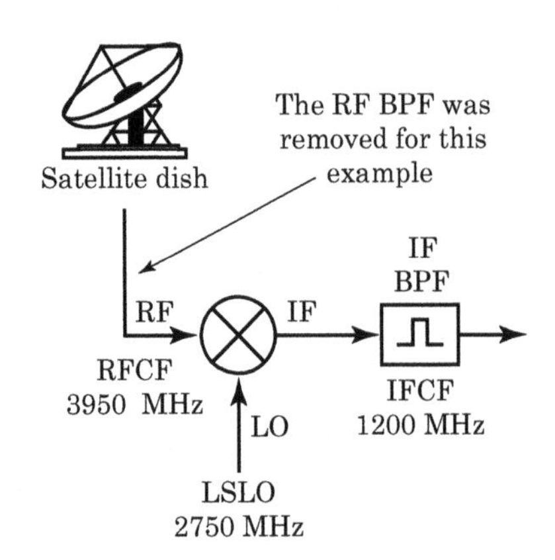

FIGURE 8-28 ■ The TVRO conversion scheme. The RF BPF has been removed to illustrate the image response of the system.

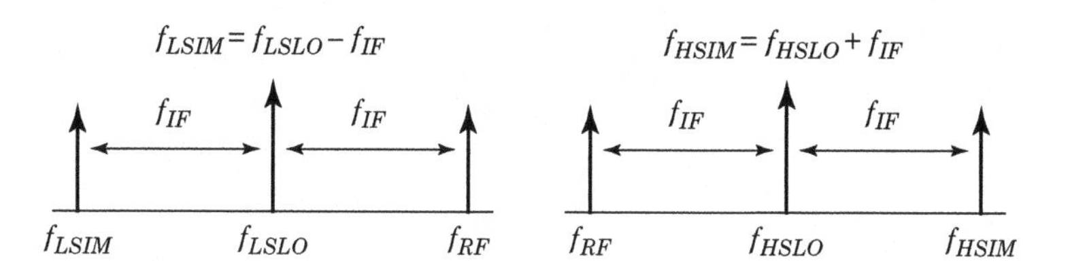

FIGURE 8-29 ■ The graphical relationship among the RF, LO, and image frequencies.

8.7.1 Locating Image Frequencies

We calculate the image frequencies of a conversion scheme using equation (8.12):

$$f_{RF} = f_{LO} \pm f_{IF} \tag{8.41}$$

Given an f_{IF} and an f_{LO}, any f_{RF} that satisfies equation (8.12) will mix down to f_{IF} at the mixer's output.

8.7.1.1 LSLO Image Frequencies

Combining equation (8.12) with equation (8.33) produces an expression for the LSLO image frequencies:

$$\begin{aligned} f_{LSIM,L} &= f_{LSLO,L} - f_{IF} \\ f_{LSIM,H} &= f_{LSLO,H} - f_{IF} \end{aligned} \tag{8.42}$$

Figure 8-29 shows this relationship graphically. This figure makes another relationship apparent:

$$f_{LSIM} = f_{RF} - 2f_{IF} \tag{8.43}$$

8.7.1.2 HSLO Image Frequencies

The expression for the HSLO image frequency comes from equations (8.12) and (8.34):

$$\begin{aligned} f_{HSIM,L} &= f_{HSLO,L} + f_{IF} \\ f_{HSIM,H} &= f_{HSLO,H} + f_{IF} \end{aligned} \tag{8.44}$$

Figure 8-29 shows this relationship graphically. We again see a second relationship among the variables:

$$f_{HSIM} = f_{RF} + 2f_{IF} \tag{8.45}$$

EXAMPLE

Commercial FM Radio

The commercial U.S. FM radio band covers an RF range of about 88 to 108 MHz. We convert each station to a 10.7 MHz IF for demodulation. Find:

a. The HSLO range
b. The range of HSLO image frequencies
c. The LSLO range
d. The range of LSLO image frequencies

Solution

a. Using equation (8.34) for the HSLO, we find

$$\begin{aligned} f_{HSLO,L} &= f_{RF,L} + f_{IFCF} & f_{HSLO,H} &= f_{RF,H} + f_{IFCF} \\ &= 88 + 10.7 & &= 108 + 10.7 \\ &= 98.7 \text{ MHz} & &= 118.7 \text{ MHz} \end{aligned} \tag{8.46}$$

b. Equation (8.44) gives us an expression for the high-side LO image frequencies:

$$\begin{aligned} f_{HSIM,L} &= f_{HSLO,L} + f_{IF} & f_{HSIM,H} &= f_{HSLO,H} + f_{IF} \\ &= 98.7 + 10.7 & &= 118.7 + 10.7 \\ &= 109.4 \text{ MHz} & &= 129.4 \text{ MHz} \end{aligned} \tag{8.47}$$

Figure 8-30(a) shows the relationships among the high-side quantities.

c. Equation (8.33) gives us f_{LSLO}:

$$\begin{aligned} f_{LSLO,L} &= f_{RF,L} - f_{IFCF} & f_{LSLO,H} &= f_{RF,H} - f_{IFCF} \\ &= 88 - 10.7 & &= 108 - 10.7 \\ &= 77.3 \text{ MHz} & &= 97.3 \text{ MHz} \end{aligned} \tag{8.48}$$

d. Equation (8.42) gives us the LSLO image frequencies:

$$\begin{aligned} f_{LSIM,L} &= f_{LSLO,L} - f_{IF} & f_{LSIM,H} &= f_{LSLO,H} - f_{IF} \\ &= 77.3 - 10.7 & &= 97.3 - 10.7 \\ &= 66.6 \text{ MHz} & &= 86.6 \text{ MHz} \end{aligned} \tag{8.49}$$

Figure 8-30(b) shows the low-side relationships.

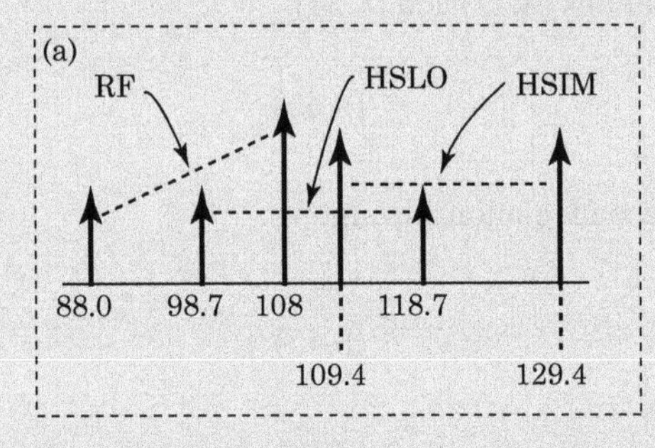

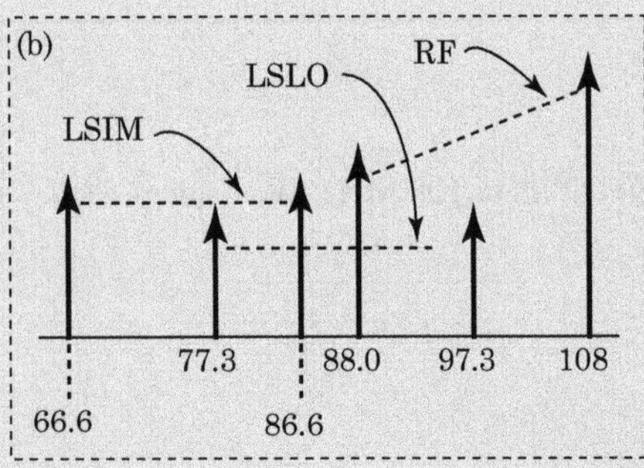

FIGURE 8-30 ■ The graphical relationship among the RF, LO, and image frequencies in a commercial FM radio using (a) an HSLO (b) an LSLO.

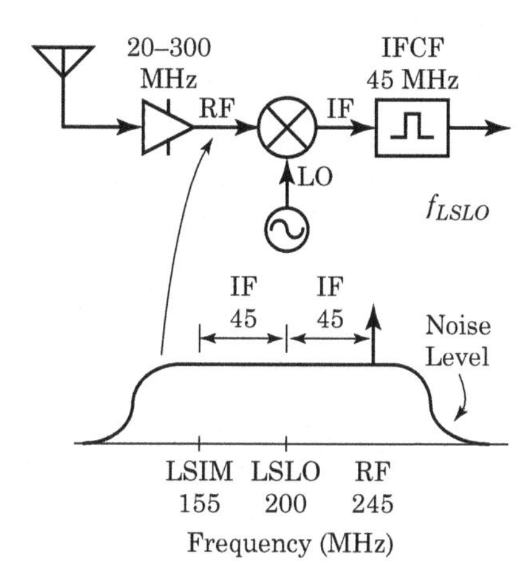

FIGURE 8-31 ■ The image response of the conversion process can translate out-of-band noise into the IF filter, increasing the noise contributed by the mixer.

8.7.1.3 Image Noise

Mixers can add extra noise to a system in a way that is not obvious. This effect, called image noise, can cause incorrect results when we measure a mixer's noise figure.

Figure 8-31 illustrates the problem. The local oscillator frequency is 200 MHz, and the center frequency of the IF filter is 45 MHz. Let us assume that the bandwidth of the RF amplifier is 20 to 1000 MHz and that the signal of interest from the antenna resides at 245 MHz.

Figure 8-31 shows the noise spectrum present at the RF port of the mixer. The architecture of the system assures that energy present on the mixer's RF port at 245 MHz will be converted to 45 MHz. However, we will also convert energy centered at 155 MHz on the mixer's RF port to the 45 MHz IF. The result is that there will be more noise present at the mixer's IF port than we anticipated and, if the noise power at 245 MHz is equal to the noise power present at 155 MHz, the noise on the mixer's IF port will be 3 dB greater than we had anticipated.

8.8 | OTHER MIXER PRODUCTS

In practice, we build a receiver by first deciding on a conversion scheme and then placing the appropriate RF and IF band-pass filters in position. The RF BPF passes more than just the signal of interest. In the case of commercial FM broadcast reception, the RF BPF allows the entire 20 MHz worth of RF spectrum through, although we want to listen to only one station at a time.

8.8.1 TVRO Example

Figure 8-32 shows the LSLO conversion scheme for our TVRO downconverter. We apply the LO and RF signals to the mixer and examine the signals present on the IF port. Equation (8.11) describes the frequency of the desired output signals:

$$\begin{aligned} f_{IF} &= f_{LO} \pm f_{RF} \\ &= 2{,}750 \pm 3{,}950 \\ &= 6{,}700 \text{ MHz and } 1{,}200 \text{ MHz} \end{aligned} \tag{8.50}$$

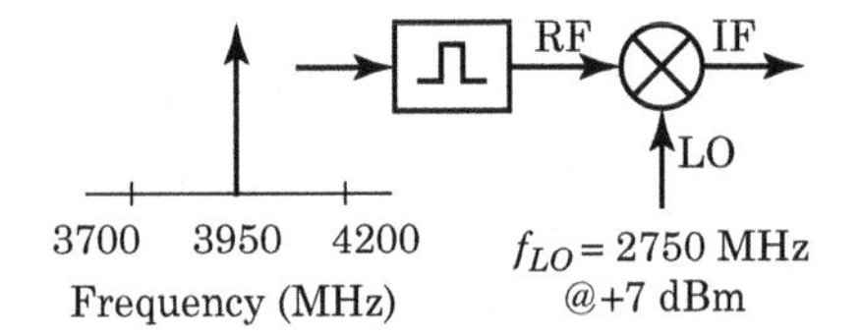

FIGURE 8-32 ■ A TVRO downconverter with one RF signal and one LO frequency.

We will observe the sum and difference signals emerging from the mixer. Due to practical considerations, signals at other frequencies will also emerge from the mixer including the following:

- The RF signal (still at f_{RF}) and its harmonics
- The LO signal (at f_{LO}) and its harmonics
- Every combination of

$$|m f_{LO} \pm n f_{RF}| \quad \text{for} \quad \begin{cases} m = 0, 1, 2, \ldots \\ n = 0, 1, 2, \ldots \end{cases} \tag{8.51}$$

including the desired signal and its image.

Table 8-1 lists some of the frequencies that will be present on the output of the mixer.

The output spectrum of the TVRO mixer will resemble Figure 8-33. There will be more signals present at the mixer's output than those we have shown. Many of these output products can be easily removed by the IF filter. Others will fall either inside the IF filter or close enough to the filter's passband that they will not experience significant attenuation. As designers, we would like to know the frequency of each mixer product and its expected power level.

8.8.2 Mixer Spur Tables

Mixer data sheets often include information regarding the strength of the spurious signals emitted by the mixer. Table 8-2 shows the data measured for one particular mixer used in

TABLE 8-1 ■ Mixer spurious components

m	n	mf_{LO} (MHz)	nf_{RF} (MHz)	$\|mf_{LO} + nf_{RF}\|$ (MHz)	$\|m_{LO} - nf_{RF}\|$ (MHz)
0	0	0	0	0	0
0	1	0	3,950	3,950	3,950
0	2	0	7,900	7,900	7,900
0	3	0	11,850	11,850	11,850
1	0	2,750	0	2,750	2,750
1	1	2,750	3,950	6,700	1,200
1	2	2,750	7,900	10,650	5,150
1	3	2,750	11,850	14,600	9,100
2	0	5,500	0	5,500	5,500
2	1	5,500	3,950	1,550	9,450
2	2	5,500	7,900	2,400	13,400
2	3	5,500	11,850	6,350	17,350
3	0	8,250	0	8,250	8,250
3	1	8,250	3,950	12,200	4,300
3	2	8,250	7,900	16,150	350
3	3	8,250	11,850	20,100	3,600

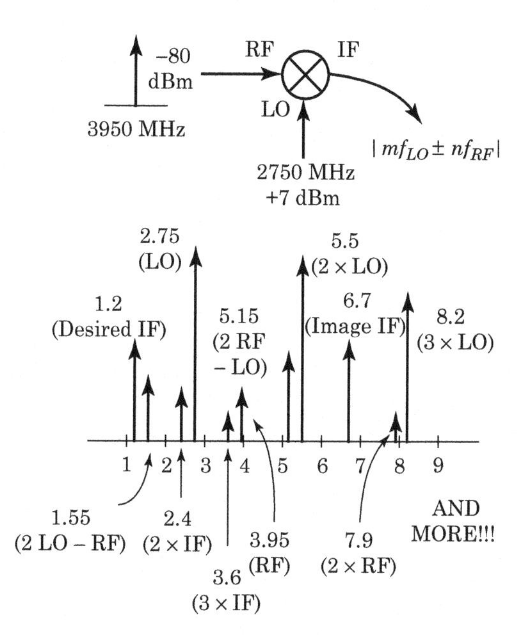

FIGURE 8-33 ■ The possible signals present at the IF port of the mixer shown in Figure 8-32.

one particular conversion scheme at specific levels of RF and LO powers. The RF and LO frequency are fixed.

These data are accurate for the specific situation under which the data were measured. However, the power of the spurious products present on the mixer's output is a strong function of the LO and RF signal power. Spurious performance also changes with RF and LO frequency. The data sheet is broadly indicative of the mixer's performance and may allow you to make modest comparisons among mixers, but the data are not applicable to a large family of conversion schemes.

Table 8-2 shows relative power levels of the various $mf_{LO} \pm nf_{RF}$ mixing products for this one particular mixer and one particular conversion scheme. The numbers in the table represent the number of dB the undesired product (at $mf_{LO} \pm nf_{RF}$) is below the desired product (at $f_{LO} \pm f_{RF}$). These data were gathered under the following conditions:

- $f_{RF} = 500$ MHz at -4 dBm

TABLE 8-2 ■ The relative power present in various mixer spurious components.

	RF Harmonic (*n*)										
LO Harmonic (*m*)	**0**	**1**	**2**	**3**	**4**	**5**	**6**	**7**	**8**	**9**	**10**
0		17	43	44	76	66	69	72	>85	>85	84
1	32	**0**	60	34	68	75	76	71	>85	83	84
2	23	39	45	49	56	67	83	73	81	84	83
3	36	17	56	35	72	53	84	74	>85	83	85
4	43	33	51	49	56	57	70	73	85	84	84
5	35	37	60	36	61	48	71	67	>85	>85	83
6	44	38	65	45	57	58	64	64	83	83	84
7	36	39	52	50	65	47	75	63	76	77	84
8	61	39	58	52	76	71	64	63	73	71	83
9	46	47	41	57	75	70	78	61	76	71	81
10	63	51	69	50	66	67	79	71	71	71	81

- $f_{LO} = 470$ MHz at 7 dBm
- $f_{IF} = 30$ MHz, measured to be -11 dBm.

The numbers in Table 8-2 are typical although the RF power is quite high.

The desired output signal of the mixer (at $f_{LO} \pm f_{RF}$) is at the intersection of RF harmonic (n) = LO harmonic $(m) = 1$. The table entry at that spot is zero.

EXAMPLE

Mixer Spurious Products

The desired signal present at the IF port of a mixer (at $f_{RF} - f_{LO}$) has a frequency of 30 MHz and an amplitude of -11 dBm. The RF is $f_{RF} = 500$ MHz, and the LO frequency f_{LO} is 470 MHz. Given the previous mixer table, find the levels of the following signals:

a. $3f_{LO} - 4f_{RF} = 590$ MHz
b. $2f_{LO} - 2f_{RF} = 60$ MHz
c. $-6f_{LO} + 7f_{RF} = 680$ MHz
d. $-4f_{LO} + 2f_{RF} = 880$ MHz

Solution

Using Table 8-2, we find

a. Looking up LO harmonic = 3, RF harmonic = 4, we find this output product is 72 dB below the desired output product at -11 dBm. The level of the signal at 590 MHz is $-11 - 72 = -83$ dBm.

b. For LO harmonic = 2, RF harmonic = 2, the output product is 45 dB below the desired output product at -11 dBm. The level of the signal at 60 MHz is $-11 - 45 = -56$ dBm.

c. The table tells us the suppression of the product with LO harmonic = 6, RF harmonic = 7 is 64 dB. The power of the signal at 680 MHz is $-11 - 64 = -75$ dBm.

d. Finally, the $m = 4$, $n = 2$ product is suppressed by 51 dB. The output power at 880 MHz is $-11 - 51 = -62$ dBm.

8.8.3 Double-Balanced Mixers

Table 8-2 characterizes a *double-balanced mixer* (DBM). This is the most common mixer used in high-end receiving equipment. We can make some general conclusions about the spurious output products of DBMs:

- *Even-by-even* products: Examples of even-by-even products are 2 by 6 and 8 by 4. Even-by-even products tend to be the weakest spurious components generated in the mixer. They are the least worry.
- *Even-by-odd* and *odd-by-even* products: Examples are 3 by 2, 4 by 5, and 1 by 7. These responses are generally higher power than the even-by-even products.
- *Odd-by-odd* products: Examples are 3 by 7 and 1 by 5. These are generally the highest-level undesired products and, consequently, the most trouble.

We worry more about low-order products than about high-order products because the power of the low-order products is usually stronger. These characteristics are a direct result of the topology of the double-balanced mixer.

The power of the spurious products generated inside of a mixer is a strong function of the input power levels and the characteristics of the particular mixer. The spurious output power is a weaker function of the frequency of operation. If you change any of the input characteristics, the mixer table probably won't be accurate.

8.9 SPURIOUS CALCULATIONS

Figure 8-34 shows a generic frequency conversion scheme complete with RF and IF filters. We know the range of RF signals we want to process, we've decided on an LO range, and we know the IF center frequency and bandwidths. We will analyze this system to determine its spurious signal performance.

The signals present on the IF port of the mixer will be

$$f_{IF} = mf_{LO} \pm nf_{RF} \tag{8.52}$$

where

f_{IF} = the frequencies of the signals present at the mixer's output port
f_{LO} = the LO frequency
f_{RF} = the RF
$m = 0, 1, 2, 3 \ldots$
$n = 0, 1, 2, 3 \ldots$

We must recognize that the RF and LO signals present at the ports of the mixer in Figure 8-34 now represent frequency bands rather than single tones. The variables f_{LO} and f_{RF} in equation (8.52) are actually the ranges of a low LO frequency ($f_{LO,L}$) to a high LO frequency ($f_{LO,H}$). We represent the low and high RFs as $f_{RF,L}$ and $f_{RF,H}$, respectively.

Our task is to determine if any of the mixer's spurious products fall within the passband of the IF filter for the possible values of LO frequency and RF. If some spurious products fall within the IF passband, what is their power level? Are they low-order signals (implying that they'll be relatively high power), or are they high-order signals (and thus tending to be low power)?

8.9.1 Assumptions

In our analysis of the system in Figure 8-34, we'll assume the following:

- The RF band-pass filter is a brick wall filter. It will pass the frequencies between $f_{RF,L}$ and $f_{RF,H}$ while completely attenuating everything else.

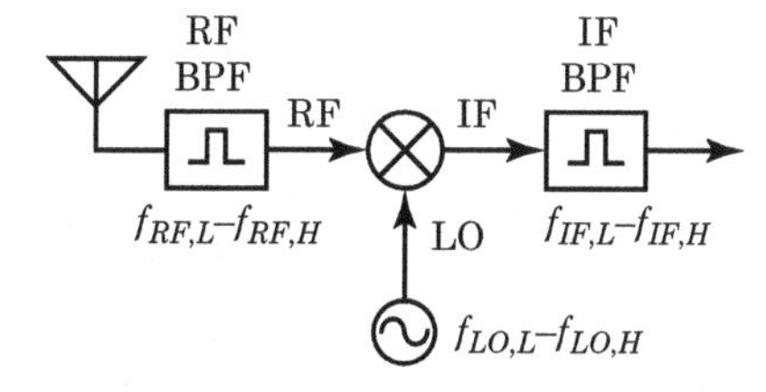

FIGURE 8-34 ■ A candidate receiver architecture, showing the definitions of frequencies used in the analysis of the receiver's spurious performance.

- Likewise, the IF band-pass filter is a "brick wall" filter. It will pass all the signals from $f_{IF,L}$ to $f_{IF,H}$ while completely attenuating signals outside of this range.
- Our conversion scheme dictates that the local oscillator will tune from $f_{LO,L}$ to $f_{LO,H}$.
- In an operational system, the RF band-pass filter will be connected to an antenna so we'll have no control over the signals entering the RF port of the mixer. We know that their frequencies lie between $f_{RF,L}$ to $f_{RF,H}$. The worst-case scenario is to assume that the entire RF band is flooded with signals, so we'll assume that every frequency in the range of $f_{RF,L}$ to $f_{RF,H}$ is present at the RF port of the mixer.
- As the receiver tunes from one channel to the next, the frequency of the LO will change. In an operational system, the LO can be constantly changing in frequency, or it could sit on one frequency for months at a time. All we really know is that the LO frequency will be somewhere between $f_{LO,L}$ and $f_{LO,H}$. The worst-case assumption is to assume that *every* frequency from $f_{LO,L}$ to $f_{LO,H}$ is present at the LO port of the mixer.

8.9.2 Analysis Procedure

Figure 8-35 shows the situation we face. We have a band of signals extending from $f_{RF,L}$ to $f_{RF,H}$ on the mixer's RF port. We also have a band of signals from $f_{LO,L}$ to $f_{LO,H}$ on the mixer's LO port. We then calculate the signals present on the output port of the mixer with special attention to those signals falling between $f_{IF,L}$ and $f_{IF,H}$. We perform this analysis for every value of m (the LO harmonic number) and n (the RF harmonic number) that we think will generate spurious signals of significant power.

8.9.3 Derivation

We seek the range of frequencies present on the mixer's IF port for each value of m and n. Equation (8.52) describes the frequencies of the spurious IF output signals given the RF and LO input frequency. As we show in Figure 8-36, the LO spectrum from $f_{LO,L}$ to $f_{LO,H}$ will mix with the RF spectrum from $f_{RF,L}$ to $f_{RF,H}$ to produce two frequency bands on the mixer's IF port. The two bands fall in the ranges of $f_{IF,-,L}$ to $f_{IF,-,H}$ (the lower sideband) and $f_{IF,+,L}$ to $f_{IF,+,H}$ (the upper sideband).

Figure 8-36 shows the situation for only one value of m and one value of n. We must recalculate for every combination of m and n.

We find expressions for $f_{IF,-,L}$, $f_{IF,-,H}$, $f_{IF,+,L}$, and $f_{IF,+,H}$ by applying equation (8.52) to the band edges of the RF and LO spectrums. For the lower sideband

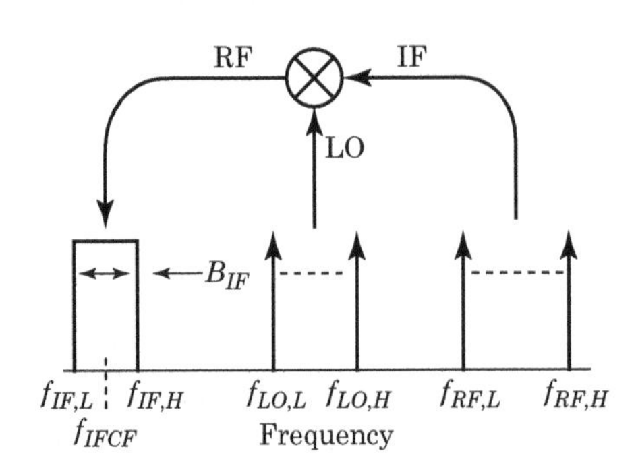

FIGURE 8-35 ■ A graphical illustration of the spurious analysis procedure. We repeat this analysis for each value of m and n.

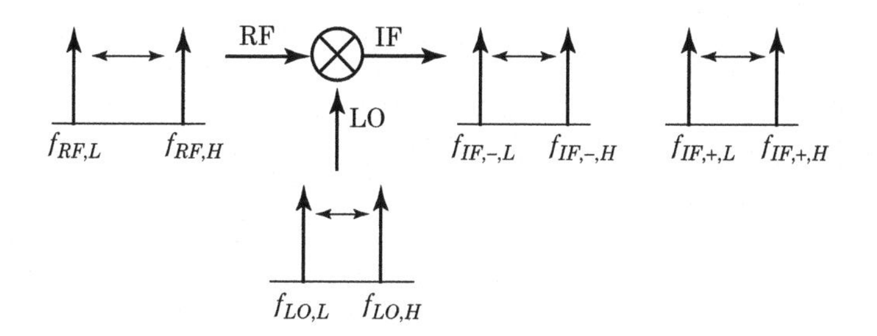

FIGURE 8-36 ■ We find the possible outputs present on the mixer's IF port given the possible RF and LO ranges and unique values of *m* and *n*.

(using $f_{IF} = mf_{LO} - nf_{RF}$), we find

$$\begin{aligned} f_{IF,-,L} &= mf_{LO,L} - nf_{RF,H} \\ f_{IF,-,H} &= mf_{LO,H} - nf_{RF,L} \end{aligned} \tag{8.53}$$

For the upper sideband, we use $f_{IF} = mf_{LO} + nf_{RF}$ to find

$$\begin{aligned} f_{IF,+,L} &= mf_{LO,L} + nf_{RF,L} \\ f_{IF,+,H} &= mf_{LO,H} + nf_{RF,H} \end{aligned} \tag{8.54}$$

EXAMPLE

Cellular Telephone

Find range of the two frequency bands emerging from a mixer's IF port for the second harmonic of the LO combined with the third harmonic of the RF. Assume the system is a cellular telephone receiver covering 825 to 890 MHz and we're using an HSLO of 870 to 935 MHz. The IF center frequency is 45 MHz, and we have a 1 MHz wide band-pass filter. Does the 2LO by 3RF spurious product fall within the IF bandwidth and hence present a possible problem?

Solution

We use equation (8.53) to find the range of the lower sideband output:

$$\begin{aligned} f_{IF,-,L} &= mf_{LO,L} - nf_{RF,H} \\ &= (2)(870) - (3)(890) = -930 \text{ MHz} \\ f_{IF,-,H} &= mf_{LO,H} - nf_{RF,L} \\ &= (2)(935) - (3)(825) = -605 \text{ MHz} \end{aligned} \tag{8.55}$$

The lower sideband will exist between 605 and 930 MHz. Equation (8.54) returns the range of the upper sideband:

$$\begin{aligned} f_{IF,+,L} &= mf_{LO,L} + nf_{RF,L} \\ &= (2)(870) + (3)(825) = 4{,}215 \text{ MHz} \\ f_{IF,+,H} &= mf_{LO,H} + nf_{RF,H} \\ &= (2)(935) + (3)(890) = 4{,}540 \text{ MHz} \end{aligned} \tag{8.56}$$

The upper sideband lives between 4,215 and 4,540 MHz. Figure 8-37 is a graphical interpretation of these results. The figure shows that neither output band overlaps the IF band-pass filter. They are both well removed from the filter's center frequency. We conclude that this particular response will not be problematic.

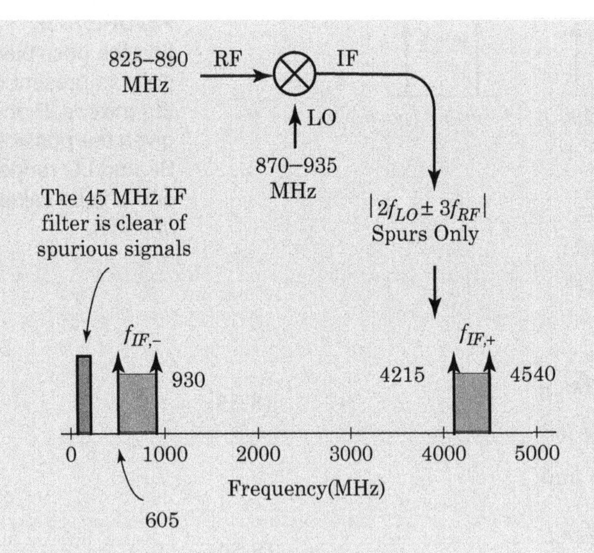

FIGURE 8-37 ■ Spurious analysis of a cellular telephone conversion scheme for $m = 2$ and $n = 3$. This combination of m and n does not produce signals that will pass through the IF filter. The $m = 2$, $n = 3$ spurious product is therefore not a concern in this conversion scheme.

EXAMPLE

Cellular Telephone

Repeat the problem for the third harmonic of the LO and the third harmonic of the RF.

Solution

We again refer to equation (8.53) to find the range of the lower sideband output:

$$\begin{aligned} f_{IF,-,L} &= mf_{LO,L} - nf_{RF,H} \\ &= (3)(870) - (3)(890) = -60 \text{ MHz} \\ f_{IF,-,H} &= mf_{LO,H} - nf_{RF,L} \\ &= (3)(935) - (3)(825) = 330 \text{ MHz} \end{aligned} \tag{8.57}$$

This band runs from −60 MHz to 330 MHz (not from 60 to 330 MHz). If we consider only positive frequencies, this response covers 0 to 330 MHz.

Equation (8.54) produces the range of the upper sideband:

$$\begin{aligned} f_{IF,+,L} &= mf_{LO,L} + nf_{RF,L} \\ &= (3)(870) + (3)(825) = 5{,}085 \text{ MHz} \\ f_{IF,+,H} &= mf_{LO,H} + nf_{RF,H} \\ &= (3)(935) + (3)(890) = 5{,}475 \text{ MHz} \end{aligned} \tag{8.58}$$

Figure 8-38 shows a graphical interpretation of these results, and it reveals that the IF passband does fall inside the range of the lower sideband. This combination of LO and RF products may produce a spurious response and our next step would be to calculate the likely power level of these signals.

When designing a conversion scheme, we would perform this calculation for every value of m and n that could produce a troublesome product. The upper limits of m and n are set by the mixer table coefficients like those shown in Table 8-2. We would calculate the results for each value of m and n that might generate noticeable power at the mixer's IF port. Computer assistance makes this analysis painless.

We'll have more to say about spurious calculations in Chapter 10.

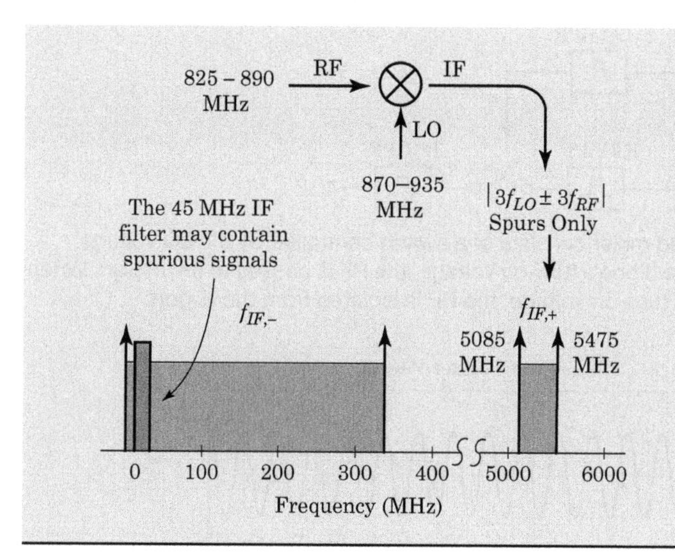

FIGURE 8-38 ■ Spurious analysis of a cellular telephone conversion scheme for $m = 3$ and $n = 3$. This combination of m and n might produce signals that will pass through the IF filter. The $m = 3$, $n = 3$ spurious product is therefore a concern for this conversion scheme.

8.10 MIXER REALIZATIONS

We will examine several mixer topologies in this section and explore the advantages and problems with each of them.

8.10.1 Single-Ended Mixers

The term *single-ended mixer* (SEM) refers to the technique of using a nonlinear device to perform frequency conversion. We seek to perform a time-domain multiplication of the RF and LO signals to generate the sum and difference output terms. We will describe a single-diode mixer, but the concepts we explore here will apply to other nonlinear devices.

8.10.1.1 SEM Operation

Figure 8-39 shows a simplified diagram of a single-ended mixer. The diode is the element that performs the time-domain multiply. The RF, LO, and IF band-pass filters increase the interport isolation.

The diode acts as a switch that is controlled by the LO (see Figure 8-40). When the instantaneous LO voltage is greater than the diode's turn-on voltage, current flows

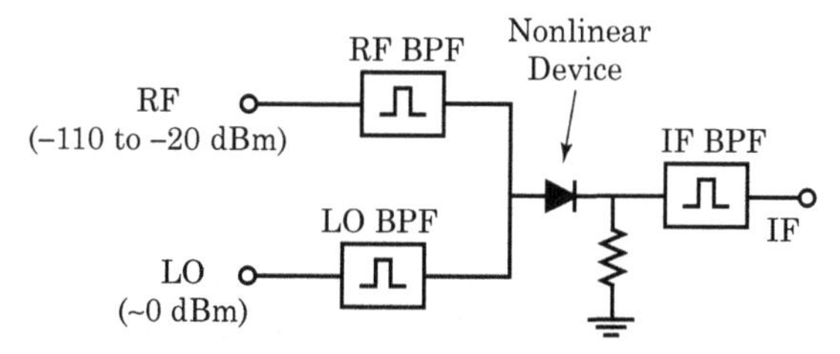

FIGURE 8-39 ■ The schematic diagram of a simple SEM. We perform the time domain multiplication of the LO waveform and the RF waveform by using the LO to control an RF switch. The LO switches the diode, gating the RF to the IF port. The band-pass filters work to improve interport isolation.

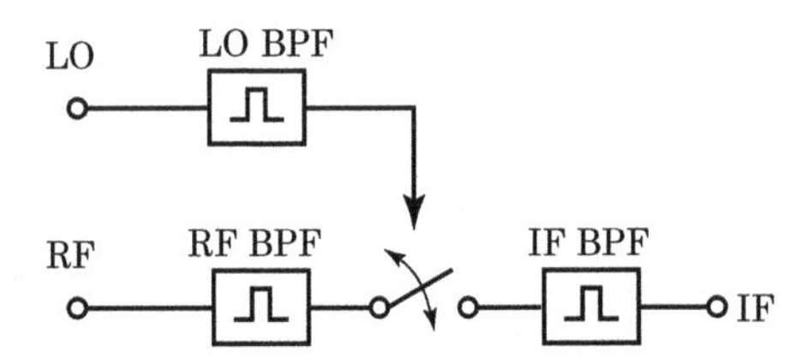

FIGURE 8-40 ■ A single-ended mixer consists of a switch controlled by the LO voltage. When the LO is greater than the diode's turn-on voltage, the RF is passed to the IF port. When the LO is less than the diode's turn-on voltage, the RF is isolated from the IF port.

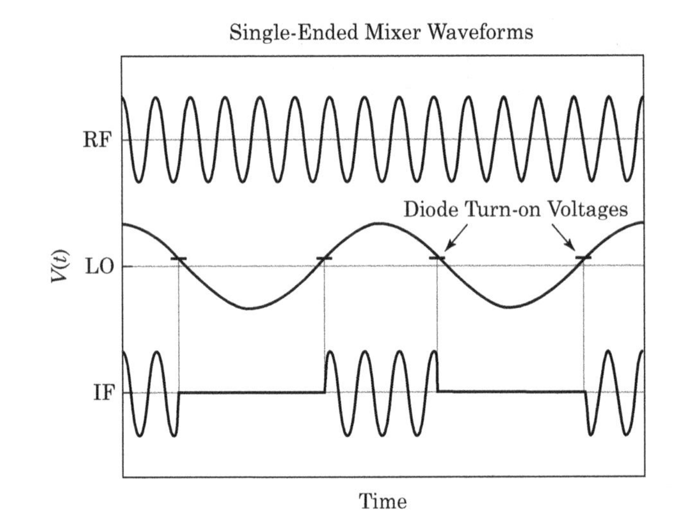

FIGURE 8-41 ■ The time-domain waveforms present in a single-ended mixer. The switching action of the diode performs an approximation of the time-domain multiply between the LO and RF signals.

through the diode, forward-biasing it and connecting the RF and LO ports together. When the instantaneous LO voltage is less than the diode's turn-on voltage, the diode is reversed biased, and the RF port is isolated from the IF port. Figure 8-41 shows the appropriate waveforms.

We immediately note that we are not performing a time-domain multiply between the LO and RF waveforms. We are multiplying the RF waveform by either zero or unity at a rate determined by the LO. When the LO voltage is greater than a diode drop, we multiply the RF by unity; when the LO voltage is less than a diode drop, we multiply the RF by zero. In equation form, the multiplying waveform, $W_{LO}(t)$, derived from the LO is

$$W_{LO}(t) = \begin{cases} 0 \text{ when } V_{LO}(t) \leq \text{ Diode Turn-on Voltage} \\ 1 \text{ when } V_{LO}(t) > \text{ Diode Turn-on Voltage} \end{cases} \tag{8.59}$$

Fourier analysis reveals an equivalent expression for $W_{LO}(t)$ is

$$W_{LO}(t) = a_0 + \sum_{k=1}^{\infty} a_k \cos\left(k\omega_{LO}t + \theta_k\right) \tag{8.60}$$

In the ideal case, we multiply the RF waveform by the LO waveform, which produces only the sum and difference products. Equation (10.87) reveals that we really multiply the RF waveform by a constant and many harmonics of the LO. This process produces not only the desired sum and difference products but also many other unwanted mixing products (also called *spurious products*).

EXAMPLE

DC and RF:IF Rejection

Show that the DC term in equation (8.60) is responsible for the poor RF:IF isolation of the single-ended and single-balanced mixers.

Solution

Equation (8.60) is the Fourier transform of the LO-derived multiplying waveform. This waveform equals +1 when the instantaneous LO voltage is greater than 0, and the waveform equals 0 when the instantaneous LO voltage is less than 0. We form a time-domain multiplication between the waveform of equations (8.59) and (8.60) with the RF signal. The result of this multiplication is

$$\begin{aligned} V_{IF}(t) &= [\cos(\omega_{RF}t)]\left[a_0 + \sum_{k=1}^{\infty} a_k \cos(k\omega_{LO}t + \theta_k)\right] \\ &= a_0 \cos(\omega_{RF}t) + [\cos(\omega_{RF}t)]\left[\sum_{k=1}^{\infty} a_k \cos(k\omega_{LO}t + \theta_k)\right] \end{aligned} \tag{8.61}$$

The first term in this equation represents the RF:IF feedthrough of the mixer.

The multiplying waveform $W_{LO}(t)$ is almost a square wave with a 50% duty cycle. This approximation approaches reality when the LO voltage is much greater than the turn-on voltage of the diode. For a large LO voltage, we can approximate $W_{LO}(t)$ as

$$W_{LO}(t) = \frac{1}{2} + \sum_{k=1}^{\infty} \frac{1}{2k-1} \sin[(2k-1)\,\omega_{LO}t] \tag{8.62}$$

This expression contains only the odd harmonics of the LO frequency. Since the even harmonics are not present in the multiplying waveform, they do not get a chance to cause spurious products. For large LO powers, the single-ended mixer suppresses the spurious products generated by the even harmonics of the LO.

Due to realization effects, the time-domain waveforms shown in Figure 8-41 are not exactly accurate. Every diode has a forward voltage drop, a forward series resistance, a reverse current, an interterminal capacitance, and a nonzero switching time. The result of these realities is that the waveform present at the input to the IF filter may be different from how we've drawn it.

The SEM can be a spurious product nightmare. However, if the RF, IF, and LO signals are widely separated, then the respective band-pass filters can provide excellent interport isolation. We can often tolerate the poor spurious performance of an SEM in narrowband systems, such as commercial FM radio receivers.

8.10.1.2 SEM LO Power

During our analysis of the single-ended mixer, we assumed the diode switch was completely controlled by the LO voltage and that the switch opened and closed exactly at the zero crossing of the LO. This is approximately true when the LO voltage is much greater than the RF voltage. As the magnitude of the RF voltage approaches the magnitude of the LO voltage, the exact switching instant of the diode will change. Equations (8.59) and (8.60) will be less accurate as the RF power increases. While the desired mixing action still occurs, the spurious output levels of the mixer will rise. In practice, we maintain the

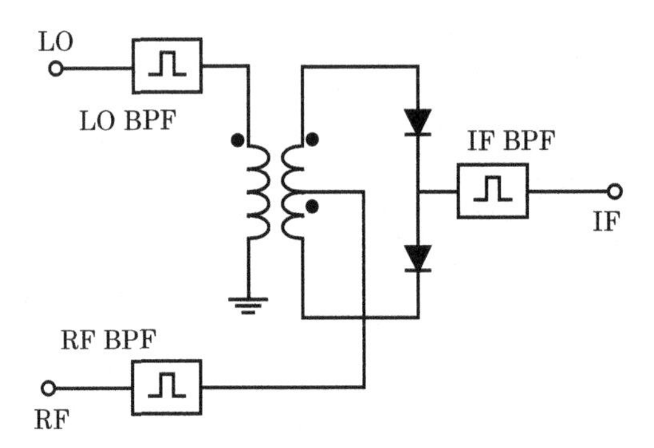

FIGURE 8-42 ■ The schematic diagram of a simple single-balanced mixer. The operation is similar to the SEM, but the balun offers improved LO:IF and the LO:RF isolation.

RF power level entering the mixer at least 20 dB below the LO power level, or

$$P_{RF,dBm} \leq P_{LO,dBm} - 20 \tag{8.63}$$

A single-ended mixer typically requires at least 0 dBm of LO power to adequately forward-bias the mixing diode. This limits the maximum RF signal power to approximately −20 dBm.

8.10.2 Single-Balanced Mixers

The single-ended mixer is inexpensive and easy to realize, but it has its problems: poor interport isolation and many spurious signals. The single-balanced mixer (SBM) is an improvement over the SEM (see Figure 8-42 for the schematic diagram of one type). This mixer contains a balanced-to-unbalanced transformer, or *balun*. We first must explore the operation of a balun.

8.10.2.1 Baluns

Baluns convert unbalanced signals to balanced signal and vice versa. An unbalanced signal is commonly carried on one wire and is referenced to a ground. A balanced signal is carried on two wires and is not referenced to ground. The information content of a balanced signal is derived from how the signal in one wire changes with respect to the other wire.

To convert an unbalanced signal to a balanced signal, we first split the signal into two equal parts and then multiply one signal by -1 or $1\angle 180°$. We put the positive signal on one wire and the negative signal on the other wire to form the balanced signal. To form an unbalanced signal from a balanced signal, we subtract the negative signal from the positive signal.

We will use the balun of Figure 8-42 to convert the unbalanced LO into a balanced signal. With the proper configuration of circuit elements, we use the presence of the positive and negative LOs to cancel the LO signal at the IF and RF ports of the SBM.

The Dot Convention The dots on the transformer terminals indicate polarity. The dot convention indicates the polarity of output windings. Figure 8-43 shows a transformer and the dot convention.

By definition, if we force current into any dotted terminal, then the balun will force current out of all the other dotted terminals as shown in Figure 8-43. If we force a current into an undotted terminal, current will flow out of all the undotted terminals of the transformer. The magnitude of the current depends on the turns ratio and the losses in the transformer.

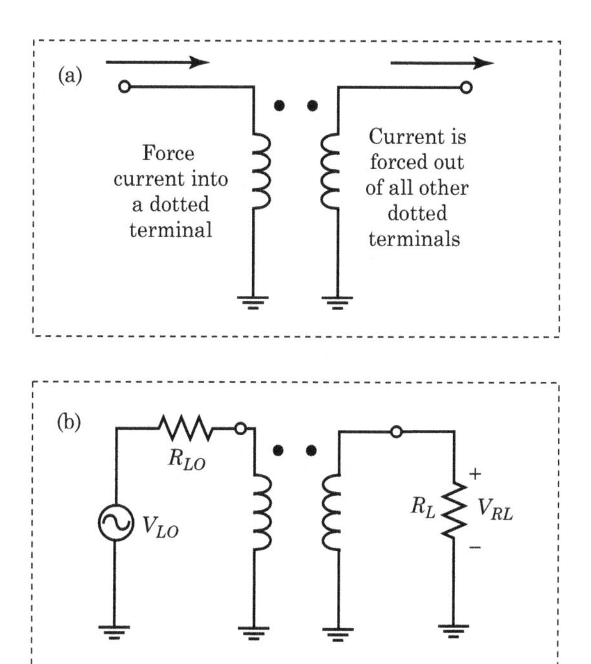

FIGURE 8-43 ■ The dot convention for a balanced-to-unbalanced transformer (balun).

Figure 8-43(b) shows the dot convention with respect to voltage. The voltage source V_{LO} forces a current into the primary's dotted terminal, so current flows out of the secondary's dotted terminal and through the load resistor. The dotted terminal will exhibit a positive voltage across the load resistor, V_{RL}, of the polarity shown in Figure 8-43(b).

Balun Voltages When we speak of mixers, the term balun often refers to a balanced transformer with a center tap, as drawn in Figure 8-44. The center tap can be grounded or can be connected to a signal source or load.

If we apply a signal to the primary side of the balun (on the left of Figure 8-44), the secondary output voltages (V_{LO1} and V_{LO2}) appear with the polarities shown. In an ideal balun, the upper and lower secondary transformers are identical and the voltage V_{LO1} exactly equals $-V_{LO2}$. In a nonideal balun, the magnitudes of the two voltages are about equal, and the two signals are approximately 180° out of phase (i.e., $V_{LO1} \approx -V_{LO2}$).

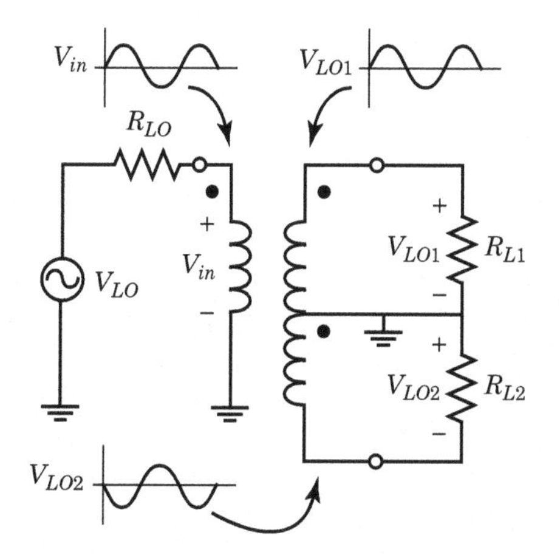

FIGURE 8-44 ■ The voltages generated in a simple balun.

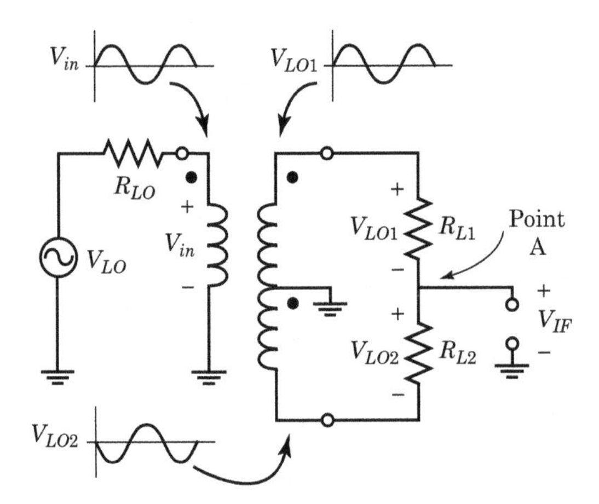

FIGURE 8-45 ■ In a balun, the transformer acts to produce equal and opposite voltages across R_{L1} and R_{L2}, canceling the LO voltage at the IF port and hence increasing the LO:IF isolation of the mixer.

Figure 8-45 shows the circuit of Figure 8-44 with a slight modification. We've removed the connection between the resistor junction and ground. Assuming that the resistors have exactly the same value (i.e., $R_{L1} = R_{L2} = R_L$) and that $V_{LO1} = -V_{LO2} = V_L$, then the current through the resistors is

$$\begin{aligned} I_{LO} &= \frac{2V_L}{2R_L} \\ &= \frac{V_L}{R_L} \end{aligned} \tag{8.64}$$

so current flows through the resistors R_{L1} and R_{L2} at a rate determined by V_{LO}.

If the resistors have *exactly* the same value and the balun is *perfectly* balanced (which means that V_{LO1} exactly equals $-V_{LO2}$), then the voltage V_{IF} at Point A will be zero. We sometimes say that Point A is a *virtual ground.* We can apply any complex waveform we like to the transformer primary and still achieve perfect cancellation at Point A if the balance remains intact over the bandwidth of the signal. If the transformers are not exactly balanced or the two components are not exactly the same, then there will be a nonzero voltage at Point A. The magnitude of the voltage at Point A is a direct measure of the system's balance.

Center Tap Drive Figure 8-46 shows yet another balun configuration. We are driving the center point of the balun with V_{RF} and are interested in the magnitude of V_{LO} and V_{IF}. We will assume the balun is perfectly balanced and that $R_{L1} = R_{L2}$.

Any current I_{RF} delivered by V_{RF} is split into I_{RF1} and I_{RF2} as it enters the transformer. With a perfectly balanced balun and with $R_{L1} = R_{L2}$, the current will split evenly as it enters the center tap of the balun so $I_{RF1} = I_{RF2} = I_{RF}/2$. Half of the RF current enters a dotted terminal, while the other half enters an undotted terminal.

On the balun primary, I_{RF1} will generate I'_{RF1} in the direction shown, while I_{RF2} will generate I'_{RF2} in the opposite direction. Since $I_{RF1} = I_{RF2}$, then $I'_{RF1} = I'_{RF2}$ and the RF current exactly cancels at the LO port. By superposition, we can also say that any current sourced from the LO port will be canceled at the RF port. The balun is responsible for improved LO:RF isolation.

The two RF currents I_{RF1} and I_{RF2} will recombine in phase at the IF port, so there will no isolation between the RF and IF ports of this mixer.

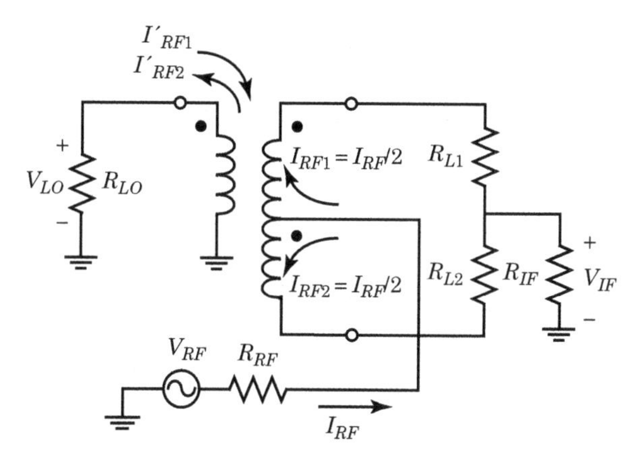

FIGURE 8-46 ■ The operation of a balanced–unbalanced transformer. An ideal balun produces equal and opposite currents through the windings in the directions shown. These currents then produce equal and opposite voltages across R_{L1} and R_{L2}, canceling the LO voltage at the IF port.

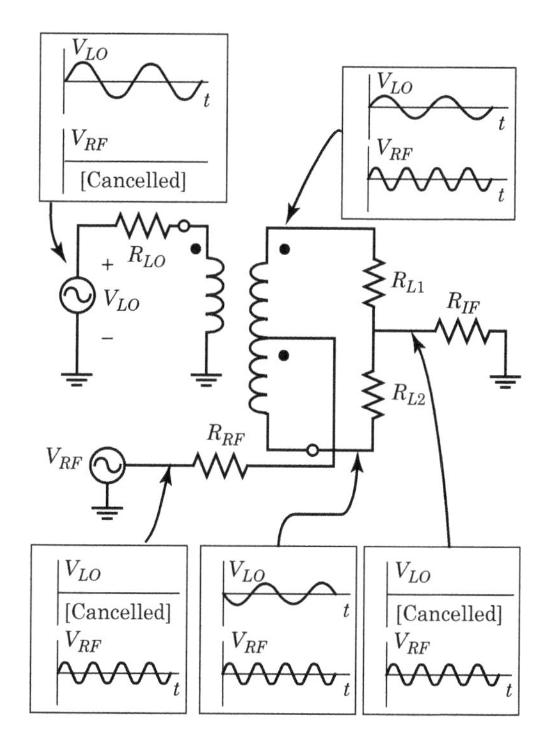

FIGURE 8-47 ■ Balun operation, illustrating the voltage cancellation that is responsible for increased LO:IF and RF:IF isolation. Similar analysis can be applied to demonstrate LO:RF isolation.

Figure 8-47 sums up the system when we apply both an LO voltage and RF voltage to the balun. The balun suppresses the LO at the two output ports. The RF port is directly connected to the IF port through transformer action.

Figure 8-48 illustrates one last point. We can we can replace the resistors of Figure 8-45 with any type of component without affecting the suppression of the balun. A capacitor, inductor, or any complex combination of the three (R, L, or C) will do, but to maintain suppression we must ensure that the two components are identical.

We can also replace the resistors with a nonlinear device such as a diode. For continued suppression, the diodes must exhibit identical V-I characteristics that must track over

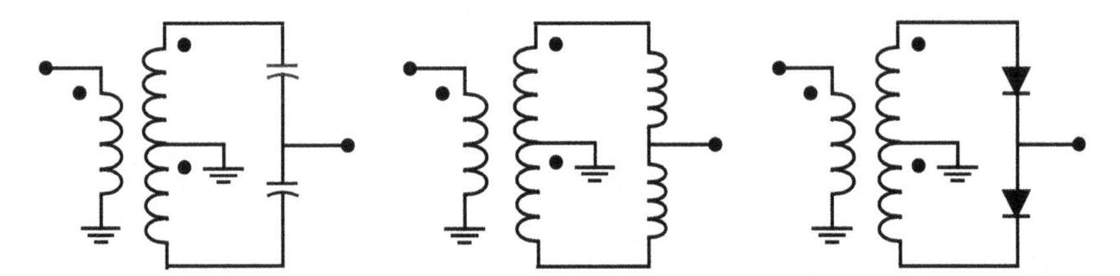

FIGURE 8-48 ■ A balun will provide isolation even when we replace the load resistors with complex, nonlinear loads. Each complex, nonlinear load must be closely matched to maintain balance and achieve high levels of cancellation.

temperature. Diodes built on the same substrate under carefully controlled conditions track quite closely.

8.10.2.2 SBM Operation

We now possess adequate tools to analyze the original single-balanced mixer circuit of Figure 8-42. The balun places the LO voltage across the two diodes, so the diodes will open and close at a rate determined by the LO. The LO will be completely canceled at both the RF and IF ports if the balun is perfectly balanced and the diodes are exactly identical.

When the two diodes are in their conducting state, the insertion loss between the RF port and the IF port of the mixer depends on the forward resistance of the diodes. Since the resistance of the diodes is a strong function of the LO power, a high LO power leads to low mixer insertion loss. When the two diodes are in their nonconducting state, the RF port is disconnected from the IF port.

Figure 8-49 shows the conceptual operation of a single-balanced mixer. The operation is identical to the conceptual operation of a single-ended mixer except that in the case of the SBM the LO must switch on two series-connected diodes rather than just one. The SBM requires more LO power than the SEM. Figure 8-50 shows the applicable waveforms.

8.10.2.3 Interport Isolation

The LO is the strongest signal present in the receiver's conversion chain, and keeping it under control is an important task. The addition of the balun and an additional diode

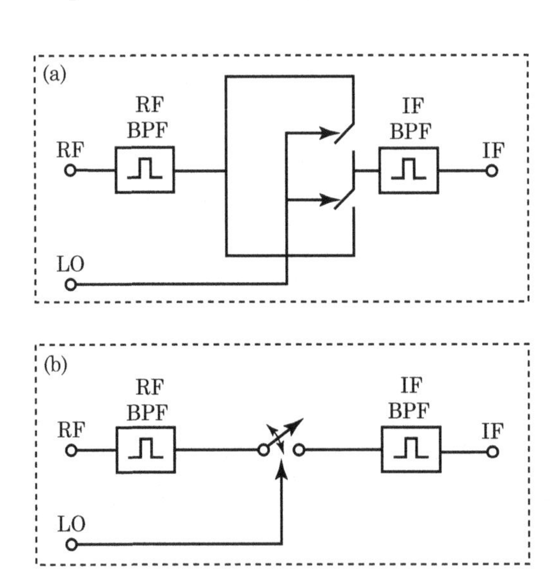

FIGURE 8-49 ■ Simplified SBM operation. The diodes have been replaced by switches controlled by the LO. The opening and closing of the switches approximates the time-domain multiply of the LO and the RF signals.

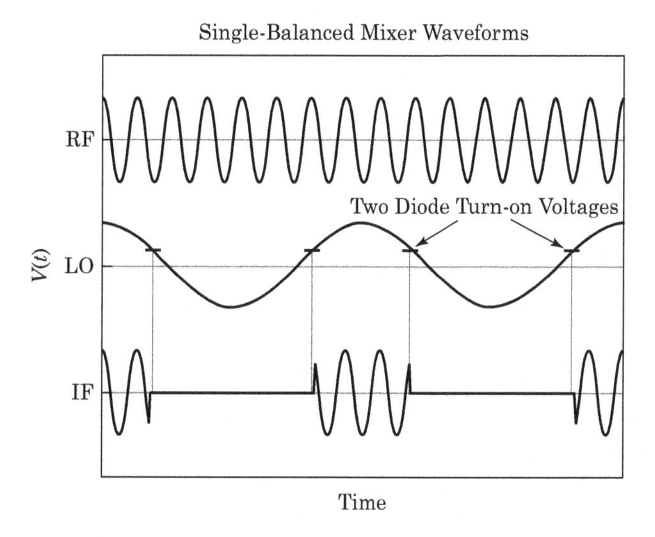

FIGURE 8-50 ■ The waveforms present in an SBM.

causes the SBM to exhibit greater LO:RF and LO:IF isolation than the SEM. This is due to the diode match and balun balance.

The spurious performance of the SBM is usually better than the SEM because the LO is not present at the IF port.

8.10.2.4 LO Power

An SBM must forward-bias two series-connected diodes so this mixer requires more LO power than the SEM. A typical rule of thumb is that the SBM requires at least 3 dBm of LO power.

8.10.3 Double-Balanced Mixers

DBMs offers improved performance over the SEM and SBM topologies. The DBM contains additional circuitry to improve the interport isolation over the single-ended and single-balanced designs. Furthermore, the design of the DBM acts to suppress the spurious products associated with the odd harmonics of both the LO and RF.

8.10.3.1 DBM Circuits

Figure 8-51 shows a common DBM topology. Both circuits are equivalent but are drawn differently. The diode topology shown in Figure 8-51 is a *diode ring*. These diodes are manufactured on a single substrate to ensure that they are nearly identical over time and temperature. In our initial analysis, we'll assume that all of the diodes are identical and that the transformers are perfectly balanced.

We've labeled the three mixer ports as the RF, LO, and IF ports. These ports are often interchangeable. The only practical difference is that each port may support a different frequency range. For example, the LO port in Figure 8-51 will extend to 0 hertz, whereas the other two ports will not.

Again, we'll consider the diodes as simple switches. Figure 8-52(a) shows an equivalent circuit for the DBM circuit of Figure 8-51 when the LO voltage is greater than the turn-on voltage of the diodes. Diodes D_2 and D_4 are in their conducting state, whereas diodes D_1 and D_3 are in their nonconducting state. In this state, the top of balun T_1 is connected to the top of balun T_2. The RF signal will pass through the mixer to the IF port with a 0° phase shift.

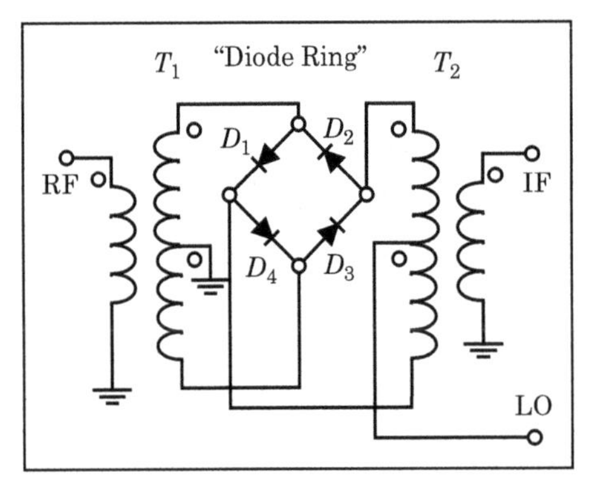

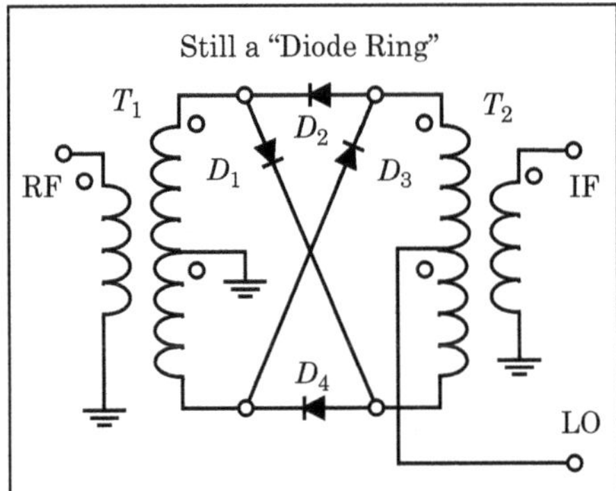

FIGURE 8-51 ■ Two equivalent circuit diagrams of a DBM.

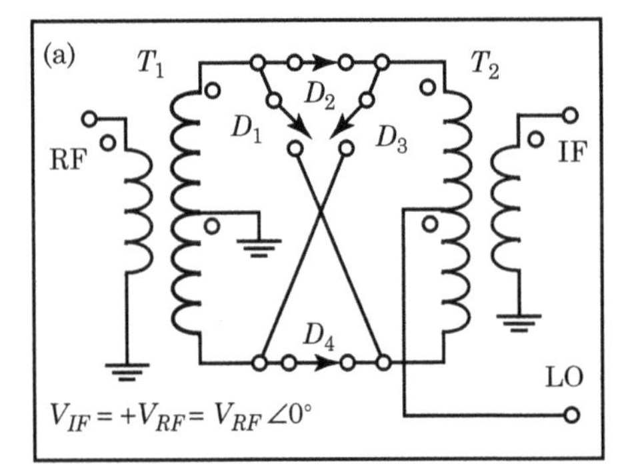

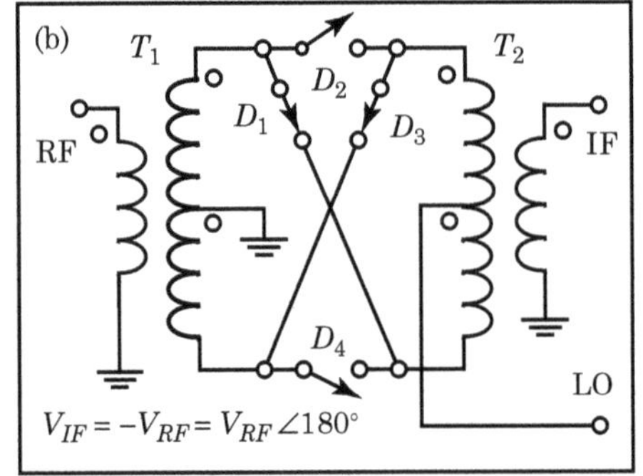

FIGURE 8-52 ■ (a) The equivalent circuit of a DBM when the LO voltage is a diode drop above 0 volts. The top of T_1 is connected to the top of T_2 and $V_{IF} = V_{RF}$. (b) The equivalent circuit of a DBM when the LO voltage is a diode drop below 0 volts. The top of transformer T_1 is now connected to the bottom of transformer T_2 and $V_{IF} = -V_{RF}$.

Figure 8-52(b) shows the DBM when the LO voltage is a diode drop below 0 volts. Diodes D_1 and D_3 are conducting, whereas diodes D_2 and D_4 are nonconducting. The top of balun T_1 is connected to the bottom of balun T_2, so the RF signal will pass through the mixer to the IF port with a 180° phase shift.

Figure 8-53 shows the waveforms over several cycles of the LO. The RF is alternately connected to the IF port with a 0° phase shift and then with a 180° phase shift. We multiply the RF signal by +1 when the LO voltage is greater than a forward diode drop and by −1 when the LO voltage is a diode drop below 0 volts. This is a change from the SEM and SBM cases when we multiplied the RF by a +1/0 waveform. In equation form, the multiplying waveform for the DBM is

$$W_{LO}(t) = \begin{cases} -1 \text{ when } V_{LO}(t) \leq \text{Diode Turn - on Voltage} \\ +1 \text{ when } V_{LO}(t) > \text{Diode Turn - on Voltage} \\ 0 \text{ Otherwise} \end{cases} \tag{8.65}$$

Fourier analysis reveals an equivalent expression for $W_{LO}(t)$ is

$$W_{LO}(t) = \frac{4}{\pi}\sum_{k=0}^{\infty} \frac{1}{2k+1} \sin\left[(2k+1)\,\omega_{LO}t\right] \tag{8.66}$$

There are two differences between equation (8.66), which describes a DBM, and equation (8.60), which describes SEMs and SBMs. The first difference is that equation (8.66) does not include a constant or DC term. The DC term in equation (8.60) is responsible

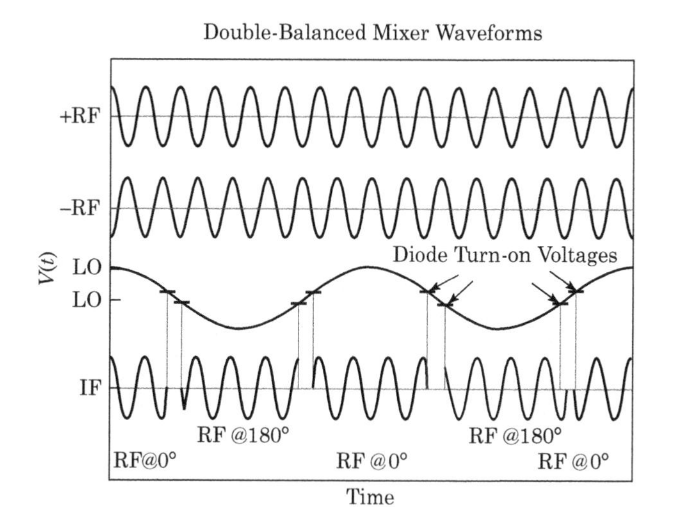

FIGURE 8-53 ■ The RF, LO, and IF voltages present in a DBM.

for the poor RF:IF isolation of the SEM and SBM. The second difference is that equation (8.66) contains only odd harmonics of the LO.

8.10.3.2 Harmonic Suppression

Like the other two mixer topologies, the DBM suppresses the spurious products associated with the even-order LO harmonics. However, the DBM makes use of additional mechanisms to suppress further the even-order LO and RF products.

Observing certain time-domain symmetries allows us to comment on the odd or even Fourier components of the wave. Fourier analysis reveals that if a time-domain function $f(t)$ satisfies the relationship

$$f(t) = -f\left(t + \frac{T_0}{2}\right) \tag{8.67}$$

for any T_0, then $f(t)$ contains only odd harmonics of $f_0 = 1/T_0$. This is a sufficient but unnecessary condition.

Figure 8-54 shows the equivalent LO-derived multiplying waveform generated by a double-balanced mixer. The six graphs represent the results for three values of LO power.

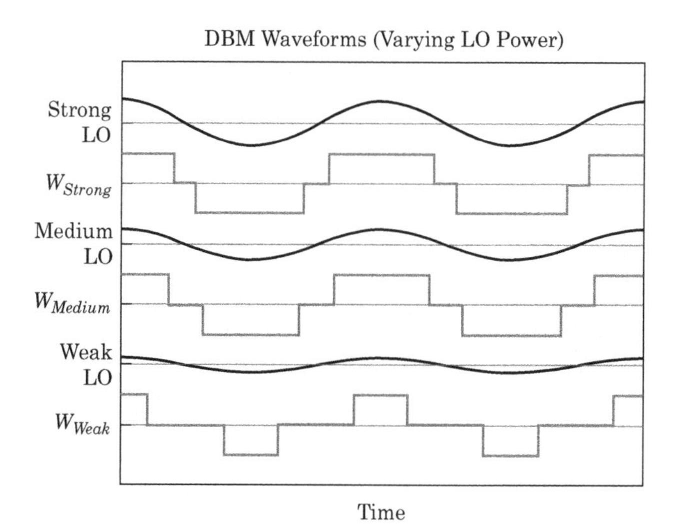

FIGURE 8-54 ■ The symmetrical waveforms generated in the DBM produce only odd LO harmonics. This effect helps suppress odd harmonic spurious products in the DBM.

In every case, the multiplying waveform satisfies equation (8.67), so it contains only the odd harmonics of the LO. The suppression of the even-order LO harmonics is a direct result of balun balance and diode match.

8.11 GENERAL MIXER NOTES

8.11.1 LO Power and Conversion Loss

In all the mixer types we've discussed, the LO turns the mixer's diodes on and off. This causes the series impedance of the mixer diodes to change from low impedance (when the diode conducts) to high impedance (when the diode is nonconducting).

An analysis of semiconductor physics tells us that the small signal resistance of a diode decreases with increasing forward voltage. The series resistance seen by the RF signal as it passes from the RF port to the IF port depends on the instantaneous voltage across the diode. A high-power LO will decrease this resistance and thus will decrease the insertion loss of the mixer. The mixer's insertion loss is a strong function of the LO power.

Figure 8-54 shows the LO-derived multiplying waveform $W(t)$ for several levels of LO power. In each case, there is a period of time when $W_{LO}(t)$ is 0. During the off time, the RF port is not connected to the IF port, so RF power is wasted. The conversion loss of the mixer is a direct function of this off time, and the length of the off time period depends on the LO power. A large LO power produces a short off time and, hence, a lower insertion loss. A small LO power produces a long off time and a higher insertion loss.

8.11.2 LO Power and Linearity

We have always made the assumption that the LO alone controls when the diodes switch. This is approximately true when the LO power is very much larger than the RF power (say by 20 dB or so). As the RF power approaches the power of the LO, the RF signal begins to affect the exact instant the diodes switch between their conducting and nonconducting states. An insufficient LO power with respect to the RF power generates a great deal of spurious signal power.

Applying a strong RF signal to a mixer similar to applying a strong RF signal to an amplifier. In an amplifier, the RF power begins to affect the DC bias levels of the transistors inside the amplifier. In a mixer, the RF signal causes the diodes to switch at the wrong moments. In both cases, we are operating the device in its nonlinear region and the results are unpredictable.

8.11.3 LO Noise

We use the zero crossings of the LO to switch the mixer diodes. To a first-order approximation, we can say that the mixer's diodes switch at the exact instant that the LO makes a zero crossing. Oscillator phase noise manifests itself as random changes in the zero crossings of the LO waveform. We can write the equation for a noisy oscillator as

$$V_{LO}(t) = A_{LO}(t) \cos\left[\omega_{LO}t + \phi_{LO}(t)\right] \tag{8.68}$$

where $A_{LO}(t)$ is the amplitude noise of the oscillator, and $\phi_{LO}(t)$ represents the phase noise of the oscillator. If we mix this noisy LO with a pure cosine wave, we obtain

$$\begin{aligned} V_{IF}(t) &= \{A_{LO}(t)\cos[\omega_{LO}t + \phi_{LO}(t)]\}\{\cos(\omega_{RF}t)\} \\ &= \frac{A_{LO}(t)}{2}\{\cos[(\omega_{LO} + \omega_{RF})t + \phi_{LO}(t)] + \cos[(\omega_{LO} - \omega_{RF})t + \phi_{LO}(t)]\} \end{aligned} \tag{8.69}$$

We can see that the amplitude noise and phase noise present on the LO are transferred to both sidebands. We can make the generalization that any signal converted from one frequency to another picks up the amplitude and phase noise of the local oscillator. A noisy oscillator can contaminate a signal so badly that it cannot be demodulated.

8.11.4 Effects of Impedance Mismatch

Like most other RF components, mixers are characterized assuming they will operate in a Z_0 environment. However, a mixer is not a device that presents a Z_0 load to the outside world. Rather, it is a device that demands that the outside world present a Z_0 load to it.

For optimum performance (e.g., low spurious output, good interport isolation), we must design our systems to present the mixer with a wideband Z_0 termination on all of its ports. This presents a problem when we must filter the RF or IF ports of the mixer. The input impedance of a filter in its stopband is usually not Z_0. The non-Z_0 filter impedance usually causes the mixer's spurious output power to increase, although we may also observe increased insertion loss and other mysterious effects.

It is very important to terminate the mixer's IF port at the sum and the difference frequencies and at the LO frequency. For example, if we're making use of the difference product from the mixer's IF port, then the difference frequency will almost naturally see a Z_0 termination. We will often install specialized circuitry to ensure good match at the sum frequency.

There are several solutions to this problem. We can follow the mixer with an amplifier whose frequency range includes the sum and difference frequencies. We can usually count on an amplifier presenting the outside world with a Z_0 termination over its operating frequency range. However, the amplifier must exhibit sufficient dynamic range to process the large range of signals present on the mixer's IF port. The LO power leaving the mixer's IF port can be very strong (0 dBm or greater). The amplifier must be able to process the weak RF signals in the presence of the strong LO signal.

We can also build special filters that are matched throughout their passband and stopband.

A third solution is to place an attenuator directly on the IF port. The worst-case return loss of an attenuator is twice the value of the attenuation (e.g., the minimum return loss of a 6 dB attenuator is 12 dB). The only downside to this solution is that we must endure the signal loss through the attenuator.

8.11.5 Mixer SOI/TOI

A mixer has second-order and third-order intercept points, which are a strong function of the LO power. The second- and third-order intercepts of a mixer decrease as the LO power decreases.

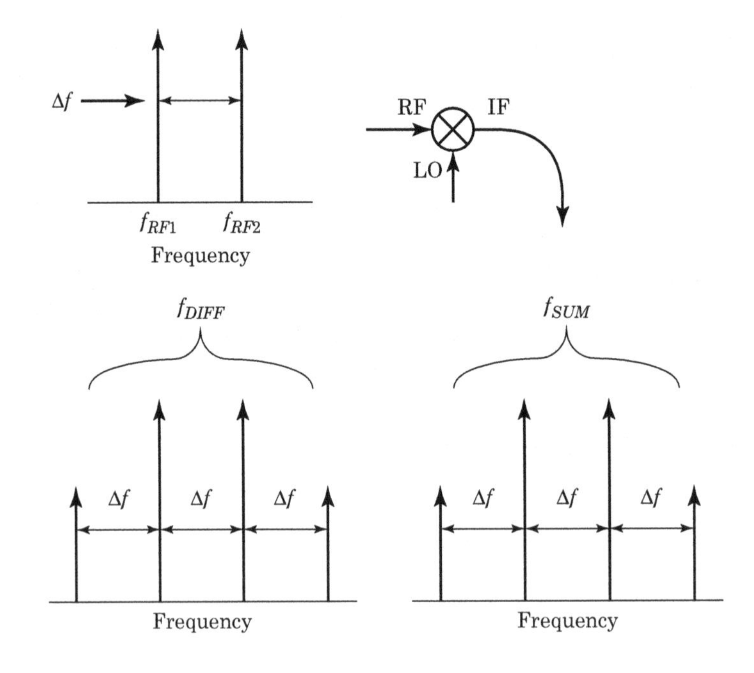

FIGURE 8-55 ■ Measuring mixer TOI. The third-order spurious signals appear at both the sum and difference frequencies.

We perform a third-order intercept test on a mixer in much the same way we perform the test on an amplifier. Figure 8-55 shows the test technique. We place two tones at the RF port (at $f_{RF,1}$ and $f_{RF,2}$). The IF port of the mixer will show two sets of four tones (each in the neighborhood of $f_{IF,SUM}$ and $f_{IF,DIFF}$). The sum and the difference frequencies will exhibit their own set of third-order distortion products. We normally select off either the sum or the difference frequencies with an IF band-pass filter, so we would examine the third-order distortion products at the desired output frequency.

We measure the mixer's second-order intercept in the same way. We apply a single tone to the RF port of the mixer and note that the sum and the difference products at the mixer's IF port each shows a second-order output product. We normally filter off either the sum or the difference frequency for use in our system, so we make our measurements at the selected frequency. Note that if the IF band-pass filter is less than an octave in bandwidth, the IF band-pass filter will attenuate the second-order signal.

One factor that affects a mixer's intercept performance is how well the mixer's ports are terminated. We can often realize a 2 dB improvement in the TOI of a mixer by properly terminating its ports.

War Story: Gilligan's Island

On a *Gilligan's Island* rerun, the professor wanted to convert that old, white transistor radio into a transmitter, but he couldn't remember precisely how. He kept muttering to himself, "It has something to do with the local oscillator."

Well, it turns out that the old professor is correct. The LO is a locally generated signal that is developed inside a receiver, and it is strong enough to be used as a weak transmitter. In commercial FM receivers, the LO is usually in the aircraft band, which is slightly higher in frequency than the commercial FM broadcast band.

8.12 | BIBLIOGRAPHY

Hayward, W.A., *Introduction to Radio Frequency Design*, Prentice-Hall, Inc., 1982.

Henderson, Bert C., "Mixers: Part 1—Characteristics and Performance," in *RF and Microwave Designers Handbook*, Watkins-Johnson Company, p. 752, 1988/1989.

Henderson, Bert C., "Mixers: Part 2—Theory and Technology," in *RF and Microwave Designers Handbook,* Watkins-Johnson Company, p. 759, 1988/1989.

Peter, Will, "Reactive Loads—The Big Mixer Menace," *Microwaves*, April 1971.

8.13 | PROBLEMS

1. **Mixer Spurious Products** The desired signal present at the IF port of a mixer (at $f_{RF} - f_{LO}$) has a frequency of 30 MHz and an amplitude of -11 dBm. The RF is $f_{RF} = 500$ MHz, and LO frequency f_{LO} is 470 MHz. Given the mixer table shown as Table 8-3, find the levels for the following:

   a. $3f_{LO} - 4f_{RF} = 590$ MHz
   b. $2f_{LO} - 2f_{RF} = 60$ MHz
   c. $-6f_{LO} + 7f_{RF} = 680$ MHz
   d. $-4f_{LO} + 2f_{RF} = 880$ MHz

2. **Spurious Output Ranges** Assume a cellular telephone receiver covering 825–890 MHz. We are using a HSLO of 870 to 935 MHz to convert the signal of interest to 45 MHz.

   a. Find range of the two frequency bands emerging from a mixer's IF port for the third harmonic of the LO combining with the second harmonic of the RF.
   b. If the IF frequency is 45 MHz and there is a 1 MHz wide band-pass filter on the mixer's IF port, do you think either of these two spurious output bands will cause a problem (i.e., interfere with the desired signal at 45 MHz) in this receiver?

3. **Gain and Noise Figure** Find the gain and noise figure of the cascade shown in Figure 8-56.
4. **HSLO and LSLO** Fill in Table 8-4 for the receiver shown as Figure 8-57. Show your work.

TABLE 8-3 ■ Mixer spurious response table.

	RF Harmonic (n)										
LO Harmonic (m)	0	1	2	3	4	5	6	7	8	9	10
0		17	43	44	76	66	69	72	>85	>85	84
1	32	**0**	60	34	68	75	76	71	>85	83	84
2	23	39	45	49	56	67	83	73	81	84	83
3	36	17	56	35	72	53	84	74	>85	83	85
4	43	33	51	49	56	57	70	73	85	84	84
5	35	37	60	36	61	48	71	67	>85	>85	83
6	44	38	65	45	57	58	64	64	83	83	84
7	36	39	52	50	65	47	75	63	76	77	84
8	61	39	58	52	76	71	64	63	73	71	83
9	46	47	41	57	75	70	78	61	76	71	81
10	63	51	69	50	66	67	79	71	71	71	81

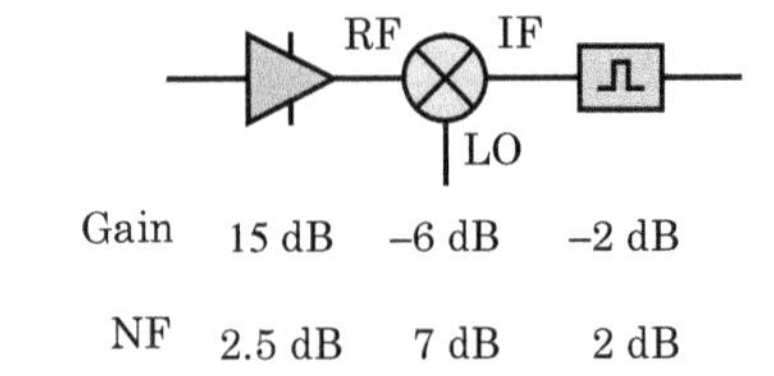

FIGURE 8-56 ■ Gain and noise figure of a cascade involving a mixer.

TABLE 8-4 ■ The LO architecture for the ISM receiver of Figure 8-57.

First Conversion		Second Conversion		Spectrum at 10.7 MHz Inverted?
LSLO		LSLO		LSLO->LSLO (Y/N) LSLO->HSLO (Y/N)
HSLO		HSLO		HSLO->LSLO (Y/N) HSLO->HSLO (Y/N)

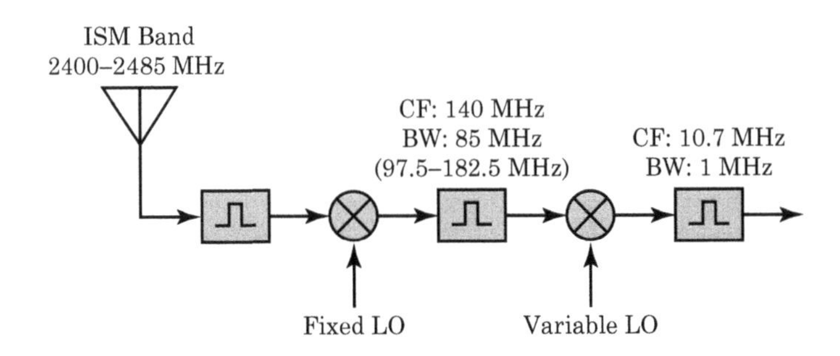

FIGURE 8-57 ■ An Industrial, Scientific and Medical (ISM) band receiver.

5. **Spurious Responses** We apply the following test signals to the receiver shown in Figure 8-58. Find the output power and output frequency for each of the following test tones. The input tone power is always 0 dBm.

   a. 904 MHz (tuned frequency)
   b. 925 MHz (a signal near the desired RF)
   c. 1044 MHz (the RF image)
   d. 70 MHz (via mixer RF:IF isolation)
   e. The LO power required by the mixer is +10 dBm.
   f. How much LO power radiates from the antenna?
      - Mixer specs: CL 7 dB, RF:IF 20 dB, RF:LO 23 dB, IF:LO 27 dB, LO power +10 dBm
      - RF BPF specs: Second-order 890-940 MHz 3 dB bandwidth, ultimate attenuation 60 dB
      - IF BPF specs: Fourth-order 68.5-72.5 MHz 3 dB bandwidth, ultimate attenuation 40 dB

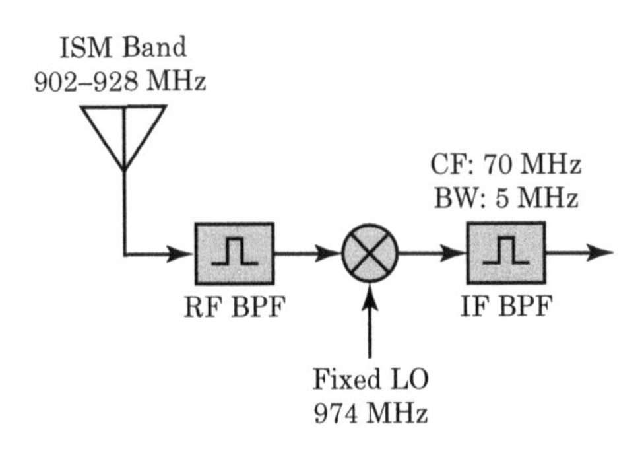

FIGURE 8-58 ■ A 900 MHz ISM band receiver.

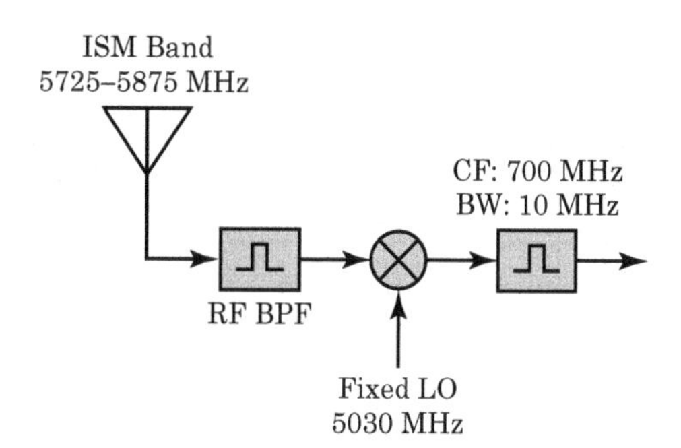

FIGURE 8-59 ■ Another ISM band receiver.

6. **Filter Specifications** Given the system of Figure 8-59, consider the following:
   a. To what frequency is the receiver tuned?
   b. The 3 dB frequencies of the RF BPF are 5,600 and 6,000 MHz. What is the required order of the RF BPF to achieve at least 30 dB of image rejection?

7. **Block Conversion** Design two single-stage Ku-band block downconverters that convert 12.20–13.00 GHz to 950–1,750 MHz.
   a. What are the HSLO and LSLO frequencies for this conversion requirement?
   b. Does either the HSLO or LSLO frequency conversion produce a spectral inversion?
   c. What are the RF image frequency ranges for the HSLO system and for LSLO system?
   d. What are the corresponding IF image frequency ranges emanating from the IF port of the mixer for the LSLO and HSLO conversion schemes?

8. **Channelized Conversion** One instantiation of the 5.8 GHz ISM band runs from approximately 5.80 to 5.85 GHz, and the band is broken down into 24 channels. Each channel is 1.8 MHz wide. Figure 8-60 shows the system. Fill in Table 8-5.

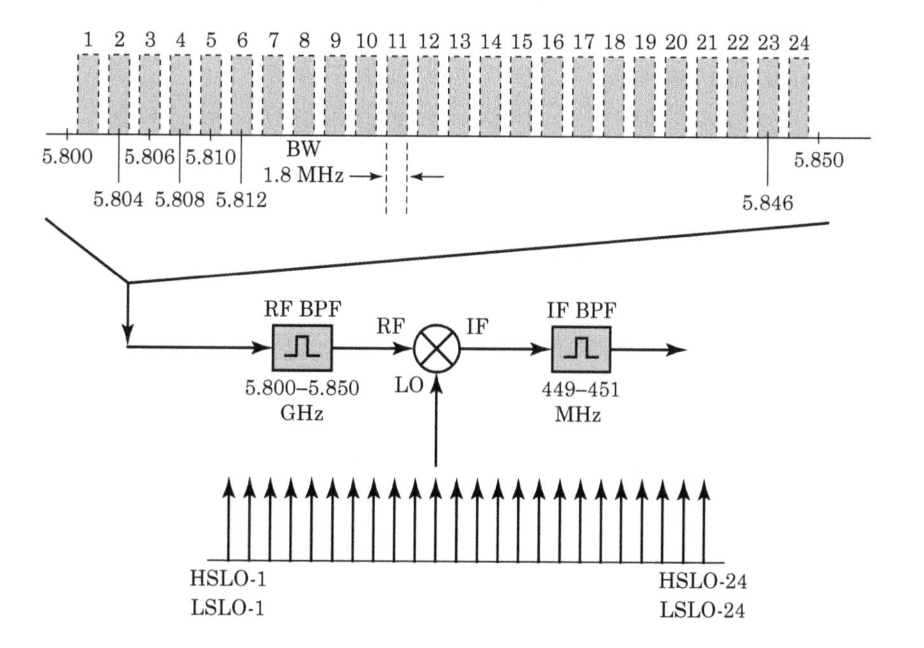

FIGURE 8-60 ■ A channelized 5.8 GHz ISM band receiver.

TABLE 8-5 ■ The mixer conversion architecture for the ISM receiver of Figure 8-60.

Channel	RF CF (MHz)	LSLO			HSLO		
		LO (MHz)	RF Image CF (MHz)	Spectrum Inverted?	LO (MHz)	RF Image CF (MHz)	Spectrum Inverted?
1							
2							
3							
4							
5							
6							
7							
8							
9							
10							
11							
12							
13							
14							
15							
16							
17							
18							
19							
20							
21							
22							
23							
24							

9. **Tuned Frequency** Answer the following questions regarding the system of Figure 8-61.
   a. To what frequency is this receiver tuned, assuming that the desired signal is centered in the 21.4 MHz band-pass filter?
   b. Is the signal at 21.4 MHz spectrally inverted with respect to the tuned RF signal?

10. **Mixer Conversion Equations**
   a. What two possible LO frequencies can you use to convert a wireless LAN signal from 5.81 GHz down to 21.4 MHz IF?
   b. If you mix a signal centered at 12.4 GHz with a 2.2 GHz local oscillator, what are the center frequencies of the two resulting signals?

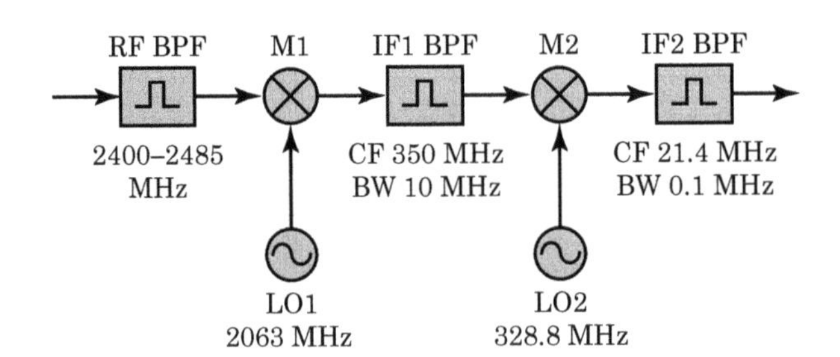

FIGURE 8-61 ■ A dual conversions 2.4 GHz ISM band receiver.

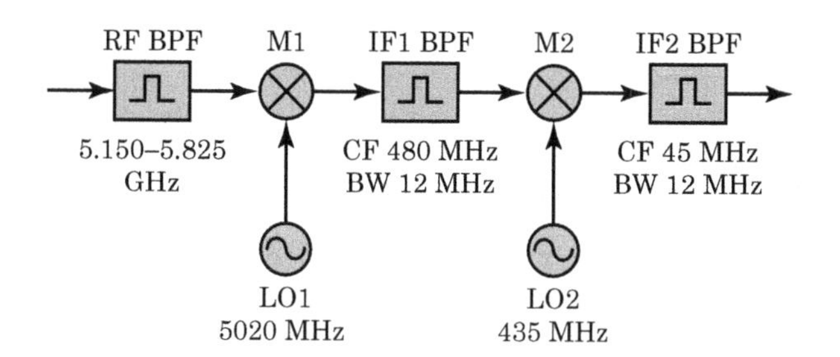

FIGURE 8-62 ■ A dual conversion receiver.

c. What are the center frequencies of the two possible signals that can be converted to an IF of 1,000 MHz using an LO frequency of 4,300 MHz?

11. Mixers and Filtering Requirements Referencing the system shown in Figure 8-62:

a. Based on the LO frequencies, to what frequency is the receiver tuned?

b. Discuss the requirements of the filters RF BPF, IF1 BPF, and IF2 BPF. Consider the effects of RF:IF, LO:RF, and LO:IF isolation of each mixer. Also consider the image of each conversion. At which frequencies do we want the filters to exhibit low and high attenuation?

12. Mixer Conversion Loss and Isolation Given the system shown in Figure 8-63 and the following mixer specifications:

- $CL_{dB} = 10$ dB
- RF:IF isolation = 17 dB
- LO:IF isolation = 22 dB
- LO:RF isolation = 25 dB
- 2 × LO power is 25 dB below the desired output
- 2 × RF power is 28 dB below the carrier

find the power level, frequency, and nature (e.g., is the signal an LO+RF, 2LO-RF) of each signal present on the IF port of the mixer.

Notes: The signal powers shown in Figure 8-63 are accurately drawn with respect to frequency but not to power level. In other words, don't make any assumptions about the power in each signal from the drawing. In the cases where you do not have enough information to find the power level of a signal, simply state that fact.

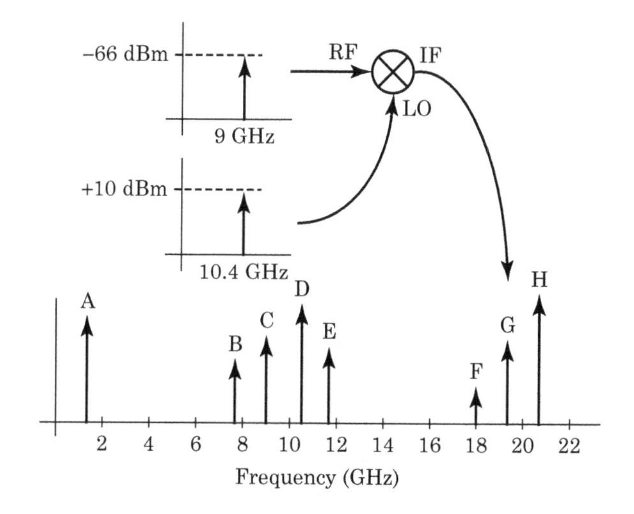

FIGURE 8-63 ■ The spurious signals produced in a mixer.

13. **LO/Frequency Inversion** We desire to convert a fixed satellite band at 4.5–4.8 GHz to an IF center frequency of 450 MHz.

    a. Find the HSLO and LSLO frequencies for this block converter system.

    b. Does either of these two LO selections produce a frequency inversion?

14. **Conversion Equations**

    a. What two possible LO frequencies can you use to convert a commercial FM radio station broadcasting at 106.9 MHz down to a 10.7 MHz IF?

    b. What are the center frequencies of the two possible signals that can be converted to an IF of 70 MHz using an LO frequency of 935 MHz?

    c. If you mix a signal centered at 215.75 MHz with 256.25 MHz local oscillator, what are the center frequencies of the two resulting signals?

15. **Mixer Conversion Loss** The power of the RF signal applied to the mixer's input port is −45 dBm. The power measured at the mixer's IF port at the sum frequency is −60 dBm, while the power measured at the difference frequency is −63 dBm. What is the mixer's conversion loss?

16. **Mixer Conversion Loss/Isolation** Given the system shown in Figure 8-64 and the following mixer specifications:

    - $CL_{dB} = 8$ dB
    - RF:IF isolation = 35 dB
    - LO:IF isolation = 32 dB
    - LO:RF isolation =28 dB

    answer the following:

    a. Find the strength of the signals at 407 MHz, 243 MHz, 82 MHz, and 325 MHz on the IF port.

    b. Find the strength of the signal at 325 MHz on the RF port.

    c. Find two third-order spurious signals that could be present on the mixer's IF port.

    d. We place a five-pole Butterworth BPF on the IF port of the mixer. The center frequency of this filter is 243 MHz, and the bandwidth is 30 MHz. Find the power levels of the signals at 407 MHz, 243 MHz, 82 MHz, and 325 MHz at the output of this filter.

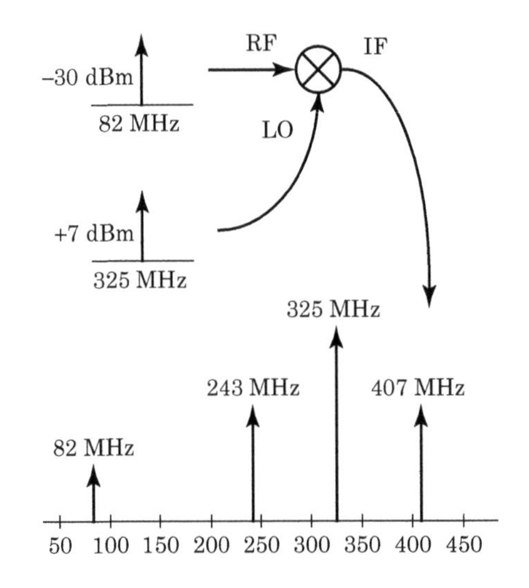

FIGURE 8-64 Spurious signals generated in a mixer.

17. **HSLO and LSLO** One U.S. cellular telephone system operates at 825 to 890 MHz. Each channel is 30 kHz apart and about 20 kHz wide.

    a. If we want to use an IF of 45 MHz, find the HSLO and LSLO frequency ranges for this channelized system.

    b. Does either of these two LO selections produce a frequency inversion?

18. **HSLO/LSLO and Images** The commercial U.S. FM radio band covers an RF range of about 88 to 108 MHz. Normally, each station is converted to a 10.7 MHz IF for demodulation. Find and draw:

    a. The HSLO range

    b. The range of HSLO image frequencies

    c. The LSLO range

    d. The range of LSLO image frequencies

19. **Spurious Responses** Answer the following questions for the system and input spectrum shown in Figure 8-65. The component parameters are as follows:

    - BPF1: 110 MHz CF, 25 MHz BW, 1 dB IL
    - BPF2: 10.7 MHz CF, 1 MHz BW, 6 dB IL
    - All filters are Butterworth
    - Mixer insertion loss: 11 dB

    a. To what frequency is the receiver tuned?

    b. Under the input conditions shown in Figure 8-65, the receiver suffers from a lack of sensitivity. In other words, we can't see the tuned signal at the output of the IF filter when the tuned signal is small but well above the receiver's MDS. What is the problem, and how would you fix it?

20. **Conversion Equations**

    a. What two possible LO frequencies can you use to convert a wireless LAN signal from 2.440 GHz down to 140 MHz IF?

    b. If you mix a signal centered at 5.5 GHz with a 4.33 GHz local oscillator, what are the center frequencies of the two resulting signals?

    c. What are the center frequencies of the two possible signals that can be converted to an IF of 510 MHz using an LO frequency of 3.30 GHz?

21. **Mixer Conversion Loss and Isolation** Given the system shown in Figure 8-66 and the following mixer specifications:

    - $CL_{dB} = 12$ dB
    - RF:IF Isolation = 22 dB
    - LO:IF Isolation = 24 dB

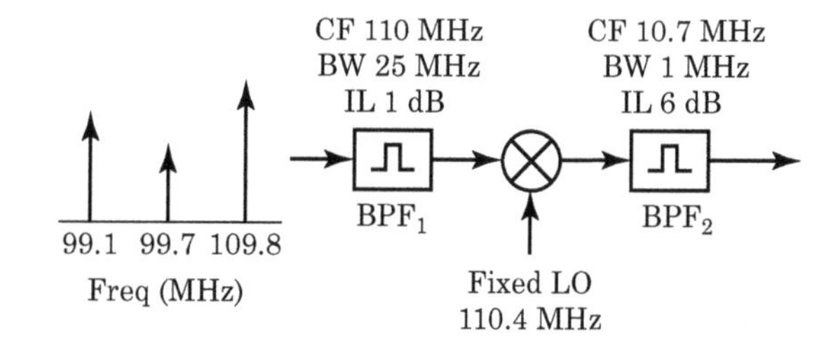

FIGURE 8-65 ■ An FM radio receiver that exhibits poor sensitivity.

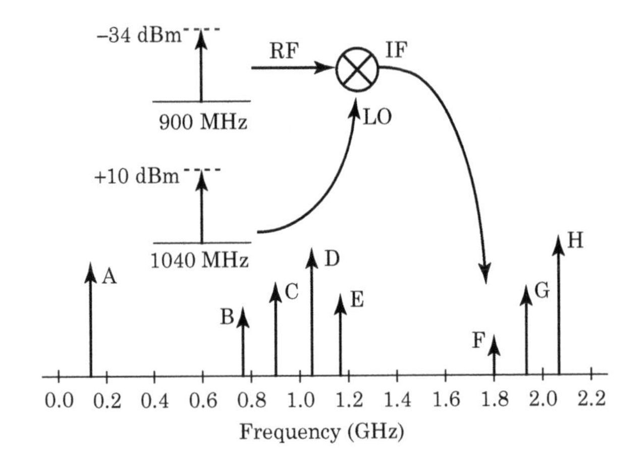

FIGURE 8-66 ▪ The spurious output signals from a mixer.

- LO:RF Isolation = 38 dB
- 2× LO Power is 22 dB below the desired output
- 2× RF Power is 30 dB below the carrier

a. Find the strength (i.e., the power level), frequency, and nature (e.g., is the signal an LO+RF, 2LO-RF) of each signal present on the IF port of the mixer.
   Notes: The signal powers shown in Figure 8-66 are accurately drawn with respect to frequency but not to power level. Don't make any assumptions about the power in each signal from the drawing. There is insufficient information to find the power level of two of the signals.

b. We place a two-pole Butterworth BPF on the IF port of the mixer to select the LO+RF frequency. The bandwidth of this filter is 200 MHz, its insertion loss is 2.5 dB, and its ultimate attenuation is 65 dB. Find the power levels present at the output of this filter.

22. HSLO and LSLO One of the industrial, scientific, and medical (ISM) bands in the United States operates at 2.400–2.485 GHz (see Figure 8-67).

a. Using an IF center frequency of 350 MHz, find the HSLO and LSLO frequency ranges for this channelized system.

b. Does either of these two LO selections produce a frequency inversion?

23. Conversion Scheme

a. Design the first mixer stage of a receiver that converts an RF signal centered in the 108–120 MHz range to a 170 MHz IF. There are two solutions (i.e., there are two possible LO ranges). Please find both LO ranges, and fill in the frequency allocation diagrams in Figure 8-68. This problem requires a tunable LO. Find:
   - $LO1_{Low}$ and $LO1_{Hi}$
   - $LO2_{Low}$ and $LO2_{Hi}$

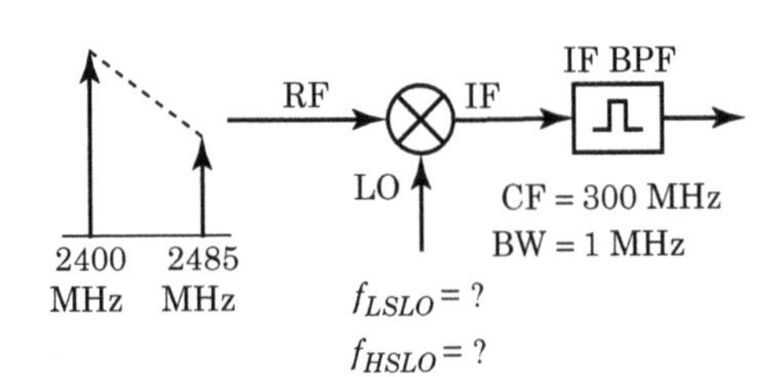

FIGURE 8-67 ▪ An ISM band first mix.

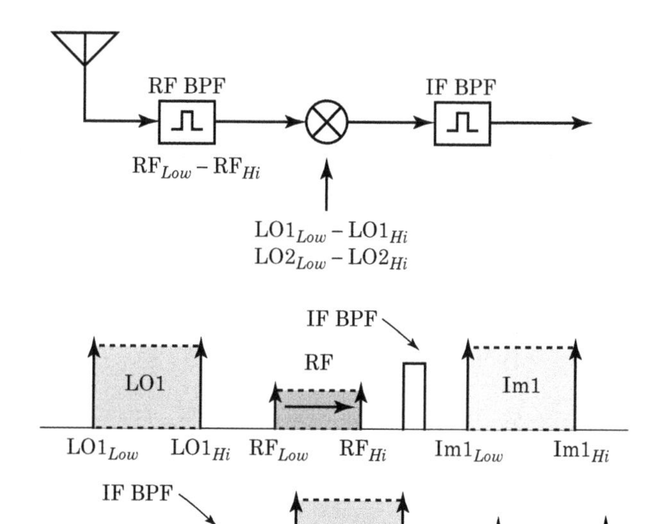

FIGURE 8-68 ■ The first mixing stage of an aircraft band receiver.

- $Im1_{Low}$ and $Im1_{Hi}$: the image frequencies associated with LO1
- $Im2_{Low}$ and $Im2_{Hi}$: the image frequencies associated with LO2

Don't assume that the drawing above is accurate with respect to frequency. In other words, the solution may require two bands to overlap when the drawing above doesn't show them overlapping.

b. Which, if any, conversion scheme results in a frequency inversion?

24. Multiple Choice Regarding a mixer's RF image, which statement is false?

a. The RF image will be converted to the IF just as efficiently as the desired signal.
b. The RF image will be converted to the IF, but with more loss than the desired signal.
c. The RF image must be heavily filtered to render its power insignificant by the time the image makes its way to the mixer's RF port.
d. The image frequency is not suppressed by garden-variety, nonspecialized mixers.

25. Mixer Spurious Table Given the system in Figure 8-69 and Table 8-6, find the power of the signals shown at the output of the mixer.

The powers shown here are relative to the desired mixer output at frequencies RF+LO or RF-LO. The conversion loss of the mixer is 10 dB.

All frequencies are in GHz. The powers may not be drawn to scale.

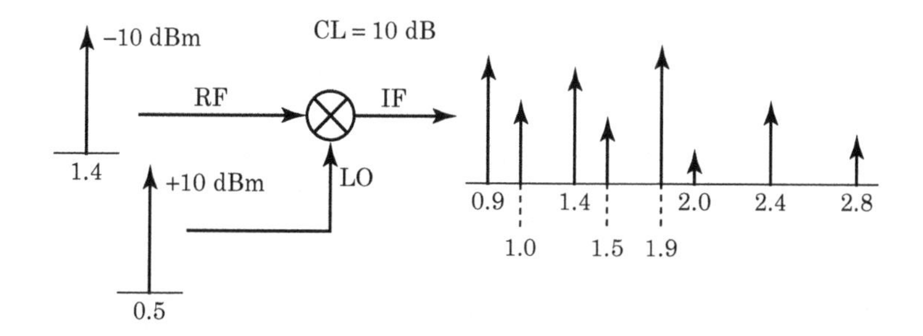

FIGURE 8-69 ■ Find the power in each of the signals at the output of the mixer using the data from Table 8-6.

TABLE 8-6 ■ A mixer spurious table.

	RF Harmonic (*n*)										
LO Harmonic (*m*)	**0**	**1**	**2**	**3**	**4**	**5**	**6**	**7**	**8**	**9**	**10**
0		17	43	44	76	66	69	72	>85	>85	84
1	32	**0**	60	34	68	75	76	71	>85	83	84
2	23	39	45	49	56	67	83	73	81	84	83
3	36	17	56	35	72	53	84	74	>85	83	85
4	43	33	51	49	56	57	70	73	85	84	84
5	35	37	60	36	61	48	71	67	>85	>85	83
6	44	38	65	45	57	58	64	64	83	83	84
7	36	39	52	50	65	47	75	63	76	77	84
8	61	39	58	52	76	71	64	63	73	71	83
9	46	47	41	57	75	70	78	61	76	71	81
10	63	51	69	50	66	67	79	71	71	71	81

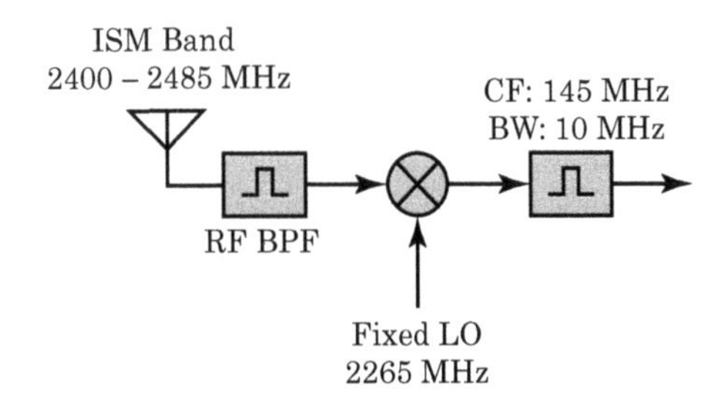

FIGURE 8-70 ■ The first conversion stage of a 2.4 GHz ISM band receiver.

26. **Mixer Characteristics** You are testing the receiver shown in Figure 8-70. The specifications of the components are as follows:

- Mixer: conversion loss 9 dB, RF:IF isolation 25 dB, RF:LO isolation 20 dB, IF:LO isolation 30 dB, LO power +14 dBm
- RF BPF: third-order Butterworth, attenuation at 2400 and 2485 is **1 dB** (not 3 dB), ultimate attenuation is 65 dB
- IF BPF: fourth-order Butterworth, center frequency is 145 MHz, 3 dB bandwidth is 10 MHz, ultimate attenuation is 50 dB

For the following questions, apply a test tone of −10 dBm to the antenna port of the receiver:

a. Apply a 2,410 MHz signal to the antenna port. What is the power level present at the output of the IF filter at 145 MHz?

b. Apply a 2,475 MHz signal to the antenna port. What is the power level present at the output of the IF filter at 210 MHz?

c. Apply a 2,120 MHz signal to the antenna port. What is the power level present at the output of the IF filter at 145 MHz?

d. Apply a 2,480 MHz signal to the antenna port. What is the power level present at the output of the IF filter at 2,265 MHz?

27. **Multiple Choice** By far, the strongest signal coming out of the IF port of a mixer is:

a. The desired signal at the mixer's IF

b. The desired signal at the signal's original RF

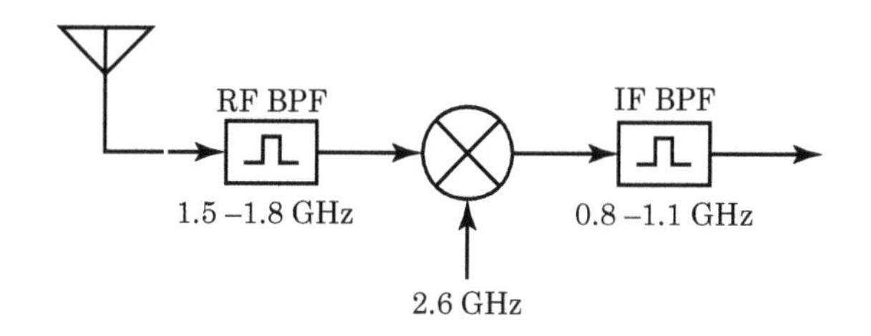

FIGURE 8-71 ■ A block downconverter.

c. The local oscillator at the LO frequency
d. The image frequency of the conversion

28. Essay on Image Reject Mixers Perform a web search on image-reject mixers. Write a short essay that answers the following questions:

a. What is their purpose?
b. Under what circumstances would you use one?
c. How do they work?
d. What are the specifications of image-reject mixers that differ from regular, nonimage reject mixers?

29. Block Converter Image Response The RF BPF of the block downconverter of Figure 8-71 has the following characteristics:

- Lower 3 dB cutoff frequency = 1.4 GHz
- Upper 3 dB cutoff frequency = 1.9 GHz
- Number of poles = 4
- Center frequency insertion loss = 0 dB

a. What is the image frequency range?
b. What is the frequency of the worst-case image response?
c. What is the worst-case image rejection of this downconverter?

30. Mixer Spurious Figure 8-72 shows a channelized conversion scheme. Fill in the spurious signal diagram for the 2RF × 1LO spurious component (Figure 8-73) and indicate if the indicated spurious combination falls within the passband of the IF BPF.

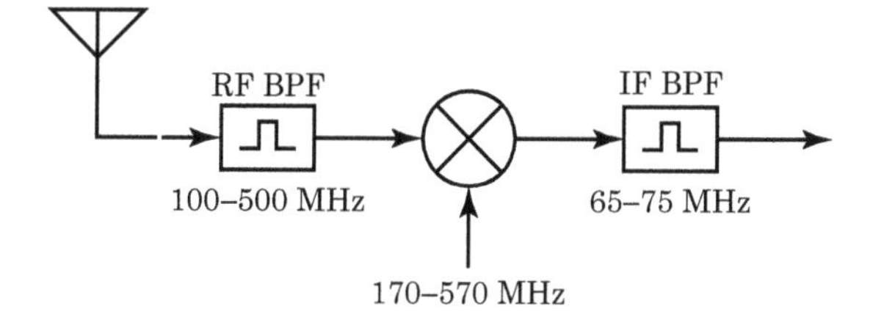

FIGURE 8-72 ■ A channelized downconverter.

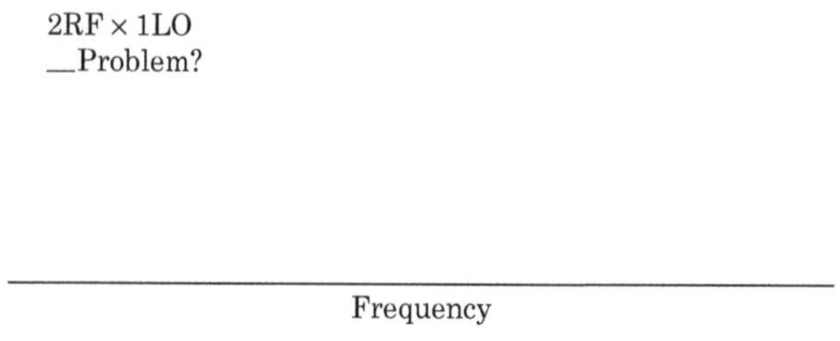

2RF × 1LO
__Problem?

Frequency

FIGURE 8-73 ■ Draw the spectrum at the output of the mixer of Figure 8-72 for the 2RF × 1LO spurious product.

Be sure to label the frequencies of the spurious signal bands coming from the mixer output and the location of the IF BPF.

31. **Multiple Choice** The most important criteria for deciding if a particular $m \times n$ mixer spurious signal will be problematic is:

    a. Does the frequency of the $m \times n$ spurious cross 0 hertz?

    b. Is the frequency of the $m \times n$ spurious signal too high?

    c. Is the estimated power of the $m \times n$ spur greater than the system's MDS + 20 dB?

    d. Does the $m \times n$ spur fall within the passband of the IF filter?

32. **Multiple Choice** The possible frequencies generated by an 3RF $\times$ 4LO mixer spur whose input frequencies are RF = 5 GHz and LO = 3.3 GHz are

    a. 29.9 and 10.1 GHz

    b. 28.2 and 1.8 GHz

    c. 8.8 and 1.7 GHz

    d. 13.3 and 6.7 GHz

33. **Conversion Scheme**

    a. Design the first mixer stage of a receiver that converts an RF signal centered in the 30–60 MHz range to a 170 MHz IF (see Figure 8-74). There are two solutions. Find:

    - $LO1_{Low}$ and $LO1_{Hi}$
    - $LO2_{Low}$ and $LO2_{Hi}$
    - $Im1_{Low}$ and $Im1_{Hi}$: the image frequencies associated with LO1
    - $Im2_{Low}$ and $Im2_{Hi}$: the image frequencies associated with LO2

    Do not assume that the frequency ranges are drawn to scale in Figure 8-74.

    b. Does either conversion scheme result in a frequency inversion?

34. **Mixer as an RF Switch** Describe how you might use a mixer as an RF switch.

35. **Mixer as a Voltage Variable Attenuator** Describe how you might use a mixer as a voltage-variable RF attenuator.

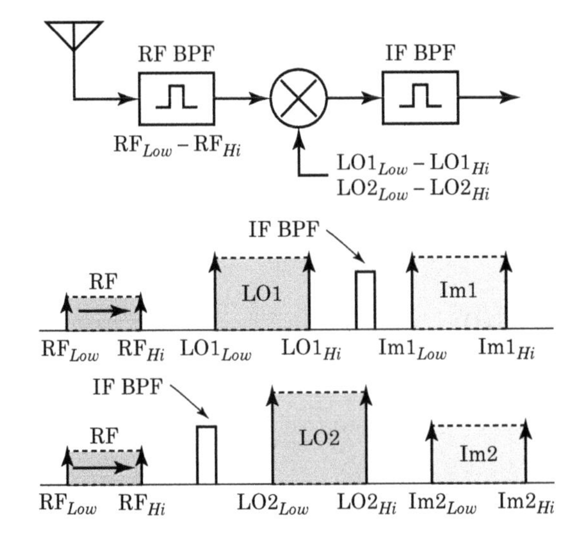

FIGURE 8-74 ■ Two different conversion schemes to convert an RF signal centered in the 30–60 MHz range to a 170 MHz IF.

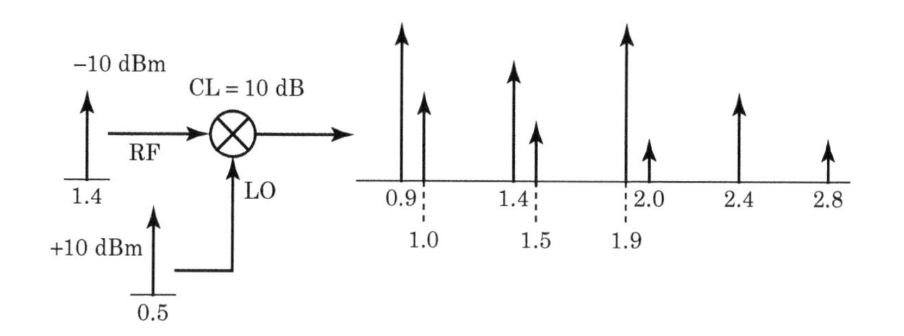

FIGURE 8-75 ■ The spurious signals present on a mixer's output.

TABLE 8-7 ■ A mixer spurious table.

LO Harmonic (*m*)	RF Harmonic (*n*)										
	0	1	2	3	4	5	6	7	8	9	10
0		17	43	44	76	66	69	72	>85	>85	84
1	32	**0**	60	34	68	75	76	71	>85	83	84
2	23	39	45	49	56	67	83	73	81	84	83
3	36	17	56	35	72	53	84	74	>85	83	85
4	43	33	51	49	56	57	70	73	85	84	84
5	35	37	60	36	61	48	71	67	>85	>85	83
6	44	38	65	45	57	58	64	64	83	83	84
7	36	39	52	50	65	47	75	63	76	77	84
8	61	39	58	52	76	71	64	63	73	71	83
9	46	47	41	57	75	70	78	61	76	71	81
10	63	51	69	50	66	67	79	71	71	71	81

36. **Mixer as a Binary Phase Shift Keying Modulator** Describe how you might use a mixer as a binary phase shift keying (BPSK) modulator.
37. **Mixer as a Phase Detector** Describe how you might use a mixer as a phase detector.
38. **Mixer Spurious Table** Given the system shown in Figure 8-75 and the mixer spurious table of Table 8-7, find the power of the signals shown at the output of the mixer. The powers shown in Table 8-7 are relative to the desired mixer output at frequencies RF+LO or RF-LO. The conversion loss of the mixer is 10 dB.

    All frequencies are in GHz. The powers may not be drawn to scale.

CHAPTER 9

Oscillators

Purity lives and derives its life solely from the Spirit of God.

David Hare (1917–1992)

Chapter Outline

9.1 INTRODUCTION

Oscillators play an important part the ultimate performance of receivers. In this chapter, we will examine oscillator phase noise, frequency accuracy, and drift and their cumulative effects on receiver performance.

9.2 IDEAL AND REAL-WORLD OSCILLATORS

The reader is invited to review Chapter 2 for a review of ideal oscillators. The first section of Chapter 2 ("A Real-Valued, Ideal Cosine Wave") describes ideal oscillators, and the section "An Ideal Sine Wave and Band-Limited AWGN" in the same chapter describes noisy, realized oscillators.

9.2.1 Phase Noise and Frequency Drift

The frequency of a nonideal oscillator will change with time, temperature, and other environmental factors. Oscillator phase noise and oscillator frequency drift are manifestations of the same effect. Both characteristics describe how the oscillator's frequency varies over

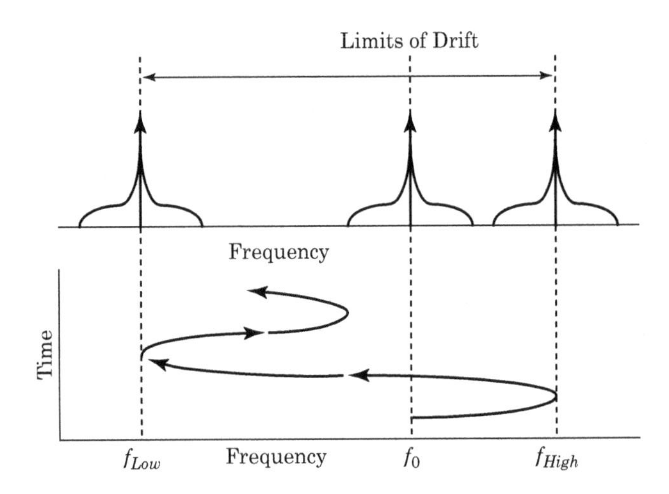

FIGURE 9-1 ■ Frequency drift and phase noise in a practical oscillator.

time. The rule contains the following two points:

- *Frequency drift* encompasses changes observed in the oscillator's frequency over time periods that are greater than about 1 second.
- *Phase noise* entails frequency changes observed over time periods that are less than 1 second or so.

Both drift and phase noise express the idea that the oscillator isn't exactly at its nominal frequency. Figure 9-1 shows oscillator drift.

9.2.1.1 Kevin's Poor Synthesizer

The spectrum of a nonperfect oscillator contains discrete unwanted signals. Most commonly, the spectrum contains harmonics of the fundamental signal, and often the output spectrum contains other discrete signals such as subharmonics and frequencies that are remnants of the oscillator realization. The spectrum will also show the oscillator's phase noise.

Figure 9-2 through Figure 9-7 show the spectral plots of a phase-locked loop (PLL) frequency synthesizer built by the author. The principal design requirements for this project

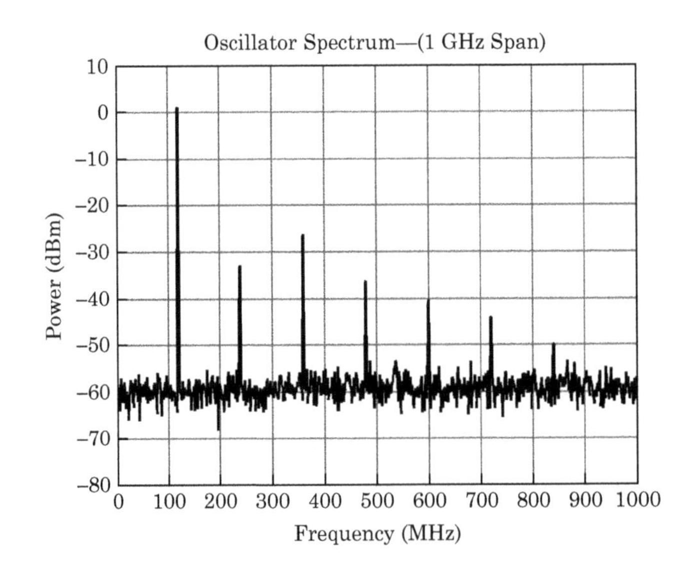

FIGURE 9-2 ■ The wideband spectrum of a PLL frequency synthesizer. The fundamental frequency is 120 MHz. Note the various harmonics.

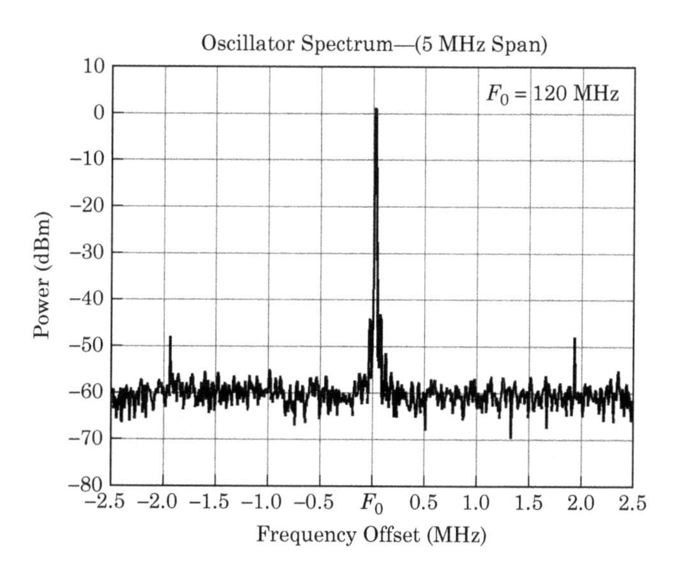

FIGURE 9-3 ■ The output of a PLL frequency synthesizer over a 5 MHz span. Note the discrete, nonharmonically related spurious signals at ±1.935 MHz.

were small physical size and low power consumption rather than low phase noise and low harmonic content. Consequently, the spectrum is not ideal. The synthesizer was tuned to 120 MHz.

Figure 9-2 shows the synthesizer's spectrum from 0 Hz to 1 GHz. Harmonics of varying strength are present at the oscillator's output. Normally, the second harmonic is strongest, followed by the third, fourth, and so forth. However, in this particular design, the second harmonic is rather small, whereas the third harmonic is the strongest (at 28 dBc). Note that the harmonics continue well into the 800 MHz range.

Figure 9-3 shows the oscillator's output over a 5 MHz bandwidth. The output spectrum suggests close-in phase noise that might be confirmed as we narrow the span. The spectrum also exhibits discrete, spurious signals about 1.935 MHz away and 50 dBc down from the carrier. These spurious products can often be traced to some waveform present in the synthesizer circuitry.

Figure 9-4 shows the synthesizer output across a 500 kHz span. The plot reveals that the phase noise we observed in Figure 9-3 contains several discrete tones about 50 kHz apart.

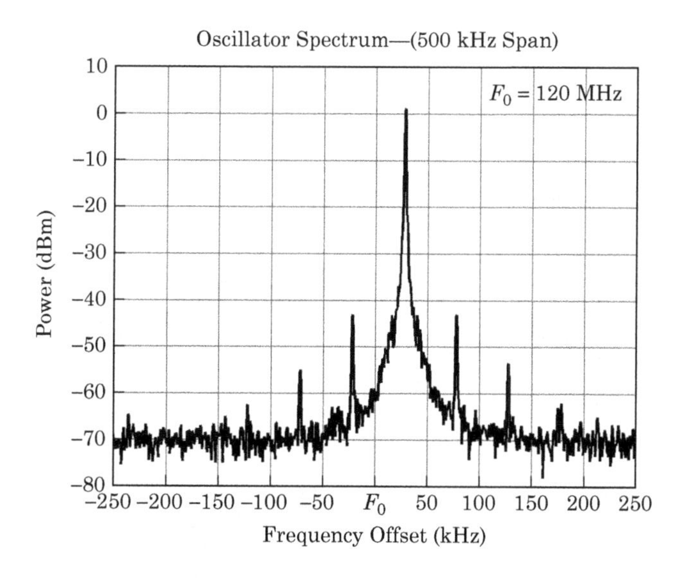

FIGURE 9-4 ■ The output of a PLL frequency synthesizer over a 500 kHz span reveals the oscillator's phase noise and several discrete components at 50 kHz intervals. The output is not exactly at its 120 MHz nominal design frequency.

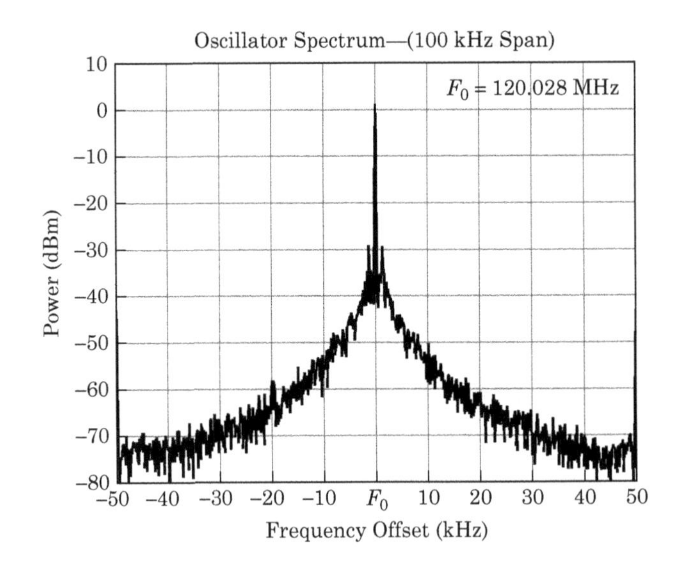

FIGURE 9-5 ■ The output of a PLL frequency synthesizer over a 100 kHz span. The oscillator's phase noise is quite apparent at this scale.

The step size of this synthesizer is 50 kHz, and tones of this nature are quite common in PLL frequency synthesizers. These spurious signals are commonly referred to as *sampling sidebands.*

Figure 9-4 shows that the oscillator is not exactly on frequency. The synthesizer frequency is about 28 kHz too high. This is a 230 parts per million (ppm) or 0.023% difference, which is quite large.

We centered up the spectrum in Figure 9-5. The oscillator's phase noise is quite apparent over the 100 kHz span.

Figure 9-6 shows the behavior of the oscillator's noise floor over a 10 kHz frequency span. The oscillator's phase noise exhibits a change in slope as we approach the carrier. As we move toward the carrier, the noise increases and then levels off about 1,200 Hz from the carrier. The noise then drops as we continue to approach the carrier. This noise profile is common in PLL frequency synthesizers. This synthesizer is particularly noisy

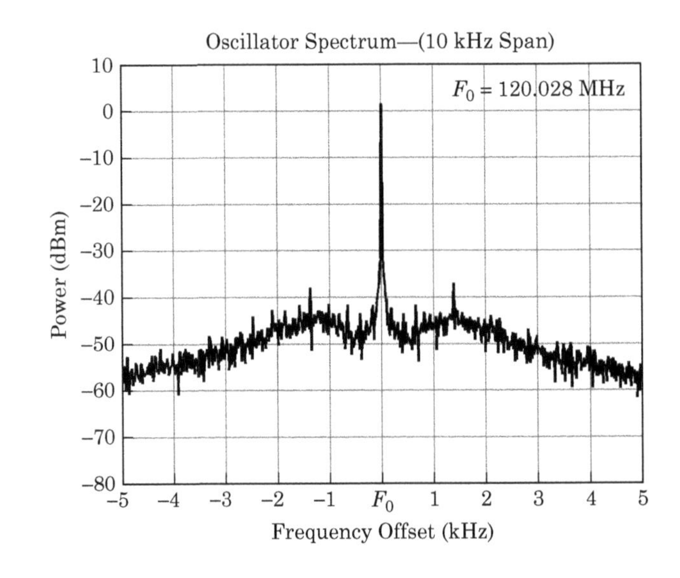

FIGURE 9-6 ■ The output of a PLL frequency synthesizer over a 10 kHz span. The change in the slope of the phase noise is due to the control loop nature of the design.

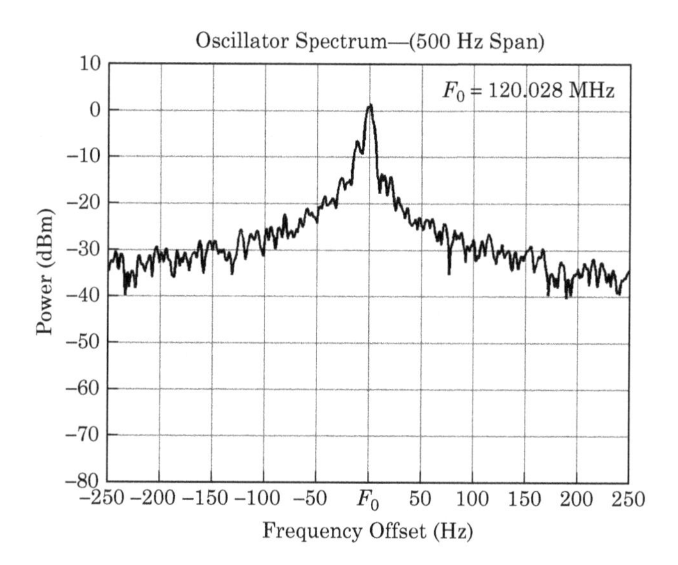

FIGURE 9-7 ■ The output of the frequency synthesizer over a 500 Hz span. The phase noise constantly increases as we approach the carrier.

due to a noisy voltage-controlled oscillator (VCO). We can also see a couple of spurious signals about 600 and 1,200 Hz from the carrier.

Finally, Figure 9-7 shows the oscillator's output over a 500 Hz span. The noise is very high close to the carrier.

9.2.2 Representations of a Noisy Oscillator

In Figure 9-8(a), we represent a perfect oscillator by an impulse function in the frequency domain. The impulse is at the oscillator's center frequency f_0. We represent a noisy oscillator in Figure 9-8(b) as the sum of independent Fourier components at $f_0 + f_m$, centered about the oscillator's nominal frequency. Each *noise sideband* of the oscillator exhibits a random amplitude and phase.

9.2.2.1 Amplitude Noise and Phase Noise

A major use an oscillator is to drive the local oscillator (LO) port of a mixer. In Chapter 8, we realized that, to a first-order approximation, we can model the action of the LO in

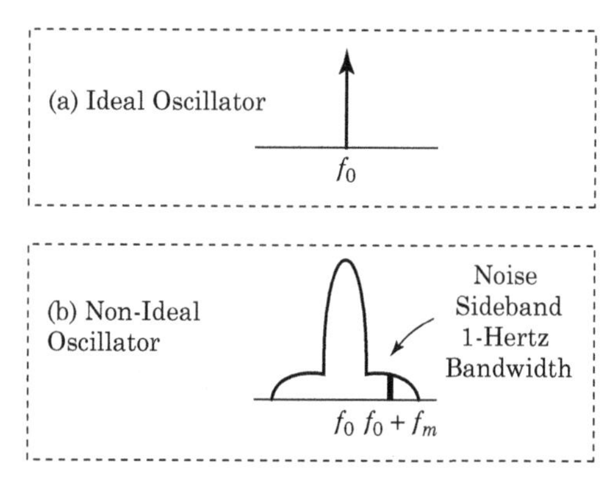

FIGURE 9-8 ■ (a) The frequency-domain representation of an ideal oscillator. (b) The frequency-domain description of an oscillator that exhibits phase noise. Each sideband (at $f_0 + f_m$) has a random amplitude and phase.

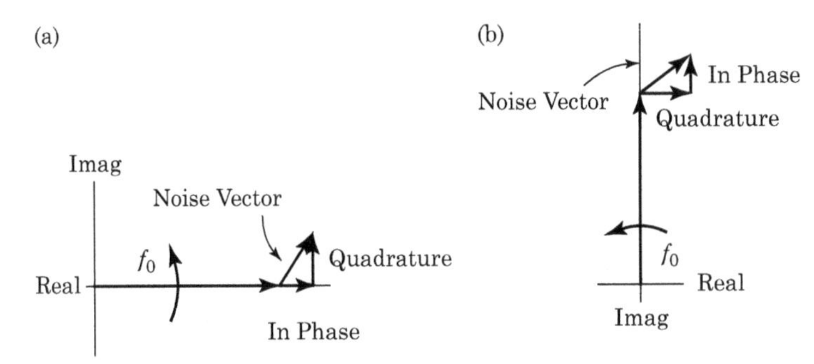

FIGURE 9-9 ■ (a) The phasor, which represents the oscillator's output, just as it passes through the real axis. The quadrature component of the noise does not strongly affect the oscillator output at this time. (b) The oscillator's output at the instant of a zero crossing. The phasor is just passing through the imaginary axis. At this instant, the quadrature component of the noise dominates the oscillator's zero crossing uncertainty. The in-phase component does not affect the zero crossing.

a mixer as controlling a diode switch. When the LO voltage is greater than a few diode drops, we route the radio frequency (RF) signal in one direction. When the LO voltage is below a few diode drops, we route the RF signal in a different direction. It is the precise instant that the LO switches from a high voltage to a low voltage (i.e., its zero crossing) that matters.

The phasor diagrams of Figure 9-9 shows the oscillator's output at various instants in time. The real part of these phasor diagrams (i.e., the representation we would observe on an oscilloscope) is the projection onto the real or horizontal axis. We have shown the noise vector at the end of the vector representing the perfect oscillator. We break the noise vector, with its random amplitude and phase, into its component parts. The in-phase or amplitude variation of the noise vector is collinear with the oscillator's phasor. The quadrature or phase variation of the noise vector is at right angles to the oscillator's phasor.

Figure 9-9(a) shows the phasor when it is passing through the real axis. The real part of the vector (and hence the instantaneous output voltage) is maximum. At this instant, the component of the quadrature component of the noise does not affect the real part of the resultant vector. The uncertainty of the output is dominated by the in-phase component of the noise.

Figure 9-9(b) shows the phasor diagram when the oscillator's output waveform is just passing through zero (i.e., at the oscillator's zero crossing). The projection of the in-phase component onto the real axis is zero, and the uncertainty in the zero crossing is totally controlled by the quadrature component of the noise.

To a first-order approximation, the behavior of an oscillator in a mixer is dominated by the quadrature component of the noise present on the oscillator. Amplitude limiting, both in the oscillator itself and in the mixer, also act to suppress amplitude noise. For these reasons, we examine only the quadrature or phase noise of the oscillator.

9.2.2.2 Phasor

Figure 9-10 shows the relationship between the power spectrum of the oscillator and the oscillator's phasor diagram for two offset frequencies f_{m1} and f_{m2}. We observe the spectral plot of the oscillator in a 1 hertz bandwidth and at various offset frequencies and examine the effect of each slice on the phasor description of the oscillator. The slice in

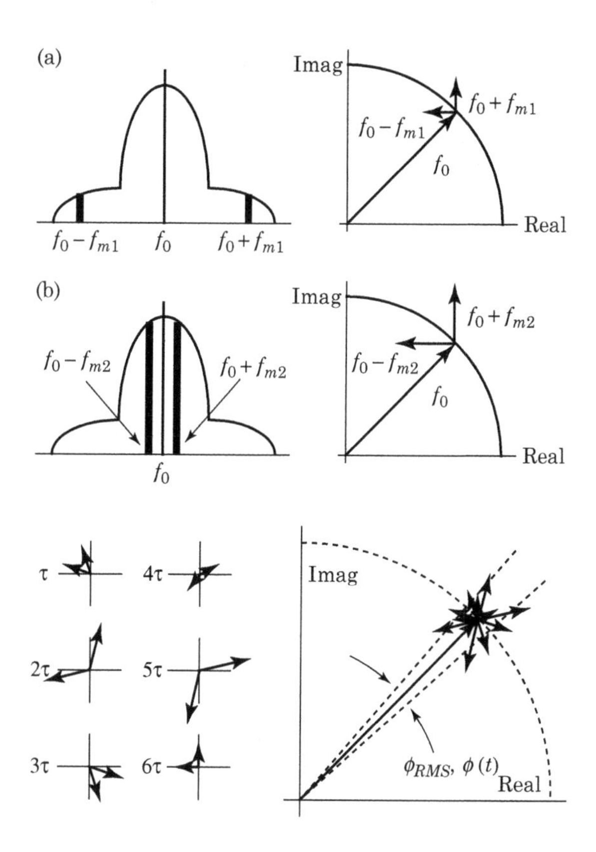

FIGURE 9-10 ■ The phase deviation in an oscillator at a particular offset frequency is related to the amount of noise power present in the oscillator's spectrum at that particular offset frequency. Higher power in the frequency spectrum results in a wider phase deviation.

FIGURE 9-11 ■ The phasor diagram of Figure 9-10 over a longer duration. The random phase variations allow us to measure $\phi(t)$ and ϕ_{RMS}.

Figure 9-10(a) (at $f_0 \pm f_{m1}$) contains only a small amount of power and causes only a small phase deviation. The larger noise power of Figure 9-10(b) (at $f_0 \pm f_{m2}$) causes a larger phase deviation.

Figure 9-10 shows the oscillator's phasor diagram at one instant in time. Since the noise present on the oscillator's phase is random, the phasor diagrams of Figure 9-10 will vary from one instant to the next. Observing the phasor diagram over a long period results in the diagram shown in Figure 9-11 and corresponding values for $\phi(t)$ and ϕ_{RMS}.

9.3 | PHASE NOISE

Phase noise is defined as the short-term variation in an oscillator's phase or frequency. Generally, phase noise represents changes in the oscillator's phase or frequency observed over time periods of 1 second or less. The term *drift* describes changes over periods of 1 second or greater.

Oscillator phase noise will limit the ultimate signal-to-noise ratio (SNR) of any signal processed by a receiver. Oscillator phase noise can cause unwanted signals to mask wanted signals as they pass through the mixing process. In radar systems, receiver phase noise enables clutter to mask the targets we want to detect. Phase noise can also limit the ultimate bit error rate (BER) of a data transmission system.

In Chapter 8, we saw that any signal that is converted from one frequency to another by a local oscillator inherits the faults of that local oscillator. If the LO possesses modulation then every then every signal converted by that LO will exhibit the same modulation.

9.3.1 Phase Modulation under Small β (or Small $\Delta\phi$) Conditions

Oscillator phase noise is often described in the literature as frequency modulation of a carrier operating *under small β conditions.* Equivalently, we can describe the oscillator as being phase modulated under *small $\Delta\phi$ conditions.* By design, modulation present on the oscillator's output is very small. The modulation is also noise-like. The instantaneous phase of a noisy oscillator is

$$\phi(t) = 2\pi f_0 t + \Delta\phi(t) \tag{9.1}$$

where

$\Delta\phi(t)$ = the time-domain representation of the noise present on the oscillator's phase
f_0 = the oscillator's center frequency

The time-domain description of the oscillator's output is

$$V_{OSC}(t) = V_{Pk}\cos\left[2\pi f_0 t + \Delta\phi(t)\right] \tag{9.2}$$

9.3.1.1 Frequency-Domain Representation

We will momentarily limit $\Delta\phi(t)$ to a sinusoid for exposition. When

$$\Delta\phi(t) = \Delta\phi_{Pk}\cos(2\pi f_m t) \tag{9.3}$$

equation (9.2) becomes

$$\begin{aligned} V_{OSC}(t) &= V_{Pk}\cos\left[2\pi f_0 t + \Delta\phi_{Pk}\cos(2\pi f_m t)\right] \\ &= V_{Pk}\cos\left[\omega_0 t + \Delta\phi_{Pk}\cos(\omega_m t)\right] \end{aligned} \tag{9.4}$$

and its frequency-domain representation will be

$$\begin{aligned} V_{OSC}(\omega_m, t) = {} & V_{Pk}J_0(\Delta\phi_{Pk})\cos(\omega_c t) \\ & + V_{Pk}J_1(\Delta\phi_{Pk})\{\sin[(\omega_0+\omega_m)t] + \sin[(\omega_0-\omega_m)t]\} \\ & - V_{Pk}J_2(\Delta\phi_{Pk})\{\cos[(\omega_0+2\omega_m)t] + \cos[(\omega_0-2\omega_m)t]\} \\ & - V_{Pk}J_3(\Delta\phi_{Pk})\{\sin[(\omega_0+3\omega_m)t] + \sin[(\omega_0-3\omega_m)t]\} \\ & + V_{Pk}J_4(\Delta\phi_{Pk})\{\cos[(\omega_0+4\omega_m)t] + \cos[(\omega_0-4\omega_m)t]\} \\ & + \cdots \end{aligned} \tag{9.5}$$

where

$J_n(\Delta\phi_{pk})$ = the Bessel function of the first kind, order n, with an argument of $\Delta\phi_{pk}$
ω_m = the radian frequency of the modulation (equation (9.3))

9.3.1.2 Small β (or Small $\Delta\phi$) Approximations

The quantities $\Delta\phi(t)$ and $\Delta\phi_{pk}$ equations (9.1) through (9.5) are small, allowing us to apply useful approximations to equation (9.5). These approximations are the small β approximations after the β from frequency modulation (FM) theory. Rather arbitrarily, we say the small β approximations are valid when

$$\Delta\phi_{pk} \le 0.2 \text{ radians} \tag{9.6}$$

For small values of $\Delta\phi_{pk}$:

1. The value of $J_0(\Delta\phi_{pk})$ in equation (9.5) is very close to unity or

$$J_0\left(\Delta\phi_{pk}\right) \approx 1 \quad \text{for } \Delta\phi_{pk} \le 0.2 \text{ radians} \tag{9.7}$$

2. The value of $J_1(\Delta\phi_{pk})$ is

$$J_1\left(\Delta\phi_{pk}\right) \approx \frac{\Delta\phi_{pk}}{2} \quad \text{for } \Delta\phi_{pk} \le 0.2 \text{ radians} \tag{9.8}$$

3. The values of the Bessel functions $J_2(\Delta\phi_{pk})$ through $J_n(\Delta\phi_{pk})$ are zero:

$$J_n\left(\Delta\phi_{pk}\right) \approx 0 \quad \text{for} \begin{cases} \Delta\phi_{pk} \le 0.2 \text{ radians} \\ n = 2, 3, 4, \ldots \end{cases} \tag{9.9}$$

Applying these approximations to equation (9.5), the Fourier spectrum for a sinusoidally phase modulated wave under small β conditions is

$$\begin{aligned} V_{Pk}(t) &= V_{Pk} J_0\left(\Delta\phi_{Pk}\right)\cos\left(\omega_0 t\right) \\ &\quad + V_{Pk} J_1\left(\Delta\phi_{Pk}\right)\left\{\sin\left[\left(\omega_0+\omega_m\right)t\right] + \sin\left[\left(\omega_0-\omega_m\right)t\right]\right\} \\ &= V_{Pk}\cos\left(\omega_0 t\right) \\ &\quad + V_{Pk}\frac{\Delta\phi_{pk}}{2}\left\{\sin\left[\left(\omega_0+\omega_m\right)t\right] + \sin\left[\left(\omega_0-\omega_m\right)t\right]\right\} \end{aligned} \tag{9.10}$$

Figure 9-12 shows the spectrum of a sinusoidally phase modulated waveform under small β conditions.

We now allow $\Delta\phi(t)$ to become an arbitrary waveform. The small β conditions force the condition that each spectral component of $V_{OSC}(t)$ is produced by only one spectral component of $\Delta\phi(t)$. The spectrum of the phase modulation (PM) $\Delta\phi(t)$ is identical to the spectrum of the oscillator's output. Figure 9-13 shows the spectrum of the modulation waveform $\Delta\phi(t)$ and the spectrum of $V_{OSC}(t)$.

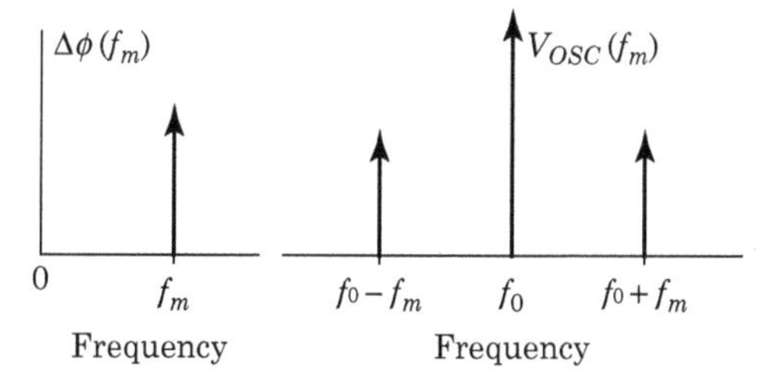

FIGURE 9-12 ■ The spectrum of $\Delta\phi(t)$ from equation (9.3) and the spectrum of $V_{OSC}(t)$ from equation (9.4) when $\Delta\phi(t)$ is a sinusoid and under small β conditions. The single spectral component of $\Delta\phi(t)$ at f_m produces spectral component in $V_{OSC}(t)$ only at offsets of $\pm f_m$.

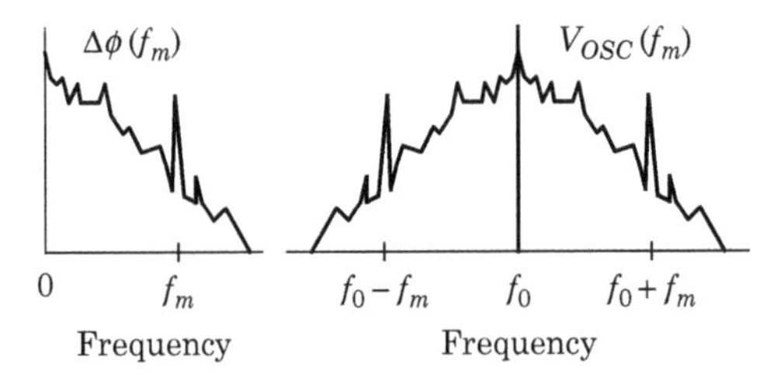

FIGURE 9-13 ■ The spectrum of $\Delta\phi(t)$ from equations (9.1) and (9.2) and the spectrum of $V_{OSC}(t)$ under small β conditions. Components in the spectrum of $\Delta\phi(f_m)$ produce spectral components in $V_{OSC}(f_m)$ only at offsets of f_m.

9.3.1.3 Small β Approximations in the Frequency Domain

We can make some approximations for the power spectrum of a phase-modulated signal under small β conditions:

- The power in the component at f_0 is

$$\begin{aligned}\frac{P_{PM}\,(f_0)}{P_{PM,Unmodulated\ Carrier}\,(f_0)} &= J_0^2\,(\Delta\phi_{Pk}) \\ &\approx 1 \end{aligned} \tag{9.11}$$

  The power in the modulated carrier approximately equals the power in the unmodulated carrier.

- The power in each of the components at $f_0 \pm f_m$ is

$$\begin{aligned}\frac{P_{PM}\,(f_0 + f_m)}{P_{PM,Unmodulated\ Carrier}\,(f_0)} &= \frac{P_{PM}\,(f_0 - f_m)}{P_{PM,Unmodulated\ Carrier}\,(f_0)} \\ &= J_1^2\,(\Delta\phi_{Pk}) \\ &\approx \frac{\Delta\phi_{Pk}^2}{4} \end{aligned} \tag{9.12}$$

  Since $\Delta\phi_{pk}$ must be less than 0.2 radians for the small β conditions to be valid, the J_2 and higher components will always be less than

$$\begin{aligned}\frac{P_{PM}\,(f_0 + f_m)}{P_{PM,Unmodulated\ Carrier}\,(f_0)} &= \frac{P_{PM}\,(f_0 - f_m)}{P_{PM,Unmodulated\ Carrier}\,(f_0)} \\ &\approx \frac{\Delta\phi_{Pk}^2}{4} \\ &\leq \frac{(0.2)^2}{4} = 0.010 = -20\ \text{dB} \end{aligned} \tag{9.13}$$

The J_2 and higher components of a phase-modulated signal under small β conditions will always be less than 20 dB below the carrier.

9.3.1.4 Phasor—Small β

Figure 9-14 shows a phasor diagram of a phase-modulated wave under small β conditions. The only nonzero Fourier components are the carrier (at f_0) and the two sidebands (at $f_0 \pm f_m$). In the left-hand diagram, the two sidebands at $f_0 \pm f_m$ are adding together

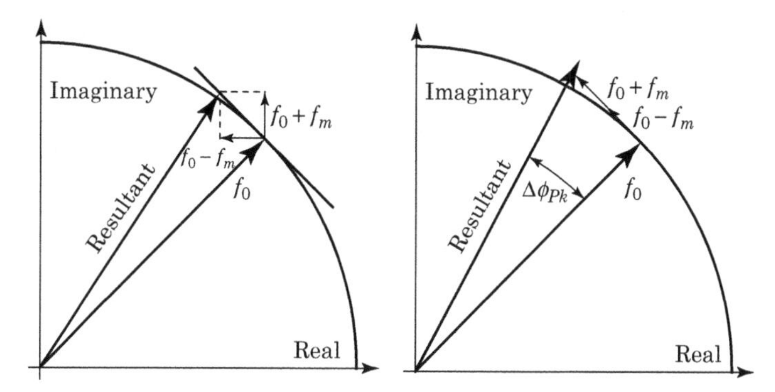

FIGURE 9-14 ■ The phasor diagram of a phase-modulated signal under small β conditions. The two sidebands at $f_0 \pm f_m$ are collinear at the instant in time shown in the right-hand diagram. The extent of these sidebands in this position allows us to calculate the peak and RMS phase deviation of the signal.

to produce the maximum phase deviation of the carrier. We can calculate the maximum and root mean square (RMS) phase deviation of the carrier when the two sidebands are in this position.

The peak phase deviation, $\Delta\phi_{pk}$, is given by

$$\Delta\phi_{Pk} = \tan^{-1}\left(\frac{2 \cdot J_1(\Delta\phi_{Pk})}{J_0(\Delta\phi_{Pk})}\right) \tag{9.14}$$

For random processes, we will discuss RMS phase deviation, $\Delta\phi_{RMS}$, where

$$\Delta\phi_{RMS} = \frac{\Delta\phi_{Pk}}{\sqrt{2}} \tag{9.15}$$

9.3.2 Measuring $\phi(f_m)$

We seek to describe the effects of a noisy oscillator on modulated signals. The first step is to measure $\phi(t)$ given $V_{OSC}(t)$. Figure 9-15 shows a phase detector consisting of a mixer and a low-pass filter (LPF). In this theoretical example, the LO has no phase noise, and the LO phase is 90° away from the phase of the oscillator under test. The voltage present at the intermediate frequency (IF) port of the mixer is

$$\begin{aligned} V_{IF}(t) &= k_{mix}\{V_{RF}(t)\}\{V_{\text{LO}}(t)\} \\ &= k_{mix}\{\cos[\omega_0 t + \phi(t)]\}\left\{\cos\left(\omega_0 t - \frac{\pi}{2}\right)\right\} \\ &= \frac{k_{mix}}{2}\cos\left[\phi(t) - \frac{\pi}{2}\right] + \frac{k_{mix}}{2}\cos\left[2\omega_0 t + \phi(t) - \frac{\pi}{2}\right] \end{aligned} \tag{9.16}$$

where k_{mix} is a constant related to the mixer's conversion loss.

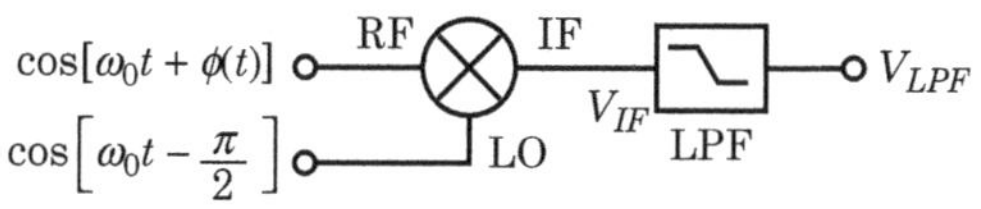

FIGURE 9-15 ■ Using a phase demodulator to measure the phase noise of an oscillator.

We set the cutoff frequency of the LPF to remove the sum term from the output of the mixer and pass the difference term. After the LPF, we are left with

$$V_{LPF}(t) = \frac{k_{mix}}{2} \cos \left[\phi(t) - \frac{\pi}{2} \right] \tag{9.17}$$

Since

$$\sin(\omega t) = \cos\left(\omega t - \frac{\pi}{2}\right) \tag{9.18}$$

then for small values of $\phi(t)$, we can write

$$\cos\left[\phi(t) - \frac{\pi}{2}\right] = \sin[\phi(t)] \approx \phi(t) \tag{9.19}$$

The output of the LPF is

$$V_{LPF}(t) \approx \frac{k_{mix}}{2} \phi(t) \tag{9.20}$$

9.3.2.1 Measurement Difficulties

Let's examine some of the problems we might encounter when trying to measure $\phi(t)$ as described.

The biggest difficulty is deriving the local oscillator. This oscillator must exhibit much lower phase noise than the device under test (DUT); otherwise, we'll just end up measuring the phase noise of our local oscillator. Also, the LO must be exactly at the same frequency as the oscillator under test. If there is any frequency difference between the two oscillators, then the results of equations (9.16) through (9.20) will be inaccurate. We can phase lock the LO to the signal we want to measure. This will produce an LO whose frequency is identical to the signal of interest. However, this introduces further measurement uncertainties. The process of phase locking one signal to a reference impresses the phase noise of the reference on the output signal. The locked LO may be as noisy as the signal we seek to measure.

9.3.2.2 $\mathcal{L}(f_m)$

We seek the quantity $\phi(f_m)$, which is mathematically useful but difficult to measure. The quantity $\mathcal{L}(f_m)$ is the power present at some offset f_m from the carrier divided by the total signal power. It is the power spectrum of an oscillator as it would appear on a suitable spectrum analyzer after normalizing the measurement bandwidth to 1 Hz (see Figure 9-16). $\mathcal{L}(f_m)$ is easy to measure, but the quantity isn't very useful from a

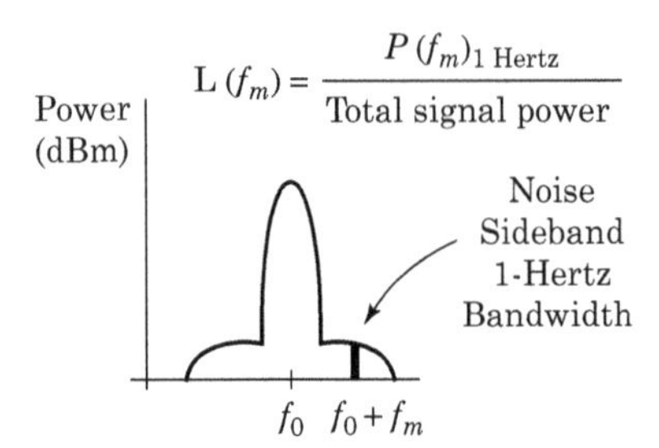

FIGURE 9-16 Single-sideband phase noise is the oscillator power present at some offset f_m from the carrier. The measurement bandwidth is 1 Hz.

mathematical perspective. However, under small β conditions, $\phi(f_m)$ is related to $\mathcal{L}(f_m)$ in a simple way. In equation form

$$\mathcal{L}(f_m) = \frac{\text{Power in a 1-Hz bandwidth measured} f_m \text{ from the carrier}}{\text{Total signal power}}$$
$$= \frac{P\,(f_m)_{1\text{ Hertz}}}{\text{Total signal power}} \tag{9.21}$$

$\mathcal{L}(f_m)$ is commonly expressed in dBc/Hz, and it is the most commonly used expression of phase noise. It's easy to measure (usually, all you need is a spectrum analyzer), and it's useful for comparing two oscillators that operate at the same frequency.

9.3.2.3 Measuring $\mathcal{L}(f_m)$

Phase noise measurement is often affected by the phase noise of the test equipment. For example, even a poorly designed quartz crystal oscillators may exhibit very low phase noise. Let us consider the results of measuring $\mathcal{L}(f_m)$ of a crystal oscillator using a spectrum analyzer. The spectrum analyzer, like any other receiver, uses an internal local oscillator to convert signals from one frequency to another. The spectrum analyzer's LO is likely to be tunable over a wide range, and its phase noise performance is likely to be poor compared to the crystal oscillator. So the phase noise of the signal appearing at the spectrum analyzer's IF will mostly represent the phase noise of the spectrum analyzer's LO rather than the phase noise of the crystal oscillator.

We can measure the phase noise of very good oscillators, but it is difficult. Scherer (n.d., April 1979, May 1979) and Hewlett-Packard (1988, 1990) are excellent sources of information on this topic.

9.3.2.4 Relating $\phi(f_m)$ to $\mathcal{L}(f_m)$

Figure 9-12, Figure 9-13, and equations (9.11) through (9.13) form the basis of our relationship between $\mathcal{L}(f_m)$ and the more mathematically useful $\phi(f_m)$. The analysis applies only under the small β conditions, which are relevant to most practical oscillators. Under the small β conditions, we know $J_0(\Delta\phi_{pk}) = 1$, $J_1(\Delta\phi_{pk}) = \Delta\phi_{pk}/2$ and $J_n(\Delta\phi_{pk}) = 0$ for n equals 2 through infinity.

The first approximation ($J_0(\Delta\phi_{pk}) = 1$) tells us that the magnitude of the carrier at f_0 is unchanged from its unmodulated value. The last approximation [$J_n(\Delta\phi_{pk}) = 0$ for $n = 2$ to infinity] tells us that the components at $f_0 \pm nf_m$ are negligible for $n >= 2$. The middle approximation describes the magnitude relationship between $\mathcal{L}(f_m)$ and $\phi(f_m)$. Under small β conditions, equation (9.12) relates $\mathcal{L}(f_m)$ and $\phi(f_m)$ as

$$\mathcal{L}\,(f_m) = \frac{\text{Power in a 1-Hz bandwidth measured } f_m \text{ from the carrier}}{\text{Total signal power}}$$
$$= \frac{P(f_m)_{1\text{ Hertz}}}{\text{Total signal power}} \tag{9.22}$$
$$= J_1^2(\Delta\phi_{Pk})$$
$$\approx \frac{\phi_{Pk}^2\,(f_m)}{4}$$

$\phi\,(t)$ is a nondeterministic, statistically described waveform, and we must describe it using RMS rather than peak values. Since much of our analysis was done assuming sinusoidal modulation and we modeled the complex modulating waveform as a combination of sinusoids, we can write

$$\Delta\phi_{RMS} = \frac{\Delta\phi_{Pk}}{\sqrt{2}} \tag{9.23}$$

and

$$\mathcal{L}(f_m) = \frac{\phi^2_{RMS}(f_m)}{2} \tag{9.24}$$

Converting this equation to decibels produces

$$\begin{aligned} \mathcal{L}(f_m)_{dB} &= 10\log\left[\frac{\phi^2_{RMS}(f_m)}{2}\right] \\ &= 20\log\left[\phi_{RMS}(f_m)\right] - 3\text{ dB} \end{aligned} \tag{9.25}$$

For small β, $\phi_{RMS}(f_m) = \mathcal{L}(f_m) + 3$ dB.

9.3.3 $S_\phi(f_m)$

The quantity $S_\phi(f_m)$ is simply the power of $\phi(f_m)$ measured in a 1 Hz bandwidth or

$$S_\phi(f_m) = \frac{\phi^2_{RMS}(f_m)}{1\text{ Hertz}}\ (\text{rad})^2 \tag{9.26}$$

$S_\phi(f_m)$ is the *spectral density of the phase fluctuations* of an oscillator and is the power of a phase discriminator's output measured in a 1 Hz bandwidth.

9.3.3.1 When the Small β Conditions Are Valid

Under what conditions does an oscillator meet the small β criteria? The -10 dB/decade line drawn on Figure 9-17 represents an RMS phase deviation of approximately 0.2 radians integrated over any one decade of offset frequency. At 0.2 radians, the power in the higher-order sidebands of the phase modulation is still small compared with the power in the first sideband. This ensures the small β criterion is satisfied. Most garden-variety oscillators

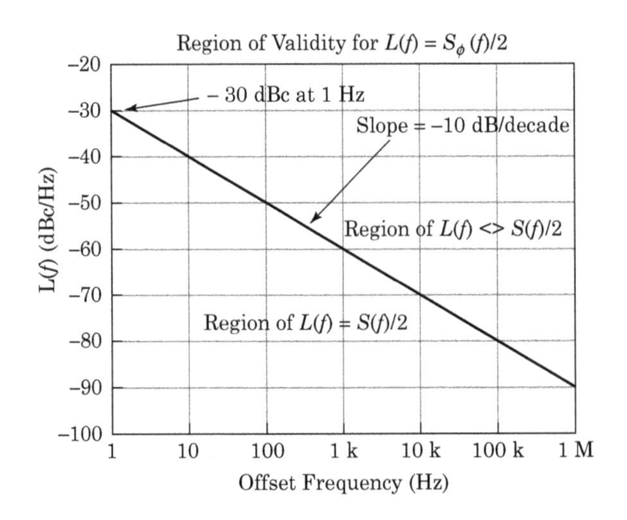

FIGURE 9-17 ■ The region of validity for the small β conditions.

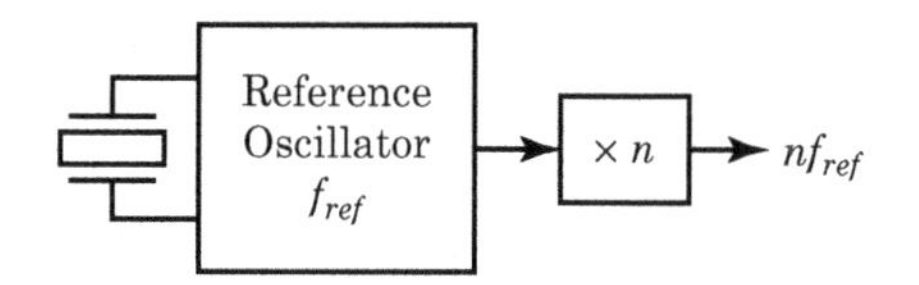

FIGURE 9-18 ■ A frequency multiplier. The output frequency is nf_{REF}.

are well within the small β area except when we are measuring very close to the carrier (i.e., f_m is very small).

9.3.3.2 Phase Noise and Multipliers

Most frequency synthesizer designs involve multiplying or dividing a stable reference oscillator. For example, to generate a 1234.567 MHz LO, we might start with a 1-MHz crystal oscillator, divide it by 1000 to get a 1 kHz oscillator, and then multiply the 1 kHz oscillator by 1,234,567 to get the required frequency.

Figure 9-18 shows the multiplication. We begin with a fixed reference oscillator (at f_{REF}). To generate a particular frequency, we multiply our *golden reference oscillator* by some number n to produce a new signal at nf_{REF}. Given the phase noise characteristics of the reference oscillator, what are the phase noise characteristics of the output of the multiplier? We'll use the expression for a noisy oscillator to describe the reference oscillator:

$$V_{REF}(t) = \cos\left[\omega_0 t + \phi(t)\right] \tag{9.27}$$

where $\phi(t)$ is a random waveform with some RMS value ϕ_{RMS}.

Passing the reference oscillator waveform through the frequency multiplier produces

$$V_{REF,\,mult}(t) = \cos\left[n\omega_0 t + n\phi(t)\right] \tag{9.28}$$

We've multiplied both the frequency and $\phi(t)$ by n. The quantity $n\phi(t)$ now has an RMS value of $n\phi_{RMS}$. Before the multiplication, the single-sideband phase noise of the reference oscillator at some arbitrary offset f_m is

$$\mathcal{L}_{REF}(f_m) = \frac{\phi_{RMS}^2(f_m)}{2} \tag{9.29}$$

After the multiplication, the single-sided phase noise at the same f_m is

$$\begin{aligned} \mathcal{L}_{REF,\,mult}(f_m) &= \frac{\left[n\phi_{RMS}(f_m)\right]^2}{2} \\ &= n^2 \frac{\phi_{RMS}^2(f_m)}{2} \end{aligned} \tag{9.30}$$

The ratio of the single-sided phase noise power of the multiplied oscillator to the unmultiplied oscillator is

$$\begin{aligned} \frac{\mathcal{L}_{REF,\,mult}(f_m)}{\mathcal{L}_{REF}(f_m)} &= n^2 \\ &= 20\log(n) \quad \text{in dB} \end{aligned} \tag{9.31}$$

If $n > 1$ (i.e., for frequency multipliers), then the $\mathcal{L}(f_m)$ of the multiplied oscillator will be higher for a given f_m. In other words, multiplying makes the phase noise of an oscillator worse. This analysis assumes the small β conditions apply for both the unmultiplied and for the multiplied carrier at the f_m of interest.

EXAMPLE

Multiplied Oscillator Phase Noise

We measure the $\mathcal{L}(f_m)$ of an oscillator at $f_m = 100$ Hz to be $\mathcal{L}(100 \text{ Hz}) = -80$ dBc. After multiplying the oscillator by 100, what is $\mathcal{L}(f_m)$ of the multiplied oscillator at

a. $f_m = 100$ Hz

b. $f_m = 10{,}000$ Hz

assuming the small β conditions apply.

Solution

Using equation (9.31), we know

$$\begin{aligned} \left[\frac{\mathcal{L}_{REF,1000x}(f_m)}{\mathcal{L}_{REF}(f_m)}\right]_{dB} &= 20\log(n) \\ &= 20\log(100) \qquad (9.32) \\ &= 40 \text{ dB} \end{aligned}$$

a. The phase noise of the unmultiplied oscillator at 100 Hz away from the multiplied carrier (i.e., at $f_m = 100$ Hz) is $-80 + 40 = -40$ dBc. Multiplying makes the phase noise worse.

b. We don't have enough information to solve this part of the problem. We would need to know $\mathcal{L}$ (10,000) of the unmultiplied oscillator to solve this problem.

Equation (9.31) tells us that each doubling of the carrier increases the phase noise measured at f_m by 6 dB. Likewise, each halving of the carrier results in a decrease in the phase noise measured at f_m by 6 dB. We didn't have to describe the multiplier in any way. It can be a nonlinear device using some high-order distortion component or a phase-locked loop.

Also note that equation (9.31) tells us what the *minimum* phase noise of the multiplier will be. If the multiplier system itself is noisy, it can introduce excess phase noise into the system, and the output of the multiplier will be noisier than equation (9.31) indicates.

9.3.3.3 Phase Noise and Dividers

Equation (9.31) also applies for frequency division. If we apply a signal to a digital flop-flop, for example, the output frequency equals one-half of the input frequency. We can divide the input signal by any number we desire using the appropriate digital techniques.

EXAMPLE

Frequency Division and Phase Noise

Given an oscillator whose $\mathcal{L}(f_m)$ is -110 dBc when $f_m = 10$ kHz. What is $\mathcal{L}$ (10,000) of the oscillator if we pass it through a divide-by-12 counter?

Solution

Using equation (9.31),

$$\left[\frac{\mathcal{L}_{REF,\div 12}(f_m)}{\mathcal{L}_{REF}(f_m)}\right]_{dB} = 20\log(n)$$

$$= 20\log\left(\frac{1}{12}\right) \tag{9.33}$$

$$= -21.6\text{ dB}$$

So, $\mathcal{L}(f_m)$ of the new oscillator at $f_m = 10{,}000$ Hz is $-110 + (-21.6) = -131.6$ dBc. For $n < 1$ (i.e., for frequency dividers), the $\mathcal{L}(f_m)$ of the divided-down oscillator is lower for a given f_m. Dividing an oscillator down improves the phase noise of the oscillator. Referring to Figure 9-19, we can view frequency division as an averaging process. Each rising edge of the input waveform exhibits some uncertainty, which contributes to the input's phase noise spectrum. The output changes only on every fourth rising edge of the input. The phase noise contributions of input transitions 1 through 3 and 5 through 7 are ignored because they don't affect the counter's output. The net effect is to spread the uncertainty of one zero crossing over the time required for eight input cycles, thus reducing the phase noise of the output waveform.

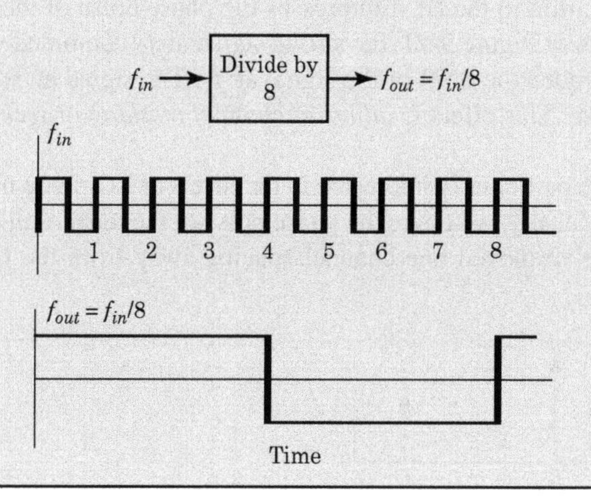

FIGURE 9-19 ■ The effects of a divider on oscillator phase noise. The uncertainty in one zero crossing of the input signal is spread across a longer time interval in the output signal. In this example, the zero crossing uncertainty of the input influences the output only at every fourth cycle.

9.3.4 The Effects of Phase Noise

A noisy LO will affect signals in a mixer in different ways. Figure 9-20 shows a noisy local oscillator combining with a pure sine wave in a mixer. The noise of the oscillator is transferred to every signal that mixes with the LO so every frequency translated signal will inherit the spectrum of the LO.

9.3.4.1 Adjacent Channel Masking

In a channelized system, our receive antenna is likely to collect many different signals at the same time. The signals will be at different frequencies, and their power levels will vary

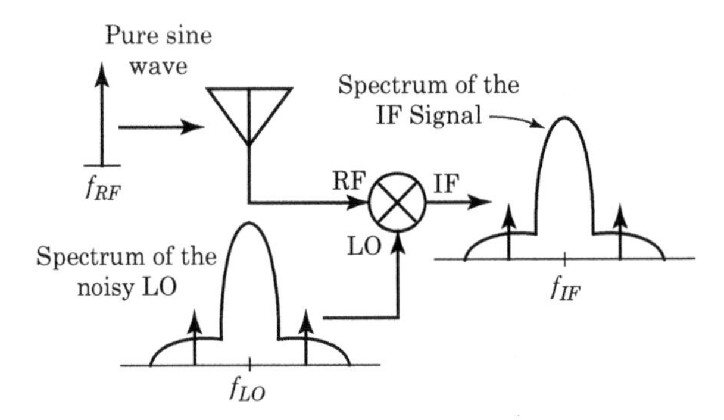

FIGURE 9-20 ■ The mixing process using a noisy LO. Noise present on the LO is transferred to every signal that is converted to a new frequency by the mixer. The noisy IF signal is the pure RF signal corrupted by the noisy LO.

dramatically. One of the most difficult receiver problems is the reception of a weak, desired signal in the presence of a strong, undesired signal. The strong signal taxes the receiver's linearity, while the weak signal taxes the receiver's noise performance. LO phase noise makes this situation even more troublesome.

Figure 9-21 shows a typical channelized spectrum entering the RF port of a mixer. The signals vary widely in power levels and are separated by the channel spacing f_{CH}. As the mixer converts the RF spectrum to the IF, it impresses the phase noise of the LO onto each signal. In the IF spectrum of Figure 9-21, the strong signal at f_2 combined with the phase noise of the LO has degraded the SNR of the signal at f_1. The signal at f_3 has been completely enveloped in noise. This effect is *adjacent channel masking* or *receiver desensitization.*

In channelized systems, we are particularly interested in the levels of LO phase noise that are f_{CH} away from the carrier. Ideally, we'd like the phase noise of the local oscillator to be insignificant by the time we've moved one channel spacing away from the LO's center frequency.

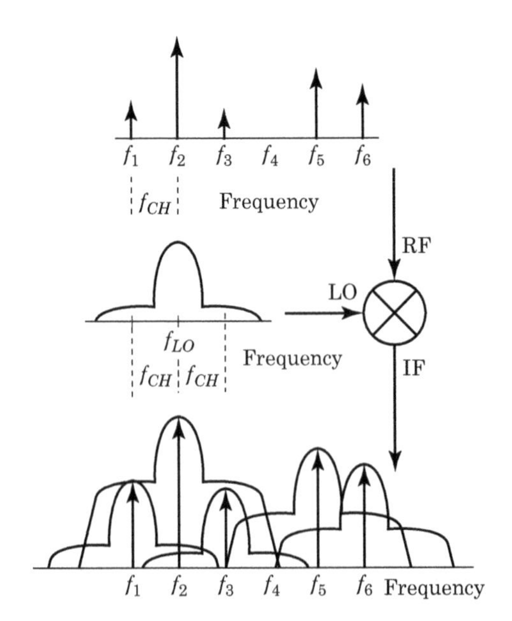

FIGURE 9-21 ■ Oscillator phase noise can cause large signals to mask weaker signals, especially in channelized systems.

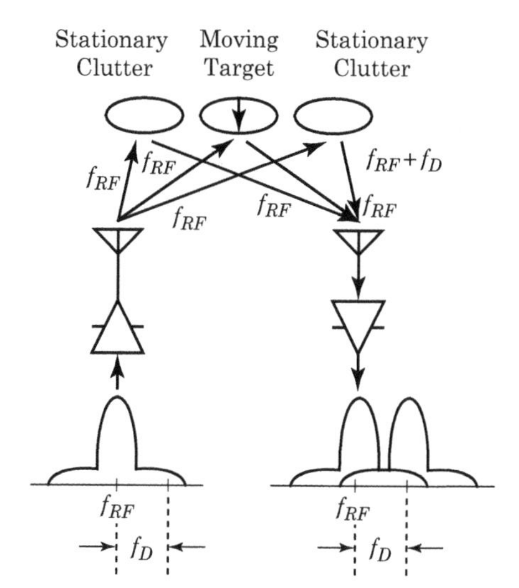

FIGURE 9-22 ■ Oscillator phase noise can cause stationary clutter to mask the returns of moving targets in radar systems.

9.3.4.2 Radar

We use Doppler radar to measure moving targets. Doppler radar works by transmitting a signal at the target and then searching for the return. The change in frequency of the returned signal is a measure of the speed of the target. The transmitter frequency is often several GHz, whereas the frequency shift caused by the Doppler effect is usually less than 10 kHz.

Figure 9-22 shows the typical return. The transmitted signal at f_{RF} (at perhaps 10 GHz) strikes both the moving target and stationary targets. The moving target shifts the frequency of f_{RF} by f_D (perhaps 10 kHz). The receiver processes the two signals at f_{RF} and $f_{RF} + f_D$. Both signals inherit the phase noise of the LO, and the SNR of the desired return from the moving target is reduced because of the nonzero phase noise of the transmitter at $f_{RF} \pm f_D$. In situations with large amounts of stationary clutter and small returns from the moving target, the phase noise of the LO can completely mask the return from the target.

9.3.4.3 Ultimate SNR or Ultimate BER

LO phase noise can limit the ultimate signal-to-noise ratio (SNR) or the ultimate BER of a receiver. Although we will cover this topic more thoroughly in Chapter 12, Figure 9-23 shows the effect. The dashed curve is the waterfall curve typical of received BER versus received signal power analyses. At higher SNR's, the BER will decrease effectively to zero errors. Since the received SNR is affected by phase noise, the SNR seen by the demodulator never increases to very large values. The effect is to flatten out the BER versus power curve as shown in Figure 9-23.

9.3.5 Sources of Phase Noise—The Leeson Model

RF engineers must choose from a large number of oscillator technologies and are interested in those that provide the lowest phase noise. The Leeson oscillator model is a reasonably simple model that supplies useful insight. Figure 9-24 shows the Leeson model, which consists of an active device with some wideband power gain G_p and some noise figure F. The model also features a resonator whose center frequency is f_0 and whose quality factor is denoted as Q.

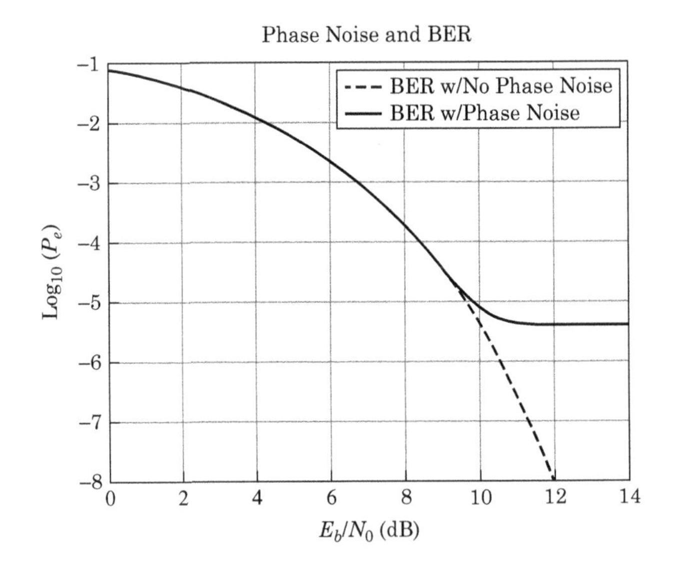

FIGURE 9-23 ■ The effect of oscillator phase noise on received BER. The phase noise of a system can limit the demodulated SNR and thus can cause the system to produce an ultimate BER below which the system cannot perform.

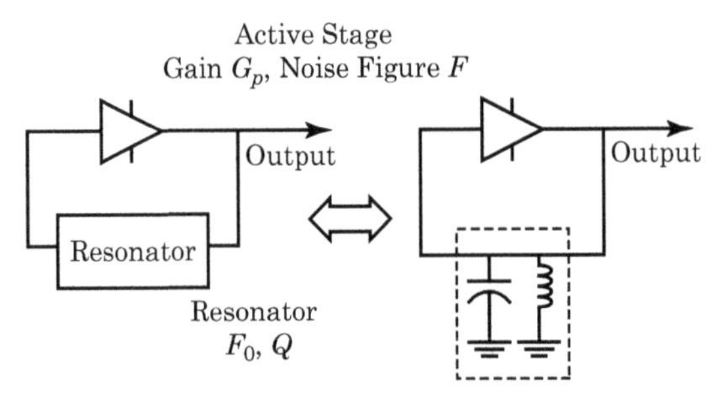

FIGURE 9-24 ■ The Leeson oscillator model consists of a gain stage with a known gain and noise figure. The resonator connects the amplifier output to its input and determines the frequency of operation.

We connect the output of the active device to its input through the resonator, creating a feedback loop. Oscillation will occur at the frequencies where the power gain around the loop is greater than unity and the phase shift around the loop is 360°. The oscillation frequency is a strong function of the magnitude and phase responses of the resonant network and the active device. The active device is wideband and exhibits a noninteresting phase response. The resonant network is the only device that plays a part in determining the oscillation frequency.

For illustration, we've shown a parallel inductor-capacitor (LC) configuration as the resonant network in Figure 9-24, but other resonators (e.g., crystals, transmission lines, surface acoustic wave [SAW] devices) are commonly used. Regardless of the resonator type, the following analysis applies almost universally.

An analysis of Leeson's model is beyond the scope of this book. The result of the analysis is an approximation for the single-sided phase noise of an oscillator:

$$\mathcal{L}(f_m) = \frac{1}{2}\frac{FkT}{P_{avg}}\left[1 + \left(\frac{1}{f_m}\frac{f_0}{2Q_L}\right)^2\right] \tag{9.34}$$

where

$\mathcal{L}(f_m)$ = the single-sideband phase noise of the oscillator
k = Boltzman's constant = 1.38E-23 watt-sec
F = the noise factor of the active device (in linear terms)
T = the physical temperature in °K
P_{avg} = the average power taken from the oscillator
f_0 = the carrier or center frequency of the oscillator
f_m = the offset from the carrier frequency
Q_L = the loaded Q of the oscillator's resonator

9.3.5.1 Resonator Q

Resonator Q is

$$Q = 2\pi \frac{\text{Maximum energy stored}}{\text{Total energy lost}} \text{ per cycle} \tag{9.35}$$

Resonator Q is a measure how well a particular circuit element stores energy. *Loaded Q* refers to how well the same element stores energy when it is placed in a circuit. A *high-Q* element stores a lot more energy than it dissipates. Different resonator technologies have different Qs, but some technologies are more suited to particular applications. For example, quartz crystals have very high Qs, but we can't change their resonant frequency. We can easily change the resonant frequency of LC networks, but their Qs are lower. Equation (9.34) tells us that phase noise and the loaded resonator Q are inversely proportional. If the resonator Q increases, the phase noise of the oscillator will tend to decrease. To maximize the loaded Q of the resonator, we must first arrange to maximize the unloaded Q of the resonator. The most common resonator choices include the following:

- Crystal: Q range is 20 k to 200 k
- Surface acoustic wave resonators: Qs of 2 k to 12 k are common
- Dielectric resonators: 500 to 5000
- LC: Qs of 20 to 300
- Voltage-controlled oscillators (VCOs): Qs of 10 to 200

This list is in the order of highest unloaded Q to the lowest unloaded Q. Equation (9.34) indicates that this list is also in the order of increasing oscillator phase noise. With regard to phase noise, even the poorest crystal oscillator will outperform the best VCO. This is almost entirely due to the very large Qs of crystal oscillator resonators. From an ability-to-tune perspective, the VCO wins easily. Experience has shown that we can change the frequency of a crystal oscillator perhaps 1000 ppm of the center frequency. Voltage-controlled oscillators can easily tune over an octave or more of frequency range if we're willing to accept poor phase noise.

9.3.5.2 Oscillator Center Frequency

Equation (9.34) tells us that phase noise of an oscillator tends to increase with the oscillator center frequency. Consequently, oscillators at lower frequencies tend to be quieter than oscillators at higher frequencies. Poor resonator Q can spoil this relationship. An RC oscillator (which has a very, very poor Q) running at 5 kHz may be much noisier than an LC oscillator running at 100 MHz.

9.3.5.3 Offset Frequency

Finally, equation (9.34) tells us that the phase noise increases as we approach the carrier. The value of $\mathcal{L}(f_m)$ increases with decreasing f_m.

9.3.6 Incidental Phase Modulation

Incidental phase modulation (IPM) is an oscillator specification that will tell us how the phase noise of a local oscillator will affect a phase-modulated signal.

Oscillator phase noise represents an uncertainty in an oscillator's phase or frequency. As we've seen, any signal converted from one frequency to another by a mixer will inherit the phase noise of the LO used to perform the conversion. The LO phase noise can mask the information present on an information-bearing signal.

For example, we can encode digital data into the phase of a sine wave. We may assign a binary 0 to 0° and a binary 1 to 180°. Anything that masks the true phase of the signal, such as poor SNR or excess LO phase noise, can cause the phase of the signal to be misread, producing a symbol error.

Incidental phase modulation is also known as β_ϕ or phase jitter.

9.3.6.1 Phase Demodulation

A phase demodulator produces an output voltage that is directly proportional to the phase difference between a reference signal (internally generated in the demodulator) and a data-bearing signal. In Figure 9-25, the information-bearing signal is

$$V_I(t) = \cos\left[\omega_0 t + \phi_I(t)\right] \tag{9.36}$$

The locally generated reference frequency is

$$V_{REF}(t) = \cos\left[\omega_0 t + \phi_{REF}(t)\right] \tag{9.37}$$

where

$\phi_I(t)$ carries the information encoded as carrier phase

$\phi_{REF}(t)$ represents the phase noise on the locally generated reference signal

The output of the phase demodulator is

$$\phi_I(t) - \phi_{REF}(t) \tag{9.38}$$

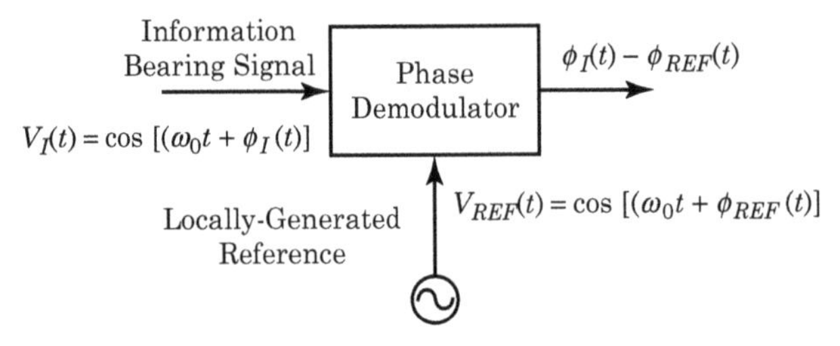

FIGURE 9-25 ■ A phase demodulator. Information is encoded in the phase of the received signal as $\varphi_I(t)$. The LO phase noise is described by $\varphi_{REF}(t)$ and $\varphi_{REF,RMS}$.

If the locally generated reference is quiet and $\phi_{I,\,RMS} \gg \phi_{REF,\,RMS}$, then the phase demodulator output represents the information-bearing $\phi_I(t)$ quite closely. If $\phi_{REF,\,RMS}$ is significant with respect to $\phi_I(t)$, the phase demodulator output will be corrupted by the local oscillator's phase noise.

Assume the phase demodulator will produce an output of 1 volt when $\phi_I(t)$ is 180°. When $\phi_I(t)$ is 0°, the demodulator produces 0 volts. The phase demodulator can by described by

$$\begin{aligned} k_\phi &= \frac{\Delta \text{ Voltage}}{\Delta \text{ Phase}} \\ &= \frac{1 \text{ Volt}}{180^\circ} \\ &= 5.6 \frac{\text{mV}}{\text{degree}} \end{aligned} \tag{9.39}$$

Since the LO phase noise, represented by $\phi_{REF}(t)$, is always present and always changing, there will always be noise present on the demodulator's output. That noise represents a nonreducible uncertainty, and hence it limits the demodulator's accuracy. Continuing the example, assume we measure 8.2 mV_{RMS} of noise on the phase demodulator's output when we apply a quiet carrier to the input of the phase demodulator in Figure 9-25; that is, $\phi_I(t) = 0$. The equivalent phase noise of the reference oscillator is

$$\begin{aligned} \Delta\text{Phase}_{RMS} &= \frac{\Delta\text{Voltage}_{RMS}}{k_\phi} \\ &= \frac{8.2 \text{ mV}_{RMS}}{5.6 \text{ mV/degree}} \\ &= 1.5 \text{ degrees}_{RMS} \end{aligned} \tag{9.40}$$

Thus, we can say our local oscillator has 1.5°_{RMS} of incidental phase modulation.

9.3.6.2 IPM Definition

The IPM of an oscillator is

$$IPM = \sqrt{\int_{fa}^{fb} S_\phi(f_m)\, df_m} \,(\text{radians}_{RMS}) \tag{9.41}$$

For small β, we can write

$$IPM = \sqrt{2\int_{f_a}^{f_b} \mathcal{L}(f_m)\, df_m} \,(\text{radians}_{RMS}) \tag{9.42}$$

where f_a and f_b represent the lower and upper frequency boundaries of the demodulated signal.

These two equations represent the amount of phase jitter present on an oscillator given its frequency spectrum $\mathcal{L}(f_m)$ or the spectral density of the phase fluctuations $S_\phi(f_m)$.

EXAMPLE

IPM

Figure 9-26 shows a local oscillator used in a 2.048 Mbps quadrature phase shift keying (QPSK) receiver. Find the IPM of this oscillator.

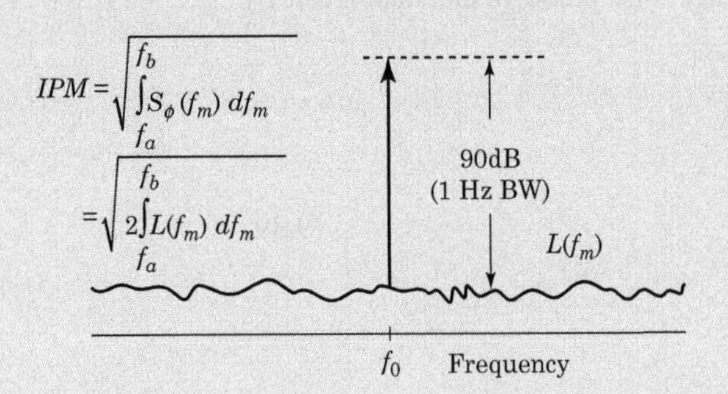

FIGURE 9-26 ■ The spectrum analyzer plot of a noisy oscillator in a 1 Hz resolution bandwidth.

Solution

We have to evaluate equation (9.42) from some lower frequency f_a to some upper frequency f_b.

Rather arbitrarily, we'll use a lower frequency limit that is equivalent to 200 symbols. In other words, every once in a while, we might expect to receive 200 of the same symbol in a row. In practice, this is a conservative number because systems purposely encode data to limit the number of consecutive identical symbols.

The lower frequency limit is $2.048 \cdot 10^6/200 = 10.24$ kHz. The peak-to-null bandwidth of a 2.048 Mbps QPSK signal is 1.024 MHz. Assuming a little excess bandwidth, we use 1.5 MHz as the upper frequency. Equation (9.42) produces

$$\begin{aligned} IPM &= \sqrt{2\int_{10.24E3}^{1.5E6} \left(10^{-9}\right) df_m} \quad (\text{radians}_{RMS}) \\ &= \sqrt{(2)\left(10^{-9}\right)\left(1.5 \cdot 10^6 - 10{,}240\right)} \\ &= 0.055 \text{ radians}_{RMS} \\ &= 3.1 \text{ degrees}_{RMS} \end{aligned} \tag{9.43}$$

The output of this demodulator will produce a noise level equivalent to $3.1°_{RMS}$ of modulation. In other words, the IPM of this oscillator will cause a $3.1°_{RMS}$ uncertainty in the phase measurements of this demodulation.

9.3.6.3 IPM and SNR

The IPM of a receiver's LO will affect the ultimate SNR of a phase-demodulated signal. The nature of the information-bearing signal will determine the affect of the receiver's IPM. Assume the RMS phase noise of a receiver is IPM_{RMS}. Figure 9-27 shows the constellation diagrams for a 4-state and an 8-state phase shift keying (PSK) signal. The phase noise of the receiver causes a phase measurement uncertainty of IPM_{RMS}, which is drawn on the diagram.

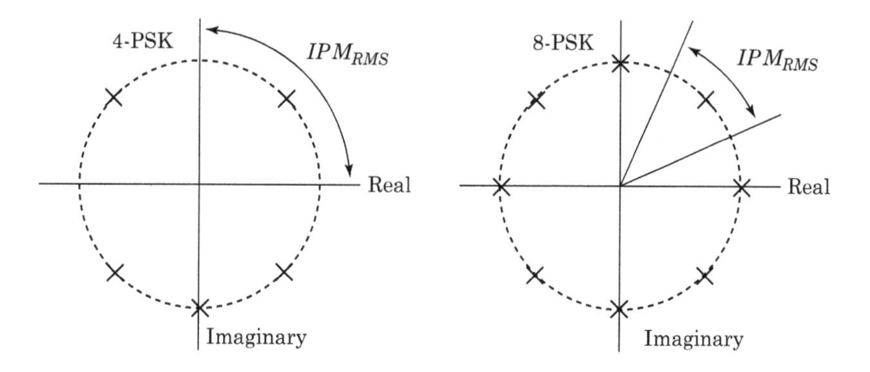

FIGURE 9-27 ▪ IPM requirements for 4- and 8-PSK signals.

The 4-state PSK demodulator will make a symbol error if the IPM of the local oscillator causes a phase error of more than $360^\circ/8 = 45^\circ$. The 8-state PSK demodulator will make a symbol error if the IPM of the local oscillator causes a phase error of more than $360^\circ/16 = 22.5^\circ$. The 8-PSK signal is more sensitive to LO phase noise than the 4-PSK signal because the 8-PSK signal requires more accuracy when it measures the phase of the received signal.

Even with a perfect input signal, oscillator phase noise generates noise on the output of a phase demodulator. We can't reduce this noise without improving our oscillator. Thus, IPM of our LO forces an ultimate limit on the SNR of a demodulated signal.

9.3.6.4 Measuring IPM

We measure IPM by applying a signal generator to the input of our DUT. We first phase modulate the signal generator at some arbitrary rate (e.g., 1 kHz) and at some known phase deviation $\Delta\phi_{pk}$, which corresponds to some RMS phase deviation $\Delta\phi_{RMS}$. When the receiver demodulates the phase-modulated signal from the generator, the RMS voltage present on the output of the demodulator will be

$$V_{out,PM,RMS} = k_\phi \Delta\phi_{RMS} \tag{9.44}$$

where k_ϕ is a constant associated with the phase demodulator.

Next, we remove the phase modulation from the signal generator and provide a quiet, noiseless carrier to the receiver. If the IPM of the signal generator is much smaller than the IPM of the receiver, the output of the phase demodulator will change to

$$V_{out,quiet,RMS} = k_\phi\,(IPM) \tag{9.45}$$

The ratio of these two voltages is

$$\begin{aligned} \frac{V_{out,PM,RMS}}{V_{out,quiet,RMS}} &= \frac{k_\phi \Delta\phi_{RMS}}{k_\phi\,(IPM_{RMS})} \\ &= \frac{\Delta\phi_{RMS}}{IPM_{RMS}} \end{aligned} \tag{9.46}$$

We know everything in this equation except the incidental phase modulation of the LO, IPM_{RMS}. Solving for IPM_{RMS}, we find

$$IPM_{RMS} = (\Delta\phi_{RMS})\,\frac{V_{out,quiet,RMS}}{V_{out,PM,RMS}} \tag{9.47}$$

Note that we didn't need to solve equation (9.44) for k_ϕ.

9.3.6.5 SNR and IPM

If the LOs in a receiver were perfect, then the smallest voltage we would ever measure from the receiver's phase demodulator port would be 0 volts. This occurs when the signal being applied to the receiver is unmodulated (i.e., a quiet carrier).

Under quiet carrier input conditions, the receiver produces noise due to the IPM of the receiver's LOs. This is the ultimate noise floor of the demodulator. It will never emit a signal smaller than that voltage. If IPM is the limiting factor in a receiver, the ultimate SNR at the output will be

$$
\begin{aligned}
SNR_{IPM} &= \left(\frac{V_{out,PM,RMS}}{V_{out,quiet,RMS}} \right)^2 \\
&= \left(\frac{k_\phi \Delta\phi_{RMS}}{k_\phi IPM_{RMS}} \right)^2 \\
&= \left(\frac{\Delta\phi_{RMS}}{IPM_{RMS}} \right)^2 \\
&= 20 \log \left(\frac{\Delta\phi_{RMS}}{IPM_{RMS}} \right)
\end{aligned} \tag{9.48}
$$

where $\Delta\phi_{RMS}$ is the RMS phase deviation of the signal we want to receive.

Signals with small RMS phase deviations are more sensitive to IPM than signals with large RMS phase deviations.

EXAMPLE

IPM and SNR

An analog signal is broadcast using phase modulation with peak phase shifts of 0° and 180°. The *IPM* of the receiver is $3°_{RMS}$. Find the ultimate *SNR* present at the output of the phase detector.

Solution

Assuming the analog signal is a sine wave, the RMS phase shift is

$$
\begin{aligned}
\Delta\phi_{RMS} &= \frac{\Delta\phi_{Pk}}{\sqrt{2}} \\
&= \frac{180°/2}{\sqrt{2}} \\
&= 63.6 \text{ degrees}_{RMS}
\end{aligned} \tag{9.49}
$$

The ultimate SNR present at the receiver's output is

$$
\begin{aligned}
SNR_{IPM} &= 20 \log \left(\frac{\Delta\phi_{RMS}}{IPM} \right) \\
&= 20 \log \left(\frac{63.6}{3} \right) \\
&= 27 \text{ dB}
\end{aligned} \tag{9.50}
$$

9.3.7 Incidental Frequency Modulation

Incidental frequency modulation (IFM) is a measure of the effects of a receiver's LO on frequency-modulated signals. IFM applies to frequency modulation in the same way that IPM applies to phase modulation.

Frequency Demodulation An FM demodulator (or FM discriminator) produces a voltage whose instantaneous value is a direct function of the instantaneous frequency of the input signal. For example, if we perform the FM demodulation at a 21.4 MHz IF, then the discriminator produces a voltage proportional to the difference between the input signal's instantaneous frequency and 21.4 MHz.

In a receiver, the phase noise of the local oscillators causes measurement uncertainty. During the frequency conversion process, the signal acquires the phase noise of the local oscillators and so we can't accurately determine the frequency of the received signal because it is contaminated with receiver phase noise. Even a pure sine wave applied to the receiver will exhibit phase noise after being processed by the receiver.

IFM Definition IFM is defined as

$$IFM = \sqrt{\int_{fa}^{fb} f_m^2 S_\phi(f_m)\, df_m} \quad (\text{Hertz}_{RMS}) \tag{9.51}$$

For small β, we can write

$$IFM = \sqrt{2\int_{fa}^{fb} f_m^2 \mathcal{L}(f_m)\, df_m} \quad (\text{Hertz}_{RMS}) \tag{9.52}$$

where f_a and f_b represent the lower and upper frequency boundaries of the demodulated waveform, respectively.

Given the frequency spectrum of an oscillator, these two equations represent the amount of variation in the instantaneous frequency of the oscillator.

EXAMPLE

IFM

Figure 9-26 shows a local oscillator to be used in a cellular telephone receiving system. Find the IFM of this oscillator.

Solution

We will evaluate equation (9.19) from a lower frequency of 300 Hz to an upper frequency of 3000 Hz since this is the bandwidth of a common telephone channel. Equation (9.19) produces

$$\begin{aligned} IFM &= \sqrt{2\int_{fa}^{fb} f_m^2 \mathcal{L}(f_m)\, df_m} \quad (\text{Hertz}_{RMS}) \\ &= \sqrt{2\int_{300}^{3000} f_m^2 \left(10^{-9}\right) df_m} \\ &= \sqrt{\left(2 \cdot 10^{-9}\right)\left(\frac{1}{3}\right)\left(3000^3 - 300^3\right)} \\ &= 4.2\ \text{Hz}_{RMS} \end{aligned} \tag{9.53}$$

Because of the phase noise on the local oscillator, it will appear as though the input signal has an uncertainty of 4.2 Hz_{RMS}.

IFM and SNR The IFM of a receiver's LO will affect the maximum SNR of the signal leaving the receiver. Consider one FM signal that carries a printer signal at 75 baud and that deviates 75 Hz. A second signal is a digital cellular telephone signal that deviates 20 kHz. To decide if the printer is sending a binary 0 or binary 1, we must measure the frequency with an accuracy of 37.5 Hz. To decode the cellular data, we have to measure the frequency of the signal with an accuracy of 10 kHz.

In the last example, our LO had an IFM of 4.2 Hz_{RMS}, and, like IPM, IFM is a statistical quantity. The instantaneous frequency of the oscillator can be larger or smaller than 4.2 Hz. The printer signal (which deviates a mere 75 Hz) will be more sensitive to oscillator phase noise than the cellular telephone signal (which deviates 20 kHz).

Measuring IFM We measure IFM using a technique similar to the IPM measurement. We measure IFM by applying a signal generator to the input of our DUT. We first frequency modulate the signal generator at some arbitrary rate (e.g., 1 kHz) and at some at some known peak frequency deviation Δf_{pk}, which corresponds to some RMS frequency deviation Δf_{RMS}. When the receiver demodulates the FM signal from the generator, the RMS voltage from the receiver will be

$$V_{out,FM,RMS} = k_{FM}\Delta f_{RMS} \tag{9.54}$$

where

k_{FM} is a constant associated with the frequency demodulator
Δf_{RMS} is the RMS frequency deviation of the test signal

Next, we remove the frequency modulation from the signal generator and provide a quiet carrier to the receiver. If the IFM of the signal generator is much smaller than the IFM of the receiver, the output of the frequency demodulator is

$$V_{out,quiet,RMS} = k_{FM} IFM_{RMS} \tag{9.55}$$

The ratio of these two voltages is

$$\frac{V_{out,FM,RMS}}{V_{out,quiet,RMS}} = \frac{k_{FM}\Delta f_{RMS}}{k_{FM}IFM_{RMS}} = \frac{\Delta f_{RMS}}{IFM_{RMS}} \tag{9.56}$$

Solving for IFM_{RMS}, we find

$$IFM_{RMS} = (\Delta f_{RMS})\frac{V_{out,quiet,RMS}}{V_{out,FM,RMS}} \tag{9.57}$$

IFM-Limited SNR If the receiver LOs were perfect, a quiet carrier would produce 0 volts at the demodulator's output. We've seen, however, that under quiet carrier input conditions the receiver produces a noise voltage due to the IFM of the receiver's LOs. We can think of this as the ultimate noise floor of the demodulator. If IFM is the limiting factor in a receiver, the ultimate SNR present at the output of a receiver will be

$$\begin{aligned} SNR_{IFM} &= \left(\frac{V_{out,FM,RMS}}{V_{out,quiet,RMS}}\right)^2 \\ &= \left(\frac{\Delta f_{RMS}}{IFM_{RMS}}\right)^2 \\ &= 20\log\left(\frac{\Delta f_{RMS}}{IFM_{RMS}}\right) \end{aligned} \tag{9.58}$$

where Δf_{RMS} is the RMS frequency deviation of the signal of interest.

Signals with small RMS frequency deviations will exhibit a smaller IFM-limited SNR than signals with large RMS frequency deviations.

EXAMPLE

IFM and SNR

A cellular telephone receiver processes an FM signal with a 20 kHz bandwidth. The signal is a voice-grade channel covering 300 to 3000 Hz. The IFM of the receiver is 5 Hz_{RMS}. Find the ultimate *SNR* present at the output of the receiver.

Solution

For this system, the RMS frequency shift is

$$\begin{aligned} \Delta f_{RMS} &= \frac{\Delta f_{PkPk}/2}{\sqrt{2}} \\ &= \frac{20k/2}{\sqrt{2}} \\ &= 7.1\ kHz_{RMS} \end{aligned} \tag{9.59}$$

The ultimate SNR present at the receiver's output is

$$\begin{aligned} SNR &= 20\log\left(\frac{\Delta f_{RMS}}{IFM_{RMS}}\right) \\ &= 20\log\left(\frac{7.1\ \text{kHz}_{RMS}}{5\ \text{Hz}_{RMS}}\right) \\ &= 32\ \text{dB} \end{aligned} \tag{9.60}$$

9.3.7.1 Comparison of IPM and IFM

Since phase and frequency are closely related, we expect the IFM and IPM of an oscillator to be related. The equations describing IPM and IFM are

$$IPM_{RMS} = \sqrt{\int_{fa}^{fb} S_{\phi}(f_m)\, df_m} \quad (\text{radians}_{RMS}) \tag{9.61}$$

and

$$IFM_{RMS} = \sqrt{\int_{fa}^{fb} f_m^2 S_{\phi}(f_m)\, df_m} \quad (\text{Hertz}_{RMS}) \tag{9.62}$$

The equations are very similar except that the IFM equation contains an f_m^2 term. In practice, we observe that if we apply white noise (which has a flat frequency spectrum) the power in the output spectrum increases parabolically with frequency. If we apply white noise to a phase demodulator, the output spectrum is flat, an observation that is validated by the equations.

9.4 | EFFECTS OF OSCILLATOR SPURIOUS COMPONENTS

Spurious signals are unwanted coherent signals generated by an oscillator that are not at the desired frequency of f_0. There are two categories of spurious signals: harmonically related spurious signals and nonharmonically related.

9.4.1 Harmonically Related Spurious Signals

Harmonically related spurious (or *harmonic spurious*) signals are coherent signals present at the output of an oscillator whose frequencies are integer multiples of the desired output signal (e.g., $2f_0$, $3f_0$). Our analysis of mixer spurious produces has assumed these signals will be present in our conversion process. We were aware of these harmonics and have considered their presence.

When the oscillator drives a transmitter, regulatory bodies are concerned with the power present in the harmonic components of the transmitter. Federal Communication Commission (FCC) part 15 requirements dictate strict guidelines for harmonic suppression,

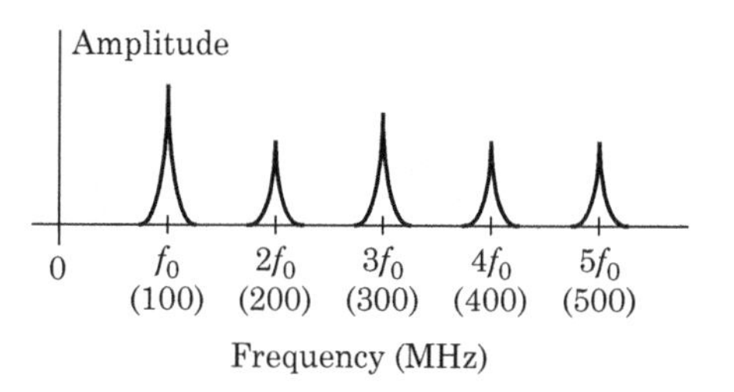

FIGURE 9-28 ■ Harmonically related spurious signals present at the output of an oscillator.

and we must often filter the output signal to meet these requirements. Figure 9-28 shows the harmonically related spurious of an oscillator. Each of the harmonic components will exhibit the multiplied phase noise of the fundamental frequency [via equation (9.31)].

9.4.2 Nonharmonically Related Spurious Signals

Nonharmonically related spurious (or *nonharmonic spurious*) signals are coherent signals present at the output of an oscillator whose frequencies are not integer multiples of the desired output signal. These signals are usually caused by realization effects within the oscillator. Nonharmonically related spurious signals also include signals that are subharmonics of the desired output signal (e.g., at $f_0/2$, $f_0/3$).

Figure 9-29(a) shows the close-in spurious outputs of an oscillator. *Close-in* is a rather arbitrary term meaning that the spurious signal falls close to the fundamental component (usually within 1% of the oscillator's fundamental frequency). The close-in spurious signals in Figure 9-29(a) are typical of a phase-locked loop frequency synthesizer.

Figure 9-29(b) shows the *far-out* spurious products that may also be present at the output of an oscillator. Far-out signals are coherent signals whose frequencies differ from the fundamental output by more than 1% and include subharmonics of the fundamental output frequency and frequencies whose relationship to the desired output frequency is not apparent.

9.4.3 Effects of Oscillator Spurious Products

Harmonically related spurious products are manageable because they may be easily filtered. Moreover, these products are likely to be generated in any system due to the system's inherent nonlinearities. We expect harmonics to be present in our systems so we plan for

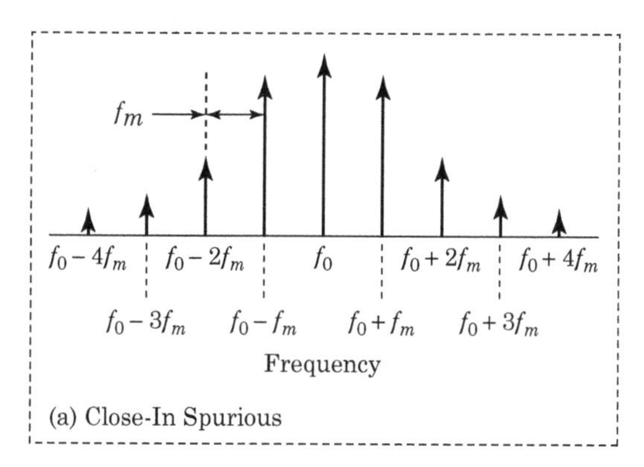

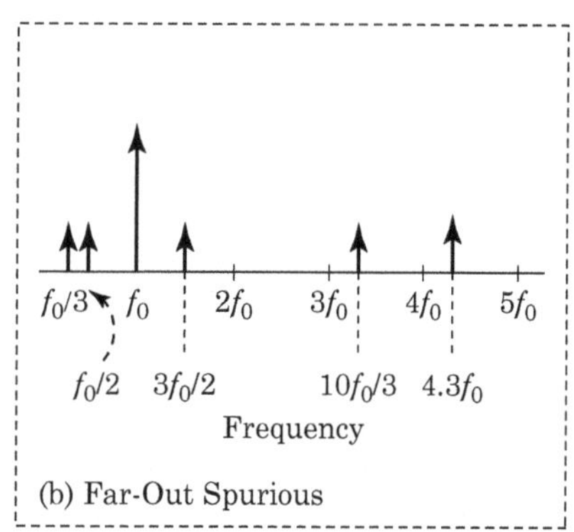

FIGURE 9-29 ■ (a) Close-in spurious signals present at an oscillator's output. (b) Far-out spurious signals present at an oscillator's output.

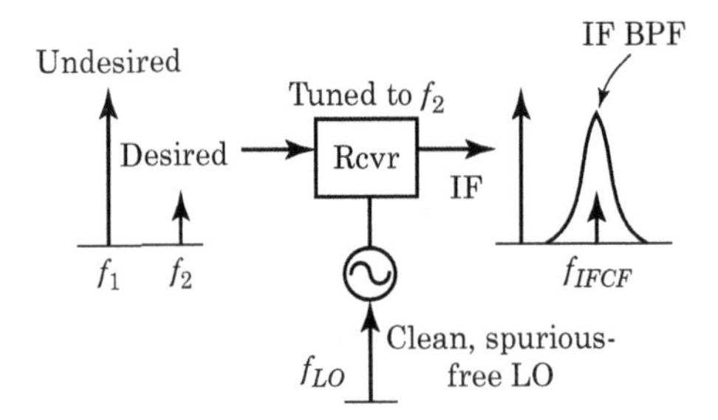

FIGURE 9-30 ■ The conversion process using a clean, spurious-free LO. The strong, undesired signal is at frequency f_1, whereas the weak desired signal is at frequency f_2.

them. Nonharmonic spurious outputs are often more problematic because we don't plan for the specific frequencies at which they appear. Designers can spend a great deal of time tracking these spurious signals to their sources.

Figure 9-30 shows a receiving system with two input signals. The strong, undesired signal is at frequency f_1, while the weak, desired signal is at frequency f_2. The receiver is tuned to f_2 and possesses a clean, spurious-free LO. The action of the receiver is to move the desired signal from f_2 to the receiver's IF center frequency. We then use the IF filter to pass only the desired signal. The IF filter also attenuates any other signals that fall outside of the IF band-pass filter's (BPF's) passband.

Figure 9-31 shows the same situation if the receiver's LO contained spurious signals at unfortunate locations. Signals that are converted from one frequency to another by a mixer acquire the characteristics of the LO. If the LO has phase noise, the output signal will have phase noise. If the LO has spurious signals present in its output, then every signal converted by the LO will contain spurious signals. If the LO exhibits spurious products that are Δf away from the desired output, the signals present at f_1 and f_2 both acquire these spurious products as they pass through the mixer. The spurious products, which are now present on the undesired signal, can mask the desired signal.

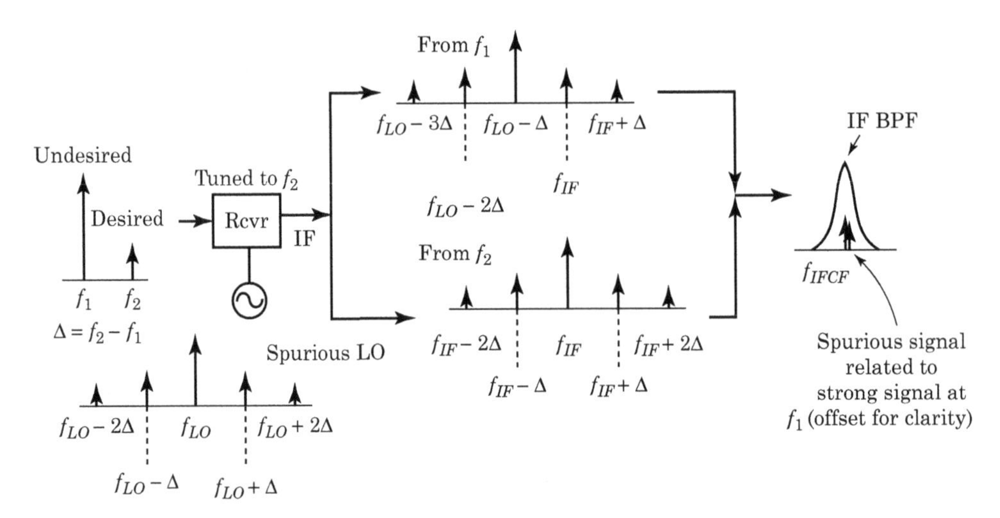

FIGURE 9-31 ■ The conversion process using an LO with close-in spurious components. The strong, undesired signal at f_1 combines with the LO's spurious components to mask the weak, desired signal at f_1.

As another example, if a receiver's LO contains a spurious product whose frequency equals the IF, then signal may bleed through from the mixer's LO port to its IF port. If that leakage occurs, every desired signal successfully converted to the IF by the mixer must compete with this spurious product. In many cases the spurious signal at the receiver's IF will seriously impede the demodulation of the desired signals. The spurious LO has made our receiver useless in demodulating weak signals.

EXAMPLE

The FM Broadcast Band and Oscillator Spurious

Figure 9-32 shows the spectral diagram associated with the U.S. commercial broadcast FM radio. The channel spacing (i.e., the distance between adjacent stations) is 200 kHz.

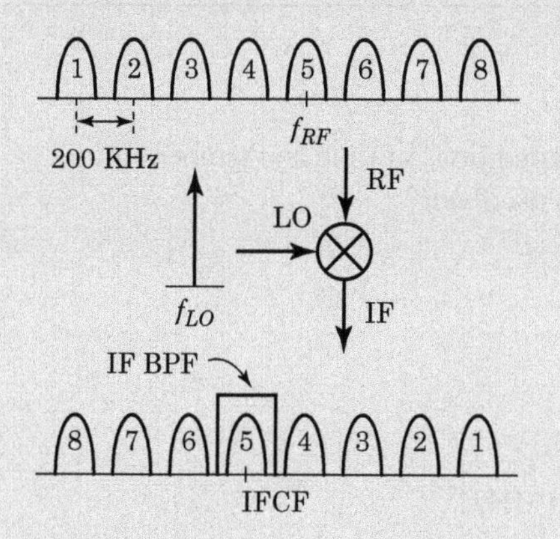

FIGURE 9-32 ■ The conversion process in a channelized system. We convert the desired RF signal to the center frequency of the IF filter.

The output of a phase-locked loop frequency synthesizer often contains spurious signals that are multiples of the channel spacing. As Figure 9-33 shows, the LO's spurious output products cause signals in adjacent channels to convert to the same IF. Now, both the desired and undesired signals must compete for the demodulator's attention.

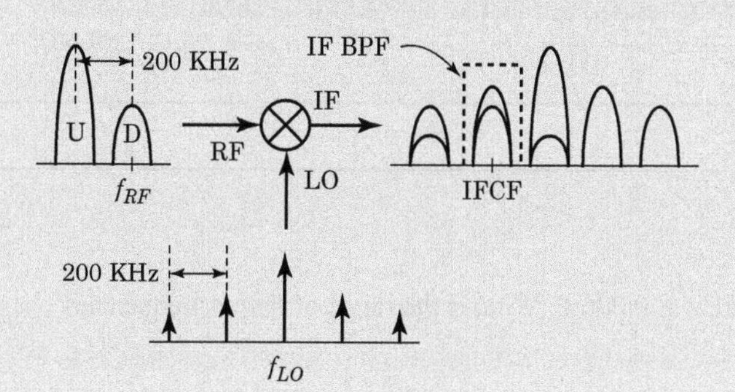

FIGURE 9-33 ■ The conversion process in a channelized system using an LO containing spurious signals. The undesired signal (annotated as U) combines with the LO spurious components to make the desired signals (annotated as D).

9.5 | FREQUENCY ACCURACY

Any signal that is converted from one frequency to another by an LO inherits the faults of that LO. If the LO frequency drifts with time and temperature, then every signal converted by that LO will drift equivalently with time and temperature. An oscillator may not produce its design frequency when we first apply power. Over time, the oscillator's frequency will change slightly depending on the temperature, power supply voltage, and component aging. Frequency changes that occur over periods longer than 1 second are called oscillator drift.

9.5.1 Quantifying Drift

Oscillator drift is often specified in ppm of the center frequency. A typical specification might read "10 MHz ± 3 ppm." If the drift is specified as "$f_0 \pm \Delta$ ppm," the frequency range of the oscillator is

$$f_{Range} = f_0 \pm \left[\frac{\Delta \text{ in ppm}}{10^6} \cdot f_0\right] \tag{9.63}$$

EXAMPLE

Oscillator Drift in ppm

An oscillator is specified to be 100 MHz ± 2.5 ppm over some conditions of time and temperature. What is the range of output frequencies we can expect from this oscillator?

Solution

Using equation (9.30), we write

$$\begin{aligned} f_{Range} &= f_0 \pm \left[\frac{\Delta \text{ in ppm}}{10^6} \cdot f_0\right] \\ &= 100\text{ MHz} \pm \left[\frac{2.5}{10^6} \cdot 100\text{ MHz}\right] \\ &= 100\text{ MHz} \pm 250\text{ Hz} \end{aligned} \tag{9.64}$$

Another common drift specification is percent of center frequency. A specification might read "40 MHz ± 0.001%." If the drift is specified as "$f_0 \pm \Delta\%$," then the oscillator's frequency will fall in the range of

$$f_{Range} = f_0 \pm \left[\frac{\Delta \text{ in }\%}{100} \cdot f_0\right] \tag{9.65}$$

EXAMPLE

Oscillator Drift in %

An oscillator has a drift specification of 25 MHz ± 0.006%. What is the range of output frequencies we can expect from this oscillator?

Solution

Equation (9.32) gives us

$$\begin{aligned} f_{Range} &= f_0 \pm \left[\frac{\Delta \text{ in }\%}{100} \cdot f_0\right] \\ &= 25\text{ MHz} \pm \left[\frac{0.006}{100} \cdot 25\text{ MHz}\right] \\ &= 25\text{ MHz} \pm 1{,}500\text{ Hz} \end{aligned} \tag{9.66}$$

Oscillator drift in ppm and percent are related by

$$(\Delta \text{ in } \%) = \frac{(\Delta \text{ in ppm})}{10{,}000} \tag{9.67}$$

9.5.2 Nomenclature

9.5.2.1 Initial Accuracy

Initial accuracy is defined as the difference between the oscillator output frequency and the specified frequency at 25°C at the time the oscillator leaves the manufacturer. If the oscillator includes a tuning feature, then the initial accuracy is specified as a range.

9.5.2.2 Temperature Stability

Oscillator frequencies vary as the temperature varies. The specification may read "±5 ppm over a −30°C to +70°C temperature range." The oscillator will be within 5 ppm of its initial frequency setting over the specified temperature range. The specification does not indicate that the oscillator's frequency will be a linear function of temperature.

Figure 9-34 shows the frequency versus temperature curves for several 1 MHz oscillators. The stability specifications on the oscillators are "X ppm over a −30°C to +70°C temperature range," where X is ±10 ppm, ±5 ppm, ±2 ppm, and ±1 ppm. The curves are fifth-order polynomials. At −30°C and +70°C, the oscillators are at the extreme ends of their tolerances. They also touch their extremes at −5°C and +45°C.

Finally, practical oscillators exhibit hysteresis. Their frequency–temperature curves don't trace the same path when the oscillator heats up as when it cools down. Hysteresis is a fundamental limit on oscillator temperature stability.

9.5.2.3 Aging (Long-Term stability)

Over longer time periods (days, months, years), the frequency of an oscillator will change even if all the external parameters such as temperature and power supply voltage are held

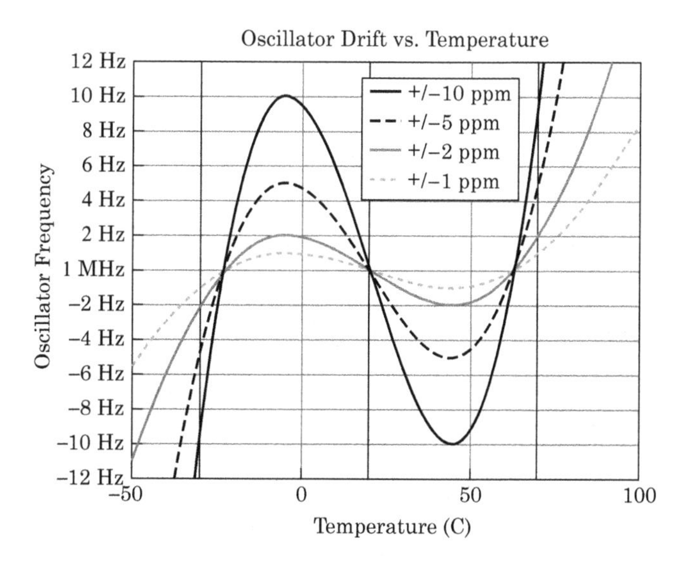

FIGURE 9-34 ■ Frequency versus temperature curves for several AT-cut crystal oscillators. The frequency is not a linear function of temperature. Measured curves may not be as "clean" as these calculated curves and may also exhibit hysteresis.

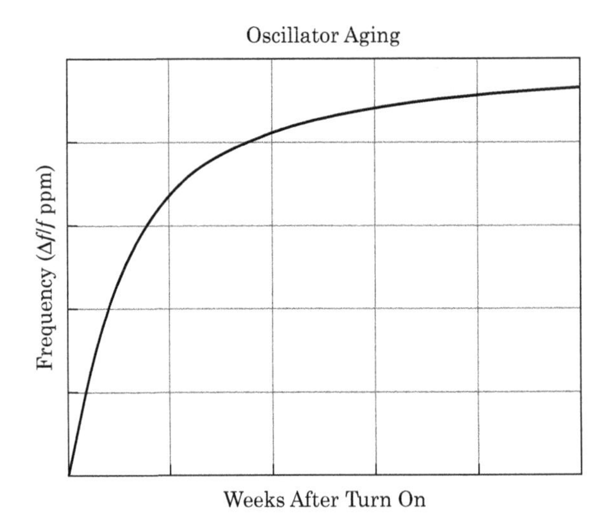

FIGURE 9-35 ■ Long-term oscillator aging. The frequency changes quickly at first and then more slowly as the stresses in the crystal relax.

constant. This is *oscillator aging*. Figure 9-35 shows a typical oscillator aging curve. When the oscillator is initially turned on, the crystal ages quickly, and stability improves rapidly with time. Most oscillators achieve their lowest aging rates within several months after turn on.

One cause of drift is the stresses generated in the crystal by the machining process when the crystal was made. These stresses affect the resonant frequency of the crystal and relax over time.

EXAMPLE

Cellular Telephones and Drift

Figure 9-36 shows a conversion scheme for a cellular telephone receiver. The RF signal is 884.460 MHz, and the oscillators have the accuracy specifications shown. Find the range of possible output frequencies at the final IF.

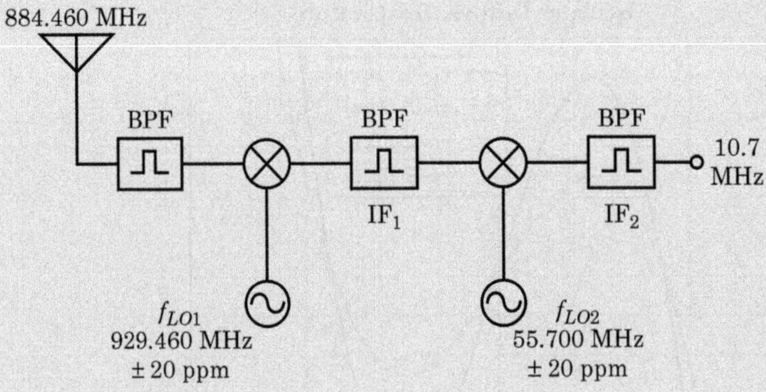

FIGURE 9-36 ■ A cellular telephone conversion scheme, with oscillator drift specifications.

Solution

Looking at the range of LO_1

$$\begin{aligned} f_{LO1} &= 929.460 \text{ MHz} \pm 20 \text{ ppm} \\ &= 929.460 \text{ MHz} \pm 18{,}589.2 \text{ Hz} \\ &= 929.441\,410 \text{ MHz to } 929.478\,589 \text{ MHz} \\ &= f_{LO1,L} \text{ to } f_{LO1,H} \end{aligned} \tag{9.68}$$

The first conversion equation is $f_{IF} = f_{LO} - f_{RF}$, so the RF signal will fall in the range of $f_{IF1,L}$ to $f_{IF1,H}$ where

$$\begin{aligned} f_{IF1,L} &= f_{LO1,L} - f_{RF} \\ &= 44.981\,410 \text{ MHz} \\ f_{IF1,H} &= f_{LO1,H} - f_{RF} \\ &= 45.018\,589 \text{ MHz} \end{aligned} \tag{9.69}$$

Next, we calculate the possible ranges of LO_2:

$$\begin{aligned} f_{LO2} &= 55.700 \text{ MHz} \pm 20 \text{ ppm} \\ &= 55.700 \text{ MHz} \pm 1114 \text{ Hz} \\ &= 55.698\,886 \text{ MHz to } 55.701\,114 \text{ MHz} \\ &= f_{LO2,L} \text{ to } f_{LO2,H} \end{aligned} \tag{9.70}$$

The second conversion equation is also $f_{IF} = f_{LO} - f_{RF}$, so the RF signal will fall in the range of $f_{IF2,L}$ to $f_{IF2,H}$, where

$$\begin{aligned} f_{IF2,L} &= f_{LO2,L} - f_{IF1,H} \\ &= 10.680\,297 \text{ MHz} \\ f_{IF2,H} &= f_{LO2,H} - f_{IF1,L} \\ &= 10.719\,703 \text{ MHz} \end{aligned} \tag{9.71}$$

The RF signal might be as much as ±19,703 Hz from the nominal 10.7 MHz IF. We now that the width of a cellular telephone signal is about 20 kHz, and we would like to ensure that the entire 20 kHz signal falls inside the final IF bandwidth.

Figure 9-37 shows the situation graphically. We've calculated that the center frequency of our signal will fall between 10.680 297 and 10.719 703 MHz. Taking the 20 kHz signal bandwidth into account means that the final IF bandwidth must accept 10.670 297 to 10.729 703 MHz. Oscillator drift has increased our required IF bandwidth from approximately 20 kHz to 40 kHz. Since our signals are on 30 kHz centers, we may end up with two signals in the IF bandwidth. Poor oscillator stability has ruined a fine bit of engineering.

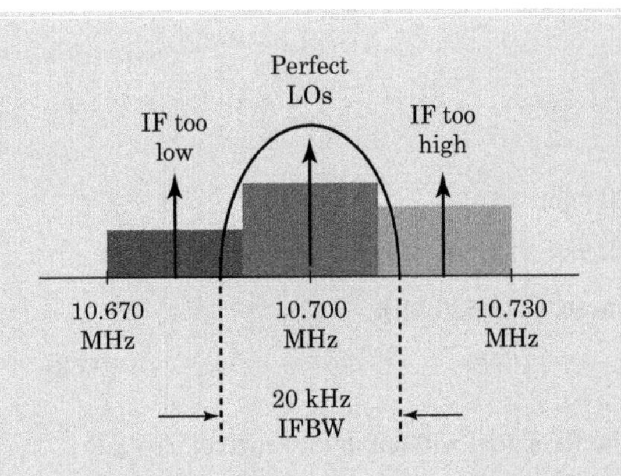

FIGURE 9-37 ■ The situation at the 10.7 MHz IF filter of Figure 9-36. The LO frequency accuracy can cause the desired signal to fall outside of the IF filter.

Typical cellular telephone reference oscillators must meet a ±2.5 ppm over a –20°C to +60°C temperature range.

9.5.3 Frequency Multiplication and Stability

If we use a multiplier to increase the frequency of an oscillator that drifts, the amount of drift (in Hz, not ppm) increases directly with n, the multiplier. The drift in ppm stays constant. For example, if we multiply a 1 MHz±1 Hz reference oscillator by 100, the output signal will be at 100 MHz ± 100 Hz. In both cases, however, the accuracy of the oscillator is still 1 ppm.

Similarly, we can see that frequency division will decrease the amount of oscillator drift (in Hz, not ppm) by the division factor n.

EXAMPLE

Frequency Multiplication

Three oscillators have stability specifications of ±4 ppm. The center frequency of the first oscillator is 1 MHz, the center frequency of the second oscillator is 250 MHz, and the center frequency of the third oscillator is 1 GHz. Find the drift of these two oscillators in Hz.

Solution

For the 1 MHz oscillator, we can write

$$1 \text{ MHz} \pm 4 \text{ ppm} = 1 \text{ MHz} \pm 4 \text{ Hz} \tag{9.72}$$

For the 250 MHz oscillator

$$250 \text{ MHz} \pm 4 \text{ ppm} = 250 \text{ MHz} \pm 1000 \text{ Hz} \tag{9.73}$$

For the 1 GHz oscillator

$$1 \text{ GHz} \pm 4 \text{ ppm} = 1 \text{ GHz} \pm 4000 \text{ Hz} \tag{9.74}$$

For a given value of temperature stability, higher-frequency oscillators exhibit greater frequency drift (in Hz) than lower-frequency oscillators. However, if we multiply or divide the frequency of an oscillator, the amount of drift expressed in ppm or percentage doesn't change.

EXAMPLE

Stability in ppm, Hz, and Percentage

A 1 MHz ± 0.5 ppm reference oscillator is multiplied by 1/5, 25, and 50. Express the drift of the input and output signals in terms of Hz, ppm, and percentage.

Solution

We'll use equations (9.30), (9.32), and (9.34).

Input oscillator: The frequency of the unmultiplied oscillator falls within the ranges of

$$\begin{aligned} 1\text{ MHz} \pm \left[\frac{0.5}{10^6} \cdot 1\text{ MHz}\right] &= 1\text{ MHz} \pm 0.5\text{ Hz} \\ &= 1\text{ MHz} \pm 0.00005\% \end{aligned} \tag{9.75}$$

$1/5\times$: The output frequency range of the divider is

$$\frac{1}{5}(1\text{ MHz} \pm 0.5\text{ Hz}) = 200\text{ kHz} \pm 0.1\text{ Hz} \tag{9.76}$$

Equation (9.30) tells us

$$\begin{aligned} 0.1\text{ Hz} &= \frac{(\Delta\text{ in ppm})}{10^6} \cdot 200\text{ kHz} \\ (\Delta\text{ in ppm}) &= 0.1 \cdot \frac{10^6}{200 \cdot 10^3} \\ &= 0.5\text{ ppm} \end{aligned} \tag{9.77}$$

$25\times$: The output frequency range of the $25\times$ multiplier is

$$25\,(1\text{ MHz} \pm 0.5\text{ Hz}) = 25\text{ MHz} \pm 12.5\text{ Hz} \tag{9.78}$$

Equation (9.30) tells us

$$\begin{aligned} 12.5\text{ Hz} &= \frac{(\Delta\text{ in ppm})}{10^6} \cdot 25\text{ MHz} \\ \Rightarrow (\Delta\text{ in ppm}) &= 12.5 \cdot \frac{10^6}{25 \cdot 10^6} \\ &= 0.5\text{ ppm} \end{aligned} \tag{9.79}$$

$50\times$: The output frequency range is

$$50\,(1\text{ MHz} \pm 0.5\text{ Hz}) = 50\text{ MHz} \pm 25\text{ Hz} \tag{9.80}$$

Equation (9.30) tells us

$$\begin{aligned} 25\text{ Hz} &= \frac{(\Delta\text{ in ppm})}{10^6} \cdot 50\text{ MHz} \\ \Rightarrow (\Delta\text{ in ppm}) &= 25\frac{10^6}{50 \cdot 10^6} \\ &= 0.5\text{ ppm} \end{aligned} \tag{9.81}$$

Frequency multiplication and division both change the oscillator's deviation from its design value when the deviation is expressed in Hz. However, the deviation in ppm or percent is fixed regardless of the multiplier or divider.

TABLE 9-1 ■ Timing inaccuracies associated with oscillator aging

Aging/Day	1 day	1 week	1 month	1 year
10^{-7}	4.30 ms	210 msec	3.90 sec	580 sec
10^{-8}	430 μsec	21.0 msec	390 msec	58.0 sec
10^{-9}	43.0 μsec	2.10 msec	39.0 msec	5.80 sec
10^{-10}	4.30 μsec	210 μsec	3.90 msec	580 msec

TABLE 9-2 ■ Timing inaccuracies associated with a fixed frequency offset

Offset	1 day	1 week	1 month	1 year
10^{-5}	860 msec	6.00 sec	26.0 sec	320 sec
10^{-6}	86.0 msec	600 msec	2.60 sec	32.0 sec
10^{-7}	8.60 msec	60.0 msec	260 msec	3.20 sec
10^{-8}	860 μsec	6.00 msec	26.0 msec	320 msec
10^{-9}	86.0 μsec	600 μsec	2.60 msec	32.0 msec

9.5.4 Timing Accuracy

We often keep track of elapsed time using nonideal oscillators. Both aging and initial offset cause errors in the perceived time lapse. Table 9-1 and Table 9-2 summarize those errors. Table 9-1 assumes that we start exactly on frequency and then our reference oscillator ages.

Table 9-2 assumes our reference oscillator is off by a fixed amount.

9.6 OTHER CONSIDERATIONS

9.6.1 Tuning Speed

Tuning speed is the time required for a receiver to tune from one frequency to another. After receiving a tune command, how long is it before the receiver is sitting at the new frequency and demodulating signals?

Tuning speed specifications vary. It can be defined as the time it takes the frequency synthesizer to tune to a new frequency, the time it takes the converted waveform to settle at the IF port of the receiver, or when the synthesizer are within 1% of their final frequency.

Tuning speed is realization dependent. Some configurations tune quickly; others tune slowly. The tuning speed of a system can also be tied into other specifications, such as LO phase noise.

9.6.2 Automatic Frequency Control

Due to oscillator drift with temperature, aging, or other factors, the desired signal may not be exactly centered in a receiver's IF filter. Each LO in the receiver gets to add its inaccuracies to the signal as the frequency conversion process proceeds.

Automatic frequency control (AFC) was developed to center a signal in the IF filter. As long as the desired signal is the largest signal in the IF passband, the receiver retunes itself slightly to compensate any frequency drift. In some systems, the transmitter includes training sequences or reference symbols used by the receiver for AFC purposes. The user initially disables the AFC and tunes the receiver to the desired signal. The operator then enables the AFC and the receiver centers the operator's signal in the IF passband. Provided the desired signal (or the receiver) drifts only slowly with time, the receiver will keep the signal center.

9.7 | OSCILLATOR REALIZATIONS

Receivers require locally generated oscillators to translate communications signals from frequency to frequency. To make intelligent design decisions, we must know a little about how frequency synthesizers are built.

9.7.1 Phase-Locked Loops

The PLL frequency synthesizer is one of the most common in use today. Several companies manufacture integrated circuits designed especially for these synthesizers. Successfully designing PLL frequency synthesizers requires an understanding of control theory.

Figure 9-38 shows a block diagram for a PLL frequency synthesizer. The PLL is a control system used to impress the good characteristics of a crystal oscillator (frequency stability and low phase noise) onto a VCO, which is characterized by poor stability and poor phase noise. The result is a digitally tuned oscillator exhibiting good phase noise and good stability.

Our conversion scheme will dictate the requirements for the frequency synthesizers. Normally, the requirements state that the synthesizer must tune from some lower frequency to some upper frequency with some resolution or step size. For example, a PLL with a 20 kHz step size can tune in frequency increments of 20 kHz, but no finer.

However, the ultimate step size of the receiver might be finer than the step size of the PLL synthesizer. Several frequency synthesizers, realized with different techniques, are often present in a receiver. For example, it is common to have a PLL and a numerically controlled oscillator (NCO) together in the same system.

9.7.1.1 PLL Nomenclature

Locked versus Unlocked A PLL synthesizer will be in one of two states: either locked or unlocked. When the PLL is locked, we say that the output has settled to the desired frequency and has been there for some time. The phase noise and accuracy of the reference are being successfully conveyed to the VCO. When locked, the PLL is in a steady-state condition.

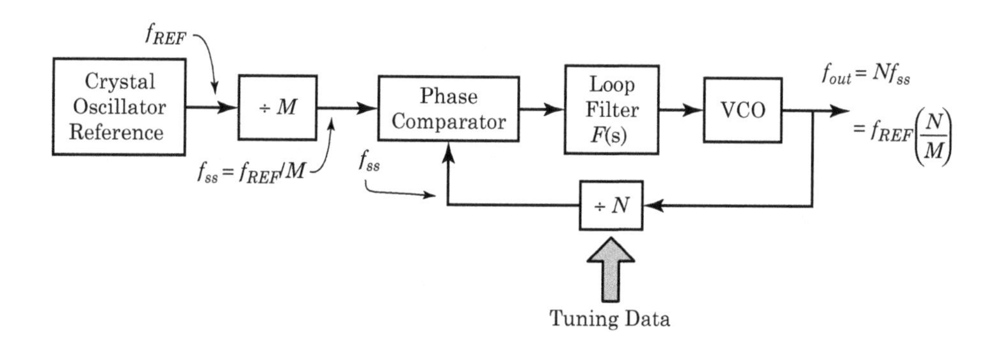

FIGURE 9-38 Block diagram for a PLL frequency synthesizer.

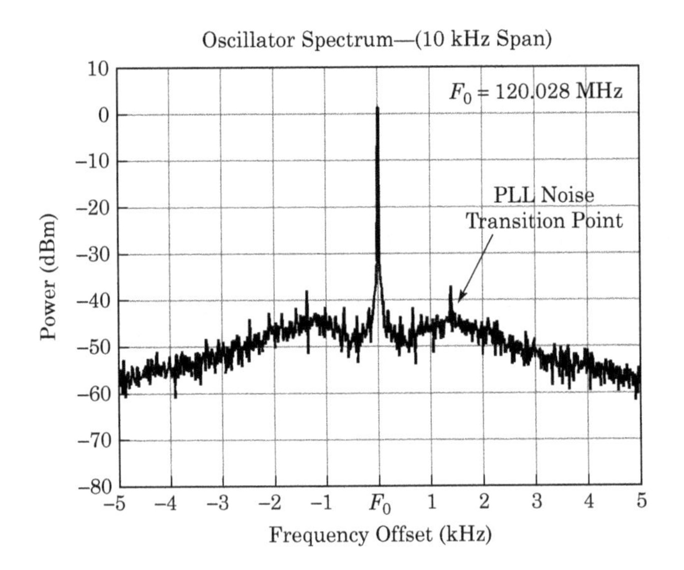

FIGURE 9-39 ■ The noise transition point for a PLL oscillator. The transition point is related to the PLL's loop bandwidth.

The PLL is in its unlocked state for some period after it has been commanded to tune to a new frequency. The output has not settled to the desired frequency, and the entire system is in a transient condition. The VCO is not being disciplined with the phase noise and accuracy of the reference.

Loop Bandwidth A PLL synthesizer is a control system. The action of a control system is to compare the output of a system with some input and then act to change the output to match the characteristics of the input. In a PLL frequency synthesizer, the input is a stable, low-phase noise reference oscillator. The output is a noisy, unstable VCO.

Two distinct regions characterized the phase noise of a PLL synthesizer: (1) phase noise close to the carrier; and (2) phase noise that is far from the carrier. Figure 9-39 shows the transition nicely. The control action of the PLL controls the noise close to the carrier to resemble that of the crystal oscillator. Far from the carrier, the phase noise resembles the uncontrolled phase noise of the VCO.

The PLL *loop bandwidth* is approximately equal to the offset from the carrier where the phase noise contributions from the VCO and reference oscillator are approximately equal. The loop controls the output over a limited range of frequencies equal to the loop bandwidth of the PLL.

9.7.1.2 Components of a PLL Frequency Synthesizer

Figure 9-38 shows a block diagram for a PLL frequency synthesizer. The functions of each block are explained in the following sections.

Reference Oscillator The reference oscillator is the most stable oscillator in the system. The output of the PLL is locked to this oscillator, so it will determine the stability and phase noise performance of the system. The crystal reference oscillator frequency often lies in the 1 to 10 MHz range; 13 and 26 MHz are common for cellular handsets. Since this oscillator is the system reference, the accuracy and stability of this oscillator should be excellent. If the entire receiver is to have a frequency accuracy of ±2 ppm, the stability of the reference oscillator must be better than ±2 ppm.

The PLL is a control system designed to impress the accuracy and phase noise of the reference onto the VCO. Fortunately, the phase noise of common reference oscillators is very good.

If we ignore the components of the PLL and consider only the input and output ports, we can see that the PLL is essentially a frequency multiplier. As such, the phase noise of the output signal is going to be at least 20log(n) times worse than the phase noise of the input. This is a lower limit. In garden-variety PLL synthesizers, the phase noise is typically much worse.

÷M Crystal oscillators operate in the low MHz range. However, inside the PLL, we need a signal whose frequency equals the step size of the receiver. The ÷M counter divides the reference oscillator's output frequency down to the step size frequency. For example, a synthesizer with a 1 MHz reference oscillator and a 20 kHz step size would require a ÷M value of

$$\frac{1\text{ MHz}}{20\text{ kHz}} = 50 \tag{9.82}$$

We would need a counter with at least 6 bits to accomplish this division. A $\div M$ counter needs at least

$$\log_2(M) \tag{9.83}$$

bits so we must choose a PLL synthesizer integrated circuit (IC) with enough capacity to perform the required frequency division. Usually, the $\div M$ counter is fixed once at power up and does not change.

Phase Comparator The phase comparator is one of the most subtle parts of the synthesizer. When the PLL is locked, its function is to compare the phase of the divided-down reference oscillator (at $f_{SS} = f_{REF}/M$) with the phase of the divided-down output frequency (at f_{VCO}/N). The phase comparator's output contains information regarding the phase noise and stability of the VCO as it compares with the reference oscillator. If the VCO's phase (or frequency) is not what it should be, the phase comparator feeds the information to the rest of the loop, and the loop acts to correct the problem. From a control system perspective, the output of the phase comparator is an error signal and the action of a control system is to drive the error signal to zero.

Frequency Acquisition The phase comparator also performs frequency acquisition. When the synthesizer tunes to a new frequency, we write new data to the $\div N$ counter, and the PLL enters its unlocked state. Initially, the VCO is still at the old frequency. Since the two signals we are now applying to the phase comparator (i.e., $f_{SS} = f_{REF}/M$ and f_{VCO}/N) are now different frequencies, a measure of the phase difference between them is meaningless. In this situation, the data from the phase comparator will be high enough in frequency that they won't pass through the PLL's loop filter. The feedback path is effectively broken, and the VCO will simply drift.

We need a method of gross frequency acquisition. We must tell the VCO to either tune up or tune down in frequency. The digital phase detectors present in modern PLL frequency synthesizer ICs contain an automatic frequency acquisition feature. When the difference between the input frequencies is large, they automatically switch to a gross frequency acquisition mode to steer the VCO frequency up or down. These digital phase detectors are often referred to as phase/frequency detectors.

The digital phase comparators are a major source of the close-in spurious products on the frequency synthesizer's output signal. These sampling sidebands are discrete signals present on the output of the PLL synthesizer. They fall at

$$f_O \pm n f_{SS} \tag{9.84}$$

where

f_O is the desired output frequency of the synthesizer
f_{SS} is the step size of the PLL synthesizer
$n = 1, 2, 3, \ldots$

The phase/frequency comparators produce digital waveforms that are filtered to provide the control voltage for the VCO. Some of the energy from the digital waveform passes through the synthesizer's loop filter, and FM modulates the VCO, producing the sampling sidebands.

Figure 9-4 shows the sampling sidebands of a PLL frequency synthesizer. This oscillator has a tuning step of 50 kHz.

Loop Filter The loop filter is the primary element controlled by the system designer. The loop filter determines the tuning speed, the settling time, the spurious suppression, and the phase noise performance of the synthesizer. Unfortunately, not all of these parameters are independently adjustable.

The design of the loop filter is not difficult for garden-variety applications. The filter is simply a LPF with carefully controlled gain and phase characteristics, but the design involves many trade-offs. Przedpelski (May 1978, September 1978a, September 1978b, March 1981) provides excellent drafts on loop filter design.

Voltage-Controlled Oscillator VCOs are tunable oscillators whose frequency is controlled by an applied voltage. The voltage-controlled oscillator determines the tuning range of the synthesizer. It also has a hand in setting the phase noise and spurious characteristics of the output.

9.7.1.3 Tuning Range

In general, a VCO that tunes over a wide frequency range has higher phase noise than a narrow-range VCO. To explain this effect, we first define a VCO characteristic. The VCO *tuning constant* or *gain* is simply the VCO's change in frequency divided by the corresponding change in input voltage, or

$$k_{VCO} = \frac{\Delta f_{VCO}}{\Delta V_{VCO}} \tag{9.85}$$

EXAMPLE

VCO Tuning Constant or Gain

If the frequency of a VCO changes from 100 MHz to 200 MHz as the input voltage changes from 10 volts to 30 volts, find the VCO gain.

Solution

Using equation (9.4), we find

$$\begin{aligned} k_{VCO} &= \frac{\Delta f_{VCO}}{\Delta V_{VCO}} \\ &= \frac{200 - 100}{30 - 10} \\ &= 5 \frac{\text{MHz}}{\text{Volt}} \end{aligned} \tag{9.86}$$

Let's assume we have two VCOs. Both require an input range of 5 to 20 volts to tune over their entire frequency range. The first oscillator, which tunes over a 10 MHz range, has a tuning constant of

$$\begin{aligned} k_{VCO1} &= \frac{\Delta f_{VCO}}{\Delta V_{VCO}} \\ &= \frac{10\text{ MHz}}{15\text{ Volts}} \\ &= 0.67 \frac{\text{MHz}}{\text{Volt}} \end{aligned} \tag{9.87}$$

The second oscillator, with the 100 MHz tuning range, has a tuning constant of

$$\begin{aligned} k_{VCO2} &= \frac{\Delta f_{VCO}}{\Delta V_{VCO}} \\ &= \frac{100\text{ MHz}}{15\text{ Volts}} \\ &= 6.67 \frac{\text{MHz}}{\text{Volt}} \end{aligned} \tag{9.88}$$

Imagine that there is a small amount of noise present on the VCO's input line. This noise can be thermal noise, or it can be an unwanted, discrete signal (e.g., the filtered digital waveform from the phase/frequency detector). For the sake of example, let us assume the undesired input voltage has an RMS value of 10 μV_{RMS}. The 10 μV_{RMS} of noise will cause a frequency deviation of

$$\begin{aligned} f_{dev1} &= (10\mu\text{Volt}_{RMS})\, k_{VCO1} \\ &= (10\mu\text{Volt}_{RMS}) \left(0.67 \frac{\text{MHz}}{\text{Volt}}\right) \\ &= 6.7\text{ Hz}_{RMS} \end{aligned} \tag{9.89}$$

on the narrow tuning VCO. That same 10 μV_{RMS} of noise will cause a frequency deviation of

$$\begin{aligned} f_{dev2} &= (10\mu\text{Volt}_{RMS})\, k_{VCO1} \\ &= (10\mu\text{Volt}_{RMS}) \left(6.67 \frac{\text{MHz}}{\text{Volt}}\right) \\ &= 66.7\text{ Hz}_{RMS} \end{aligned} \tag{9.90}$$

The wider the VCO tunes, the more susceptible it is to undesired signals on its tuning line. If the undesired signal is noise, then the oscillator exhibits excess phase noise. If the undesired signal is a discrete waveform, then the oscillator will exhibit discrete sidebands. While it helps to decrease the amount of thermal noise present on the control line of a VCO, some of the noise is due to the components internal to the VCO itself.

Phase Noise A PLL frequency synthesizer is a control system that is capable of controlling the phase noise of the output only within the loop bandwidth. The loop bandwidth covers

$$f_0 - f_{LBW} \text{ to } f_0 + f_{LBW} \tag{9.91}$$

where

f_0 is the current output frequency of the synthesizer
f_{LBW} is the loop bandwidth of the PLL synthesizer

Well inside the loop bandwidth, the PLL disciplines the VCO phase noise with the phase noise of the reference oscillator. Inside the loop bandwidth, the phase noise of the VCO is suppressed. At frequencies well outside of the loop bandwidth, the control loop doesn't affect the VCO phase noise. The phase noise at these offsets approaches the phase noise of the basic VCO. The PLL phase noise in the neighborhood of the loop bandwidth is a composite of the multiplied reference noise and the undisciplined VCO noise. If we want low phase noise outside of the loop bandwidth, we must choose a VCO with low phase noise at the offsets of interest.

$\div N$ The frequency of the reference oscillator, the $\div N$ counter, and the $\div M$ counter act in concert to set the output frequency of the PLL. In normal operation, both the reference and the $\div M$ counter are fixed. To change the output frequency, the user simply writes a new number to the $\div N$ counter.

When we change the $\div N$ counter, the loop unlocks for a period and then settles back into a steady-state condition after the synthesizer reaches the correct output frequency. This is the tuning speed of the synthesizer.

9.7.1.4 Spectrum

Figure 9-2 though Figure 9-7 show various views of the spectrum of a PLL frequency synthesizer.

9.7.1.5 PLL Design Trade-Offs

The design of a PLL synthesizer is full of trade-offs. The following discussion focuses on defining and quantifying the various trade-offs and their interaction.

Rules of Thumb The following list provides general guidance on the design of PLL frequency synthesizers:

- A VCO with a large tuning range is noisier than a similar VCO with a small tuning range.
- A PLL frequency synthesizer containing a VCO with a large tuning range will tend to have stronger close-in spurious signals than a synthesizer containing a VCO with a small tuning range.
- A PLL frequency synthesizer with a large loop bandwidth will tune faster than a synthesizer with a small loop bandwidth.
- Synthesizers whose loop bandwidths are a large percentage of the step size will exhibit higher close-in spurious signals than an equivalent synthesizer with a small percentage bandwidth.

- Because of stability concerns and other realizations factors, we often limit the loop bandwidth to less than 10% of the synthesizer's step size. This limit also suppresses reference spurs.
- The phase noise of the output is determined by the VCO for offsets well outside of the loop bandwidth. The phase noise of the output is determined by the reference oscillator for offsets well inside the loop bandwidth. In the neighborhood of the loop bandwidth, the phase noise is a combination of both.
- Synthesizers with small step sizes must have small loop bandwidths. Therefore, these synthesizers will tend to tune slowly, and their VCO's must have good close-in phase noise.

Phase Noise To realize a low-phase noise synthesizer, we would like either:

- A wide-loop bandwidth (which will let us discipline the phase noise of the VCO out to large offset frequencies) or
- A narrow-loop bandwidth and a VCO whose phase noise is acceptable for offsets greater than the loop bandwidth.

A wide-loop bandwidth implies a large step size (because the loop bandwidth is limited to $\approx$10% of the step size). This tells us that PLL frequency synthesizers with small step sizes, low phase noise, and a wide tuning range are difficult to build. It's easier to build one of the following:

- Small step size and low phase noise but narrow tuning range
- Small step size and wide tuning range but poor phase noise performance
- Low phase noise and wide tuning range with large step size

Well-designed communications receivers recognize these limits in their conversion schemes. The first LO of a wideband receiver will often tune over a very large range but with a 1 MHz step size. This is the *coarse-step* synthesizer. The large step size allows for a large loop bandwidth and thus good phase noise performance. The coarse-step synthesizer places the signal of interest somewhere in the first IF filter. Of course, the first IF filter is wide enough to handle this situation—usually not a problem.

The second LO has a small step size (perhaps at the ultimate step size of the receiver), but it must tune over only 1 MHz. The small step size forces a narrow loop bandwidth. The phase noise of the raw VCO will be low since the VCO doesn't have to tune over a very large range. This synthesizer is often called a *fine-step* synthesizer.

Spurious Signals The magnitude of the sampling sidebands is determined by the digital phase/frequency comparator, the loop filter, and the VCO tuning characteristics. Commonly used digital phase/frequency detectors produce discrete signals at their output ports even when the synthesizer is locked. These signals contain strong components at the step size frequency and its harmonics. So if the synthesizer has a step size of 1 kHz, the signals present at the phase/frequency detector outputs will have strong components at, for example, 1 kHz, 2 kHz, and 3 kHz.

The larger the VCO gain, the more sensitive the VCO is to noise or other unwanted signals on its control line. The digital signals from the phase/frequency detector will make a larger impact on a device with a large VCO gain than a device with a small VCO gain.

Figure 9-38 shows that the loop filter lies between the digital phase/frequency detector and the VCO. The loop filter is a low-pass filter whose cutoff frequency is directly proportional to the loop bandwidth of the synthesizer, and we use the loop filter to suppress the digital waveforms from the phase/frequency detector. The lower the cutoff frequency of the loop filter, the more suppression the filter provides for the digital waveforms.

The cutoff frequency of the loop filter is directly proportional to the loop bandwidth. Synthesizers whose loop bandwidth is a smaller percentage of the step size will tend to have lower spurious output power than synthesizers with larger percentage bandwidths. For example, if two PLL synthesizers are identical except that one has a 1 kHz loop bandwidth while the other has a 10 kHz loop bandwidth, we would expect to see a lower spurious output power from the synthesizer with the 1 kHz loop bandwidth.

The fundamental conflict is that we want a large loop bandwidth because it provides us with better phase noise, but the narrow loop bandwidth offers lower close-in spurious signal power. We often add separate filters to the loop whose sole purpose is to suppress the sampling sidebands. Of course, we have to carefully evaluate the phase and magnitude response of these sideband suppression filters, lest they introduce instability.

Tuning Speed Usually, the synthesizer with the widest loop bandwidth will tune the fastest. A wide-loop bandwidth allows information to travel quickly around the loop, so the loop can react quickly to changing conditions. In receivers that contain several PLL frequency synthesizers, the system tuning speed will be determined by the PLL with the narrowest loop bandwidth.

One method commonly used to improve the tuning speed is to employ a coarse-step PLL synthesizer as the first LO. This synthesizer will have a large step size, will offer low phase noise, and will tune quickly. The second LO will be provided digitally by an NCO, which offers a very fine-step size and a very quick tuning speed but with a only fair spurious output and limited frequency range.

9.7.2 Numerically Controlled Oscillator

An NCO generates sine waves using digital counters, ROM tables, and digital-to-analog converters (DACs). The NCO offers very small step sizes (< 1 Hz, in some cases) and very fast tuning speeds. However, the output exhibits many nonharmonically related spurious signals at unusual (but predictable) frequencies. The maximum output frequency of NCO is limited, but technology will advance the speed.

Often, we use the output of an NCO entirely in its digital form. We apply the NCO to our downconverted and digitized signal in the digital domain, thus avoiding the issues associated with converting the waveform back to analog form via DACs. Figure 9-40(a) shows an analog architecture while 10.10.1.2(b) shows a more modern equivalent.

Figure 9-41 shows a basic NCO basic configuration. The NCO uses an n-bit phase accumulator driven at a reference clock at f_{REF}. At each rising edge of the reference clock, the phase accumulator output increases by the phase increment, P_i. The output of the phase accumulator is fed into a lookup table that converts the phase information into a digitized sine wave of b-bits.

9.7.2.1 Operation

For a simple example, we'll assume a 4-bit data path ($n = 4$). Typically, NCOs use a larger number of bits (32 bits and higher). The frequency resolution of an NCO is related to the number of bits in its data path. The output of the *binary adder* is the modulo-16

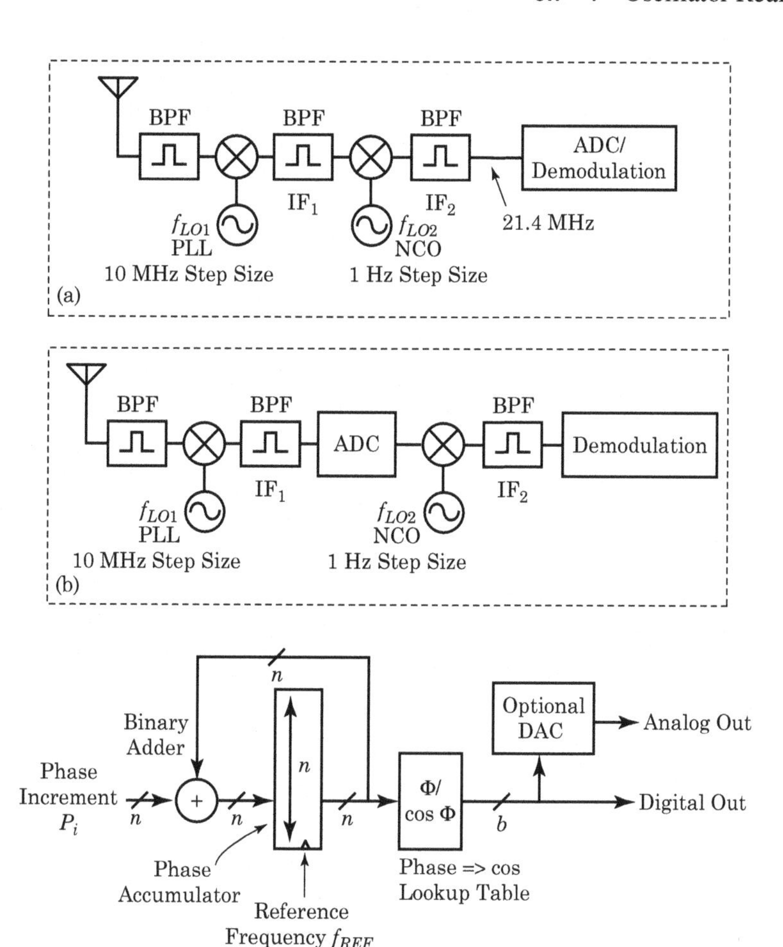

FIGURE 9-40 ▪ (a) A conversion scheme in which we convert the output of an NCO into analog form (b) A conversion scheme in which we use an NCO entirely in its digital form, bypassing the issues associated with converting the NCO output into analog.

FIGURE 9-41 ▪ Simplified block diagram of an NCO. Additional circuits allow the user to modulate the signal.

sum of the two inputs. The user applies the phase increment at the P_i port, which controls the output frequency f_{out}.

The "Φ/cosΦ Table" is a lookup table containing a digitized sine wave. The output of the phase accumulator is zero to 2^{n-1}, which corresponds to 0 to 359.999... degrees of the sine wave. The number of bits we apply to the Φ/cosΦ table is the *phase resolution* of the NCO. In our example, the NCO's phase resolution is 4 bits.

Set the phase accumulator initially to zero, and set P_i to 01_H $(= 0001_B)$. At every rising edge of f_{REF}, the phase accumulator copies the data present on its input port to its output. The phase accumulator holds this data on its output port until f_{REF} produces another rising edge.

On successive rising edges of the clock line, the phase accumulator output becomes 02_H (0010_B), 03_H (0011_B), 04_H (0100_B), 05_H (0101_B), 06_H (0110_B), 07_H (0111_B), ..., $0F_H$ (1111_B), 00H (0000_B) ... The phase accumulator output is increasing incrementally by 1 for each rising edge of the f_{REF} line. When the phase accumulator reaches 15 $(0F_H = 1111_B)$, the output rolls over and starts at zero again. We have a modulo-16 counter that increases incrementally by 1 on each cycle of the clock.

Imagine that P_i is 02_H (0010_B). A little thought will convince us that we have a modulo-16 counter that increases incrementally by 2s. When P_i is 3, we have a counter that increases incrementally by 3s, and so on. We've built a user-programmable frequency divider.

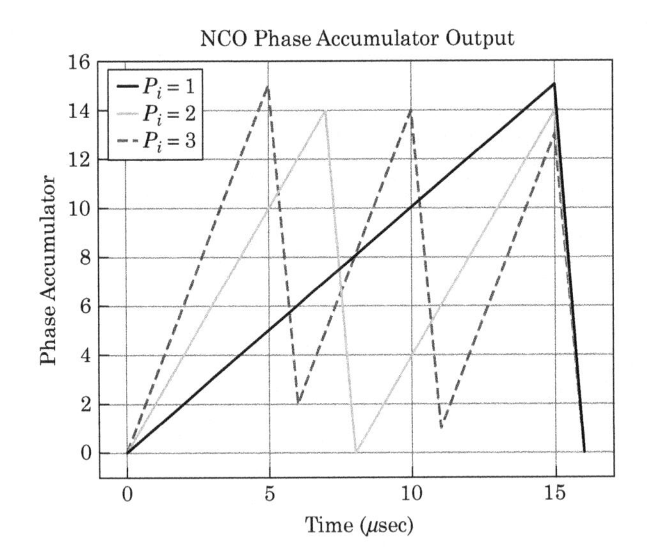

FIGURE 9-42 ■ The phase accumulator output of Figure 9-41 for different values of P_i. The data are shown in Table 9-3.

For f_{REF} of 1 MHz, the clock line provides a rising edge every microsecond. Figure 9-42 and Table 9-3 show the output of the phase accumulator over time for various values of P_i.

9.7.2.2 Frequency of Output

The value of P_i determines the output frequency. For our example using the 1 MHz clock rate, the output frequency is

$$f_{out} = \frac{P_i}{16} (1 \text{ MHz}) \tag{9.92}$$

TABLE 9-3 ■ The output of the phase accumulator of Figure 9-41 over time for various values of P_i. Figure 9-42 shows these data graphically.

	Phase Increment P_i				
T (μsec)	1	2	3	4	5
0	0	0	0	0	0
1	1	2	3	4	5
2	2	4	6	8	10
3	3	6	7	12	15
4	4	8	12	0	20 = 4
5	5	10	15	4	9
6	6	12	18 = 2	8	14
7	7	14	5	12	19 = 3
8	8	16 = 0	8	0	8
9	9	2	11	4	13
10	10	4	14	8	18 = 2
11	11	6	17 = 1	12	7
12	12	8	4	16 = 0	12
13	13	10	7	4	17 = 1
14	14	12	10	8	6
15	15	14	13	12	11
16	16 = 0	16 = 0	16 = 0	16 = 0	16 = 0

For any P_i and f_{REF}, the output frequency is

$$f_{out} = \frac{P_i}{2^n} f_{REF} \tag{9.93}$$

where

n = the number of bits carried by the phase accumulator and binary adder
f_{REF} = the reference frequency
P_i = the phase increment supplied by the user

9.7.2.3 More Bits

We can increase the number of discrete frequencies available from the NCO by increasing the number of bits carried through the NCO. The number of discrete frequencies generated by an NCO is given by

$$\text{Number of Frequencies} = 2^{n-1} \tag{9.94}$$

The number of frequencies is not 2^n because of sampling effects in the DAC and because of rollover effects in the phase accumulator. A 32-bit NCO can produce over 2 billion discrete frequencies.

EXAMPLE

NCO Rollover

Examine the behavior of the 4-bit NCO example when $P_i = 0F_H$.

Solution

Assuming the clock occurs at a 1 MHz rate and P_i is $0F_H$, the NCO will produce the output stream shown in Table OSC-4.

TABLE 9-4 NCO rollover example.

T (μsec)	Phase Acc Output	T (μsec)	Phase Acc Output
0	15 ($0F_H$)	10	5 (05_H)
1	14 ($0E_H$)	11	4 (04_H)
2	13 ($0D_H$)	12	3 (03_H)
3	12 ($0C_H$)	13	2 (02_H)
4	11 ($0B_H$)	14	1 (01_H)
5	10 ($0A_H$)	15	0 (00_H)
6	9 (09_H)	16	15 ($0F_H$)
7	8 (08_H)	17	14 ($0E_H$)
8	7 (07_H)	18	13 ($0D_H$)
9	6 (06_H)	19	12 ($0C_H$)

The frequency of the output waveform is the same as if the phase increment was 01_H. This is an effect of the modulo-16 adder and of the Nyquist theorem.

Nyquist states that you have to sample a waveform at least twice as fast as its bandwidth. In the NCO case, we have to provide at least two output samples at the highest frequency we generate. This limits the theoretical maximum frequency of the NCO to $f_{CLK}/2$. We're limited to a maximum output frequency of about 40% of f_{CLK} after we address filtering and other realization concerns.

9.7.2.4 NCO Phase Noise

Since the output frequency of an NCO is less than the input frequency, the output can have better phase noise performance than the input by the $20\log(n)$ rule.

9.7.2.5 Tuning Speed

The value of the phase increment P_i controls the NCO frequency. Changing P_i changes the frequency instantaneously and with coherent phase. There is no abrupt, transient waveform present at the output of the NCO. The time-domain waveform is smooth and continuous.

The output of a PLL frequency synthesizer is not phase continuous as it changes frequency. When we change the $\div N$ counter, the loop unlocks and begins its sweep to the new frequency. The output is a transient, complex waveform that is not phase continuous.

9.7.2.6 Modulation

The digital nature of the NCO provides precise control over the output waveform. We can change the frequency very quickly, and the frequency changes are phase continuous. With some minor changes to the architecture of Figure 9-41, we can accommodate phase, amplitude modulation, and frequency modulation. We can also configure the NCO to easily produce both sine and cosine waves. The two signals, always 90° apart, are useful in complex basebanding.

9.7.2.7 The Real World

NCOs generate spurious signals along with the desired output tone. The source of these signals lies in the realization details of Figure 9-41. Figure 9-43 shows the n-bit phase accumulator divided into two parts: one part is p bits wide, and the other is $q = n - p$ bits wide. A typical commercial NCO IC contains a 32-bit phase accumulator. A 32-bit phase/cos table would require over 4 billion addresses. We continue to carry the full n bits in the phase accumulator and binary adder to produce a small step size, but we present only the p most significant bits (MSBs) of the phase accumulator to the phase/cos table.

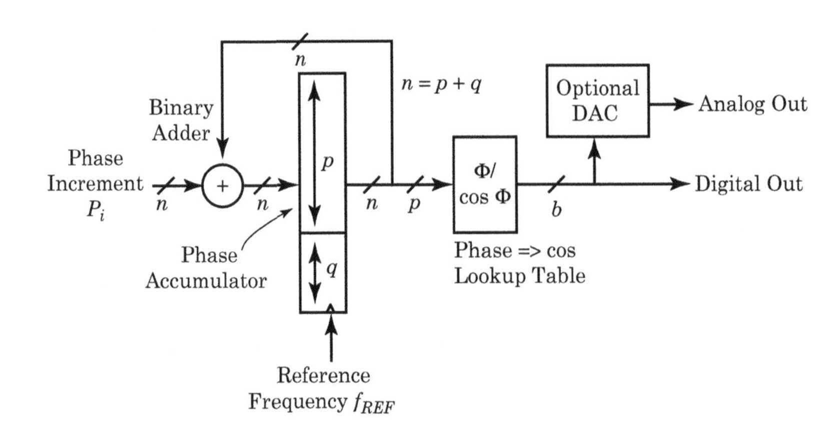

FIGURE 9-43 ■ A more practical block diagram of an NCO. Only the most significant bits of the phase accumulator are passed to the Φ/cos Φ lookup table.

TABLE 9-5 ■ Phase truncation example.

Clock #	Phase Accumulator	Output to the Phase/Cos Lookup Table (Top 3 MSBs)	Clock #	Phase Accumulator	Output to the Phase/Cos Lookup Table (Top 3 MSBs)
0	0	0	16	16	4
1	3	0	17	19	4
2	6	1	18	22	5
3	9	2	19	25	6
4	12	3	20	28	7
5	15	3	21	31	7
6	18	4	22	2	0
7	21	5	23	5	1
8	24	6	24	8	2
9	27	6	25	11	2
10	30	7	26	14	3
11	1	0	27	17	4
12	4	1	28	20	5
13	7	1	29	23	5
14	10	2	30	26	6
15	13	3	31	29	7
			32	0	0

The value p is often in the range of 8 to 12 bits. This *phase truncation* is one source of spurious NCO signals.

Phase Truncation Referencing Figure 9-43, assume a 7-bit phase accumulator ($n = 7$), and feed the 4 MSBs into the phase/cos table ($p = 4$). The phase accumulator knows the output phase to 7 bits, but it is not passing that complete knowledge on to the phase/cos table lookup table. Consequently, the lookup table always produces a slightly wrong phase value, and the output is slightly incorrect.

When P_i is three, Table 9-5 shows the state of the NCO for each clock cycle.

The first two cycles (clocks 0 through 10 and clocks 11 through 21) are both 11 cycles long. The third cycle (clocks 22 through 31) is 10 cycles long. At clock #32, the state of the NCO is identical to its state at clock #0 so this cycle will repeat. The NCO will continually produce this string of [11,11,10,11,11,10,11,11,10 . . .]-long cycles (two long cycles, then a short cycle) forever.

The average frequency is correct, but the instantaneous frequency is always incorrect. The net effect is that the NCO output always exhibits a little frequency modulation, and that modulation brings output spurs. The exact level of the spurs and their frequencies depend on the number of bits carried through the phase accumulator and on how many bits we use in the lookup table.

Some "perfect" numbers produce no FM at all because they do not affect the bits not presented to the lookup table. As an exercise, work though Figure 9-43 when P_i is four. Figure 9-44 through Figure 9-47 show NCO spectra generated under the following conditions: 10-bit phase accumulator ($n = 10$) and a 5-bit phase/cos table lookup ($p = 5$). The clock, f_{REF}, runs at 100 MHz.

- Figure 9-44: Phase increment P_i set to 32 (frequency is 3,125,000 Hz). This is a "perfect" P_i number that generates a spectrum with no FM and no spurious outputs.

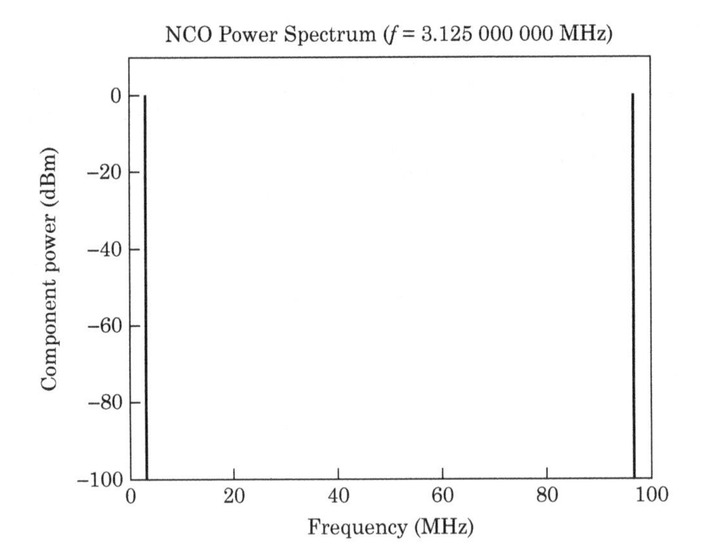

FIGURE 9-44 ■ NCO output for $P_i = 32$. This is a "perfect" P_i that generates no unwanted spurious signals.

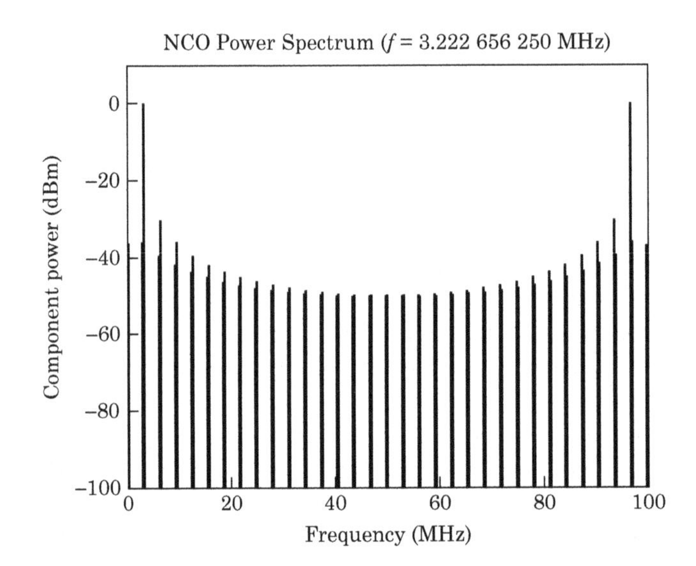

FIGURE 9-45 ■ NCO output for $P_i = 33$. Although we are only one step size away from a "perfect" P_i of 32, the spurious output of the NCO has changed significantly.

- Figure 9-45: 10-bit phase increment set to 33 (frequency is 3,222,656.25 Hz). This is one step size away from the perfect P_i setting. A 1-bit change in the phase increment has changed the output spectrum from almost no spurs to spurs that are only 30 dB below the desired output. This is a very common effect in NCOs.
- Figure 9-46: P_i is set to 29 (frequency is 2,832031.25 Hz). Spurious products are still only 30 dB below the desired signal, but their frequencies have changed.
- Figure 9-47: P_i is set to 38 (frequency is 3,710,937.5 Hz).

Figure 9-48 and Figure 9-49 show the effect of increasing p, the number of bits sent to the phase/cos lookup table. In both cases, P_i is ten bits and the value is 17. The f_{REF} clock is 100 MHz. This configuration produces an output signal at 1.660156250 MHz.

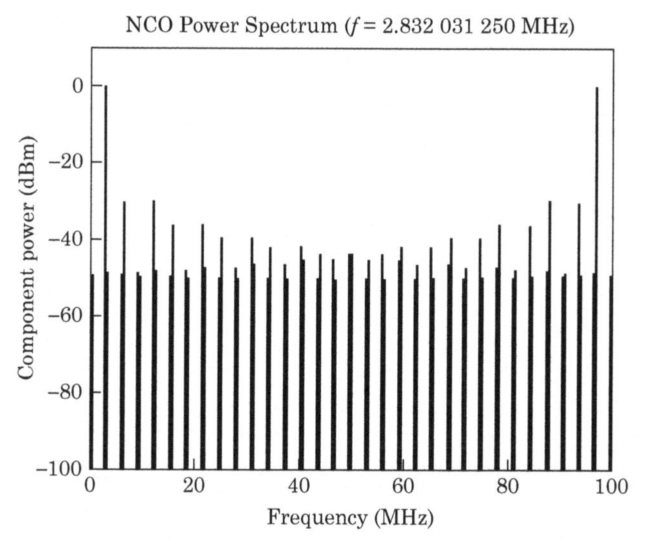

FIGURE 9-46 NCO output for $P_i = 29$. The spectrum is similar to Figure 9-45, but the frequencies and amplitude of the spurious signals have changed.

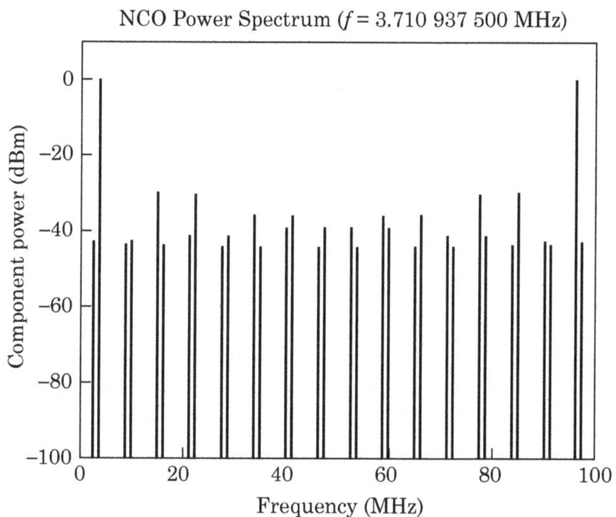

FIGURE 9-47 NCO output for $P_i = 38$. The amplitudes and frequencies of the spurious signals differ radically from those of Figure 9-45 and Figure 9-46.

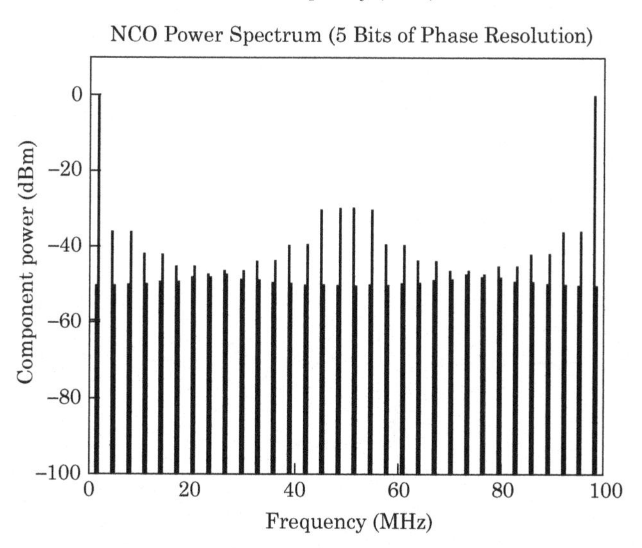

FIGURE 9-48 NCO output when sending 5 bits of phase resolution to the $\Phi/\cos\Phi$ lookup table. $P_i = 17$, $f_{Ref} = 100$ MHz and the output frequency is 1.660156250 MHz. The worst-case products are 30 dBc.

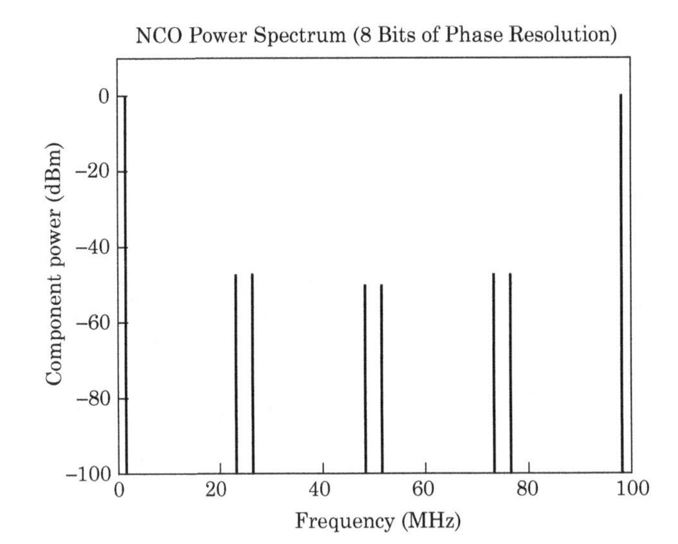

FIGURE 9-49 ■ The NCO from Figure 9-48 except we are sending 8 bits of phase resolution to the Φ/cos Φ lookup table. The 3 extra bits have significantly decreased the number of spurious products in the NCO output and have decreased their magnitude to 50 dBc.

We're sending 5-bit phase resolution to the sine lookup table in Figure 9-48. In Figure 9-49, we're using 8 bits of phase resolution. The more bits of phase resolution we pass to the lookup tables, the fewer spurious signals we generate.

To a rough approximation, the NCO spur level due to phase quantization is

$$\text{Spur Level (due to PM)} = -6p + 5.17 \quad \text{(dB)} \tag{9.95}$$

where p is the number of bits into the input of the phase/cos table.

Phase Dithering The spurious signals present on an NCO's output are due to repetitive events occurring in the NCO. If we can force those events to become more random, we can change the coherent spurious products into a more noise-like phenomenon. Figure 9-50 shows one method of randomizing the output of the phase accumulator.

The PN generator is a pseudo-random sequence generator that produces semirandom numbers. We add a small amount of this random noise to the output of the phase accumulator and then pass the randomized data onto the phase/cos table. Figure 9-51 and Figure 9-52 show the output of the NCO with and without phase dithering. We have reduced most of the spurs below 50 dBc. We have also reduced the strength of the two strongest spurs about 10 dB. We did not remove the spurious energy; we simply spread it out over the full bandwidth of the output.

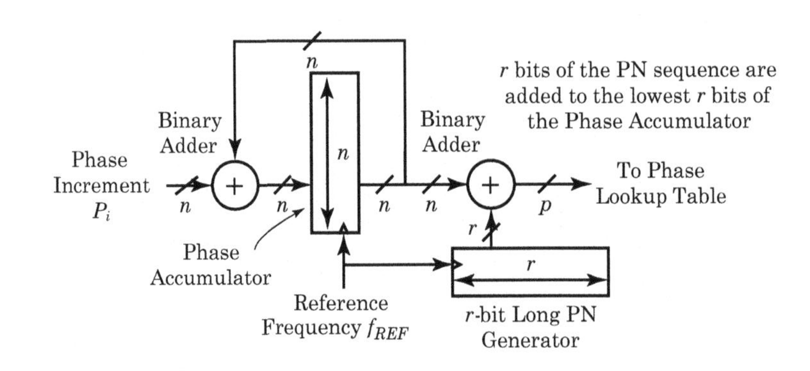

FIGURE 9-50 ■ One method of randomly dithering the phase of an NCO using a pseudo-random PM sequence.

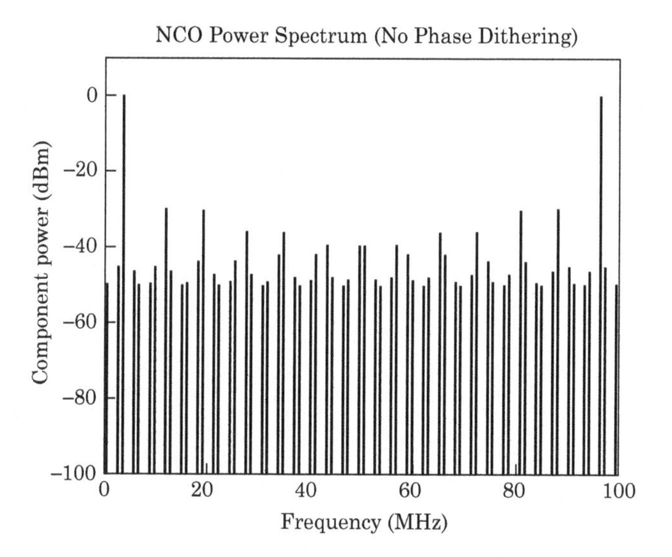

FIGURE 9-51 ■ The output of an NCO without the phase dithering mechanism of Figure 9-50. The spurious signals are coherent tones that are 30 dBc.

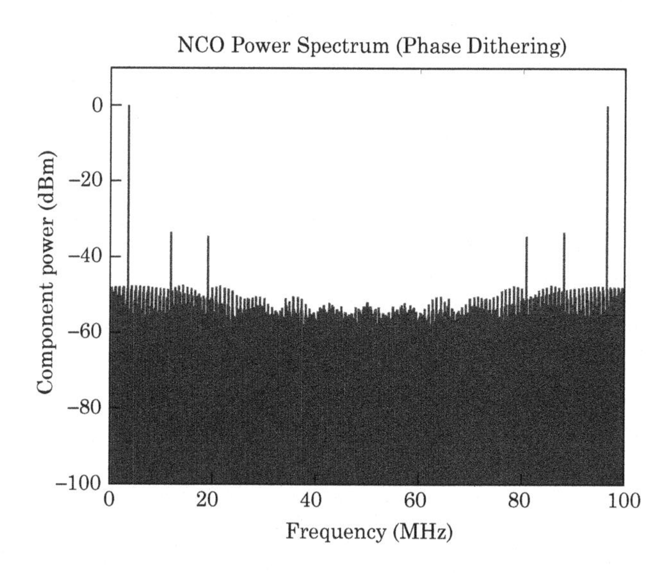

FIGURE 9-52 ■ The output of an NCO when we dither the oscillator's phase using the circuitry of Figure 9-50. Many coherent spurious signals have been converted into pseudo-random noise. The worst-case spurious signals are 35 dBc.

DACs The distortion in the figures shown thus far is due only to phase truncation. The NCO output signals are represented to full floating-point precision, and we have not yet considered DAC distortion. DAC distortion occurs only when we must convert the NCO into analog form. Most modern architectures never convert the NCO into analog and thus are immune from DAC issues.

The DAC converts the digital output of the sine lookup table into an analog waveform. The NCO can express the instantaneous value of a sine wave only with finite precision, and the output waveform will always be quantized to some accuracy. There will always exist some nonzero amount of error between an ideal sine wave and the output waveform of the NCO.

There are two nonlinear effects at work in a practical DAC. One effect is the mathematically derivable distortion the DAC introduces as it attempts represent a continuous analog waveform as discrete steps. The second source of distortion arises from DAC

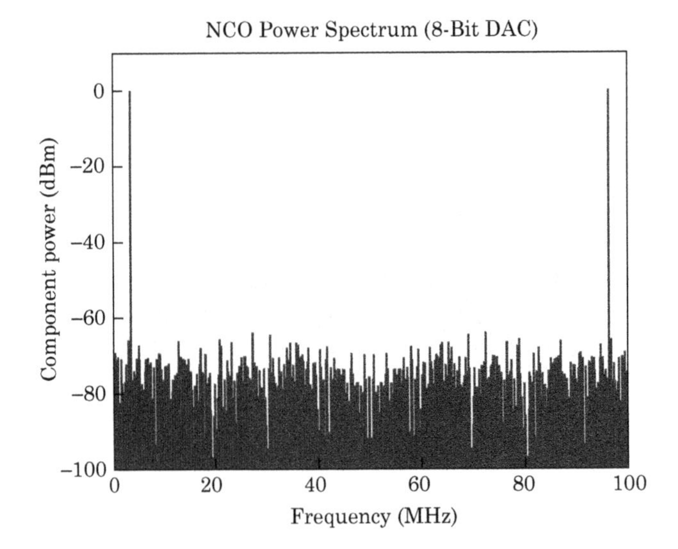

FIGURE 9-53 ■ NCO output for an ideal 8-bit DAC. The spurious products are about 70 dBc.

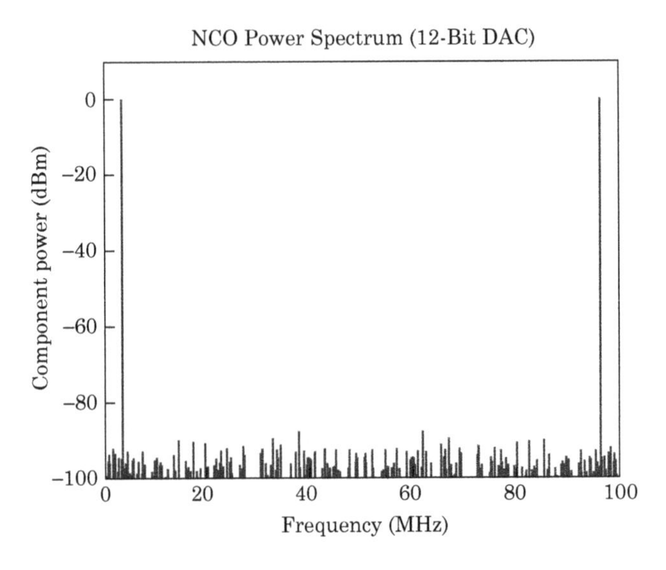

FIGURE 9-54 ■ NCO output of Figure 9-53 when we increase the number of DAC bits to 12. The spurious products are now about 90 dBc.

nonlinearities. DACs exhibit missing codes, nonuniform step sizes, and a host of other parameters that act to generate nonlinear distortion. These are fully covered in Chapter 11.

Amplitude Quantization Let us assume our NCO does not exhibit phase truncation so we are free to concentrate on the amplitude quantization. Amplitude quantization behaves similarly to phase truncation. The frequency and amplitude of the nonlinear products are strong functions of P_i, the phase increment, and of b, the number of bits we feed to the DAC. If the DACs are perfect, then larger quantities of b result in lower NCO output noise. Figure 9-53 and Figure 9-54 show two identical NCOs; one uses an 8-bit DAC, whereas the other uses a 12-bit DAC. The nonlinear products shown in these figures are due entirely to DAC nonlinearities.

Compare Figure 9-55 and Figure 9-56. Both NCOs use 8-bit DACs but are tuned to slightly different frequencies. The character of the spurious signals is very different. Again, the distortion presented here is caused entirely by the DAC nonlinearities.

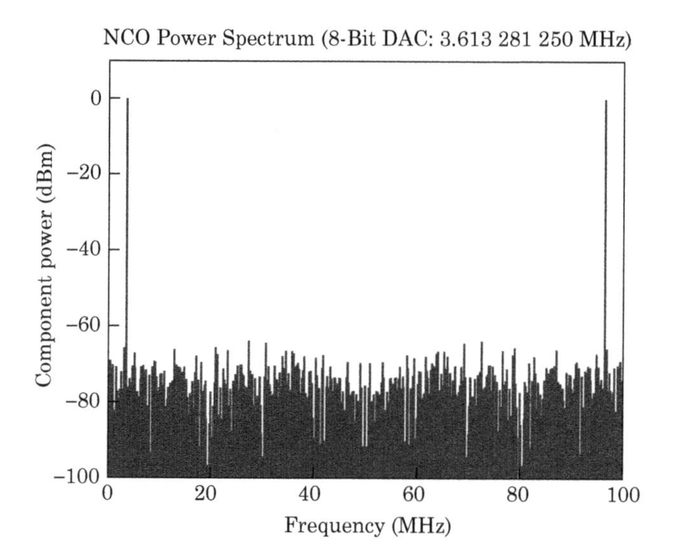

FIGURE 9-55 ■ NCO output tuned to one frequency. Note the quantity and character of the spurious signals compared with those in Figure 9-56.

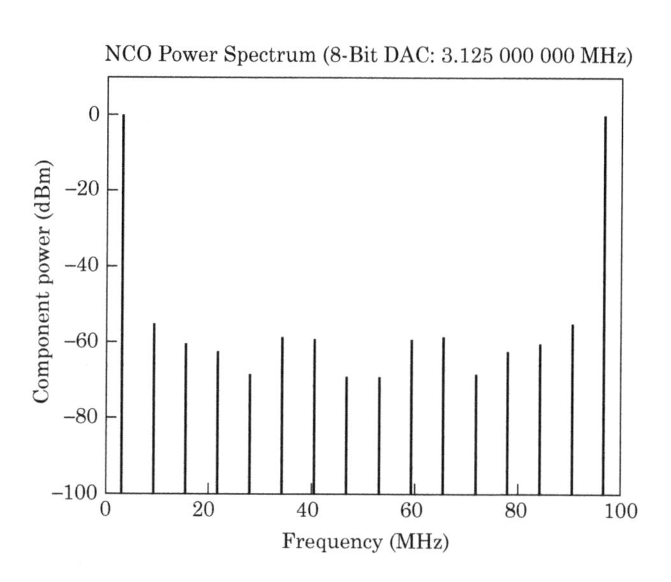

FIGURE 9-56 ■ The NCO from Figure 9-55 tuned to a slightly different frequency. The quantity and strength of the spurious products can change dramatically with only slight changes to output frequency.

We can separate the NCO's output waveform into two components: the ideal sine wave and an error signal. The NCO output signal is the time-domain sum of the ideal sine wave and the error signal. Likewise, the frequency spectrum of the NCO's output is the sum of the frequency-domain plot of the ideal sine wave plus the frequency-domain plot of the error function.

The nature of the error function depends on the exact frequency the NCO is trying to generate and the number of bits in the DAC. For example, at some frequency settings, the NCO will generate a sine wave whose exact values just happen to coincide with the values that the DAC can produce. The error function is zero, and the output spectrum is a pure, ideal sine wave.

With other frequency settings, the exact values of the ideal sine wave and the NCO hardly ever correspond, and the result is a complex, nonzero error function with a complex, nonzero frequency spectrum.

The error functions are repetitive, and any repetitive signal present in the error waveform translates into unwanted spurious products in the frequency domain.

The repetition rate of the error function will be related to the desired NCO output signal; however, that doesn't mean the repetition rate of the error function is the same as the desired output signal. There is really no simple way to determine the spurious output frequencies given the architecture of an NCO and the desired output frequency.

The NCO spur level due to DAC quantization is roughly

$$\text{Spur Level (due to AM)} = -6.02b + 1.76 \quad \text{(dB)} \tag{9.96}$$

where b is the number of amplitude bit in the DAC.

9.8 BIBLIOGRAPHY

Analog Devices Corp., "A Technical Tutorial on Digital Signal Synthesis," Analog Devices Corp, 1999.

Cercas, Francisco A.B., Tomlinson, M., and Albuquerque, A.A., *Designing with Digital Frequency Synthesizers*, Proceedings of RF Expo East, 1990.

Cordesses, Lionel, "Direct Digital Synthesis: A Tool for Periodic Wave Generation (Part 1)," *IEEE Signal Processing Magazine*, July 2004.

Cordesses, Lionel, "Direct Digital Synthesis: A Tool for Periodic Wave Generation (Part 2)," *IEEE Signal Processing Magazine*, September 2004.

Engleson, Morris, and Breaker, Ron, "Interpreting Incidental FM Specifications," *Frequency Technology Magazine,* p. 13, February 1969.

Gentile, Ken, Brandon, David, and Harris, Ted, "Direct Digital Synthesis Primer," PowerPoint presentation, Analog Devices Corp., May 2003.

Grebenkemper, C. John, "Local Oscillator Phase Noise and Its Effect on Receiver Performance," *Watkins-Johnson Tech-Note,* Vol.8, No. 6, Gaithersburg, MD, November/December 1981.

Hewlett-Packard Corp., "Low Phase Noise Applications of the HP-8662A and 8663A Synthesized Signal Generators," Application Note 283-2, December 1986.

Hewlett-Packard, "RF and Microwave Phase Noise Measurement Seminar," Hewlett-Packard Corp., March 1988.

Hewlett-Packard, "Signal Generator Spectral Purity," Application Note 388, Hewlett-Packard Corp., 1990.

Leeson, D.B., "A Simple Model of Feedback Oscillator Noise Spectrum," *Proceedings of the IEEE,* Vol. 54, No. 2, February 1966.

McCune, Earl W., Jr., "Control of Spurious Signals in Direct Digital Synthesizers," Digital RF Solutions Application Note AN1004. 1988.

Payne, John B., "Measure and Interpret Short-Term Stability," *Microwaves*, July 1976.

Piezo Technology, Inc., *Frequency Control Products—Handbook for the Design and Component Engineer,* Orlando, FL, March 1990.

Przedpelski, Andrzej, "Analyze, Don't Estimate, Phase-Lock-Loop," *Electronic Design,* May 10, 1978.

Przedpelski, Andrzej, "Optimize Phase-Lock Loops to Meet Your Needs—or Determine Why You Can't," *Electronic Design*, September 13, 1978a.

Przedpelski, Andrzej, "Suppress Phase-Lock-Loop Sidebands without Introducing Instability," *Electronic Design*, September 13, 1978b.

Przedpelski, Andrzej, "Programmable Calculator Computes PLL Noise, Stability," *Electronic Design,* March 31, 1981.

Scherer, Dieter, "Design Principles and Test Methods for Low Phase Noise RF and Microwave Sources," RF and Microwave Measurement Symposium and Exhibition. n.d.

Scherer, Dieter, "Today's Lesson—Learn about Low-Noise Design—Part I," *Microwaves,* April 1979.

Scherer, Dieter, "Today's Lesson—Learn about Low-Noise Design—Part II," *Microwaves,* May 1979.

Upham, Art, "Low-Noise Oscillator Paper," Hewlett-Packard, Inc., 1620 Signal Drive, TAF C-34, Spokane, WA 99220, n.d.

Terman, Frederick Emmons, *Radio Engineering*; 2d ed., McGraw-Hill, 1937.

Vectron Laboratories, Inc., *Crystal Oscillators Catalog*, Norwalk, CT, 1990–1991.

Zavrel, Robert J., Jr., "Digital Modulation Using the NCMO," *RF Design Magazine*, March 1988.

9.9 | PROBLEMS

1. **Frequency Accuracy** For the receiver shown in Figure 9-57:

   a. What is the nominal tuned frequency of the receiver? In other words, to what frequency, when presented to the antenna of this receiver, will be converted down to the 45 MHz IF? Assume that both oscillators are perfectly accurate.

   b. If all of the oscillators drift in the worst-case direction, what is the frequency of the signal leaving the 140 MHz filter? In other words, what is the farthest the output signal can deviate from 140 MHz? Give your answer in hertz.

   c. For the worst-case condition, how much frequency offset (in Hz) is each oscillator causing?

   d. Draw the worst-case frequency drift for all the signals present in the various parts of the receiver.

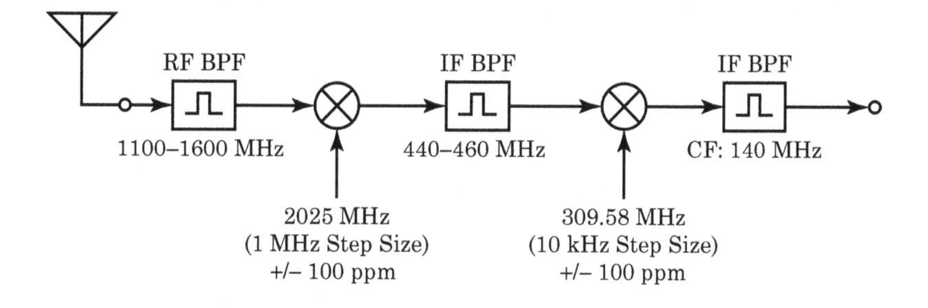

FIGURE 9-57 ■ A 1,100–1,600 MHz dual conversion receiver.

2. **Frequency Accuracy** For the receiver shown as Figure 9-58:

   a. What is the nominal tuned frequency of the receiver? In other words, to what frequency, when presented to the antenna of this receiver, will be converted down to the 45 MHz IF? Assume that both oscillators are perfectly accurate.

   b. If all of the oscillators drift in the worst-case direction, what is the frequency of the signal leaving the 45 MHz filter? In other words, what is the farthest the output signal can deviate from 45 MHz? Give your answer in hertz.

   c. For the worst-case condition, how much frequency offset (in Hz) is each oscillator causing?

   d. Is the final output spectrum (at 45 MHz) inverted?

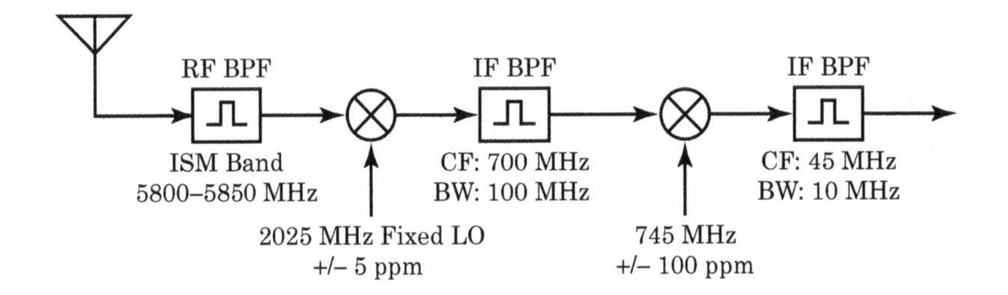

FIGURE 9-58 ■ A 5,800 MHz ISM band, dual conversion receiver.

3. **Multiple Choice** Oscillator phase noise:

    a. Is a measure of the uncertainty of an oscillator's zero crossings.
    b. Increases the spectral width of an oscillator.
    c. Can cause multiple signals to interfere with each other in ways that wouldn't happen if the oscillator were pure.
    d. Causes problems when we demodulate phase modulated signals.
    e. All of the above.

4. **Multiple Choice** Oscillator spurious components are unwanted sinusoidal components present at the output of an oscillator. Oscillator spurious components are:

    a. Always at a harmonic frequency of the fundamental oscillator frequency.
    b. Increase the number of possible spurious signals present in a system.
    c. Are measured as incidental phase modulation.
    d. Are measured as incidental frequency modulation.
    e. All are true

5. **Oscillator Phase Noise** Figure 9-59 shows the constellation plot of a received QPSK signal. The signal has 30 dB of SNR and the receiver's LO has 0° of IPM or phase noise. Ideally, there should be four infinitesimally small dots in each corner of the plot. The fuzziness of the dots in this diagram is due to the 30 dB SNR of the received signal.

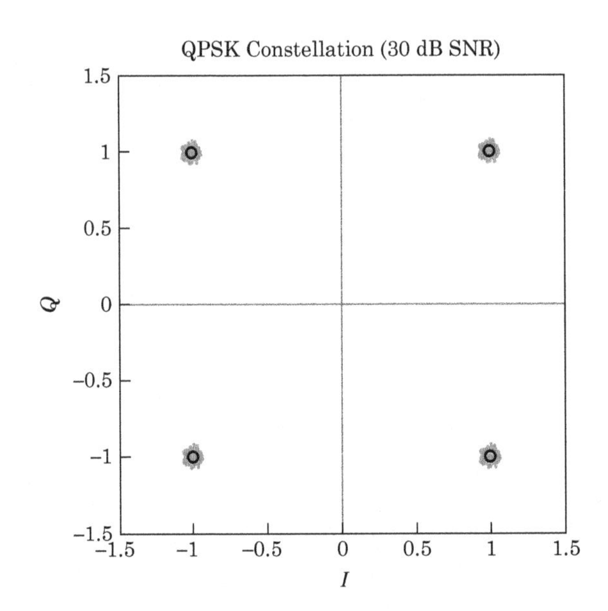

FIGURE 9-59 ■ The constellation patter of a demodulated QPSK signal. This signal has 30 dB of SNR.

Figure 9-60 shows the same received signal (QPSK with 30 dB of SNR) with oscillator phase noise present in the system. The more phase noise, the larger the arcs.

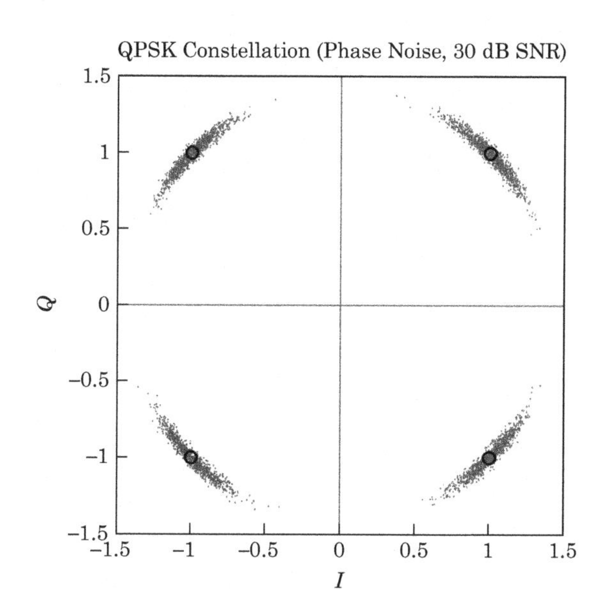

FIGURE 9-60 ■ The constellation patter of a demodulated QPSK signal. This signal has 30 dB SNR and is also contaminated with phase noise.

Estimate the amount of oscillator phase noise present on the signal of Figure 9-60, in degrees RMS.

Hint: Oscillator phase noise causes the phase of the received signal to vary in a random manner. That randomness in the phase is directly measurable from this diagram.

6. **Oscillator Spurious** Complete parts a. and b. for the conversion schemes shown in Figure 9-61 through Figure 9-64.

   a. For each conversion scheme, identify the conversion scheme. In other words, is the conversion scheme RF − LO = IF, LO − RF = IF, and so forth? All frequencies are in MHz.

   b. Figure 9-61 through Figure 9-64 each shows an input spectrum applied to their respective RF systems. Each system exhibits a spurious problem that masks the desired signal. In other words, some combination of the input signals and LO is producing a spurious response at or near the IF. Explain the problem and how you might fix it.

   Part A

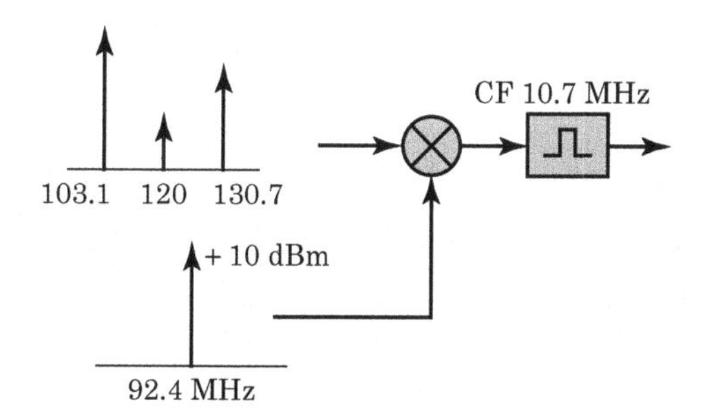

FIGURE 9-61 ■ A conversion architecture with a problematic input spectrum.

Part B

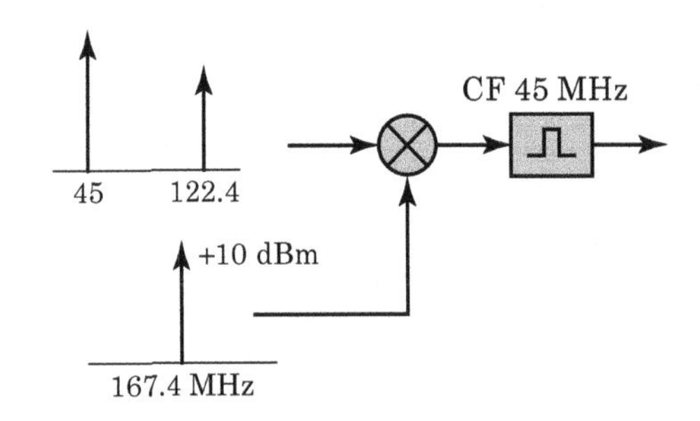

FIGURE 9-62 ■ A conversion architecture with a problematic input spectrum.

Part C

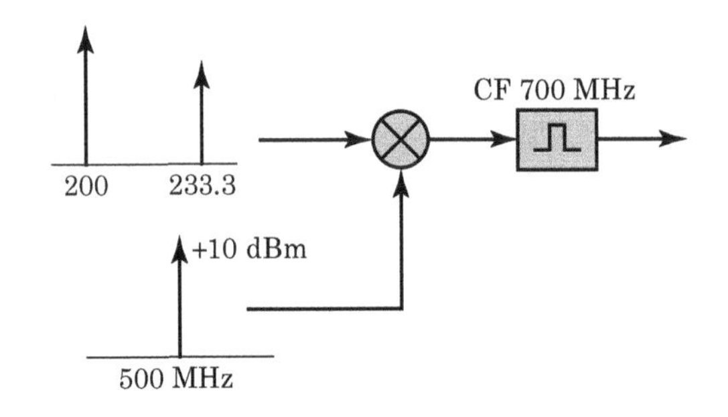

FIGURE 9-63 ■ A conversion architecture with a problematic input spectrum.

Part D

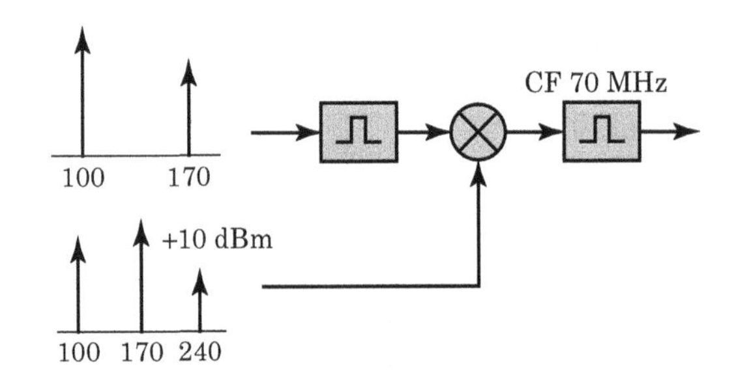

FIGURE 9-64 ■ A conversion architecture with a problematic input spectrum.

7. **Oscillator Accuracy** For the system of Figure 9-65, what is the oscillator accuracy required to ensure that the RF signal will fit completely in the IF filter? Give your answer in ppm (i.e., find X in ppm).

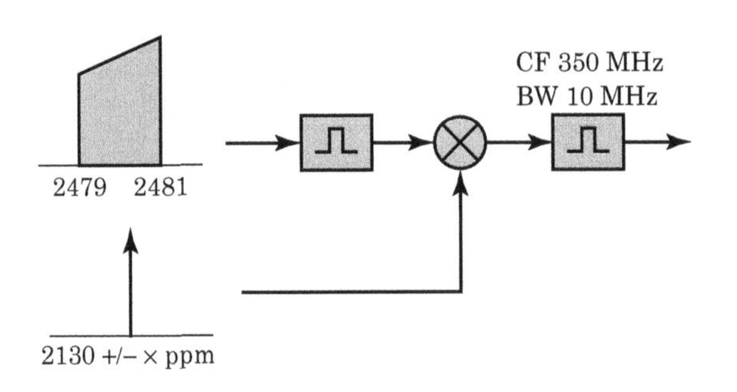

FIGURE 9-65 ■ Find the worst case oscillator accuracy.

8. **Phase Noise and Digital Modulation** Research the effects of phase noise on the BER of quadrature amplitude modulated (QAM) signals. Describe the degradation in received BER when the receiver exhibits poor phase noise. How does phase noise affect the BER/received SNR waterfall curve?

9. **Frequency Accuracy** For the receiver shown as Figure 9-66:

   a. What is the nominal tuned frequency of the receiver? In other words, to what frequency is the receiver tuned if both oscillators were perfectly accurate?

   b. If all of the oscillators drift in the worst-case direction, what is the frequency of the signal leaving the 10.7 MHz filter? In other words, what is the farthest the output signal can deviate from 10.7 MHz? Give your answer in ppm.

   c. For the worst-case condition, how much frequency offset (in Hz) is each oscillator causing?

   d. Is the final output spectrum (at 10.7 MHz) inverted?

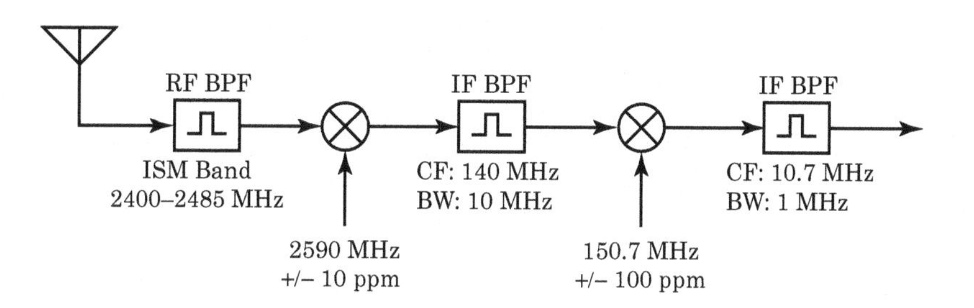

FIGURE 9-66 ■ A dual-conversion, 2.4 GHz ISM band receiver.

10. **Multiple Choice** Which statement regarding a frequency-modulated signal under small β conditions is false?

    a. The amplitude of the carrier and sideband components are given by Bessel functions.

    b. Significant energy is present only at the carrier frequency and at the first sideband frequencies.

    c. The bandwidth of the FM signal is about twice the frequency of the modulating signal.

    d. At some points in time, the magnitude of the carrier is completely suppressed.

    e. All statements are true.

11. **Oscillator Accuracy** In the downconverter schematic shown in Figure 9-67, the labels on the oscillators include their individual frequency accuracies over temperature. The downconverter is tuned to the L_1 frequency of the global positioning system (GPS), 1,575.42 MHz. The L_1 signal contains a mix of navigational messages, coarse acquisition (C/A) codes, and encrypted precision P(Y) codes.

    a. What are the minimum and maximum frequencies of the 2,025 MHz coarse-step oscillator? Answer in hertz.

    b. What are the minimum and maximum frequencies (in Hz) of the 309.58 MHz fine-step oscillator will exhibit over frequency?

    c. What is the maximum frequency range (in Hz) of the output signal (i.e., the signal at the nominal IF of 140 MHz)?

    Don't assume the two oscillators will drift in the same direction. For example, when the 2,025 MHz has drifted to +100 ppm, the 309.58 MHz oscillator may have drifted to −100 ppm.

12. **Output Spectrum** Given the input spectrum and the local oscillator spectrum shown in Figure 9-68, draw the indicated output spectrum. Show frequencies and approximate relative power levels.

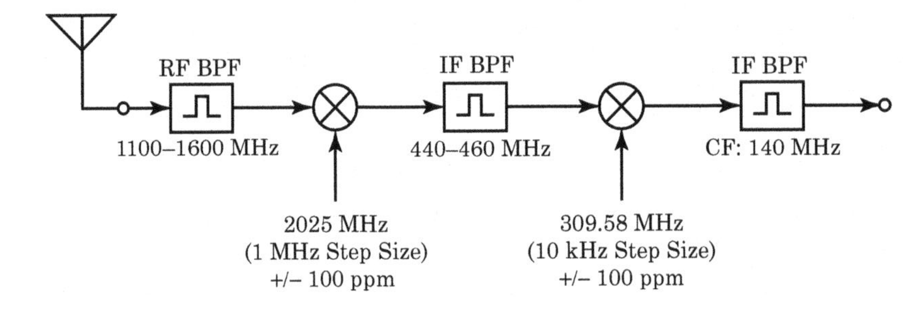

FIGURE 9-67 ■ A wide band receiver used to process GPS signals.

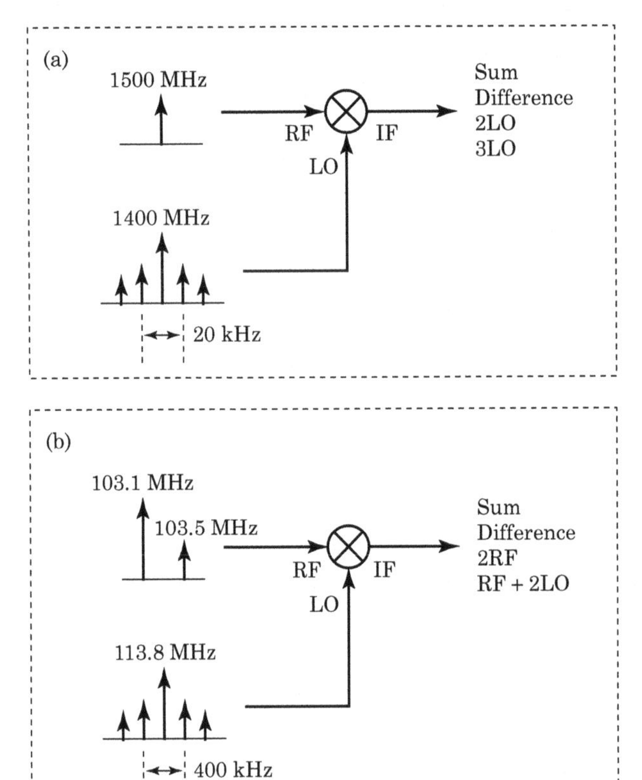

FIGURE 9-68 ■ Two conversions schemes and their input spectra.

13. **Leeson Model** Referring to the Leeson model of oscillator phase noise and to the equation relating IPM to oscillator spectrum $S_\phi(f)$:

$$IPM = \sqrt{\int_{fa}^{fb} S_\phi(f_m)\,df_m} \quad (\text{radians}_{RMS}) \tag{9.97}$$

Which of the following actions will drive the phase noise (or IPM) of an oscillator lower, given all other factors are equal:

- Lower noise figure of the active device
- Higher noise figure of the active device
- Lower oscillator center frequency
- Higher oscillator center frequency

- Lower value of IPM integration limit f_a
- Higher value of IPM integration limit f_a
- Lower value of IPM integration limit f_b
- Higher value of IPM integration limit f_b
- Higher oscillator output power
- Lower oscillator output power
- Higher resonator quality factor
- Lower resonator quality factor

14. Oscillator Phase Noise

a. Find the IPM present on the oscillator shown in Figure 9-69 assuming that small β conditions are valid. The integration limits are 5 KHz to 50 kHz. Answer in degrees_{RMS}.

b. Which bin contributes the most phase noise to the oscillator?

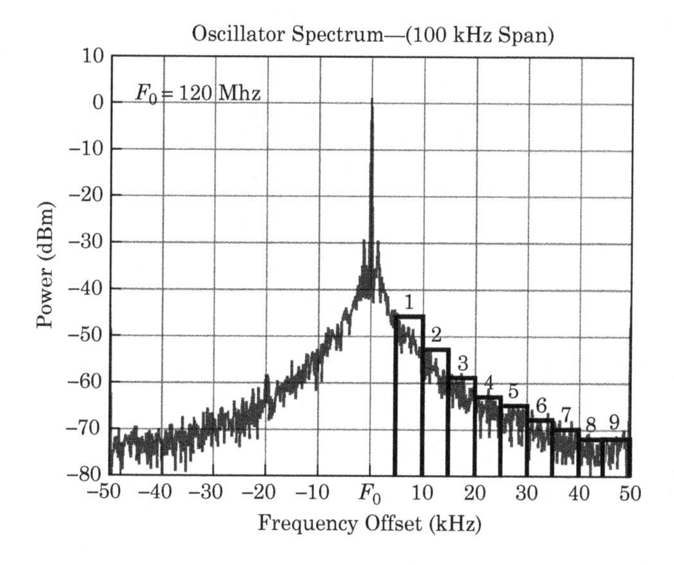

FIGURE 9-69 ■ Oscillator spectrum, broken down into individual bins for integration.

Each bin, labeled 1-9, is 5 kHz wide and is shown in Table 9-6.

TABLE 9-6 ■ The oscillator spectrum of Figure 9-21 in tabular form.

Bin Number	Width (Hz)	Amplitude (dBm)
1	5,000	−46
2	5,000	−53
3	5,000	−59
4	5,000	−63
5	5,000	−65
6	5,000	−68
7	5,000	−70
8	5,000	−72
9	5,000	−72

15. **Oscillator Accuracy** Your MP3 player is driven by a 1.0 MHz crystal oscillator whose frequency tolerance is ±1000 ppm. With a perfect oscillator (i.e., an oscillator exactly on frequency), Led Zepplin's *Stairway to Heaven* is 8 minutes (or 480 seconds) long. What is the minimum and maximum play time for this song using the imperfect oscillator? Give your answer in seconds.

    Hint: Think in terms of the number of zero crossings required to play the song. The MP3 player will clock out the full song after the oscillator has generated 480 million cycles.

16. **Oscillator Accuracy and Conversion Scheme** Given the FM radio receiver shown in Figure 9-70:

    a. If the desired RF is 88.1–107.9 MHz and the final IF is centered at 10.7 MHz, what are the two possible ranges of the LO?
    b. Given the LO shown, to what nominal frequency is the receiver tuned?
    c. If we apply a signal into the receiver at the frequency determined in part b), what are the maximum and minimum center frequencies of the tuned signal at the receiver's IF due to LO drift?

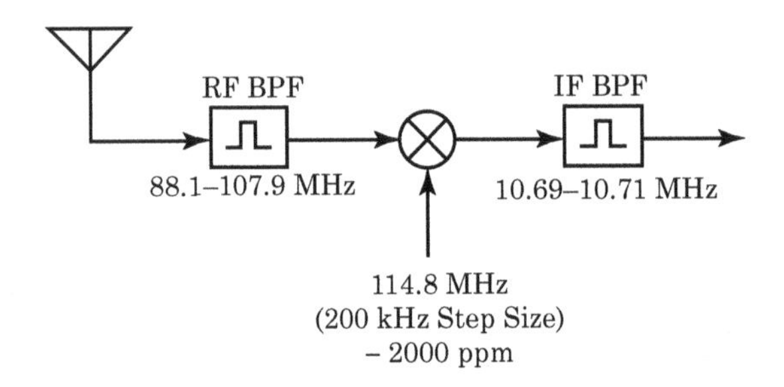

FIGURE 9-70 ■ An FM radio receiver.

17. **Multiple Choice: Oscillator Spectrum on a Spectrum Analyzer** If we display the spectrum of an oscillator on a spectrum analyzer, the function diplayed is

    a. $S(\phi)$ if the oscillator can be assumed to be operating under the large β conditions.
    b. $\mathcal{L}(f_m)$ if the oscillator can be assumed to be operating under the small β conditions
    c. $\mathcal{L}(f_m)$, which is easy to measure but not mathematically useful
    d. $S(\phi)$, which is both easy to measure and mathematically useful

18. **Phasor Diagrams** Draw a phasor diagram for an FM waveform with $\beta = 0.1$. Assume the small β conditions are valid. Show the diagram at one instant in time, being sure to show the relative positions of the sidebands with respect to the carrier. Label the sidebands and their amplitudes.

19. **Oscillator Phase Noise** In the following paragraphs, explain which system would be more problematic in terms of oscillator phase noise. Justify your answers:

    a. System 1: QPSK at 2 kbps. System 2: 8PSK at 3 kbps.
    b. System 1: QPSKat f_0. System 2: QPSK at $2f_0$.
    c. System 1: frequency shift keying (FSK) deviating at 50 Hz. System 2: FSK deviating at 5 kHz.
    d. System 1: on–off keying (OOK) with a data rate of 1 kbps. System 2: OOK with a data rate of 10 kbps.

20. **Oscillator Spurious** The receiver of Figure 9-71 is tuned to 101.1 MHz by the 111.8 oscillator $(111.8 - 101.1 = 10.7$ MHz). The oscillator exhibits the spurious signals shown.

a. The signal at 101.1 MHz is not detectable at the output of the 10.7 MHz IF filter? Explain.
b. How much bigger does the signal at 101.1 MHz have to be before detection is possible?
c. Is the signal at 100.1 MHz detectable if the signal at 100.9 MHz goes off the air?

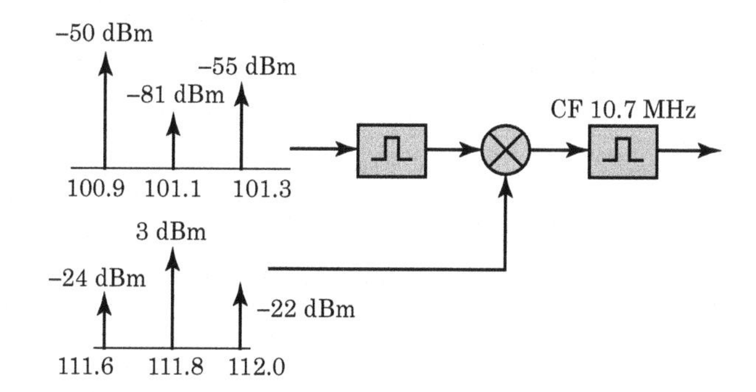

FIGURE 9-71 ■ A single-stage conversion scheme and its input spectrum.

21. Oscillator Accuracy Fill in Table 9-7 with the missing data. Assume we want the entire signal to be present somewhere in the final IF filter. Give oscillator accuracies in ppm and filter bandwidths in hertz.

TABLE 9-7 ■ Fill in the missing sections for the given conversion architectures.

	FM Radio	Wireless LAN	CB Radio	Cellular Telephone
Center Frequency	101.1 MHz	2400 MHz	27 MHz	824 MHz
Signal Bandwidth	150 kHz	11 MHz	10 kHz	30 kHz
Oscillator Accuracy		±10 ppm	±100 ppm	
IF Filter Bandwidth	200 kHz			50 kHz

22. Limits of Integration You are evaluating a candidate QPSK system that works at a 10 kbps data rate. The data is Manchester encoded. What limits of integration would you use in the IPM equation for this system? Why?

The following equation describes IPM:

$$IPM = \sqrt{\int_{fa}^{fb} S_\phi (f_m)\, df_m} \text{ (radians}_{RMS}\text{)} \tag{9.98}$$

CHAPTER 10

Cascade Design

What are you able to build with your blocks? Castles and palaces, temples and docks.

Robert Louis Stevenson

Most receiver designers pick their conversion schemes the same way that men pick their hairstyles. They choose one early in life, then stick with it.

KJ McClaning

Chapter Outline

10.1 INTRODUCTION

In this chapter, we discuss the art of receiver design. This chapter is largely about engineering trade-offs. We discuss the interaction of filters, component noise, component linearity, mixers, oscillators, and transmission lines in cascade. We will first examine gain distribution in a cascade of components. We will then discuss frequency planning and conversion schemes in detail. Finally, we'll look at some example designs.

10.1.1 Our Task

Consider the receiving system of Figure 10-1, which contains three pieces: the antenna, the analog-to-digital converter/demodulator (ADC/Demodulator), and the electronics between the two.

The antenna port presents the receiver with various signals. The signals will be at different frequencies, with varying power levels and with different types of modulation. The ADC expects its input signal to be at a single frequency—the intermediate frequency (IF)—and it expects that input signal be at a particular power level.

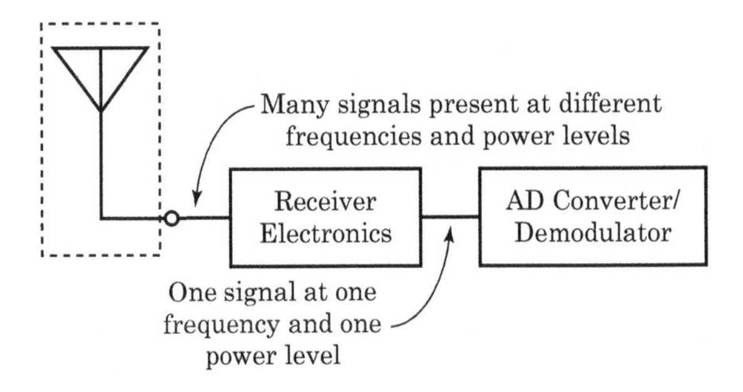

FIGURE 10-1 ■ A generalized receiving system, illustrating our design goals.

We desire to convert the signal of interest from its on-the-air frequency to the IF. We must also adjust the gain of the receiver so the ADC always sees a suitable power level. Oscillator drift (both in the transmitter and receiver), phase noise, interfering signals, multipath, and a host of other difficulties complicate our task.

10.1.2 Input/Output Requirements

Figure 10-2 shows some of the specifications of the antenna (or signal source) and the signal sink (or the ADC/demodulator).

10.1.2.1 Signal Source

The signal source provides signals with the following characteristics:

- Different center frequencies
- Different bandwidths
- Different signal power levels
- A noise floor that changes with frequency
- A nominal impedance environment

10.1.2.2 Signal Sink

The signal sink may demand that the signal of interest:

- Be centered at one particular frequency
- Exist in a filter large enough to pass the signal without distortion yet narrow enough not to let excess noise or interfering signals through
- Be at a particular power level
- Exist in a specific impedance level

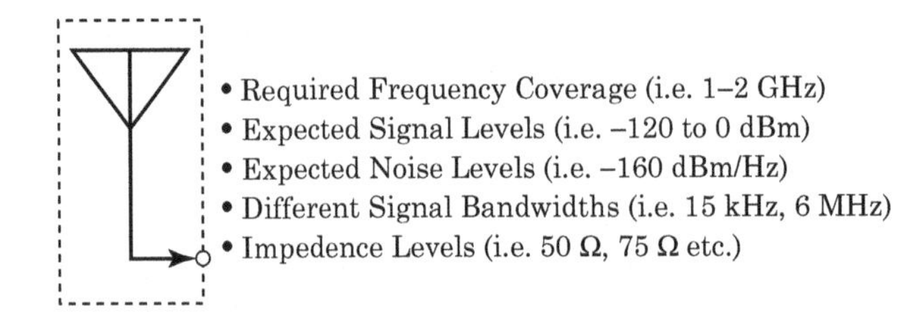

FIGURE 10-2 ■ The input and output conditions of a signal source (i.e., an antenna) and signal sink (i.e., the ADC/demodulator block).

10.1.2.3 System Specifications

The receiver converts signals from the relatively messy antenna port to the relatively clean and controlled environment required by the demodulator. The receiver electronics of Figure 10-1 must perform this conversion under a variety of system specifications, including:

- Required frequency range
- Maximum allowable added noise (relates to the system's noise temperature or noise figure)
- Maximum allowable signal distortion—relates to the system's linearity => second-order intercept (SOI), third-order intercept (TOI)
- Allowed power range of the input signals that the system must process (very small to very large)

10.2 | MINIMUM DETECTABLE SIGNAL

The minimum detectable signal (MDS) of a system is the minimum signal power we can apply to a receiver and still be able to detect the signal's presence at the output. With this specification, we are not interested in demodulating the signal; we just want to be able to tell whether the signal is present. We define the *MDS* power as input signal whose signal power equals the equivalent input noise power, or

$$S_{MDS} = N_{in} \tag{10.1}$$

where N_{in} is the total noise power dissipated in the system's input resistor R_{sys}. N_{in} includes the noise from the antenna and the receiver's own internally generated noise (see Figure 10-3).

When the input noise temperature is T_0, the total noise power dissipated in R_{sys} is

$$\begin{aligned} N_{in} &= F_{sys} k T_0 B_n \\ &= S_{MDS} \end{aligned} \tag{10.2}$$

so

$$S_{MDS} = F_{sys} k T_0 B_n \tag{10.3}$$

Converting to dBm

$$S_{MDS,dBm} = F_{sys,dB} - 174 \frac{\text{dBm}}{\text{Hz}} + 10 \log (B_n) \tag{10.4}$$

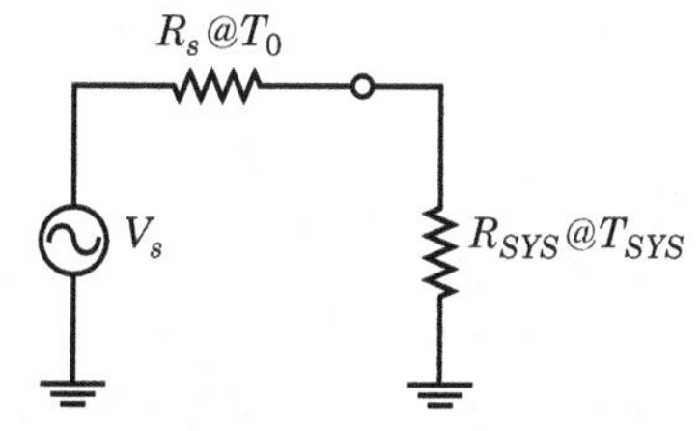

FIGURE 10-3 ■ Internal and external noise sources in a receiving system.

EXAMPLE

A large, parabolic antenna feeds satellite signals into a receiver with a 12 MHz bandwidth and a 1.5 dB noise figure.

a. What is the *MDS*?

b. Assuming the system requires 20 dB of signal-to-noise ratio (SNR) to demodulate these signals, what is the minimum signal power the receiver can process?

Solution

a. The input noise temperature of a satellite system is usually lower than T_0 or 290°K. Because we do not have any other information, we will assume that the input noise temperature is T_0 and proceed. Using equation (10.4), we know

$$\begin{aligned} S_{MDS,dBm} &= F_{sys,dB} - 174\frac{\text{dBm}}{\text{Hz}} + 10\log(B_n) \\ &= 1.5 - 174\frac{\text{dBm}}{\text{Hz}} + 10\log\left(12\cdot 10^6\right) \\ &= -101.7\ \text{dBm} \end{aligned} \tag{10.5}$$

b. If the system requires a 20 dB SNR, the minimum input signal power is

$$\begin{aligned} S_{min} &= S_{MDS,dBm} + SNR_{dB} \\ &= -101.7 + 20 \\ &= -81.7\ \text{dBm} \end{aligned} \tag{10.6}$$

10.3 DYNAMIC RANGE

The term *dynamic range* describes to us how well a system handles small signals in the presence of large signals. System designers have developed many ways to specify this characteristic, but we will discuss only the three most common specifications.

10.3.1 Linear Dynamic Range

Linear dynamic range is the difference between the *MDS* and the 1 dB input compression point (*ICP*) of a system:

$$DR_{lin,dB} = ICP_{dBm} - S_{MDS,dBm} \tag{10.7}$$

Linear dynamic range is useful when comparing systems, but this specification does not provide insight into the amount of distortion generated by a system.

10.3.2 Gain-Controlled Dynamic Range

Gain-controlled dynamic range, or *noninstantaneous dynamic range*, describes the ability of a system to process signals of different power levels—one at a time. We present only one signal to the system, and then let the system adjust itself to that one signal. For example, when first entering a dark room, our eyes see little. After waiting for some time, our pupils

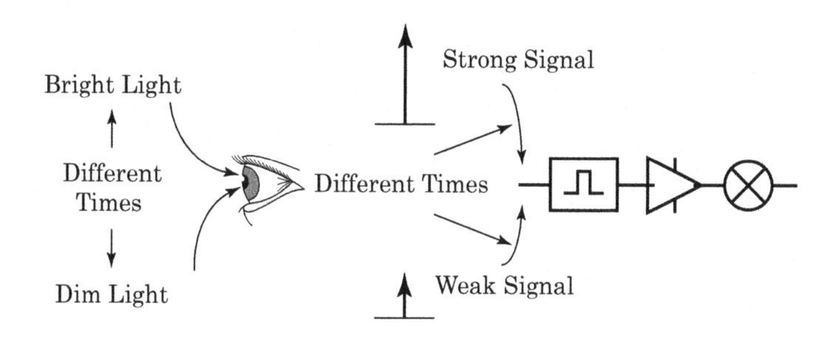

FIGURE 10-4 ■ Gain-controlled dynamic range is similar to the behavior of the eye in dim and brightly lit conditions.

open and our eyes become more sensitive. Similarly, we can also function in a bright, well-lit room after allowing our eyes to adapt (see Figure 10-4).

Gain-controlled dynamic range also describes the range of input signals the receiver is capable of processing. The lower end is set by the noise performance of the system. The upper end is usually set by some system nonlinearity.

10.3.3 Spur-Free Dynamic Range

Spur-free dynamic range (SFDR), or *instantaneous dynamic range*, describes the ability of a receiver to process a small, desired signal in the presence of a large, unwanted signal. For example, imagine driving down the road late at night, and a car approaches in the other lane. The approaching driver is using his high beams, and they are shining directly into your eyes. Spur-free dynamic range describes your ability to see the relatively faint light coming off the lines on the road (the desired signal) in the presence of the other driver's high beams (the undesired signal). The equivalent receiver situation is processing a very small signal in the presence of a very large signal (see Figure 10-5).

We are interested in the range of input signals the system can process simultaneously without:

- Losing the desired signal in the noise. This is a noise problem principally controlled by the noise temperature of the receiver and the noise gathered by the receiving antenna.
- Creating undesirable spurious products that can mask or distort the small, desired signal. This is a linearity problem described principally by the system's SOI and TOI.

Spur-free dynamic range is the most used, and most useful, dynamic range specification. Frequently, maximizing dynamic range is the designer's major goal. Spur-free dynamic range is the difference between the MDS and the input power that will cause the third-order spurious products to be equal to the MDS. The low-level input condition is

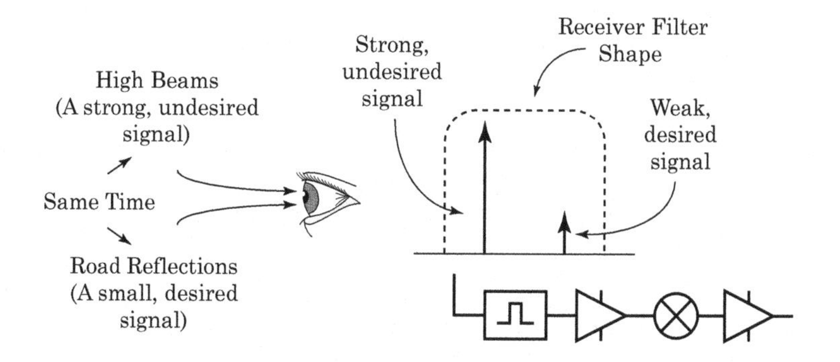

FIGURE 10-5 ■ Instantaneous dynamic range in a receiver is analogous to the behavior of the human eye when it views dim and bright objects at the same time.

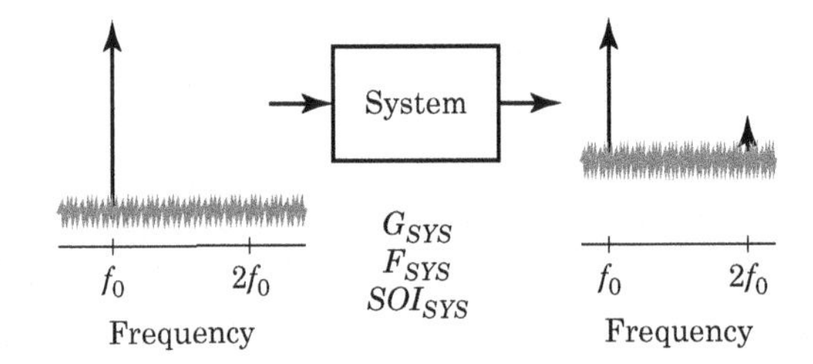

FIGURE 10-6 ▪ The high signal level input criterion when second-order distortion limits the dynamic range.

clear: The input signal equals the MDS. There are two high-level input conditions—one for second-order limited systems and one for systems limited by third-order distortion.

10.3.3.1 Second-Order Limited Dynamic Range

Figure 10-6 illustrates the high-level input conditions when second-order distortion is the major concern. The input consists of a signal tone and a noise floor. We increase the signal power until the second-order spurious product just peaks out of the noise. At this point, the power in the spurious tone equals the system MDS. The power in the second-order tone is

$$P_{in,2f} = S_{MDS} \tag{10.8}$$

and our discussions of linearity in Chapter 7 revealed

$$P_{in,2f,dBm} = 2P_{in,f,dBm} - ISOI_{dBm} \tag{10.9}$$

When the second-order distortion power just equals the system's MDS, the fundamental input power is

$$P_{in,f,dBm} = \frac{S_{MDS,dBm} + ISOI_{dBm}}{2} \tag{10.10}$$

Second-order spur-free dynamic range presents problems only in wideband systems (e.g., some RADARs, ELINT receivers). Most of the world is interested in communications systems, which are inherently narrowband and thus are limited by third-order distortion.

10.3.3.2 Third-Order Limited Dynamic Range

Figure 10-7 illustrates the high-level input conditions in systems limited by third-order distortion. We increase the signal power of the two input tones until the third-order spurious products are just discernable in the output noise. At this point, the power in each spurious tone equals the system MDS. This is the highest input signal we will allow.

Given a system with a minimum detectable signal of S_{MDS}, a noise figure of F_{sys}, and an input third-order intercept (*ITOI*), we desire to find the fundamental input power when

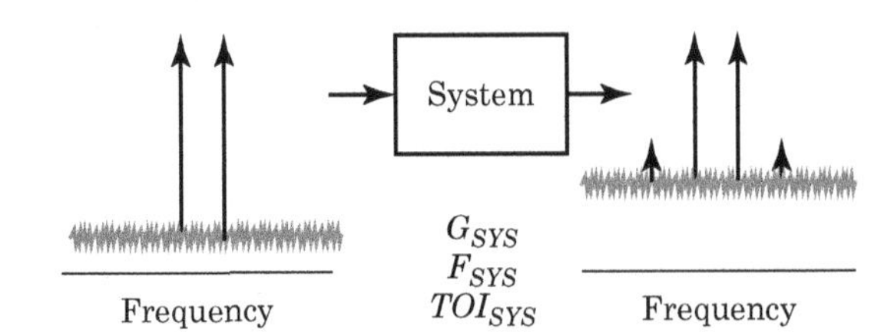

FIGURE 10-7 ▪ The high signal level input criterion when third-order distortion limits the dynamic range.

the third-order power equals the S_{MDS} or

$$P_{in,2f-f} = S_{MDS} \tag{10.11}$$

Our discussion of linearity in Chapter 7 revealed that

$$P_{in,2f-f,dBm} = 3P_{in,f,dBm} - 2ITOI_{dBm} \tag{10.12}$$

Combining equation (10.12) with equation (10.11) produces

$$P_{in,f,dBm} = \frac{1}{3}S_{MDS,dBm} + \frac{2}{3}ITOI_{dBm} \tag{10.13}$$

This is the maximum input signal power that the system can process without producing detectable third-order spurious signals.

10.3.3.3 Spur Free Dynamic Range

Using the third-order criteria, the *SFDR* is the difference between the maximum allowable input signal [$P_{in,f,dBm}$ of equation (10.13)] and the system's minimum detectable signal, or

$$\begin{aligned} SFDR_{dB} &= P_{in,f,dBm} - S_{MDS.dBm} \\ &= \frac{1}{3}S_{MDS,dBm} + \frac{2}{3}ITOI_{dBm} - S_{MDS.dBm} \\ &= \frac{2}{3}[ITOI_{dBm} - S_{MDS.dBm}] \\ &= \frac{2}{3}\left[ITOI_{dBm} - F_{sys}kT_0B_n\right] \end{aligned} \tag{10.14}$$

Designers seek a large spur-free dynamic range. To increase the *SFDR*, we must increase the *ITOI* (increase the system's linearity), decrease the noise figure, or decrease noise bandwidth.

Spur-free dynamic range applies only to spurious signals produced by the nonlinearities in the receiver's cascade chain. Spurious signals can arise in a receiver through other mechanisms such as poor local oscillator (LO) rejection, mixing products, or electromagnetic interference generated within the receiver itself.

EXAMPLE

SFDR

Given the cascade shown as Figure 10-8, find the *MDS* and the *SFDR* in a 30 kHz bandwidth.

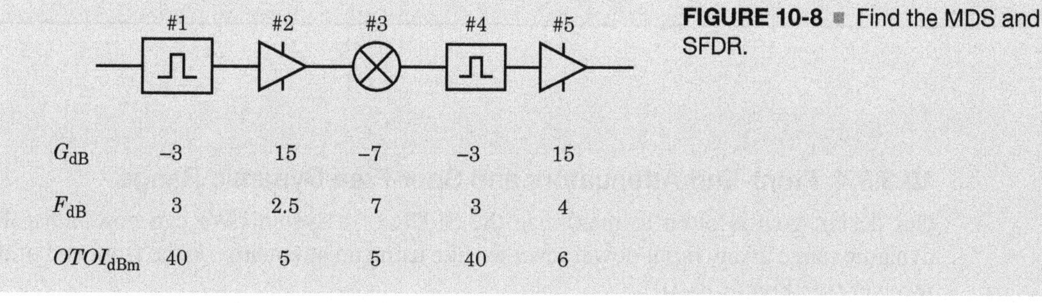

FIGURE 10-8 ■ Find the MDS and SFDR.

Solution

The cascade power gain $G_{p,cas}$ is 17 dB, the cascade noise figure F_{sys} is 7.05 dB, and the cascade's *ITOI* is -12.7 dBm. Using equation (10.4), the *MDS* is

$$\begin{aligned} S_{MDS,dBm} &= F_{sys,dB} - 174\frac{\text{dBm}}{\text{Hz}} + 10\log(B_n) \\ &= 7.05 - 174\frac{\text{dBm}}{\text{Hz}} + 10\log(30{,}000) \\ &= -122.2 \text{ dBm} \end{aligned} \tag{10.15}$$

Equation (10.14) gives us the *SFDR*

$$\begin{aligned} SFDR_{dB} &= \frac{2}{3}[ITOI_{dBm} - S_{MDS,dBm}] \\ &= \frac{2}{3}[-12.7 - (-122.2)] \\ &= 73 \text{ dB} \end{aligned} \tag{10.16}$$

Figure 10-9 shows the results of these calculations graphically. The output noise floor is

$$\begin{aligned} N_{out,dBm} &= G_{p,dB} + F_{sys,dB} - 174\frac{\text{dBm}}{\text{Hz}} + 10\log(B_n) \\ &= 17 + 7 - 174 + 10\log(30{,}000) \\ &= -105.2 \text{ dBm} \end{aligned} \tag{10.17}$$

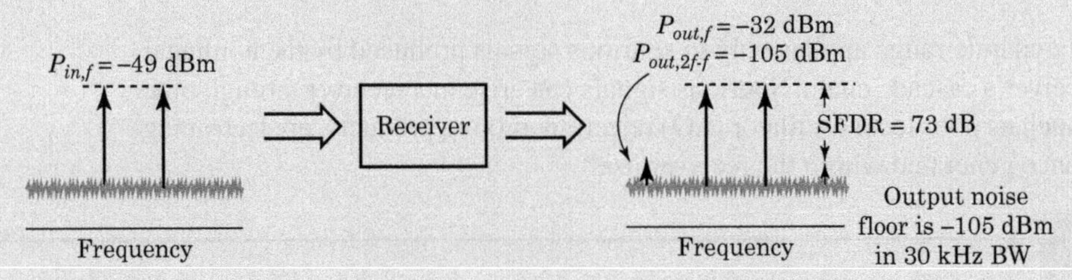

FIGURE 10-9 ■ Graphical depictions of the MDS and SFDR.

We can just detect the third-order distortion products when the fundamental input power is −49 dBm.

10.3.3.4 Front-End Attenuators and Spur-Free Dynamic Range

Our design goal is often to maximize the SFDR of a system. We can now *move* this dynamic range to any input power level we like using an attenuator on the front end of the receiver (see Figure 10-10).

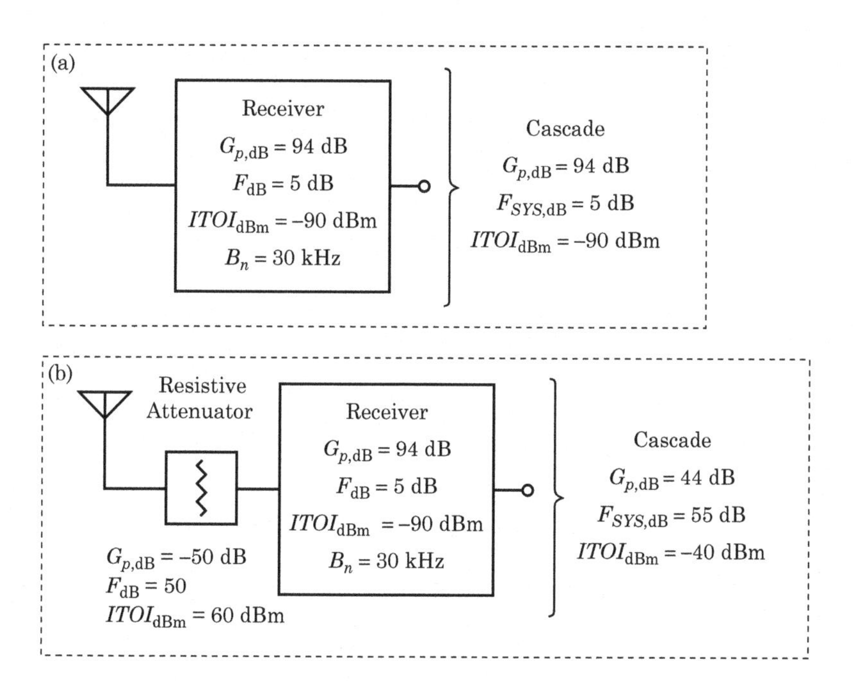

FIGURE 10-10 System SFDR with and without a resistive attenuator on the system front end.

Without Attenuator Figure 10-10(a) shows a receiving system with a 5 dB noise figure, a 30 kHz noise bandwidth, and a −90 dBm ITOI. The system has an MDS of

$$\begin{aligned} S_{MDS,dBm} &= F_{sys,dB} - 174\frac{\text{dBm}}{\text{Hz}} + 10\log(B_n) \\ &= 5 - 174\frac{\text{dBm}}{\text{Hz}} + 10\log(30{,}000) \\ &= -124 \text{ dBm} \end{aligned} \tag{10.18}$$

The *SFDR* is

$$\begin{aligned} SFDR_{dB} &= \frac{2}{3}[ITOI_{dBm} - S_{MDS,dBm}] \\ &= \frac{2}{3}[-90 - (-124)] \\ &= 22.7 \text{ dB} \end{aligned} \tag{10.19}$$

With Attenuator The system with the attenuator is shown in Figure 10-10(b). When we add the attenuator, the gain of the system drops to 44 dB, the noise figure rises to 55 dB, and the system's ITOI rises to −40 dBm. The *MDS* is

$$\begin{aligned} S_{MDS,dBm} &= 55 - 174\frac{\text{dBm}}{\text{Hz}} + 10\log(30{,}000) \\ &= -74 \text{ dBm} \end{aligned} \tag{10.20}$$

The attenuator has made the system very insensitive. The *SFDR* is

$$\begin{aligned} SFDR_{dB} &= \frac{2}{3}[ITOI_{dBm} - S_{MDS.dBm}] \\ &= \frac{2}{3}[-40 - (-74)] \\ &= 22.7 \text{ dB} \end{aligned} \tag{10.21}$$

The *SFDR* is the same. The receiver simply performs over a different range of input power levels. In general, a receiver can process signal power levels in the range of

$$MDS_{dBm} \text{ to } (MDS_{dBm} + SFDR_{dB}) \tag{10.22}$$

Without the attenuator, the system can process signals over the input power range of

$$\begin{aligned} &-124 \text{ to } (-124 + 83) \\ &= -124 \text{ dBm to} - 41 \text{ dBm} \end{aligned} \tag{10.23}$$

With the attenuator, the system can process signals over the input power range of

$$\begin{aligned} &-74 \text{ to } (-74 + 83) \\ &= -74 \text{ dBm to} + 9 \text{ dBm} \end{aligned} \tag{10.24}$$

An input attenuator affects not the *SFDR* of a receiver but only the range of input signals. These calculations assume we're using a resistive or other highly linear attenuator.

10.3.4 Dynamic Range Notes

10.3.4.1 What's Being Specified

There are many dynamic range specifications, so it is prudent to ensure which dynamic range specification is being discussed. Designers of fixed-gain systems usually specify the SFDR. Designers of receiving systems often quote the gain-controlled dynamic range. Moreover, we must consider the second-order/third-order ambiguity.

10.3.4.2 Other Sources of Spurious Signals

There are many sources of spurious signals in a receiver. The dynamic range quantities we have specified here address only the nonlinearity of the system (i.e., the second- and third-order intercepts of the components making up the system).

Dual Conversion Systems Figure 10-11 shows one source of additional spurious products. In a dual conversion receiver, we convert the signal from the RF to the first IF. After filtering,

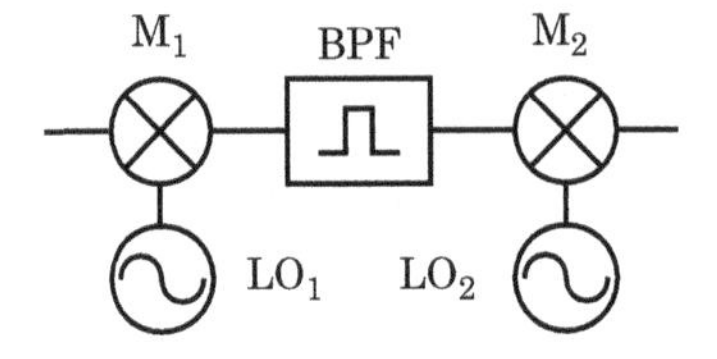

FIGURE 10-11 ■ One source of spurious signals in a receiving system. LO_1 can pass through the band-pass filter (BPF) and participate in M2's mixing process. Similarly, LO_2 can pass through the BPF and mix with the signals present in mixer M1.

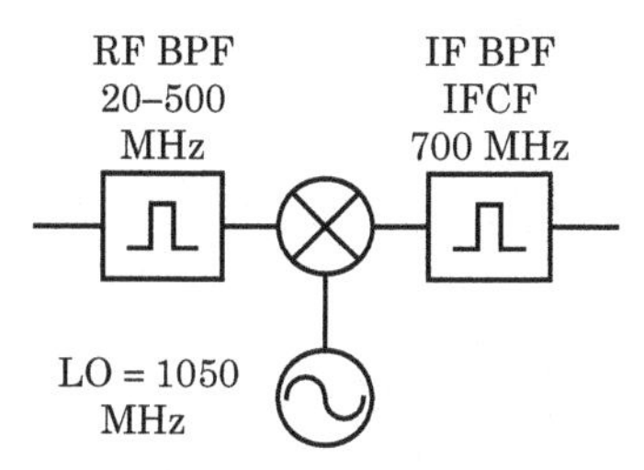

FIGURE 10-12 ■ Another source of spurious signals in a receiving system. A signal at 350 MHz will pass through the RF BPF, producing a second harmonic of 700 MHz. The 700 MHz second harmonic is centered at the IF BPF center frequency, so the spurious signal will pass into the receiver unencumbered.

we then convert the signal from the first IF to a second IF. The dual conversion requires that two separate LOs be simultaneously active in the receiver. Due to leakage effects, LO_1 can find its way into mixer M_2, and LO_2 can find its way into mixer M_1. These two signals will mix and produce spurious products at

$$|mf_{LO1} \pm nf_{LO2}| \text{ for } \begin{cases} n = 0,1,2,3\ldots \\ m = 0,1,2,3\ldots \end{cases} \tag{10.25}$$

When one of these products falls inside any one of the IF filters in the receiver, the result is a spurious signal. This type of spurious response (termed *internally generated spurious*) is unrelated to the linearity of the amplifiers and mixers that make up the system. The amount of spurious signal power is purely a function of how well we've separated the two LOs. Our choice of LO frequencies and IFs in our designs is extremely important when we seek to minimize these spurs.

Other Nonlinearities Figure 10-12 shows an example architecture that will generate spurious products. A 350 MHz RF signal will pass through the RF band-pass filter (BPF) and generate second-order distortion in the mixer. This effect produces a 700 MHz signal at the mixer's output regardless of the LO frequency. This second-order effect is not addressed by equation (10.14), so a spurious product generated in this way won't be predicted by the equation. This spurious product will be very large and will almost certainly reduce the measured spur-free dynamic range of the receiver.

Digital Logic Another serious source of receiver spurious products is the digital control system. Figure 10-13 shows a typical microprocessor-controlled receiving system. Although designers take great pains to isolate the sensitive receiving circuits from the digital logic, the digital signals occasionally find their way onto the receiver's signal path. Consider a receiver with a 60 MHz IF and a 10 MHz microprocessor crystal. Since the microprocessor is probably a 0 to 5 volt square wave, it is rich in strong harmonics. The sixth harmonic of a 10 MHz microprocessor clock will fall directly on top of the 60 MHz IF center frequency.

The receiver's control logic often contains functional counters, gating, and other functions that produce many nonobvious frequency components. A microprocessor clocked at 10 MHz might produce components at 45.6 kHz, 22.3 MHz, and other frequencies. Digital control lines that must pass into sensitive RF compartments are especially dangerous. Such lines should be heavily filtered.

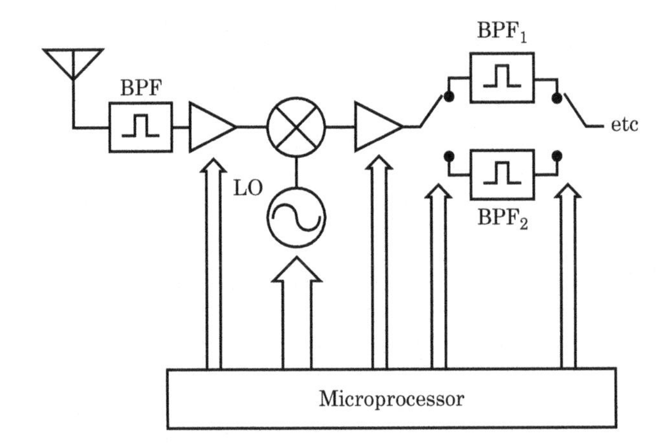

FIGURE 10-13 ■ Noise and other coherent signals can leak into the sensitive RF path of a receiving system. These coherent signals can show up as coherent spurious products on the receiver's output.

10.3.4.3 Measuring the SFDR

Receivers users, as opposed to receiver designers, are not concerned with the mechanisms that produce a spurious signal. Users are interested only in whether a particular signal is from the antenna or has been generated internally by the receiver. Testing should reflect the concerns of the receiver user.

The SFDR is measured by tuning the receiver with the antenna port terminated in a matched load. Any signals present in the receiver have been generated internally. The equivalent input power level of the largest of these internally generated spurious products is then used in place of S_{MDS} in equation (10.14). Internally generated spurious signals, compared with second- or third-order spurious signals, are generally the limiting factor in receiver dynamic range.

10.4 | GAIN DISTRIBUTION—NOISE AND LINEARITY IN CASCADE

The design goal is to appropriately convert signals present at the output of our antenna into a format compatible with some downstream signal processor. This task requires we change the power level and frequency of the input signal. We also strive to remove unwanted signals that lie close to our signal of interest.

10.4.1 Required Gain

Figure 10-14 shows a receiving system in its initial design stages. We assume that the noise figure of the completed receiver must be 5 dB and that its ITOI will be 0 dBm. The receiver must support frequencies of 500 to 1000 MHz.

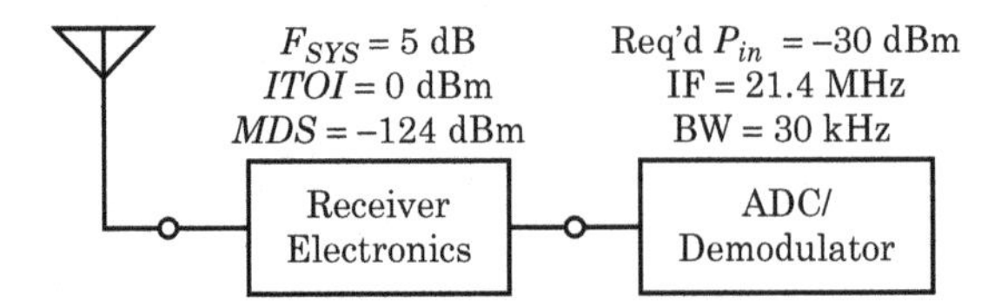

FIGURE 10-14 ■ A receiving system in its initial design stages. We have shown only the high-level design specifications.

The ADC/demodulator block requires a signal with a 30 kHz bandwidth. The signal must be centered at 21.4 MHz and must be at least −30 dBm. The minimum detectable signal of this system will be

$$\begin{aligned} S_{MDS,dBm} &= F_{sys,dB} - 174\frac{\text{dBm}}{\text{Hz}} + 10\log(B_n) \\ &= 5 - 174\frac{\text{dBm}}{\text{Hz}} + 10\log(30{,}000) \\ &= -124\ \text{dBm} \end{aligned} \tag{10.26}$$

Since we need −30 dBm at the output of the receiver even under MDS input conditions, the receiver must supply a power gain of

$$\begin{aligned} G_{p,dB} &= -30 - (-124) \\ &= 94\ \text{dB} \end{aligned} \tag{10.27}$$

We must also translate any signal in the 500 to 1000 MHz range to a 21.4 MHz IF center frequency. We seek the best architecture to distribute the gains and losses of the system to achieve maximum SFDR.

10.4.2 Three Gain Distributions

Figure 10-15 shows three ways to distribute the 94 dB of power gain:

- In Figure 10-15(a), we have placed all of the gain at the front of the cascade. After the signal passes through the lossy components, the cascade gain is 94 dB.
- In Figure 10-15(b), we have placed all of the power gain at the very end of the cascade. After accounting for the lossy elements, we have a total cascade gain of 94 dB.
- Figure 10-15(c) shows a distributed system. We apply gain in various places throughout the cascade and in concert with the losses. Total cascade gain is still 94 dB.

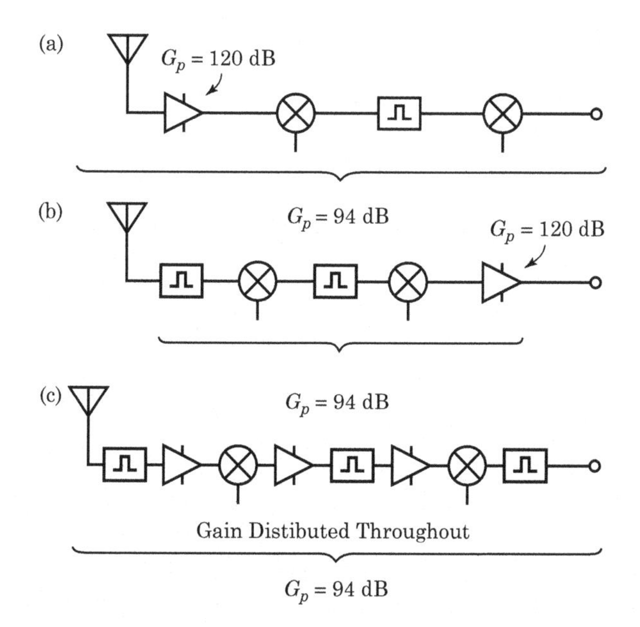

FIGURE 10-15 Three ways to distribute power gain in a receiving system. (a) Place all of the gain at the front end of the receiver. (b) Place all of the gain at the end of the receiver. (c) Distribute the gain throughout the receiver using as yet unspecified rules.

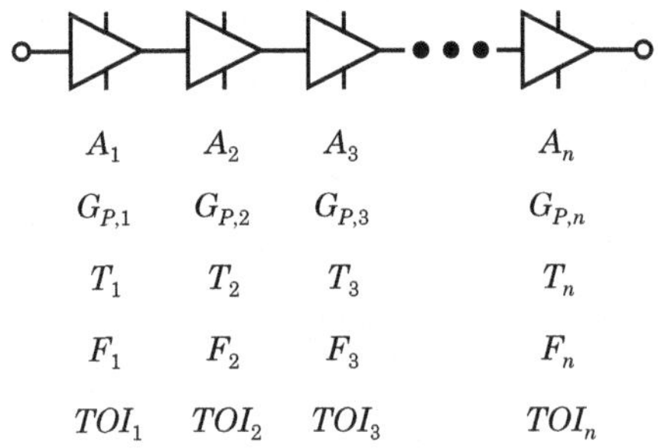

FIGURE 10-16 ■ A general n-element cascade.

A_1	A_2	A_3	A_n
$G_{P,1}$	$G_{P,2}$	$G_{P,3}$	$G_{P,n}$
T_1	T_2	T_3	T_n
F_1	F_2	F_3	F_n
TOI_1	TOI_2	TOI_3	TOI_n

We seek the architecture that will produce the largest dynamic range. We gain insight by examining several of the cascade equations developed over the last few chapters. For the n-element cascade of Figure 10-16, we can write:

1. The noise temperature of a cascade is

$$T_{cas} = T_1 + \frac{T_2}{G_{p,1}} + \frac{T_3}{G_{p,1}G_{p,2}} + \cdots + \frac{T_n}{G_{p,1}G_{p,2}G_{p,3}\ldots G_{p,n-1}} \tag{10.28}$$

Similarly, the noise figure of a cascade given the noise figure and gain of its components

$$F_{cas} = F_1 + \frac{F_2 - 1}{G_{p,1}} + \frac{F_3 - 1}{G_{p,1}G_{p,2}} + \cdots + \frac{F_n - 1}{G_{p,1}G_{p,2}G_{p,3}\ldots G_{p,n-1}} \tag{10.29}$$

2. The ITOI of a cascade (assuming coherent addition) is

$$\begin{aligned}\frac{1}{ITOI_{cas}} = \frac{1}{ITOI_1} &+ \frac{1}{ITOI_2/G_{p,1}} \\ &+ \frac{1}{ITOI_3/G_{p,1}G_{p,2}} \\ &+ \cdots \\ &+ \frac{1}{ITOI_n/G_{p,1}G_{p,2}\ldots G_{p,n-1}}\end{aligned} \tag{10.30}$$

We will discuss only the ITOI equation in this chapter, but the following arguments are valid for other linearity specifications.

We will begin by examining some common themes in equations (10.28), (10.29), and (10.30).

10.4.3 Excess Gain

The three cascade equations contain expressions regarding *excess gain*. Excess gain describes the amount of power gain between the cascade input and a particular component. Figure 10-17 shows an example of excess gain calculation. The first row of numbers, placed under each component, is the power gain of the component. The row of numbers centered between components is the excess gain of the cascade up to that point.

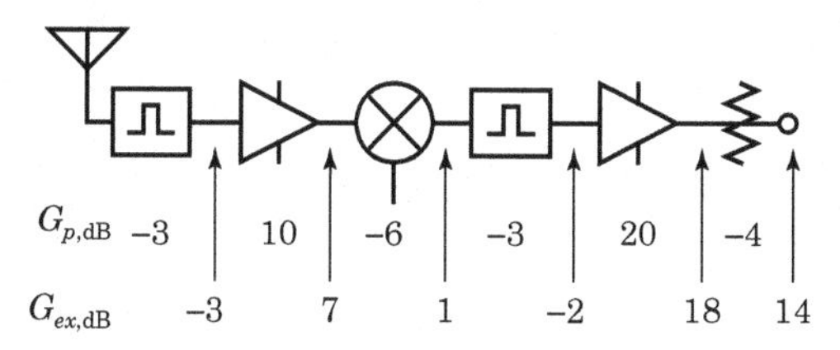

FIGURE 10-17 ■ An example of excess gain calculation. The excess gain is the amount of gain between the antenna and any point in the cascade.

To find the excess gain of any point in a cascade, simply add up the power gains (in dB) of all the components before the point of interest. Remember that gains are associated with components and that excess gain is associated with the points between components.

The excess gain appears in equations (10.28), (10.29), and (10.30) as

$$G_{excess} = G_{p,1} G_{p,2} G_{p,3} \ldots G_{p,n-1} \tag{10.31}$$

10.4.4 Translation of Component Specifications

We observe that the individual terms of the cascade equations mathematically translate the component specifications to either the input or output port.

10.4.4.1 Linearity

For example, to determine the linearity contribution of the third element in a cascade, equation (10.30) causes us to compute the term:

$$\frac{1}{ITOI_3 / G_{p,1} G_{p,2}} \tag{10.32}$$

The term $ITOI_3$ represents a power in watts. The term

$$ITOI_3 / G_{p,1} G_{p,2} \tag{10.33}$$

mathematically moves the power ITOI_3 to the input of the cascade. Every term in equation (10.30) operates similarly. For example, Figure 10-18 shows three amplifiers in cascade.

It is not obvious which amplifier is limiting the cascade's ITOI until we move each component's ITOI to the input port of the cascade:

Amplifier A_1: This amplifier's TOI is already referenced to the cascade's input, so

$$\begin{aligned} A_1\text{'s TOI referenced to cascade input} &= \text{ITOI}_1 \\ &= -10 \text{ dBm} \end{aligned} \tag{10.34}$$

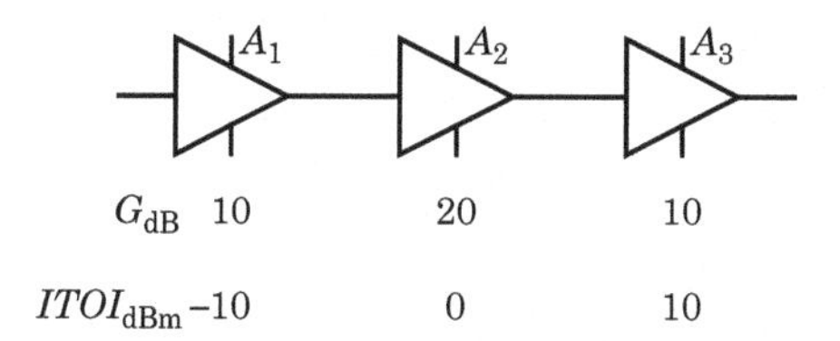

FIGURE 10-18 ■ Three amplifiers in cascade. Determine which amplifier limits the cascade TOI.

Amplifier A_2: The relevant part of equation (10.30) is

$$ITOI_2/G_{p,1} \tag{10.35}$$

so

$$\begin{aligned} A_2\text{'s TOI referenced to cascade input} &= \mathrm{ITOI_2}/\mathrm{G_{p,1}} \\ &= 0 - 10 \text{ dBm} \\ &= -10 \text{ dBm} \end{aligned} \tag{10.36}$$

Amplifier A_3: The relevant part of equation (10.30) is

$$ITOI_3/G_{p,1}G_{p,2} \tag{10.37}$$

so

$$\begin{aligned} A_3\text{'s TOI referenced to cascade input} &= \mathrm{ITOI_3}/\mathrm{G_{p,1}G_{p,2}} \\ &= 10 - 30 \text{ dBm} \\ &= -20 \text{ dBm} \end{aligned} \tag{10.38}$$

After moving the ITOI of all the amplifiers to a common point, we find two amplifiers both produce a -10 dBm equivalent input TOI, whereas the third amplifier produces a -20 dBm value. Since the ITOI of a device is linearly related to the amount of power the device can handle before it *goes nonlinear,* we see that the third amplifier is limiting the TOI of the cascade. The ITOI for this cascade, using equation (10.30), is -20.8 dBm.

10.4.4.2 Noise Temperature

The concept of translation to a common point is also suitable for the noise temperature and noise figure. Figure 10-19 shows three amplifiers in cascade. Which one limits the noise performance of the cascade? We turn to equation (10.28).

Amplifier A_1: The noise temperature contribution of this amplifier is already referenced to the cascade's input, so

$$\begin{aligned} A_1\text{'s noise temperature at the cascade input} &= T_1 \\ &= 290°\text{K} \end{aligned} \tag{10.39}$$

Amplifier A_2: Equation (10.28) indicates

$$\begin{aligned} A_2\text{'s noise temperature at the cascade input} &= T_1/G_{p,1} \\ &= 8{,}881/10 \\ &= 888.1°\text{K} \end{aligned} \tag{10.40}$$

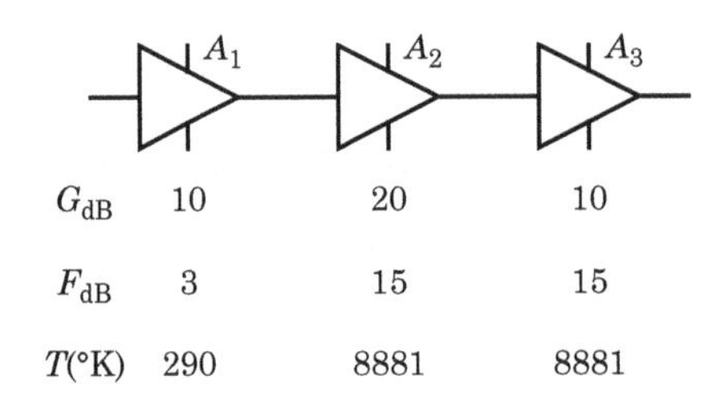

FIGURE 10-19 ■ Three amplifiers in cascade. Determine which amplifier limits the cascade's noise performance.

Amplifier A_3: Equation (10.28) reveals

$$\begin{aligned} A_3\text{'s noise temperature at the cascade input} &= T_3/G_{p,1}G_{p,2} \\ &= 8{,}881/1{,}000 \\ &= 8.9^\circ\text{K} \end{aligned} \tag{10.41}$$

Translating the noise to a common port reveals that amplifier A_2 limits the noise performance.

10.4.5 Gain Distribution and Noise Temperature

Figure 10-16 and equation (10.28) describe an n-element cascade:

$$\begin{aligned} T_{cas} &= T_1 && \text{[Component \#1]} \\ &+ \frac{T_2}{G_{p,1}} && \text{[Component \#2]} \\ &+ \frac{T_3}{G_{p,1}G_{p,2}} && \text{[Component \#3]} \\ &+ \cdots \\ &+ \frac{T_n}{G_{p,1}G_{p,2}G_{p,3}\ldots G_{p,n-1}} && \text{[Component \#n]} \end{aligned} \tag{10.42}$$

- **Component #1:** This term describes the noise temperature contribution of amplifier A_1. T_1 appears in no other term of equation (10.42) so this expression represents the complete contribution of amplifier A_1. To achieve a low cascade noise temperature, the first amplifier should have a low noise temperature. Any change in the noise temperature of the first element directly affects the cascade's noise temperature. The cascade noise temperature will never be lower than the noise temperature of the first component.
- **Component #2:** This term describes the entire noise temperature contribution of amplifier A_2 and, like T_1, T_2 appears in no other term of equation (10.42). The noise temperature contribution of amplifier A_2 is reduced by the gain of amplifier A_1. This term will be small when A_1 has a large gain or when A_2 has a low noise temperature.

  For example, if A_1 has 20 dB of gain and A_2 has a 290°K noise temperature, then the noise temperature contribution of A_2 is

$$\frac{T_2}{G_{p,1}} = \frac{290}{100} = 2.9^\circ\text{K} \tag{10.43}$$

  The gain of A_1 has greatly reduced the noise contribution of A_2.
- **Component #3:** This term represents the noise contribution of A_3 to the cascade. In this case, the noise temperature of A_3 is reduced by $G_{p,1}G_{p,2}$, the gains of A_1 and A_2. Large values of gain for amplifiers A_1 and A_2 reduce the noise contribution of A_3.
- **Component #*n*:** The noise contribution of the n-th component is reduced by the gain of the $(n-1)$ components preceding the n-th component.

Relative Levels Figure 10-20 shows the noise levels present in a cascade when the excess gain of the cascade is large. Most of the cascade's noise power comes from the antenna

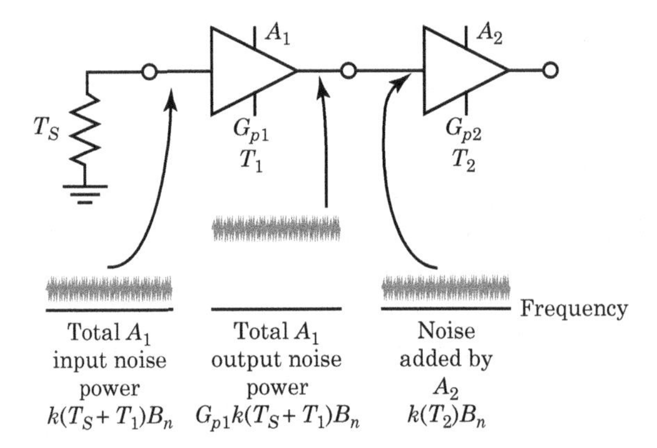

FIGURE 10-20 ■ Relative noise levels in a cascade when the excess gain is large. The output noise of amplifier A_1 is much greater than the internal noise generated by amplifier A_2's input resistor. Thus, most of the system noise originates from the cascade's input.

and the noise generated by the first amplifier because the large excess gain reduces the contribution of the subsequent amplifiers. After the input noise experiences the gain of A_1, the noise power from A_1's output is much larger than the noise added by A_2.

Figure 10-21 shows the relative noise levels when the excess gain of the cascade is small. When the excess gain preceding a particular amplifier is small, the noise contribution of that amplifier is large. For example, in Figure 10-21, the noise added by amplifier A_2 is significant because of the small excess gain prior to A_2.

If we want amplifier A_2 to add only a small amount of noise to the cascade, Figure 10-20 and equation (10.28) tell us that we would like

$$\begin{aligned} &\text{Noise at A}_1\text{'s output} \gg \text{Noise added by A}_2 \\ &G_{p,1}k\,(T_s + T_1)\,B_n \gg kT_2B_n \\ &G_{p,1}\,(T_s + T_1) \gg T_2 \end{aligned} \tag{10.44}$$

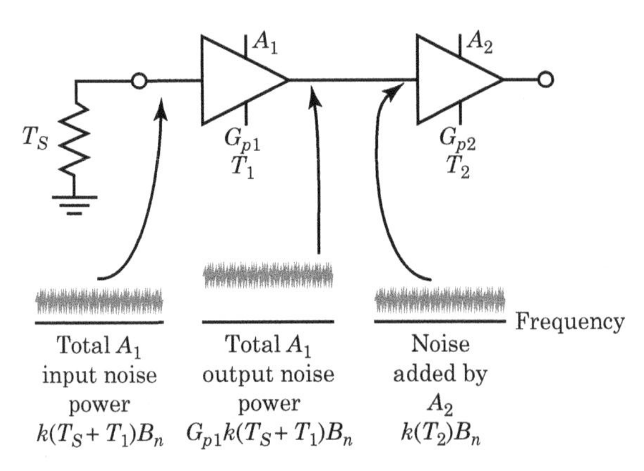

FIGURE 10-21 ■ Relative noise levels in a cascade when the excess gain is small. The input noise of amplifier A_2's input resistor is significant with respect to the output noise of amplifier A_1. Most of the system noise arises from the source temperature T_S, the internally generated noise of the first amplifier (noise temperature of T_1) and the internally generated noise of the second amplifier (noise temperature of T_2).

If we let the source noise temperature $T_s = T_0$, we can speak in terms of noise figure. The equation becomes

$$\begin{aligned} G_{p,1}(T_0 + T_1) &\gg T_2 \\ G_{p,1}[T_0 + T_0(F_1 - 1)] &\gg T_0(F_2 - 1) \\ \Rightarrow G_{p,1}F_1 &\gg F_2 - 1 \\ \Rightarrow G_{p,1}F_1 + 1 &\gg F_2 \end{aligned} \tag{10.45}$$

If the $G_{p,1}F_1$ product is large, we can make one final approximation:

$$\begin{aligned} G_{p,1}F_1 &\gg 1 \\ \Rightarrow G_{p,1}F_1 &\gg F_2 \end{aligned} \tag{10.46}$$

Using decibels, we can write

$$G_{p,1,dB} + F_{1,dB} \gg F_{2,dB} \tag{10.47}$$

In words, add the gain of the first amplifier to its noise figure (all in dB). If that number is much greater than the noise figure of the second amplifier, then the noise added to the cascade by the second amplifier will be insignificant.

10.4.5.1 A Noise Rule of Thumb

A successful strategy is to keep the excess gain plus the noise figure the first amplifier at least 15 dB greater than the noise figure of the second amplifier:

$$G_{p,1,dB} + F_{1,dB} > F_{2,dB} + 15 \tag{10.48}$$

Under the worst conditions, this rule ensures that the second component will contribute less than 0.1 dB of noise figure to the cascade. Greater gain before the second amplifier results in the second device adding smaller amounts of noise to the cascade. However, while high excess gains are very good for low-noise cascades, they degrade linearity. The goal of cascade design is to maximize the SFDR, which depends on both low cascade noise temperature and on high cascade linearity.

EXAMPLE

Noise Cascades

You're considering adding another component to the end of an existing cascade. The new component has a 7.5 dB noise figure. How much excess gain do you need to be sure this new component will not significantly alter the noise figure of the existing cascade?

Solution

Figure 10-22 shows the situation. Equation (10.48) reveals that the most gain is required from the existing cascade when the existing cascade's noise figure is very low. For a worst-case analysis, we'll assume the noise figure of the existing cascade is 1 dB. Thus,

$$\begin{aligned} G_{p,1,dB} + F_{1,dB} &> F_{2,dB} + 15 \\ G_{p,1,dB} + 1 &> 7.5 + 15 \\ \Rightarrow G_{p,1,dB} &> 21.5 \text{ dB} \end{aligned} \tag{10.49}$$

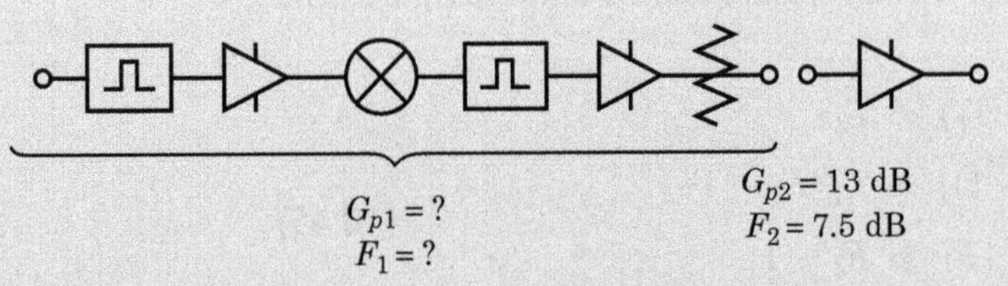

FIGURE 10-22 ■ What is the effect of adding a component to the end of the cascade?

Assuming the worst-case noise figure for the existing cascade, we require about 21.5 dB of excess gain prior to the last component. We can reduce the amount of required excess gain by decreasing the noise figure of the last amplifier.

10.4.5.2 Front-End Attenuators and Noise

In Chapter 6, we showed that the noise figure of a resistive attenuator equals its attenuation if the attenuator is at room temperature. In equation form,

$$\text{Attenuator Noise Figure} = \text{Attenuator Loss} \quad \text{for attenuators at } T_0 \tag{10.50}$$

Under the special conditions where the noise figure of the first device is inversely related to its gain, the cascade noise figure becomes

$$\begin{aligned} F_{cas} &= F_1 + \frac{F_2 - 1}{1/F_1} \\ &= F_1 + F_1(F_2 - 1) \\ &= F_1 F_2 \\ F_{cas,dB} &= F_{1,dB} + F_{2,dB} \end{aligned} \tag{10.51}$$

In words, we can simply add the value of the attenuator to the noise figure of the amplifier to produce the noise figure of the cascade (see Figure 10-23).

FIGURE 10-23 ■ The effect of adding a lossy component to the end of the cascade.

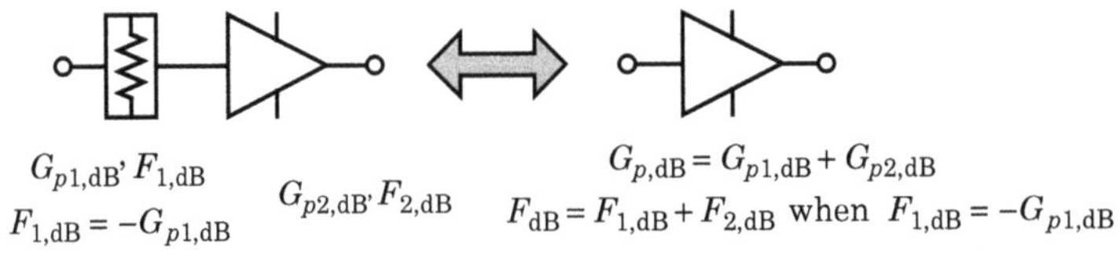

EXAMPLE

TV Antenna Amplifier

Figure 10-24(a) shows a typical household television receiving system. An outdoor TV antenna sits on a mast on the roof of the house. The antenna is connected to the television by 50 feet of RG-59 coaxial cable, which has a nominal loss of 6.5 dB at 500 MHz. The noise figure of the television is 15 dB. We are considering the addition of an amplifier to our system to bring in new channels. The amplifier has a 4 dB noise figure and 20 dB of gain. We have three choices:

1. We can attach the amplifier directly to the television. This is the least troublesome solution because the amplifier is inside the house, and we can plug it directly into the wall socket by the TV.

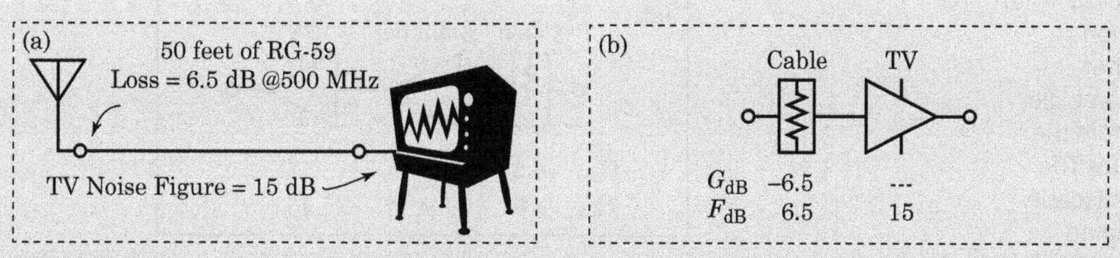

FIGURE 10-24 ■ (a) A typical household television system with antenna, cable, and television set. (b) The RF cascade equivalent of the initial receiving system.

2. We can place the amplifier inside, right where the cable enters the house. This places the amplifier about equidistant between the antenna and TV. It's more difficult to provide power to the amplifier than in (1) and the location is less convenient.
3. Finally, we can place the amplifier right on the mast, next to the antenna. This is a lot of trouble because we have to protect the amplifier against the weather and provide adequate power.

Examine the noise consequences of these three situations.

Solution

Cable Alone: Figure 10-24 shows the system as it currently stands, without the amplifier. Table 10-1 shows the relevant information about the cascade.

TABLE 10-1 Initial performance of the television receiving system

	Cable	TV
G_{dB}	−6.5	–
G	0.224	–
F_{dB}	6.5	15
F	4.47	31.6
T (°K)	1005	8,881
F (referenced to cascade input)	4.47	$\frac{F_2-1}{G_1}=\frac{31.6-1}{0.224}=136.8$
T (referenced to cascade input −°K)	1,005	$\frac{T_2}{G_1}=\frac{8{,}881}{0.224}=39{,}688\ °\text{K}$

The cascade noise figure and temperature are

$$\begin{aligned} F_{cas} &= F_1 + \frac{F_2-1}{G_1} & T_{cas} &= T_1 + \frac{T_2}{G_1} \\ &= 4.47 + 136.8 & &= 1005 + 39{,}688 \\ &= 141.3 & &= 40{,}675\ °\text{K} \\ &= 21.5\ \text{dB} \end{aligned} \tag{10.52}$$

In this simple case, we could use equation (10.51) to find the noise figure of the cascade as $6.5 + 15 = 21.5$ dB. The television is responsible for the bulk of the noise produced in the cascade although any reduction in cable loss will directly affect the noise figure of the cascade.

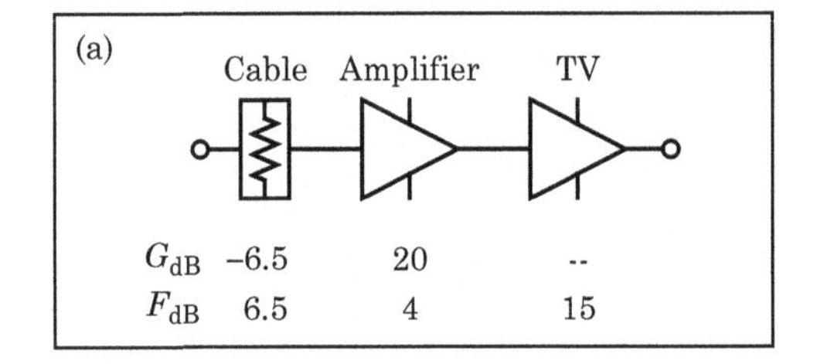

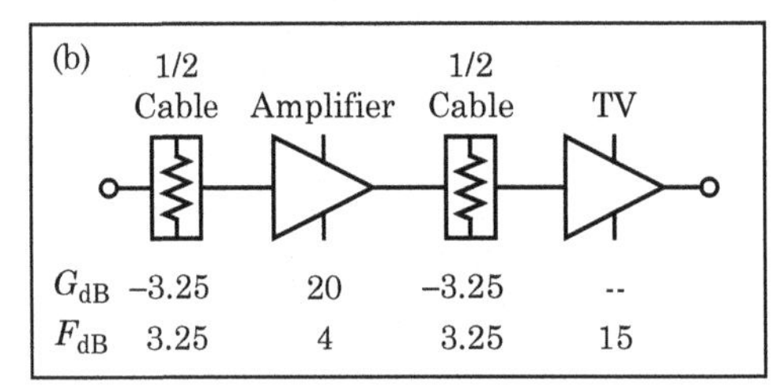

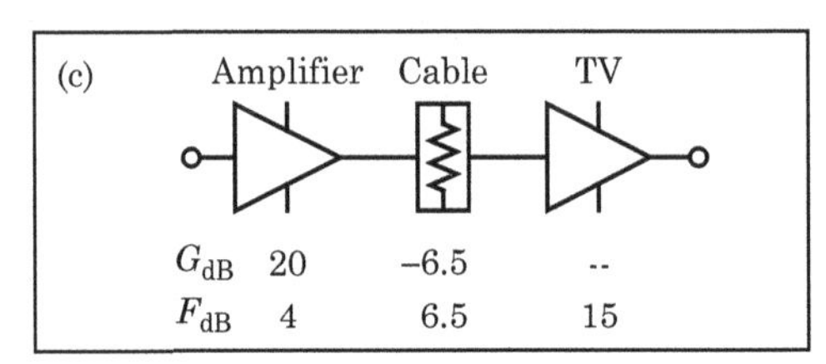

FIGURE 10-25 (a) The television cascade after adding an amplifier directly before the television. (b) The television cascade after adding an amplifier midway between the television and the antenna. (c) The television cascade with an amplifier placed directly after the antenna.

Amplifier Next to the TV: Figure 10-25(a) shows the system when we place the amplifier immediately before the television. Table 10-2 describes the cascade when the amplifier is directly connected to the television.

The cascade noise figure and temperature are

$$\begin{aligned} F_{cas} &= F_1 + \frac{F_2 - 1}{G_1} + \frac{F_3 - 1}{G_1 G_2} & T_{cas} &= T_1 + \frac{T_2}{G_1} + \frac{T_3}{G_1 G_2} \\ &= 4.47 + 6.74 + 1.37 & &= 1{,}005 + 1{,}955 + 397 \\ &= 12.6 & &= 3{,}357\ ^\circ\mathrm{K} \\ &= 11.0\ \mathrm{dB} \end{aligned} \tag{10.53}$$

TABLE 10-2 Performance of the television receiving system cascade when we place an amplifier directly before the television

	Cable	Amplifier	TV
G_{dB}	−6.5	20	–
G	0.224	100	–
F_{dB}	6.5	4	15
F	4.47	2.51	31.6
T(°K)	1005	438	8881
F (referenced to cascade input)	4.47	$\frac{F_2 - 1}{G_1} = \frac{2.51 - 1}{0.224} = 6.74$	$\frac{F_3 - 1}{G_1 G_2} = \frac{31.6 - 1}{(0.224)(100)} = 1.37$
T (referenced to cascade input °K)	1005	$\frac{T_2}{G_1} = \frac{438}{0.224} = 1955\ ^\circ\mathrm{K}$	$\frac{T_3}{G_1 G_2} = \frac{8{,}881}{(0.224)(100)} = 397\ ^\circ\mathrm{K}$

TABLE 10-3 ■ Performance of the television receiving system with an amplifier placed at the cable midpoint

	1/2 Cable	Amplifier	1/2 Cable	TV
G_{dB}	−3.25	20	−3.25	–
G	0.473	100	0.473	–
F_{dB}	3.25	4	3.25	15
F	2.11	2.51	2.11	31.6
T(°K)	323	438	323	8881
F (referenced to cascade input)	2.11	$\frac{F_2 - 1}{G_1} = \frac{2.51 - 1}{0.473} = 3.19$	$\frac{F_3 - 1}{G_1 G_2} = \frac{2.11 - 1}{(0.473)(100)} = 0.02$	$\frac{F_4 - 1}{G_1 G_2 G_3} = \frac{31.6 - 1}{(0.473)(100)(0.473)} = 1.37$
T (referenced to cascade input, °K)	323	$\frac{T_2}{G_1} = \frac{290}{0.473} = 926$ °K	$\frac{T_3}{G_1 G_2} = \frac{323}{(0.473)(100)} = 7$ °K	$\frac{T_3}{G_1 G_2 G_3} = \frac{8881}{(0.473)(100)(0.473)} = 397$ °K

Adding the amplifier before the TV has decreased the system noise figure to 11.0 dB. The amplifier and cable contribute approximately equal amounts of noise and the noise from the TV is now insignificant.

Amplifier in the Middle of Cable: Figure 10.25(b) shows the system when the amplifier is halfway between the antenna and the television. Table 10-3 describes the cascade when the amplifier is at the cable midpoint.

The noise figure and noise temperature are

$$\begin{aligned} F_{cas} &= F_1 + \frac{F_2 - 1}{G_1} + \frac{F_3 - 1}{G_1 G_2} + \frac{F_4 - 1}{G_1 G_2 G_3} \\ &= 2.11 + 3.19 + 0.03 + 1.37 \\ &= 6.7 \\ &= 8.3 \text{ dB} \end{aligned} \qquad \begin{aligned} T_{cas} &= T_1 + \frac{T_2}{G_1} + \frac{T_3}{G_1 G_2} + \frac{T_4}{G_1 G_2 G_3} \\ &= 323 + 926 + 7 + 397 \\ &= 1{,}653°\text{K} \end{aligned} \tag{10.54}$$

The system noise figure is down to 8.3 dB, which is a 13.2 dB decrease over the original system. The second term of the cascade equation, involving the amplifier and its excess gain, contributes the lion's share of the noise to the system.

Amplifier on the Antenna Mast: Figure 10-25(c) shows the system when we place the amplifier on the antenna mast as close to the antenna as possible. Cascade analysis is shown in Table CAS-4.

In this case, the cascade noise figure and temperature are

$$\begin{aligned} F_{cas} &= F_1 + \frac{F_2 - 1}{G_1} + \frac{F_3 - 1}{G_1 G_2} \\ &= 2.51 + 0.03 + 1.37 \\ &= 3.91 \\ &= 5.9 \text{ dB} \end{aligned} \qquad \begin{aligned} T_{cas} &= T_1 + \frac{T_2}{G_1} + \frac{T_3}{G_1 G_2} \\ &= 438 + 10.1 + 397 \\ &= 845°\text{K} \end{aligned} \tag{10.55}$$

The cable contributes only 10°K of noise temperature to the cascade, whereas the TV supplies 397°K and the front-end amplifier supplies 438°K. The cascade noise figure is down to 5.9 dB. Table CAS-5 offers a summary.

TABLE 10-4 ■ Performance of the television receiving system cascade with an amplifier placed immediately after the antenna

	Amplifier	Cable	TV
G_{dB}	20	−6.5	–
G	100	0.224	–
F_{dB}	4	6.5	15
F	2.51	4.47	31.6
T(°K)	438	1005	8881
F (referenced to cascade input)	2.51	$\frac{F_2 - 1}{G_1} = \frac{4.47 - 1}{100} = 0.03$	$\frac{F_3 - 1}{G_1 G_2} = \frac{31.6 - 1}{(100)(0.224)} = 1.37$
T (referenced to cascade input – °K)	438	$\frac{T_2}{G_1} = \frac{1005}{100} = 10.1°\text{K}$	$\frac{T_3}{G_1 G_2} = \frac{8881}{(100)(0.224)} = 397°\text{K}$

TABLE 10-5 ■ Summary of the television receiving system noise performance

Amplifier Position	System Noise Figure	System Noise Temperature
None	21.5 dB	40,675°K
At TV	11.0 dB	3,357°K
Halfway	8.3 dB	1,653°K
At Antenna	5.9 dB	845°K

The best noise figure occurs when we place the amplifier on the mast next to the antenna. However, we achieve a significant improvement when we attach the amplifier right to the TV input terminals. We could further decrease the system's noise figure by increasing the gain of the first amplifier. However, high excess gain can reduce the cascade linearity.

10.4.5.3 Conclusions

For a low-noise cascade, we want to use an amplifier with a low-noise temperature as the first amplifier in the cascade. If we care only about noise temperature, Equation (10.28) indicates that we should to keep the excess gain in front of every component as large as possible. But noise figure is only half of the story. We are dearly concerned about linearity, which suffers when we build a cascade with high excess gain.

10.4.6 Gain Distribution and Linearity

In this section, we examine the effects of gain distribution on the linearity of a cascade. Although we'll speak chiefly about TOI, the arguments apply to SOI as well. Figure 10-26 shows a general n-element cascade that we will use for analysis.

10.4.6.1 OTOI versus ITOI

Most design engineers are interested in the port that interfaces to the outside world (i.e., the antenna port). In transmitter design, we are interested in the level of the unwanted signals that our system will radiate into free space. Since the output of the cascade is attached to the antenna, transmitter designers pay more attention to output specifications (i.e., the OTOI or OSOI) rather than input specifications.

In receiver design, we are interested in the smallest and largest signals our system will reliably process. Since the input of the cascade is attached to the antenna, receiver designers are most interested in the cascade's input specifications (i.e., the ITOI or ISOI).

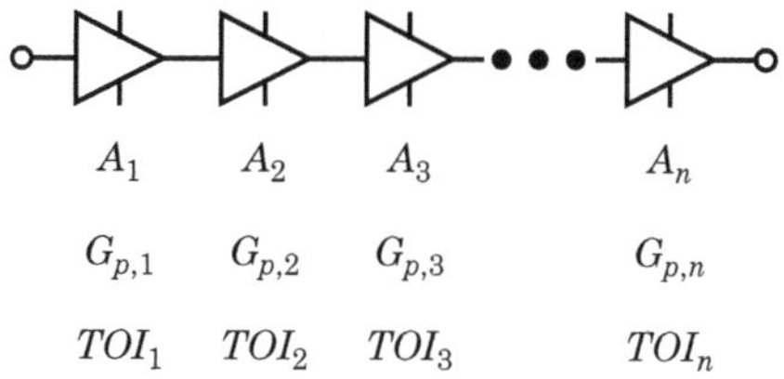

FIGURE 10-26 ■ An n-element cascade.

This book concentrates primarily on receiver design so we will concentrate our efforts on input specifications. Of course, these arguments will be valid for the output specifications. We will assume that coherent distortion products add coherently since that situation constitutes the worst-case scenario.

10.4.6.2 Linearity Cascade Equation

Equation (10.30) describes the linearity behavior of the cascade in Figure 10-26:

$$\begin{aligned}\frac{1}{ITOI_{cas}} = & \frac{1}{ITOI_1} && \text{[Component \# 1]} \\ & + \frac{1}{ITOI_2/G_{p,1}} && \text{[Component \# 2]} \\ & + \frac{1}{ITOI_3/G_{p,1}G_{p,2}} && \text{[Component \# 3]} \\ & + \cdots \\ & + \frac{1}{ITOI_n/G_{p,1}G_{p,2}\ldots G_{p,n-1}} && \text{[Component \# } n\text{]}\end{aligned} \tag{10.56}$$

The concepts of excess gain and translation apply. As with noise figure, equation (10.56) breaks apart nicely and each term expresses the linearity contribution of a single component.

Resistors in Parallel Equation (10.56) resembles the equation that describes adding resistors in parallel:

$$\frac{1}{R_p} = \frac{1}{R_1} + \frac{1}{R_2} + \frac{1}{R_3} \tag{10.57}$$

Figure 10-27 shows the similarities. The value of each resistor represents the TOI of each component when it is referenced to the input port. Adding a new device to the cascade effectively adds another parallel resistor to the equivalent circuit; lowering the TOI (resistance) of the cascade. We cannot improve the linearity of the cascade by adding another component; we can only make the linearity worse.

Pieces of the Linearity Cascade Equation Remembering the resistors in parallel analogy, each term of equation (10.56) signifies:

- **Component #1:** The first term

$$\frac{1}{ITOI_1} \tag{10.58}$$

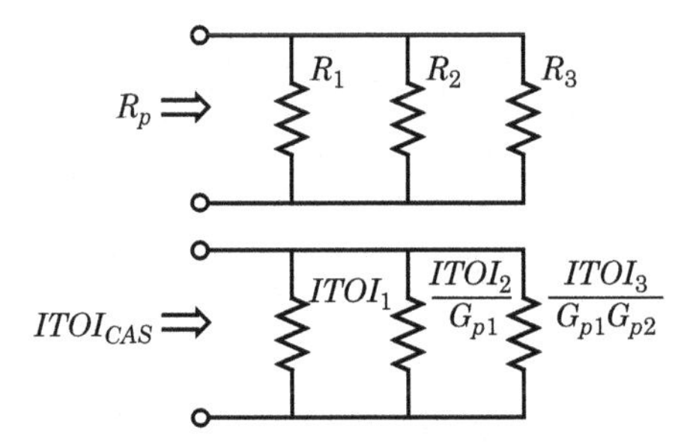

FIGURE 10-27 ■ The equation describing resistors in parallel is similar to the equation describing the linearity of a cascade. Adding a component to a cascade will always decrease the linearity of the cascade.

represents the distortion the first component contributes to the cascade. Any change in the ITOI of the first element directly affects the cascade's linearity. ITOI_1 appears in no other term of equation (10.56) so this expression represents the complete contribution of amplifier A_1.

- **Component #2:** This term

$$\frac{1}{ITOI_2/G_{p,1}} \tag{10.59}$$

represents the entire ITOI contribution of amplifier A_2 to the cascade. The ITOI contribution of amplifier A_2 is reduced (i.e., the cascade is made less linear) by the gain of amplifier A_1. In the discussion of noise temperature, gain preceding a stage was helpful. With linearity, excess gain is harmful.

Remember the parallel resistor analogy of Figure 10-27. If we add a low-value resistor (a low ITOI to the cascade) in parallel, then we limit the parallel resistance to a low value (a low cascade ITOI). The cascade linearity will always be limited by component whose translated ITOI is the smallest.

- **Component #3:** This term represents the linearity contribution that amplifier A_3 makes to the cascade. In this case, the ITOI of A_3 is reduced by $G_{p,1}G_{p,2}$, the excess gain up to the input of A_3.

If we want A_3 to contribute only a small amount of distortion to the cascade, we would like the combination of $ITOI_3/G_{p,1}G_{p,2}$ to be as large as practical.

- **Component #*n*:** The ITOI contribution of the n-th term is reduced by the excess gain from the $(n-1)$ components before the n-th component.

EXAMPLE

Three-Element Cascade

Figure 10-28 shows a three-element cascade. Examine the cascade.

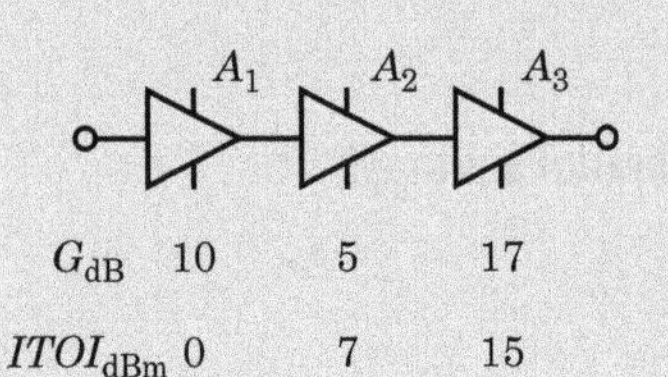

FIGURE 10-28 ■ Simple three-element cascade.

Solution

Table CAS-6 describes the cascade.

TABLE 10-6 Linearity contributions in the three-element cascade of Figure 10-28

	A_1	A_2	A_3
G_{dB}	10	5.0	17
G	10	3.16	50.1
$ITOI_{dBm}$	0.0	7.0	15.0
$ITOI_{mW}$	1.0	5.0	31.6
$ITOI_{mW}$ (referenced to cascade input)	1.0	$\frac{ITOI_2}{G_1} = \frac{5.0}{10} = 0.50$	$\frac{ITOI_3}{G_1 G_2} = \frac{31.6}{(3.16)(10)} = 1.0$
$ITOI_{dBm}$ (referenced to cascade input)	0.0	−3.0	0.0

Note: The reference point is the cascade's input port.

Table 10-6 reveals that the first and third amplifiers contribute equally to the cascade ITOI. The second component contributes the most nonlinearity to the cascade because this amplifier exhibits the lowest linearity when we compare the amplifier at a common point. Complete evaluation of equation (10.56) tells us the *ITOI* of the cascade is

$$\begin{aligned}\frac{1}{ITOI_{cas}} &= \frac{1}{ITOI_1} + \frac{1}{ITOI_2/G_{p,1}} + \frac{1}{ITOI_3/G_{p,1}G_{p,2}} \\ &= \frac{1}{1.0} + \frac{1}{0.5} + \frac{1}{1.0} \\ &= 4.0 \\ ITOI_{cas} &= 0.25 \text{ mW} \\ &= -6.0 \text{ dBm}\end{aligned} \tag{10.60}$$

We can perform this identical process on the output of the cascade. The equation for cascade *OTOI* in terms of component *OTOI* is

$$\frac{1}{OTOI_{cas}} = \frac{1}{G_{p,2}G_{p,3}OTOI_1} + \frac{1}{G_{p,3}OTOI_2} + \frac{1}{OTOI_3} \tag{10.61}$$

We also know that component *ITOI* is related to component *OTOI* via

$$OTOI_n = ITOI_n G_{p,n} \tag{10.62}$$

Thus, we can write equation (10.61) as

$$\frac{1}{OTOI_{cas}} = \frac{1}{G_{p,1}G_{p,2}G_{p,3}ITOI_1} + \frac{1}{G_{p,2}G_{p,3}ITOI_2} + \frac{1}{G_{p,3}ITOI_3} \tag{10.63}$$

Again, this equation demonstrates the concept of moving the TOI specification of a component to a particular port and then comparing these translated TOIs. Table CAS-7 is also applicable to equation (10.63).

TABLE 10-7 Linearity contributions in the three-element cascade of Figure 10-28

	A_1	A_2	A_3
G_{dB}	10	5.0	17
G	10	3.16	50.1
$ITOI_{dBm}$	0.0	7.0	15.0
$ITOI_{mW}$	1.0	5.0	31.6
$ITOI_{mW}$ (referenced to cascade output)	1580*	792**	1,580***
$ITOI_{dBm}$ (referenced to cascade output)	32.0	29.0	32.0

* $G_{p,1}G_{p,2}G_{p,3}ITOI_1 = (10)\,(3.16)\,(50.1)\,(1.0) = 1580$

* $G_2G_3ITOI_2 = (3.16)\,(50.1)\,(5.0) = 792$

** $G_3ITOI_3 = (50.1)\,(31.6) = 1580$

Note: The reference point is the cascade's input port.

We again see that components #1 and #3 contribute equally to the cascade TOI. We also see that component #2 is the weak link, contributing the most nonlinearity to the cascade.

10.4.6.3 Strength of Distortion Products

Equation (10.12), repeated here, indicates that the strength of the distortion products produced by a nonlinear device is a strong function of the fundamental input power:

$$P_{out,2f-f,dBm} = 3P_{out,f,dBm} - 2OTOI_{dBm} \tag{10.64}$$

In a cascade, the strength of the fundamental signal into a particular device is determined by the gain preceding the device. To keep the distortion products low (and the linearity high), we try to keep the excess gain through the cascade low.

For example, if we apply a −80 dBm signal to the cascade of Figure 10-29, then amplifier #1 will see a −80 dBm signal. Amplifier #2 will see a −60 dBm signal because of the 20-dB power gain of amplifier #1. Although amplifiers #1 and #2 have the same TOI, amplifier #2 will generate higher levels of distortion because it sees a larger signal. Likewise, amplifier #3 will see a −40 dBm signal and will generate the most distortion power of all the devices in the cascade. This is the reason that amplifier #3 dominates the TOI of the cascade.

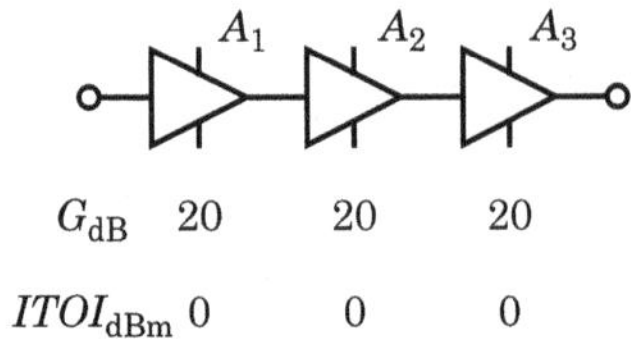

A_1 A_2 A_3

G_{dB} 20 20 20

$ITOI_{dBm}$ 0 0 0

FIGURE 10-29 ■ A three-element cascade. Amplifier A_3 will always see a larger signal than either A_1 or A_2 so A_3 will limit this cascade's linearity.

10.4.6.4 Conflict

Good noise performance dictates that we should keep the excess gain high. Good linearity performance forces us to keep the excess gain low. We have a fundamental compromise between low noise temperature and high linearity.

10.4.6.5 Rule of Thumb

When we translate the TOI of every component to a common port, we would like the translated TOI of each component to be about equal. If all the translated TOIs are equal, then all of the elements in the cascade will go nonlinear at the same input power level. If one translated TOI is much smaller than all the rest, that component will set the linearity of the cascade.

EXAMPLE

Component TOI and Excess Power Gain

Figure 10-30 shows three identical amplifiers in cascade. Each amplifier has a power gain of 20 dB (or 100x), an OTOI of +10 dBm (or 10 mW), and an ITOI of −10 dBm (or 0.10 mW). Analyze the cascade. Find which component dominates the linearity.

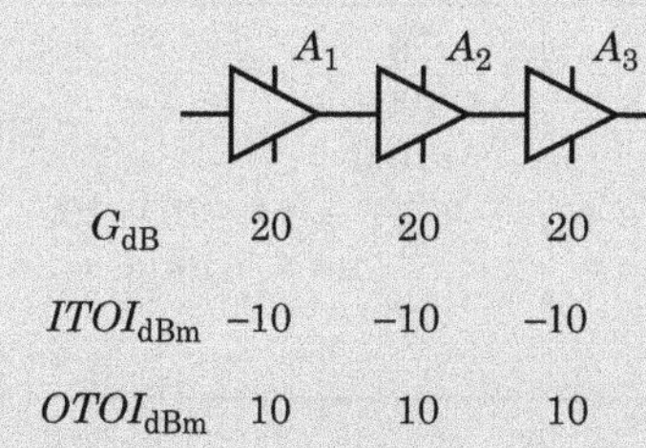

FIGURE 10-30 ■ A three-element cascade. The component TOIs are poorly matched to the cascade's excess gain.

Solution

We first translate the ITOI of each amplifier to the cascade's input (Table CAS-8).

TABLE 10-8 The linearity contributions of each element in the cascade of Figure 10-30. Amplifier A3 limits the cascade linearity

	A_1	A_2	A_3
G_{dB}	20	20	20
G	100	100	100
$ITOI_{dBm}$	−10.0	−10.0	−10.0
$ITOI_{dBm}$ (referenced to cascade output)	$ITOI_1 = -10$	$ITOI_2 - G_{p,1} = -10 - 20 = -30$ dBm	$ITOI_3 - G_{p,1} - G_{p,1} = -10 - 20 - 20 = -50$ dBm

Clearly, amplifier #3 dominates the cascade TOI. The third amplifier goes nonlinear at a much lower cascade input power level than the other two amplifiers. The 40 dB power gain preceding amplifier #3 reduces its effective linearity when we translate its TOI to the cascade's input.

EXAMPLE

Matching Component TOI to Excess Power Gain

Figure 10-31 shows another three-component cascade. This time, we have selected the TOI of each component based on the gain preceding the component. Analyze the cascade. Find which component dominates the cascade TOI.

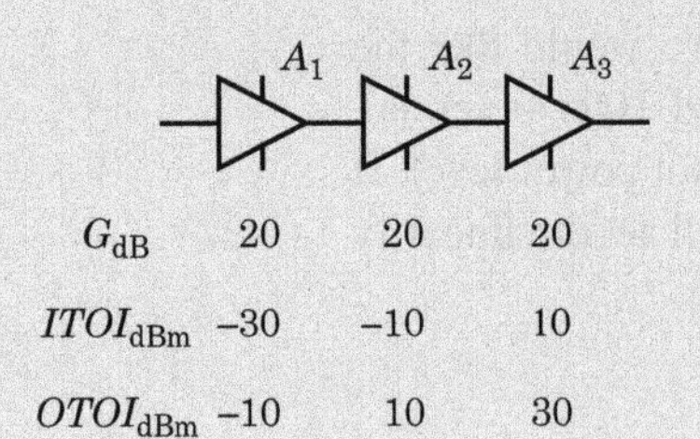

FIGURE 10-31 ■ A three-element cascade. The component TOIs are properly matched to the cascade's excess gain.

Solution

We construct Table CAS-9.

TABLE 10-9 The linearity contributions of each element in the cascade of Figure 10-31

	A_1	A_2	A_3
G_{dB}	20	20	20
G	100	100	100
$ITOI_{dBm}$	−30.0	−10.0	+10.0
$ITOI_{dBm}$ (referenced to cascade output)	$ITOI_1 = -30$	$ITOI_2 - G_{p,1} = -10 - 20 = -30$ dBm	$ITOI_3 - G_{p,1} - G_{p,1} = 10 - 20 - 20 = -30$ dBm

Note: Each amplifier contributes equally to the cascade's TOI.

The translated TOI of each component is the same. No one component dominates the linearity of the cascade.

10.4.6.6 Coherent versus Noncoherent Addition

In Chapter 7, we made two different assumptions and derived two different equations that describe the linearity of a cascade given the linearity of its components. One equation assumed that the distortion products were coherent and in phase, which is the worst-case assumption. The second equation assumed that the phases of the distortion products were noncoherent and random. The power of the distortion components add.

In a well-designed system (where the equivalent intercept points of all the devices are equal), the difference between coherent and noncoherent summation is 4 to 5 dB.

In our experience, we've found that well-designed cascades usually behave as though the distortion products are adding up noncoherently. For the most part, these systems followed the noncoherent equations plus 1 or 2 dB. With wideband systems, the cascade SOI or TOI will stay at noncoherent levels over most of the frequency range of the system. However, over narrow frequency ranges, the SOI and TOI will spike up to coherent summation levels.

When designing a system, the authors perform the calculations for both the coherent and noncoherent cases. These calculations express the variation we're likely to expect over time and frequency.

For reference, the coherent and noncoherent cascade equations are as follows:

Third-order intercept (coherent):

$$\begin{aligned}\frac{1}{ITOI_{cas}} = &\frac{1}{ITOI_1}\\ &+\frac{1}{ITOI_2/G_{p,1}}\\ &+\cdots\\ &+\frac{1}{ITOI_n/G_{p,1}G_{p,2}\ldots G_{p,n-1}}\end{aligned} \tag{10.65}$$

Third-order intercept (noncoherent):

$$\begin{aligned}\frac{1}{ITOI^2_{cas}} = &\frac{1}{(ITOI_1)^2}\\ &+\frac{1}{(ITOI_2/G_{p,1})^2}\\ &+\cdots\\ &+\frac{1}{(ITOI_n/G_{p,1}G_{p,2}G_{p,3}\ldots G_{p,n-1})^2}\end{aligned} \tag{10.66}$$

Second-order intercept (coherent):

$$\begin{aligned}\frac{1}{\sqrt{ISOI_{cas}}} = &\frac{1}{\sqrt{ISOI_1}}\\ &+\frac{1}{\sqrt{ISOI_2/G_{p,1}}}\\ &+\cdots\\ &+\frac{1}{\sqrt{ISOI_n/G_{p,1}G_{p,2}\ldots G_{p,n-1}}}\end{aligned} \tag{10.67}$$

Second-order intercept (noncoherent):

$$\begin{aligned}\frac{1}{ISOI_{cas}} = &\frac{1}{ISOI_1}\\ &+\frac{1}{ISOI_2/G_{p,1}}\\ &+\cdots\\ &+\frac{1}{ISOI_n/G_{p,1}G_{p,2}\ldots G_{p,n-1}}\end{aligned} \tag{10.68}$$

EXAMPLE

TOI Cascade (Coherent and Noncoherent Addition)

Figure 10-32 shows three devices in cascade. Find the ITOI of the cascade using both coherent and noncoherent summation.

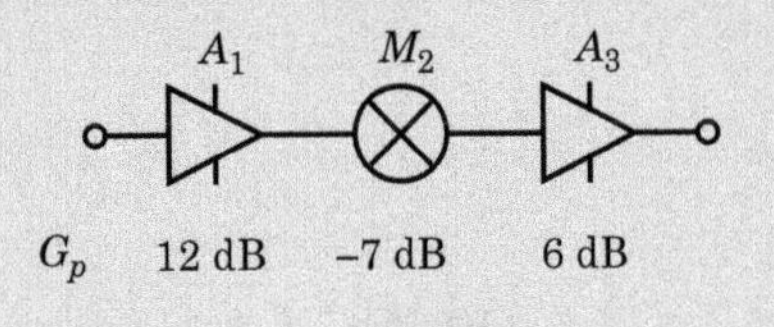

G_p	12 dB	−7 dB	6 dB
ITOI	20 dBm	---	28 dBm
OTOI	---	32 dBm	---

FIGURE 10-32 ■ A three-element cascade. Find the coherent and noncoherent TOI of the cascade.

Solution

We move the TOI of each device to the cascade's input and then apply the appropriate equations:

Amplifier #1: We know the $ITOI_{1,dBm}$ at the input of the cascade so

$$ITOI_{ampl,input} = 20 \text{ dBm} = 100 \text{ mW} \tag{10.69}$$

Mixer #2: Moving the *OTOI* of the mixer to the input of the cascade produces

$$\begin{aligned} ITOI_{mix2,input} &= 32 + 7 - 12 \\ &= 27 \text{ dBm} \\ &= 501 \text{ mW} \end{aligned} \tag{10.70}$$

Amplifier #3: The equivalent input *TOI* of amplifier #3 is

$$\begin{aligned} ITOI_{amp3,input} &= 28 + 7 - 12 \\ &= 23 \text{ dBm} \\ &= 200 \text{ mW} \end{aligned} \tag{10.71}$$

Using this method, we can see that amplifier #1 is the lowest *ITOI*, so it will dominate the result. Amplifier #3 is the next lowest. The coherent TOI equation above reveals that the coherent summation *ITOI* is

$$\begin{aligned} \frac{1}{ITOI_{cas}} &= \frac{1}{100 \text{ mW}} + \frac{1}{501 \text{ mW}} + \frac{1}{200 \text{ mW}} \\ ITOI_{cas} &= 58.8 \text{ mW} \\ &= 17.7 \text{ dBm} \end{aligned} \tag{10.72}$$

The noncoherent assumption produces

$$\begin{aligned} \frac{1}{ITOI_{cas}^2} &= \frac{1}{(100 \text{ mW})^2} + \frac{1}{(501 \text{ mW})^2} + \frac{1}{(200 \text{ mW})^2} \\ ITOI_{cas} &= 88.1 \text{ mW} \\ &= 19.4 \text{ dBm} \end{aligned} \tag{10.73}$$

In this particular case, the noncoherent assumption produces a result that is 1.7 dB better than the coherent assumption.

EXAMPLE

Cascade TOI (Coherent and Noncoherent Cases)

Find the ITOI of the cascade shown as Figure 10-33 for both the coherent and noncoherent cases.

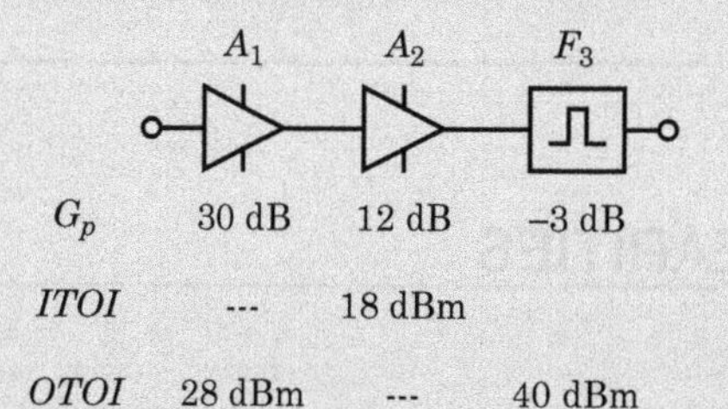

	A_1	A_2	F_3
G_p	30 dB	12 dB	–3 dB
ITOI	---	18 dBm	
OTOI	28 dBm	---	40 dBm

FIGURE 10-33 ■ A three-element cascade. Find the coherent and noncoherent TOI of the cascade.

Solution

We first move the TOI specification for each component to the input of the cascade:
Amplifier #1: Moving the OTOI of amplifier #1 to the cascade's input produces

$$\begin{aligned} ITOI_{amp1,input} &= 28 - 30 \\ &= -2 \text{ dBm} \\ &= 0.631 \text{ mW} \end{aligned} \tag{10.74}$$

Amplifier #2: The TOI of this amplifier referenced to the cascade's input is

$$\begin{aligned} ITOI_{amp2,input} &= 18 - 30 \\ &= -12 \text{ dBm} \\ &= 63.1 \cdot 10^{-3} \text{ mW} \end{aligned} \tag{10.75}$$

Filter #3: The filter's equivalent cascade input TOI is

$$\begin{aligned} ITOI_{filter3,input} &= 40 + 3 - 12 - 30 \\ &= 1 \text{ dBm} \\ &= 1.26 \text{ mW} \end{aligned} \tag{10.76}$$

Amplifier #2 is the weak link in the cascade. For coherent addition, we find

$$\begin{aligned} \frac{1}{ITOI_{cas}} &= \frac{1}{0.631} \quad \text{(Amp1 contribution)} \\ &+ \frac{1}{63.1\text{E} - 3} \quad \text{(Amp2 contribution)} \\ &+ \frac{1}{1.26} \quad \text{(Filter3 contribution)} \\ ITOI &= 54.9 \cdot 10^{-3} \text{ mW} = -12.6 \text{ dBm} \end{aligned} \tag{10.77}$$

The noncoherent power summation produces

$$\begin{aligned} \frac{1}{ITOI_{cas}^2} &= \frac{1}{(0.631)^2} \quad \text{(Amp1 contribution)} \\ &+ \frac{1}{(63.1\text{E}-3)^2} \quad \text{(Amp2 contribution)} \\ &+ \frac{1}{(1.26)^2} \quad \text{(Filter3 contribution)} \\ ITOI &= 62.7 \cdot 10^{-3} \text{ mW} = -12.0 \text{ dBm} \end{aligned} \tag{10.78}$$

In this case, the difference between coherent and noncoherent addition is only 0.6 dBm. Since one component is contributing most of the distortion to the cascade, it does not matter how we add up the distortion products. If all of the components in the cascade contributed equally to the cascade TOI, there would be a greater difference between the coherent and noncoherent summations.

10.5 SYSTEM NONLINEARITIES

10.5.1 Detection

How can we discern if a receiving system is suffering from linearity problems? There are several clues:

- Signals are not where they are supposed to be. For example, we might observe FM radio stations at 500 MHz when they are supposed to be at 100 MHz.
- We observe signals appearing and disappearing abruptly at the same time. This effect can be caused by continuous signals combining with push-to-talk (PTT) signals. If either signal is very strong, the intermodulation products developed by the combination will also exhibit the on/off rate of the push-to-talk.
- Signals seem to have more than one type of modulation present. A signal with both AM and FM characteristics can be a result of intermodulation.
- The noise floor of the receiver bounces up and down, especially at push-to-talk rates.
- Finding a signal with the same information on it at several different frequencies. Although this occasionally occurs intentionally, it can also result from intermodulation.
- When tuning a receiver suspected of intermodulation distortion, watch the spectrum at the IF port on a spectrum analyzer. As we tune up in frequency, intermodulation products will move through the IF at different rates than signals that are not due to intermodulation. We may also see distortion products increase in frequency as we tune down.

EXAMPLE

Receiving Problems

While tuning through the spectrum, we observe an FM radio station (which normally broadcasts at 101.1 MHz) at 247 MHz. This is a result of the FM broadcast at 101.1 MHz combining with some other signal in a nonlinear device according to

$$f_{out} = |\pm n \cdot f_1 \pm m \cdot f_2| \text{ where } \begin{cases} n = 0, 1, 2, 3, \ldots \\ m = 0, 1, 2, 3, \ldots \end{cases} \tag{10.79}$$

where $m + n$ is the order of the spurious product. If f_1 is the FM broadcast station at 101.1 MHz, find all of the possible values of f_2 which will result in an output frequency of 247 MHz. Assume $m + n \leq 5$.

Solution

For this particular problem, we can write

$$247 = n \cdot 101.1 \pm m \cdot f_2 \tag{10.80}$$

or

$$f_2 = \left| \frac{247 \pm n \cdot 101.1}{m} \right| \tag{10.81}$$

We generate Table 10-10 using equation (10.81).

TABLE 10-10 The possible spurious products generated in the receiver

n,m	f_2 (MHz) (+, −)	*n,m*	f_2 (MHz) (+, −)	*n,m*	f_2 (MHz) (+, −)
0,0	*,*	1,1	348.1, 145.9	2,3	149.7, 14.9
0,1	247, 247	1,2	174.1, 73.0	3,0	*,*
0,2	123.5, 123.5	1,3	116, 48.6	3,1	550.3, 56.3
0,3	82.3, 82.3	1,4	87, 36.5	3,2	275.2, 28.2
0,4	61.8, 61.8	2,0	*,*	4,0	*,*
0,5	49.4,49.4	2,1	449, 44.8	4,1	651.4, 157.4
1,0	*,*	2,2	224.6, 22.4	5,0	*,*

Table 10-10 shows all of the frequencies that can combine with 101.1 MHz to produce a signal at 247 MHz (up to order 5). The * indicates that there is no solution for these particular values of n and m.

10.5.2 Corrective Action

There are two major corrective actions we can take when faced with a system exhibiting nonlinear behavior. We can increase the linearity of the system or remove the signals that are causing the system to distort. You improve the system's linearity by using more linear components. Limiting the input signal power often requires some trade-offs.

We first examine the complete spectrum being applied to the system being careful to observe both in-band and out-of-band signals in the search. Very strong signals that are far removed from the center frequency of the system can still cause problems. Ideally, we should examine all of the signals present within the bandwidth of all the components in the system.

PTT systems often cause intermittent nonlinear behavior in receiving systems because the transmitters can be very close to the receiving antenna. The intermittent nature of the problem can also make this type of interference difficult to track down and fix. Be sure to examine the input spectrum over a length of time.

Airport communications, taxi cab transmitters, FM radio, and television stations can all be sources of strong signals that are likely to cause intermodulation problems.

We can often remove the offending signals with a low-pass, high-pass, or notch filter if the filter will not also remove the signals in which we are interested. We can sometimes just reorient the receive antenna to reduce the amplitude of the interferer.

10.6 TOI TONE PLACEMENT

Figure 10-34 shows a typical receiving system. The system consists of amplifiers, filters, mixers, attenuators, and an ADC. The last filter in the system (labeled BPF_4 and realized in the digital domain) is typically the narrowest filter in the system. This filter is just large enough to pass the entire signal of interest. This filter sets the system noise bandwidth.

By definition, there is only one signal present in the system after it passes through BPF_4. This single signal is the signal we want to demodulate.

Third-order distortion results from unwanted signals combining to form phantom signals that might obscure our wanted signal. Since BPF_4 passes only our one signal of interest, we can safely state that there are no unwanted signals present in the system beyond BPF_4. In other words, the components following BPF_4 do not contribute to the nonlinearity of the cascade. We can consider the TOI of the components beyond BPF_4 to be infinite.

10.6.1 TOI Input Tones

Suppose we wanted to measure the TOI of the receiver in Figure 10-34. The third-order test requires us to apply two tones to the receiver and look for third-order distortion products. We measure the power of all four tones present at the output of the receiver, and then we calculate the receiver's TOI. To maximize sensitivity, we set the frequency of the two tones such that the nonlinear distortion will appear at the receiver's tuned frequency.

10.6.1.1 100 kHz Tone Spacing

For the first example, we run the TOI test with the two input tones spaced 100 kHz apart. Figure 10-35(a) shows the relationship between the two tones, the receiver's tuned frequency and the filter bandwidths of Figure 10-34.

We've tuned the receiver to the upper distortion product. The two input signals pass through the entire receiver until they are finally stopped by BPF_4. Every amplifier and mixer in the cascade has an opportunity to generate distortion power since all the components experience the two large input signals. However, components beyond BPF_4 do not contribute to the cascade's distortion power because BPF_4 attenuates the two large input signals. Components downstream from BPF_4 never see the input signals.

10.6.1.2 1 MHz Tone Spacing

Figure 10-35(b) shows the second example. We space the two large input signals 1 MHz apart. Since the receiver is tuned to the upper third-order distortion product, the two input

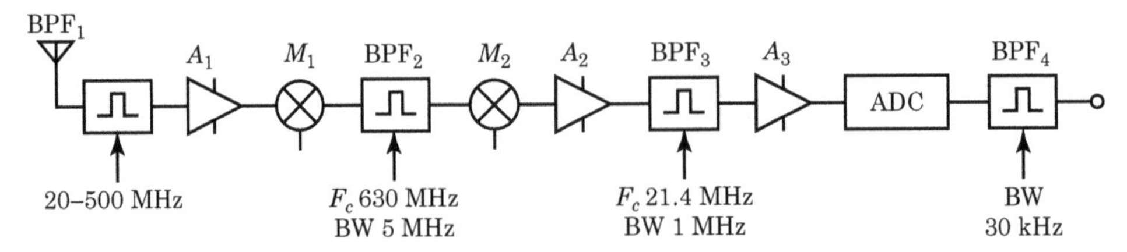

FIGURE 10-34 ■ A practical receiving system consisting of amplifiers, filters, mixers, attenuators and an ADC. Filter BPF_4 is a digital filter, realized in the digital signal processing (DSP).

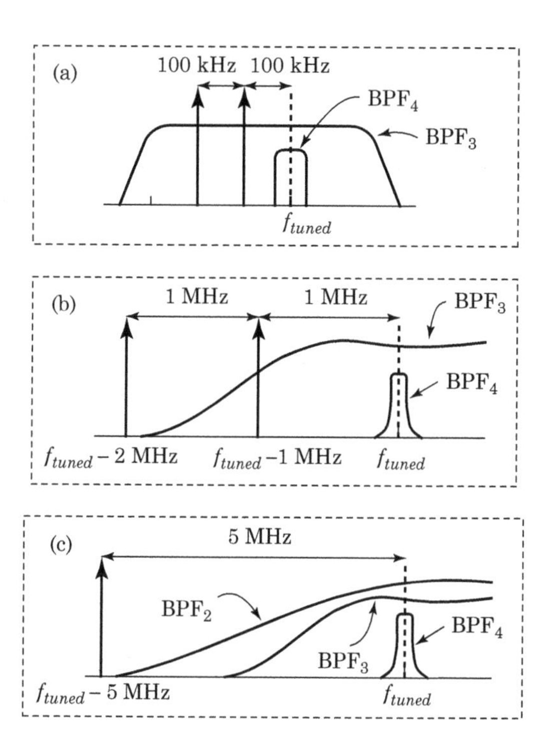

FIGURE 10-35 ■ Testing receiver TOI with two input signals. The relationship between the spacing of the two test tones, the receiver's tuned frequency and the bandwidth of the filters affects the results of the TOI measurement.

signals are 1 and 2 MHz below the tuned frequency. Both input signals will experience some attenuation due to BPF_3 (because of its 1 MHz bandwidth). The components before BPF_3 contribute to the cascade's distortion (and its TOI), but the components after BPF_3 will contribute less than they did when the tones were only 100 kHz apart. When we increase the input tone spacing, the cascade appears to have a higher TOI because fewer components experience the large input signals.

10.6.1.3 5 MHz Tone Spacing

Figure 10-35(c) shows the results when the input tones are 5 MHz apart. Only components between the antenna and the input of BPF_2 contribute distortion to the cascade. The larger tone spacings have made the cascade appear more linear than it did with smaller tone spacings.

Figure 10-36 shows the relationship between the measured TOI of a receiver and the measurement tone spacing. The tone spacing relative to the bandwidth of the band-pass filters in the system is the important metric.

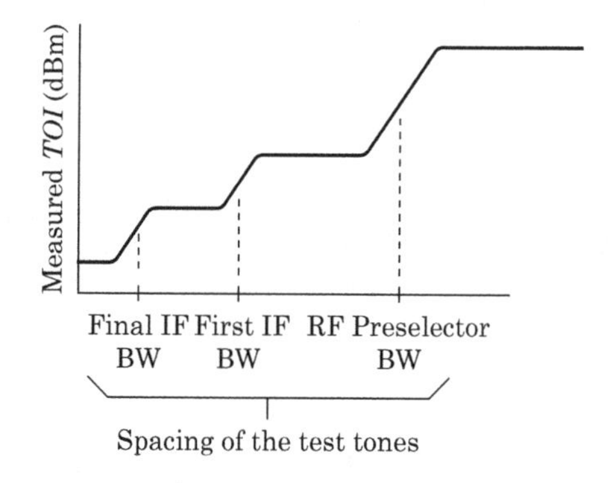

FIGURE 10-36 ■ When measuring the TOI of a receiver, the apparent linearity will change with the spacing of the test tones used to perform the test.

10.6.1.4 So, Where?

The measured linearity of our cascade changes depending on how we perform the test. When we perform a linearity test on a receiver, how far apart do we want to place the input tones? What constitutes a useful test? It's common procedure to specify the tones so that they're stopped only by the final IF filter. This is the worst-case situation because every component up to the final IF filter contributes to the system's nonlinearity.

Consider the receiver's operational environment. In a channelized system, such as a commercial FM radio receiver, place the tones on the two adjacent channels. For example, tune the receiver to 99.1 MHz and place the two tones at 98.7 and 98.9 MHz (or 99.3 and 99.5 MHz). The TOI specifications of a receiver should also include:

- The IF bandwidths used
- The tuned frequency of the receiver
- Frequencies of the two tones

Given this information, others will be more able to judge the receiver's performance in a particular signal environment.

10.6.1.5 Application to Gain Distribution

Given that every device prior to the final IF filter will contribute to the system's nonlinearity, we desire to keep the excess gain low prior to the final IF filter. After the final IF filter, only one signal is present in the receiver, and we can apply gain without a linearity penalty. In other words, we can apply all of the gain we like, after the final IF filter, without incurring a degradation in the cascade's third-order intercept.

10.7 AUTOMATIC GAIN CONTROL

So far, we have far assumed that the signal present at the input port of our receiver is very weak. In our example, we applied the MDS to the system and calculated the amount of power gain we would need to bring this signal up to the level required by the downstream processor. Our MDS was −124 dBm, and we needed 94 dB of gain to make this signal compatible with the ADC. Our receiver, as designed, will process a signal whose power is the MDS power.

However, the signals arriving at the input of a real-world receiver will not all be weak. Some signals could be very strong, which will cause our high-gain system to distort and impair our ability to process the signals. What do you do when faced with a very bright day? You attenuate the light using sunglasses.

EXAMPLE

SNR of Large Signals

What is the equivalent input SNR of a −30 dBm signal presented to a receiver with a 5 dB noise figure and a 30 kHz noise bandwidth?

Solution

We know

$$S_{in,dBm} = -30 \text{ dBm} \tag{10.82}$$

If we assume a 290°K antenna noise temperature, the receiver's input noise power is

$$\begin{aligned} N_{in,dBm} &= F_{dB} - 174\frac{\text{dBm}}{\text{Hz}} + 10\log(B_n) \\ &= 5 - 174\frac{\text{dBm}}{\text{Hz}} + 10\log(30{,}000) \\ &= -124 \text{ dBm} \end{aligned} \tag{10.83}$$

So the SNR is

$$\begin{aligned} \left(\frac{S}{N}\right)_{dB} &= -30 - (-124) \\ &= 94 \text{ dB} \end{aligned} \tag{10.84}$$

10.7.1 Noise/Linearity Trade-Offs

Strong signals exhibit high SNR. We are often in the position to exchange abundant signal power (and, hence, SNR) for linearity. Earlier we found that an attenuator placed at the receiver's front end performs exactly that trade-off. The attenuator preserves the receiver's dynamic range while allowing operation at higher signal powers.

Figure 10-37 shows an example. We design a receiver with enough gain to process an input signal at −124 dBm. If our signal arrives with −30 dBm of power, the receiver will go into compression and the demodulator will see a severely distorted signal [Figure 10-37(a)]. Let us assume we need 50 dB of SNR on the signal, which is suitable SNR for even the most finicky demodulator. We know that the equivalent receiver input noise is −124 dBm. For an input SNR of 50 dB, we know

$$\begin{aligned} \left(\frac{S}{N}\right)_{dB} &= 50 \text{ dB} \\ &= S_{in,dBm} - (-124) \\ S_{in,dBm} &= -74 \text{ dBm} \end{aligned} \tag{10.85}$$

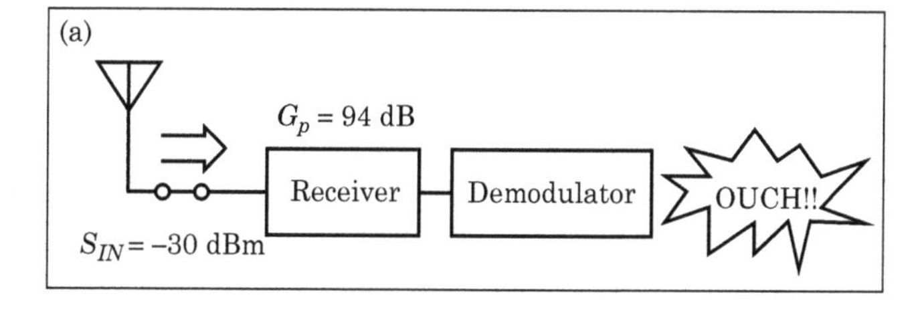

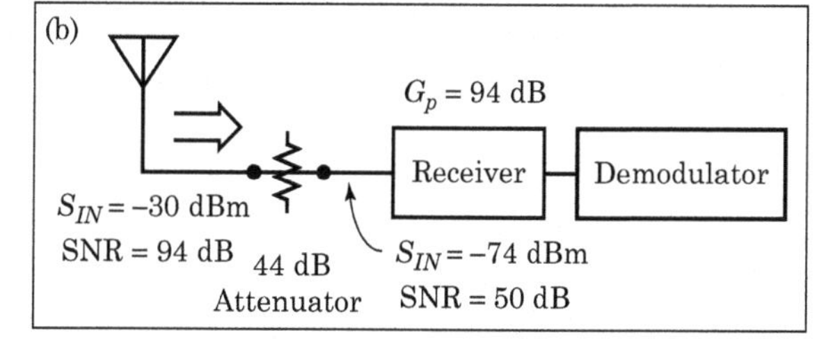

FIGURE 10-37 ■ Trading noise performance for linearity. A large signal can drive a high-gain system into distortion. We adjust the receiver's gain based on the input power to process a wide range of signal powers.

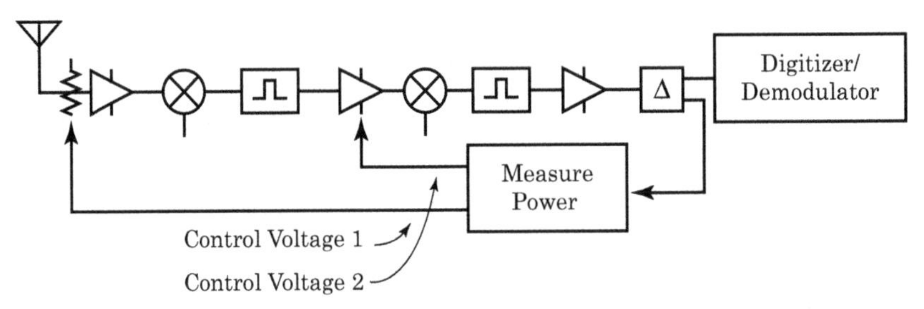

FIGURE 10-38 ■ Automatic gain control (AGC) applied to separate stages in the receiver. As the signal power increases, gain is often first reduced in the sections of the cascade lying closest to the antenna. This technique is commonly referred to as delayed AGC.

This equation tells us that the output SNR will be 50 dB if the receiver sees an input signal of −74 dBm. We can use a 44 dB attenuator that lies between the antenna and the receiver to reduce the −30 dBm antenna signal to −74 dBm [Figure 10-37(b)]. The signal will still possess 50 dB of SNR, and the receiver will not suffer the strong −30 dBm signal.

10.7.2 AGC Action

In a well-designed receiver, we place devices with variable gains at carefully-chosen locations in our cascades. These variable gain devices can be amplifiers or attenuators whose gain is controlled by an external voltage. We might also switch different amplifiers or attenuators into the cascade depending on the input signal.

We then measure the signal power at one, or several, points in the cascade and use that information to adjust the cascade gain. We change the gain of each amplifier to maximize the SFDR at every input power level and to produce a constant signal level at some point in the cascade (usually at the input to the demodulator). Figure 10-38 shows a diagram.

Most commonly, the gain reduction is performed in two stages. As the input signal level increases, the gain reduction is performed first at the sections of the receiver that are farthest from the antenna. When the gain reduction has reached its maximum value, we start to apply a second gain reduction or delayed automatic gain control (AGC) to the sections of the receiver that are closest to the antenna.

10.7.3 AGC Bandwidth

Successful AGC action requires us to pay attention to the bandwidth of the AGC control system (or, equivalently, the AGC attack and decay times). For example, a receiver to work with an AM signal containing human speech must demodulate signals that extend from 300 to 3000 hertz. The information we seek to process is present in the amplitude variation of the speech signal. If the AGC bandwidth is too high, the AGC will change the gain of the receiver as the amplitude of the amplitude modulation (AM) waveform changes. The fast AGC will remove the AM modulation we seek to process. We must set the AGC bandwidth to be much less than the lower frequency of the signal we wish to process. In the speech system, we might set the AGC bandwidth to as small as 3 Hz.

10.7.3.1 Attack and Decay Times

AGC bandwidth is often specified in terms of attack and decay times rather than bandwidth. Attack time refers to how quickly a receiver will adjust to a signal whose amplitude suddenly increases. Decay time refers to how quickly the receiver adjusts to a signal whose amplitude suddenly decreases.

10.7.4 Cascade Gain Distribution Rules

We are finally in a position to answer the question we posed earlier in this chapter. Given a requirement for 94 dB of gain, which configuration of Figure 10-15 will produce the system with the largest dynamic range? The answer is that none of the candidate systems is fully adequate. From a noise perspective, Figure 10-15(a) is ideal because it applies a great deal of gain at the very beginning of the cascade. This system will result in a minimum noise figure. From a linearity perspective, Figure 10-15(a) is a very poor choice. Since we are asking every component beyond the first amplifier to process a very large signal, each component will produce a lot of distortion and the linearity of the cascade will suffer. Figure 10-15(b) is a very linear solution because all of the components process very small signals. However, Figure 10-15(b) is a poor performer in term of noise because each lossy component robs our receiver of precious signal power while contributing noise to the system.

Figure 10-15(c) is a reasonable trade-off between noise and linearity. We seek to distribute the gain in an intelligent manner, keeping sight of both our noise and linearity goals. We have arrived at the following design rules of thumb.

10.7.4.1 Excess Gain

Keep the excess gain between 15 and 25 dB. At low levels of excess gain, individual components contribute too much noise to the cascade. At high levels of excess gain, individual components add distortion because they are forced to process signals that are too large.

Follow the 15 and 25 dB rule until you pass through the narrowest filter in the system. This filter determines the noise bandwidth and only one signal is present in this filter. Distortion, which results from unwanted signals combining to form phantom signals, is no longer a problem since we are processing only one signal. After passing through this filter, add enough gain to ensure that the system's MDS will experience enough power gain to bring it up to the levels required by further processing.

10.7.4.2 IF Bandwidth

To avoid second-order nonlinearities, keep all the bandwidths in the system much less than an octave. This might mean choosing a high center frequency for one of your IFs if the system must process signals with large bandwidths. You should have a very good reason for building a system containing a multioctave bandwidth. Second-order distortion is typically a lot stronger than third-order distortion.

10.7.4.3 Stability

Stability is an added benefit to placing most of the gain after the final IF filter. The possibility of oscillation exists whenever a system exhibits power gain. Oscillation is less likely in systems with low IFs.

10.7.4.4 AGC

Supply enough gain to process the system's MDS. Rely on the receiver's AGC to reduce the gain when processing larger signals. When dealing with a high SNR signal, we can afford to trade noise performance for linearity.

10.7.4.5 Limit Bandwidth

It is beneficial to limit bandwidth at every opportunity, almost without exception. Limiting bandwidth will:

- Decrease the possibility of producing spurious signals in nonlinear devices. A device processing a large number of unwanted signals is more likely to generate an in-band spurious response than a device processing a small number of unwanted signals.
- Limit stability problems. Oscillation occurs because a signal finds its way from a system's output back to its input. Limiting bandwidth limits the number of signals that can pass through the system with gain. This reduces the chance of oscillation.
- Limit external pickup of unwanted signals. There are many sources of spurious signals in a real-world receiver. Limiting the bandwidth reduces the chance that any spurious energy will experience gain through the RF system.

10.7.4.6 Matching

Most components we use to build receivers are designed, built, and tested assuming they will be operated in a wideband Z_0 impedance environment. If we do not terminate the ports, the devices will often not perform as we expect them to. This is especially true of mixers and filters.

We can occasionally count on amplifiers to provide a wideband match to the outside world, although this is not good design practice. If misterminated, a poorly designed amplifier can oscillate or change its gain and noise characteristics. However, there are many examples in functioning receivers that have poorly terminated amplifiers providing a match to external devices. That doesn't make it good design practice.

Attenuators are useful matching tools. Recall that the maximum return loss of a resistive attenuator is twice its attenuation value (i.e., a 5 dB attenuator will always present at least a 10 dB return loss to anything connected to it). Although we have to suffer some signal loss, the matching provided by the attenuator is often worth it.

War Story: Amplifier Stability and Terminal Loading

The author was working with a system that contained a parabolic dish antenna. Using good design practices, the antenna feed was followed immediately by a wideband amplifier. After working for about an hour, the author noted that the RF environment was behaving strangely. The noise floor, as viewed on a spectrum analyzer, would rise about 40 dB as he moved the antenna past one particular azimuth. Finally, he found a huge signal present at the same azimuth. The author reasoned that the large signal was causing the front-end amplifier to go nonlinear, which can raise the system's noise floor and cause other nonlinear effects.

On closer examination, the large signal disappeared when the antenna was moved just a few degrees. This was unusual considering the power in the signal and the sidelobe levels of the antenna. Closer examination revealed that the frequency of the signal changed slightly as the antenna changed azimuth. In this particular facility, the antenna was some distance from the receiver. The positioning was done by remote control. When the author physically viewed the antenna, he found that it faced a piece of steel supporting structure when the antenna was receiving the large, troublesome signal.

When the antenna was pointing at the metal, the front-end amplifier was exposed to a poor impedance match on its input, and the amplifier broke into self-oscillation. Since the reflection coefficient of the poor match changed slightly as the author moved the antenna, the frequency of oscillation changed the antenna moved. When the antenna was viewing perfectly matched free space, the oscillation stopped all together. Swapping the defective front-end amplifier solved the problem.

10.7.5 Linearity and Power Consumption

An amplifier with a high TOI will usually require more direct current (DC) power than an amplifier with a low TOI. High DC power indicates the amplifier will generate more heat. We can build a very linear system if we don't care about power consumption and have no problem getting rid of the heat generated. Similarly, a mixer with a high TOI will usually require a higher LO drive level than a mixer with a low TOI. This increases the complexity of the LO as well as complicates the receiver's internal isolation problems.

10.7.5.1 Balancing TOI with Excess Gain

When we translate the linearity specifications of each component in the cascade to a common port, the translated specifications of every component should be about equal. If all the translated TOIs are equal, then all of the elements in the cascade go nonlinear at the same input power level. If one element's TOI is too small, the TOI of the entire cascade will be dominated by that one weak link.

10.7.6 Cascades, Bandwidth, and Cable Runs

In some situations, we are forced to place an antenna a long distance from the receiving system. In these cases, the attenuation of the cable connecting the antenna and the receiver is a significant factor in system's gain distribution. The cable attenuation is more problematic when we are handling wideband signals because the cable's attenuation changes with frequency. Table 10-11 lists the attenuation of three types of coaxial cable over frequency. Figure 10-39 shows this data graphically.

TABLE 10-11 ▪ The attenuation of three types of cables (100 foot runs)

Frequency (MHz)	RG-58A	RG-223	1/4-inch Heliax
10	1.4	1.3	0.5
100	4.9	4.0	2.0
200	7.3	5.7	2.8
300	9.1	7.1	3.3
400	11.0	8.3	4.0
500	13.5	9.4	4.3
600	15.0	10.4	5.0
700	17.0	11.3	5.2
800	18.0	12.2	6.0
900	19.0	12.9	6.1
1,000	20.0	13.3	6.8
3,000	41.0	36.0	12.0
5,000	–	51.0	18.0
8,000	–	74.0	23.0
10,000	–	85.0	29.0

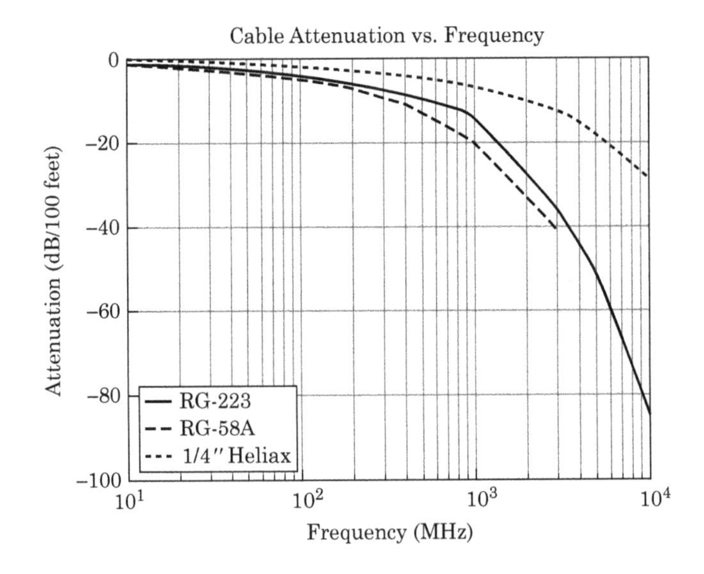

FIGURE 10-39 Cable attenuation versus frequency for three cable types.

EXAMPLE

Narrowband Cable Runs

We must move a signal that is centered at 8 GHz and is 30 MHz wide. The percentage bandwidth is 100 * 30E6/8E9 = 0.375%, which is very small. Figure 10-39 reveals that the attenuation across the bandwidth of the signal is approximately constant so we can work this problem at a single frequency of 8 GHz. The attenuation of 100 feet of z inch Heliax cable is about 23 dB at 8 GHz. We simply insert amplifiers at the proper places in the cable to keep the excess gain between 15 and 25 dB.

EXAMPLE

Wideband Cable Runs

Our problem now changes to transporting signals in the range of 10 to 1000 MHz over a distance of 450 feet. Figure 10-40 shows the cable attenuation for a 450 foot run over a 10 to 1000 MHz frequency range. The loss of the RG-58A cable varies from 6 dB at 10 MHz to about 90 dB at 1000 MHz; a difference of 84 dB. Low-frequency signals will suffer little attenuation through the cable, while high-frequency signals will suffer much more attenuation. At the end of the cable, we will observe many strong, low-frequency signals and fewer, weaker high-frequency signals.

It will be difficult for the amplifiers in the signal path to remain linear in this situation because the gain of the system changes with frequency. If we place an amplifier in the cable to increase the strength of the high-frequency signals, then the amplifier will have to process the strong, low-frequency signals that didn't experience much attenuation. Returning to Figure 10-40, we note that the differential attenuation for the RG-223 cable is 58 dB, while the attenuation difference for the 3-inch Heliax is about 28 dB. The low-loss cable eases the wideband problem.

The usual solution to the wideband problem is to use low-loss cable—not because of its low attenuation but because low-loss cable usually exhibits lower differential attenuation with frequency. We can also design amplitude-compensating networks to flatten the cable's gain profile.

The hardest cable run to realize is a wideband (several octaves) link between an antenna and a remote receiver. The primary design factor is the cable's attenuation characteristics over frequency.

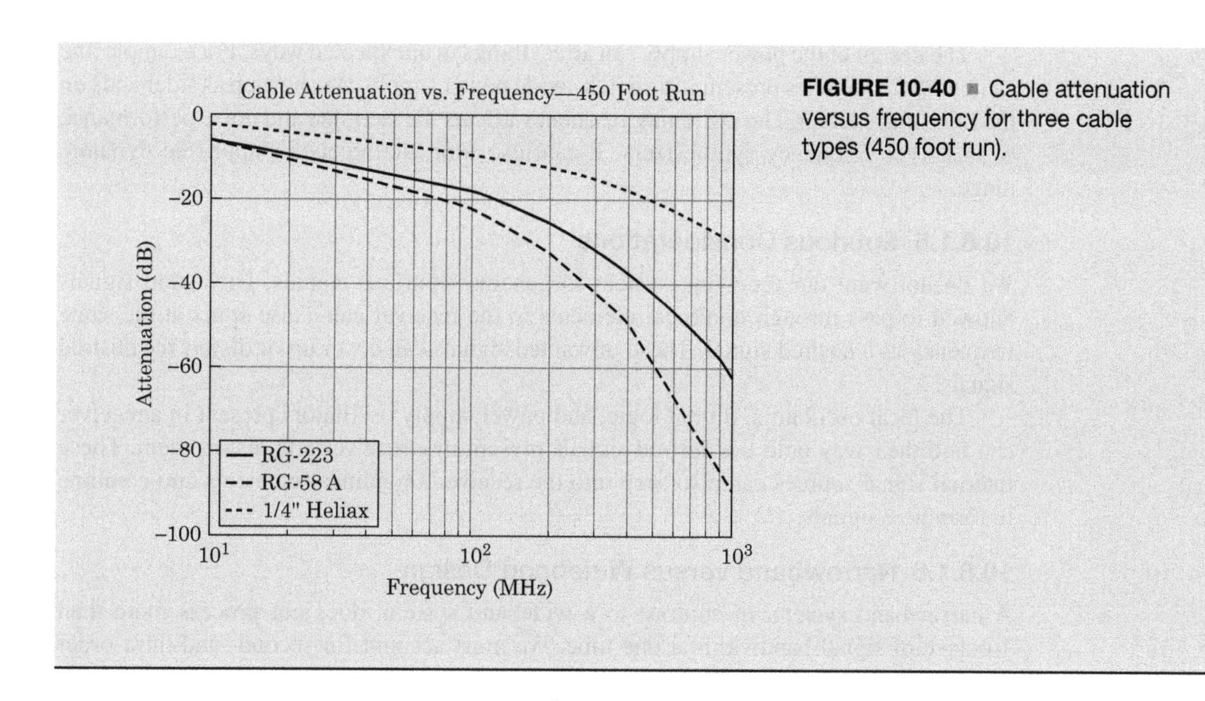

FIGURE 10-40 ■ Cable attenuation versus frequency for three cable types (450 foot run).

10.8 | FREQUENCY PLANNING AND IF SELECTION

10.8.1 Introduction

We'll discuss design strategies for frequency conversion schemes in this section. Design considerations include the following.

10.8.1.1 Cost

We desire the most inexpensive system possible, to include the cost of mixers, filter, amplifiers, LOs and their support circuitry, demodulator, and microprocessor. For example, some cellular telephones use a first IF of 45 MHz. Most televisions use this IF so there are plenty of cheap, small and available components that work in that frequency range. Common IFs include 455 kHz, 10.7 MHz, 21.4 MHz, 70 MHz, and 140 MHz.

10.8.1.2 Inertia

Designing a conversion scheme for a high-performance receiver can be quite involved once you consider all the trade-offs, so there is some validity to designing a receiver using a previously fielded design.

10.8.1.3 Physical Size

Small physical size, commercial components, low power, and common IFs go hand in hand.

10.8.1.4 Power Consumption

Small physical size makes it difficult to remove heat. Small systems require that the power supply be very efficient and that the receiver itself require little power. We often must contend with battery life.

The design of the power supply can affect things in unexpected ways. For example, the switching frequencies present in a switch-mode power supply can show up as sidebands on the LOs of a receiver. The switching frequency affects the step size and noise performance of PLL-type frequency synthesizers. It can also limit the receiver's spur-free dynamic range.

10.8.1.5 Spurious Considerations

We do not want our receiving system to generate spurious signals. Unwanted signals allowed to pass through nonlinear elements in the receiver can cause spurs at the same frequency as a desired signal. These unwanted signals can cover up or distort the desired signals.

The local oscillators, digital logic, and power supply oscillators present in a receiver can find their way onto the desired signals present in a receiver and distort them. These internal signal sources can also seep into the receiver's nonlinear elements and combine to form new signals.

10.8.1.6 Narrowband versus Wideband Design

A narrowband system, in contrast to a wideband system, does not process more than 1 octave of signal bandwidth at one time. We must account for second- and third-order distortion in a wideband system but only for third-order distortion in a narrowband system. In short, a narrowband design is easier than a wideband design.

10.8.2 Our Task

Figure 10-41 shows a simple conversion scheme and the labels we will use in the following discussions. The goal is to convert a swatch of spectrum centered at the RF to some IF.

10.8.2.1 Observations

Our analysis will assume the following conditions:

- At any time, any signal in the range of $f_{RF,L}$ to $f_{RF,H}$ may pass through the RF BPF to the RF port of the mixer.
- Since we don't know to which particular frequency the receiver will be tuned, we will assume that any signal in the range $f_{LO,L}$ to $f_{LO,H}$ can be present on the mixer's LO port.
- We are interested in signals that pass through the IF BPF. Any mixing component, other than the desired sum or difference frequencies, that passes through the IF BPF is undesired.

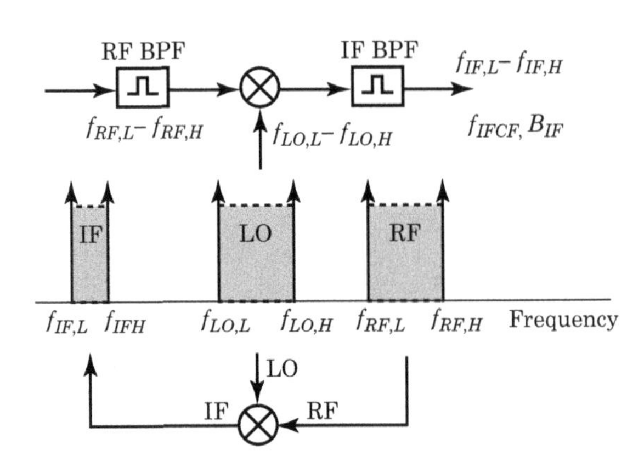

FIGURE 10-41 ■ A simple frequency conversion scheme defining the variables we'll use for the analysis of the frequency conversion.

10.8.2.2 Conversion Equation

The conversion equation is

$$f_{IF} = \left|mf_{LO} \pm nf_{RF}\right| \tag{10.86}$$

where

f_{LO} = any frequency from $f_{LO,L}$ to $f_{LO,H}$
f_{RF} = any frequency from $f_{RF,L}$ to $f_{RF,H}$
$m = 0, 1, 2, 3 \ldots$
$n = 0, 1, 2, 3 \ldots$

For each value of m and n, the band of frequencies from $f_{LO,L}$ to $f_{LO,H}$ will mix with the band of frequencies from $f_{RF,L}$ to $f_{RF,H}$ to produce two frequency bands on the mixer's IF port. The two bands fall in the ranges of $f_{IF,-,L}$ to $f_{IF,-,H}$ for the lower sideband and from $f_{IF,+,L}$ to $f_{IF,+,H}$ for the upper sideband. Figure 10-41 shows the operation for only one value of m and one value of n. If we change either m or n, the upper and lower IF sidebands will shift in frequency.

To calculate $f_{IF,-,L}$, $f_{IF,-,H}$, $f_{IF,+,L}$ and $f_{IF,+,H}$, we evaluate the conversion equations at the band edges of the RF and LO ports. For the upper and lower sidebands, we write

$$\begin{aligned} f_{IF,+,L} &= \left|mf_{LO,L} + nf_{RF,L}\right| & f_{IF,-,L} &= \left|mf_{LO,L} - nf_{RF,H}\right| \\ f_{IF,+,H} &= \left|mf_{LO,H} + nf_{RF,H}\right| & f_{IF,-,H} &= \left|mf_{LO,H} - nf_{RF,L}\right| \end{aligned} \tag{10.87}$$

EXAMPLE

Cellular Telephone

Figure 10-42 shows a conversion scheme for a cellular telephone receiver. We can write

- $f_{RF,L} = 825$ MHz
- $f_{RF,H} = 890$ MHz
- $f_{LO,L} = 870$ MHz
- $f_{LO,H} = 935$ MHz
- $f_{IF,L} = 42.5$ MHz
- $f_{IF,H} = 47.5$ MHz
- $f_{IFCF} = 45$ MHz
- $B_{IF} = 5$ MHz

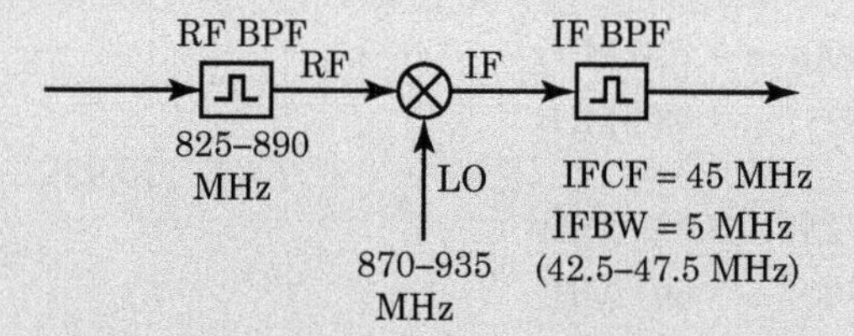

FIGURE 10-42 ■ A simplified frequency conversion scheme for a cellular telephone receiver.

Draw the output spectrums for the $mf_{LO} \times nf_{RF}$ combinations of 1×1, 1×2, 2×1, 2×2, 3×1, and 3×3 products.

Solution

We find the frequencies using equation (10.87).

1 × 1:

$$\begin{aligned} f_{IF,-,L} &= (1)(870) - (1)(890) = -20 \text{ MHz} \\ f_{IF,-,H} &= (1)(935) - (1)(825) = 110 \text{ MHz} \\ f_{IF,+,L} &= (1)(870) + (1)(825) = 1{,}695 \text{ MHz} \\ f_{IF,+,H} &= (1)(935) + (1)(890) = 1{,}825 \text{ MHz} \end{aligned} \tag{10.88}$$

The 1 × 1 output contains frequencies which overlap the 45 MHz IF, as it should.

1 × 2:

$$\begin{aligned} f_{IF,-,L} &= (1)(870) - (2)(890) = -910 \text{ MHz} \\ f_{IF,-,H} &= (1)(935) - (2)(825) = -715 \text{ MHz} \\ f_{IF,+,L} &= (1)(870) + (2)(825) = 2{,}520 \text{ MHz} \\ f_{IF,+,H} &= (1)(935) + (2)(890) = 2{,}715 \text{ MHz} \end{aligned} \tag{10.89}$$

The 1 × 2 output contains no frequencies which overlap the 45 MHz IF. This product is harmless to our conversion scheme.

2 × 1:

$$\begin{aligned} f_{IF,-,L} &= (2)(870) - (1)(890) = 850 \text{ MHz} \\ f_{IF,-,H} &= (2)(935) - (1)(825) = 1{,}045 \text{ MHz} \\ f_{IF,+,L} &= (2)(870) + (1)(825) = 2{,}565 \text{ MHz} \\ f_{IF,+,H} &= (2)(935) + (1)(890) = 2{,}760 \text{ MHz} \end{aligned} \tag{10.90}$$

The 2 × 1 output contains no frequencies which overlap the 45 MHz IF. This product is also harmless.

2 × 2:

$$\begin{aligned} f_{IF,-,L} &= (2)(870) - (2)(890) = -40 \text{ MHz} \\ f_{IF,-,H} &= (2)(935) - (2)(825) = 220 \text{ MHz} \\ f_{IF,+,L} &= (2)(870) + (2)(825) = 3{,}390 \text{ MHz} \\ f_{IF,+,H} &= (2)(935) + (2)(890) = 3{,}650 \text{ MHz} \end{aligned} \tag{10.91}$$

The 2 × 2 products contains frequencies which overlap the 45 MHz IF. These signals are in-band spurious products, and they might be troublesome.

3 × 1:

$$\begin{aligned} f_{IF,-,L} &= (3)(870) - (1)(890) = 1{,}720 \text{ MHz} \\ f_{IF,-,H} &= (3)(935) - (1)(825) = 1{,}980 \text{ MHz} \\ f_{IF,+,L} &= (3)(870) + (1)(825) = 3{,}435 \text{ MHz} \\ f_{IF,+,H} &= (3)(935) + (1)(890) = 3{,}695 \text{ MHz} \end{aligned} \tag{10.92}$$

The 3 × 1 components are harmless.

3 × 3:

$$f_{IF,-,L} = (3)(870) - (3)(890) = -60 \text{ MHz}$$
$$f_{IF,-,H} = (3)(935) - (3)(825) = 330 \text{ MHz}$$
$$f_{IF,+,L} = (3)(870) + (3)(825) = 5{,}085 \text{ MHz} \qquad (10.93)$$
$$f_{IF,+,H} = (3)(935) + (3)(890) = 5{,}475 \text{ MHz}$$

The 3 × 3 component also contains signals that overlap the 45 MHz IF.

Figure 10-43 shows this data in graphical format. There are many products we did not consider. Note that the 1 × 1, the 2 × 2, and the 3 × 3 products all can produce signals that fall within the 45 MHz IF band-pass filter.

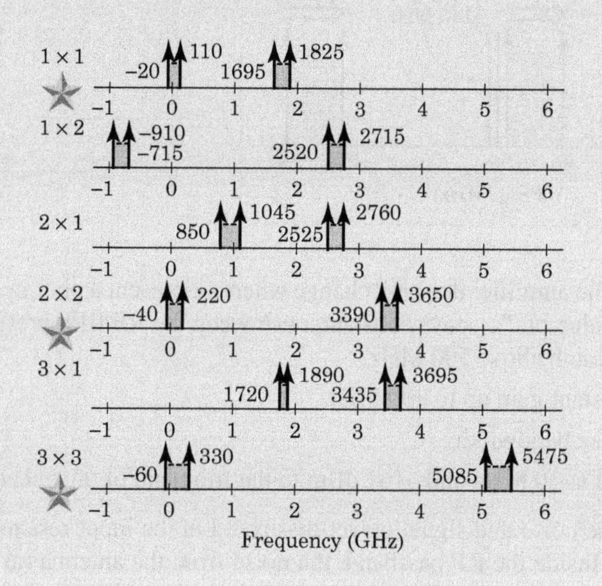

FIGURE 10-43 ■ A partial list of the $m \times n$ mixer products generated by the conversion scheme of Figure 10-42.

10.8.3 Image Noise

In a poorly designed system, the conversion process will move out-of-band noise to our IF along with the desired signal. Figure 10-44(a) shows the front end of a receiving system. We have modeled the antenna as a voltage source in series with a noisy resistor. The noisy resistor models the antenna noise, and we will assume the temperature of R_{ant} is 290°K (or T_0).

The RF BPF passes 10 to 50 MHz. We follow the RF BPF with a low-noise amplifier ($F = 3$ dB; $T_{sys} = 290$°K) whose gain is 25 dB. A mixer converts a signal from the RF to 145 MHz using a low-side LO (LSLO). Figure 10-44(b) shows where the LO, RF, and image frequencies fall.

We make several simplifying yet reasonable assumptions:

- The insertion loss and the noise figure of the RF BPF are both 3 dB.

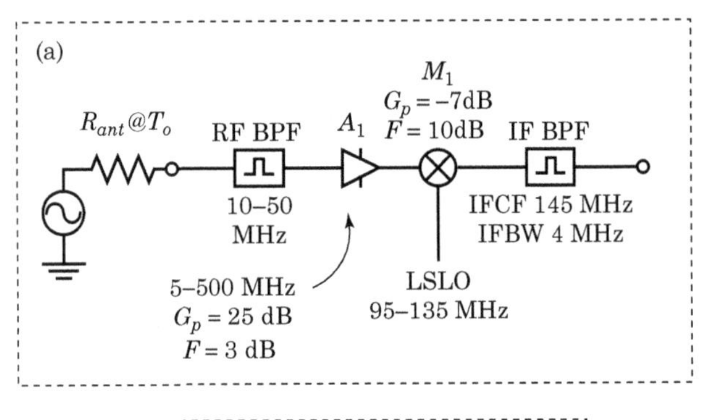

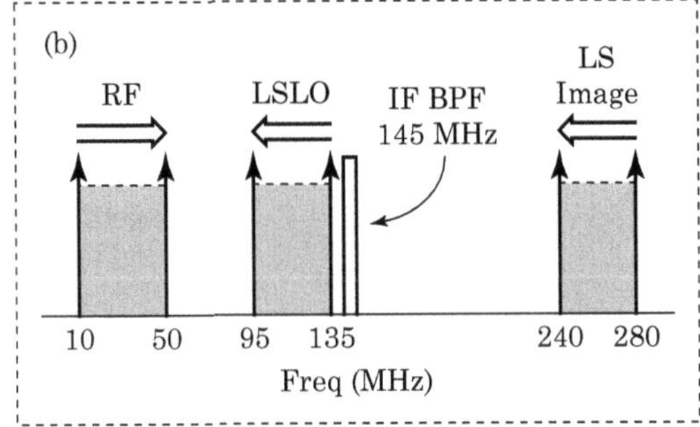

FIGURE 10-44 ■ A conversion scheme exhibiting image noise problems. (a) The conversion scheme. (b) A graphical view of the conversion showing the RF, LSLO, IF, and image frequencies.

- The noise performance of the amplifier does not change when we present a poor match to its input terminal. The value of T_{sys} stays constant, even when the RF BPF presents the amplifier with a poor match above 500 MHz.
- Amplifier A_1 provides constant gain up to 300 MHz.
- We will assume a 1 Hz noise bandwidth.
- We will apply an RF signal at 30 MHz and –161 dBm to the front end of this receiver.

Figure 10-45(a) shows the noise and signal power dissipated in the input resistor of amplifier A_1 over frequency. Inside the RF passband, the noise from the antenna (at T_0) adds to the internally generated noise of amplifier A_1 to produce the noise hump from 10 to 50 MHz. In the RF passband, the noise power dissipated in A_1's input resistor is

$$\begin{aligned} N_{in,in\text{-}band} &= k\left(T_{sys} + T_{ant}\right) B_n \\ &= k\,(290 + 290)\,(1) \\ &= 8.00 \cdot 10^{-21} \text{ Watts} \\ &= -171 \text{ dBm} \end{aligned} \tag{10.94}$$

The input SNR is 10 dB.

The antenna noise is severely attenuated in the stopband of the RF BPF, effectively reducing the antenna noise temperature to 0°K. In the stopband of the RF BPF, the noise dissipated in A_1's input resistor is

$$\begin{aligned} N_{in,out\text{-}of\text{-}band} &= k\left(T_{sys} + T_{ant}\right) B_n \\ &= k\,(290 + 0)\,(1) \\ &= 4.00 \cdot 10^{-21} \text{ Watts} \\ &= -174 \text{ dBm} \end{aligned} \tag{10.95}$$

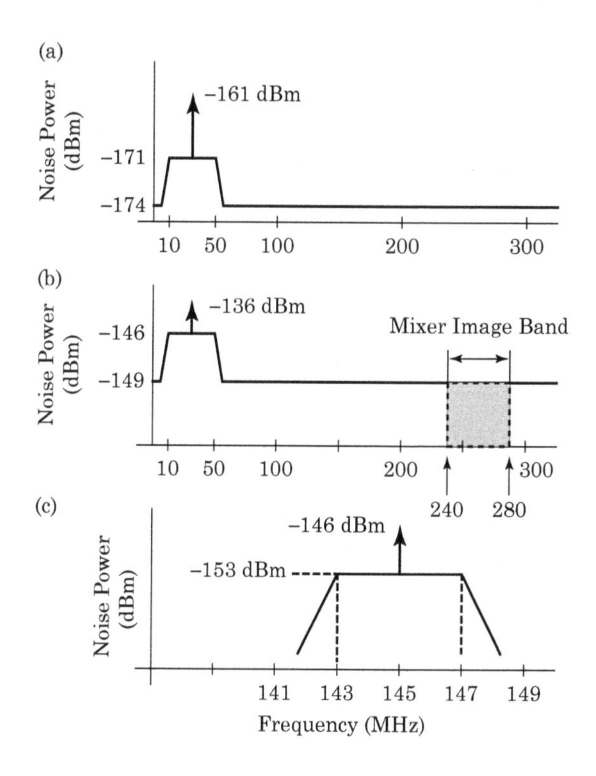

FIGURE 10-45 The noise and signal powers at various points in Figure 10-44. (a) The noise power present at the input of amplifier A_1. (b) The spectrum present at the input of mixer M_1. (c) The spectrum present at the output of the 145 MHz IF band-pass filter.

Figure 10-45(b) shows the noise and signal power at the output of amplifier A_1. Note that we have a significant amount of noise up to and beyond 300 MHz due to the amplifier's gain and noise figure. However, the SNR (measured in a 1 hertz bandwidth around the signal) is still 10 dB.

The noise figure of mixer M_1 is 10 dB. M_1's noise temperature is

$$\begin{aligned} T_{mix,in} &= T_0 \left(F_{mix} - 1\right) \\ &= 290\,(10 - 1) \\ &= 2610\ ^\circ\mathrm{K} \end{aligned} \tag{10.96}$$

The mixer's input resistor dissipates

$$\begin{aligned} N_{mix} &= kT_{mix}B_n \\ &= k\,(2610)\,(1) \\ &= 6.0 \cdot 10^{-21}\ \text{Watts} \\ &= -164.4\ \text{dBm} \end{aligned} \tag{10.97}$$

due to its own internally generated noise. At the mixer's image frequency band (240 to 280 MHz), there are two sources of input noise: (1) the noise from A_1 (at -149 dBm); and (2) the noise due to the mixer's own internal noise (at -164 dBm). The external noise is 15 dB larger than mixer's internal noise. The major source of noise at this node is the amplified noise supplied by A_1. At the mixer's input port, signals present at both the RF and at the image frequencies are both converted to the IF center frequency with equal efficiency.

Figure 10-45(c) shows the spectrum on the output of M_1. The signal's SNR has changed from 10 dB to only 7 dB. We lost 3 dB in SNR because the noise at the image

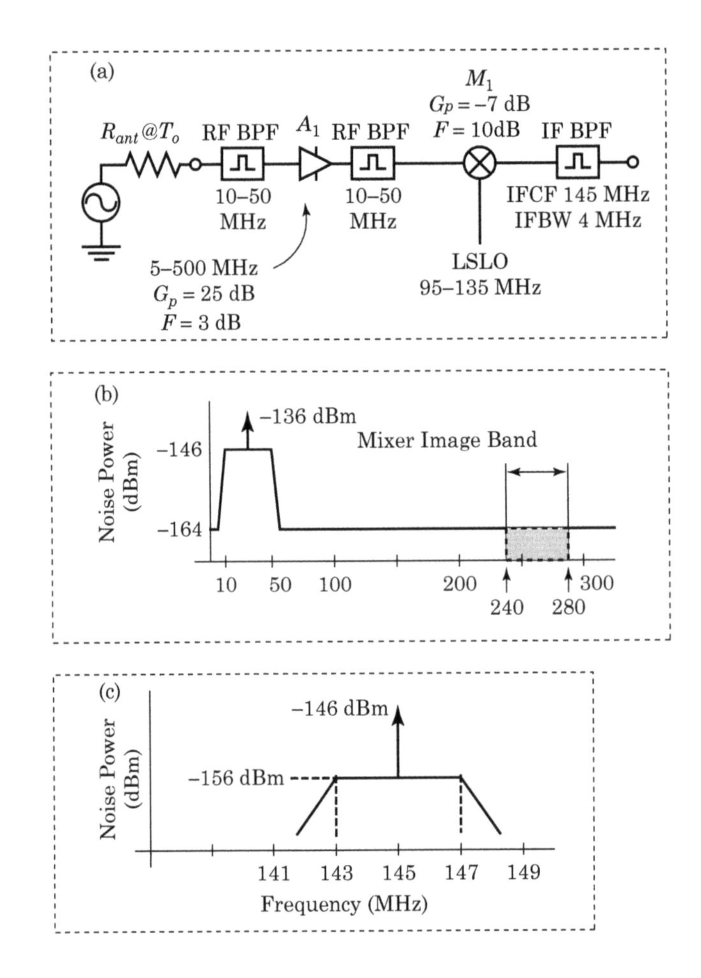

FIGURE 10-46 Reducing the image noise problem of Figure 10-44 and Figure 10-45 (a) The modified cascade. Note the addition of the second 10 to 50 MHz band-pass filter between the amplifier and the mixer. (b) The spectrum present at the input to the mixer. (c) The spectrum present at the output of the 145 MHz band-pass filter. Note the improved SNR over the signal shown in Figure 10-45(c).

frequency of the mixer was converted to the IF along the signal and noise present at the desired frequency. Since the noise powers at the RF and its image was about equal, the net result is a loss of 3 dB of SNR (or an increase in the effective noise temperature of the mixer).

Figure 10-46(a) shows the solution to this problem. Adding a filter between amplifier A_1 and mixer M_1 attenuates the noise generated by the amplifier at the image frequency and preserves the received SNR. Figure 10-46(b) shows the spectrum applied to the RF port of the mixer. We could use either a BPF or a LPF between A_1 and M_1 to attenuate the noise at 240 to 280 MHz. Figure 10-46(c) shows the output spectrum at 145 MHz. The signal's SNR is preserved.

10.8.4 Upconversion versus Downconversion

When we move a signal from one frequency to another, we make one of the four choices shown in Figure 10-47(a-d). We can:

- Upconvert with an high-side local oscillator (HSLO)
- Upconvert with an low-side local oscillator (LSLO)
- Downconvert with an HSLO
- Downconvert with an LSLO

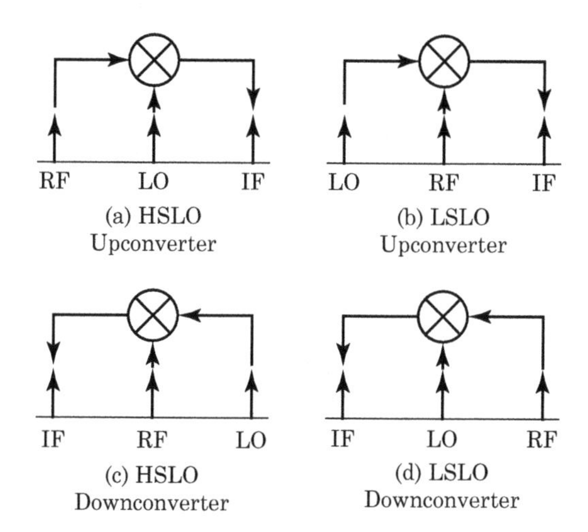

FIGURE 10-47 Four possible types of conversion schemes. (a) HSLO upconverter. (b) LSLO upconverter. (c) HSLO downconverter. (d) LSLO downconverter.

where

- *Upconvert* means to move a signal from its original frequency to a higher frequency
- *Downconvert* means to move a signal from its original frequency to a lower frequency
- *LSLO* = the LO frequency is less than the RF
- *HSLO* = the LO frequency is greater than the RF

In equation form, we can write

$$\begin{gathered}\text{Low - Side LO} \Rightarrow f_{LO} < f_{RF} \\ \textit{and} \\ \text{High - Side LO} \Rightarrow f_{LO} > f_{RF}\end{gathered} \tag{10.98}$$

Recall that our signal will experience a spectrum inversion when we use a HSLO and we're selecting the lower sideband.

10.8.4.1 Example Architecture for Upconversion with a HSLO

The receiver architecture of Figure 10-48 shows a system that will convert a signal from the 20 to 500 MHz RF range to an IF of 700 MHz using a HSLO. The directions of the horizontal arrows on the spectrum plot of Figure 10-48 show the relative directions of the frequency changes. As we tune the receiver from 20 MHz upward, the LO starts at 720 MHz and moves upward also. The HS image frequency starts at 1420 MHz and moves upward in frequency.

Strengths This architecture has several strengths and one major weakness. Strengths include:

- The IF and image rejection as well as the LO radiation at the antenna port can all be effectively serviced with a low-pass filter as the RF BPF. We can achieve very high rejection with low complexity and small physical size.
- It is a very cheap and simple solution to the many spurious problems present in a typical receiver.
- The simplicity of this system makes it very easy to realize a complex yet precise measurement system. The architecture of Figure 10-48 is typical of most spectrum analyzers.

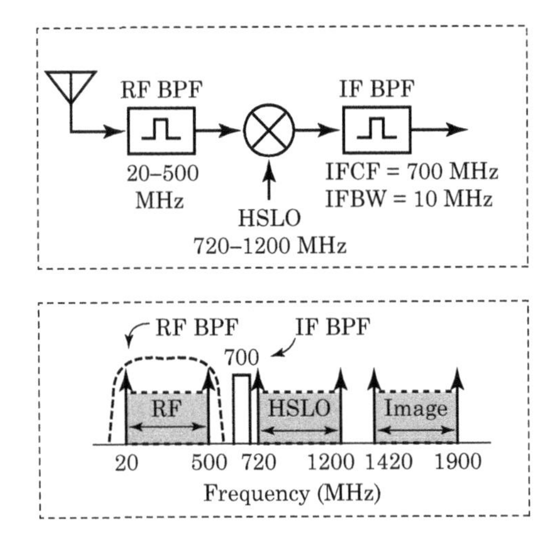

FIGURE 10-48 The architecture and frequency plan for an HSLO upconverter.

Weaknesses The major weakness of this system is that its dynamic range is limited due to harmonics of the RF falling within the IF passband. Imagine the receiver is tuned to 80 MHz. The microprocessor tunes the LO to 780 MHz to convert the 80 MHz RF signal to a signal centered at 700 MHz. We now apply a 350 MHz signal to the front end of the system (while it's still tuned to 80 MHz). The second harmonic of 350 MHz is 700 MHz. The second-order nonlinearity of the mixer will generate a 700 MHz signal from both the 350 MHz signal and from the 80 MHz signal. The signal powers will compete with each other for control of the IF.

If the 700 MHz energy from the 350 MHz second harmonic is greater than the 700 MHz energy generated from the 80 MHz signal, the distortion energy will cover up the wanted signal energy at 80 MHz. The receiver is tuned to 80 MHz, yet is will be processing the signal at 350 MHz. We can improve this design by selecting an IF which is higher in frequency. If we set the first IF to 1100 MHz, then second-order distortion is no longer a problem because the highest second-order distortion product produced by the system will be 1000 MHz.

Even with a higher IF, the receiver is vulnerable to higher-order distortion products. If we apply $700/3 = 233.33$ MHz to the receiver, the third harmonic of this RF signal falls within the IF passband. The same argument hold for 175 MHz ($=700/4$), 140 MHz ($=700/5$), 116.67 MHz ($=700/6$), etc.

An IF at 1100 MHz prevents second-order distortion, but higher-order products are still a problem. For an upconverter with an HSLO, it's generally necessary to place the IF at least 2.1 times the higher than the highest RF signal, or

$$f_{IFCF} \geq (2.1)\, f_{RF,H} \tag{10.99}$$

10.8.4.2 Example Architecture for Upconversion with a LSLO

Figure 10-49 shows an upconverting receiver using an LSLO and an IF of 700 MHz. The receiver tunes from 450 to 600 MHz. Note the directions of the arrows on the spectrum plot of Figure 10-49. As we tune the receiver from 450 MHz upward, the LO starts at 250 MHz and moves downward. The low-side image frequency starts at 950 and moves downward in frequency.

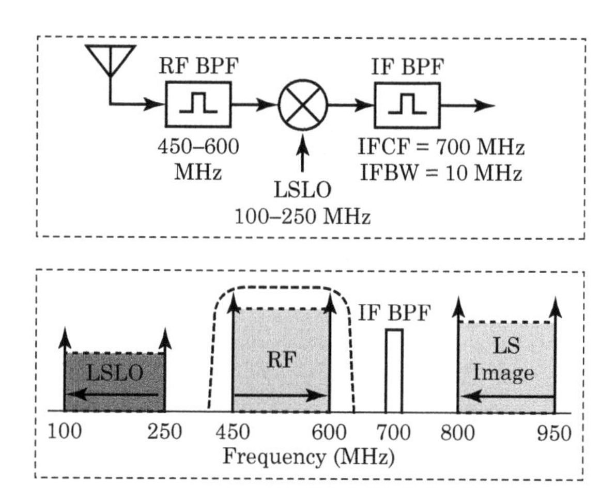

FIGURE 10-49 ■ The architecture and frequency plan for an LSLO upconverter.

Strengths This is a simple, straightforward architecture. We need an RF BPF to achieve our LO, IF, and image rejection.

Weaknesses The harmonics of the LO can cross both the RF and IF, resulting in LO leakage from the antenna port and a possible jamming of the IF by harmonics of the LO. Depending on the precise frequencies involved, harmonics of the RF can appear in the IF. We place the IF at a high frequency to mitigate this effect.

10.8.4.3 Example Architecture for Downconverting with a HSLO

Figure 10-50 shows a system to convert signals in the 100 to 500 MHz range to 40 MHz. As we increase the tuned frequency, the HSLO and HS image frequencies also increase.

Strengths Since the RF and LO are both above the IF, there is no possibility of harmonics of either signal getting into the IF. This increases the dynamic range of the system considerably.

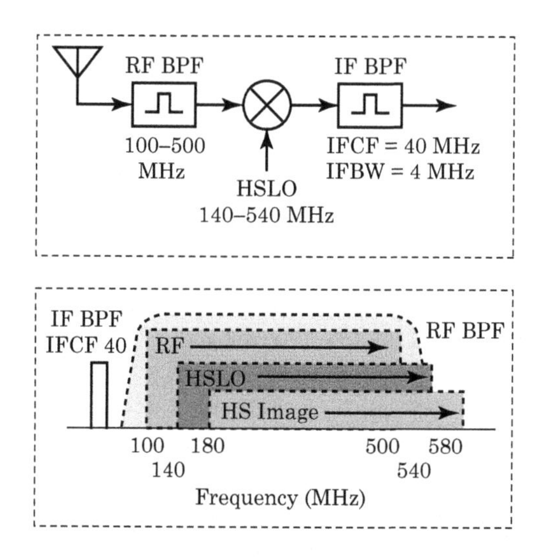

FIGURE 10-50 ■ The architecture and frequency plan for an HSLO downconverter.

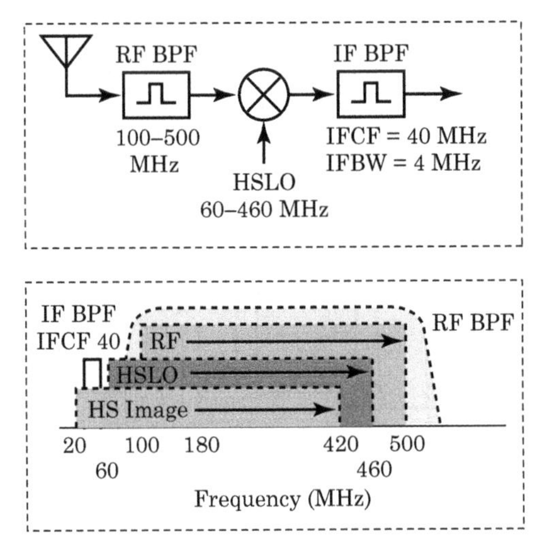

FIGURE 10-51 The architecture and frequency plan for an LSLO downconverter.

Weaknesses The IF, image, and LO rejection problems have increased dramatically over the upconversion cases. The most common solution to these rejection problems is to use a tracking RF BPF as the preselection filter. This will provide the IF and image rejection as well as suppressing the LO leakage to the antenna port, however it will significantly increase the cost of the receiver.

10.8.4.4 Example Architecture for Downconverting with a LSLO

Figure 10-51 shows a system used to convert signals in the 100 to 500 MHz range to 40 MHz using a LSLO. As we increase the tuned frequency, the HSLO and HS image frequencies also increase.

Strengths The RF and LO are still above the IF, so there is no possibility of harmonics of either signal getting into the IF. This increases the dynamic range of the system considerably.

Weaknesses The IF, image, and LO rejection problems are still difficult to deal with over the upconversion cases. You almost have to use a tracking RF BPF as the preselection filter to provide adequate IF and image rejection as well as suppressing the LO leakage to the antenna port. Again, this will add significant cost and complexity to the receiver.

10.8.4.5 Simplicity

Upconverters are generally simpler and less complicated than downconverters. Upconverters provide LO, IF, and image rejection with a simple low-pass front-end filter. Downconverters usually require tracking band-pass filters or switched filters on the RF port to accomplish these goals. However, upconverters have limited dynamic range because harmonics of RF signals can fall into the IF passband.

10.8.4.6 LO Rejection

The HSLO upconverter usually has a smaller amount of LO power present on the antenna port than the LSLO upconverter because the HSLO always falls outside of the RF passband. In an LSLO upconverter, the LO can fall inside the RF passband.

10.8.4.7 Oscillator Tuning Range

While the HSLO upconverter relies on a higher LO frequency (which may be difficult to realize), the synthesizer has to tune over a smaller fractional bandwidth. This usually reduces the phase noise of the oscillator and makes for a cleaner demodulation.

10.8.4.8 When to Upconvert and When to Downconvert

Eventually, we must bring the signal to some common frequency to demodulate it and, more often than not, the final IF is at a relatively low frequency (e.g., 10.7 MHz). We must eventually use a downconverter in some part of our receiver.

In a narrowband environment, the difference between upconversion and downconversion schemes is small. There are several reasons:

- In both cases, the LO must tune over only a very small percentage bandwidth, so LO design considerations even out.
- If we design carefully, we can achieve all of the LO, image, and IF rejection we need in a downconverter using band-pass filters.

In summary, it's desirable to upconvert in a wideband environment and can weather the limited dynamic range. Once the signal has passed through the first IF filter, we have considerably simplified the problem and can downconvert to place the signal to the final IF for demodulation or digitization.

War Story: Unintended Effects in Conversion Schemes

The effects of conversion schemes can cause unexpected problems. The author once designed a voltage-controlled oscillator (VCO). These devices are nonlinear and not easy to characterize theoretically, so designers spend much time on the bench verifying the oscillator over variations in, for example, power supply voltages, temperature, and load impedances. One of the characteristics of a poor VCO design is that, at some frequency, the oscillator's output signal will widen. The oscillator will exhibit modulation and the noise floor near the oscillator enter frequency will increase.

Your author spent several days debugging one particular VCO. It worked well but always misbehaved at one single frequency, although the design was familiar and benign. We built similar oscillators in the past. However, none of the usual debugging techniques solved the problem. In desperation, the author placed a known good oscillator in the test fixture, and it exhibited the same problem at the exact same frequency range. Panic set in. This design had been fielded. Changing the oscillator now would be very expensive.

We then connected a commercial signal generator to the test fixture and observed the same problem at the same frequency. The problem was in the test fixture. Figure 10-52 shows the block diagram of the spectrum analyzer used in the test configuration. The problem occurred when the VCO was tuned to 405 MHz, which is $f_{IF}/5$. The fifth harmonic of the oscillator fell directly into the passband of the IF. The problem was with

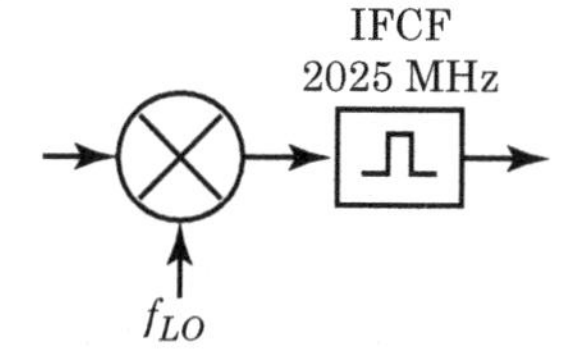

FIGURE 10-52 ■ A simplified block diagram of the input stage of a spectrum analyzer.

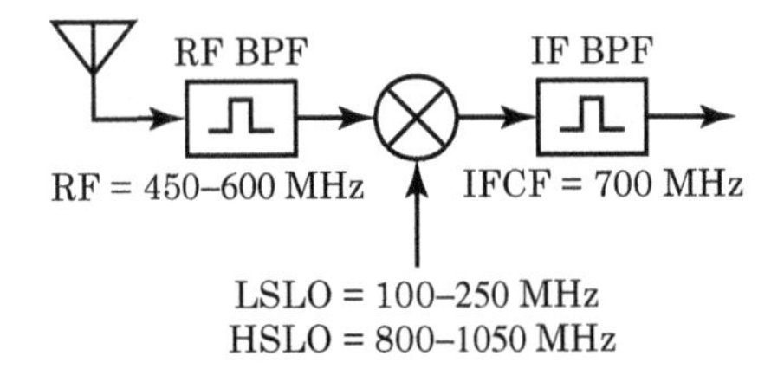

FIGURE 10-53 ■ Block diagram of a receiver showing the two possible choices for the LO range. The HSLO is 800–1050 MHz, while the LSLO is 100–250 MHz.

the conversion scheme of the spectrum analyzer (and with the author for believing the spectrum analyzer).

10.8.5 LOs, Tuning Range, and Phase Noise

In Chapter 9, we found that a local oscillator that must tune over a large percentage bandwidth will be noisy.

10.8.5.1 HSLO versus LSLO Upconverters

Figure 10-53 shows the conversion architecture for a 450–600 MHz receiver. The LSLO range covers 100–250 MHz, which is a 150 MHz tuning range on a geometric center frequency of 158 MHz. This is a 95% tuning range. The HSLO tunes from 800–1050 MHz, which is the same 150 MHz tuning range centered on 917 MHz or a 16% tuning range.

For a given conversion scheme, the HSLO is always at a higher center frequency than the LSLO so the HSLO will tune over a smaller percentage bandwidth than does the LSLO converter. As we discussed in Chapter 9, reducing the tuning range of an oscillator (in percentage of center frequency) will often improve its phase noise. However, higher-frequency oscillators tend to be noisier than lower-frequency oscillators so there is no clear trade-off here.

10.8.5.2 Large Tuning Range, Small Step Size

Figure 10-54(a) shows one possible architecture for a wideband receiver. The receiver must tune from 20 to 500 MHz in 1 kHz steps (i.e., the receiver must tune to 20.000 MHz, 20.001 MHz, 20.002 MHz . . . 499.999 MHz, 500.000 MHz). The first IF is 700 MHz with a HSLO. LO_1 must tune over a 720 to 1200 MHz range (a percentage bandwidth of 52%) in 1 kHz steps.

A phase-locked loop oscillator is ideal for this receiver because it can easily realize the step size and the tuning range. However, we expect phase noise problems due to the large tuning range. A PLL will exhibit improved phase noise when we observe the LO within a loop bandwidth from the carrier. Practical considerations limit the loop bandwidth of simple PLL synthesizers to less than 10% of the step size or 100 Hz in this case.

10.8.5.3 Two Synthesizers

Figure 10-54(b) shows the common solution to the phase noise issue that uses two synthesizers. The first synthesizer tunes from 720 to 1200 MHz in 1 MHz steps. The coarse frequency step size of 1 MHz allows the loop bandwidth of the first synthesizer to be approximately 100 kHz, quieting the phase noise of this oscillator significantly.

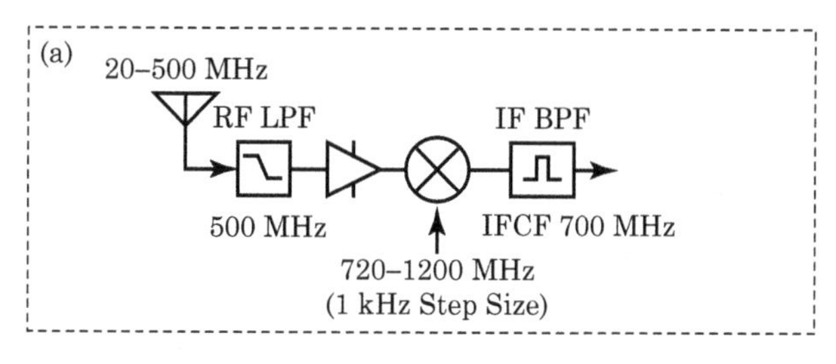

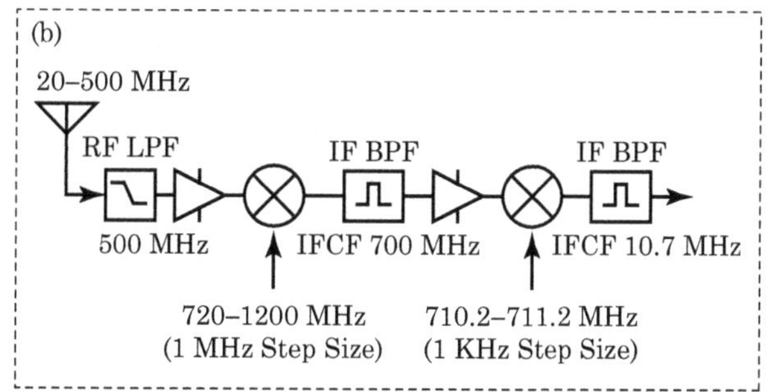

FIGURE 10-54 ■ Two possible architectures for a wideband receiver with a small tuning resolution. (a) One synthesizer tunes over the entire LO range with a small step size. (b) Two-synthesizer solution. One synthesizer tunes over a wide range with a large step size. The second synthesizer tunes over a small range with a small step size.

The signal of interest now lies within 500 kHz of the center frequency of the first IF filter. The second LO tunes with a 1 kHz step size, but over only 1 MHz. Since the tuning range of this oscillator is relatively small, the phase noise of the basic oscillator will be small, and we can suffer the small loop bandwidth.

10.8.6 IF Selection Guidelines

Designing a conversion scheme can be difficult because the final performance of the receiver is determined by quantities that cannot be readily measured or calculated. Such quantities include the ultimate attenuation of the filters and the isolation achievable between various compartments of the receiver. Long, hard experience has taught us a few useful rules of thumb to use in the initial stages of designing a conversion scheme.

10.8.6.1 Suboctave Preselection

Use suboctave preselection filters to reduce the effects of second-order distortion. Figure 10-55(a) shows a system with a single 100 to 500 MHz RF preselection filter. Instead of a single filter, it is wise to use several filters as shown in Figure 10-55(b). Each filter should be less than an octave in bandwidth.

10.8.6.2 Image Noise

Image noise can seriously reduce the sensitivity of your system. Be sure to provide filtering to eliminate it.

10.8.6.3 Upconverting

When upconverting, select the IF to ensure the harmonics of the RF lie well outside of the IF filter. A rough rule of thumb is to place the center frequency of the IF passband at least 2.1 times the highest RF.

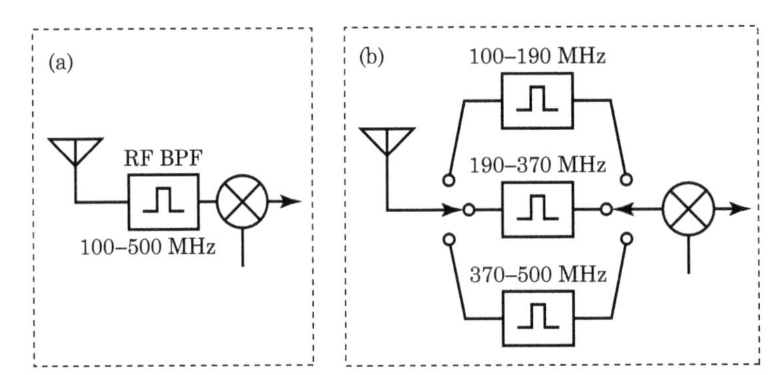

FIGURE 10-55 ■ Suboctave preselection reduces the effects of second-order distortion. (a) The wideband RF BFP is greater than an octave wide and can cause second-order problems to occur. (b) Replacing the wideband RF BFP with several suboctave filters prevents second-order distortion from becoming a problem.

10.8.6.4 Butterworth Approximations

Use the Butterworth filter approximations (from Chapter 5) to find the approximate attenuation of the filters in your system. Remember, though, that the Butterworth approximations will not predict the filter's ultimate attenuation.

10.8.6.5 Eventually Downconvert

The insertion loss of a band-pass filter tends to increase as its percentage bandwidth (B_{IF}/f_{IFCF}) decreases. A filter with a small percentage bandwidth will tend to be lossier than a filter with a large percentage bandwidth. We must eventually downconvert to perform filtering with a small insertion loss and a narrow IF bandwidth. For example, a narrowband signal of 30 kHz will require that we eventually move the signal to a low frequency to properly filter it.

10.8.6.6 Oscillator Center Frequency

High-frequency oscillators are noisier than low-frequency oscillators. Oscillators that must tune over a wide percentage bandwidth are noisier than oscillators that tune over a small percentage bandwidth.

10.8.6.7 Common IFs

Use common IFs to take advantage of the plethora of cheap, small, commercial parts available on the open markets. Some common IFs are as follows:

- 21.4 MHz (high-end receivers)
- 10.7 MHz (commercial receivers)
- 45 MHz (television, cellular telephones)
- 70 MHz (satellite television, military gear)
- 455 kHz (commercial equipment)
- 160 MHz (commercial satellite equipment)

10.8.6.8 Bad IFs

Avoid placing IFs in sections of the spectrum where there are many strong signals. The receiver's IF rejection has to be that much better to function well. For example, a poor designer places the first IF falls within the commercial FM band. The receiver is likely to encounter very strong signals in this frequency range, which puts a heavy burden on the rejection of the RF BPF at the IF center frequency. Leakage around the filter will also be problematic. It is wise to avoid the commercial aircraft bands, the cellular telephone band and common amateur radio frequencies.

10.8.6.9 LO Harmonics

Since the local oscillator is the strongest signals present in a receiver, you should never design a conversion scheme that allows harmonics of the LO to overlap the IF passband. The harmonics of the LO will likely be larger than most of the RF signals the receiver will process.

10.8.7 Practical Design Considerations

10.8.7.1 Separate Compartments

If possible, provide separate, isolated compartments for each subsystem of the receiver. Ideally, supply a separate compartment for each of the following receiver subsystems:

- Switching power supply: The switching transients will get everywhere, especially the voltage-controlled oscillators used in the frequency synthesizers. Avoid switchers on critical supplies if at all possible.
- Microprocessor: The high-level digital signals present in clocked logic are deadly to any high gain system. The harmonics of the digital waveforms can be a problem to several hundred MHz.
- Voltage-controlled oscillators: The VCOs used in PLL frequency synthesizers are especially sensitive to power supply, radiated, and conducted noise. The clever designer will provide sensitive reference oscillators with their own linear, low-noise power supply regulators.
- Digital PLL components: In the ideal case, we would provide a separate compartment for the PLL circuitry other than the VCOs. This circuitry is a unique combination of digital and analog systems working together. However, we often package these components in the same compartment as the VCOs with only minor problems.
- Each compartment should house only one LO. We do not want two separate oscillators present in the same compartment at the same time. The interface between compartments provides a unique opportunity to increase the absolute attenuation of the filter that separates the two mixers.

10.8.7.2 Single Printed Circuit Board Systems

Cost and manufacturability are important considerations. These factors often force us to make unpleasant design choices. Although we'd like to place all of our sensitive circuitry in a nice, quiet compartment, we are often forced to place them on the same physical printed circuit board as the noisy circuitry.

With careful layout techniques, you can achieve reasonably high levels of isolation on a single circuit board. See Dexter (1989) for more details.

10.8.7.3 Ground Planes

The purpose of a ground plane is to provide a low-impedance return path for power supply currents and signal currents. The key term is "low impedance." We can also call on a properly designed ground plane to provide isolation between components (to increase the ultimate attenuation of a filter, for example) and to form printed-circuit transmission lines directly on the circuit board.

10.8.7.4 Power Supplies

Many of the subsystems of a receiver are sensitive to power supply variations. Voltage-controlled oscillators, for example, exhibit "frequency pushing." Frequency pushing describes the variation of oscillator frequency with changes in its power supply voltage. If the power supply of the oscillator contains a component at 60 Hz, for example, the oscillator will exhibit sidebands at ± 60 Hz from the carrier.

Amplifiers are also affected by power supply variations.

Switching Power Supplies Switching power supplies accommodate the changing input voltage and generate other voltages required by the various receiver subsections. However, these power supplies are a notorious source of high-level, hard-to-filter transients.

Switching power supplies operate using fast, high-current switches to gate the input voltage into AC waveforms. Then, the AC waveforms are rectified back into noisy DC voltages. The noisy DC voltages are then filtered to produce the still-noisy DC voltages used throughout the receiver. Switching power supplies currently operate from 20 kHz to beyond 200 kHz. These relatively low frequencies require large inductors and large capacitors.

We often use the switcher to generate a voltage that is three or four Volts higher than required and then sub-regulate with a linear regulator. Linear regulators are quiet and, if properly designed, they will remove most of the switching transients from the switcher's output voltage. See **Linear Regulators** below for more details.

The high-speed switching transients produce much of the noise. These switching transients can have peak values of several Amps and their magnitudes are directly related to the DC current supplied by the power supply. The large transient currents are present on the system's ground plane which cause voltage spikes on the ground plane. Designing your system to draw as little power as possible will help suppress the noise of the switcher.

Keep the sensitive and low-level circuitry as physically far away from the switching supply as practical. Imagine the path that the transient currents from the switcher must take and place the sensitive circuitry accordingly.

Linear Regulators Linear regulators are always quieter than switching regulators. Linear regulators are simple control systems. They contain a pass transistor (or FET) and an internal voltage reference. The idea is to monitor the output of the regulator and change the base (or gate) conditions on the pass transistor so that the output voltage equals some multiple of the reference voltage. The control process generates internal noise and some regulators are quieter than others are.

Linear regulators possess a control bandwidth. Inside this bandwidth, they will apply some measure of attenuation to noise at the regulator's input. However, we will experience little or no attenuation of V_{noise} components beyond the regulator's control bandwidth. In Figure 10-56, the first voltage source, labeled "V_{DC}", represents the pure DC applied to

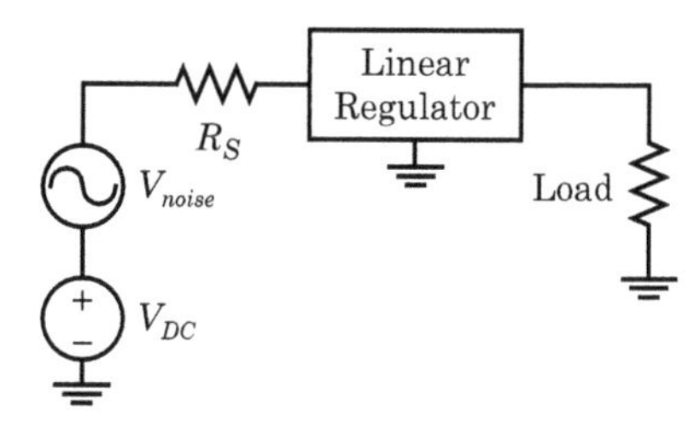

FIGURE 10-56 ■ Using a linear regulator to reduce power supply noise. The regulator will reduce power supply noise if the frequency of the noise is lower than the control bandwidth of the regulator. Noise outside of the regulator's control bandwidth may not be suppressed.

the regulator. The second voltage source, labeled V_{noise}, represents noise present on the unfiltered power supply. This noise may exhibit a 60 Hz mains component or it could be a 40 kHz component from a switching power supply. We will assume the noise component is a simple sinusoid.

When V_{noise} is at a very low frequency, the control loop inside the regulator may have no trouble following this variation and suppressing it at the output. If we increase the frequency of V_{noise}, we will reach a frequency at which the loop will have trouble following the variation on the regulator's input and some amount of V_{noise} will be present on the regulator's output. As we further increase the frequency of V_{noise}, we will ultimately find a point where the regulator is ineffective and V_{noise} will pass unaffected through the regulator.

10.8.7.5 The Love–Hate Relationship with Digital Logic

The high-level digital signals derived from the microprocessor's clock present horrendous isolation problems to the receiver designer. The harmonics of the digital waveforms can extend to several hundred MHz—well into the passband of most radio receivers. However, we must get information from the digital portion of the receiver to the rest of the system, so we must run conductors from the microprocessor portion of the receiver to the frequency synthesizers, the demodulator and to the various IF and RF subsections.

Continuously Clocked versus Unclocked Logic From a noise perspective, we can place digital logic into two categories: continuously clocked and unclocked. Continuously clocked logic is logic that requires a continuous clock to operate. For example, most microprocessors require a continuous clock. Unclocked logic does not require a continuous clock to operate. Flip-flops, shift registers, and static memory are examples.

Considerations to Receiver Design Continuously clocked logic is always a threat to the performance of a receiver because the digital waveforms are always present. Unclocked logic is not a source of noise because both its inputs and outputs are not changing unless we specifically clock the logic and command it to change. We can place unclocked logic in proximity to sensitive electronic components without a noise penalty.

Number of Conductors We are very concerned about the number of wires that transverse the boundaries between the separate subsections of a receiver. We seek to minimize these interconnections:

- Electromagnetic interference: Each conductor passing through the boundary between a noise-generating piece of the system and a noise-sensitive part of the system presents

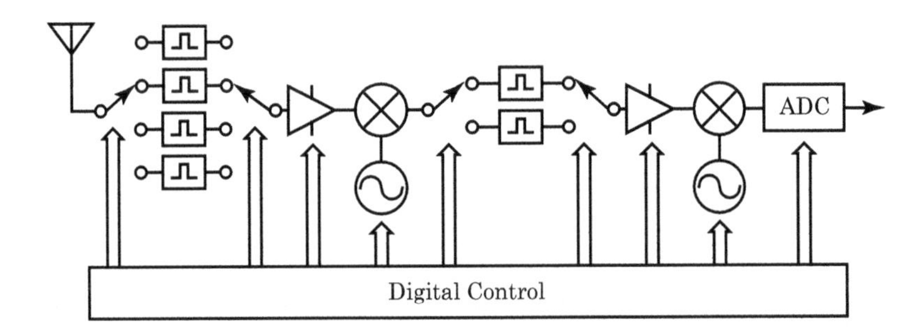

FIGURE 10-57 ■ Portions of a receiver that require digital control.

an opportunity for an interfering signal to escape from the noisy subsection and contaminate the quiet subsection.

- Ease of manufacture: Minimizing the number of interconnections will decrease number of operations required to build the receiver, reducing the complexity and cost of the receiver.
- Cos: Decreasing the interconnections reduces the assembly cost and the cost of associated components.
- Miniaturization: The interconnections between separate subsystems can occupy a surprising amount of physical space. Fewer required connections frees up space for more worthwhile functions.

Digital Control of Receiver Subsystems Figure 10-57 shows an example of the control structure in a garden-variety receiver. Functions we wish to control include preselection filters, IF filters, PLL synthesizers, and the gain of individual components.

10.9 A TYPICAL SYSTEM

Figure 10-58 shows the architecture of an idealized receiving system.

10.9.1 Filter BPF_1

The RF preselection filter immediately limits the bandwidth of the system prior to any nonlinear component. We take every opportunity to limit bandwidth because allowing the receiver to view signals outside of our range of interest will not help the receiver's performance and permits out-of-band signals to drive our system into its nonlinear range.

This filter should be as narrow as possible without excessive loss. Any loss at this point in the receiver directly adds to the receiver's noise figure and limits the receiver's MDS.

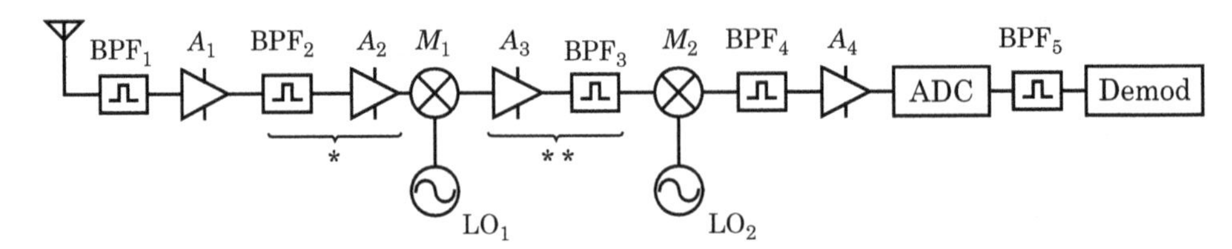

FIGURE 10-58 ■ The architecture of an ideal receiving system. Many systems will omit parts of the architecture shown here due to cost, physical size constraints, or other trade-offs.

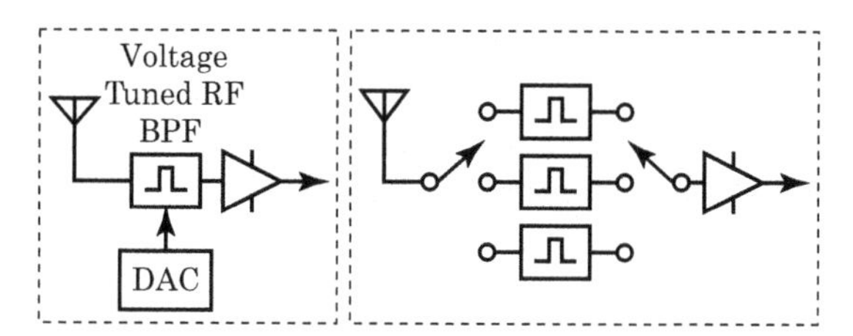

FIGURE 10-59 ■ Voltage-tuned preselection filters and an equivalent filter configuration. The first filters in the system are often voltage tuned preselection filters or a group of switched RF filters. The purpose is to reduce second-order effects (note that each filter is less than an octave in bandwidth) and to remove any large out of band signals.

BPF_1 and BPF_2 (if present) are responsible for the image and IF rejection of the receiver.

In high performance, wideband receivers, this BPF is often a set of voltage-tunable band-pass filters. Figure 10-59 illustrates this concept. The complete receiver tunes over a 500 to 1000 MHz frequency range. BPF_1 and BPF_2 are voltage-tuned band-pass filters with a 5% bandwidth. A microprocessor inside the receiver changes the center frequency of the band-pass filter to the receiver's tuned frequency.

10.9.2 Amplifier A_1

This is the first amplifier in the system, so it should have a low noise figure. This amplifier should posses enough gain to get the signal to the next amplifier without the excess gain dropping below 15 dB.

The ITOI of this amplifier is not a great concern. Since there is no gain before this amplifier, it will always see relatively small signals.

The reverse isolation of amplifier A_1 keeps internally generated signals (e.g., LO_1, LO_2 and any noise from the microprocessor) from leaving the receiver.

10.9.3 Filter BPF_2

This is an optional component. If amplifier A_1 has significant gain at mixer M_1's image frequency, the wideband noise from A_1 present at the image frequency will be converted to the IF center frequency by M_1. This will degrade the effective noise figure of mixer M_1.

This filter is partially responsible for the mixer's image and IF rejection as well as suppressing internally generated signals on their way out of the receiver.

10.9.4 Amplifier A_2

Amplifier A_2 is an optional component. Amplifier A_2 provides gain (if the signal path through BPF_2 was particularly lossy) and helps provide a wideband match to M_1's RF port. If we leave A_2 out, then mixer M_1 does not see a wideband match on its RF port and M_1's linearity may suffer.

10.9.5 Mixer M_1

Mixer M_1 converts the RF to the first IF. The finite LO:RF isolation of this mixer is partially responsible for the presence of LO_1 at the receiver's antenna port.

The finite LO:IF isolation of M_1 makes it possible for LO_1 to find its way to M_2. Two large signals present in a mixer at the same time is a major cause of intermodulation products. If the sum or difference of any of the harmonics of LO_1 and LO_2 fall inside any IF filter, we have a spur. The greater the LO:IF isolation of mixer M_1, the less LO power leaks through the mixer into the receiver's first IF and the lower the danger of a detectable spur.

The RF:IF rejection of mixer M_1 improves the IF rejection of the receiver.

10.9.6 Amplifier A_3

Amplifier A_3 provides a wideband match to M_1 and provides gain in the receiver chain if we need it at this particular point. If we do not place a filter between M_1 and A_3, then A_3 must be able to process the LO_1 leakage from M_1. Since this is a fairly strong signal (≈ -30 dBm), amplifier A_3 must have good linearity.

The reverse isolation of A_3 helps keep the local oscillator from mixer M_2 out of mixer M_1.

If we place BPF_3 between M_1 and A_3, then amplifier A_3 does not have such a strict linearity requirement. Since linearity usually comes at the cost of more DC power, we can sometimes save power with this filter.

10.9.7 Filter BPF_3

We select the first IF with the characteristics of BPF_3 in mind. The roll-off, ultimate attenuation, and bandwidth of this band-pass filter are important in determining the spurious responses of the receiver. BPF_3 rejects the out-of-band spurious products generated by mixer M_1 including the image, all of the $|mLO \pm nRF|$ products and LO_1. The ultimate rejection of BPF_3 sets the lower limit to the strength of the unwanted signals passing between M_1 and M_2. Since LO_1 will experience gain through amplifier A_3, BPF_3 should reject LO_1. We may add a notch in BPF_3 at the frequency of LO_1.

10.9.8 Mixer M_2

Mixer M_2 is much less critical to the operation of the receiver than mixer M_1. LO_2 is much less likely to find its way to the antenna port than is LO_1. Also, at this point in the cascade, the band-pass filters will be considerably narrowed so spurious problems caused by unwanted signals have been significantly reduced.

10.9.9 Filter BPF_4

This is the narrowest analog BPF in the system. It protects the ADC from large signals and from image noise present in unwanted Nyquist bands. This filter should be wide enough to pass the largest signal of interest without significant distortion.

10.9.10 Amplifier A_4

After BPF_4, we must apply enough gain to bring the signals contained in BPF_4 to a level suitable for the ADC, which is in the neighborhood of one Volt at 1 kΩ. Achieving this level requires a great deal of gain, so this amplifier must be very linear. It is also important that BPF_4 is as small as possible to reduce the number of signals present in this amplifier.

10.9.11 ADC

We transition the signal into the digital domain at this point (see Chapter 11). After we digitize, gain is available with no linearity penalty. We must not allow our signal to fall close to the noise floor, which is dictated by the number of bits with which we digitized the signal.

10.9.12 Filter BPF_5

Techniques diverge widely after the ADC, depending on the nature of the follow on processing. However, we will mention just one more component.

Filter BPF_5 is a digital filter. This filter sets the noise bandwidth, and thus the sensitivity, of the receiver. This filter must be wide enough to pass the signal of interest without significant distortion. If it is wider, it will pass more noise than is absolutely necessary (thus decreasing the SNR of the signal presented to the demodulator). If BPF_5 is too wide, it may also pass strong, adjacent signals that can ruin the performance of the demodulator.

This filter is responsible for the adjacent channel rejection of the receiver.

10.10 DESIGN EXAMPLES

You can learn much from both good and bad examples.

10.10.1 Design Example #1

This example is an expensive, high-performance receiver. The most important design parameters are performance, low power, and relatively small size. The specifications are as follows:

- Tuning range: 20 to 500 MHz
- Tuning plan: Frequency synthesized local oscillators locked to an internal reference.
- Noise figure: 8.5 dB maximum.
- Input third-order intercept: -10 dBm, 10 kHz IF bandwidth. Tones placed at $f_{tuned} \pm$ 100 kHz and at $f_{tuned} \pm$ 200 kHz
- Reference accuracy: ± 3.5 ppm over -10 to $+55°$C temperature range
- Tuning step size/tuning speed: 2.0 kHz/25 ms
- Front-end RF preselectors: The bandwidth of the RF preselection filters will be less than 15% of the tuned frequency
- Detector modes: AM, FM
- IF bandwidths: 10, 50, 200 and 1000 kHz, selectable. IF Filter Shape Factors: 3:1 nominal (6 to 60 dB)
- AM sensitivity: For an input signal level of -110 dBm, a 1-kHz, 60% modulated AM tone, the audio SINAD will be greater than 10 dB. Measured in a 10 kHz IF bandwidth and a 3 kHz audio bandwidth.
- FM sensitivity: For an input signal level of -110 dBm, a 1 kHz FM tone which deviates over 80% of the 50 kHz IF bandwidth, the signal-to-noise-and-distortion (SINAD) ratio will be greater than 10 dB. Measured with a 3 kHz audio BW.

- Third-order intercept: The ITOI of the receiver will be greater than −10 dBm with test signals at 0.5 and 1.0 MHz from the tuned frequency with the 50 kHz IF bandwidth selected.
- Image rejection: The power difference between a test signal at the image frequency and at the tuned frequency will be greater than 65 dB.
- IF Rejection: The power difference between a test signal at the either of the first IFs and at the tuned frequency will be greater than 65 dB.
- Local oscillator radiation: The power of any local oscillator present at the antenna port of the receiver shall be less than −100 dBm
- Local oscillator phase noise: The combined incidental FM of all the local oscillators is less than 100 Hz RMS when measured in an audio bandwidth of 3 kHz.
- Internally generated spurious signals: All spurious signals shall be less than −110 dBm relative to the antenna input. No more than six spurious responses shall be greater than −120 dBm.
- Automatic gain control: The AGC shall act to keep less than 2 dB of variation in the IF signal for an input signal variation of 80 dB.
- Computer control via RS-232 with hardware reset
- Operating/storage temperature: −10 to +55°C, operating −55 to 70°C, storage
- Power requirement: 6-16 V_{DC}, 1.25 watts maximum

10.10.1.1 Analysis of the Specifications

This is a digitally controlled receiver, capable of autonomous action. It requires a microprocessor, which can perform sophisticated controlling functions (e.g., switching of preselectors, demodulator selection, user interfaces). The microprocessor is also a source of noise and we will experience RF/digital isolation issues that we must address from the start of the design.

10.10.1.2 Overall Block Diagram

Figure 10-60 shows the block diagram of the receiver. We first pass the antenna output into a set of preselection filters. These filters perform initial filtering and contain enough

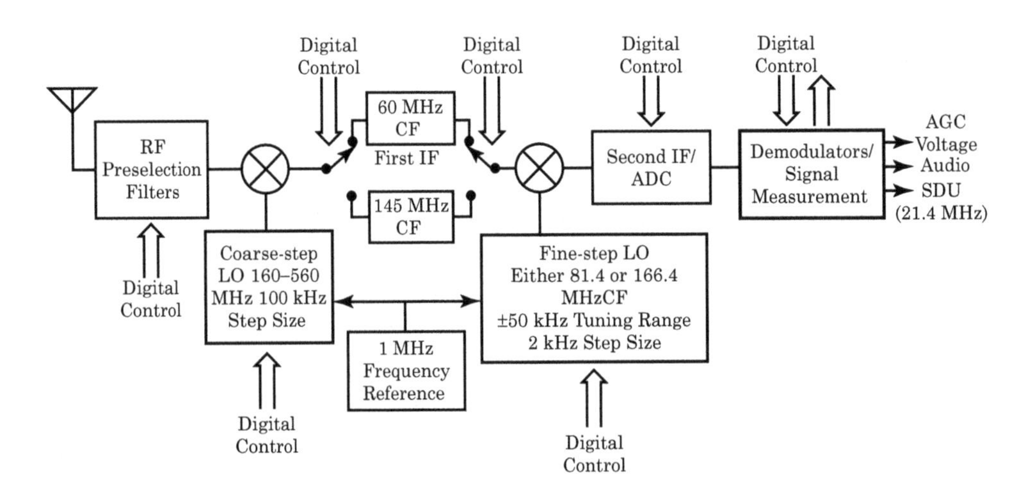

FIGURE 10-60 ■ Block diagram of a candidate receiving system.

gain to set the noise figure of the receiver. We then convert the RF to an IF of either 60 or 145 MHz, depending on the tuned frequency. The receiver must generate a first LO of 160 to 560 MHz, with a 100 kHz step size.

The 60/145 MHz IF signal is then converted to the final IF of 21.4 MHz using either a 81.4 MHz LO or a 166.4 MHz LO. Finally, we perform the ADC at 21.4 MHz. We perform the demodulation digitally.

RF Preselection Filters The front-end preselection filters enable the receiver to operate in high-interference (i.e., urban or military) environments. These filters protect the receiver from direct IF feedthrough and from generating intermodulation products caused by large interfering signals because the probability of two signals producing a third-order distortion product is directly proportional to the preselector bandwidth.

The preselectors should also provide high rejection at the LO frequencies to keep the LO from radiating from the receiver's antenna terminal. The minimization of the LO leakage is particularly important in multiple receiver configurations, where the LO of one receiver can appear as a valid signal to another receiver.

Figure 10-61 shows a block diagram of the voltage-tuned 20–500 MHz preselector filters. The microprocessor switches the 20–500 MHz antenna input to the appropriate set of voltage-tuned band-pass filters. The preselector tuning voltage comes from a 12-bit DAC. The receiver contains a ROM-based calibration table to look up the proper DAC value, given the tuned frequency. The ROM table accommodates the nonlinear tuning characteristics of the preselectors. The processor also switches power to the individual amplifiers (V_{CC1} through V_{CC5}).

The losses in the input switch and the first two-pole band-pass filter add directly to the receiver's noise figure. The amplifier isolates the two filters and establishes the receiver noise figure. The gain of the first amplifier is fairly low (12 dB) because we have not performed significant filtering.

The output of the preselector module goes to the first mixer.

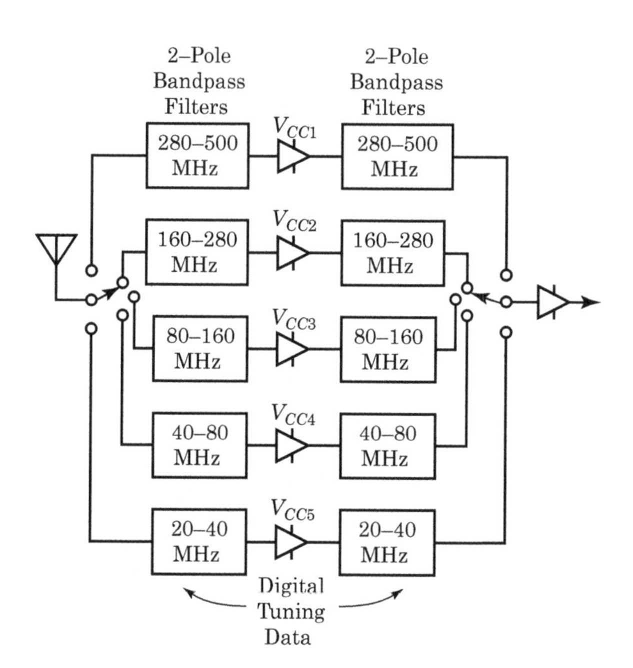

FIGURE 10-61 ■ Structure of the voltage-tuned preselection filters (20–500 MHz) at the receiver's front end.

TABLE 10-12 ■ The band breaks of the RF preselection filters

Preselector	Frequency Range
Band 1	20–40 MHz
Band 2	40–80 MHz
Band 3	80–160 MHz
Band 4	160–280 MHz
Band 5	280–500 MHz

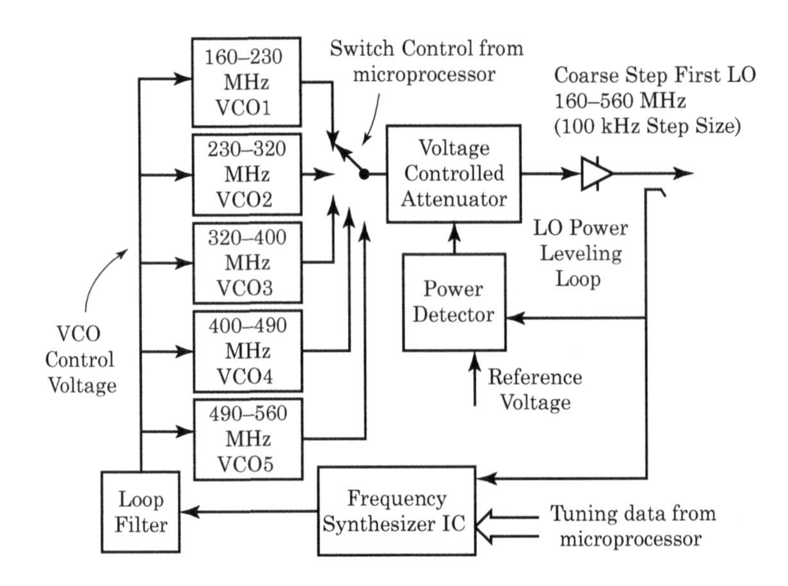

FIGURE 10-62 ■ Block diagram of a candidate LO architecture. Several switched VCOs are used to reduce phase noise.

Table 10-12 lists the five subbands that tune from 20 to 500 MHz.

Coarse-Step Synthesizer/First LO The first LO tunes from 160 MHz to 560 MHz, which is almost two octaves. Since oscillator phase noise is a strong function of percentage tuning range, we used five separate oscillators that are power-switched as required (Figure 10-62). Since each oscillator tunes much less than an octave, we reduce the system phase noise while maintaining the required tuning range.

The microprocessor provides power control for each separate VCO. Only one oscillator is powered up at any one time, which greatly reduces the likelihood of internally generated spurious signals.

Table 10-13 shows the breakdown of the VCO frequencies.

TABLE 10-13 ■ The band breaks of the frequency synthesizer voltage-controlled oscillators

VCO Band	Frequency
VCO1	160–230 MHz
VCO2	230–320 MHz
VCO3	320–400 MHz
VCO4	400–490 MHz
VCO5	490–560 MHz

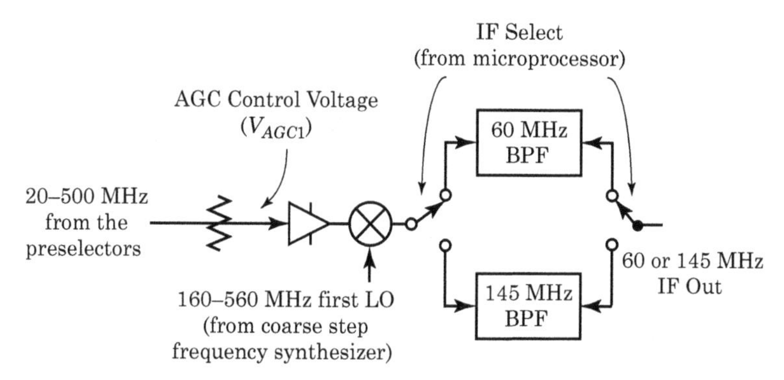

FIGURE 10-63 ■ Block diagram of the first conversion and the first IF stages. The IF is switched to reduce intermodulation distortion and to allow the receiver to reuse some of its LO range. The first LO has a 100 KHz step size to reduce phase noise.

The step size of the first synthesizer is 100 kHz, which allows for a large synthesizer loop bandwidth and hence reduced phase noise. The large loop bandwidth also enables fast tuning speed. One effect of the 100 kHz step size is that the first mix will not exactly center the IF signal in the first IF band-pass filter. The signal will be ±50 kHz removed from the first IF center frequency. We remedy this situation in the second mix.

The LO leveling loop ensures that the LO power remains constant as we tune the receiver. Constant LO power preserves the system's performance over the receiver tuning range.

First IF The first IF center frequencies are 60 and 145 MHz (see Figure 10-63). Filter selection for the tuned frequency is based on an analysis of the spurious signals that might be generated. The microprocessor operates the IF switches and controls the power of each amplifier.

The first synthesizer upconverts the 20–100 MHz RF to a 145 MHz IF with a first LO of 165–245 MHz. The 100-500 MHz band is downconverted to 60 MHz with an LO of 160–560. This conversion scheme allows some of the first LO tuning range to be reused.

Table 10-14 shows the IF selection and LO frequency with respect to tuned frequency.

Figure 10-63 shows the first instance of gain control in the receiver. One of the AGC voltages developed in the demodulator is fed to a voltage-controlled attenuator. When the signal of interest is large, the AGC acts to reduce the receiver gain through V_{AGC}.

Fine-Step Synthesizer/Second LO The fine-step synthesizer/second LO converts the first IF (at either 60 or 145 MHz) down to 21.4 MHz. Figure 10-64 shows architecture of the

TABLE 10-14 ■ IF and LO frequencies

Tuned Frequency	First IF Frequency	First LO Frequency (100 kHz Step Size)
20–85 MHz	145 MHz	165–230 MHz (VCO 1)
85–100 MHz	145 MHz	230–245 MHz (VCO 2)
100–170 MHz	60 MHz	160–230 MHz (VCO 1)
170–260 MHz	60 MHz	230–320 MHz (VCO 2)
260–340 MHz	60 MHz	320–400 MHz (VCO 3)
340–430 MHz	60 MHz	400–490 MHz (VCO 4)
430–500 MHz	60 MHz	490–560 MHz (VCO 5)

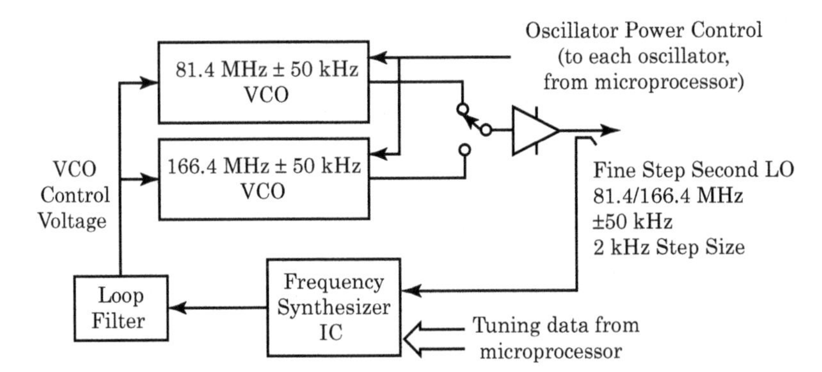

FIGURE 10-64 ■ Block diagram of the second LO. Two separate oscillators are required because the receiver contains two radically different first IFs. The second LO tunes over a narrow range that allows it to have a 2 kHz step size, good phase noise, and adequate tuning speed.

fine-step synthesizer, which tunes in 2 kHz steps. These LOs are centered about 81.4 and 166.4 MHz, depending on the first IF selection. When the first IF is 60 MHz, the second LO tunes 81.4 MHz ± 50 kHz. When the first IF is 145 MHz, the second LO tunes from 166.4 MHz ± 50 kHz. The second LO tunes ±50 kHz because we must center the first IF in the final IF bandwidth. The narrow tuning range coupled with the fine-step size preserves the tuning speed and phase noise of the receiver.

Second IF/ADC Figure 10-65 shows the second IF/ADC, where the fine-step frequency synthesizer converts the 60/145 MHz first IF to the 21.4 MHz second IF. The microprocessor selects the proper band-pass filter based on the user's choice of IF bandwidth. We then digitize the signal. We also generate a second AGC voltage, V_{AGC2}, to control the receiver gain.

Digital Signal Processing The DSP develops the housekeeping information for the receiver. We measure the signal power to generate several AGC voltages for the receiver. These voltages are designed to intelligently trade off noise performance for linearity. The AM detector also generates a squelch signal that turns off the receiver's audio when there

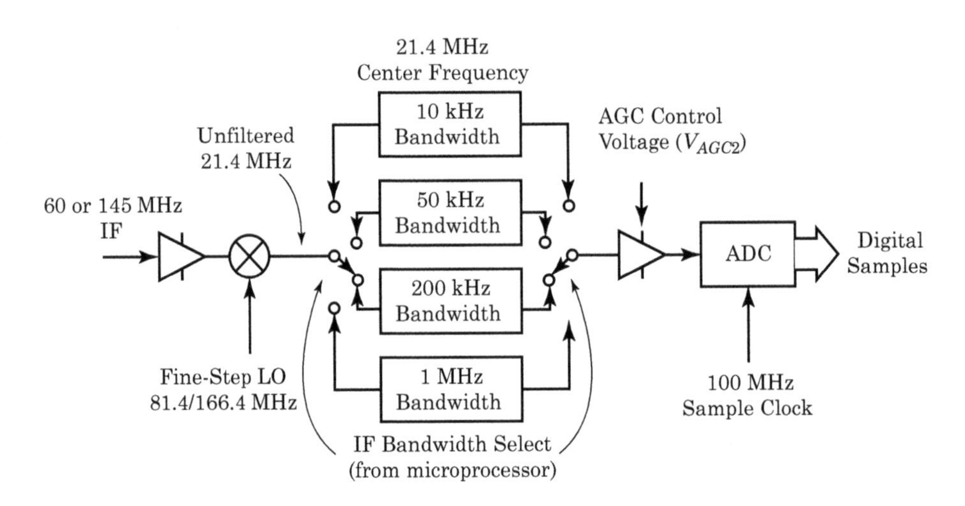

FIGURE 10-65 ■ Block diagram of the second IF at 21.4 MHz. Several IF filters can be selected to limit signal bandwidth before the ADC. Finally, we sample the signal at 100 MHz and present those digital samples to the follow-on ADC.

TABLE 10-15 ■ The aggregate receiver control architecture

Tuned Frequency (MHz)	LO Frequency (MHz)	IF (MHz)	Preselector (MHz)
20–40	165–185 (VCO 1)	145	20–40 (RF 1)
40–80	185–225 (VCO 1)	145	40–80 (RF 2)
80–85	225–230 (VCO 1)	145	80–160 (RF 3)
85–100	230–245 (VCO 2)	145	80–160 (RF 3)
100–160	160–220 (VCO 1)	60	80–160 (RF 3)
160–170	220–230 (VCO 1)	60	160–280 (RF 4)
170–260	230–320 (VCO 2)	60	160–280 (RF 4)
260–280	320–340 (VCO 3)	60	160–280 (RF 4)
280–340	340–400 (VCO 3)	60	280–500 (RF 5)
340–430	400–490 (VCO 4)	60	280–500 (RF 5)
430–500	490–560 (VCO 5)	60	280–500 (RF 5)

is no signal present. When a signal arrives, the squelch is released, the audio amplifier is enabled, and the user can hear the signal. The FM detectors generate a signal used for automatic frequency control (AFC). If the receiver is slightly off-tuned, we can measure the frequency error and center the signal in the IF filter.

Aggregate Plan Table CAS-15 describes the operation of the receiver over its tuned frequency range.

10.10.2 Design Example #2

At the time of its manufacture, this receiver achieved a state-of-the-art size, power, and performance specifications:

- Tuned frequency: 20–500 MHz
- Tuning step size: 5 kHz
- Noise figure: 8 dB maximum
- IF bandwidths: 30 kHz, 200 kHz, and 1 MHz
- Input third-order intercept point, out-of-band: Two tones: $F_0 + 3.5$ MHz and $F_0 + 7.0$ MHz with 1 MHz bandwidth selected: -11 dBm, minimum
- Input third-order intercept point, in-band: Two tones: $F_0 + 0.5$ MHz and $F_0 + 1.0$ MHz with 30 kHz IF bandwidth selected: $+20$ dBm minimum
- Internally generated spurious signals: -120 dBm maximum, referenced to the antenna input
- Image rejection: >60 dB
- IF rejection: >60 dB
- Reference accuracy: ±5 ppm, maximum, over temperature for 3 years
- Incidental frequency modulation: <300 Hz_{RMS}, measured over a 100 Hz to 3.0 kHz bandwidth.
- Demodulation: FM, AM
- FM sensitivity: Signal deviation of IFBW/2, audio SINAD of 10 dB (in a 3 kHz audio bandwidth). The FM sensitivities are -108 dBm (30 kHz IFBW), -100 dBm (200 kHz IFBW) and -93 dBm (1 MHz IFBW).

- AM sensitivity: Modulation index of 60%. audio SINAD of 10 dB (in a 3 kHz audio bandwidth). The AM sensitivities are −104 dBm (30 kHz IFBW), −96 dBm (200 kHz IFBW) and −89 dBm (1 MHz IFBW).
- IF bandwidths: 30 KHz, 200 kHz and 1 MHz.
- IF filter shape factors (60dB/3dB): 30 kHz Bandwidth: 2:1, 200 kHz: 2.5:1, 1 MHz: 2.0:1
- AGC characteristics: 10 ms attack time, 80 ms decay time.
- IF output (10.7 MHz): −30 dBm ± 2.5 dB. 80 dB AGC range.
- Maximum power dissipation less than 1.2 watts (with prime power at 6 VDC)

For this receiver, the prioritized characteristics are (with the most important listed first):

1. Size
2. Power consumption
3. Performance

The cynical engineer might argue that a very small bag of sand meets the first two requirements perfectly.

10.10.2.1 Analysis of the Specifications

This receiver requires a digital control system, but it does not have to be very sophisticated. The design does not include a microprocessor. After the receiver is tuned, there is no active digital circuitry. This quiets the inside of the receiver and greatly alleviates the isolation demands.

The small physical size of this receiver dictates that our design is a simple as we can make it yet still provides the needed functionality. We'll use the smallest number of frequency conversions that we can. We now discuss the issues associated with this decision:

Two Candidate Architectures Figure 10-66 shows a single-conversion candidate architecture. We simply move the desired signal immediately down to 15 MHz to perform the analog-to-digital conversion. Cursory examination of the filter characteristics reveal that this architecture cannot provide suitable image rejection and the low LO radiation specification will be difficult to achieve. We readily abandon this architecture in favor of an upconvert then downconvert architecture.

When upconverting, the RF tuning range sets most of the constraints for the first IF. Filter roll-off considerations place the minimum center frequency of the IF at 600 MHz. Ideally, we would like the first IF to be at least 2.1 times the highest RF, which places the first IF above 1050 MHz. We desire the first IF bandwidth to be as small as practical. The 1 MHz final bandwidth specification sets the minimum bandwidth of the first IF filter to 2 MHz.

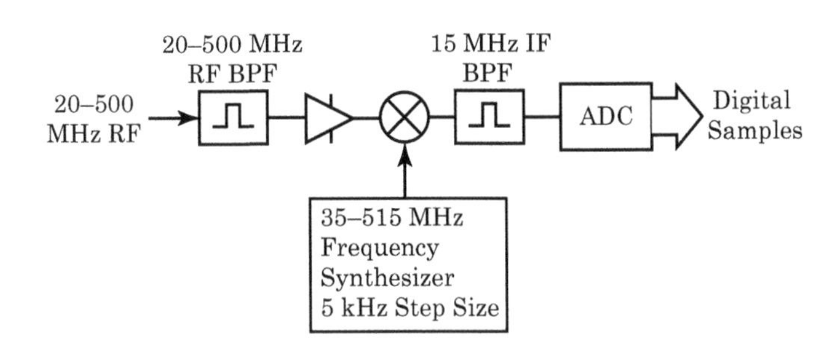

FIGURE 10-66 ■ The high-level block diagram of a single-conversion architecture. This approach has no hope of achieving the required specifications.

At the time of this design, the technology most suitable for the first IF filter (center frequency >600 MHz with a 2 MHz bandwidth) was a surface acoustic wave (SAW) filter. A session with the filter catalogs revealed that SAW filters with center frequencies from 500 MHz to 725 MHz were available with 2 MHz bandwidths.

Image rejection is determined by the RF preselection filters, the first IF and the LO selection. We used an HSLO because that architecture would attenuation the LO on its way to the antenna port. The HSLO also makes for easier filtering of LO:IF leakage using the first IF filter.

Physical size and phase noise considerations also dictated a SAW-based second LO. SAW oscillators are not very stable, but there was room for a compensation scheme to partially correct for this shortcoming.

Selection of the Second IF In this design, physical size was a crucial specification. This forced the consideration of standard, commercially available and physically small parts. Second IFs of 10.7 MHz or 21.4 both meet these requirements. Detailed research indicated that a 10.7 MHz second IF was the most feasible. The final design included a first IF of 710.7 MHz using an HSLO for the first conversion. The second IF was 10.7 MHz using an LSLO to convert the 710.7 MHz to 10.7 MHz.

The Design of the First LO The characteristics of the LOs are as follows:

- First LO frequency coverage: 730.7 to 1210.7 MHz
- Second LO frequency: 700 MHz

Figure 10-67 shows a candidate LO architecture using a frequency-synthesized first LO. A crystal oscillator and multiplier generate the second LO. The multiply-and-filter scheme used to generate the second LO requires a lot of filtering and shielding to develop a clean, spurious-free second LO. This architecture was abandoned.

Figure 10-68 shows a different architecture that uses two frequency synthesizers. The first LO tunes in 5 kHz step sizes, while the second LO is fixed at 700 MHz. This scheme achieves good frequency stability because both local oscillators are referenced to one reference. However, two complete synthesizers require more surface area and power. After some investigation, the architecture of Figure 10-68 was accepted.

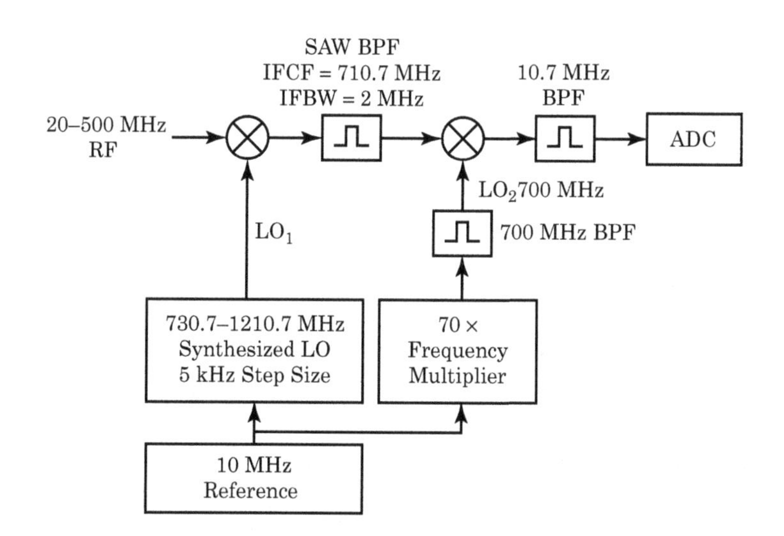

FIGURE 10-67 ■ The LO architecture using a single synthesized first LO and a ×70 multiplier to generate the fixed second LO. We have ignored gain distribution in this figure to concentrate on the frequency conversion scheme.

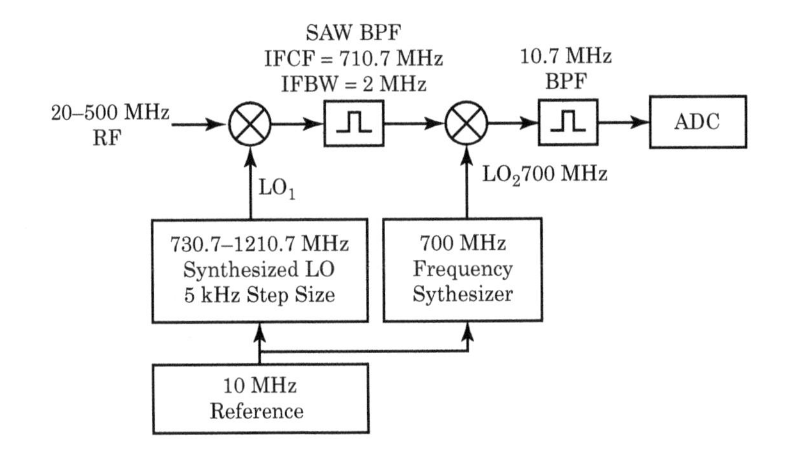

FIGURE 10-68 The LO architecture using a two-synthesized LOs.

10.11 BIBLIOGRAPHY

Dexter, Charles E., "Design Considerations for Miniaturized Receivers," Watkins-Johnson Tech Notes, Vol. 16, No. 5, September–October 1989.

Gross, Brian P., "Calculating the Cascade Intercept Point of Communications Receivers," *Ham Radio Magazine*, August 1980.

Ha, Tri T., *Solid-State Microwave Amplifier Design*, John Wiley and Sons, 1981.

Hayward, W.H., *Introduction to Radio Frequency Design*, Prentice-Hall, Inc., 1982.

Hewlett-Packard Corp., "Fundamentals of RF and Microwave Noise Figure Measurement," Application Note 57-1, July 1983.

Horowitz, Paul, and Hill, Winfield, *The Art of Electronics*, Cambridge University Press, 1980.

Perlow, Stewart M., "Basic Facts about Distortion and Gain Saturation," *Applied Microwaves Magazine*, p. 107, May 1989.

Rohde, Ulrich L., and Bucher, T.T.N., *Communications Receivers—Principles and Design,* McGraw-Hill Book Company, 1988.

Sabin, William E., and Schoenike, Edgar O., *Single-Sideband Circuits and Systems,* McGraw-Hill Book Company, 1987.

Terman, Frederick Emmons, *Electronic and Radio Engineering*, McGraw-Hill, 1955.

Watson, Robert, "Receiver Dynamic Range, Part 1—Guidelines for Receiver Analysis," *Microwaves and RF Magazine*, p. 113, December 1986.

Watson, Robert, "Receiver Dynamic Range, Part 2—Use One Figure of Merit to Compare All Receivers," *Microwaves and RF Magazine*, p. 99, January 1987.

Williams, Richard A., *Communications Systems Analysis and Design—A Systems Approach,* Prentice-Hall, Inc., 1987.

10.12 PROBLEMS

1. **Block Converter RF Image Rejection** For the block downconverter shown in Figure 10-69:

   a. Find the desired RF and RF image.

   b. What is the worst-case RF image of this downconverter? What is the image frequency of the worst-case image rejection?

   c. Name one change we could make to this downconverter to increase the RF image rejection.

d. Find the two bands of IF spectrum leaving the IF port of the mixer. Which is the IF image band?

e. What frequency in the IF image band will be attenuated the least by the IF filter? To what RF does this frequency correspond?

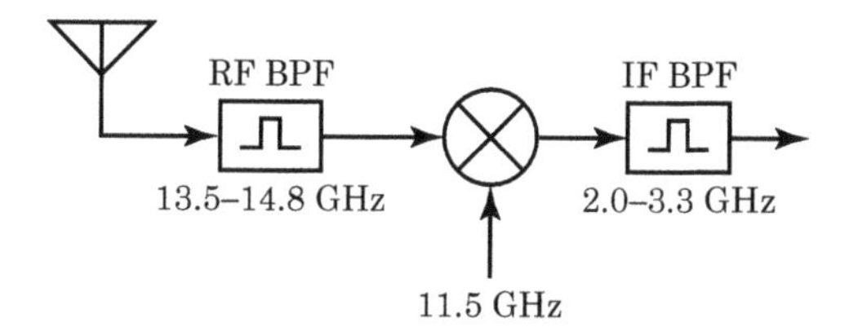

FIGURE 10-69 Block downconverter.

2. **Multiple Choice: Amplifiers and Mixers** Amplifiers placed after mixers are often required to process large signals because

   a. Large signal powers are present after the mixer because of the characteristically high excess gain present at this point in the cascade.
   b. The LO of the mixer leaks from the LO to the IF port of the mixer.
   c. This is a false statement. Amplifiers placed after mixers don't have to process large signals.
   d. Multiple signals from the antenna are present at this point in the cascade. This situation dictates an amplifier with high linearity.

3. **Multiple Choice: Cascade Gain Distribution** Regarding the gain distribution in a cascade, which statement is false?

   a. Noise effects push us toward high excess gains.
   b. Linearity effects push us toward low excess gains.
   c. Both noise and linearity effects push us toward high excess gains.
   d. We carefully design our systems to balance linearity and noise issues to maximize spurious-free dynamic range (SFDR).

4. **Effects of Bad Design** Is high noise figure or poor linearity responsible for the following effects (or is it something else)? Explain your answers:

   a. You pull up to a guard post. As one of the guards brings his walkie-talkie up to his face to talk, you notice that the radio station you're listening to disappears.
   b. Your favorite FM radio station is veiled behind a haze of static.
   c. Your favorite FM radio station appears at a frequency assigned to other users (e.g., a signal originally broadcast at 103.1 MHz appears at 304.8 MHz).
   d. Your cable TV signal is distorted only on certain channels. You notice that the channels in which the distortion appears are the same channels broadcast over the air (i.e. channel 2 on cable TV is distorted in an area where channel 2 is a broadcast channel).
   e. Your cable picture is snowy at the higher channel numbers. The lower channel numbers are clear. Hint: Higher channel numbers correspond to higher-frequency signals.
   f. You're using your cellular telephone. Suddenly, a meeting adjourns and several other people light up their cellular phone. Your call disconnects.

5. **Noise Power Ratio** We've discussed noise figure and linearity as separate specifications for RF systems. We've also discussed minimum detectable signal and dynamic range as figures of merit that combine noise and linearity specifications.

There are other dynamic range type system specifications. One is NPR. Write a 500-word essay on noise power ratio. Answer the following questions:

a. Describe the systems in which this test is useful and how to measure NPR.
b. Why is NPR a combination of noise and linearity specifications?

6. **MDS and SFDR** A large, parabolic antenna feeds satellite signals into a receiver with an 18 MHz bandwidth and a 0.7 dB noise figure. The antenna noise temperature is 290°K, and the ITOI of the receiver is −40 dBm.

   a. What is the MDS?
   b. What is the third-order limited SFDR?
   c. Assuming the system needs about 35 dB of SNR to demodulate these signals, what is the minimum signal power the receiver can process?

7. **HSLO Upconverter** Design a HSLO upconverting receiver that converts an RF signal centered in the 100–200 MHz range to a 450 MHz IF. Your circuit should contain two filters, a mixer and a local oscillator. The bandwidth of the IF filter is 30 MHz, and the signal bandwidth is 1 MHz (see Figure 10-70).
   Derive the frequency allocation plan and include

   - Filter frequencies
   - RFs
   - LO frequencies
   - Image frequencies

   Show the orientation of the image and LO frequency ranges (i.e., if the receiver tunes from 100 MHz up to 200 MHz, does the LO frequency increase or decrease?). Similarly, if the receiver tunes from 100 to 200 MHz, does the image frequency increase or decrease?

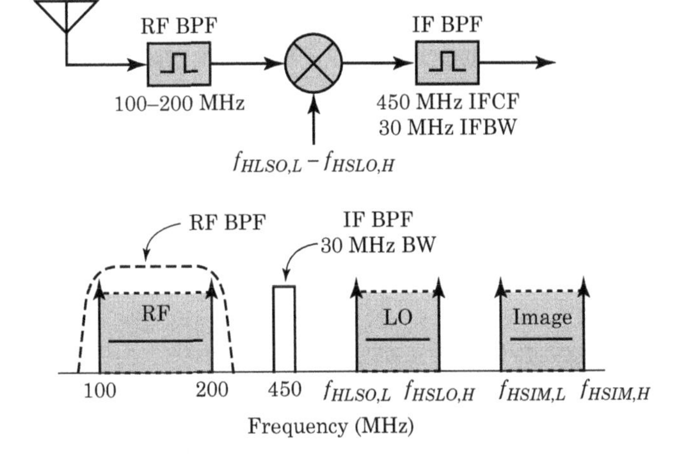

FIGURE 10-70 ■ A 100–200 MHz upconverter.

8. **HSLO Downconverter** Design an HSLO downconverting receiver that converts an RF signal centered in the 210–270 MHz range to a 130 MHz IF. Your circuit should contain two filters, a mixer and a local oscillator. The bandwidth of the IF filter is 1 MHz and the signal bandwidth is 10 kHz (see Figure 10-71).
   Derive the frequency allocation plan and include

   - Filter frequencies
   - RFs

- LO frequencies
- Image frequencies

Show the orientation of the image and LO frequency ranges (i.e., if the receiver tunes from 210 MHz up to 270 MHz, does the LO frequency increase or decrease?). Similarly, if the receiver tunes from 210 to 270 MHz, does the image frequency increase or decrease?

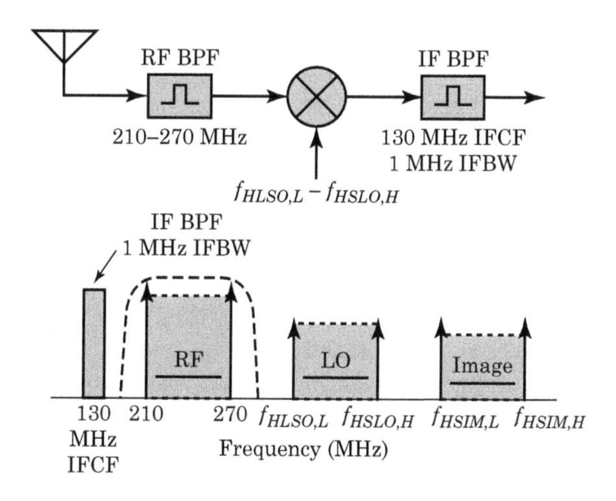

FIGURE 10-71 ■ A 210–270 MHz downconverter.

9. **LSLO Downconverter** Design an LSLO downconverting receiver that converts an RF signal centered in the 70–110 MHz range to a 10.7 MHz IF. Your circuit should contain two filters, a mixer and a local oscillator. The bandwidth of the IF filter is 200 kHz, and the signal bandwidth is 100 kHz (see Figure 10-72).

   Derive the frequency allocation plan and include

- Filter frequencies
- RFs
- LO frequencies
- Image frequencies

Show the orientation of the image and LO frequency ranges (i.e., if the receiver tunes from 70 MHz up to 110 MHz, does the LO frequency increase or decrease?). Similarly, if the receiver tunes from 70 to 110 MHz, does the image frequency increase or decrease?

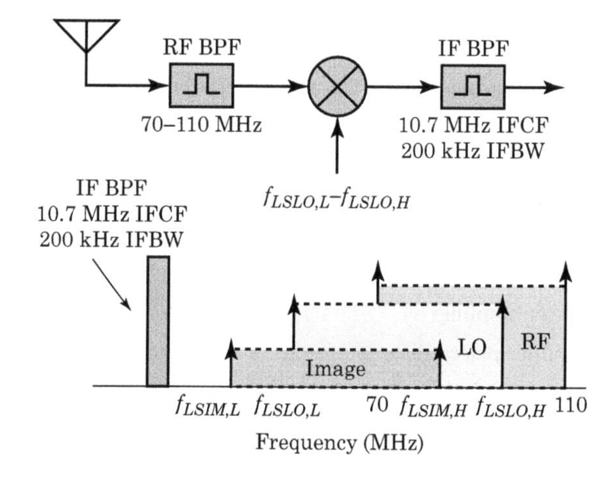

FIGURE 10-72 ■ A 70–110 MHz downconverter.

10. **Compare and Contrast** Figure 10-73 shows two candidate architectures for a receiver design. We need about 17 dB of gain between the two mixers, and we must filter the output of mixer M1 before we apply it to the input of mixer M2. Which architecture requires a more linear RF amplifier? Why?

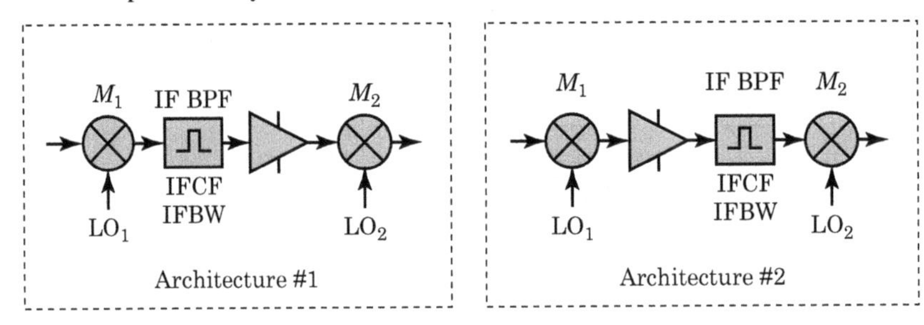

FIGURE 10-73 Two receiver architectures.

11. **Superconducting Filters** The main use of a filter is to limit bandwidth (i.e., reject strong, out-of-band signals) while not inflicting too much loss on in-band signals. Important filter metrics include out-of-band rejection (which we'd like to be as high as possible) and insertion loss (which we'd like to be as low as possible).

    Write a 500-word description (approximately one typewritten page) detailing why engineers believe that these filters will enhance their systems. This isn't an exercise in financial analysis. Describe, in technical terms based on what you know about filters and amplifiers, why these filters might be useful.

    Hint: This isn't about only insertion loss.

12. **Preselection Filter** In a receiver, we universally place a band-pass filter between the antenna and the first amplifier of the receiver.

    a. What is the purpose of this filter?

    b. What are the attenuation characteristics we would like to achieve at the receiver's tuned frequency?

    c. What are the frequencies at which we evaluate the attenuation of this filter?

13. **Noise Figure versus Linearity Essay** When designing a receiver, we must make trade-offs between designing a quiet (i.e., a low noise figure) receiver and a linear receiver. The trade-off largely depends on the RF environment in which the receiver will operate.

    Under what environmental conditions is linearity more important than noise figure? Under what environmental conditions is noise figure more important than linearity?

14. **Noise Figure versus Linearity** Answer the following questions with either *linearity* or *noise figure* indicating whether linearity or noise figure is the more important consideration. Explain your answer.

    Generally, linearity is important when there are large, undesired signals present in the environment. Noise figure is more important when the received signal strength is likely to be small and when there is a low likelihood of large numbers of strong signals.

    a. An urban environment

    b. A desert environment

    c. A cable TV repeater amplifier

    d. A TV station power amplifier

    e. A television amplifier placed directly after the TV antenna

    f. A probe of the planet Saturn

15. **HSLO Upconverter** Design an HSLO upconverting receiver that converts an RF signal centered in the 300–400 MHz range to a 900 MHz IF. Your circuit should contain two filters,

a mixer and a local oscillator, as shown in Figure 10-74. The bandwidth of the IF filter is 40 MHz and the signal bandwidth is 1 MHz.

Derive the frequency allocation plan and include:

- Filter frequencies
- RFs
- LO frequencies
- Image frequencies

Show the orientation of the image and LO frequency ranges (i.e., if the receiver tunes from 300 MHz up to 400 MHz, does the LO frequency increase or decrease?). Similarly, if the receiver tunes from 300 to 400 MHz, does the image frequency increase or decrease?

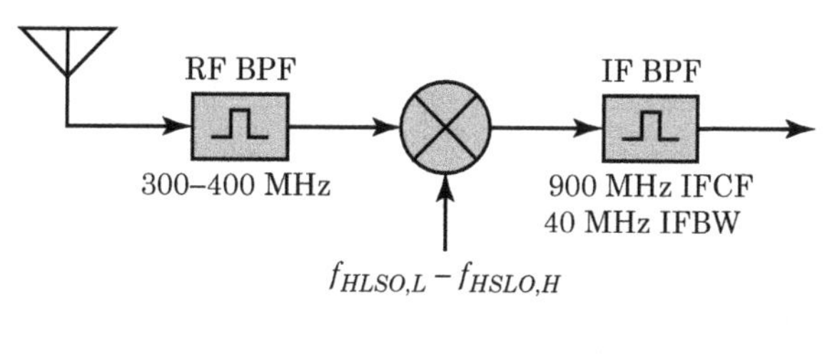

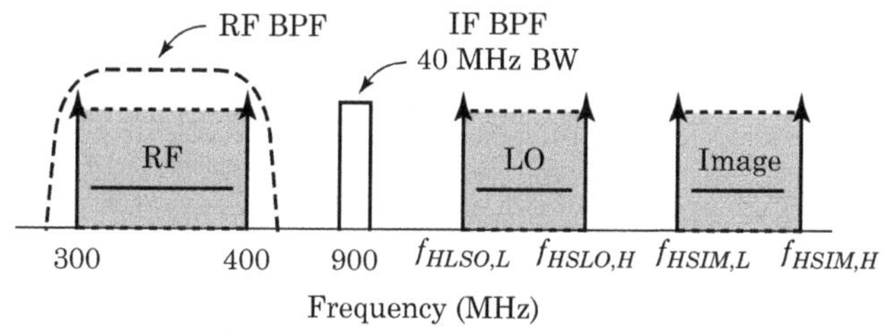

FIGURE 10-74 ■ A HSLO upconverter.

16. HSLO Downconverter Design an HSLO downconverting receiver that converts an RF signal centered in the 110–170 MHz range to a 30 MHz IF. Your circuit should contain two filters, a mixer and a local oscillator, as shown in Figure 10-75. The bandwidth of the IF filter is 1 MHz, and the signal bandwidth is 10 kHz.

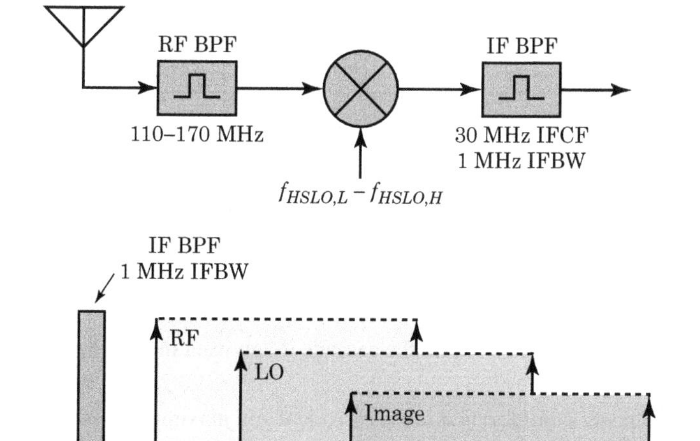

FIGURE 10-75 ■ A HSLO downconverter.

Derive the frequency allocation plan and include

- Filter frequencies
- RFs
- LO frequencies
- Image frequencies

Show the orientation of the image and LO frequency ranges (i.e., if the receiver tunes from 110 MHz up to 170 MHz, does the LO frequency increase or decrease?). Similarly, if the receiver tunes from 110 to 170 MHz, does the image frequency increase or decrease?

17. **LSLO Downconverter** Design an LSLO downconverting receiver that converts an RF signal centered in the 170–210 MHz range to a 70 MHz IF. Your circuit should contain two filters, a mixer and a local oscillator. The bandwidth of the IF filter is 8 MHz and the signal bandwidth is 100 kHz. See Figure 10-76.

    Derive the frequency allocation plan and include

    - Filter frequencies
    - RFs
    - LO frequencies
    - Image frequencies

    Show the orientation of the image and LO frequency ranges (i.e., if the receiver tunes from 170 MHz up to 210 MHz, does the LO frequency increase or decrease?). Similarly, if the receiver tunes from 170 to 210 MHz, does the image frequency increase or decrease?

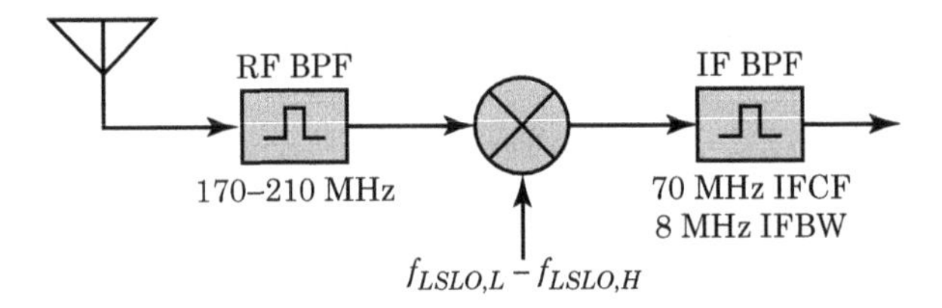

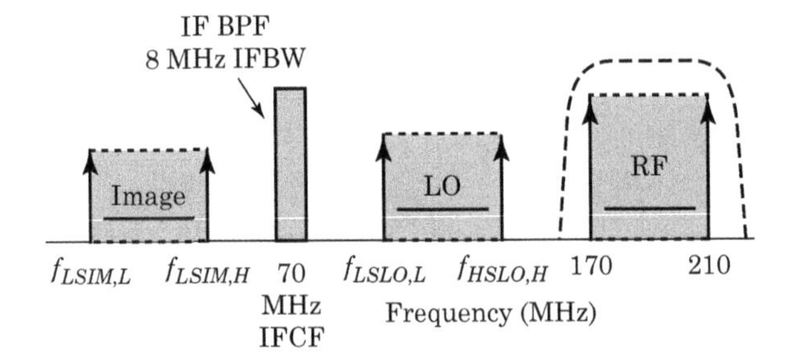

FIGURE 10-76 ■ A HSLO downconverter.

18. **Spur-Free Dynamic Range** Given the cascade shown in Figure 10-77, find the MDS and the third-order limited SFDR in a 1 MHz bandwidth and in a 6 MHz bandwidth.

19. **Order in Cascade** Table 10-16 shows the components available to build a receiver.

    a. Which component should go first in the cascade if you want minimum noise figure? Justify your answer.

    b. Which components should go first in the cascade if you maximum noise figure? Justify your answer.

    c. Which arrangement of components will produce the highest gain?

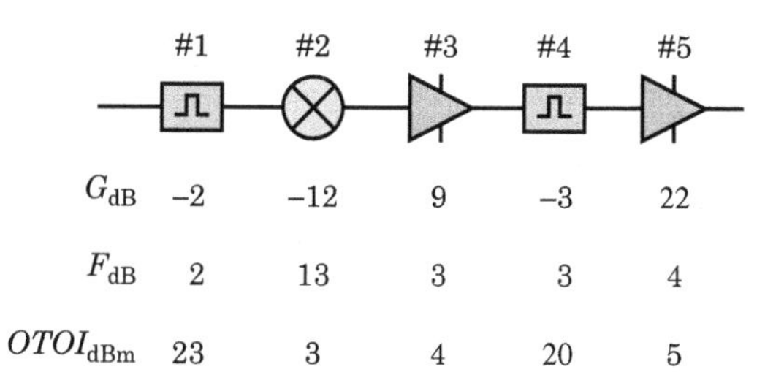

FIGURE 10-77 ■ An RF cascade.

TABLE 10-16 ■ Shows the components available to build a receiver

	Attenuator	Amplifier #1	Amplifier #2	Filter
Gain (dB)	−12.0	15.0	25.0	−2.0
Noise Figure (dB)	13.0	7.0	2.0	2.5

20. **Identification of Spurious Signals** Table 10-17 shows the result of observing six signals on the output of a mixer. We suspect some of the components on the mixer's output are spurious signals so we perform the following tests and record the following data:

    a. Increase the value of the input attenuator by 3 dB.
    b. Reset the input attenuator and then shift the frequency of the LO by 1 MHz.

    Identify the source of the signals in the table and the order of the response. Justify your answers.

TABLE 10-17 ■ Possible spurious signals

Signal Number	1	2	3	4	5	6
Change in output level with 3 dB attenuator (dB)	−9	−3	−6	−6	−3	−9
Change in frequency with 1 MHz LO shift (MHz)	0	−1	1	0	−2	0

21. **Never Helping** Your antenna is seeing two signals (a desired signal and an undesired signal). The two signals are at different frequencies but close enough to each other that the undesired signal is interfering with the reception of the desired signal. What action will never help suppress the unwanted signal?

    a. Change the filters in your front end with filters of a different frequency.
    b. Swap the front end amplifier with another amplifier whose only difference is lower noise figure.
    c. Change the polarization of your antenna.
    d. Point your antenna in a slightly different direction.

22. **Linearity or Noise Figure** Answer the following questions with either *linearity* or *noise figure,* indicating whether linearity or noise figure is the more important consideration in the selection of an amplifier operating in the following environments. Explain your answer.

    a. An RF anechoic chamber
    b. At a military base
    c. A cable TV amplifier
    d. A desert environment
    e. A TV station power amplifier

f. A television amplifier placed directly after the TV antenna

g. A probe around the planet Mercury

23. **Mixer IF Filter** Referring to the filter placed on the IF port of a mixer:

a. What is the purpose of this filter?

b. Identify four frequencies at which we normally evaluate the attenuation of this filter.

24. **Received SNR** Figure 10-78 shows the layout of a reliable, 22 Mbit data link. To achieve the required system reliability, the signal power at the receiver must be 50 dB above the receiver's MDS. The following characteristics apply:

- Transmitter. 3 watts, 16QAM modulation, 4.5 GHz. The transmitter is physically located at base of Tx tower.
- Receiver: 6 MHz noise bandwidth, 7 dB noise figure. The receiver is physically located at base of Rx tower.
- Cable: loss is 4.5 dB/100 feet
- Antennas: The distance between the transmit and receive towers is 6 km. The gain of the transmitting antenna and of the receiving antenna is 32 dBi. The receiver antenna noise temperature, viewed at the input of the receiver, is 140°K. Both antennas should be the same height.
- Assume free-space path loss.
- The antenna tower heights are 150 feet.

Find the *SNR* at the receiver's input.

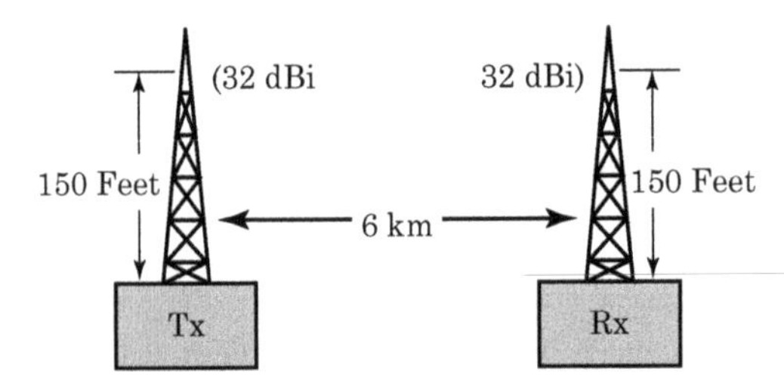

FIGURE 10-78 ■ A transmitter/receiver pair.

25. **Filter Performance Frequencies** For the typical architecture of Figure 10-79, give four frequencies at which we're interested in the attenuation of the IF BPF. Explain why we're interested.

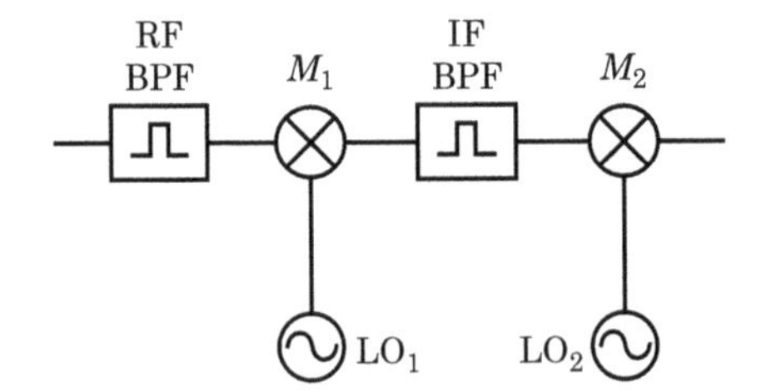

FIGURE 10-79 ■ A dual conversion with two local oscillators.

26. **Multiple Choice: Image Rejection** The image rejection of a receiver:

a. Depends solely on the number of poles in the RF BPF placed between the antenna and the first amplifier.

b. Is the limiting factor in determining the measured third-order limited spur-free dynamic range

c. Is determined largely by the characteristics (i.e., attenuation profile) of the filters between the antenna and the first mixer

d. Is a significant problem only in wideband systems.

27. Multiple Choice To evaluate the IF rejection of a receiver, we must know the conversion scheme of the receiver and which two other parameters?

a. The attenuation of the RF BPF at the IF and the RF:IF rejection specification of the mixer.

b. The characteristics of IF BPF (i.e., the number of poles, the insertion loss, the center frequency and the bandwidth) and the RF:IF rejection specification of the mixer.

c. An analysis of the spurious signals arising from the mRF × nLO combinations to see if one of them falls within the passband of the IF BPF.

d. The image frequency of the first conversion and the attenuation of the RF filter at the image frequency.

28. Conversion Scheme Design Design a conversion scheme to convert a swatch of spectrum at 4–6 GHz to 1–3 GHz with no 2LO, 2RF, 3LO, 3RF, (2RF-LO), or (2LO-RF). The following LOs are available to you: 10 GHz, 13 GHz, 14 GHz, 15 GHz, 16 GHz, and 18 GHz.

Be sure to show all IF bandwidths and all LO frequencies.

29. Identification of Spurious Signals Table 10-18 shows the results of observing six signals on the output of a mixer. We suspect some of the components on the mixer's output are spurious signals so we perform the following tests and record the following data:

a. Increase the value of the input attenuator by 3 dB.

b. Reset the input attenuator and then shift the frequency of the LO by 1 MHz.

Identify the source of the signals in the table and the order of the response. Justify your answers.

TABLE 10-18 ■ Spurious signals and their behavior

Signal Number	1	2	3	4	5	6
Change in output level with 3 dB attenuator (dB)	−6	−3	−3	−6	−3	−9
Change in frequency with 1 MHz LO shift (MHz)	0	1	1	0	−2	0

30. SETI Write a paper on SETI receivers. Comment on the following issues (and any others you think might be interesting and applicable):

a. What are the major problems with detecting a signal of extraterrestrial origin?

b. What about Doppler and minimizing noise bandwidth?

c. What frequencies would you transmit/listen on?

d. How could you tell that a signal is extraterrestrial?

e. How could you tell the signal is due to an intelligent entity and to some physical process?

Try to concentrate on receiver/signal detection issues, not little green men issues.

31. Creating an Equalized Wideband Cable Run . . . in which we design an equalized wideband cable run.

Statement of the Problem We've been tasked to design a wideband cable run from a shipboard tower to the ship's radio room. This cable run is 150 feet long, must support 2–18 GHz, and, to overcome the noise figure of the follow-on receiver, must deliver a signal with about

25 dB of gain at the end of the cable run. The gain must be relatively flat with respect to frequency.

The major problem we face in this design is the variation of cable attenuation with frequency. In general, RF cable has lower loss at low frequencies than at high frequencies. If we were to simply run the cable directly from the antenna to the receiver, signals at 2 GHz will experience very little loss, whereas the signals at 18 GHz will suffer a great deal of loss. If we allow this situation, then the gain, noise figure, and linearity of the complete system will be a strong function of frequency.

We are using a cable with solid outer and inner conductors, which is very good cable but exhibits much variation in its attenuation over the 2–18 GHz frequency range. The attenuation of 150 feet of this cable is 13 dB at 2 GHz and 46.5 dB at 18 GHz. The attenuation, in dB/foot plotted with respect to frequency, is shown in Figure 10-80.

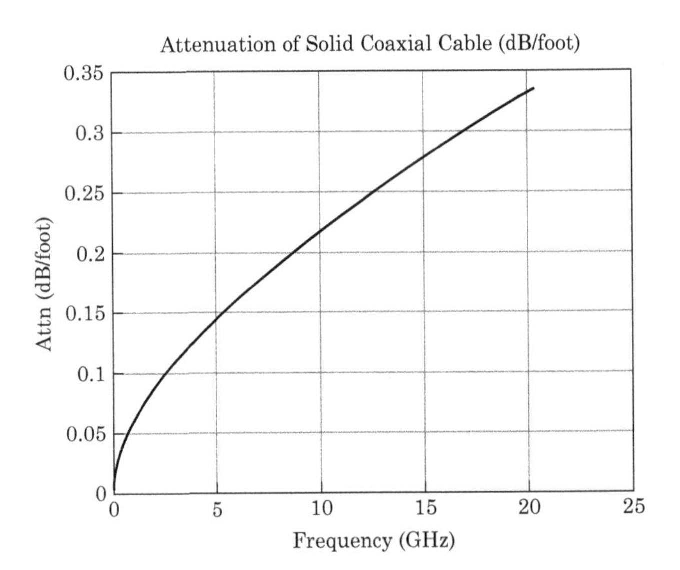

FIGURE 10-80 ■ The attunation of solid coaxial cable.

We often solve the variation of cable loss with frequency problem with cable equalizers. Equalizers are designed to have higher loss at low frequencies and lower loss at high frequencies—just the opposite of the cable attenuation curve. We insert equalizers into various places in our system cascade to flatten the gain. We have three different types of equalizers available to us in our design:

- 10 dB equalizer: −12.0 dB of gain at 2 GHz and −2.0 dB of gain at 18 GHz
- 15 dB equalizer. −17.5 dB of gain at 2 GHz and −2.5 dB of gain at 18 GHz
- 20 dB equalizer: −23.5 dB of gain at 2 GHz and −3.5 dB of gain at 18 GHz

The attenuation of these equalizers, with respect to frequency, is shown in Figure 10-81.

Figure 10-82 shows the architecture of the system, which is due to the physical layout of the ship.

Our task is to get any signal in the 2–18 GHz range to the radio room with about 25 dB of gain. The gain should be relatively flat with respect to frequency. You can use the following parts to build your system, and you have an unlimited number of these parts in your toolbag:

Amplifiers

- 2–18 GHz low-noise amplifier (LNA) with 2 dB of noise figure and a gain of 48 dB

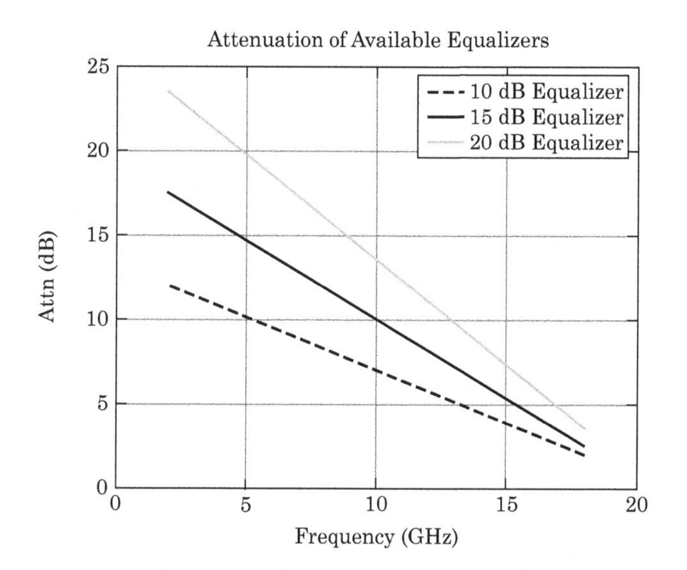

FIGURE 10-81 ■ The attenuation of cable equalizers that are available for our design.

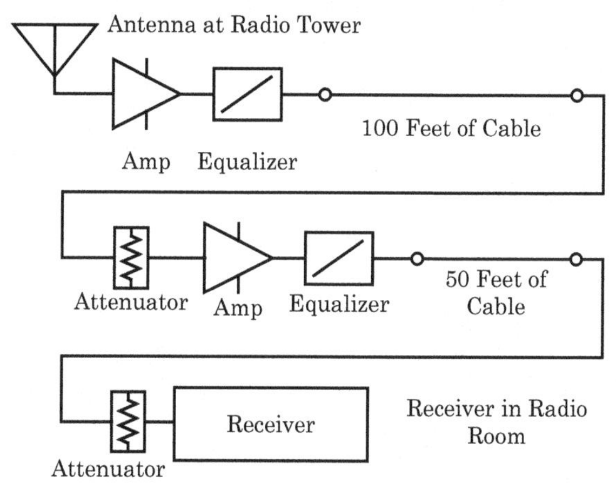

FIGURE 10-82 ■ A diagram of the system to equalize.

DC-18 GHz Attenuators

- 3 dB: −3.0 dB of gain, flat across the 2–18 GHz frequency band
- 6 dB: −6.0 dB of gain, flat across the 2–18 GHz frequency band
- 10 dB: −10.0 dB of gain, flat across the 2–18 GHz frequency band
- 15 dB: −15.0 dB of gain, flat across the 2–18 GHz frequency band
- 20 dB: −20.0 dB of gain, flat across the 2–18 GHz frequency band

Equalizers

- 10 dB equalizer: −12.0 dB of gain at 2 GHz and −2.0 dB of gain at 18 GHz
- 15 dB equalizer. −17.5 dB of gain at 2 GHz and −2.5 dB of gain at 18 GHz
- 20 dB equalizer: −23.5 dB of gain at 2 GHz and −3.5 dB of gain at 18 GHz

Receiver

- The noise figure of the receiver in the radio room is 15 dB. This is a typical number for these devices.

Rules

- The entire frequency band must arrive at the input port of every amplifier with at least 15 dB of excess gain at all frequencies. In other words, the gain between the antenna and the input port of the amplifier must be at least 15 dB. This rule ensures that the noise figure of the cascade is maintained throughout the system.
- Your signals must arrive at the receiver with about 25 dB of excess gain. This rule also ensures that the noise figure of the cascade is maintained, despite the relatively high noise figure of the receiver.

Deliverables

- A drawing of your system with the attenuator and equalizer values labeled.

32. **Maximum Antenna Height** Figure 10-83 shows the layout of a reliable, 65 Mbit data link. For reliability, the signal power at the receiver must be 50 dB above the receiver's MDS. The components have the following properties:
    - Transmitter. 1 watt, QPSK modulation, 6 GHz. Physically placed at base of Tx tower
    - Receiver: 35 MHz noise bandwidth, 6 dB noise figure. Physically placed at base of Rx tower
    - Cable: loss is 5 dB/100 feet
    - Antennas: Distance between towers is 4 km. Antenna gains are both 30 dBi. Rx antenna temperature is 290°K. Both antennas should be the same height. Assume free-space path loss.

    Find h, the maximum height of the antennas.

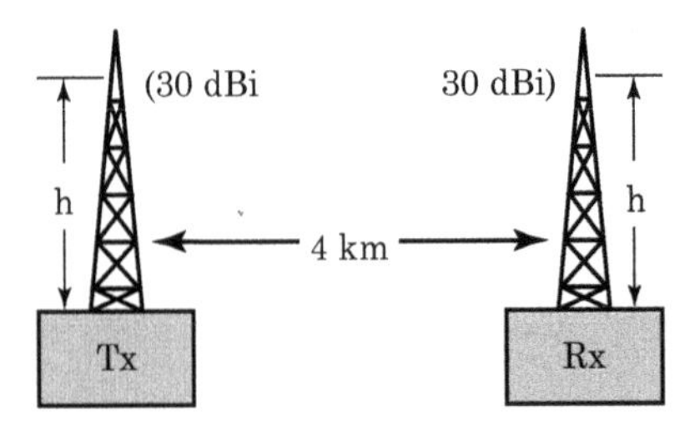

FIGURE 10-83 ■ A data link.

33. **Conversion Scheme**
    a. Design the first mixer stage of a receiver that converts an RF signal centered in the 108-120 MHz range to a 170 MHz IF. There are two solutions (i.e., there are two possible LO ranges). Please find both LO ranges and fill in the frequency allocation diagrams of Figure 10-84. This problem requires a tunable LO. Find:
    - $LO1_{Low}$ and $LO1_{Hi}$
    - $LO2_{Low}$ and $LO2_{Hi}$
    - $Im1_{Low}$ and $Im1_{Hi}$: the image frequencies associated with LO1
    - $Im2_{Low}$ and $Im2_{Hi}$: the image frequencies associated with LO2

    Don't assume that Figure CASPS-31 is accurate with respect to frequency. In other words, the solution may require two bands to overlap when the figure doesn't show them overlapping.

    b. Which, if any, conversion scheme results in a frequency inversion?

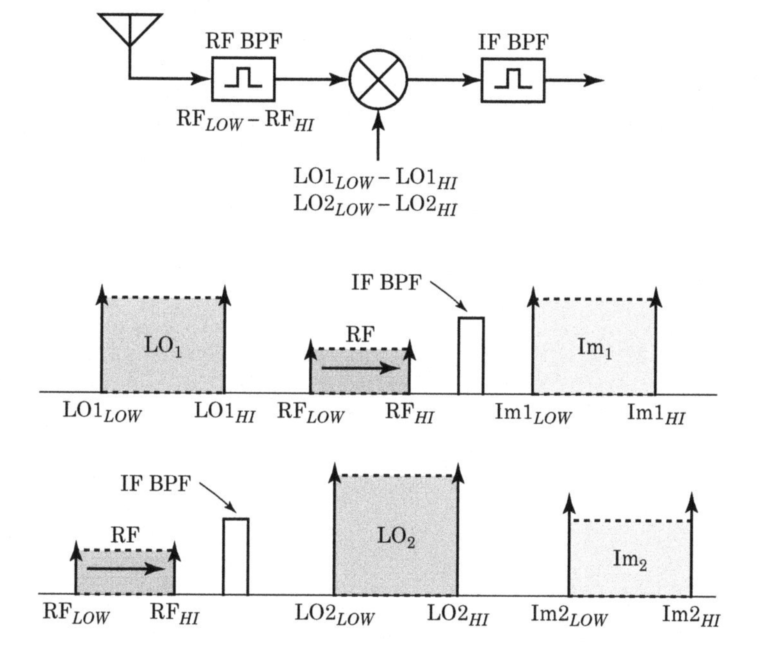

FIGURE 10-84 ■ Two possible solutions for a conversion scheme given the RF and IF ranges.

CHAPTER 11

Digitizing

Well, here goes nothing...

Dr. Lora Baines, preparing to digitize an orange in Tron, 1982, Walt Disney Productions

However, as every parent of a small child knows, converting a large object into small fragments is considerably easier than the reverse process.

Andrew Tanenbaum

Chapter Outline

11.1 INTRODUCTION

An analog-to-digital converter (ADC) converts a signal from its continuous analog representation into a series of finite-resolution samples taken at discrete instances in time. In an ideal receiver, we would place the ADC directly at the antenna to minimize the effects of receiver noise, nonlinearities, and spurious signals. This architecture is not yet practical, so we must first translate the frequency band of interest to a frequency compatible with the ADC input. We must also amplify the antenna output to make use of the ADC's full range.

Sampling resolution (i.e., number of bits in the ADC) and sampling rate are the primary ADC parameters we specify. There are also the issues of ADC saturation, nonlinear ADC response, and sampling jitter.

11.2 NYQUIST-SHANNON THEOREM

The original Nyquist-Shannon sampling theorem states:

> *If a function $f(t)$ contains no frequencies higher than W cps, it is completely determined by giving its ordinates at a series of points spaced $1/(2W)$ seconds apart.*

The Nyquist-Shannon theorem tells only part of the story. If our signal is band-limited and we choose our sampling frequency carefully, we need to sample our signal only at twice its bandwidth rather than twice its highest frequency. This process is *undersampling*.

11.3 SAMPLING AT DISCRETE INSTANTS IN TIME

An ADC samples an analog waveform at discrete instants in time. Figure 11-1 shows a sine wave generator feeding an ADC through a band-pass filter (BPF).

11.3.1 Aliasing

Figure 11-2 shows the samples present at the output of the ADC of Figure 11-1 when the ADC sample rate is 10 MHz. The series of samples shown will be generated when the signal source is a 1 MHz oscillator and BPF center frequency is 1 MHz. The same series of samples will be generated when the signal source and BPF are tuned to 9 MHz. Although not shown in Figure 11-2, input signals of 11, 19, 21, 29, . . . MHz will also produce the same series of samples. Investigating further reveals that the samples shown will match any one of an infinite number of sine waves that meet certain conditions. The determining element of the true frequency present at the input to the system is the BPF.

11.3.2 Undersampling

Figure 11-3 shows that the frequency domain consequences of sampling a signal. The ADC process replicates the spectrum of the input signal at frequencies related to an

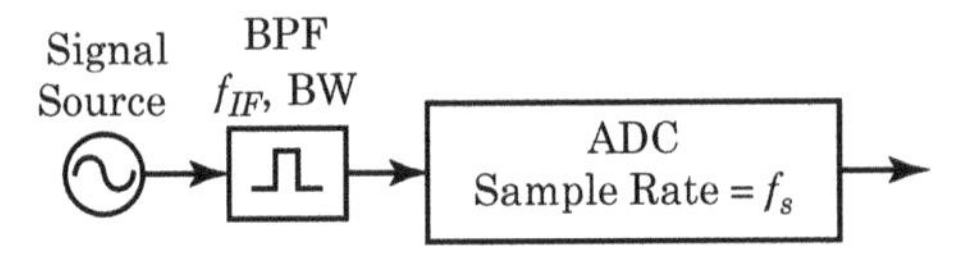

FIGURE 11-1 ▪ ADC architecture with an input BPF. The BPF center frequency is F_{IF}, and the ADC sample rate is f_S.

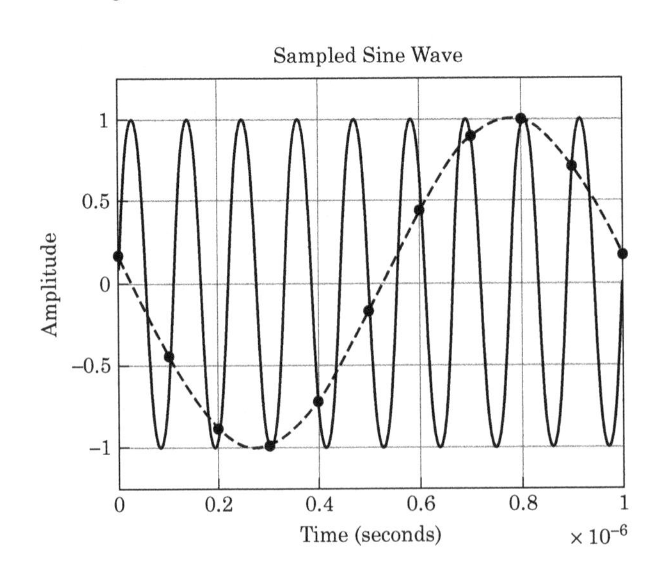

FIGURE 11-2 ▪ The samples at the output of the ADC of Figure 11-1. In this diagram, the sample rate is 10 MHz. The samples taken by the ADC could describe either a 1 MHz or a 9 MHz sine wave. There are an infinite number of sine waves that match the sample stream shown.

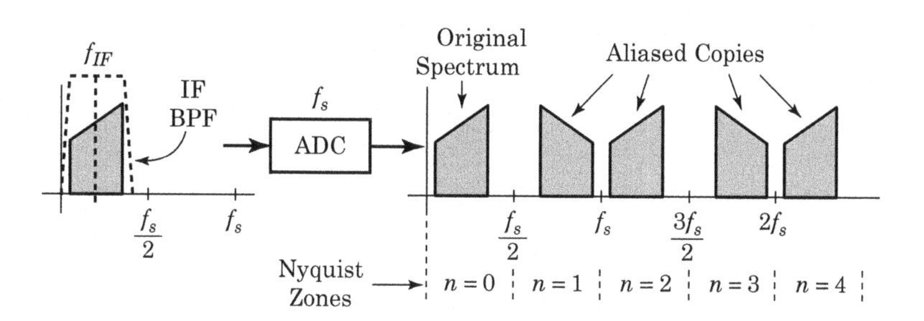

FIGURE 11-3 ■ The spectrum of the sampled signal is replicated by the ADC process. Each piece of spectrum between $nf_S/2$ and $(n+1)\,f_S/2$ is a Nyquist zone.

integer multiple of the sampling clock frequency divided by 2. The piece of spectrum between $nf_S/2$ and $(n+1)f_S/2$ is the n-th Nyquist zone.

We can extend the spectrum duplication concept of Figure 11-2 and Figure 11-3 to include signals that entirely lie within a Nyquist zone on the analog side of an ADC. Figure 11-4 illustrates the concept.

The undersampling phenomenon is very similar to the mixing process, with the sampling clock and its harmonics serving as the local oscillator (LO). If the sample rate exceeds twice the signal bandwidth, the Nyquist criteria is not violated even though the input frequency, f_{IF}, exceeds the sampling frequency. The analog input port of the ADC must still provide sufficient bandwidth and dynamic range to successfully sample a signal at f_{IF}.

If we choose an inappropriate sample rate for a signal center frequency and bandwidth, the signals at the output of the ADC will overlap in the frequency domain. Figure 11-5 shows the consequences of choosing the wrong sampling rate. The various spectra overlap, which irretrievably ruins the signal.

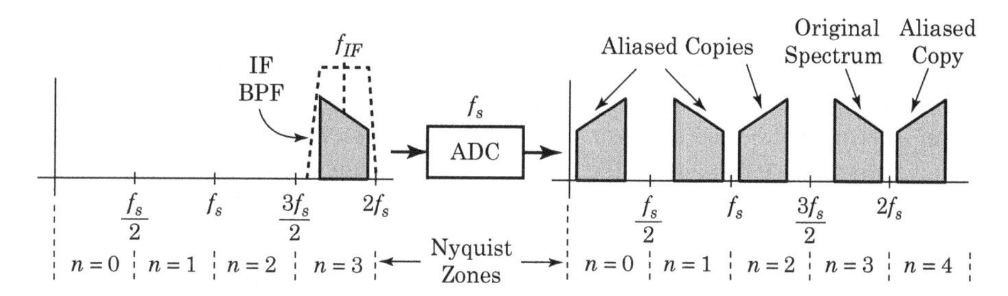

FIGURE 11-4 ■ The spectrum of the sampled signal is replicated by the ADC process. The spectrum present in any Nyquist zone is replicated by the ADC process. Undersampling allows us to sample a signal below its Nyquist rate and still obtain useful data.

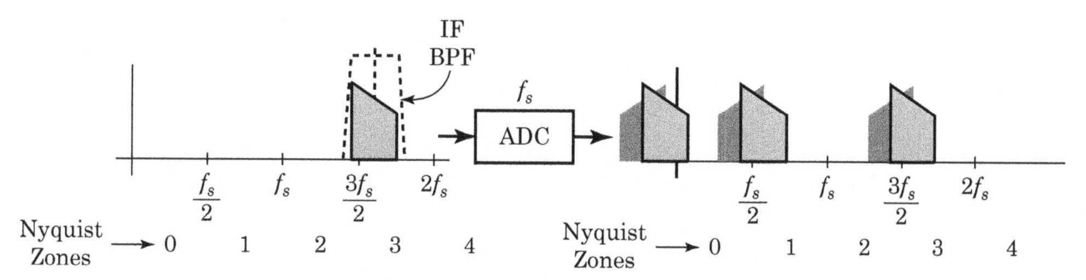

FIGURE 11-5 ■ The consequences of choosing the wrong sampling frequency for a signal. The spectra of various aliasing components overlap, ruining the signal.

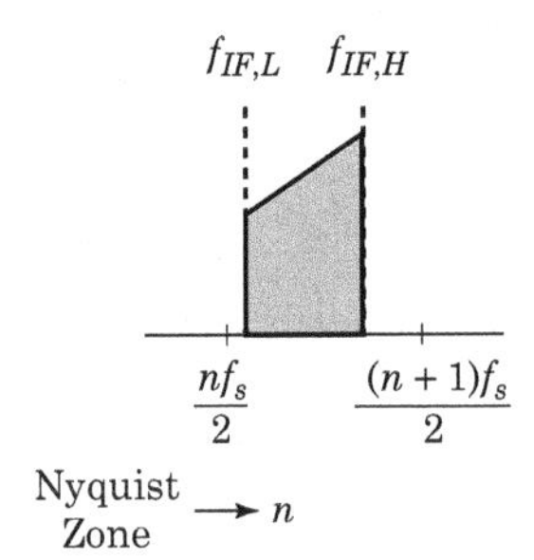

FIGURE 11-6 Signal bandwidth compared with sampling rate for an undersampled system. This drawing is for Nyquist zone n.

11.3.2.1 Nyquist Zones

Figure 11-6 illustrates the various frequencies associated with the subsampling process. Given an IF signal whose extent runs from $f_{IF,L}$ to $f_{IF,H}$, we seek the allowable sampling rates. We can write

$$f_{IF,L} > \frac{nf_S}{2} \quad \text{and} \quad f_{IF,H} < \frac{(n+1)f_S}{2} \tag{11.1}$$

where

n is the integer Nyquist zone, $n = 0, 1, 2, 3, \ldots$
f_S is the sampling rate
$f_{IF,L}$ and $f_{IF,H}$ span the spectral bandwidth to digitize

Conventional sampling occurs within the 0-th Nyquist zone (i.e., $n = 0$), and the input signal frequency ranges from DC to one-half of the sampling frequency or $f_S/2$. Signals present in the other Nyquist zones ($n \geq 1$) up to the bandwidth of the ADC analog input will also be converted, or aliased, into the same output frequency range.

The *anti-aliasing* filter before the ADC input serves a similar function as the image rejection filter placed before a mixer. The mixing process will convert both the desired frequency range and its image to the intermediate frequency (IF). The image rejection filter selects the desired frequency range and rejects the undesired range. In the case of an ADC, the digitizing process will convert a range of signals presented to the ADC's input port to the same range of frequencies in the digitized representation. The anti-aliasing filter selects the one frequency band of interest and presents it to the ADC for conversion. In the 0-th Nyquist range, the anti-aliasing filter can be a low-pass filter. In all other cases, the anti-aliasing filter must be a band-pass filter centered on the desired frequency range.

Figure 11-7 shows the effects of converting from even- and odd-order Nyquist zones. Odd-order Nyquist zones produce a frequency inversion. A signal present in the n-th Nyquist zone will convert to f_0 in the 0-th Nyquist zone where

$$f_0 = \begin{cases} \left| n\frac{f_S}{2} - f_{IF} \right|; & n \text{ even} \\ \left| (n+1)\frac{f_S}{2} - f_{IF} \right|; & n \text{ odd} \end{cases} \tag{11.2}$$

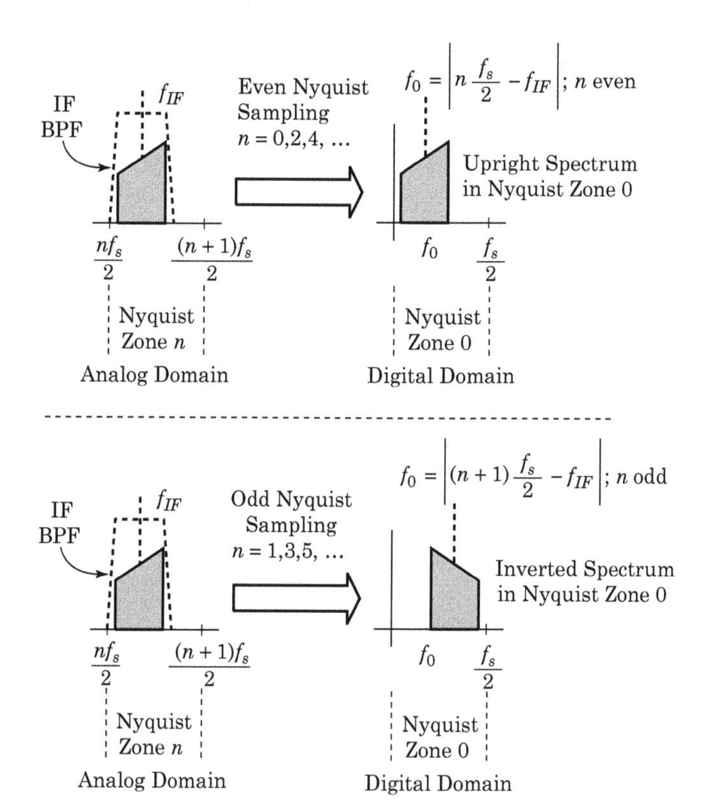

FIGURE 11-7 ■ Odd- and even-order subsampling. Subsampling from odd Nyquist zones results in a frequency inversion.

EXAMPLE

Subsampling a Common IF

A standard IF is centered at 140 MHz and extends over a finite bandwidth. What is the center frequency of the signal in the digital domain when the sampling frequency is 205 MHz? Is the spectrum inverted? Figure 11-8 shows the system.

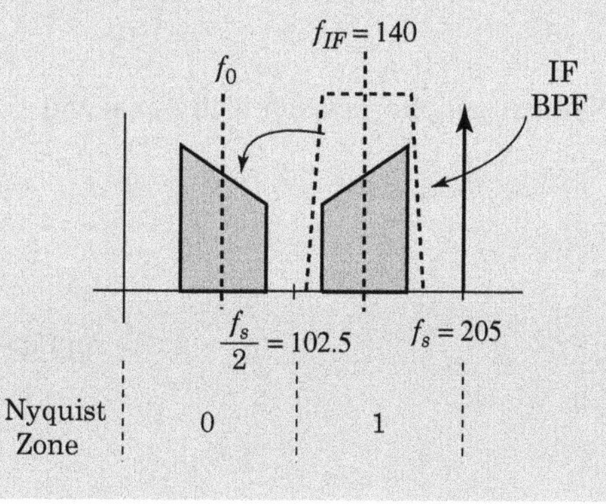

FIGURE 11-8 ■ Using a 205 MHz sampling frequency to convert a 140 MHz IF to 65 MHz. This is an odd Nyquist zone system with $n = 1$.

Solution

This system is an odd Nyquist zone system with $n = 1$ and $f_S = 205$ MHz. Equation (11.2) tells us the 140 MHz IF signal will convert to

$$\begin{aligned} f_0 &= \left|(n+1)\frac{f_S}{2} - f_{IF}\right|;\ n \text{ odd} \\ &= \left|(2)\frac{205}{2} - 140\right| \\ &= 65\,\text{MHz} \end{aligned} \tag{11.3}$$

Since the signal starts in an odd Nyquist zone, the equivalent signal in 0-th Nyquist zone is inverted. Figure 11-9 shows the solution.

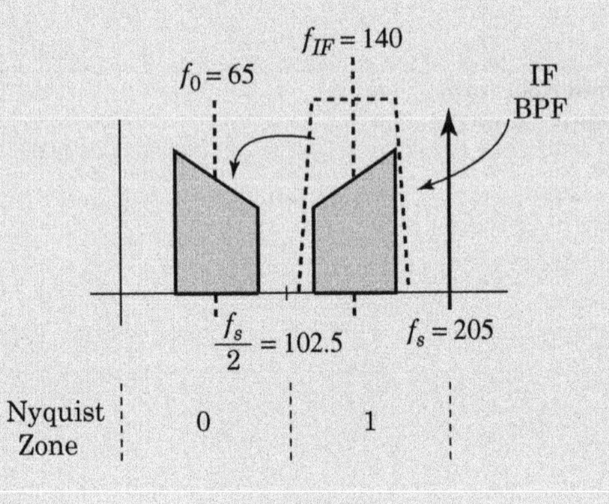

FIGURE 11-9 ■ Example solution.

EXAMPLE

Subsampling Another IF

A signal exists at an IF of 255 MHz. If this signal is sampled at 110 MHz, what is the center frequency of the signal in the 0-th Nyquist zone? Is the signal in the 0-th Nyquist zone inverted or upright?

Solution

Examining Figure 11-10, we realize the system is an even Nyquist sampling system with $n = 4$ and $f_S = 110$ MHz. The 255 MHz IF signal will convert to

$$\begin{aligned} f_0 &= \left|n\frac{f_S}{2} - f_{IF}\right|;\ n \text{ even} \\ &= \left|4\frac{110}{2} - 255\right| \\ &= 35\,\text{MHz} \end{aligned} \tag{11.4}$$

with no frequency inversion.

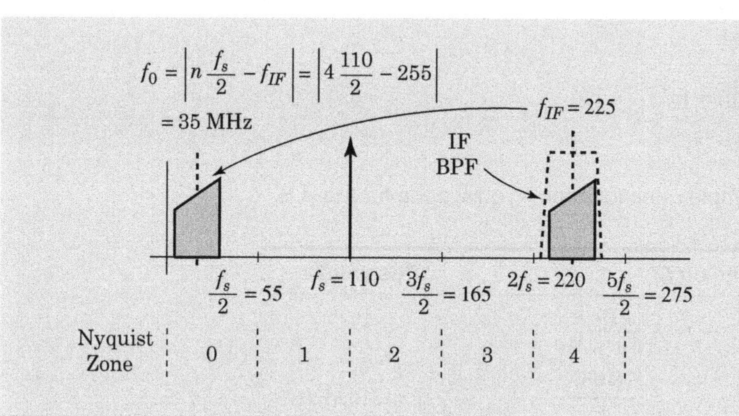

FIGURE 11-10 ■ Using a 110 MHz sampling frequency to convert a 255 MHz IF to 35 MHz. This is an even Nyquist zone with $n = 4$.

11.3.2.2 Calculating Sampling Rates

Given a signal that extends from some $f_{IF,L}$ to $f_{IF,H}$, what are the possible sampling rates that can convert the signal to the 0-th Nyquist band when subsampling is an option? The bandwidth and center frequency of the signal places restrictions on our selection of sampling frequency f_S. Figure 11-11 shows the problem graphically.

Equation (11.1) shows the relationship between signals in the n-th Nyquist zone and their possible sample rates. We can rewrite these equations to produce possible sample rates for a given frequency plan:

$$\begin{aligned} f_{IF,L} &> \frac{nf_S}{2} & f_{IF,H} &< \frac{(n+1)f_S}{2} \\ \Rightarrow f_S &< \frac{2f_{IF,L}}{n} & \Rightarrow f_S &> \frac{2f_{IF,H}}{(n+1)} \end{aligned} \tag{11.5}$$

The two conditions for f_S in equation (11.5) must both be true for each value of Nyquist zone number n.

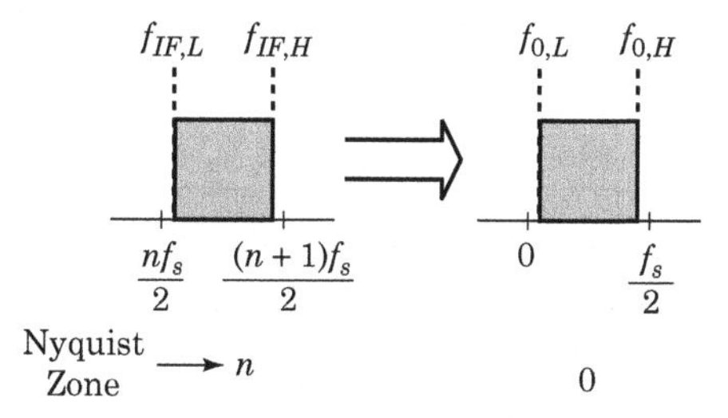

FIGURE 11-11 ■ Converting a signal from the n-th Nyquist zone to the 0-th Nyquist zone.

EXAMPLE

Subsampling the FM Broadcast Band

In the United States, the commercial FM broadcast band covers 88–108 MHz. Find the sample rates useful to digitize this section of spectrum.

Solution

Using equation (11.5), we can build Table 11-1.

TABLE 11-1 Calculating the valid sample rates for each Nyquist zone for the U.S. commercial FM broadcast band

Nyquist Zone (n)	Limits on f_S		Possible?
0	$f_S < \frac{2f_{IF,L}}{n}$ $< \frac{2(88)}{0}$ $< \infty$	$f_S > \frac{2f_{IF,H}}{(n+1)}$ $> \frac{2(108)}{1}$ $> 216\,\text{MHz}$	Yes $f_s > 216\,\text{MHz}$
1	$f_S < \frac{2f_{IF,L}}{n}$ $< \frac{2(88)}{1}$ $< 176\,\text{MHz}$	$f_S > \frac{2f_{IF,H}}{(n+1)}$ $> \frac{2(108)}{2}$ $> 108\,\text{MHz}$	Yes $108 < f_s < 176\,\text{MHz}$
2	$f_S < \frac{2f_{IF,L}}{n}$ $< \frac{2(88)}{2}$ $< 88\,\text{MHz}$	$f_S > \frac{2f_{IF,H}}{(n+1)}$ $> \frac{2(108)}{3}$ $> 72\,\text{MHz}$	Yes $72 < f_s < 88\,\text{MHz}$
3	$f_S < \frac{2f_{IF,L}}{n}$ $< \frac{2(88)}{3}$ $< 58.7\,\text{MHz}$	$f_S > \frac{2f_{IF,H}}{(n+1)}$ $> \frac{2(108)}{4}$ $> 54\,\text{MHz}$	Yes $54 < f_s < 58.7\,\text{MHz}$
4	$f_S < \frac{2f_{IF,L}}{n}$ $< \frac{2(88)}{4}$ $< 44\,\text{MHz}$	$f_S > \frac{2f_{IF,H}}{(n+1)}$ $> \frac{2(108)}{5}$ $> 43.2\,\text{MHz}$	Yes $43.2 < f_s < 44\,\text{MHz}$
5	$f_S < \frac{2f_{IF,L}}{n}$ $< \frac{2(88)}{5}$ $< 35.3\,\text{MHz}$	$f_S > \frac{2f_{IF,H}}{(n+1)}$ $> \frac{2(108)}{6}$ $> 36\,\text{MHz}$	No $36 < f_s < 35.3\,\text{MHz}$

Figure 11-12 shows the data of Table 11-1 in graphical form.

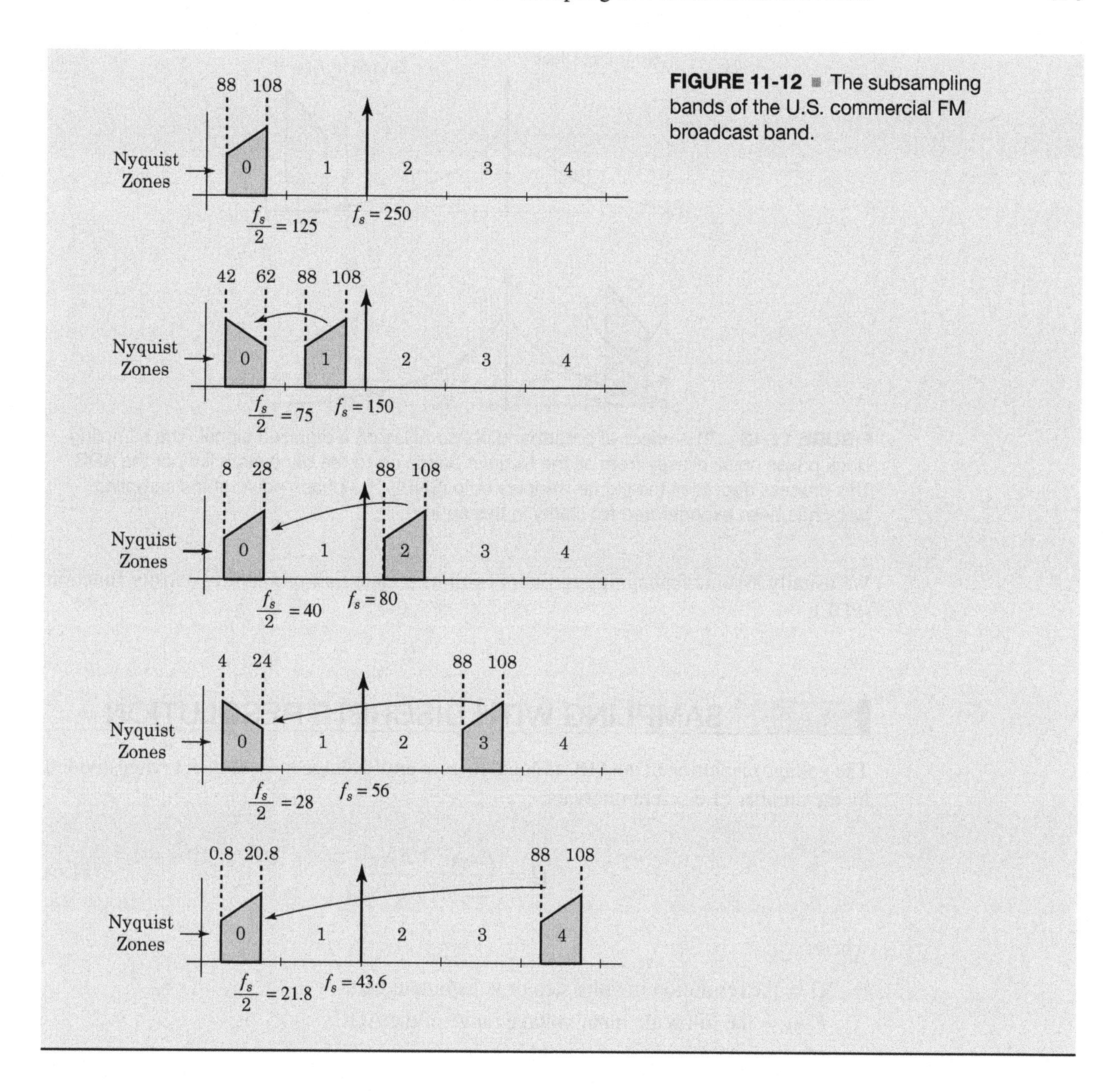

FIGURE 11-12 ■ The subsampling bands of the U.S. commercial FM broadcast band.

11.3.3 Sampling Jitter

Sampling jitter is a measure of the uncertainty of the precise time at which a digital sample is taken. Sampling jitter arises from internal noise present in the ADCs sample-and-hold circuitry and from phase noise present on the digital clock.

As in the case with analog mixers, any wideband noise present on the ADC sample clock affects all the signals generated by the ADC. The noise present in every Nyquist band within the ADC bandwidth limit is folded into all of the other Nyquist bands, including the 0-th band (see Figure 11-13).

Sampling jitter, τ_J, has the units of RMS seconds. Sampling jitter behaves much like phase noise in that the result is undesired phase modulation present on the sampled signal.

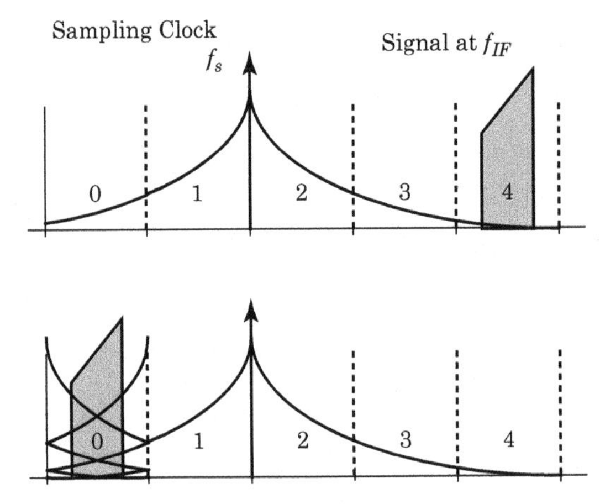

FIGURE 11-13 ■ The effect of broadband phase noise on a digitized signal. The sampling clock phase noise aliases from all the Nyquist bands up to the bandwidth limit of the ADC. This process degrades the signal-to-noise ratio (SNR). The phase noise of the sampling clock has been exaggerated for clarity in this figure.

We usually assume that sampling jitters exhibits a Gaussian probability density function (PDF).

11.4 | SAMPLING WITH DISCRETE RESOLUTION

The voltage resolution of an ADC is equal to its overall voltage measurement range divided by the number of discrete intervals:

$$Q = \frac{E_{FSR}}{2^M} = \frac{E_{FSR}}{N} \tag{11.6}$$

where:

Q = the resolution in volts/step or volts/output code
E_{FSR} = the full scale input voltage range of the ADC
M = the ADC's resolution in bits
N = the number of available output codes

EXAMPLE

ADC Ranges

A 12-bit ADC has a full-scale input range of -1 V to $+1$ V. What is the resolution?

Solution

A 12-bit ADC can distinguish $2^{12} = 4096$ levels. The voltage resolution at 2 volts full-scale input is 2 V/4096 levels = 488 μvolts/level.

11.4.1 Quantization Error

Quantization error is caused by the finite resolution of the ADC. A continuous analog signal arriving at the ADC input will be reported by the ADC to have arrived at one of N discrete levels. The maximum error associated with this effect is $\pm Q/2$.

The major questions are as follows:

- How does the quantization error correlate with the input signal?
- What is the spectrum of the quantization error?

11.4.2 Low Input Levels

At low input signal levels, the ADC can misbehave in a highly nonlinear fashion. The top section of Figure 11-14 shows the ideal situation. The input signal is centered about the threshold between two consecutive output codes. The zero crossings of the signal are preserved, and, hence, the frequency and phase of the input are available at the output of the ADC. However, the amplitude information is lost. The middle section of Figure 11-14 shows the effect when the input signal is slightly above the threshold between two output codes. The output resembles a pulse train whose duty cycle depends on the DC level of the input signal. The bottom section of Figure 11-14 shows the situation when the input signal is centered directly on the threshold level between two output codes. The output from the ADC is completely flat, erasing all traces of the input signal. The three examples of Figure 11-14 are difficult to model and exhibit wildly different harmonic output spectra.

11.4.3 Large Input Levels

If the input signal is much larger than one quantization level, the quantization error is more suitable for analysis. The magnitude of the quantization error will always fall between $-Q/2$ and $+Q/2$.

11.4.3.1 Random Input Signals

We define a random signal as a signal that is uncorrelated with the sample clock. For such a random signal, the quantization error of each sample will be a uniformly distributed random process over $[-Q/2 + Q/2]$ with a uniform probability density function. The

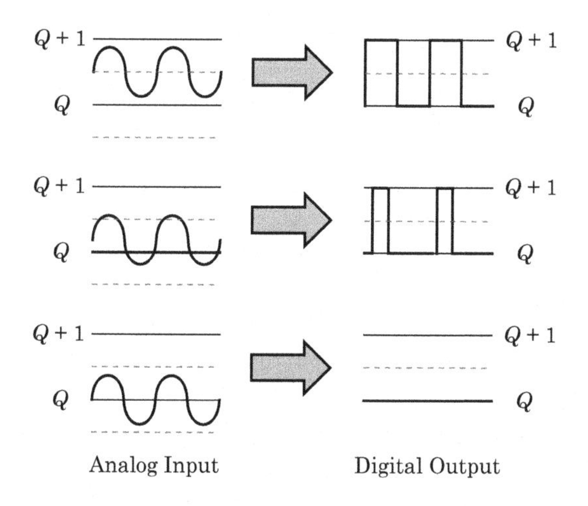

FIGURE 11-14 ■ Effects of DC offset on the output of an ADC.

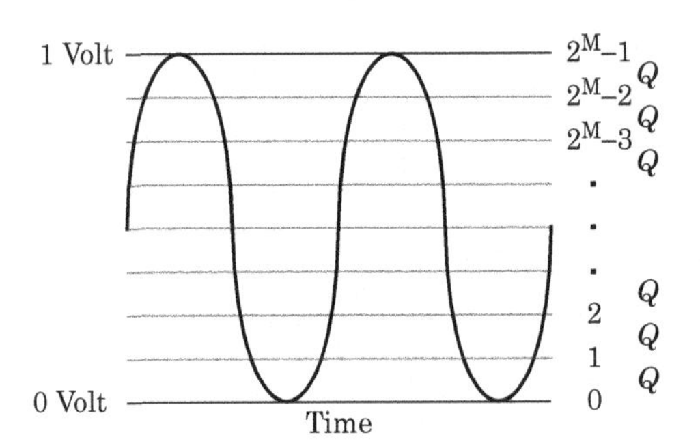

FIGURE 11-15 ■ A full-scale sine wave as applied to an ADC. The E_{FSR} of this signal is 1 volt.

power available from a signal generated via a random process equals the variance of the process. Over many samples, the mean, μ, of our uniform distribution is zero and the variance, σ^2, is $Q^2/12$ where Q is the resolution of the ADC [equation (11.6)].

Figure 11-15 shows a full-scale sine wave applied to an ADC. We seek an expression for the SNR of this waveform. The peak-peak and RMS values of this sine wave are $V_{PkPk} = 1$ volt and

$$\begin{aligned} V_{RMS} &= \frac{V_{PkPk}/2}{\sqrt{2}} \\ &= \frac{V_{PkPk}}{\sqrt{8}} \\ &= \frac{1}{\sqrt{8}} \end{aligned} \tag{11.7}$$

In an impedance environment of 1 Ω, the power available from the sine wave is

$$\begin{aligned} P_{Signal} &= \frac{V_{RMS}^2}{R} \\ &= \frac{\left(1/\sqrt{8}\right)^2}{1\Omega} \\ &= \frac{1}{8}\text{ W} \end{aligned} \tag{11.8}$$

The SNR of the digitized signal is

$$\begin{aligned} \frac{P_{Signal}}{P_{Noise}} &= \frac{1/8}{Q^2/12} \\ &= \frac{1.5}{Q^2} \\ &= 1.5/\left(2^{-M}E_{FSR}\right)^2 \\ &= 1.5(2^{2M}) \\ \left[\frac{P_{Signal}}{P_{Noise}}\right]_{dB} &= 1.76 + 6.02\,M \end{aligned} \tag{11.9}$$

If the sample clock is uncorrelated with the input signal, the SNR increases by 6 dB for every bit of resolution we add to the ADC.

11.4.3.2 Spectral Extent of the Sampling Noise

When deriving equation (11.9), we looked only at the power in the signal and the expected power in the noise. We assumed only that the quantization noise followed a uniform distribution, but we made no assumptions regarding the spectrum of the noise or its extent.

However, if the signal is noise-like and uncorrelated with the sample clock, we often make the assumption that the spectrum of the noise is flat and extends evenly over the Nyquist bandwidth of the digitizing process.

11.4.3.3 Assuming Spectrally Flat Sampling Noise

If we assume that the noise is spectrally flat, then we can increase the SNR by restricting bandwidth, much as we filter to reduce thermal noise. If the noise is spread evenly over the Nyquist bandwidth and we examine only a fraction of the available bandwidth, B_{Max}, the SNR is

$$SNR_{dB} = 1.76 + 6.02\,M + 10\log\left(\frac{f_{Nyquist}}{B_{Max}}\right) \tag{11.10}$$

EXAMPLE

1-Bit ADCs

Consumer grade electronics often boast of 1-bit ADC technology. Find the sample rate if these devices must deliver 40 dB of SNR over 10 KHz of bandwidth.

Solution

Using equation (11.10), we can write

$$SNR_{dB} = 1.76 + 6.02\,M + 10\log\left(\frac{f_{Nyquist}}{B_{Max}}\right) \tag{11.11}$$

If we perform the digitization in the 0-th Nyquist zone, then $f_{Nyquist} = f_S/2$ and we can write;;;

$$\begin{aligned} SNR_{dB} &= 1.76 + 6.02\,M + 10\log\left(\frac{f_{Nyquist}}{B_{Max}}\right) \\ &= 1.76 + 6.02\,M + 10\log\left(\frac{f_S}{2B_{Max}}\right) \\ &\Rightarrow 40 = 1.76 + 6.02(1) + 10\log\left(\frac{f_S}{2(10k)}\right) \\ &\Rightarrow f_S = 33.34\,\text{MHz} \end{aligned} \tag{11.12}$$

11.5 | SOURCES OF SPURIOUS SIGNALS

The act of digitizing a signal to a finite number of states produces spurious signals, even with a perfect ADC. Figure 11-16 shows a continuous sine wave and its corresponding quantized samples. The figure also shows the error signal, which we define as the difference between the continuous sine wave and its samples.

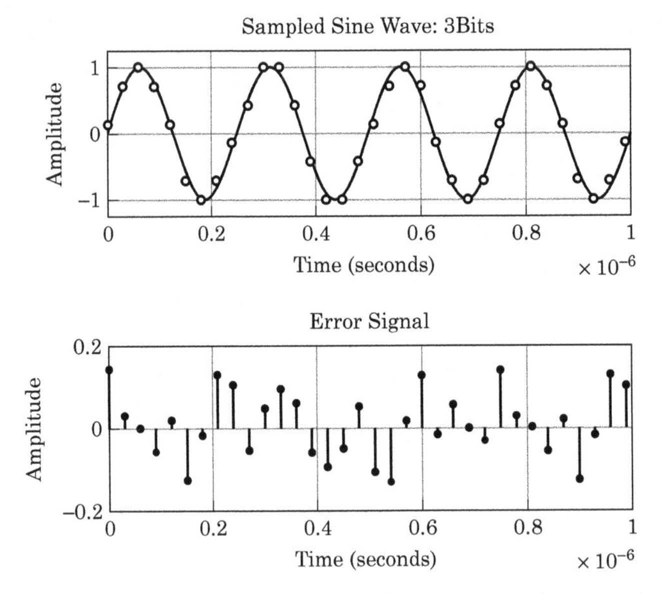

FIGURE 11-16 ■ (top) A continuous sine wave and its corresponding samples. Note that the samples are not exactly collocated on the continuous sine wave. (bottom) The difference between the continuous sine wave and its digitized representation. Note that the error signal for this particular configuration repeats every 750 μsec.

We can interpret the quantized samples of the sine wave of Figure 11-16 as the sum of two signals. One signal consists of the sine wave samples expressed to infinite precision. These samples will generate no spurious signals because they perfectly describe the input sine wave. The second signal is the error between the perfect samples and the samples from our quantized ADC. If the sample clock is uncorrelated with the input signal, the error signal will be spectrally flat. The error signal will be responsible for the noise floor present on output of the digitizer.

If the input signal is correlated with the sample clock, then the error signal will exhibit repetition. Some of the energy in the error signal will express itself as coherent spurious signals on the ADC output.

EXAMPLE

Spurious Frequencies

The signal of Figure 11-16 is a 4 MHz sine wave sampled at 33.33333 MHz. Over what duration will the error signal repeat itself? What does this duration indicate about a possible spurious frequency?

Solution

A sampling frequency of 33.33333 MHz corresponds to a sample time of 30 ns. The 4 MHz sine wave has a period of 250 ns. The least common multiple of 250 ns and 30 ns is 750 ns so we can expect the error signal to exhibit a periodicity of 750 ns or 1.3333 MHz. We also expect spurious signals to exist at harmonics of 1.3333 MHz.

Examination of Figure 11-16 shows the error signal repeats every 750 ns. Figure 11-17 shows the power spectral density of the error waveform from Figure 11-16.

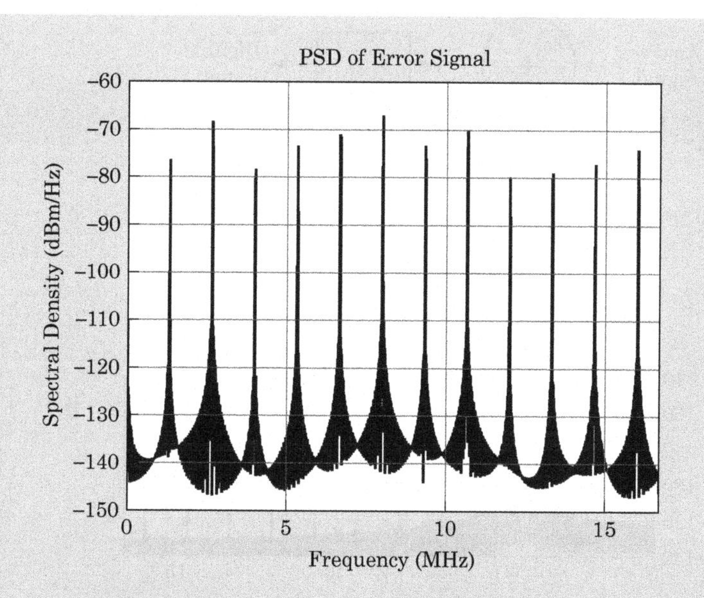

FIGURE 11-17 ■ The PSD of the error signal from Figure 11-16. Note the strong spurious signals at the harmonics of 1.3333 MHz.

Figure 11-18 shows the spectrum of the oscillator. The dynamic range is approximately 30 dB and is limited by the ADC spurious characteristics. It is not possible suppress the spurious spectral components by averaging or other techniques that are often of use against wideband noise.

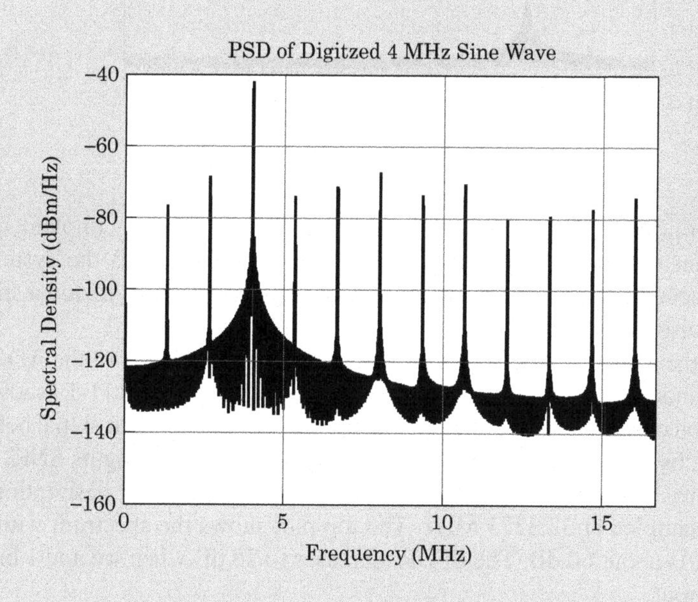

FIGURE 11-18 ■ The power spectral density of the digitized sine wave of Figure 11-16. The signal-to-spurious ratio is about 30 dB.

11.5.1 Dithering

The correlation of the input signal with the sample clock causes the coherent spurious signals of Figure 11-18. If we can disturb that correlation, we will spread the spurious energy over a larger frequency range and increase the spur-free dynamic range (SFDR) of the signal. One method of defeating the sample clock/input signal correlation is by

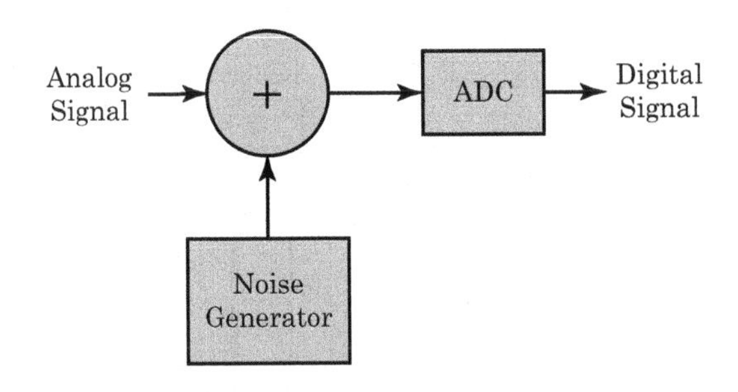

FIGURE 11-19 One implementation of dithering. This is destructive dithering.

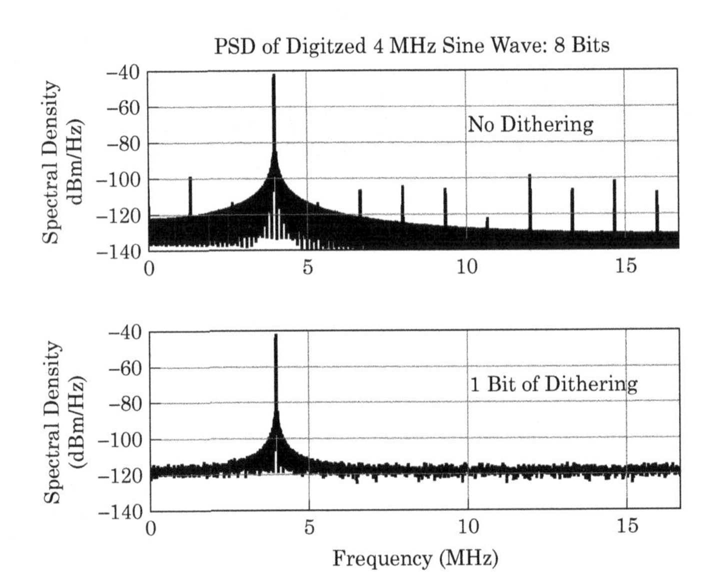

FIGURE 11-20 The PSD of a digitized sine wave without (top) and with (bottom) dithering. The SFDR of the signal is about 60 dB when there is no dithering and increases to about 75 dB when we add 1 bit of noise to the analog input.

introducing a random jitter on the sample clock. However, unless carefully implemented in the analog domain, this method may cause unrecoverable damage to the system's fundamental timing accuracy. The second method of dithering requires us to add a small amount of random noise to the input signal.

Adding noise to the analog input signal causes the least significant bits of the ADC to toggle based on the random noise rather than on the coherent signal. Figure 11-19 shows a simple implementation of dithering. We simply add noise to the analog input signal before digitizing. Of course, by adding noise to the input signal, we are decreasing its SNR.

Figure 11-20 shows the results of dithering. This figure shows an 8-bit description of a 4 MHz sine wave sampled at 33.3333 MHz. The top plot shows the spectrum without dithering. The SFDR is about 60 dB. The SFDR increases to 75 dB when we add 1 bit of noise to the analog input.

11.5.2 Nondestructive Dithering

If the decrease in SNR caused by destructive dithering is not acceptable, we can nondestructively dither the input signal as shown in Figure 11-21. We generate pseudo-random noise in our DSP, then convert that noise into an analog form and add the noise to our incoming signal. We then subtract off the noise in the digital domain, after the signal has passed through the ADC.

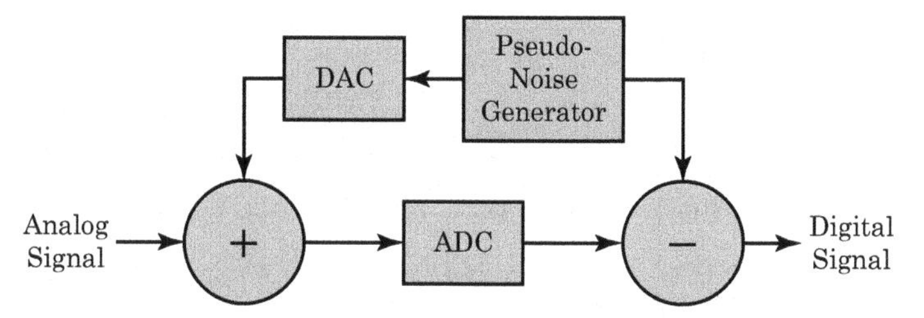

FIGURE 11-21 ■ A second implementation of dithering. We reduce the effect of the dithering noise on the input signal by subtracting the noise we added from the signal after digitizing. This is nondestructive dithering or subtractive Dither.

11.6 | ANALOG-TO-DIGITAL CONVERTERS

ADCs convert an analog input signal into its corresponding digital representation. Practical ADCs deviate from the ideal linear transfer function, and there is a plethora of ADC specifications to quantify those deviations. There are two broad classes of performance deviations: (1) DC accuracy specifications; and (2) dynamic specification. In the signals world, we are more concerned with the dynamic specifications such as SNR, harmonic distortion, and noise figure. We also want to ensure that the samples are taken at evenly spaced intervals.

In some cases, we'll use an ADC to measure temperature, pressure, or some other parameter in which the absolute accuracy is important. The ADC specifications that describe this type of accuracy are offset error, full-scale error, differential nonlinearity error, and integral nonlinearity error.

11.6.1 An Ideal ADC Transfer Function

Figure 11-22 shows the ideal transfer function of a 3-bit ADC. The input is a continuous function but the output is a discrete set of 2^M codes, where M is the number of bits in the ADC. Ideally, the transfer function would be a straight line. A line drawn through the points at each discrete output boundary will begin at the origin. There are no missing output codes, and each output code is equally spaced.

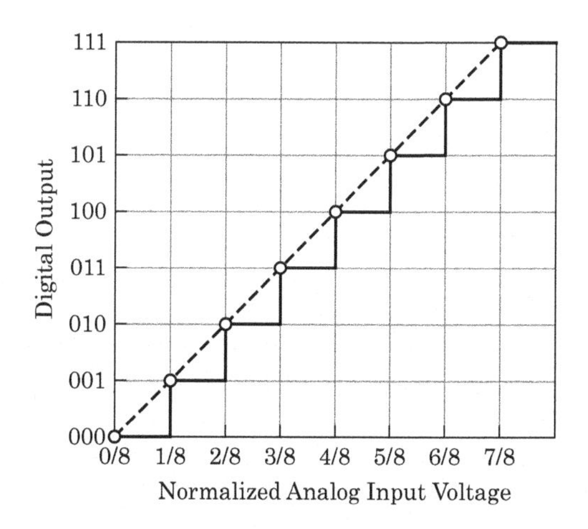

FIGURE 11-22 ■ The ideal voltage transfer function for a 3-bit ADC.

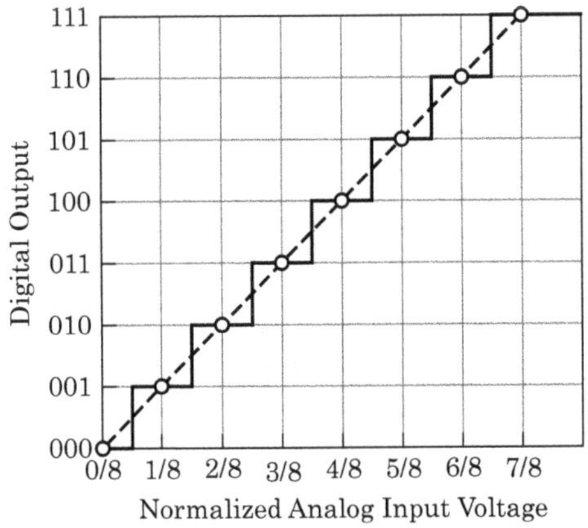

FIGURE 11-23 The ideal voltage transfer function for a 3-bit ADC, with an offset of 1/2 LSB. This offset forces the error between the input and output to lie between −1/2 LSB and +1/2 LSB. Real-world ADCs won't exhibit this perfect transfer function.

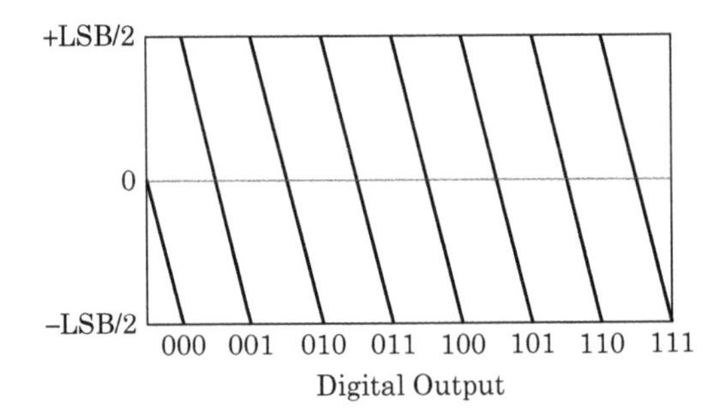

FIGURE 11-24 Quantization error versus ADC output code.

Figure 11-22 shows the transfer function of an ideal ADC with reference points at the output transition boundaries. The error between the ADC input and output falls between 0 and 1 lower sideband (LSB). We can redraw Figure 11-22 to place the error between −1/2 LSB and +1/2 LSB (see Figure 11-23).

The 1/2 LSB offset shown in Figure 11-23 is intentional, but it is often incorporated into the data sheet of an ADC IC as part of the offset error.

11.6.1.1 Quantization Error (Again)

Figure 11-24 shows the quantization error versus ADC output.

11.6.1.2 Offset Error

As Figure 11-25 shows, offset error is represented by a left or right shift of the linear approximation to the ADC transfer function.

11.6.1.3 Full-Scale Error

Full-scale error is equivalent to specifying a change in slope of the realized ADC transfer function from the ideal ADC transfer function.

Full-scale error is the difference between the ideal code transition to the highest ADC output code and the measured transition to the highest output code when the offset error is zero (see Figure 11-26).

Gain error is a specification similar to full-scale error, but the latter accounts for the deviation of both gain and offset from the ideal case.

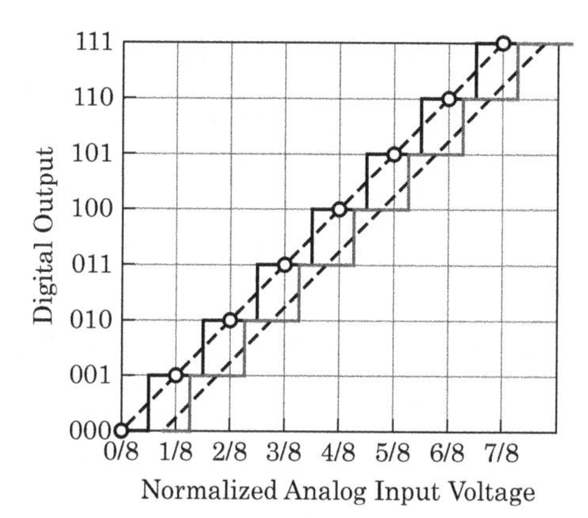

FIGURE 11-25 ■ Offset error is represented by a left or right shift of the linear approximation to the ADC transfer function. This offset error is error to the right.

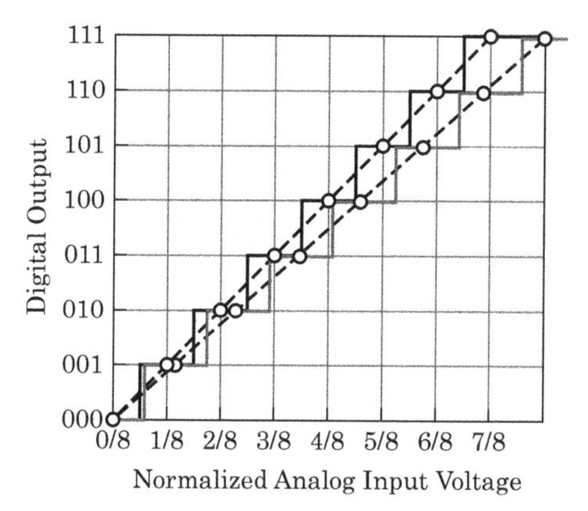

FIGURE 11-26 ■ Full-scale error is represented by a change in slope from the ideal ADC transfer function.

11.6.1.4 Differential Nonlinearity

In our ideal model, a change of 1 least significant bit (LSB) on the output of the ADC represents a uniform change in the input signal. In our 3-bit ADC example, each LSB change in the output corresponds to an input change of $(^1\!/\!_2)^3 = 1/8$ of the ADC full-scale input range. In equation form, we can write

$$V_{LSB} = \frac{V_{FullScale}}{2^M} \tag{11.13}$$

The measured difference in input signal changes between consecutive output code changes is the Differential Nonlinearity of the ADC. Differential nonlinearity presents itself as uneven spacing in the output codes or as uneven transition boundaries in the ADC's transfer function (see Figure 11-27).

For a particular transition, we can express differential nonlinearity (*DNL*) as

$$DNL = \left| \frac{V_{n+1} - V_n}{V_{LSB}} \right| \tag{11.14}$$

11.6.1.5 Integral Nonlinearity

Integral nonlinearity (*INL*) is a measure of the deviation of an ADC transfer function from a straight line. There are several methods by which we can characterize the deviation including best fit, end-point fit, etc. *INL* is expressed in units of LSB. See Figure 11-28.

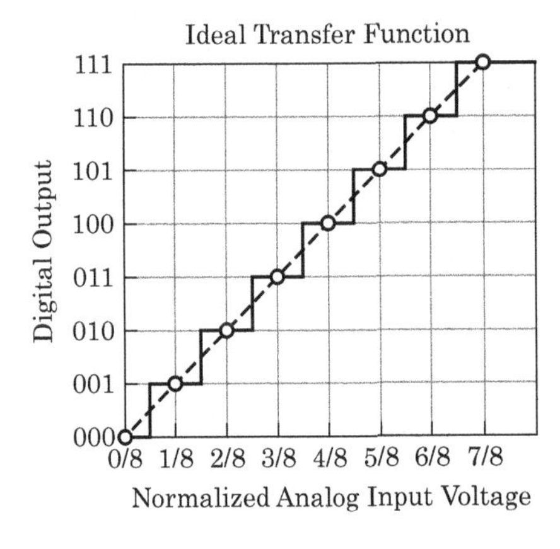

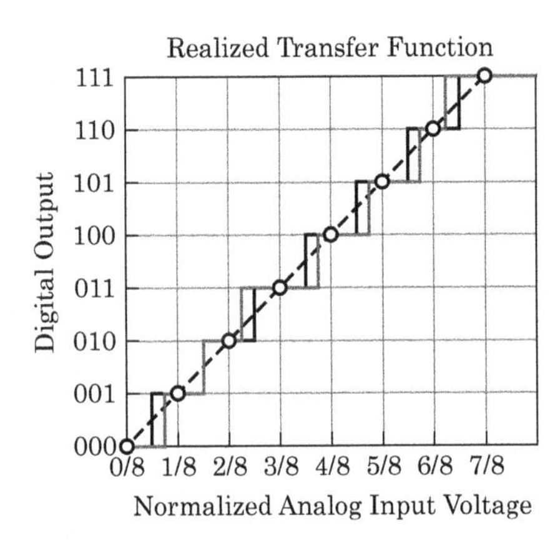

FIGURE 11-27 ■ Differential nonlinearity in an ADC.

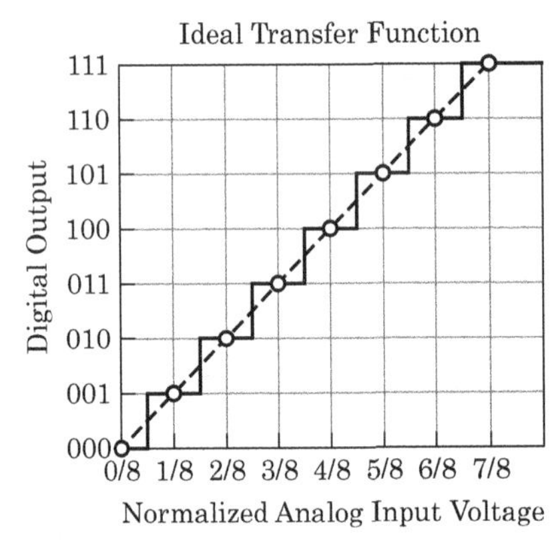

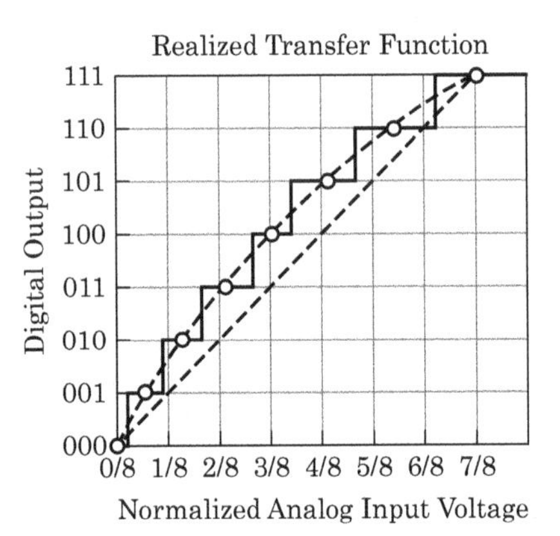

FIGURE 11-28 ■ Integral nonlinearity in an ADC. The INL represents a deviation from a straight-line transfer function.

As we discussed in Chapter 7, the amount of distortion developed by an ADC is a strong function of the exact nature (i.e., curviness) of the transfer function.

11.6.1.6 Absolute Error

In a DC measurement, the absolute error is the sum of all the ADC errors we've discussed so far; the offset, full-scale, *DNL*, and *INL* errors. We normally don't include quantization errors in the absolute error because it's an inherent property of the ADC. Most of these errors can be calibrated out at some cost.

11.6.2 Sample Histograms

The histograms of the digital output codes are useful debugging tools.

11.6.2.1 Saturation

The effect is very similar to amplifier limiting, but ADC saturation is much more abrupt. There is no graceful transition from linear to nonlinear behavior. Figure 11-29 shows ADC saturation.

11.6.2.2 Input Too Low

Figure 11-30 shows the results if the gain prior to the ADC is low. The input signal is not filling up the ADC, and, consequently, we are wasting dynamic range. Since the

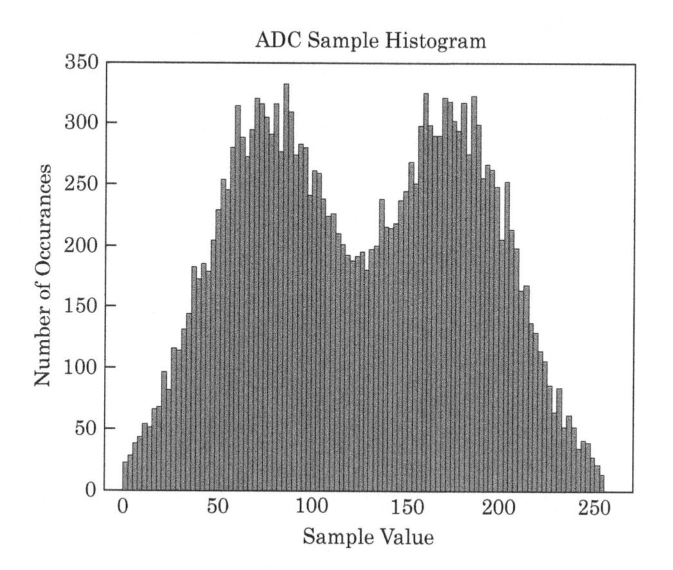

FIGURE 11-29 Histogram of an 8-bit ADC samples when the ADC is undergoing saturation. The sample histogram exhibits a nonzero number of samples beyond the minimum and maximum sample values.

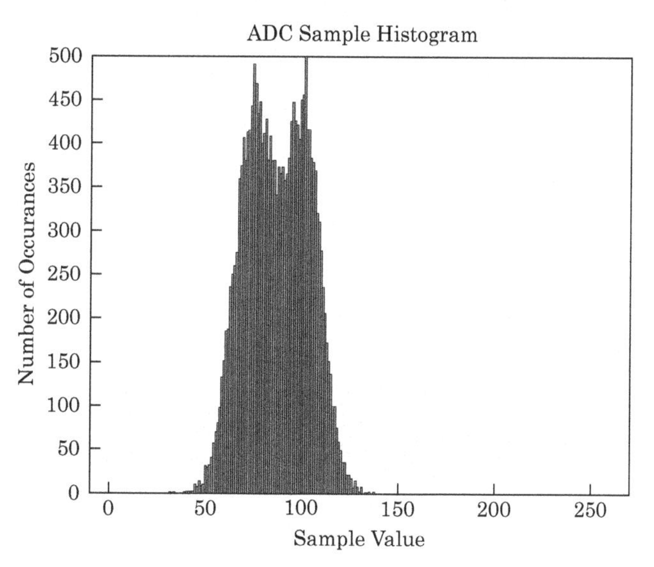

FIGURE 11-30 Histogram of ADC samples when the ADC is driven by a nonfull-scale signal. This histogram also shows a nonzero DC offset.

quantization noise is only a function of the voltage resolution, the SNR also suffers. Figure 11-30 also shows that the input signal exhibits a DC offset because the samples are not centered about the midscale. DC offset is usually harmless in an RF system.

11.6.2.3 Missing Codes

Because of realization issues in the ADC, the ADC may not ever produce particular output codes. Figure 11-31 shows the effects of missing output codes.

11.6.3 Spur-Free Dynamic Range

SFDR is the measured power difference between the desired (or full-scale) signal and the highest spur in the band. Unlike our discussions of amplifier nonlinearity, the nonlinearity exhibited by an ADC is not well behaved, even at small input levels (see Figure 11-32).

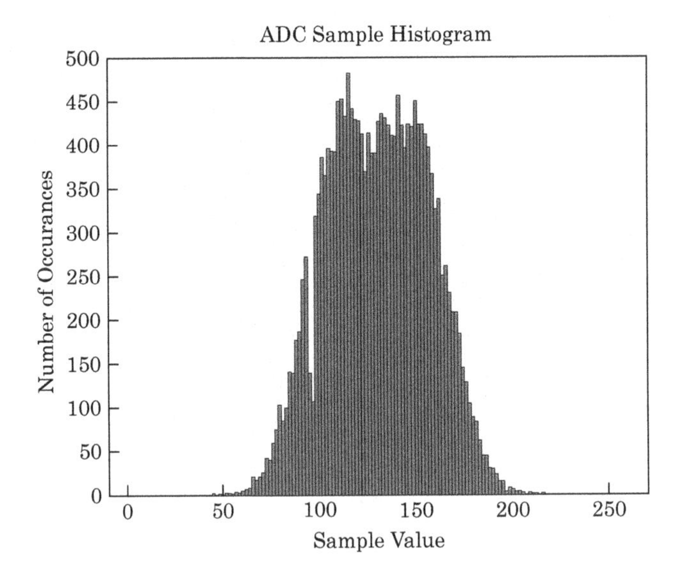

FIGURE 11-31 ■ Histogram of ADC samples when the ADC exhibits missing codes.

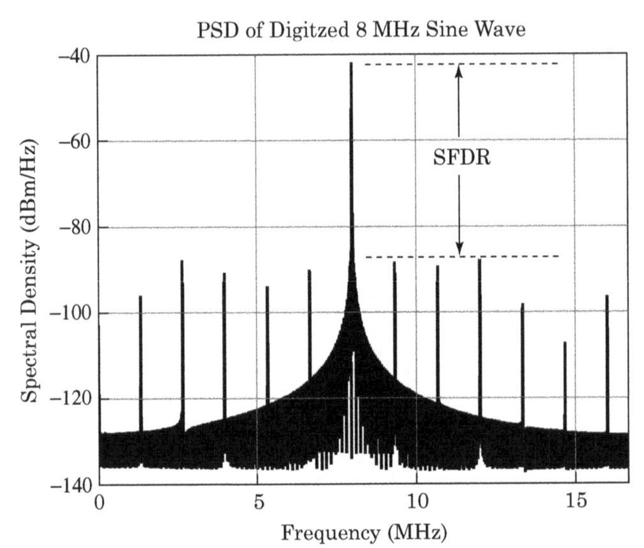

FIGURE 11-32 ■ Spur-free dynamic range in an ADC is the difference between the desired signal and the highest spur.

11.6.3.1 Effective Number of Bits

We can express the measured SFDR as the number of effective bits of resolution offered by a particular ADC. Given a measured *SFDR* as shown in Figure 11-32, we can calculate an effective number of bits (*ENOB*) as

$$ENOB = \frac{SFDR_{dB} - 1.76}{6.02} \tag{11.15}$$

11.7 USING AN ADC IN AN RF SYSTEM

ADC parameters of interest to a receiving system include the following:

- IF center frequency
- IF bandwidth

TABLE 11-2 ■ The required SNR for various modulation formats, assuming a P_{error} of 1E-8

Modulation	Required SNR (dB) for Prob(error) = 1E-8
BPSK	12
QPSK	15
8PSK	20.5
16QAM	22.5
64QAM	28.5
128QAM	31.5

- ADC sampling rate
- The desired SFDR
- The type of signal we'll be digitizing. We'll have to be more careful with high-order quadrature amplitude modulation (QAM) than we might be with quadrature phase shift keying (QPSK).
- The peak (or crest) factor of the signal

SFDR is the most important driving specification. The ADC must exhibit sufficient dynamic range to support the SNR requirements of the desired signal types. Table 11-2 shows the SNR required for various modulation types for a symbol error rate of 1E-8. Even very complex signals (e.g., 128 QAM) require only 32 dB of SNR for error-free reception.

The SNR required in practice is, of course, higher than that indicated by Table 11-2 indicates. For example, the minimum SNR required for a 64QAM signal is about 28.5 dB in an ideal system. The required SNR will be higher in the real world, of course, after phase noise, clock jitter, and other real-world effects impair the signal.

Assume, for example, we've taken measurements and have decided our 64QAM signal requires 32 dB of SNR for tolerable reception. Our underlying assumption is that the noise is of Gaussian character, is uncorrelated with the digitized signal, and exhibits a flat frequency spectrum. In a practical system, the noise associated with the digitizer is not Gaussian, may be correlated with the input signal and it may not exhibit a flat frequency spectrum.

11.7.1 Required ADC Resolution

Equation (11.9) tells us we need at least 5 bits of resolution to support an SNR of 32 dB. In practice, we typically specify more bits. Equation (11.9) assumes that the input signal voltage equals the ADC's full-scale value. It is usually not possible or wise to constantly apply a full-scale signal to the ADC because we might overdrive it if the signal strength changes and the receiver AGC cannot adjust quickly enough. As a rule of thumb, designers often underdrive the ADC by a 10 dB margin. The cost of underdriving the ADC is that we are not using the full number of bits supplied by the ADC.

When we apply very small signals to our receiver, we would like the least significant bits of the ADC to toggle due to avoid the effect shown Figure 11-14. A more realistic assumption for the number of bits required in an ADC is

$$M \geq \frac{(SNR_{dB} + HR_{dB}) - 1.8}{6.02} + LOBits \tag{11.16}$$

where

M = the ADC's resolution in bits

SNR_{dB} = the required SNR in dB

HR_{dB} = the required headroom in dB (often set to 10 dB)

$LOBits$ = the number of ADC bits kept moving by the equivalent input noise of the receiver (often 2 bits).

Using these estimates, we rewrite equation (11.16) as

$$M \geq \frac{SNR_{dB} + 20.2}{6.02} \tag{11.17}$$

Our 64QAM signal needs about 9 bits of resolution under these assumptions.

11.7.2 ADC Equivalent Noise Temperature

ADC manufacturers do not directly specify noise temperature, but we can characterize the noise performance an ADC based on other items commonly listed in the data sheets.

11.7.2.1 Histogram of the Output Codes

Figure 11-33 shows a histogram of the output samples taken from a 14-bit ADC when the input is shorted. If the ADC were noiseless, we would expect a constant output consisting only of the mid-scale output code. The variation we see in the output is a measure of the noise performance of the device.

Curve fitting the histogram of Figure 11-33 to a normal distribution reveals the standard deviation is 2.5 bit_{RMS}. Suppose the ADC of Figure 11-33 has a full-scale input voltage of 2.2 V. The resolution of the ADC is

$$\begin{aligned} \text{Resolution} &= \frac{2.2V}{2^{14}} \\ &= \frac{2.2V}{16384} \\ &= 134\,\mu V \end{aligned} \tag{11.18}$$

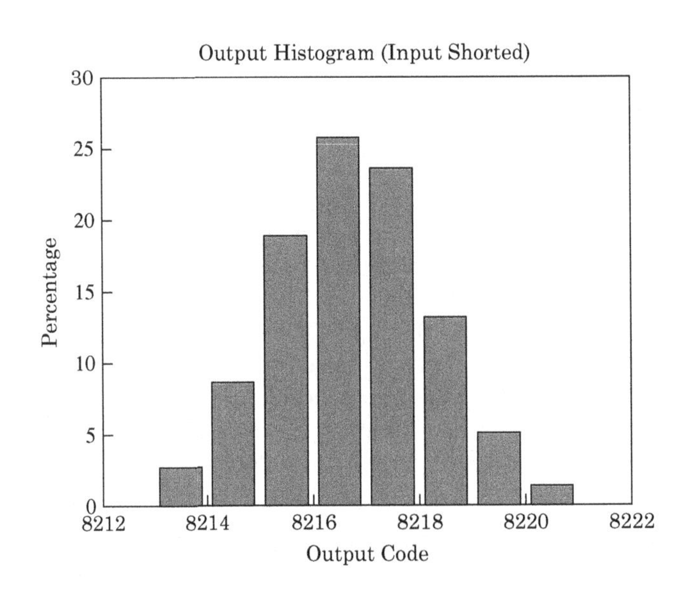

FIGURE 11-33 ■ The histogram of the output samples of a 14-bit ADC when the input is shorted. The nonconstant output is a measure of the noise performance of the ADC.

The standard deviation of the noise is $2.5(134\ \mu V) = 335\ \mu V_{RMS}$. In Chapter 6, we found that noise temperature and noise voltage were related by

$$V_{n,OC,RMS} = \sqrt{4kTB_n R} \tag{11.19}$$

Solving for noise temperature produces

$$T_{ADC} = \left(\frac{V_{n,OC,RMS}}{4kB_n R}\right)^2 \tag{11.20}$$

where R is the input impedance of the ADC, and B_n is the noise bandwidth of the filter preceding the ADC. The other variables are defined in Chapter 6. Assuming that the bandwidth of the filter preceding the ADC is 100 MHz and the ADC input impedance is 500 Ω, the ADC noise temperature is approximated by

$$\begin{aligned} T_{ADC} &= \left(\frac{V_{n,OC,RMS}}{4kB_n R}\right)^2 \\ &= \left(\frac{335\ \mu V}{4(1.38E-23)(100E6)(500)}\right)^2 \\ &= 121E6°\text{K} \end{aligned} \tag{11.21}$$

The noise figure is

$$\begin{aligned} F_{ADC} &= \frac{T_0 + T_{ADC}}{T_0} \\ &= \frac{290 + 121E6}{290} \\ &= 418k \\ F_{ADC,dB} &= 56\,dB \end{aligned} \tag{11.22}$$

11.7.2.2 Required Excess Gain

In the cases in which the data sheet does not supply a figure like Figure 11-33, we can estimate the ADC noise figure from system requirements. Our system design requires us to drive our ADC with at least 2 bits of noise voltage, which is $4(134\ \mu\text{V}) = 536\ \mu\text{V}$. Using the procedure in the previous section, the noise temperature of the ADC is

$$\begin{aligned} T_{ADC} &= \left(\frac{V_{n,OC,RMS}}{4kB_n R}\right)^2 \\ &= \left(\frac{536\ \mu V}{4(1.38E-23)B_n R}\right)^2 \\ &= 194\ M°\text{K} \end{aligned} \tag{11.23}$$

and the noise figure is

$$\begin{aligned} F_{ADC} &= \frac{T_0 + T_{ADC}}{T_0} \\ &= \frac{290 + 194E6}{290} \\ &= 670k \\ F_{ADC,dB} &= 58\,dB \end{aligned} \tag{11.24}$$

11.7.2.3 An Example

A perfect ADC would transform the receiver's analog input signal into its digital equivalent without degradation. This situation is not realistic, so we design the cascade before the ADC to minimize signal degradation due to the digitizing process.

We will present 20 MHz of bandwidth to our 10-bit ADC. This statement does not mean our final IF or noise bandwidth will be 20 MHz; it means only that we are presenting 20 MHz of bandwidth to the ADC. The final bandwidth of the signal may be much narrower.

Let's also assume that our antenna noise is $T_0 = 290°$ K and that our system's noise figure is 10 dB. The system's minimum detectable signal (*MDS*) in a 20 MHz bandwidth is

$$\begin{aligned} MDS &= F_{sys} k T_0 B_n \\ MDS_{dBm} &= F_{sys,dB} - 174 + 10\log(B_n) \\ &= 10 - 174 + 10\log(20E6) \\ &= -91.0\, dBm \end{aligned} \tag{11.25}$$

Despite the processing in the DSP, which occurs after the ADC, our system needs sufficient gain to convert the MDS from the input to a level high enough to toggle the lower bits of the ADC. If the full-scale voltage input range of the 10-bit ADC is 0–1 V, the voltage represented by a single bit is

$$\begin{aligned} Q &= \frac{E_{FSR}}{2} \\ &= \frac{1.0}{2^{10}} \\ &= 977\, \mu V \end{aligned} \tag{11.26}$$

Toggling the bottom 2 bits means that we must bring the noise voltage up to $4(977\, \mu V) =$ 3.91 mV. This voltage is equivalent to −35.2 dBm in a 50 Ω system. Our system needs 91.0 − 35.2 = 55.8 dB of gain.

EXAMPLE

14-Bit ADC

We present 80 MHz of bandwidth to an ADC in a system with the following parameters. What's the gain required between the antenna and the ADC?

System parameters:

- Noise figure = 20 dB
- Antenna noise temperature = 290°K
- ADC bits = 14
- ADC full-scale voltage range = −1 to +1 Volt

Solution

The system's *MDS* in an 80 MHz bandwidth is

$$\begin{aligned} MDS &= F_{sys} k T_0 B_n \\ MDS_{dBm} &= F_{sys,dB} - 174 + 10\log(B_n) \\ &= 20 - 174 + 10\log(80E6) \\ &= -75.0\, dBm \end{aligned} \tag{11.27}$$

A 14-bit ADC can distinguish $2^{14} = 16{,}384$ levels. The voltage resolution at 2 volts full-scale input is $2/16{,}384 = 122\ \mu$ volts. Toggling the bottom 2 bits means that we must bring the noise voltage up to $4(122\ \mu V) = 488\ \mu V = -53.2$ dBm in a 50 Ω system. Our system needs $75.0 - 53.2 = 21.8$ dB of gain.

11.8 | BIBLIOGRAPHY

Kester, Walt, "Taking the Mystery Out of the Infamous Formula, 'SNR = 6.02N + 1.76dB,' and Why You Should Care," Analog Devices Tutorial MT-001, 2009.

Kester, Walt, "What the Nyquist Criterion Means to Your Sampled Data System Design," Analog Devices Tutorial MT-002, 2009.

Kester, Walt, "ADC Noise Figure—An Often Misunderstood and Misinterpreted Specification," Analog Devices Tutorial MT-006, 2009.

Lyons, R., *Understanding Digital Signal Processing*, 2nd ed., Prentice Hall, 2004.

Proakis, John G., *Digital Signal Processing—Principles, Algorithms and Applications*, 4th ed., Prentice Hall.

Viniotis, Y., *Probability and Random Processes for Electrical Engineers*, McGraw-Hill, 1998.

Wannamaker, R., "The Theory of Dithered Quantization," Ph.D. Thesis, Department of Applied Mathematics, University of Waterloo, ON, Canada, July 1997.

Wannamaker, R., et al., "A Mathematical Theory of Non-Subtractive Dither," *IEEE Transactions on Signal Processing*, Vol. 48, pp. 499–516, February 2000.

11.9 | PROBLEMS

1. **Subsampling the ISM Band** The US ISM band covers 2,400–2,485 MHz. This 85 MHz chunk of spectrum has been converted to 400–485 MHz by an external mixer.
   a. Find the possible Nyquist zones and sampling frequency ranges for each zone.
   b. Pick a sampling frequency in each possible Nyquist zone and draw the spectrum of the conversion in each possible case.

2. **Nyquist Ranges** Given the digitizing system shown in Figure 11-34:
   a. Is the digitized spectrum inverted?
   b. What is the minimum center frequency at which the signal resides after it has passed through the ADC? In other words, what is f_C?

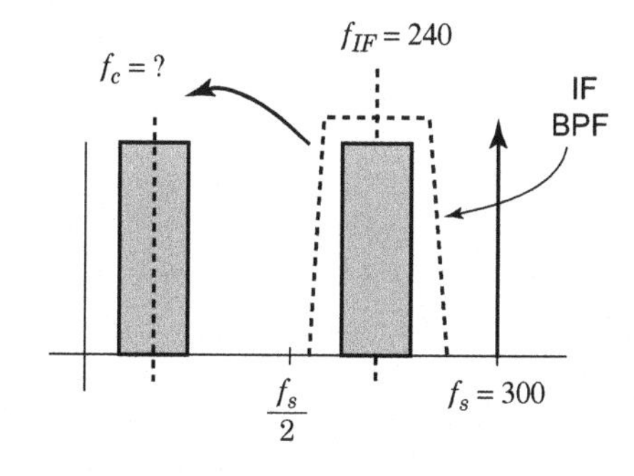

FIGURE 11-34 ■ A digitizing system that performs frequency conversion.

3. **Multiple choice** When designing a digitizing downconverter, we must account for which of the following before deciding the number of bits required in the ADC?

   a. Signal peak factor
   b. SNR required for the particular modulation
   c. The noise figure of the downconverter before the ADC and gain before the ADC
   d. All of the items listed above

4. **Multiple Choice: ADC Bits** When designing a digitizing downconverter, we must account for which of the following before deciding the number of bits required in the ADC?

   a. Signal peak factor
   b. SNR required for the particular modulation
   c. The noise figure of the downconverter before the ADC and gain before the ADC
   d. All of the items listed above

CHAPTER 12

Demodulation

Digital signal processing: That discipline which has allowed us to replace a circuit previously composed of a capacitor and a resistor with two anti-aliasing filters, an A-to-D converter and a D-to-A converter, and a general purpose computer (or array processor) so long as the signal we are interested in does not vary too quickly.

Thomas P. Barnwell, 1974

Chapter Outline

12.1 INTRODUCTION

Demodulation is the process of recovering the transmitter's symbol stream from the received signal. There are entire books devoted to demodulation, so a single chapter in this book has no hope of covering the topic adequately. Our goal is describe, in broad strokes, the nature of the received signal and some introductory methods we might use to compensate for the corruption of the signal by the propagation environment, oscillator inaccuracies, and other effects. We will examine only continuous quadrature amplitude modulation (QAM)-type signals in this chapter.

The propagation path through which the signal travels exhibits a variable time delay, multipath, and signal power loss. These parameters will change with time. The oscillators in the receiver and transmitter exhibit frequency drift that causes the received signal to appear at an incorrect frequency and baud rate. The time delay of the propagation path and the oscillator inaccuracies cause the phase of the received signal to be indeterminate and varying.

At the receiver, we wish to measure the precise phase and amplitude of a constellation point. That is our received symbol. The (considerable) task of the demodulator is to remove the ambiguities (amplitude, frequency, and phase) of the arrived signal and allow the symbol decision to proceed with accuracy.

12.2 A TRANSMITTER MODEL

Figure 12-1 shows the general framework for a data transmission system. This system absorbs a single-user data stream, a symbol clock, and a radio frequency (RF) carrier and produces a modulated output signal.

12.2.1 User Data Source

The user data source supplies the data to send over the channel. We place no particular restrictions on the user data source. The data may be highly repetitive or highly random.

12.2.2 Scrambler (Randomizer)

Allowing data that exhibit nonrandom characteristics to pass into a modulator will produce problems when we attempt to perform timing recovery and channel equalization at the receiver. Problematic data patterns include long strings of ones or zeroes or long strings of repetitive patterns (e.g., 1010... or 1110011100...). The scrambler or randomizer converts nonrandom data into pseudo-random data exhibiting qualities that aid timing recovery and equalization. Such qualities include an approximate 50% ratio of zeros to ones and limited run lengths of zeroes or ones.

12.2.3 Bit to Symbol Mapping

The bit to symbol mapping function converts binary data into the appropriate alphabet for the modulator. For example, four user bits will produce one symbol in a 16QAM system.

12.2.4 Pulse-Shaping Filter

Pulse shaping is done primarily to limit the bandwidth of the signal that we apply to the modulator. Proper selection of the pulse-shaping filter allows the receiver to apply a matched filter resulting in zero intersymbol interference (ISI) at the output of the matched filter in the receiver.

12.2.5 Modulator

The modulator absorbs the pulse shaper output and the RF carrier to produce the modulated RF waveform that we will apply to our antenna. For this discussion, the modulator will be a quadrature modulator of the type described in Chapter 2.

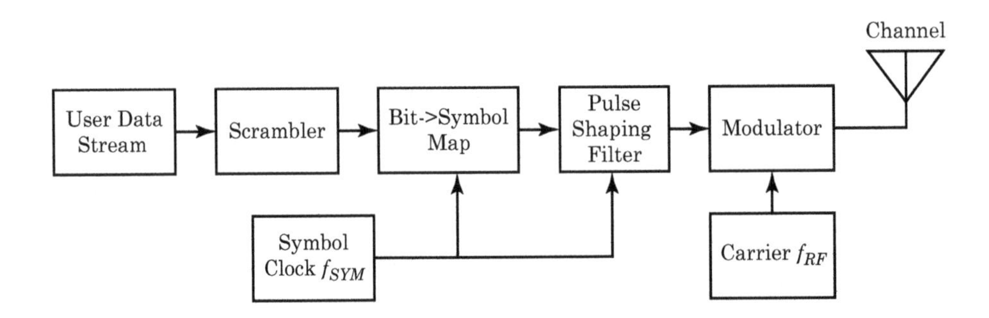

FIGURE 12-1 ■ A data transmission system, emphasizing the conversion of user data into a transmitted waveform.

12.3 | THE PULSE-SHAPING FILTER

The major function of the pulse-shaping filter is to limit the bandwidth of the RF signal so that the signal occupies only a carefully controlled amount of spectrum. Proper selection of the transmitter's pulse shaping filter requires that we accommodate the needs of the receiver. At some point in the receiver, we must filter the received signal with a matched filter to achieve maximum signal-to-noise ratio (SNR) before we make a symbol decision. Therefore, we must choose a transmitter pulse-shaping filter that can adequately limit the RF bandwidth. When we pass the received signal through the receiver's matched filter, the resulting waveform must exhibit no ISI. A final filter requirement is that we allow enough information through the channel to aid in carrier and baud clock acquisition in the receiver.

Most designers of data transmission systems have settled on the root raised cosine (RRC) filter for both the transmitter pulse-shaping filter and for the receiver's matched filter. After passing through two RRC filters, the signal will exhibit zero ISI and will be suitable for the follow-on decision process. We can easily accommodate the addition of excess bandwidth to aid the receiver's carrier and baud clock acquisition.

12.3.1 Root Raised Cosine Spectral Shaping

The root raised cosine is the square root of the raised cosine waveform. The square root is performed in the frequency domain. The equation describing the impulse response of the root raised cosine is

$$f_{RRC}(t) = 4\beta\sqrt{R_S}\frac{\cos\left[(1+\beta)\pi t R_S\right] + \sin\left[(1-\beta)\pi t R_S\right]/4\beta R_S t}{\pi\left[1-(4\beta R_S t)^2\right]} \tag{12.1}$$

where

R_S = the symbol rate and

β = the roll-off factor ($0 <= \beta <= 1$), which describes how quickly the filter rolls off in the frequency domain and is directly responsible for allowing enough excess bandwidth to aid carrier and baud clock acquisition

Figure 12-2 shows the time and frequency domain plots of a RRC filter. The RRC filter does not exhibit zero ISI (i.e., the waveform is not zero at multiples of the baud rate). However, when we filter the RRC waveform of Figure 12-2 with its matched filter (identically the same RRC filter we used as a pulse-shaping filter in the transmitter), we will produce a waveform with zero ISI (see Figure 12-3).

12.3.2 RRC Filter Roll-Off Factor β

The roll-off factor, β, of a raised cosine waveform is a measure of the waveform's excess bandwidth. A system that processes R_S symbols/second requires a theoretically minimum bandwidth of $R_S/2$ hertz. The excess bandwidth of the filter is the bandwidth occupied by the signal beyond the Nyquist bandwidth of $R_S/2$. The receiver needs some amount of excess bandwidth to acquire carrier lock to the received signal. Figure 12-4 shows the frequency response of several RRC filters with varying value of roll-off factor β.

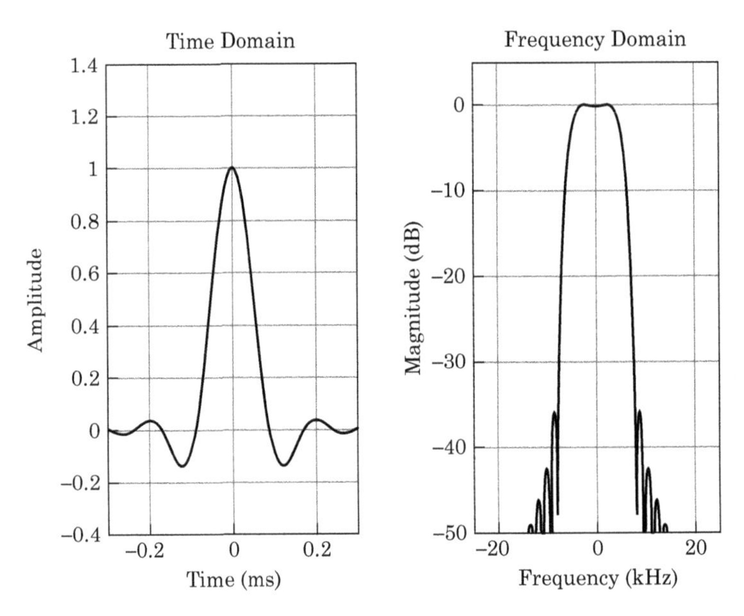

FIGURE 12-2 ■ The time- and frequency-domain response of a root raised cosine waveform, applicable to a 10 k symbols/second system. The roll-off factor, β, for this filter is 0.5.

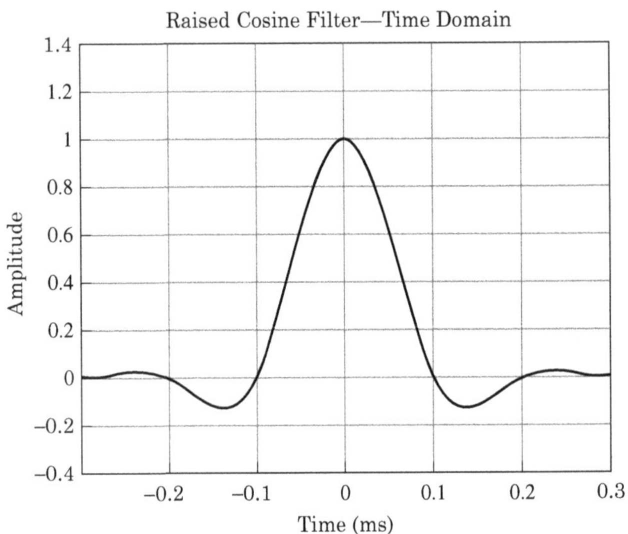

FIGURE 12-3 ■ The time-domain response of a raised cosine waveform, applicable to a 10 k symbols/second system. This waveform is the impulse response of two RRC filters in series. This is the waveform present in the receiver directly after its matched filter. This waveform exhibits zero ISI.

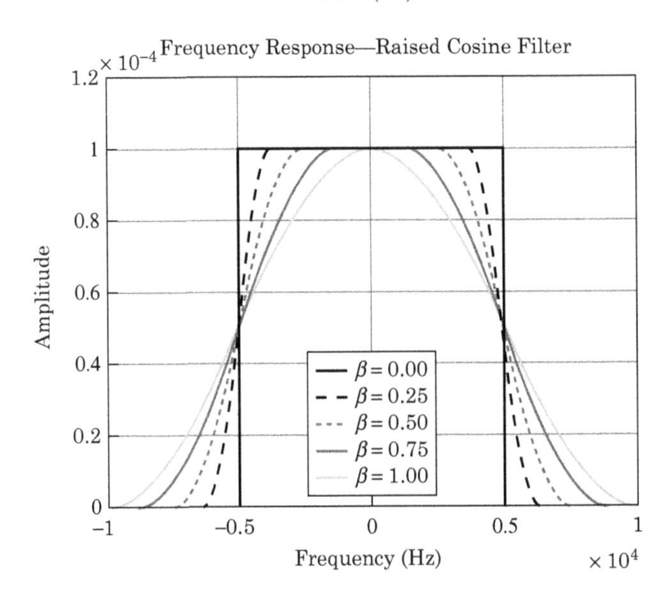

FIGURE 12-4 ■ The frequency-domain response of a raised cosine waveform, applicable to a 10 k symbols/second system. This waveform is the frequency response of two RRC filters in series. A waveform passing through this filter will exhibit zero ISI.

12.4 | A 16QAM MODULATOR

We now explore the waveforms present in the transmitter of Figure 12-1. We will ignore the forward error correction (FEC) portion of the transmitter and point the reader to the many excellent books available on that subject. We will assume the user's data have been previously randomized. Figure 12-5 shows the parts of the modulator we will examine.

In a 16QAM modulator, our goal is to encode the user's symbol stream into one of 16 possible combinations of amplitude and carrier phase. Thus, the symbol rate is one-fourth of the bit rate, or $R_S = R_b/4$. We desire to place the constellation points on an equally spaced rectangular grid consisting of 16 points as shown in Figure 12-6.

We will assume that the user's data (after FEC and scrambling) are presented to the transmitter as a series of 4-bit symbols. We use each symbol to select one of the 16 possible constellation points, and we will use a gray code to map the user's data to the modulator symbols. The gray code causes the most likely symbol errors to translate into only 1 bit error. Figure 12-7 and Table 12-1 show one possible mapping.

We will use a digital signal processing (DSP) finite impulse response (FIR) filter to perform the $\sin(x)/x$ pulse shaping. Table 12-1 shows the magnitude and phase angles

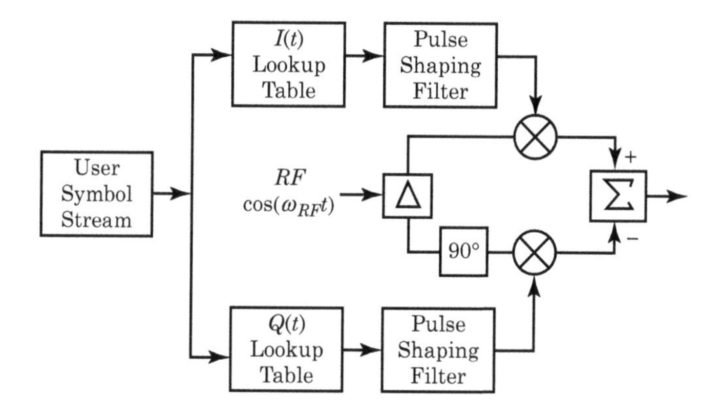

FIGURE 12-5 ■ A generic modulator consisting of $I(t)$ and $Q(t)$ lookup tables, two identical pulse-shaping filters, and a generic IQ modulator from Chapter 2.

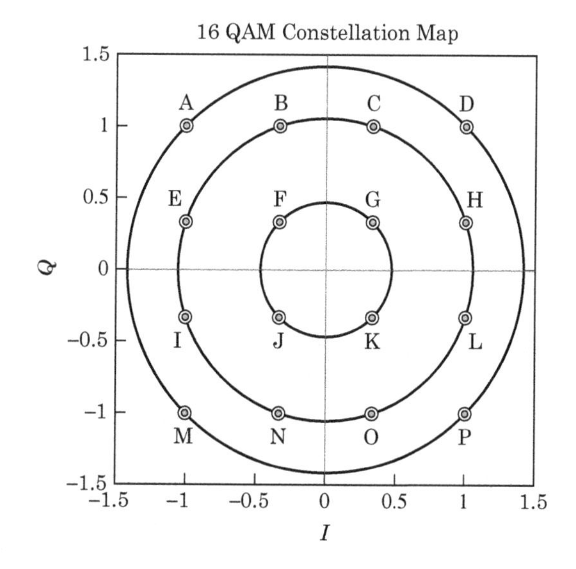

FIGURE 12-6 ■ The constellation of a 16QAM signal. The symbols have been mapped into the amplitude and phase of a cosine. There are three distinct amplitudes and three distinct, rotationally invariant phases.

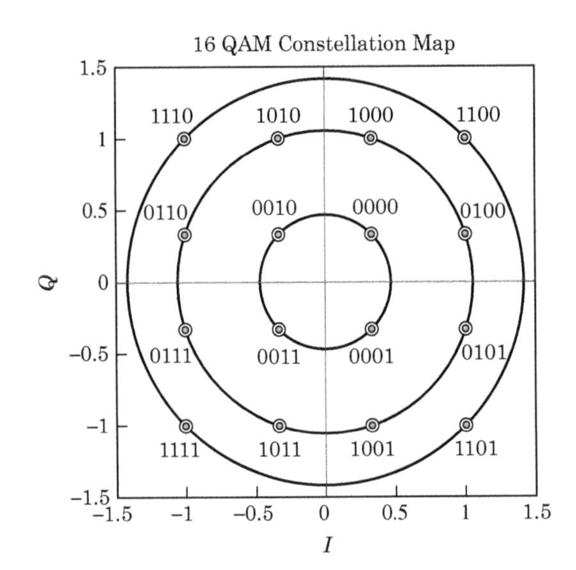

FIGURE 12-7 ■ The constellation of a 16QAM signal. The symbols have been mapped into the amplitude and phase of a cosine. There are three distinct amplitudes and three distinct, rotationally invariant phases. The bit –> symbol mapping here is a gray code and maps the most likely symbol errors to cause only 1 bit error.

TABLE 12-1 ■ The amplitude, phase, and assigned symbol value for every constellation point in a 16QAM constellation

Symbol	Transmitted Amplitude and Phase	*I* and *Q*
A = 1110 (14)	$1.414\angle 135^\circ$	$I = -1.000;\ Q = 1.000$
B = 1010 (10)	$1.054\angle 108.4^\circ$	$I = -0.333;\ Q = 1.000$
C = 1000 (8)	$1.054\angle 71.6^\circ$	$I = 0.333;\ Q = 1.000$
D = 1100 (12)	$1.414\angle 45^\circ$	$I = 1.000;\ Q = 1.000$
E = 0110 (6)	$1.054\angle 161.6^\circ$	$I = -1.000;\ Q = 0.500$
F = 0010 (2)	$0.471\angle 135^\circ$	$I = -0.333;\ Q = 0.500$
G = 0000 (0)	$0.471\angle 45^\circ$	$I = 0.333;\ Q = 0.500$
H = 0100 (4)	$1.054\angle 18.4^\circ$	$I = 1.000;\ Q = 0.500$
I = 0111 (7)	$1.054\angle 198.4^\circ\ (1.054\angle -161.6^\circ)$	$I = -1.000;\ Q = -0.500$
J = 0011 (3)	$0.471\angle 225^\circ\ (0.471\angle -135^\circ)$	$I = -0.333;\ Q = -0.500$
K = 0001 (1)	$0.471\angle 315^\circ\ (0.471\angle -45^\circ)$	$I = 0.333;\ Q = -0.500$
L = 0101 (5)	$1.054\angle 341.6^\circ\ (1.054\angle -18.4^\circ)$	$I = 1.000;\ Q = -0.500$
M = 1111 (15)	$1.414\angle 225^\circ\ (1.414\angle -135^\circ)$	$I = -1.000;\ Q = -1.000$
N = 1011 (11)	$1.054\angle 251.6^\circ\ (1.054\angle -108.4^\circ)$	$I = -0.333;\ Q = -1.000$
O = 1001 (9)	$1.054\angle 288.4^\circ\ (1.054\angle -71.6^\circ)$	$I = 0.333;\ Q = -1.000$
P = 1101 (13)	$1.414\angle 315^\circ\ (1.414\angle -45^\circ)$	$I = 1.000;\ Q = -1.000$

of the complex pulses we will apply to the FIR for each symbol. Figure 12-8 shows the results of passing the user's symbol stream through the in-phase (I) and quadrature (Q) lookup tables of Figure 12-5. The output of the lookup tables is a series of pulses whose magnitude and angle reflect the data of Table 12-1.

For this example transmitter, we will assume an RRC pulse-shaping filter with a roll-off factor, β, of 0.5. Figure 12-9 shows the RRC waveform stored in the pulse-shaping filter of Figure 12-5. We can place any waveform we desire into this filter although the RRC shape is the most commonly used.

Figure 12-10 shows the waveforms present on the outputs of the I and Q pulse-shaping filters of Figure 12-5. In a zero ISI system, each sample point, measured at the

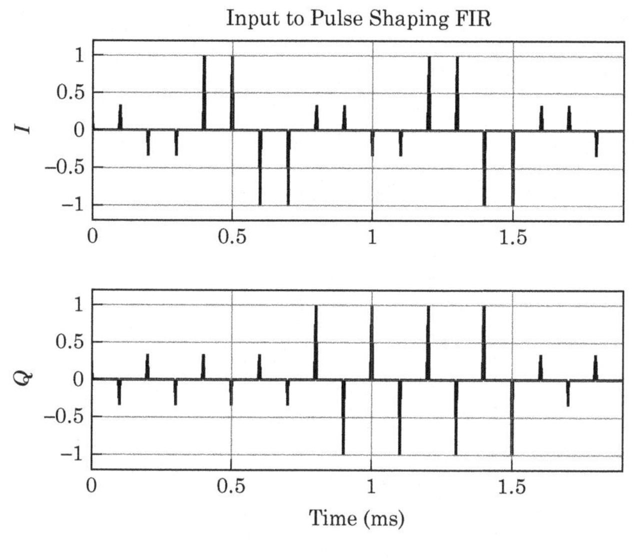

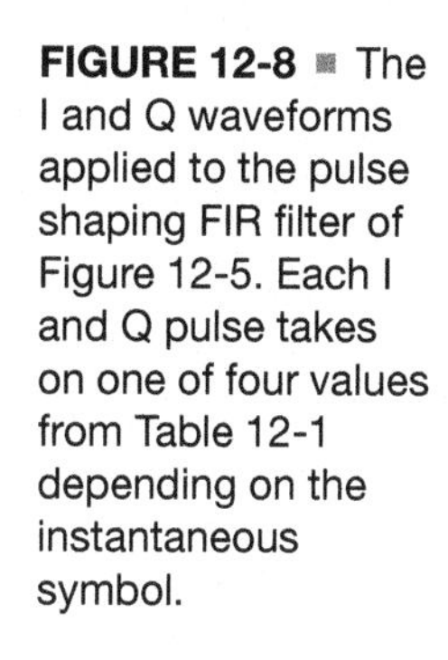
FIGURE 12-8 ■ The I and Q waveforms applied to the pulse shaping FIR filter of Figure 12-5. Each I and Q pulse takes on one of four values from Table 12-1 depending on the instantaneous symbol.

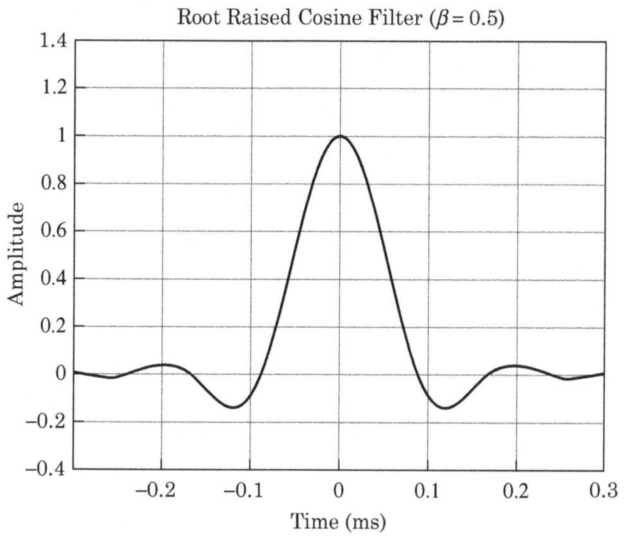

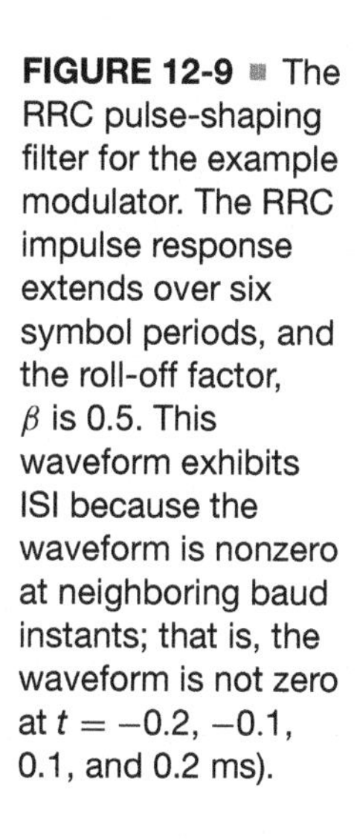
FIGURE 12-9 ■ The RRC pulse-shaping filter for the example modulator. The RRC impulse response extends over six symbol periods, and the roll-off factor, β is 0.5. This waveform exhibits ISI because the waveform is nonzero at neighboring baud instants; that is, the waveform is not zero at $t = -0.2$, -0.1, 0.1, and 0.2 ms).

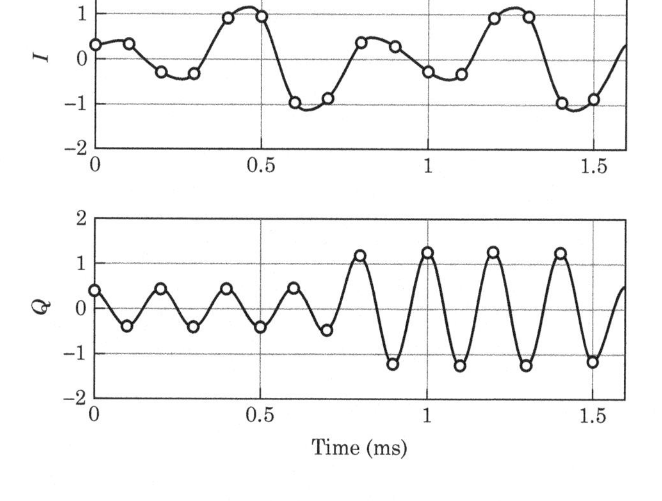

FIGURE 12-10 ■ The $I(t)$ and $Q(t)$ waveforms after the impulse train of Figure 12-8 has been applied to the FIR shown in Figure 12-5. The circles indicate the baud clock. In a zero ISI system, each point at the baud clock would lie on -1.0, -0.33, 0.33, or $+1.0$.

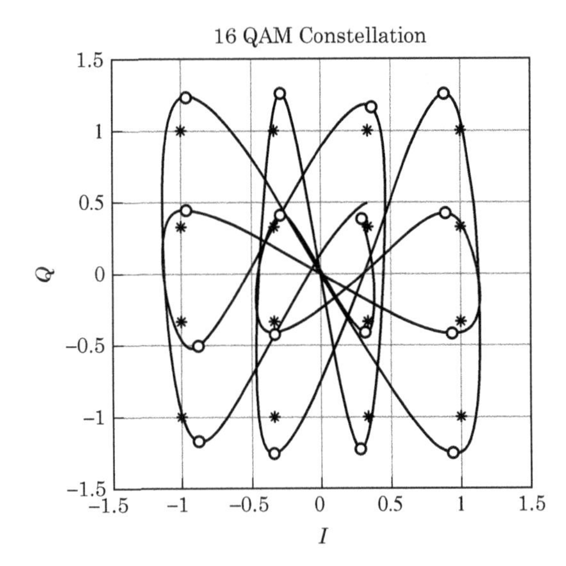

FIGURE 12-11 ■ The resulting constellation after applying the $I(t)$ and $Q(t)$ waveforms of Figure 12-10 to a the quadrature modulator of Figure 12-5. The circles indicate the symbol generated by the modulation, while the * indicate the ideal constellation points. This constellation exhibits ISI due to the root raised cosine pulse-shaping filter.

baud instants, would lie on $-1.0, -0.33, 0.33$, or $+1.0$. Figure 12-11 shows the IQ diagram of the waveforms of Figure 12-10. This waveform is sent to the antenna, and it contains significant ISI.

Figure 12-12 shows the transmitted constellation averaged over many symbols. The on-the-air waveform exhibits a large amount of ISI.

We have generated a 16QAM signal. We now seek to reverse the process and recover the user's data stream.

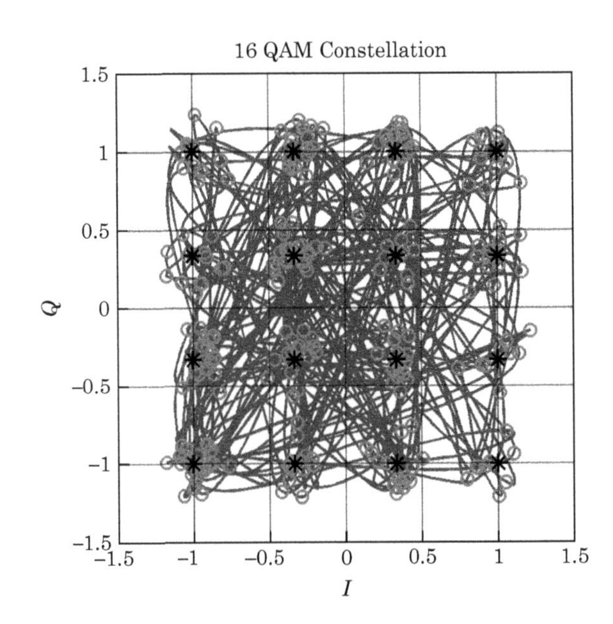

FIGURE 12-12 ■ The constellation of Figure 12-11 after many symbols. The ideal constellation points are shown as the black *, while the constellation points generated by the transmitter are shown as circles. Note the ISI.

12.5 | A RECEIVER MODEL

Figure 12-13 shows a functional block diagram of a generic data receiving system and demodulator. A demodulator is more complicated than a modulator because the demodulator must recover the RF carrier and timing signal from the incoming modulated signal. The demodulator often acts in two distinct modes: an initial acquisition mode and a decision directed mode.

The initial acquisition mode is characterized by the receiver measuring various signal parameters in an open loop fashion. During the initial acquisition mode, the receiver derives an initial estimate for the signal power, RF carrier, and the baud clock frequency. The receiver derives the initial gain estimate by measuring the power in the received signal and adjusting its internal gain until the signal power matches the expected power in the constellation. The receiver performs initial carrier recovery via some nonlinear process such as squaring the signal and feeding the resultant into a DSP phase-locked loop (PLL). Symbol timing is usually recovered from the received signal via filtering and then performing an amplitude modulation (AM) detection and feeding that result into a DSP phase-locked loop. The transmitter's data scrambler, coupled with the excess bandwidth of the pulse-shaping filter, ensures there is sufficient information in the transmitted waveform to derive these initial estimates of signal parameters.

Once the receiver is satisfied that it has derived sufficiently accurate first estimates of the signal power, carrier frequency, and baud clock frequency, the receiver switches into decision-directed mode. In the decision-directed mode, the receiver examines its symbol decisions and adjusts the receiver to compensate for any anomalies it observes. For example, if the receiver observes that the symbols are tending to rotate clockwise, it will use this information to adjust the carrier phase to compensate. Similarly, the receiver may observe that its decoded constellation points are all slightly smaller in magnitude than the optimum constellation points. The receiver will use this information to increase its gain.

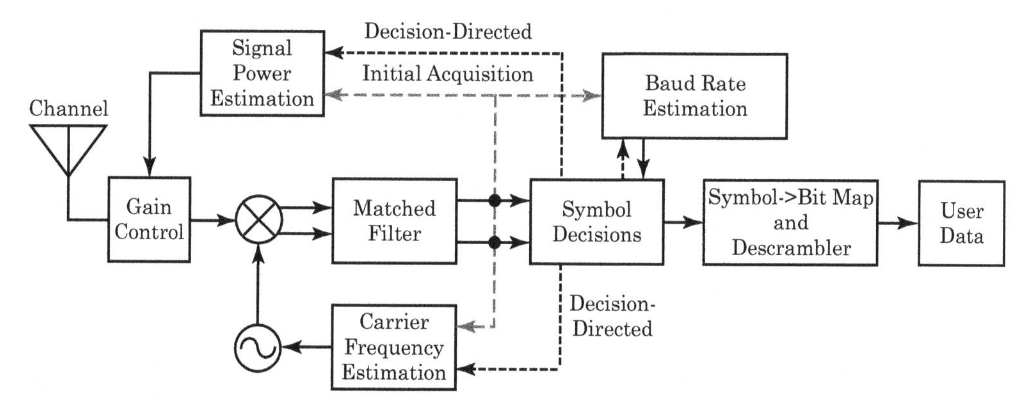

FIGURE 12-13 ■ A data-receiving system. The red lines indicate the feedback paths required for initially estimating received signal power, carrier frequency/phase, and baud clock/phase. The blue lines show the decision-directed information feedback paths. We use information from the symbol decision engine to further refine our parameter estimates once the receiver has achieved lock. We have not shown FEC or an equalizer in this diagram.

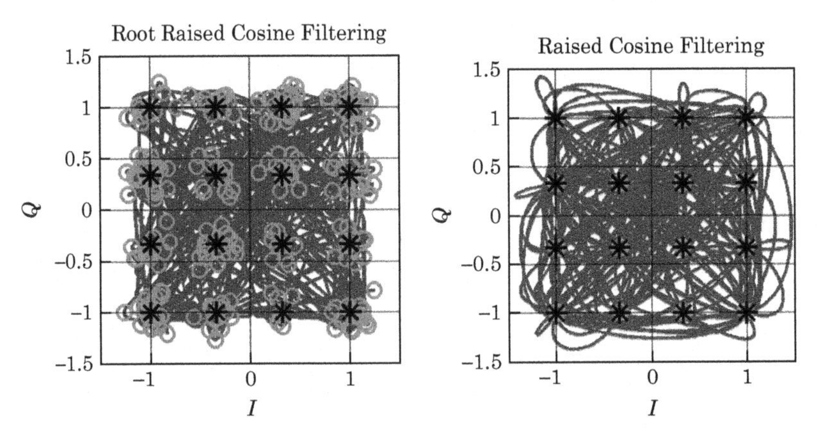

FIGURE 12-14 The constellation of a received signal before and after the RRC matched filter. Before the matched filter, the constellation shows significant ISI. After the matched filter, much noise has been removed from the signal and the zero ISI condition has been restored.

12.5.1 Matched Filtering

The purpose of the matched filter in the receiver is to maximize the SNR of the received signal and to restore a zero ISI condition to the received constellation. This effect is most clearly described if we assume the receiver has already achieved gain, carrier, and baud clock lock. Figure 12-14 shows the received constellation before and after the matched filter. After the matched filter, the SNR has improved, and the zero ISI waveform has been restored.

12.5.2 Gain Control

The gain control block adjusts the gain of the system so that the RMS power level of the incoming signal equals the RMS power level of an ideal constellation. A practical system usually has gain blocks in both its analog and digital sections. The primary purpose of the analog gain section is to ensure that the incoming signal fills up the ADC. Figure 12-15 shows the received IQ diagram when the gain of the system is incorrect.

Figure 12-16 shows the signal at the correct gain but before we have achieved carrier or baud lock. The constellation points selected by the demodulator do not correspond with the ideal constellation points of the demodulator.

Initially, measure the power in the unlocked constellation cloud and set the gain appropriately to fill up the IQ constellation plot. As carrier and baud rate tracking routines lock, we use decision-directed feedback to manage gain control. The signal power estimates we generate using these open loop methods provide the seed values to the gain tracking control loops.

12.5.3 Signal Power Estimation

The power present in a complex signal described by a set of points $s(k)$ is given by S_{RMS}^2 where

$$S_{RMS}^2 = \frac{1}{N}\sum_{k=1}^{N} s(k)s^*(k) \tag{12.2}$$

where N is the total number of points, and $s^*(k)$ is the complex conjugate of the sample.

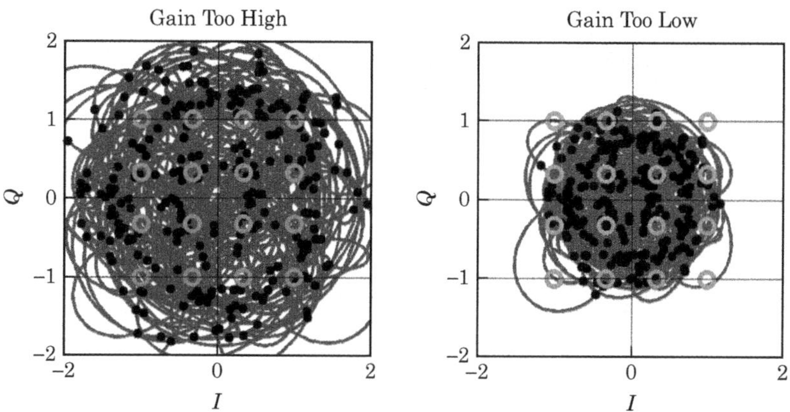

FIGURE 12-15 ■ The IQ diagram of the received signal when the gain is not appropriate to the normalized ideal constellation diagram. We first filter the signal, measure the power, and compare the measured power with the constellation power. Finally, we adjust the gain so that the two power levels match. These plots show the signal with incorrect gain and before carrier and baud lock has been achieved.

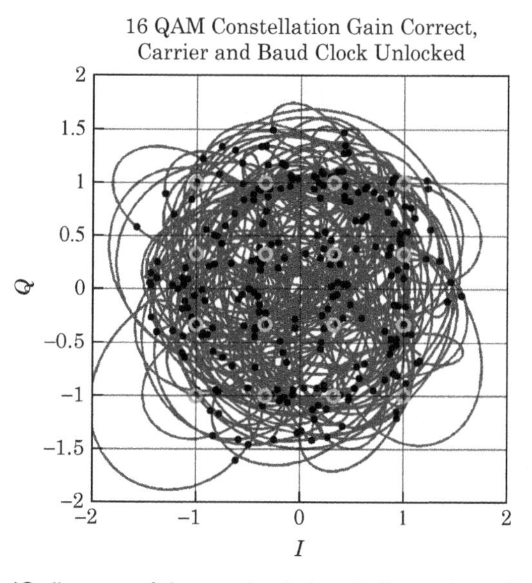

FIGURE 12-16 ■ The IQ diagram of the received signal after we've adjusted the gain of the signal to correspond with the ideal constellation power. However, we have not yet achieved carrier or baud lock. The constellation points selected by the demodulator (the dots) do not line up with the ideal constellation points (the circles). More corrections will be required before we can adequately demodulate this signal.

EXAMPLE

Power of a Sampled Signal

Find the power present in the following series:

$$0.333 + j0.333, 0.333 - j0.333, -0.333 + j0.333, -0.333 - j0.333,$$
$$1.000 + j0.333, 1.000 - j0.333, -1.000 + j0.333, -1.000 - j0.333,$$
$$0.333 + j1.000, 0.333 - j1.000, -0.333 + j1.000, -0.333 - j1.000,$$
$$1.000 + j1.000, 1.000 - j1.000, -1.000 + j1.000, -1.000 - j1.000$$

This is the constellation cloud for a 16QAM signal.

Solution

Equation (A.2) produces

$$\begin{aligned} S_{RMS}^2 &= \frac{1}{N}\sum_{k=1}^{N} s(k)s^*(k) \\ &= \frac{1}{16}\sum_{k=1}^{16} \begin{bmatrix} (0.333 + j0.333)\,(0.333 - j0.333) + \\ (0.333 - j0.333)\,(0.333 + j0.333) + \\ (-0.333 + j0.333)\,(-0.333 - j0.333) + \\ \cdots \\ (1.000 - j1.000)\,(1.000 + j1.000) + \\ (-1.000 + j1.000)\,(-1.000 - j1.000) + \\ (-1.000 - j1.000)\,(-1.000 + j1.000) \end{bmatrix} \\ &= 1.111 \end{aligned} \tag{12.3}$$

12.5.4 The Effects of Oscillator Accuracy in a Sampled System

The reference oscillators of the transmitter and receiver are never exactly synchronized. In the receiver, we must measure features in the yet unsynchronized signal to estimate both the carrier frequency and baud rate as seen by the receiver. The particular method that will be most successful depends on amount of frequency difference between the transmitter and receiver, the baud rate, and the carrier frequency. After the initial estimation, we will switch to decision-directed tracking.

An example is useful to describe the effects of oscillator accuracy on our system. We will examine a QPSK system operating at 2.4 GHz and 12 MBaud. The oscillators in the transmitter and receiver are ±100 ppm devices. In the worst case, the transmitter's oscillator will be 100 ppm above the nominal frequency, whereas the receiver is 100 ppm below the nominal frequency (or vice versa). At an RF of 2.4 GHz, a ±100 ppm difference is ±240 kHz, while the ±100 ppm difference at the baud rate of 12 MBaud is ±1.2 kHz. Under the worst-case conditions, the receiver will see a signal that is 480 kHz too high with a baud rate of 12.0024 MBaud (2.4 kBaud too fast). Figure 12-17 shows the system.

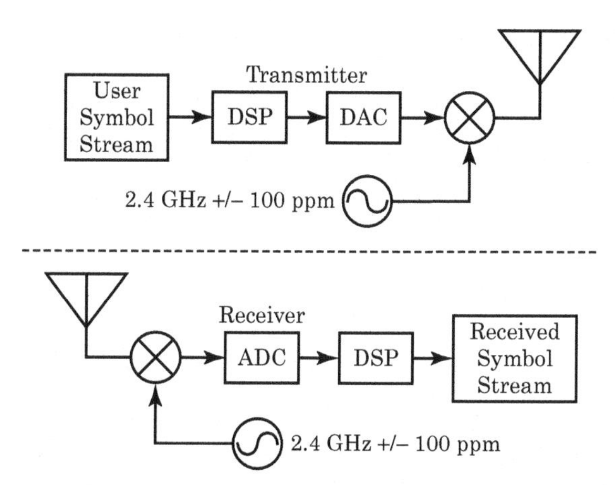

FIGURE 12-17 ■ Our example system uses a QPSK system operating at 2.4 GHz and 12 MBaud. The transmitter and receiver oscillator accuracies are both ±100 ppm. We'll assume one of the worst-case conditions where the transmit oscillator is 100 ppm too high, while the receive oscillator is 100 ppm too low. The DSP sees a signal that is 480 kHz too high in center frequency and 2.4 kBaud too fast.

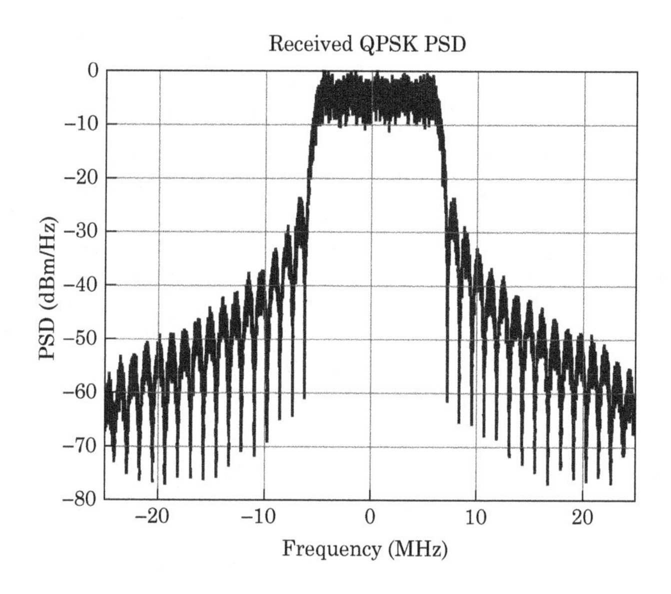

FIGURE 12-18 ■ The spectrum of our example received signal after it has been downconverted and digitized by the receiver. The received signal is centered at 480 kHz, and the baud rate is 12.0024 MBaud.

We would normally digitize this signal at slightly greater than 2 samples/symbol (e.g., at 25M samples/second), but for this example we will digitize the IF at approximately 4 samples/symbol or 50M samples/second. Figure 12-18 shows the spectrum of the received, digitized signal.

12.6 | ESTIMATION OF CARRIER FREQUENCY

Often, we begin our carrier-tracking journey by estimating the offset frequency of the received signal. We then use the estimate to seed a tracking loop.

12.6.1 Average Phase Advance per Sample

A powerful carrier estimation technique is to examine the accumulated phase of each sample. If the signal were a 480 kHz quiet carrier sampled at 50 MHz, the phase of each sample would increase by $\Delta\phi$ where

$$\begin{aligned}\Delta\phi &= \frac{2\pi/T}{fs}\\ &= \frac{2\pi f}{fs}\\ \Rightarrow f &= \frac{\Delta\phi}{2\pi} fs\end{aligned} \tag{12.4}$$

Applying our signal parameters to equation (A.4) produces

$$\begin{aligned}\Delta\phi &= \frac{2\pi f}{fs}\\ &= \frac{2\pi\ (480E3)}{(50E6)}\\ &= 0.06\,\text{rad}\\ &= 3.5^\circ\end{aligned} \tag{12.5}$$

The random data impressed on our signal will cause the signal's phase to change rapidly over short periods. For our 12 MBaud QPSK signal sampled at 50 MHz, the phase can change by 180° in four samples. However, if we observe the signal over many symbols, we expect the phase transitions due to the user's data to average to zero (because the data have been randomized). We also expect to observe a cumulative phase advance of 3.5° per sample due to the carrier offset. Figure 12-19 shows the accumulated phase of the QPSK

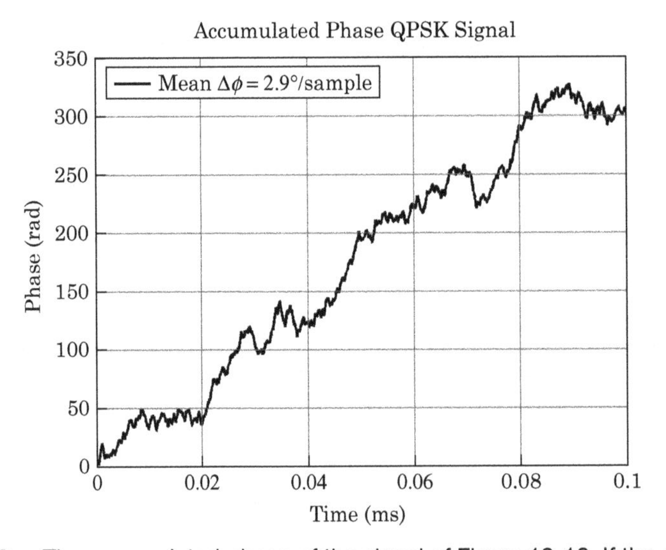

FIGURE 12-19 ■ The accumulated phase of the signal of Figure 12-18. If the data are random, we expect a long-term bias in the phase difference of each sample that is proportional to the center frequency of the signal. This signal's center frequency is 480 kHz, indicating an aggregate phase advance of 3.5°/sample. The measured phase advance of this signal is 2.9°/sample, measuring over 0.1 msec or 1,200 symbols. We expect this estimate to improve with a longer observation time.

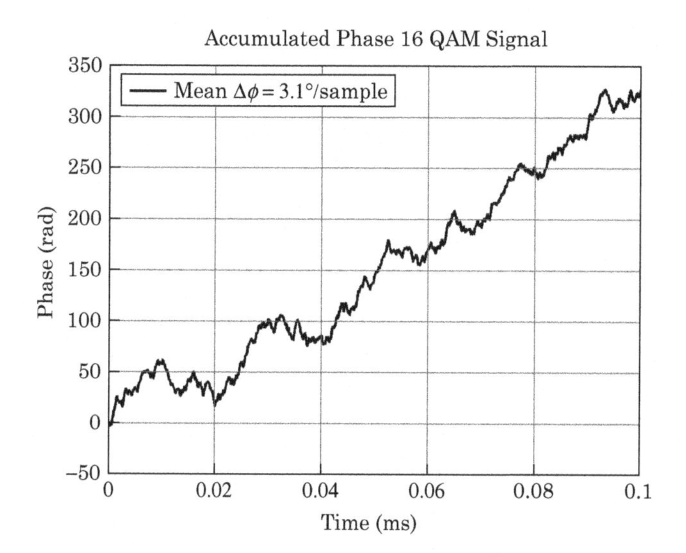

FIGURE 12-20 ■ The accumulated phase of a 16QAM signal whose center frequency is 480 kHz, exhibiting the same accumulated phase bias of Figure 12-19.

signal of Figure 12-18. For comparison, Figure 12-20 shows the accumulated phase of a 16 QAM signal. The long-term phase advance does not depend on the signal's modulation.

We use the information of Figure 12-19 and Figure 12-20 to drive a numerically controlled oscillator (NCO) control loop as shown in Figure 12-21. The action of this loop will drive the center frequency of the received signal to 0 hertz.

If the underlying user data were not random, it would be more difficult to average out the phase deviations due to data and observe the continuous per-sample phase advance useful to carrier acquisition.

12.6.2 Using Nonlinearities to Estimate Carrier Offset

A second common method of measuring the carrier offset uses nonlinear operations. For example, applying a second-order nonlinearity to a binary phase shift keying (BPSK) signal will generate a signal containing an artifact of the carrier offset. Applying a fourth-order nonlinearity to a QPSK or higher complexity QAM signal will produce another signal that contains a tone at four times the signal's carrier frequency.

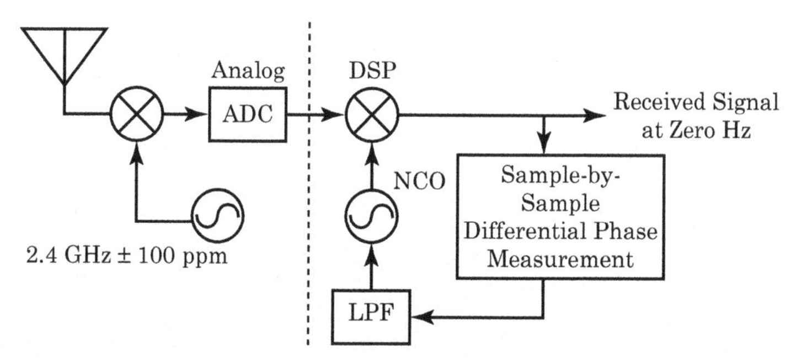

FIGURE 12-21 ■ The receiving system of Figure 12-17 modified to correct the center frequency of the received signal using the sample-to-sample differential phase driving a low-pass filter (LPF) and then an NCO. The action of the control loop is to bring the signal's center frequency to 0 hertz.

12.6.2.1 BPSK

The time-domain equation for BPSK with rectangular pulse shaping is

$$\begin{aligned} V_{BPSK}(t) &= A(t)\cos(\omega_{RF}t + n\pi); \quad n = 0, 1 \\ &= \pm A(t)\cos(\omega_{RF}t) \\ &= A(t)\cos(\omega_{RF}t) \; or \; A(t)\cos(\omega_{RF}t + \pi) \end{aligned} \tag{12.6}$$

Squaring the BPSK signal produces

$$\begin{aligned} V_{BPSK}^2(t) &= A^2(t)\cos^2(\omega_{RF}t + n\pi); \quad n = 0, 1 \\ &= \frac{A^2(t)}{2}[1 + \cos(2\omega_{RF}t + 2n\pi)]; \quad n = 0, 1 \\ &= \frac{A^2(t)}{2}[1 + \cos(2\omega_{RF}t)] \; or \; \frac{A^2(t)}{2}[1 + \cos(2\omega_{RF}t + 2\pi)] \\ &= \frac{A^2(t)}{2}[1 + \cos(2\omega_{RF}t)] \end{aligned} \tag{12.7}$$

The output of the squaring operation produces a signal at twice the RF. It is necessary only to lock to this frequency and divide the frequency by 2 to obtain an estimate of the RF carrier. Figure 12-22 graphically shows the effect of squaring a BPSK signal with a rectangular pulse-shaping filter. The squaring process destroys the information in the BPSK signal.

Applying a pulse-shaping filter to the BPSK signal does not eliminate the effect. Figure 12-23 shows the BPSK signal present at the output of the matched filter of Figure 12-13. The signal is off-tuned by 3,142 Hz and has been shaped by a $\sin(x)/x$ filter. Figure 12-24 shows the same signal after it has been passed through a second-order nonlinear process. The spike at twice the offset frequency is clearly visible.

The strength of the carrier component increases with increasing integration time.

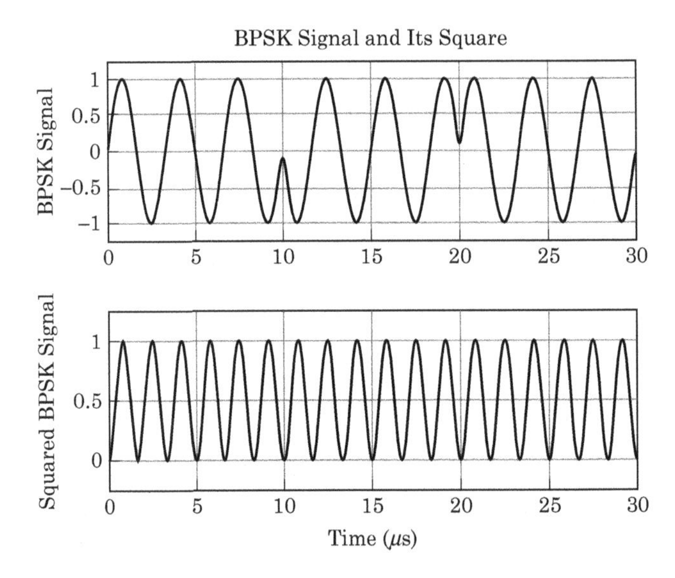

FIGURE 12-22 ■ The effect of squaring a BPSK signal with a rectangular pulse-shaping filter. The baud rate and user information are completely removed from the signal, but a strong carrier estimate is present. The symbol rate is 100 kHz.

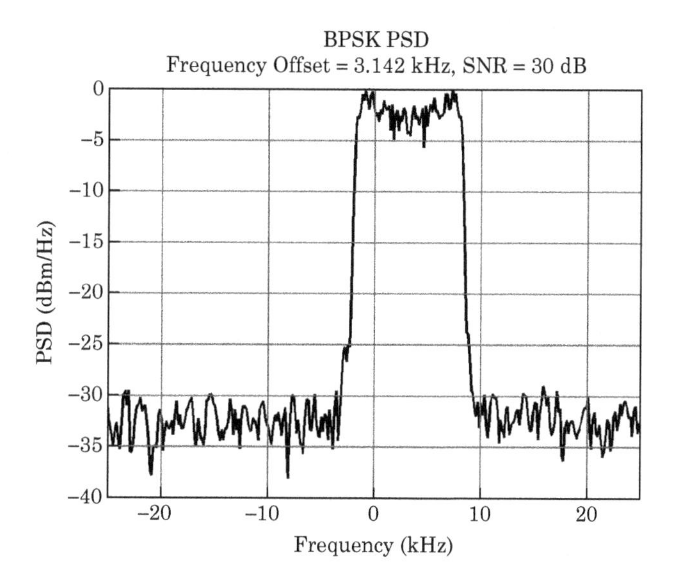

FIGURE 12-23 ■ A BPSK signal with 30 dB of SNR with a center frequency of 3142 Hz. This signal is present at the output of the matched filter of Figure 12-13 and thus exhibits a $\sin(x)/x$ pulse shape. The symbol rate is 10 kHz.

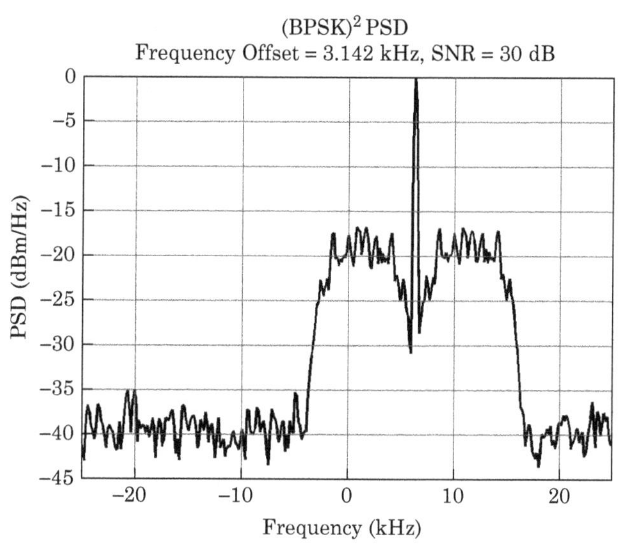

FIGURE 12-24 ■ The effect of squaring the BPSK signal of Figure 12-23. Note the pronounced signal spike at 3(3,142) = 6,284 Hz, which is twice the offset frequency of the signal of Figure 12-23. We would normally follow this process with a narrow filter and PLL to track the carrier of the BPSK signal.

12.6.2.2 Phasor Interpretation of Second-Order Spectra on BPSK

A BPSK signal containing nonbiased random data will not exhibit a carrier artifact. The signal consists of $-\cos(\omega t)$ for the $(-1 + j0)$ symbol and $\cos(\omega t)$ for the $(1 + j0)$ symbol. Since the data are nonbiased, the signal will contain equal parts of $-\cos(\omega t)$ and $\cos(\omega t)$ so the carrier cancels out.

Figure 12-25 shows the effect of squaring the constellation points of a BPSK signal. The two constellation points at $(-1 + j0) = -\cos(\omega t)$ and $(1 + j0) = \cos(\omega t)$ collapse to the single point $(1 + j0) = \cos(\omega t)$ and the squared signal will contain an artifact of the carrier.

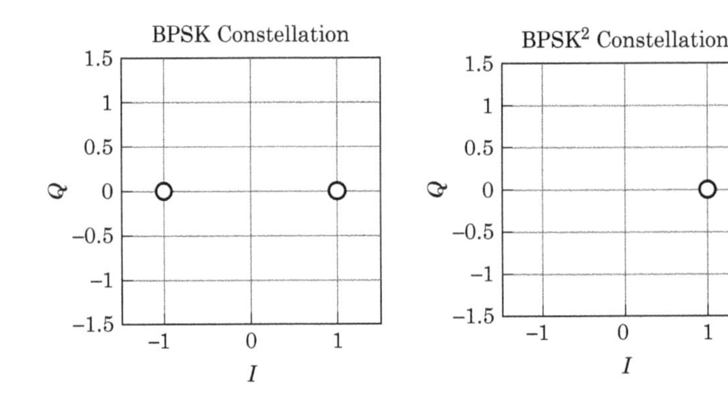

FIGURE 12-25 The constellation diagram of BPSK and BPSK^2. The two BPSK points at $(-1 + j0)$ and $(1 + j0)$ collapse to a single point $(1 + j0)$ in the BPSK^2 constellation, resulting in a carrier artifact.

12.6.2.3 QPSK

A QPSK signal, described in the time domain, can be expressed as

$$V_{QPSK}(t) = A(t)\cos\left(\omega_{RF}t + \frac{n\pi}{4}\right); \quad n = 0, 1, 2, 3 \tag{12.8}$$

Squaring the QPSK signal produces

$$\begin{aligned} V^2_{QPSK}(t) &= A^2(t)\cos^2\left(\omega_{RF}t + \frac{n\pi}{4}\right); \quad n = 0, 1, 2, 3 \\ &= \frac{A^2(t)}{2}\left[1 + \cos\left(2\omega_{RF}t + 2\frac{n\pi}{4}\right)\right]; \quad n = 0, 1, 2, 3 \\ &= \frac{A^2(t)}{2}[1 + \cos(2\omega_{RF}t)] \ or \ \frac{A^2(t)}{2}[1 + \cos(2\omega_{RF}t + \pi)] \end{aligned} \tag{12.9}$$

which is a BPSK signal. We've seen that squaring a BPSK signal will produce a 2× carrier component, so squaring $V^2_{QPSK}(t)$ will produce a coherent component at the fourth harmonic of the QPSK signal.

Figure 12-26 shows a QPSK signal with $\sin(x)/x$ pulse shaping and an offset frequency of 3,142 Hz. Figure 12-27 shows the spectrum of the signal after it has been

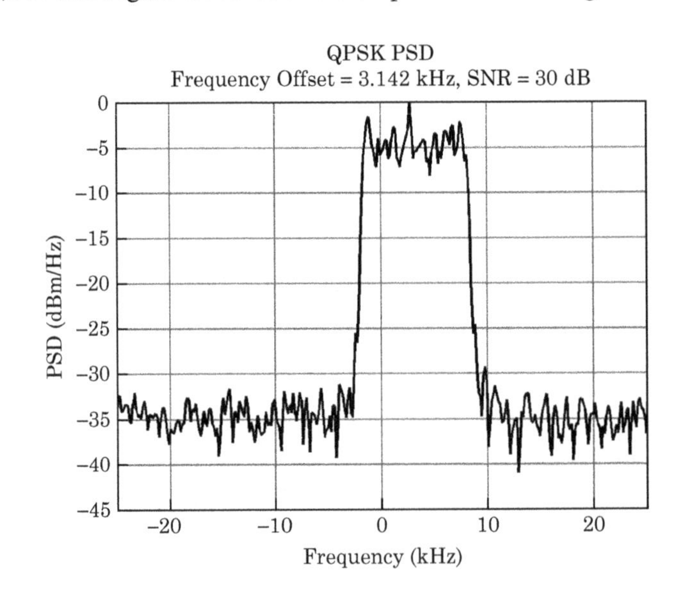

FIGURE 12-26 A QPSK signal with 30 dB of SNR with a center frequency of 3,142 Hz. This signal is present at the output of the matched filter of Figure 12-13 and thus exhibits a $\sin(x)/x$ pulse shape. The symbol rate is 10 kHz.

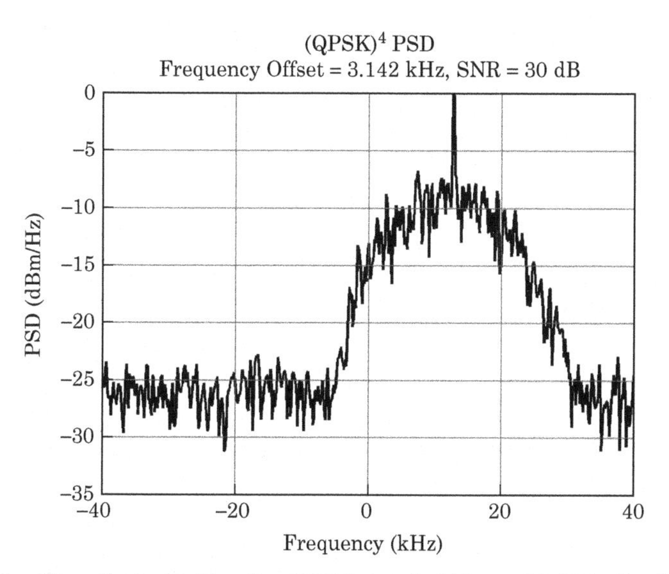

FIGURE 12-27 ■ The effect of taking the QPSK signal of Figure 12-26 to the fourth power. This process generates a quiet carrier at 4(3,142) = 12,568 Hz or four times the offset frequency of the signal of Figure 12-26. The SNR of the quiet carrier is not as great as that of the squared spectrum of the BPSK signal of Figure 12-24.

processed by a fourth-order nonlinearity. The spike at four times the offset frequency is clearly visible, but the SNR is decreased from the BPSK signal of Figure 12-24.

The SNR of the carrier component increases with increasing integration time.

12.6.2.4 Phasor Interpretation of Fourth-Order Spectra on QPSK

Like BPSK, a QPSK signal generated using nonbiased random data will exhibit no carrier component. The average of the four constellation points $[(0.707 + j0.707), (0.707 - j0.707), (-0.707 + j0.707)$ and $(-0.707 - j0.707)]$ is zero, indicating no carrier component. The squared constellation points are $[(0 + j), (0 - j), (0 - j)$ and $(0 + j)]$ and also exhibit zero mean. Hence, the squared QPSK signal contains no carrier. The fourth power of the constellation points $[(-1 + j0), (-1 + j0), (-1 + j0)$ and $(-1 + j0)]$ exhibit a nonzero mean, and hence the fourth power signal contains a carrier component. Figure 12-28 shows the constellation points of a QPSK signal after raising the signal to the second and fourth powers.

12.6.2.5 8PSK

The time-domain description of an 8PSK signal is

$$V_{8PSK}(t) = A(t)\cos\left(\omega_{RF}t + \frac{n\pi}{8}\right); \quad n = 0, 1, 2, 3, 4, 5, 6, 7 \tag{12.10}$$

Raising this equation to the eighth power (i.e., generating $V_{8PSK}{}^8(t)$) produces a signal containing an artifact of $8\omega_{RF}$. Figure 12-29 and Figure 12-30 show the spectrum of the signal and the signal raised to the eighth power.

12.6.2.6 Phasor Interpretation of Eighth-Order Spectra on 8PSK

Applying an eighth-order nonlinearity to an 8PSK signal collapses the eight constellation points into one single point, leaving a carrier component (see Figure 12-31).

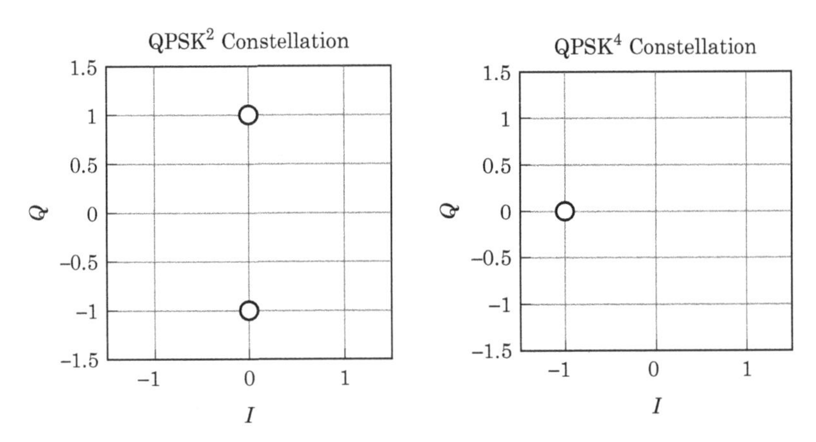

FIGURE 12-28 ■ The constellation diagram of QPSK2 and QPSK4. The four QPSK constellation points collapse into two points with zero mean when the signal is raised to the second power. The four constellation points converge to a single point upon raising the signal to the fourth power.

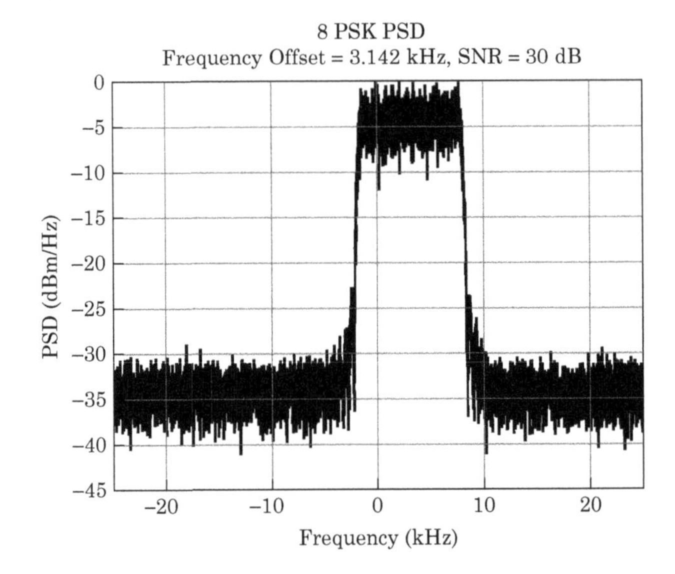

FIGURE 12-29 ■ An 8PSK signal with 30 dB of SNR with a center frequency of 3,142 Hz. This signal is present at the output of the matched filter of Figure 12-13 and thus exhibits a $\sin(x)/x$ pulse shape. The symbol rate is 10 kHz.

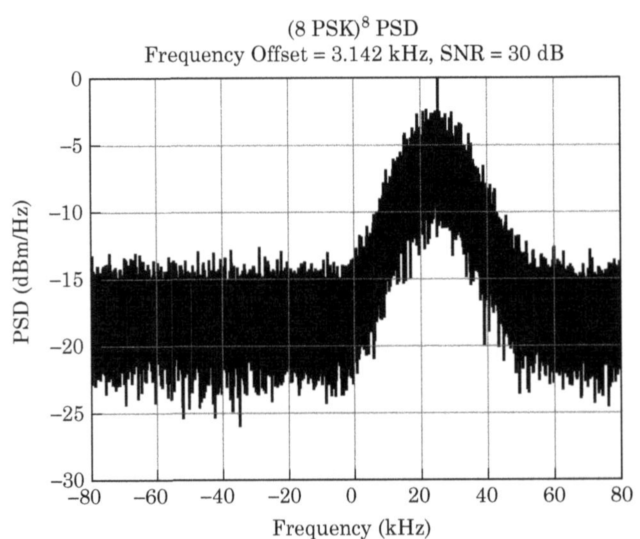

FIGURE 12-30 ■ The effect of raising the 8PSK signal of Figure 12-29 to the eighth power. This process generates a quiet carrier at $8(3{,}142) = 25{,}136$ Hz or eight times the offset frequency of the signal of Figure 12-29. The SNR of the quiet carrier is smaller than the carrier spectrums of Figure 12-24 and Figure 12-27.

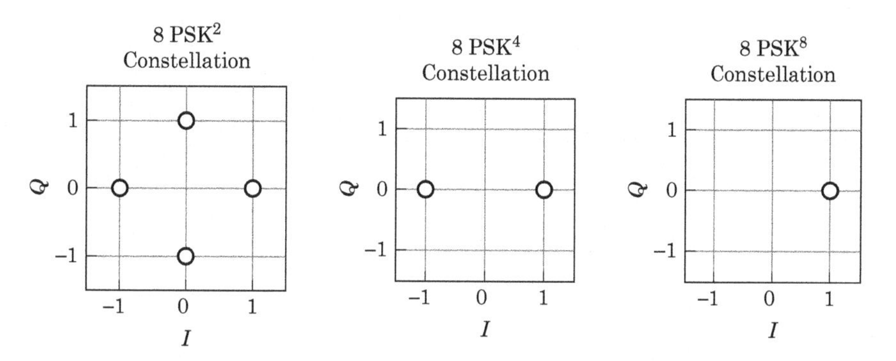

FIGURE 12-31 ■ The constellation diagram of 8PSK^2, 8PSK^4, and 8PSK^8. The eighth-order nonlinearity is the lowest-order nonlinearity that will collapse the 8PSK constellation points into a single point, leaving a carrier component.

12.6.3 Fourth Power Spectra

With the exceptions of BPSK and 8PSK, all QAM constellations exhibit a 90° rotationally invariant symmetry. A 90° rotation does not affect the appearance of the constellation. The practical effect of this observation is that most QAM constellations will reveal a carrier artifact when we raise the time-domain waveform to the fourth power. The frequency of the artifact will be four times the carrier frequency of the signal, and we must often observe a large number of samples to allow this artifact to build up. Figure 12-32 through Figure 12-34 shows the fourth power constellations of three QAM-type signals.

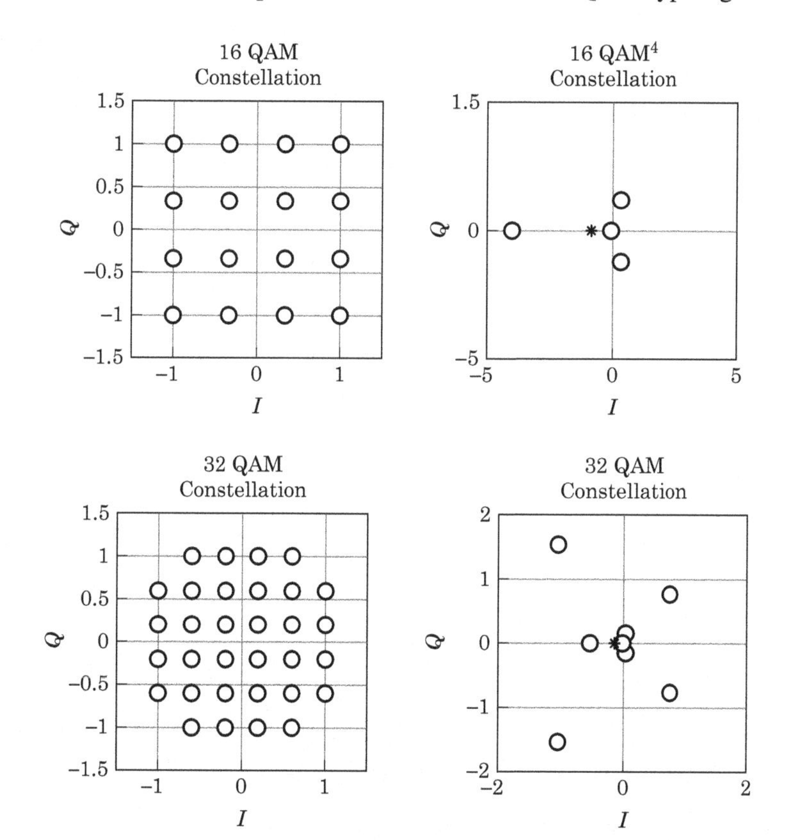

FIGURE 12-32 ■ A 16QAM constellation and the constellation of the signal after it has been raised to the fourth power. The mean of the fourth power constellation is nonzero, which corresponds to a carrier artifact.

FIGURE 12-33 ■ A 32QAM constellation and the constellation of the signal after it has been raised to the fourth power. The fourth power constellation exhibits a nonzero mean.

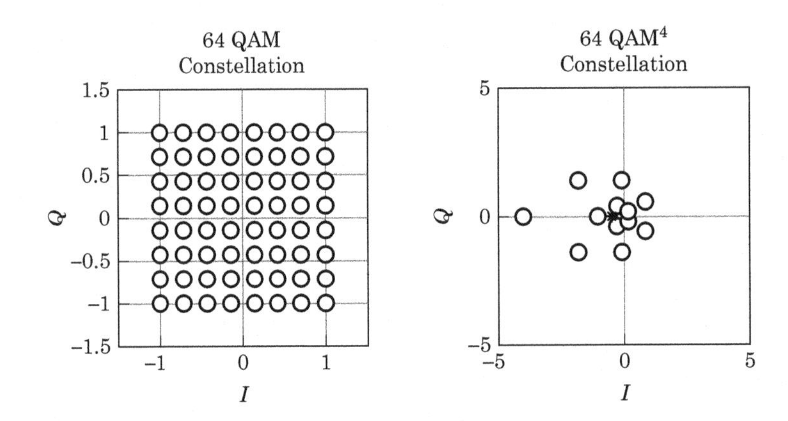

FIGURE 12-34 ■ A 64QAM constellation and the constellation of the signal after it has been raised to the fourth power. The nonzero mean of the fourth power constellation is evident.

12.6.3.1 Carrier Recovery Block Diagram

Figure 12-35 shows the block diagram of a BPSK, QPSK, or 8PSK carrier recovery system.

12.6.4 32QAM—A Comparison of Carrier Estimation Parameters

Figure 12-36 shows the spectrum of an off-tuned 32QAM signal. We seek an estimate of the signal's carrier frequency and phase.

Figure 12-37 shows the fourth-power spectrum of the received signal. The carrier artifact at −7,100 Hz is clearly visible. The SNR of this tone will increase as we increase the integration time.

Figure 12-38 shows the differential phase between each time-domain sample of the 32QAM signal of Figure 12-36. The average differential phase is −0.962° (or −0.0168 rad) and the sample rate is 672 kHz. Equation (A.4) then reveals the offset frequency to be

$$\begin{aligned} f &= \frac{\Delta\phi}{2\pi} f_S \\ &= \frac{-0.0168}{2\pi}(672{,}000) \\ &= -1{,}797\,\text{Hz} \end{aligned} \tag{12.11}$$

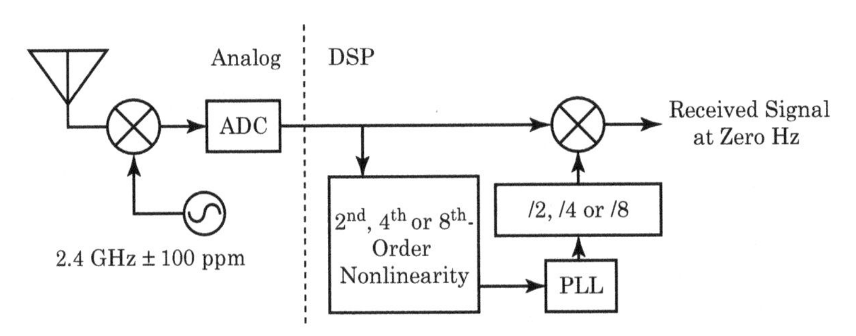

FIGURE 12-35 ■ The block diagram of a system using a second-, fourth-, or eighth-order nonlinearity to generate the carrier frequency of a BPSK, QPSK, and 8PSK signal, respectively. The PLL locks to a signal at 2×, 4×, or 8× the carrier of the input signal. We use the fourth-order nonlinearity for most constellations above 8PSK.

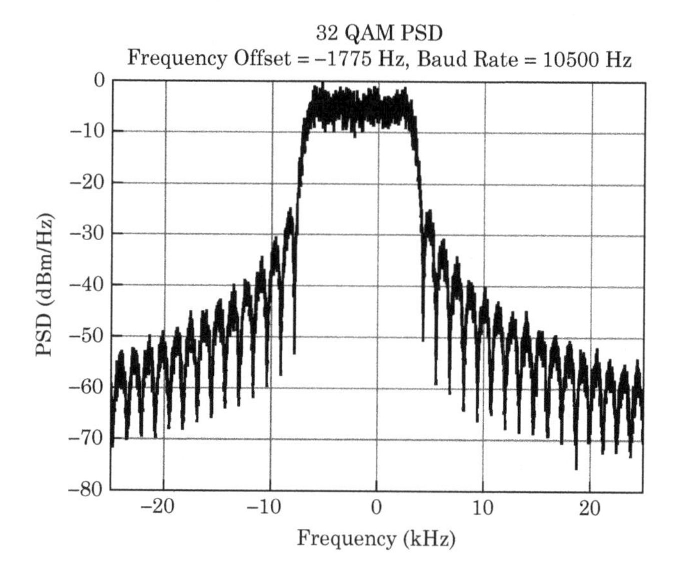

FIGURE 12-36 The spectrum of a frequency-offset 32QAM signal. We would like to measure the frequency offset of this signal.

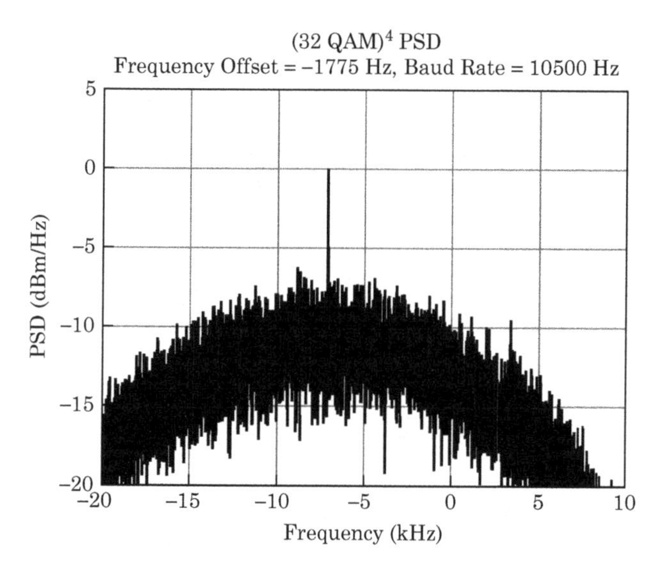

FIGURE 12-37 The spectrum of a frequency-offset 32QAM signal raised to the fourth power. The spectrum shows a carrier artifact at −7.1 kHz, which is four times the true carrier frequency of −1,775 Hz.

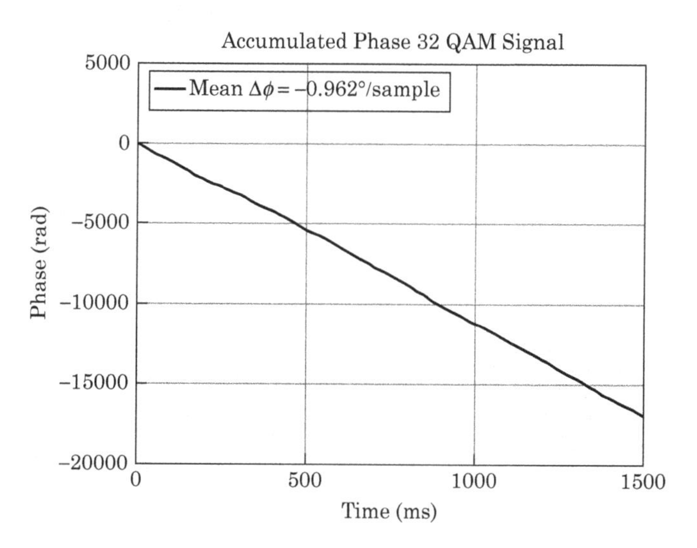

FIGURE 12-38 The differential phase between each sample of the 32QAM signal of Figure 12-36.

The measured offset frequency is $-1{,}797$ Hz, which compares favorably with the true offset frequency of $-1{,}775$ Hz.

12.7 | ESTIMATION OF BAUD RATE

The most common method of estimating the baud rate of a signal is to filter the signal with sufficient bandwidth to generate amplitude modulation at the baud rate and then AM detect. We set the filter's cutoff frequency to something in the neighborhood of the signal's baud rate. We seek to reduce the amplitude of the signal in the neighborhood of signal transitions to produce a strong baud rate artifact.

12.7.1 Filter, then AM Detect—BPSK

We will illustrate this technique using the BPSK signal of Figure 12-22. This signal was generated using a rectangular baseband filter, so it exhibits a large excess bandwidth. Figure 12-39 shows the effect of filtering on the BPSK signal and then performing the amplitude detection. The filter attenuates the sharp edges of the rectangularly windowed BPSK and exposes a baud clock artifact.

The effect of Figure 12-39 is strongest when the signal has a large excess bandwidth (i.e., when the signal exhibits an abrupt change at the baud boundaries). The power of the baud rate artifact is reduced with decreasing excess bandwidth.

We envelope (or AM) detect the signal of Figure 12-39 to generate the baud clock (see Figure 12-40).

For comparison, Figure 12-41 shows the baud tone recovered from the 32QAM signal of Figure 12-36 via filtering and then AM detection.

12.7.2 Envelope (AM) Detection

The strength of the baud tone generated by the nonlinearity depends on the pulse-shaping filter, the channel characteristics, the constellation, and the particular nonlinearity used to

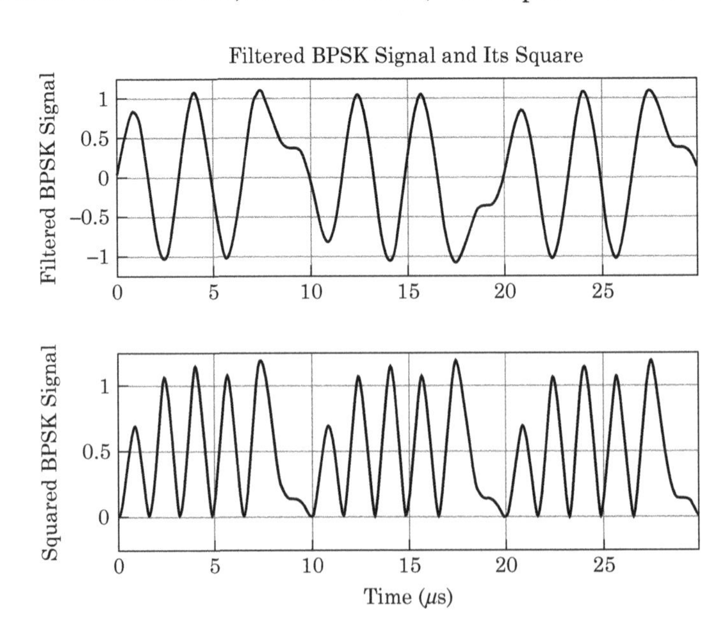

FIGURE 12-39 The effect of filtering and then squaring the BPSK signal of Figure 12-22. Squaring a signal performs a crude AM detection. The resulting waveform exhibits artifacts of both the carrier frequency and of the baud clock. The symbol rate is 100 kHz, and the LPF's 3 dB point is approximately 400 kHz.

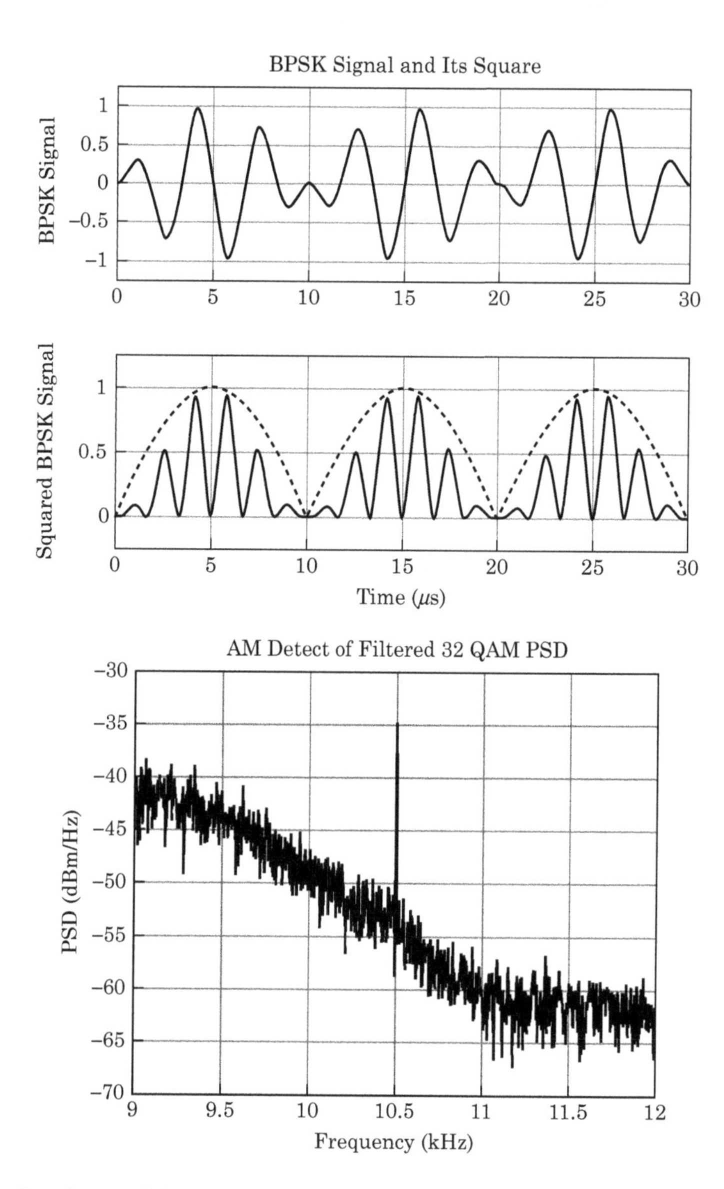

FIGURE 12-40 The effect of squaring a BPSK signal with a nonrectangular pulse-shaping filter. The squaring process removes the user's data, but the outline of the each symbol remains, allowing the estimation of the baud clock. The symbol rate is 100 kHz.

FIGURE 12-41 The result of filtering and then AM detecting the 32QAM signal of Figure 12-36. The 10.5 kHz baud spike is clearly visible.

generate the baud tone. We will now examine the Fourier transform of the envelope of a general QAM signal. Our description of a QAM signal is

$$\begin{aligned} s(t) &= \mathrm{Re}\left\{ \sum_{n=-\infty}^{\infty} \left[(x_n + jy_n)\, g\,(t - nT)\, e^{j\omega_0 t} \right] \right\} \\ &= \mathrm{Re}\{\hat{s}(t)\} \end{aligned} \tag{12.12}$$

where

$\hat{s}(t) =$ the complex representation of the general QAM signal
$x_n + jy_n =$ the n-th transmitted symbol (or constellation point)
$g(t) =$ the pulse shaping filter
$T =$ the baud period
$\omega_0 =$ the radian carrier frequency

If we allow $z_n = x_n + jy_n$ (for brevity's sake in writing the equations), we describe the envelope of a generic QAM signal as

$$
\begin{aligned}
\hat{s}(t) \cdot \hat{s}'(t) &= \left\{ \sum_{n=-\infty}^{\infty} \left[z_n g\left(t - nT\right) e^{j\omega_0 t} \right] \right\} \left\{ \sum_{m=-\infty}^{\infty} \left[\hat{z}_m g\left(t - mT\right) e^{-j\omega_0 t} \right] \right\} \\
&= \left\{ \sum_{n=-\infty}^{\infty} z_n g(t - nT) e^{j\omega_0 t} \right\} \left\{ \sum_{m=-\infty}^{\infty} \hat{z}_m g(t - mT) e^{-j\omega_0 t} \right\} \\
&= \sum_{n=-\infty}^{\infty} z_n \hat{z}_n g^2(t - nT) + \sum_{n,m=-\infty|(n \neq m)}^{\infty} [z_n g(t - nT)][\hat{z}_m g\left(t - mT\right)] \\
&= \sum_{n=-\infty}^{\infty} \left(x_n^2 + y_n^2\right) g^2(t - nT) + \text{ cross terms}
\end{aligned}
\tag{12.13}
$$

where $\hat{s}'(t)$ is the complex conjugate of $\hat{s}(t)$. Let's look at the Fourier transform of

$$
\sum_{n=-\infty}^{\infty} \left(x_n^2 + y_n^2\right) g^2\left(t - nT\right) \tag{12.14}
$$

Equation (A.14) is the convolution of the square of the amplitude of our symbol stream, $(x_n^2 + jy_n^2)$, with the square of the impulse response of our pulse-shaping filter, $[g^2(t - nT)]$. The symbol stream has zero mean and changes at the baud rate of the QAM signal, so it possesses large spectral components at the baud rate and the harmonics of the baud rate. The Fourier transform of $g^2(t)$ is the convolution of the Fourier transform of our pulse shaping filter, $g(t)$. If the pulse shape produces a total spectral bandwidth B, then the squared response will have a total spectral bandwidth of $2B$.

The Fourier transform of the complex envelope will be the product of the Fourier transforms of $(x_n^2 + jy_n^2)$ and $g^2(t)$. So the Fourier transform of the complex envelope will exhibit spectral components at the baud frequencies and all of its integer multiples. The magnitude of these components will be modified by the Fourier transform of the squared pulse-shaping filter. In the case of the Nyquist pulse-shaping filters discussed in Chapter 2, the bandwidth of $g(t)$ is larger than the baud rate frequency so the bandwidth of $g^2(t)$ is larger than twice the baud rate.

This effect causes spectral lines at $+/-$ the baud rate frequency. We use these spectral components to achieve an initial phase lock with the baud rate of the transmitted signal. The signals and their associated spectra are shown in Figure 12-42.

The *noise* level in Figure 12-42 arises from the *cross terms* of equation (A.13). These components form a noise-like spectrum with no notable coherent spectral components if the complex symbols (z_n's) are zero mean and the symbol stream is random.

The amplitude baud rate discrimination degrades and may fail on perfect, constant modulus signals (e.g., frequency shift keying [FSK] or carefully filtered staggered quadrature phase shift keying [SQPSK]); however, it may perform well in multipath scenarios.

Other nonlinear operators will also bring out a baud rate artifact. These operators include squaring the signal, performing frequency discrimination, and performing a delay-conjugate-multiply operation. Figure 12-43 shows the baud recovery of a typical demodulator.

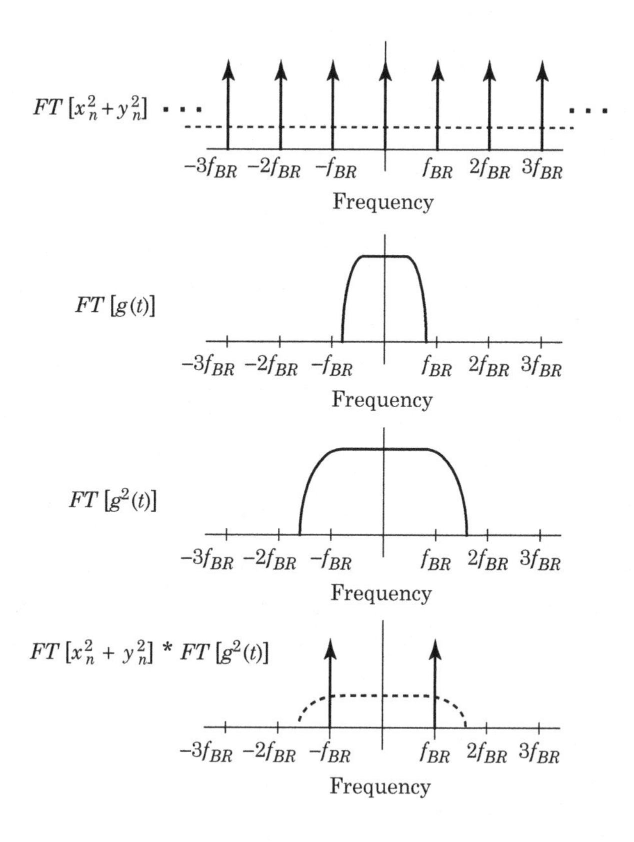

FIGURE 12-42 The development of Equation (12.14) in graphical format.

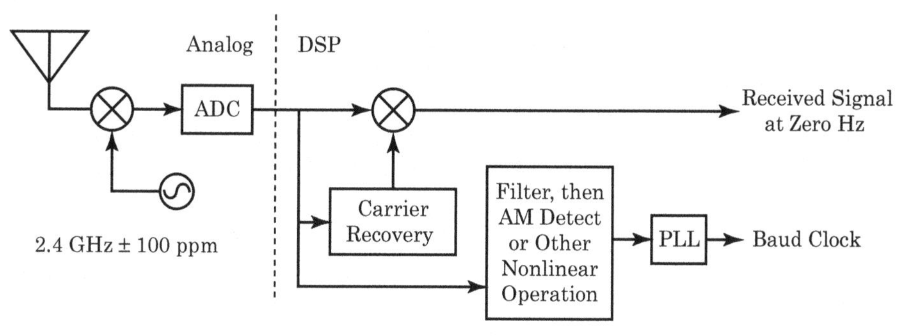

FIGURE 12-43 The recovery of the baud timing requires that we filter the signal and then perform an AM detection. We use a PLL to filter the clock component.

12.7.3 The Requirement for Excess Bandwidth

The strength of the baud rate tone we are able to generate is a strong function of the shape of the pulse-shaping filter we use in the transmitter and of the matched filter we use in the receiver. A large roll-off factor of the RCC filter (i.e., the β of equation (A.1)) will allow more bandwidth through the channel and will make the recovery of the carrier easier.

12.8 CONSTELLATION IMPAIRMENTS

There is much diagnostic information present in the constellation plot of a received signal. For this discussion, we will examine 16QAM constellations, but the underlying concept

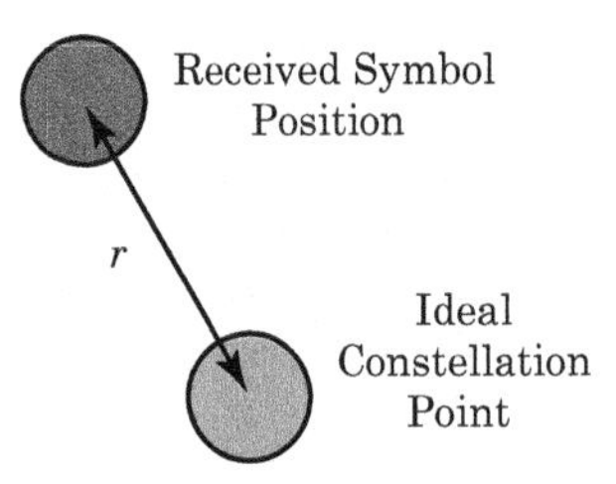

FIGURE 12-44 ■ Cluster variance is a measure of how accurately the received, decoded constellation points align with the ideal constellation points.

applies to most constellations. However, more complex the modulation schemes are less tolerant to noise, nonlinear behavior, and misalignment in either the modulator or demodulator.

12.8.1 Cluster Variance

Cluster variance is the average deviation of the received constellation points from the ideal constellation points. When the receiver is perfectly synchronized to the transmitter, cluster variance is a measure of the average SNR of the input signal.

To compute cluster variance, we examine each constellation point individually. The every symbol decision the receiver makes, we examine the vector difference between the ideal constellation point and the measured detection point (see Figure 12-44). For each demodulated data point, we determine an error power that is the square of the distance between the input data point and the nearest ideal constellation point. We define cluster variance to be the average error power normalized by the ideal average input signal power, or

$$CV_{dB} = 10\log\left(\frac{\text{Average Error Power}}{\text{Average Ideal Signal Power}}\right) \tag{12.15}$$

The *average ideal signal power* is the average of the squares of the distances of the ideal constellation points from the center (0,0) of the constellation. Using Figure 12-44 as a guide, we can express equation (12.15) as

$$CV_{dB} = -10\log\left(\frac{(\overline{r})^2}{\text{P}}\right) \tag{12.16}$$

where P is the mean power in the constellation.

The average SNR is the reciprocal of the cluster variance. Table 12-2 shows the cluster variance requirements for various modulated signals.

12.8.2 Normal, Healthy Constellation

The symbol points are placed on a 4 × 4 rectangular grid. The cluster size (or "dot size") is small, which indicates low values of thermal noise, interference, nonlinearity, and ISI. The SNR in Figure 12-45 is 35 dB.

12.8.3 Degraded SNR Constellation

Figure 12-46 shows the 16QAM constellation with degraded SNR. The constellation points change from tight dots into "fuzzy balls." The SNR in Figure 12-46 is 17 dB.

TABLE 12-2 ■ Cluster variance ranges for particular modulation types. Cluster variance is similar to received SNR

Modulation Type	Low Signal Quality. CV is less than:	High Signal Quality. CV is greater than:
BPSK	7	10.5
QPSK	10	13.5
8PSK	15	19
16QAM	16.5	20.5
32QAM	19.5	23.5
64QAM	22.5	26.5
SQPSK	10	16.1
2LFSK	14	18.1
3LFSK	14	18.1
4LFSK	14	18.1

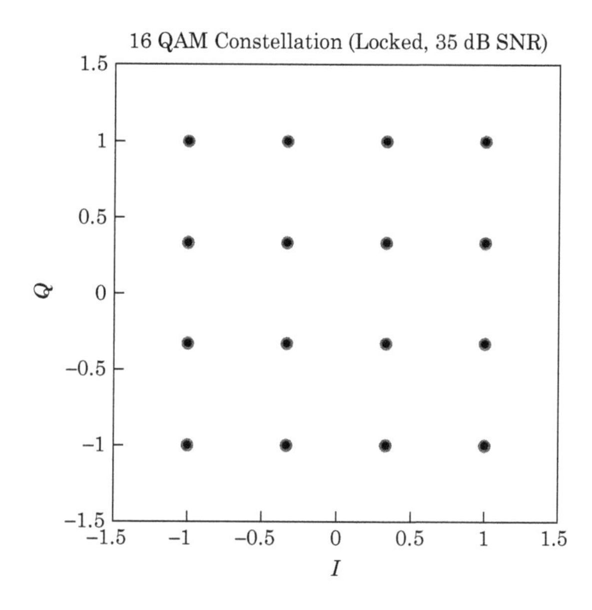

FIGURE 12-45 ■ A healthy 16QAM constellation with no anomalies. The SNR is 35 dB.

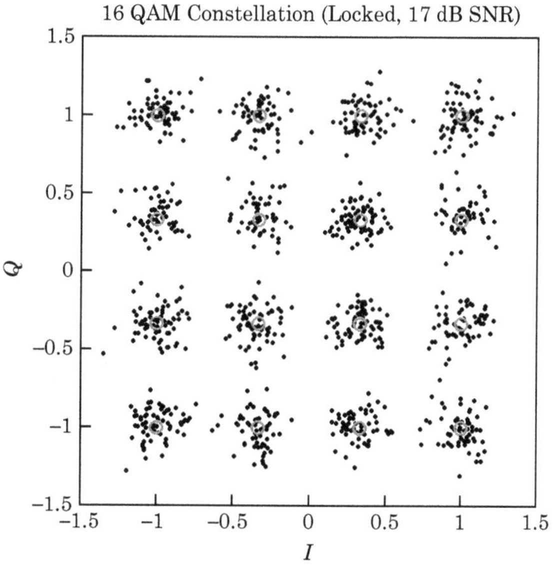

FIGURE 12-46 ■ A 16QAM constellation suffering from poor SNR. The SNR of this signal is 17 dB, which produces a symbol error rate of 2.4E-3.

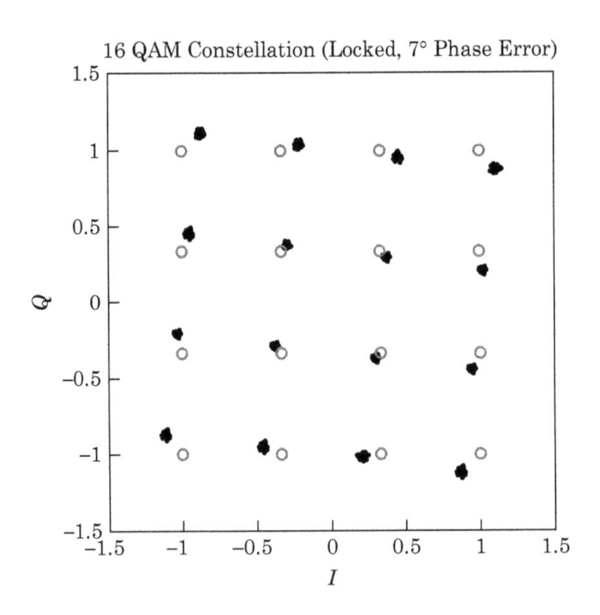

FIGURE 12-47 ■ A 16QAM constellation suffering from poor carrier lock. The carrier lock is offset by 7°, creating a rotation of the received constellation points about the nominal value.

12.8.4 Carrier Recovery Loop Phase Offset

The rotation of the constellation in Figure 12-47 is due to a phase offset present in the receiver's carrier recovery loop. The carrier offset in Figure 12-47 is 7°.

12.8.5 Carrier Recovery Loop Unlocked

Figure 12-48 shows the 16QAM constellation when the receiver has lost carrier lock but has achieved symbol rate lock.

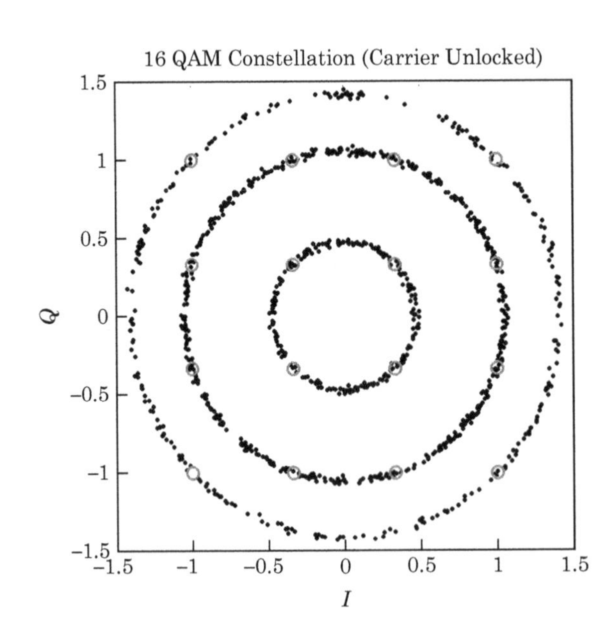

FIGURE 12-48 ■ The appearance of a 16QAM constellation when the receiver has not achieved carrier lock but has achieved symbol clock lock. The loss of carrier lock causes the constellation points to rotate about the center of the constellation.

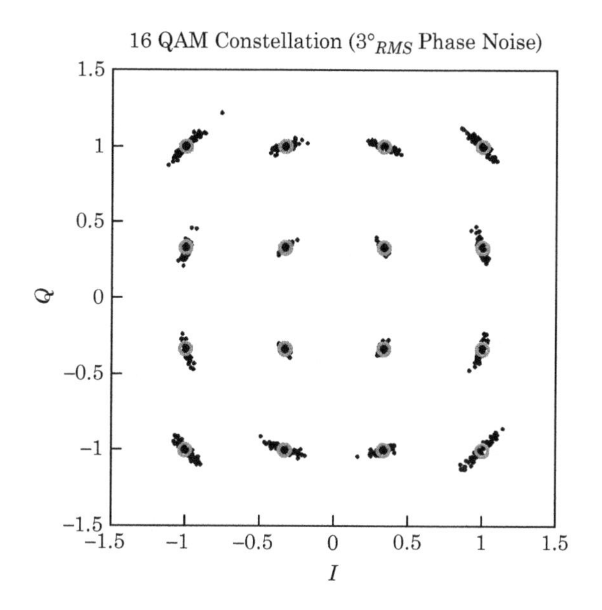

FIGURE 12-49 ■ A 16QAM constellation suffering from poor phase noise. The receiver phase noise is only 3° RMS but is sufficient to creating the characteristic arcing of the constellation points.

12.8.6 Poor Phase Noise

Figure 12-49 shows a constellation suffering from poor phase noise. The phase noise impressed upon the 16QAM constellation of Figure 12-49 is $3°_{RMS}$.

12.8.7 Coherent Carrier Interference

The received constellation shown as Figure 12-50 is suffering from a coherent interfering signal. The sinusoidal vector generates a doughnut shape caused by the rotating vector of the interfering signal.

The coherent interference causes a net decrease in the Euclidean distance between constellation points.

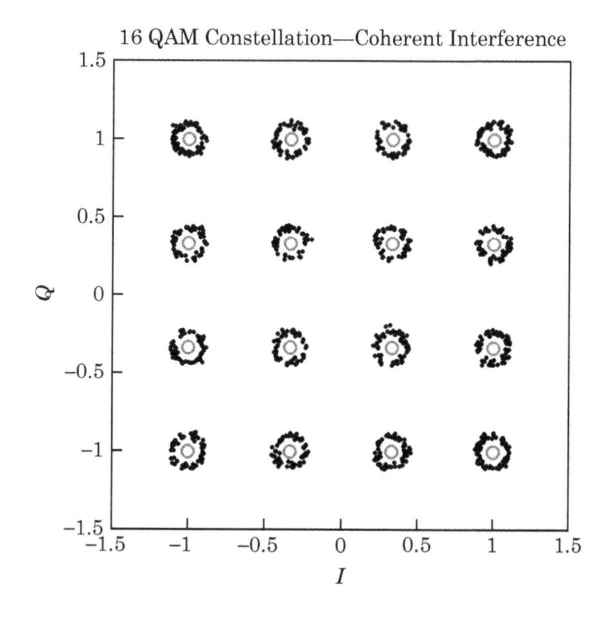

FIGURE 12-50 ■ A 16QAM constellation suffering from interference with a coherent, in-band carrier. The phasor addition of the coherent carrier with the received signal causes the constellation points to form a circle around the ideal constellation points.

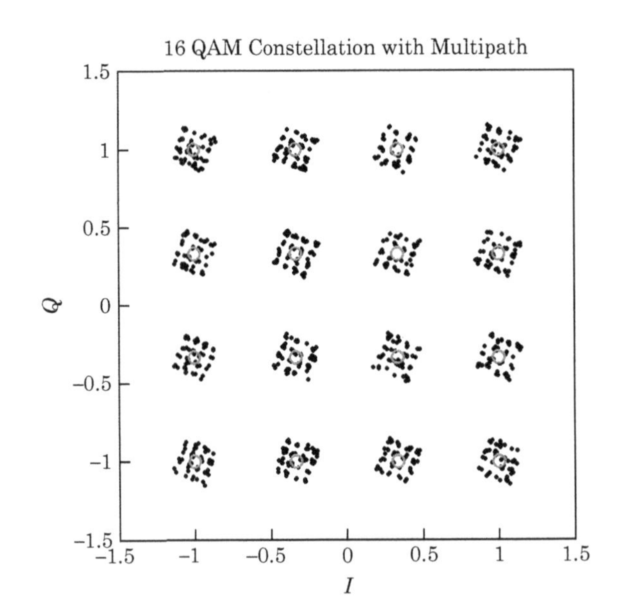

FIGURE 12-51 ■ A 16QAM constellation suffering from multipath. Each constellation point contains the entire constellation of the multipath signal. The multipath amplitude is one-tenth of the amplitude of the main signal. Each "tiny" constellation will rotate as the time difference of arrival between the main signal and the multipath signal changes.

12.8.8 Multipath

Multipath occurs when a delayed, attenuated version of the signal arrives at the antenna with the desired signal. The demodulator will lock to the larger signal, but the constellation will show the delayed, attenuated multipath signal along with the desired signal. Figure 12-51 shows the received constellation of a signal containing multipath.

12.8.9 IQ Phase Imbalance

IQ phase imbalance points to a modulator or demodulator circuit impairment. The phase between the I and Q channels of the quadrature modulator (at the transmitter) or the quadrature downconverter (at the receiver) are not exactly 90°. This effect causes the nominally square constellation to become rhombic. The phase imbalance between the I and Q channels of Figure 12-52 is 10°.

12.8.10 I/Q Amplitude Imbalance

IQ amplitude imbalance is another transmitter/receiver circuit problem. The gains of the I and Q channels of the quadrature modulator (at the transmitter) or the quadrature downconverter (at the receiver) are not identical. This effect causes the nominally square constellation to become rectangular. See figure 12.53.

12.8.11 Gain Compression

The outer constellation points are at a higher instantaneous level than the inner constellation points. As the system is driven into gain compression, the outer constellation points are compressed (or experience a different amount of gain) with respect to the inner constellation points. See figure 12.54 for an example constellation.

A system under compression is more vulnerable to noise because the Euclidean distances between points have been reduced.

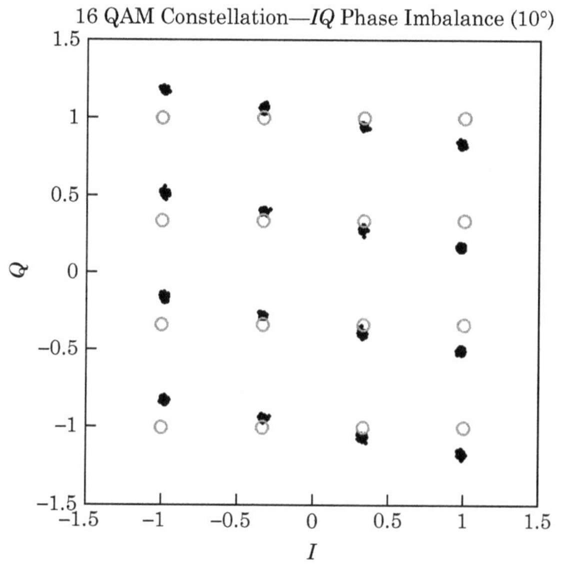

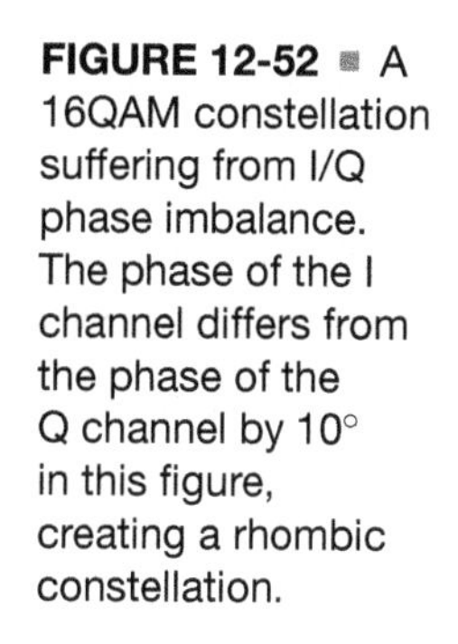

FIGURE 12-52 ■ A 16QAM constellation suffering from I/Q phase imbalance. The phase of the I channel differs from the phase of the Q channel by 10° in this figure, creating a rhombic constellation.

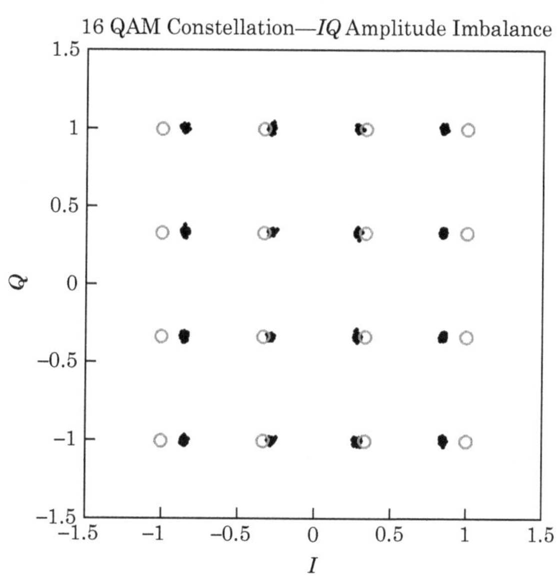

FIGURE 12-53 ■ A 16QAM constellation suffering from IQ amplitude imbalance. The amplitude of the I channel differs from the amplitude of the Q channel by 1.0 dB in this figure, creating a rectangular constellation.

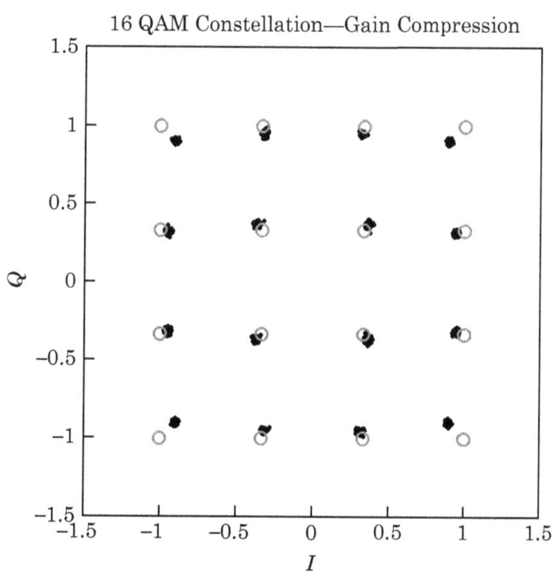

FIGURE 12-54 ■ A 16QAM constellation suffering from compression. The outer constellation points are experiencing less again than the inner points, producing the pincushion effect shown here. The outer corners move in while the inner corners move out.

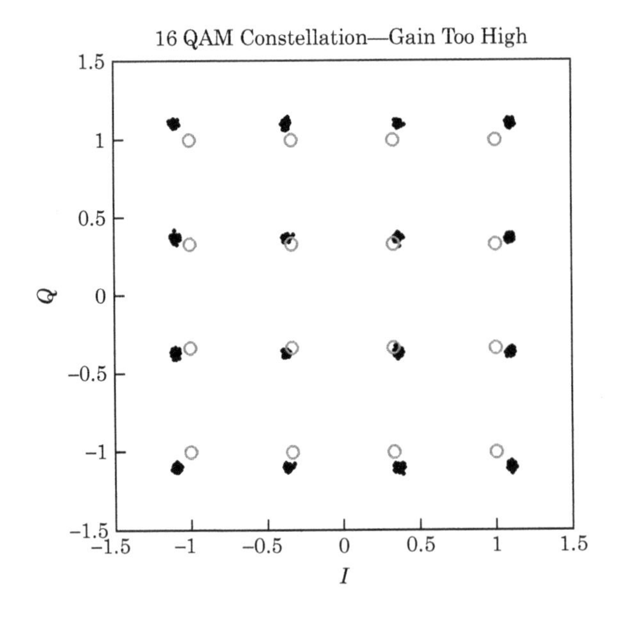

FIGURE 12-55 ■ A 16QAM constellation exhibiting too much gain.

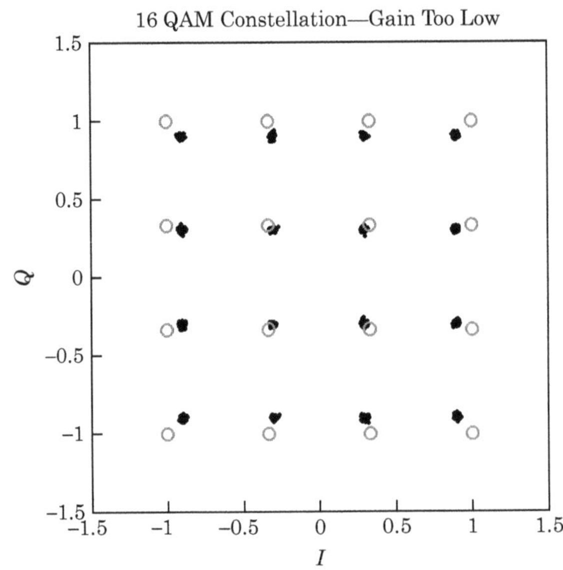

FIGURE 12-56 ■ A 16QAM constellation exhibiting too little gain.

12.8.12 Improper Amounts of Gain

Figures 12-55 and 12-56 show the results of an improperly set receiver gain. The constellation of figure 12-55 is the result of too much gain before the demodulator. Figure 12-56 shows the results of too little gain.

12.8.13 Amplitude Variation beyond Automatic Gain Control Bandwidth

Figure 12-57 shows a 16QAM constellation suffering from wild amplitude variation. The receiver automatic gain control (AGC) bandwidth is not fast enough to remove the variation. All the constellation clusters point linearly toward the center, indicating that the

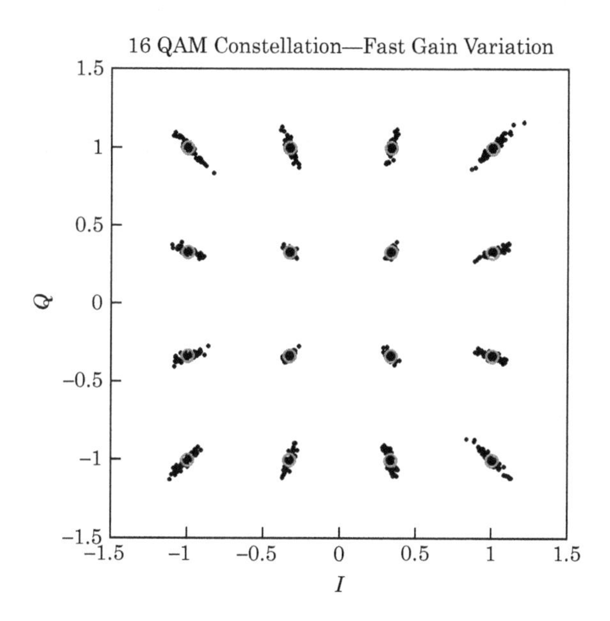

FIGURE 12-57 ■ A 16QAM constellation after passing through a system that is unable to compensate for channel's rapidly varying propagation loss.

demodulator has good baud clock phase lock. The outer groups are longer than the inner groups.

12.8.14 AM/PM Conversion

In amplitude modulation (AM)/phase modulation (PM) distortion, the phase shift experienced by a signal depends on its instantaneous amplitude. In Figure 12-58, the inner constellation points (at low amplitude) experience a $+5^\circ$ phase shift while the outer points (at a higher amplitude) experience a -5° phase shift. The result is the twisted, corkscrew constellation of Figure 12-58.

The outer constellation points rotate in one direction while the inner constellation points rotate in the opposite direction.

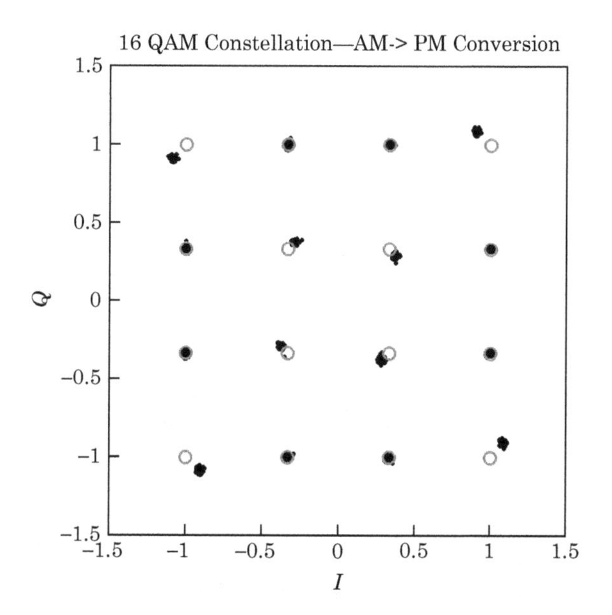

FIGURE 12-58 ■ A 16QAM constellation showing the effects of AM -> PM conversion. Constellation points of differing amplitudes experience different amount of phase shift.

12.9 BIBLIOGRAPHY

Bingham, John A.C., *The Theory and Practice of Modem Design*, Wiley-Interscience, 1988.

Frerking, Marvin E., *Digital Processing in Communications Receivers*, Von Nostrand Reinhold, 1994.

Harada, Hiroshi, and Prasad, Ramjee, *Simulation and Software Radios for Mobile Communications*, Artech House Press, 2002.

Mengali, Umberto, and D'Andrea, Aldo N., *Synchronization Techniques for Digital Receivers,* Pelnum Press, 1997.

Rappaport, Theodore S., *Wireless Communications Principles and Practice*, 2d ed., Prentice-Hall, 2002.

Snedecor, George W., and Cochran, William G., *Statistical Methods*, 7th ed., Iowa State University Press, 1980.

Webb, W.T., and Hanzo, L., *Modern Quadrature Amplitude Modulation*, IEEE Press, 1994.

12.10 PROBLEMS

1. **Multiple Choice: Estimation at the Receiver** Three items that we must first estimate and then lock to before successfully demodulating a received digital signal are:

   a. Signal peak factor, signal SNR, signal distortion

   b. Signal power, baud rate frequency and phase, carrier (or center) frequency and phase

   c. Multipath delay, oscillator tolerances, antenna size in wavelengths

   d. Transmitting antenna type (e.g., Yagi, aperature), receiver noise figure, receiver linearity

2. **Multiple Choice: Problematic Constellation** What is wrong with the constellation shown in Figure 12-59?

   a. The receiver has not properly adjusted its gain.

   b. The receiver is undergoing gain compression.

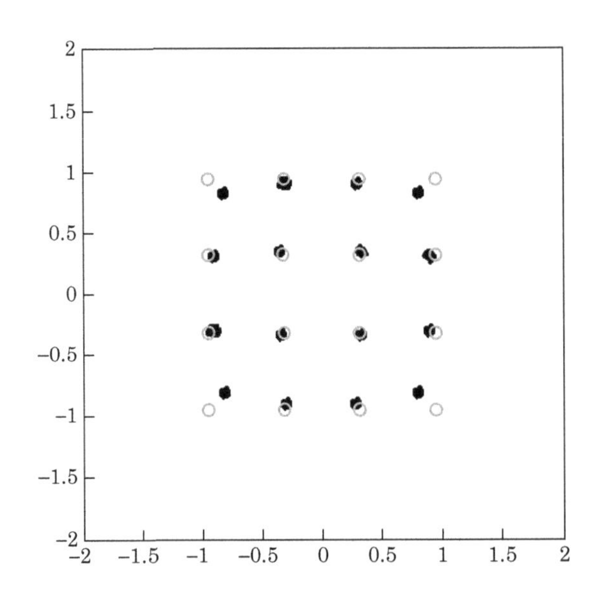

FIGURE 12-59 ■ Impared constellation.

c. Nothing.

d. The signal power is varying beyond the receiver's ability to compensate.

3. **Multiple Choice: 16QAM and QPSK** Figure 12-60 shows a QPSK and a 16QAM constellation. Which statement is false?

   a. 16QAM is more sensitive to phase noise than QPSK.
   b. 16QAM is more sensitive to gain compression than QPSK.
   c. 16QAM has a higher peak to average ratio than QPSK.
   d. All are true.

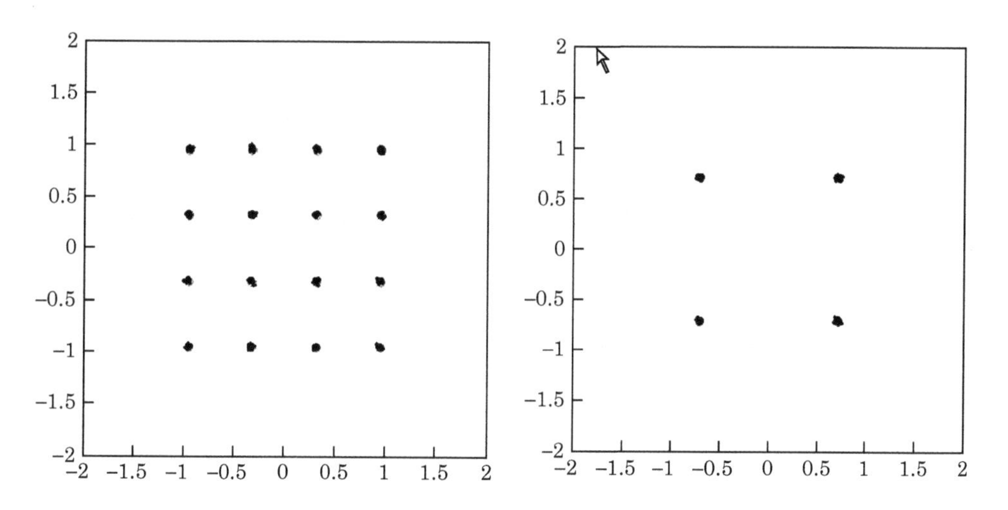

FIGURE 12-60 ■ Two constellations.

4. **Multiple Choice: Digitizer Headroom** A QPSK signal needs at least 18 dB of SNR to demodulate reliably. If we need 2 bits of headroom (to accommodate the peak/average ratio of the signal) and we need the 2 lower bits to always be driven by the noise floor of the receiver, how many bits do we need in the ADC?

   a. 2.7 bits
   b. 3 bits
   c. 7 bits
   d. 9 bits

Appendix

Outside of a dog, a book is man's best friend. Inside of a dog, it's too dark to read.

Groucho Marx

A.1 | MISCELLANEOUS TRIGONOMETRIC RELATIONSHIPS

This section lists several useful trigonometric identities.

Product of Sine and Cosines

$$\sin(\alpha)\sin(\beta) = \frac{1}{2}\cos(\alpha-\beta) - \frac{1}{2}\cos(\alpha+\beta) \tag{A.1}$$

$$\cos(\alpha)\cos(\beta) = \frac{1}{2}\cos(\alpha-\beta) + \frac{1}{2}\cos(\alpha+\beta) \tag{A.2}$$

$$\cos(\alpha)\sin(\beta) = \frac{1}{2}\sin(\alpha+\beta) - \frac{1}{2}\sin(\alpha-\beta) \tag{A.3}$$

$$\sin(\alpha)\cos(\beta) = \frac{1}{2}\sin(\alpha+\beta) + \frac{1}{2}\sin(\alpha-\beta) \tag{A.4}$$

Squares of Trigonometric Functions

$$\cos^2(\alpha) = \frac{1+\cos(2\alpha)}{2} \tag{A.5}$$

$$\sin^2(\alpha) = \frac{1-\cos(2\alpha)}{2} \tag{A.6}$$

Sum and Differences of Angles

$$\sin(\alpha \pm \beta) = \sin(\alpha)\cos(\beta) \pm \cos(\alpha)\sin(\beta) \tag{A.7}$$

$$\cos(\alpha \pm \beta) = \cos(\alpha)\cos(\beta) \mp \sin(\alpha)\sin(\beta) \tag{A.8}$$

$$\sin\left(\alpha \pm \frac{\pi}{2}\right) = \pm\cos(\alpha) \tag{A.9}$$

$$\cos\left(\alpha \pm \frac{\pi}{2}\right) = \mp\sin(\alpha) \tag{A.10}$$

A.2 EULER IDENTITIES

Various forms of Euler's equation include

$$e^{-j\omega t} = \cos(\omega t) - j\sin(\omega t) \quad e^{j\omega t} = \cos(\omega t) + j\sin(\omega t) \tag{A.11}$$

and, equivalently,

$$\cos(\omega t) = \frac{e^{j\omega t} + e^{-j\omega t}}{2} \quad \sin(\omega t) = \frac{e^{j\omega t} - e^{-j\omega t}}{2j} \tag{A.12}$$

A.3 LAW OF COSINES

Referring to Figure A-1, the relationship between the length of the sides and the angles of the triangle is

$$c^2 = a^2 + b^2 - 2ab\cos(\gamma) \tag{A.13}$$

$$\cos(\alpha) = \frac{b^2 + c^2 - a^2}{2bc} \quad \cos(\beta) = \frac{a^2 + c^2 - b^2}{2ac} \quad \cos(\gamma) = \frac{a^2 + b^2 - c^2}{2ab} \tag{A.14}$$

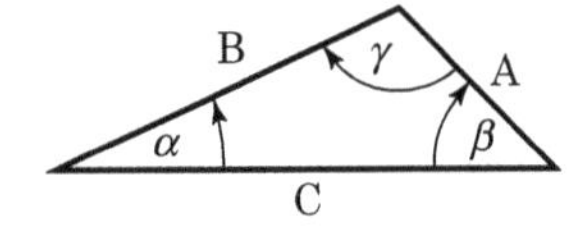

FIGURE A-1 ■ A graphical illustration of the law of cosines.

Selected Answers

Chapter 1

1. 10 db
3. The gain is

$$\begin{aligned} G_{p,cas,\text{dB}} &= G_{p,1,\text{dB}} + G_{p,2,\text{dB}} + G_{p,3,\text{dB}} + \cdots + G_{p,n,dB} \\ &= 10 + 15 + 20 \\ &= 45\text{ dB} \end{aligned}$$

5. The actual gain is −6 dB, which is less than the advertised gain of −5 dB, so the advertised claim is false.
7. 4 dB
9. 157.0 dB
11. 5.62
13. The voltage in the right channel will be 0.0316 times the magnitude of the voltage in the left channel.
15. **a.** In order: 0.191, 158, 0.100, 0.0794
    b. −6.2 dB
    c. In order (dBm): −87.2, −65.2, −75.2, −86.2
17. **a.** In order: 0.355, 15.849, 0.631, 0.20, 50.119, 0.398, 0.251
    b. 5.5 dB
    c. 3.55
    d. $-87.5\text{ dBm} = 1.78 \times 10^{-9}\text{ mW}$
21. **e**

Chapter 2

1. **a.** 1
   b. 1.53

5. a.

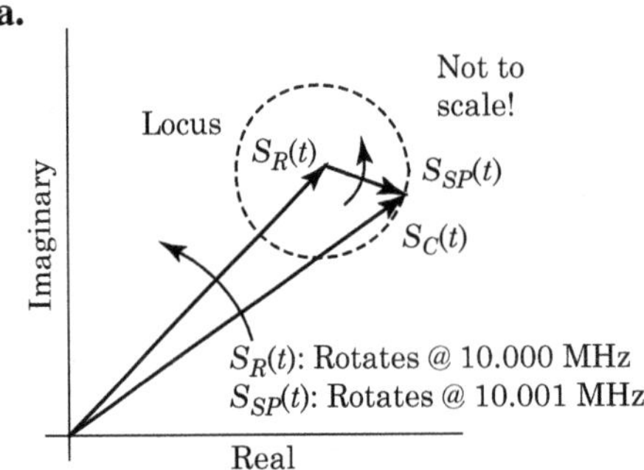

b. The maximum phase deviation is $2.3° = 40.1\text{E} - 3$ rad. The amplitude deviation is $\Delta A/2 = 0.080$.

c. 318.4 psec

Chapter 3

1. $D = 10^{\frac{A_{dB}-63.8}{34}}$

3. a. 2540 m
$\phi_0 = 0$

b. $\phi_1 = -\pi = -180°$
$\phi_2 = -115.2°$

c. $\tau_1 = 0.42\,\text{ns} \quad \tau_2 = 0.0\,\text{ns} \quad \tau_3 = 0.56\,\text{ns}$

Chapter 4

1. **a** or **b**

3. a. 38.8 dB

5. a. $A_{eff}(0°) = 7.16 \times 10^{-3}\ \text{m}^2$; $A_{eff}(45°) = 2.26 \times 10^{-7}\ \text{m}^2$

b. 53 dB

c. 56 dB

7. a. $f = 9.524\,\text{GHz}$

b. $G = 63.4\,\text{dBi}$

c. $G = 90.8\,\text{dBi}$

d. $T_{Mars} = 137\dot{\text{K}}$

e. If the antenna is polarized (either horizontal or vertical), then the antenna will see only the horizational or vertical energy of both the background radiation and of the planet Mars. The final answer will remain unchanged.

9. 1200-mile high satellite: $A = 4223\ \text{m}^2$; 22300-mile high satellite: $A = 464{,}000\ \text{m}^2$

Chapter 5

1. **d**

2. **c**

5.

Frequency (MHz)	LPF Attn (dB)	Amplifier Gain (dB)	BPF Attn (dB)	System Gain (dB)
4200	1.4	22	5.0	15.6
4400	2.4	22	5.0	14.6
3140	0.0	22	88.7 => 85	−63.0
5460	12.0	22	76.5	−66.5
5470	12.1	22	76.8	−66.9
3300	0.1	22	81.0	−59.1
3500	0.1	22	69.9	−48.0

7. a. $P_{out,\text{dBm}} = -32.03$ dBm
b. $P_{out,\text{dBm}} = -101.1$ dBm

8. a. 3.16
b. 4.67

11. a. 1.5 MHz
b. −105.6 dBm
c. −107.0 dB

13. a. Ultimate rolloff = 6 dB/octave (6 poles − 5 zeroes) = 6 dB/octave,
b. Ultimate rolloff = 6 dB/octave (7 poles − 6 zeroes) = 6 dB/octave,
c. Ultimate rolloff = 6 dB/octave (7 poles − 2 zeroes) = 30 dB/octave,
d. Ultimate rolloff = 6 dB/octave (3 poles − 1 zero) = 12 dB/octave.

15. a. The three shunt inductors and the series capacitor next to RL indicate that this is a highpass filter.
b. Since the stopband of a high-pass filter is at low frequencies, we have to examine the behavior of the filter for low frequencies. The element closest to the source is a shunt inductor, so this filter looks like a low impedance in the stopband.
c. At the load end of the first, the first element is a series capacitor. At low frequencies, a capacitor looks like a high impedance, so the filter presents a high impedance to the load in the stopband.

17. a. $f_{c,arith} = 700.2$ MHz. The difference between f_c and $f_{c,arith}$ is 0.2 MHz = 0.029%.
b. $f_{c,arith} = 700.9$ MHz. The difference between f_c and $f_{c,arith}$ is 0.9 MHz = 0.13%.
c. $f_{c,arith} = 703.5$ MHz. The difference between f_c and $f_{c,arith}$ is 3.5 MHz = 0.50%.
d. $f_{c,arith} = 782.8$ MHz. The difference between f_c and $f_{c,arith}$ is 82.8 MHz = 11.8%.

19. a. 2.89
b. 1.97

Chapter 6

1. $P_{Net\,Flow} = k\,(T_{Hot} - T_{Cold})\,B_n$

3. 3 dB

5. **a.** 27.1×10^{-15} W $= -105.7$ dBm
   b. 29.5%
   c. 46.7×10^{-15} W $= -103.3$ dBm
   d. 59.1%
7. Noise temperature: 827K; Noise figure: 5.9 dB
9. System gain: 18.8 dB; Noise figure: 7.2 dB; Noise temperature: 1220 K
11. **a.** 26.8 μV_{RMS}
   b. 9.65 μV_{RMS}
   c. 1.39%
   d. 5.5%
13. **e**
15. **c**, **a**, **b**
17. **a.** -104 dBm
   b. 61%
   c. 38%
19. 6 dB
21. **a**
23. **a.** -101.7 dBm
   b. -81.7 dBm
25. **a.** -99 dBm
   b. 73%
   c. 27%

Chapter 7

1. **a.** -199.2 dBm, -30.2 dBm, -185.8 dBm
   b. The possible frequencies present at the output of the system are: 0, 90, 180, 270, 360, 470, 560, 650, 740, 830, 920, 1010, 1100, 1190, 1280, 1370, 1480, 1570, 1660, 1750, 1840, 1930, 2020, 2110, 2200, 2310, 2400, 2490, 2580, 2670, 2760, 2850, 2940, 3030, 3120, 3210, 3320, 3410, 3500, 3590, 3680, 3770, 3860, 3950, 4040, 4150, 4240, 4330, 4420, 4510, 4600, 4690, 4780, 4870, 4960 and 5050 MHz.
5. **c**, **b**, **a**
7. **b**
9. **a** and **c**
11. $f_1 = 490$ MHz, $f_2 = 510$ MHz
   $f_A = 1960$ MHz $= 4f_1$
   $f_B = 1980$ MHz $= 3f_1 + f_2$
   $f_C = 2000$ MHz $= 2f_1 + 2f_2$
   $f_D = 2020$ MHz $= f_1 + 3f_2$
   $f_E = 2040$ MHz $= 4f_2$

16. **a.** $G_{\mathrm{dB}} = 17.0\,\mathrm{dB}$; $F_{cas} = 5.9\,\mathrm{dB}$; $ITOI_{cas} = -17.6\,\mathrm{dBm}$; $OTOI_{\mathrm{dBm}} = -0.6\,\mathrm{dBm}$; $ISOI_{cas} = -30.3\,\mathrm{dBm}$; $OSOI_{\mathrm{dBm}} = -13.3\,\mathrm{dBm}$

b.

	(1)	(2)	(3)	(4)	(5)
$G_{P,\mathrm{dB}}$	18	−7	−4	19	−9
Input $ISOI_{\mathrm{dBm}}$	−3	−26	29	−14	−16
Input $ITOI_{\mathrm{dBm}}$	2	−17	44	−5	−6

17. By ITOI: #3, #1, #2; BY OTOI: #3, #2, #1

19. 54.05 MHz and 176.2 MHz

22. **a**, **c**, and **e**

25.

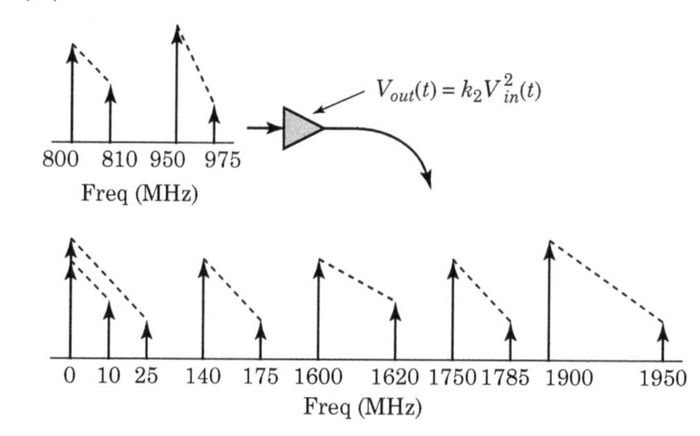

27. **a.** Component (e) contributes the most to the noise figure of the cascade because there is insufficient gain before the amplifier to overcome its internally generated noise.

b. Component (c), the mixer, is limiting the cascade's second-order intercept because the combination of excess gain before the mixer and the mixer's ISOI causes the mixer to develop nonlinear products at a small input power level.

32. **a.** 1, 2, 5, 4, 3

b. 5, 2, 3, 4, 1

c. $MDS_{\mathrm{dBm}} = -107\,\mathrm{dBm}$;
$SFDR_{\mathrm{dB}} = 62.9\,\mathrm{dB}$

33. $f_1 = 910\,\mathrm{MHz}$, $f_2 = 930\,\mathrm{MHz}$
$f_A = 2690\,\mathrm{MHz} = 5f_1 - 2f_2$
$f_B = 2710\,\mathrm{MHz} = 4f_1 - f_2$
$f_C = 2730\,\mathrm{MHz} = 3f_1$
$f_D = 2750\,\mathrm{MHz} = 2f_1 + f_2$
$f_E = 2770\,\mathrm{MHz} = f_1 + 2f_2$
$f_F = 2790\,\mathrm{MHz} = 3f_2$
$f_G = 2810\,\mathrm{MHz} = -f_1 + 4f_2$
$f_H = 2830\,\mathrm{MHz} = -2f_1 + 5f_2$
The highest order is seven.

Chapter 8

1. **a.** −83 dBm
   b. −56 dBm
   c. −75 dBm
   d. −62 dBm
3. $G_{cas,\mathrm{dB}} = 7\,\mathrm{dB}$; $F_{cas,\mathrm{dB}} = 3.0\,\mathrm{dB}$
5. **a.** $P_{out} = -7.16\,\mathrm{dBm}@70\,\mathrm{MHz}$
   b. $P_{out} = -47.11\,\mathrm{dBm}@49\,\mathrm{MHz}$
   c. $P_{out} = -34.41\,\mathrm{dBm}@70\,\mathrm{MHz}$
   d. $P_{out} = -80\,\mathrm{dBm}@70\,\mathrm{MHz}$
   e. $P_{out} = -27\,\mathrm{dBm}@974\,\mathrm{MHz}$
7. **a.** $LSLO = 11.25\,\mathrm{GHz}$
   $HSLO = 13.95\,\mathrm{GHz}$
   b. The HSLO produces a frequency inversion.
   c. $f_{RF,LSLO} = \{10.30 \text{ to } 9.50\,\mathrm{GHz}\}$ and $\{12.20 \text{ to } 13.00\,\mathrm{GHz}\}$;
   $f_{RF,HSLO} = \{13.00 \text{ to } 12.20\,\mathrm{GHz}\}$ and $\{14.90 \text{ to } 15.70\,\mathrm{GHz}\}$
   d. $f_{IF,LSLO} = \{0.95 \text{ to } 1.75\,\mathrm{GHz}\}$ and $\{23.45 \text{ to } 24.25\,\mathrm{GHz}\}$;
   $f_{IF,HSLO} = \{1.75 \text{ to } 0.95\,\mathrm{GHz}\}$ and $\{26.15 \text{ to } 26.95\,\mathrm{GHz}\}$
9. **a.** 2413.2 MHz
   b. No
11. **a.** 5500 MHz
13. **a.** $f_{LO} = 4200 \;\; or \;\; 5100\,\mathrm{MHz}$
    b. The 5100 MHz LO will produce a frequency inversion.
15. $CL_{LSB} = 18$ dB
17. **a.** $f_{LSLO,L} = 780$ MHz; $f_{LSLO,H} = 845$ MHz; $f_{HSLO,L} = 870$ MHz;
    $f_{HSLO,H} = 935$ MHz
    b. The RF signal at 825 MHz goes to 77.5 MHz and the RF signal at 890 MHz goes to 12.5 MHz. We do get a frequency inversion.
19. **a.** 99.7 MHz
    b. There are several possible causes for poor sensitivity.
    - The 109.8 MHz signal could mix with the 99.1 MHz signal to produce an undesired response at 10.7 MHz at the mixer's IF port.
    - This receiver has almost no image rejection because the attenuation of the RF BPF at 121.1 MHz is minimal. If there happens to be an RF signal present at 121.1 MHz, the receiver will be jammed.
    - The IF filter might be too wide. This receiver might be a commercial FM radio receiver and so, the channel bandwidth and IF filter bandwidth should both be 200 kHz.
    - In general, the filters are too wide. To fix the problem, we would examine the system specifications more closely and use narrower filters where we could.

21. a. $f_{IF} = 140$ MHz and 1940 MHz
$P_{IF,140} = -46$ dBm
$P_{IF,900} = -56$ dBm
$P_{IF,1040,dBm} = -14$ dBm
$P_{IF,1800} = -76$ dBm
$P_{IF,2080} = -68$ dBm

23. a. By inspection, using the sum and difference formulas, we find:

- RF = 108 – 120, LO_1 = 62 – 50 MHz, $Image_1$ = 232 – 220 MHz
- RF = 108 – 120, LO_2 = 278 – 290 MHz, $Image_2$ = 448 – 460 MHz

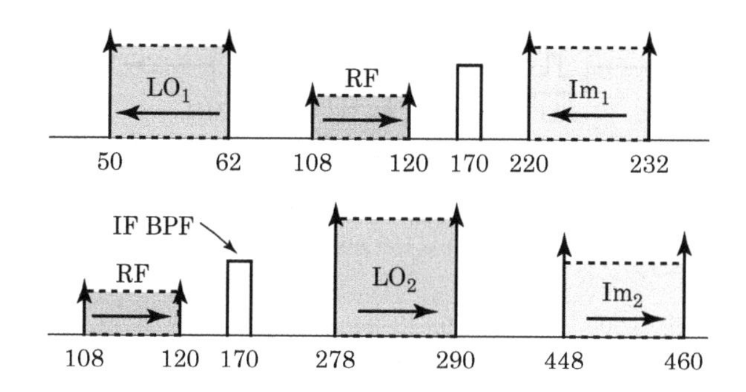

b. LO_2 produces a frequency inversion.

25. –20 dBm, –43 dBm, –37 dBm, –56 dBm, –59 dBm, –63 dBm

27. c

29. a. 3.4 – 3.7 GHz

b. 3.4 GHz

c. 57.5 dB

31. d

33. a. By inspection, using the sum and difference formulas, we find:

- LO_1 = 110 – 140 MHz, $Image_1$ = 280 – 310 MHz
- LO_2 = 200 – 230 MHz, $Image_2$ = 370 – 400 MHz

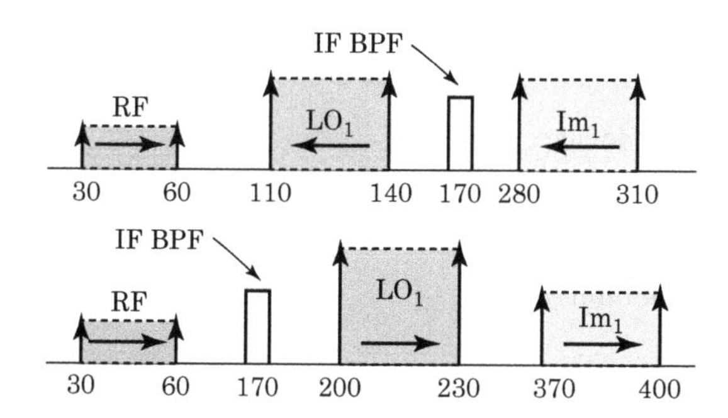

b. LO_2 produces a frequency inversion.

Chapter 9

1. a. ~1575 MHz

b. ±233.458 kHz

c. The 2025 MHz oscillator is responsible for 202.5 kHz of frequency drift while the 309.58 MHz oscillator is responsible for 30.958 kHz of drift.

d.

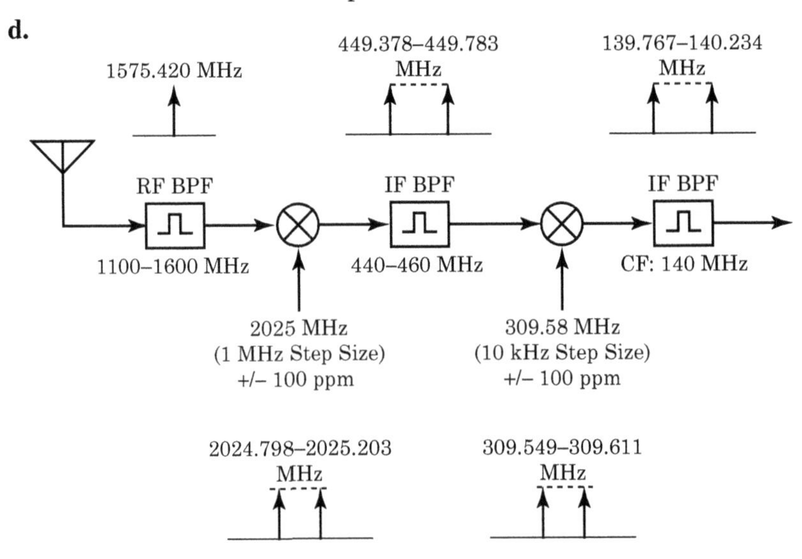

3. e

5. about 10° RMS

7. 2130 +/− 1878 ppm

9. a. 2450 MHz

b. 3829 ppm

c. From part a) A, the 2590 MHz oscillator is responsible for 25.9 kHz of frequency drift while the 150.7 MHz oscillator is responsible for 15.07 kHz of drift.

d. No

11. a. $f_{Low} = 2024.798\,\text{MHz}$ $\quad f_{High} = 2025.203\,\text{MHz}$

b. $f_{Low} = 309.549\,\text{MHz}$ $\quad f_{High} = 309.611\,\text{MHz}$

c. 139.767 to 140.234 MHz

13. Factors which will decrease the oscillator's phase noise are:

- Lower noise figure of the active device
- Lower oscillator center frequency
- Higher value of IPM integration limit f_a
- Lower value of IPM integration limit f_b
- Lower oscillator output power
- Higher resonator quality factor

15. Minimum time = 479.52 secs, Maximum time = 480.48 secs

17. c

21.

	FM Radio	Wireless LAN	CB Radio	Cellular Telephone
Center Frequency	101.1 MHz	2400 MHz	27 MHz	824 MHz
Signal Bandwidth	150 kHz	11 MHz	10 kHz	30 kHz
Oscillator Accuracy	±247 ppm	±10 ppm	±100 ppm	±12 ppm
IF Filter Bandwidth	200 kHz	11.048 MHz	15.4 kHz	50 kHz

Chapter 10

1. a. $f_{RF} = 13.5$ *or* $9.5\,\text{GHz}$; $f_{RF} = 14.8$ *or* $8.2\,\text{GHz}$

b. The worst-case image frequency is the image frequency that is closest to the passband of the RF BPF. In this problem, the worst-case RF image is the image at 9.5 GHz.

c. To increase the image rejection, we could use a larger number of poles in the input RF BPF to reject the image response.

d. $f_{RF} = 25.0$ *or* $2.0\,\text{GHz}$; $f_{RF} = 26.3$ *or* $3.3\,\text{GHz}$

e. $f_{RF} = 36.5$ *or* $13.5\,\text{GHz}$; The equivalent RF frequency is 13.5 GHz.

3. c

6. a. $S_{MDS,\text{dBm}} = -100.8\text{ dBm}$

b. $SFDR_{\text{dB}} = 40.5\,\text{dB}$

c. $S_{min} = -70.8\text{ dBm}$

7. LO frequencies: 550 MHz, 650 MHz
Image frequencies: 1000 MHz, 1100 MHz

9. LO frequencies: 59.3 MHz, 99.3 MHz
Image frequencies: 48.6 MHz, 88.6 MHz

14. a. Linearity

b. Noise Figure

c. Linearity

d. Linearity

e. Both

f. Noise Figure

15. LO frequencies: 1200 MHz, 1300 MHz
Image frequencies: 2100 MHz, 2200 MHz

17. LO frequencies: 100 MHz, 140 MHz
Image frequencies: 30 MHz, 70 MHz

19. a. Amplifier #2

b. Attenuator

c. Component placement has no effect on the overall gain.

21. **b**

24. 64 dB

26. **c**

27. **a**

32. 126 ft

Chapter 11

1. **a.** Zone 0: $f_s > 970$ MHz; Zone 1: $485 > f_s > 800$ MHz;
Zone 2: $194 > f_s > 200$ MHz

3. **d**

4. **d**

5. **c**

Chapter 12

1. **b**

2. **b**

3. **d**

4. **c**

Index

D

Q

R

T

U

V

W

Z

Author Biography

Kevin McClaning is an electrical engineer with more than 30 years of experience in technical design and management positions with the Department of Defense. He has served on the faculty of the Johns Hopkins University Whiting School of Engineering since 1990 and as a technical consultant. Mr. McClaning is coauthor (with Tom Vito) of the textbook *Radio Receiver Design* (Noble Publishing, 2000).

Lightning Source UK Ltd.
Milton Keynes UK
UKOW07n1346180417
299382UK00002B/28/P